THE EPIDEMIOLOGY OF PLANT DISEASES

Second Edition

The Epidemiology of Plant Diseases

Second Edition

Edited by

B.M. COOKE
University College Dublin, Ireland

D. GARETH JONES

and

B. KAYE
University College Dublin, Ireland

A C.I.P. Catalogue record for this book is available from the Library of Congress.

ISBN-10 1-4020-4580-8 (PB)
ISBN-13 978-1-4020-4580-6 (PB)
ISBN-10 1-4020-4579-4 (HB)
ISBN-13 978-1-4020-4579-0 (HB)
ISBN-10 1-4020-4581-6 (e-books)
ISBN-13 978-1-4020-4581-3 (e-books)

Published by Springer,
P.O. Box 17, 3300 AA Dordrecht, The Netherlands.

www.springer.com

Printed on acid-free paper

CONTENTS

Part Two: Case Examples

CONTRIBUTORS

James K.M. Brown
Department of Disease and Stress Biology
John Innes Centre
Colney Lane
Norwich NR4 7UH
UK

Valerie Cockerell
Official Seed Testing Station for Scotland
Scottish Agricultural Science Agency
1 Roddinglaw Road
Edinburgh EH 12 9FJ
Scotland
UK

Robert J. Cook
18 Green End
Comberton
Cambridge
CB3 7DY
UK

B.M. Cooke
School of Biology and Environmental Science
College of Life Sciences
University College Dublin
Belfield
Dublin 4

M.L. Deadman
Department of Crop Sciences
College of Agricultural and Marine Sciences
P O Box 34
Sultan Qaboos University
Al Khod 123
Sultanate of Oman

Maria R. Finckh
Department of Ecological Plant Protection
University of Kassel
Nordbahnhofstrasse 1a
D-37 213 Witzenhausen
Germany

B.D.L. Fitt
Plant Pathogen Interactions
Rothamsted Research
Harpenden
Hertfordshire AL5 2JQ
UK

R.T.V. Fox
School of Biological Sciences,
The University of Reading
2 Earley Gate
Reading RG6 6AU
UK

William E. Fry
Plant Pathology Department
Cornell University
334 Plant Science Building
Ithaca NY 14853
USA

Nigel V. Hardwick
Central Science Laboratory
Sand Hutton
York Y041 1LZ
UK

Bernhard Hau
Institut fur Pflanzenkrankheiten und Pflanzenschutz
Universitat Hannover
Herrenhauser Str. 2
D-30419 Hannover
Germany

Laurent Huber
Unité Mixte de Recherche
INRA/INAP-G
Environnement et Grandes Cultures
78850 Thiverval-Grignon
France

Gareth Hughes
School of Biological Sciences
University of Edinburgh
Kings Buildings
Edinburgh EH9 3JG
Scotland
UK

Alan L. Jones
Department of Plant Pathology
Michigan State University
East Lansing MI 48824-1311
USA

D. Gareth Jones
10 Clos Ffawydden
Ystrad Owen
Cowbridge
Vale of Glamorgan CF71 7SE
Wales
UK

Bernard Kaye
School of Biology and
Environmental Science
College of Life Sciences
University College Dublin
Belfield
Dublin 4
Ireland

Philippe Lucas
INRA – Agrocampus Rennes
UMR BiO3P
(Biologie des Organismes et des
Populations appliquée à la
Protection des Plantes)
BP 35327
35653 Le Rheu Cedex
France

L.V. Madden
Plant Pathology Department
Ohio State University
Wooster OH 44691-4096
USA

Robert B. Maude
The Lindens
Lenchwick
Evesham
Worcestershire WR11 4TG
UK

H.A. McCartney
Plant Pathogen Interactions
Rothamsted Research
Harpenden
Hertfordshire AL5 2JQ
UK

Neil McRoberts
Crop and Soil Research Group
Scottish Agricultural College
Edinburgh EH9 3JG
Scotland
UK

Eduardo S.G. Mizubuti
Depto. de Fitopatologia
Universidade Federal de Viçosa
36570-000 Viçosa
MG
Brazil

H.P. Narra
School of Biological Sciences,
The University of Reading
2 Earley Gate
Reading RG6 6AU
UK

Adrian C. Newton
Scottish Crop Research Institute
Invergowrie
Dundee DD2 5DA
Scotland
UK

G.W. Otim-Nape
National Agricultural Research Organization
PO Box 295
Entebbe
Uganda

William J. Rennie
1 St Fillans Grove
Aberdour
Fife
KY3 OXG
Scotland
UK

Michael W. Shaw
School of Biological Sciences
The University of Reading
2 Earley Gate
Reading RG6 6AU
UK

Christine Struck
University of Rostock
Faculty of Agricultural and Environmental Sciences
Institute for Land Use
Crop Health
Satower Strasse 48
18059 Rostock
Germany

George W. Sundin
Department of Plant Pathology
Michigan State University
East Lancing MI 48824-1311
USA

John M. Thresh
Natural Resources Institute
University of Greenwich
Chatham Maritime
Kent ME4 4TB
UK

Claude de Vallavieille-Pope
UMR INRA/INA-PG
d'Epidémiologie Végétale et Ecologie des Populations
BP01
F-78850 Thiverval–Grignon
France

Jonathan S. West
Plant Pathogen Interactions
Rothamsted Research
Harpenden
Hertfordshire AL5 2JQ
UK

Martin S. Wolfe
Wakelyns Agroforestry
Fressingfield
Suffolk IP21 5SD
UK

Xiangming Xu
East Malling Research
New Road
East Malling
Kent ME19 6BJ
UK

David J. Yarham
Mill House
Croxton
Fakenham
Norfolk NR21 0NP
UK

PREFACE

Since the first edition of this book was published in 1998 by Kluwer Academic Publishers, I have been inspired to produce a second edition of the text but in a format that would make it much more affordable to research workers and senior research students studying and working in plant disease epidemiology. The inspiration for the first edition came solely from D. Gareth Jones who edited the volume around the time of his retirement from the University of Wales, Aberystwyth, where I studied for both degrees in the Department of Agricultural Botany, and where as an undergraduate was taught almost everything about plant pathology, and carried out postgraduate research on the epidemiology of Septoria diseases of wheat, under the supervision of DGJ. This research was probably one of the first attempts to evaluate the importance of the two diseases under UK conditions; the initial work was greatly inspired by the late Ellis Griffiths.

DGJ agreed late in 2003 that a second edition of the book should be put in train, and that a softbound version should be produced in addition to a hardbound version, so reducing the cost compared to the first edition published in hardbound only. The book publication proposal was rapidly approved by Springer, the present publishers; dealing with this publisher made the task easier, as being the Editor-in-Chief of the *European Journal of Plant Pathology,* also published by Springer, I had already established the necessary publishing contacts within the organization in Dordrecht, The Netherlands - Zuzana Bernhart and Ineke Ravesloot - who have been most helpful at all times.

The next task was to persuade the original authors to revise their chapter contributions in the light of the rapid modern developments in plant disease epidemiology that have occurred since the first edition was published, such as in molecular diagnostics and information technology (see Chapters 1 and 12). I need not have worried. The response has been truly amazing. Although this second edition follows largely the pattern of the first, all chapters have been extensively updated by the original authors, and significant contributions added by new authors who are at the cutting edge of their respective fields (see Chapters 1, 6, 8, 12, 13 and 14). Again the text is divided into two parts: Principles and Methods, and Case Examples. The result is a comprehensive text on all aspects of the epidemiology of plant diseases that should serve as an invaluable reference work for those involved in this dynamic and fascinating science of crop plants.

I could not have edited this second edition without the considerable expertise of my colleague Bernard Kaye who re-formatted all the chapters (several times) and electronically enhanced the figures using Apple Macintosh computer software. Finally, my thanks go to D. Gareth Jones for his kind permission to revise and update the original text of the first edition.

B.M. Cooke
November 2005

PART ONE

Principles and Methods

CHAPTER 1

PLANT DISEASE DIAGNOSIS

R.T.V. FOX AND H.P. NARRA

1.1 INTRODUCTION

The diagnosis of disease, during the course of studies on epidemiology or more commonly as a prelude to control, differs in many ways from the identification and taxonomy of the causal microorganisms or other pathogenic agents. A plant is said to be diseased when its normal functions are disturbed and harmed (Holliday, 1989). This is a complex condition and provides many more opportunities for detection and diagnosis than the pathogen alone. For example, in human medicine, diminished glucose levels form the diagnosis of diabetes and human hormone imbalances can indicate several problems. However, in plant pathology, conditions analogous to these are regarded as disorders, not diseases. Therefore, in line with current practice, this review does not include such disorders, as in plants these are only indirectly caused by pathogenic agents but can directly result from genetic defects, mineral imbalance, environmental pollution and a number of non-biotic reasons.

With the possible exception of prions, all known biotic pathogens depend on nucleic acids for their reproduction and hence diversity, including their virulence. Recognizing various aspects of this biodiversity at different levels provides the basis, directly or indirectly, for all of the existing methods of detection of pathogens, but not the diseases that they cause. The latter can be distinguished additionally by their effects on the host plant, including the macro- and micro-symptoms that are visible (or smelled in some cases) or the underlying biochemical changes; this topic is covered more fully elsewhere (Fox, l993a). However, in general, apart from a few exceptions, the diagnosis of the early onset of plant disease, unlike much human disease, depends on much more invasive techniques to detect and identify the pathogen. The main exceptions to this rule are the immunoassays to specific microbial toxins (Candlish *et al.,* 1992). Although most commercial kits to assay mycotoxins have primarily been developed for food safety, wilt phytotoxins have been monitored in epidemiological studies (Benhamou *et al.,* 1984, 1985a,b).

It is also useful to start a review of plant disease diagnosis by attempting to differentiate those distinctive features that characterize diagnosis from traditional identification techniques, as these are governed by well defined precepts. Conventional keys employ only the principal distinctive differences between related individuals. Classical taxonomic descriptions for cladistics consist of a collation of as many characters as possible, usually morphological and anatomical features, to ensure that the differentiation of taxa is as comprehensive as possible. For this

B. M. Cooke, D. Gareth Jones and B. Kaye (eds.), The Epidemiology of Plant Diseases,
2nd edition, 1–42.

reason, taxonomists have increasingly been aided by the huge advances in information technology. By contrast, for diagnosis of whatever sort, it is the end that justifies the means. Like crop protection, or even agriculture itself, the important feature of diagnosis is that it is essentially just the means used to effect an end result: one diagnostic test is merely one tool among many that could be effective. This is as true if the diagnosis is for a scientific study of epidemiology and ecology of species, strains or other intraspecific variants, such as those resistant to fungicides (Martin *et al.*, 1992b) (Fig. 1.1), as when diagnosis is simply used to justify the application of a pesticide. Often direct methods of diagnosis are now being adopted in the latter situation. The problems involved with the detection of the cereal disease, eyespot, highlight the superiority of diagnostic methods that rely on the direct detection of the pathogen to those that merely record its effects. In this case, an extremely wide range of symptoms other than 'typical eyespots' can be seen on the stem bases of wheat from which *Pseudocercosporella* spp. can nonetheless still be isolated. Just as in medical general practice, the control of disease is usually the most pressing immediate objective of plant disease diagnosis, although the means vary.

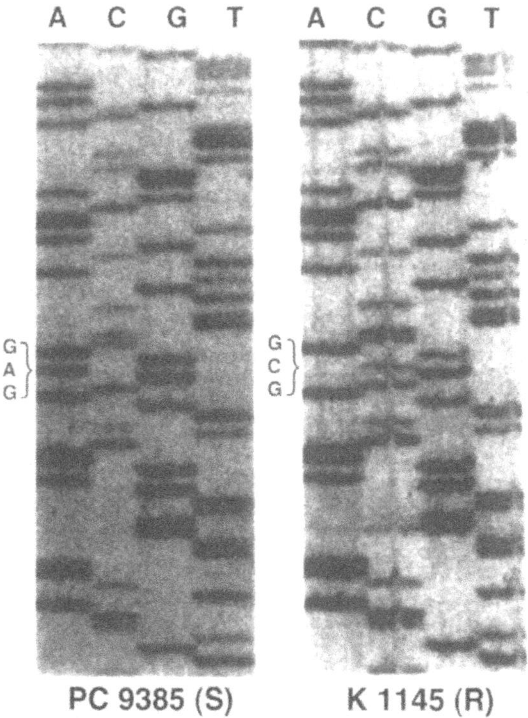

Figure 1.1. Autoradiograph of pair sequencing gels showing the region of the point mutation to carbendazim (MBC) resistance at nucleotide 293 of the sequences. This single base-pair change from A in the sensitive strain PC 9385 (5) to C in the resistant strain K 1145 (R) results in a GAG codon (glutamic acid) being changed to GCG (alanine) at amino acid 198 in the resistant isolate. (Photograph from Martin et al., 1992b).

1.2 CHOICE OF DIAGNOSTIC

Plant pathologists have gained access to a variety of different types of diagnostic techniques, along with rapid electronic methods of reporting, storing and analysing the results (Fox, 1993a). Unless the disease is unknown or under a specialized scientific study, the form of diagnosis that is chosen (and there may be several contenders) is generally the quickest, easiest, cheapest and the most accurate method that can confirm the correct identification of the causal agent (Fox, 1990a). The multiplicity of methods used to study the behaviour of plant pathogens in their natural environments largely reflects the different diagnostic problems. The skill of the diagnostician and the epidemiologist is to know the limitations of each method and to choose the most appropriate. An effective diagnostic test should be as simple, accurate, rapid and safe to perform as the technology allows, yet also be sufficiently sensitive to avoid 'false positives'. Many farmers and growers now need to be more self-sufficient as a consequence of the reduction in advisory support resulting from the worldwide political support for free market forces. Such general agreements on tariff and trade tend to result in the loss of subsidized technical advice. Therefore, commercial diagnostic kits are increasingly marketed for the identification of plant pathogens in the field (Miller *et al.*, 1988), as well as in the laboratory, so improvements in portability are important. As a spin-off, some of the miniaturized kits already developed for farm use should also benefit epidemiological research, as their accuracy is as vital for monitoring the onset of serious epidemics.

Many of the different methods of diagnosing plant disease have proved to be complementary. It is doubtful that there will ever be one single, universal, easy yet dependable method of identifying pathogens or the diseases that they cause, either for studies of plant disease epidemiology or directly for crop protection. An assortment of diagnostic methods will probably continue to be used in epidemiology. The extent to which routine recognition of the pathogen by unique features of its molecular biochemistry, such as the ability of its nucleic acid to hybridize or its proteins to bind immunologically, will displace the traditional visual and microscopic examination of pathogens, both *in planta* in the natural environment or *in vitro* after the isolation of cultures, is less clear. To a great degree, this depends on whether epidemiologists choose to benefit from the technological advances designed to provide rapid solutions to problems in the field, or whether they decide instead to continue to devote the time necessary for specialist diagnostic laboratory studies. If we can learn anything from what has already happened in medicine, both the general practitioner and specialist diagnostic laboratories coexist to their mutual advantage and that of the patient. Perhaps cooperation between practitioners based in the field and specialist diagnosticians will also increasingly be a feature of integrated crop production and maybe from this to epidemiology.

Speed is naturally a prime consideration whenever healthy planting material or a control agent for a disease outbreak has to be selected urgently. While efficient diagnosis is also important for academic investigations and can easily be measured, the other advantages of one method over another as practical diagnostics are less immediately apparent. When a critical comparison has been attempted (Fox, 1993a), some crucial differences may be exposed. From an intuitive perspective, few diagnostic tests seem as quick and as straightforward as the routine visual inspection of a specimen for symptoms. However, the differences are only readily noticeable when the progress of infection of leaves is already well advanced. Since soil obscures the symptoms of diseased roots, they take longer to examine than foliage (Fox, 1990b; Fox *et al.*, 1994). Wilted plants can be rather troublesome to diagnose as the pathogen is deep seated. In such cases, less tangible techniques of diagnosis are required.

The most significant criterion for any test is reliability. Hence, for traditional diagnostics to be feasible, the specimen must be in a reasonably good condition. An absence of other pathogens or saprophytes as well as a minimal level of varietal and phenotypic variation between samples is vital. In practice, although potentially simple, albeit at a late stage, in many cases the assessment of symptoms can be inherently unreliable. The value of any associated microscopical techniques also depends on the quality of the specimens. Plant material that is provided for examination may be contaminated with a mixture of microorganisms, which greatly hinders the isolation of pure cultures – especially if several pathogens cause similar symptoms – thus delaying any subsequent investigations.

By their nature, symptoms are easily influenced by minor environmental changes, such as an unusually dry summer or cold spring. If this is likely to be the case, it can again be prudent to choose a method based on fundamental biochemical differences, such as the immunology of proteins and the hybridization of nucleic acids which are relatively unaffected by external factors. Neither of these types of molecular diagnosis requires an extensive knowledge of plant pathogens nor expertise to operate. Although there is an increased opportunity to acquire such information and advice via the Internet, some basic expertise, talent and more time are still required. The molecular methods have the potential to achieve even greater cost-effectiveness if automated. The larger the scale of the investigation, the sooner can the initial costs of equipment and reagents be written off against the need for highly trained personnel. Nonetheless, when a protocol is being developed, care should be taken to ensure that any time saved in performing the diagnosis is not squandered by inefficient sample collection or assessment of results. Again, electronic communication and information systems are likely to play an increasingly important role in the future interpretation, dissemination and storage of all plant disease diagnoses.

1.3 DIAGNOSIS BY CONVENTIONAL TECHNIQUES

In practice, there is a range of diagnostic tests that are traditionally regarded as conventional. These techniques are generally considered to include the recognition of symptoms and the isolation and examination of plant pathogens using microscopy (Fox, 1993a). Most epidemiological investigations involve the detection of outbreaks of disease in the field, which means that at least some symptoms must already be present. In this situation it is usual to walk through the field in order to assess the level of disease, but where the terrain is unsuitable or too extensive, or extra speed is required, aerial surveys by manned or remote-controlled aircraft have been used. This approach is possible for some plant diseases but for many pathogens it is not practical.

The early stages of many diseases are inconspicuous and it is not feasible to make a rapid visual assessment until the level of disease is sufficiently high, which may be so late in the life-cycle that timing may become critical. In this case, alternative methods may have to be adopted such as the detection of airborne spores by the use of traps or the collection of samples for laboratory examination by other methods.

Symptoms that are observed are the accumulated result of numerous changes at the cellular level generally as a result of damage to the host tissues. Sometimes the most obvious indication that a plant is diseased is the appearance of the vegetative or fruiting bodies of the pathogen. The latter are often termed the 'signs' of infection. In some diseased plants, the only symptoms or signs are internal and so assessment requires dissection, thus destroying the host.

Another difficulty is that few symptoms are so completely distinctive that they can be used by themselves to distinguish the presence of a particular pathogen unequivocally on every occasion. In most cases, the identification of a pathogen relies on the presence of a characteristic syndrome or specific combination of symptoms, some of which might not appear very distinctive. This raises the difficulty that more than one pathogen might be present. Since plants are covered in epiphytic fungi and bacteria, it can sometimes be surprisingly difficult to be sure that the microorganism that is most obviously present on or in plant material is actually the pathogen responsible for those symptoms that are observed. In many cases a vigorously growing saprophyte is far more noticeable than a more deep-seated pathogen. For this reason, unless an organism is familiar, it may be necessary to satisfy Koch's postulates by testing its pathogenicity. Even in a routine investigation the identity of pathogen should be confirmed. In many cases, this may only require the use of a hand lens but frequently some microscopy or even isolation in the laboratory may be needed.

For the majority of fungi, an extra stage involving microscopy is not such a significant problem as it would be for bacteria; for viruses such an examination would be still more complicated. For this reason, in the latter cases, the recognition of clear symptoms in field assessments is more important than for most fungi. Fortunately, there is a profusion of well illustrated literature in the areas of symptomology and fungal morphology to aid identification of the pathogens of major crops but this help is less readily available for minor or exotic hosts.

Even though their use is so widespread and generally acceptable, the use of symptoms in practice is beset with a few other problems. Their main drawback is that both pathogens and a number of other causes – such as insects and other pests, as well as abiotic agents – result in damage that originates due to very similar cellular or physiological injuries. Partly in order to avoid such confusion, the types of symptom that occur have been classified into several major groups and sub-divisions. For example, a number of pathogens disrupt the growth of their hosts. Some pathogens stunt the host, causing symptoms such as dwarfing, atrophy, suppression of growth or differentiation. Others cause excessive elongation (etiolation). If the meristematic activity of the host is affected, the result may be abnormal cell enlargement causing deformations such as curling, or excessive cell division causing proliferations such as galls, tumours and callus. The life of storage organs can be shortened. The behaviour of other organs, such as petioles, may be affected by epinasty caused by pathogens and leaves may be lost by abscission and chlorosis. A more subtle disruption to growth results when there is simply a loss of yield without any other associated symptoms.

Other pathogens, such as seedling blights, disrupt the mobilization of stored food, whereas root and foot rots disrupt the absorption of water and minerals. Vascular wilts disrupt water transport. Phyllody and decline follow the disruption of translocation. Many pathogens cause disruption to many functions and it is these that are most commonly mistaken for the consequences of abiotic agents. Among the most common symptoms are yellowing, chlorosis caused by the break down of chlorophyll, water soaking and exudation as well as necrotic reactions ranging from minute spotting to completely blighted plants. More specialized pathogens disrupt the secondary metabolism of the plants and it is these that result in pigmentation change – sometimes with colourful results, as for example the viruses that cause colour breaking.

Since reliance on symptomology alone is such an imperfect means of identifying any other than the most familiar pathogens, in many cases it is necessary to confirm the identity of the causal agent by the use of microscope techniques, isolation or both. If the putative pathogen that is isolated is unfamiliar then its pathogenicity should be checked by satisfying Koch's postulates by reinoculating the isolate into the host plant. This process is both time-consuming and fraught with problems. The initial isolation of microorganisms from plants is beset with difficulties, due to contamination during and after the putative agent has been targeted. Numerous pathogens are biotrophs which cannot be grown *in vitro*; many others are so fastidious that a suitable medium is too troublesome to find and in practice isolation is not feasible; and others grow so slowly or weakly that they are lost too readily for reliable identification. Even before isolation, it is necessary to prepare the host material appropriately to induce sporulation. Where sporulation is recalcitrant, progress may be stifled at the onset.

Where conventional isolation methods are used, a number of specialized techniques have had to be perfected in order to cope with microorganisms in specific substrates such as soil, roots, aerial parts and seeds. Successful isolation from most of these substrates involves considerable expertise in the special problems involved and the normal structure and functioning of the plant host. Where such knowledge is absent, it is all too easy for a novice to make serious mistakes. Selective media offer one potential source of aid for the inexperienced, but in practice, suitable products are rare. Nonetheless, where they are available, selective media and selection techniques can greatly increase the chances of successful isolation, subsequent purification and conservation of fungi and bacteria. Routines for the extraction and purification of viruses require equally professional proficiency and skill. Once isolation has been successful, other sorts of expertise are needed to confirm the identity of the pathogenic agents.

Among the methods available once a pathogen has been isolated are a number of biochemical and physiological ones which traditionally have been used mainly with bacteria but may be used more widely in future. However, in most cases microscope techniques are preferred as the most conventional method to distinguish fungi using morphological differences. To be effective, such methods need to combine an extensive knowledge of fungal taxonomy with a practical understanding of microscopy. Even when used by an expert, access to a reasonably substantial collection of reference books and other publications is essential, either directly or via electronic sources. Fox (1993a) gives details of a number of such books. The choice of material sampled for examination is also critical and since the position of the pathogen is not always obvious from the location of the symptoms, a reasonable background in plant pathology is necessary. Often specialized techniques are required – for example, when woody tissues or the embryos from seeds need to be examined. With the exception of those based on specific immunological properties, fewer of the impressive array of microscopic stains that are currently available are sufficiently selective by themselves to be valuable as diagnostics for fungi than bacteria.

In summary, conventional plant pathological techniques need substantial expertise and experience of a wide array of specialized technologies. All these features tend to make them less attractive for routine use than the techniques based on the intrinsic microbial biochemistry of the pathogens, described below.

Despite most common established preconceptions, although the assessment of symptoms has the potential to be simple, albeit usually at a late stage, in practice it is often inherently unreliable.

1.4 USE OF IMMUNOLOGICAL REACTIONS

Immunological methods were amongst the first, and the simplest to use and interpret, of the techniques based on the measurement of molecular rather than the observation of visible criteria. The myriad of different molecular modifications that occur at a cellular level are also responsible for the development of the symptoms that are discernible by the naked eye or microscope. Due to the longer time for any change to become apparent, visual methods based on the much more substantial scale of the reaction of the host to infection are inevitably slower, in contrast to the subtle changes that are detected by selective molecular methods for directly detecting the presence of pathogens. Hence, in traditional pathology it is often necessary to concentrate on the pathogen but instead of the days or even weeks required if cultures are made, the detection of a characteristic antibody binding site (epitope) on a molecule (antigen) can be completed in a few hours at most. As well as being usually accurate and always quick, immunological methods are invaluable for diseases with inconsistent or undeveloped symptoms. For this reason, many commercial kits are now used on a large scale by plant breeders and propagators when disease-indexing propagating material. Most diagnose important viral pathogens, nearly all of which are invisible under the light microscope and morphologically indistinct even under the electron microscope. Although there are a number of similar tests for bacterial diseases, fewer commercial kits have so far been developed for fungal diseases. Fungi are usually obvious as they may grow on the plant tissue culture medium used during disease-indexing. Away from the tissue culture laboratory, other commercial kits are used in field situations by non-specialists for rapid diagnosis of fungal pathogens on high-value crops such as sports turf (Miller *et al.*, 1988).

1.4.1 Principles of immunoassay

There are numerous different diagnostic immunoassay procedures, but all ultimately depend on the efficacy of a biochemical response of living animals, particularly mammals. Animals manufacture antibodies that react to some of the characteristic substances that comprise pathogens and other environmental agents that potentially challenge their integrity. Antibodies that otherwise share a basic common structure (two heavy and two light chains held together by disulphide bonds) differ in the ability of their tips to bind to sites on different 'foreign' protein, glycoprotein, lipoprotein, lipopolysaccharide and carbohydrate molecules. Provided such antigenic substances are not already present in the body of that animal, they stimulate its immune system to produce an assortment of specific antibodies, each of which specifically recognizes and will link almost exclusively to its complementary epitope on the antigen. This interaction between an antibody and an antigen depends on an extraordinary pairing of an individual epitope of the latter in three dimensions between the contours of the tips of the side chains of a particular antibody, since the molecular configuration of each of these binding site regions varies subtly in both affinity and avidity. While the overall capacity of the system is truly astounding, nearly all of this enormous potential is normally latent. Nonetheless, once an animal has been stimulated by inoculation, in theory a specific antibody can be generated that possesses the ability to bind with any of a virtually unlimited variety of different antigens. Afterwards, the immune system of the animal can produce more copies of an antibody from 'memory'. In practice, as well as several constituents of the walls of pathogens, the antigens recognized by antibodies include a diversity of other innocuous substances. Unfortunately, many fungi and other microorganisms consist of many non-specific antigenic substances that are immunodominant. As a result, polyclonal serum may be insufficiently reliable for certain diagnoses so monoclonal antibodies are necessary (Dewey, 1992). A uniform population of a particular monoclonal antibody is formed by each hybridoma (Fox, 1993a). Although each monoclonal antibody generally reacts with its own particular antigenic determinant (epitope), this site could be restricted to a particular strain or it may be common to an entire species, genus or even more widely (Hardham *et al.,* 1985). Barring mishaps in culture due to contamination by noxious yeasts and mycoplasmas or unexpected thawing, a hybridoma chosen for its valuable properties can be kept almost indefinitely at low temperature. This makes it possible to maintain a plentiful supply of a highly specific monoclonal antibody (MCA). Monoclonal antibodies are valuable for numerous applications such as increasing the specificity of detection of the antigen – for example, after electrophoresis (Western blotting), and to purify yet more antigen (affinity chromatography).

1.4.2 Classical immunoassay tests

Several immunoidentification techniques, including agglutination and precipitation tests, have a long and successful history and have proved very useful in the epidemiology of viruses as they are portable and the results are clearly visible to the eye. These simple serological tests depend on two similar properties: agglutination and precipitation. In **agglutination** tests the antigen must be particulate, as the positive reaction to the antiserum is the clumping of the suspension. The antigen is precipitated out of solution by the antiserum in **precipitation** tests; when mixed with its respective antiserum, the initially opalescent suspension containing antigen particles settles out, producing a clear supernatant by granulation or flocculation. A microscope is used in slide agglutination tests if agglutination is indistinct to the naked eye. Drops of antigen and dilute antiserum are mixed together on a glass microscope slide and compared with antigen/saline and antigen/normal serum controls. When greater sensitivity is necessary, the agglutination can be carried out in tubes rather than on slides.

Although used less frequently in plant pathology than in human medicine, **complement fixation** can assist some difficult immunoassays. Complement, a non-specific non-immunogenic thermolabile substance in blood, is only able to combine with genuine antigen-antibody complexes, not unsuccessful mixtures. Some antigens and antibodies require the addition of complement before they are able to complex properly after they have already been 'sensitized' by being mixed together.

These reactions can also be enhanced by absorbing the antibodies or antigens onto latex beads coated with protein A from *Staphylococcus aureus* or the intact cocci themselves (Chirkov *et al.,* 1984). The outer membranes of Gram-negative bacteria contain **lipopolysaccharide (LPS)** which is highly antigenic and easily extracted and purified. Antiserum raised against LPS may be so specific that cultivar-specific strains of *Pseudomonas syringae* may be detected which are not differentiated by antiserum raised to intact bacteria.

Bacterial extracellular polysaccharides (EPS) may also be used to raise antisera with a high level of specificity. Cross-reactivity caused by contaminant bacteria and the host plant hinder the detection of pathogenic bacteria in infected plant material as they usually occur there in such relatively low numbers. Provided the bacteria are not among those phytoplasms restricted to the phloem and other prokaryotes that have so far not been isolated in culture, detection can be improved if the population of the bacteria can be increased by initially incubating the infected plant extracts with an enrichment medium. Although it is very difficult, antisera can be raised which are specific to the small numbers of obligate phytoplasms present in infected plants (Bove, 1984). Monoclonal antibody preparation and selection techniques avoid the necessity to prepare obligate pathogen immunogens free from host contamination (Lin and Chen, 1986; Clark *et al.,* 1989; Clark, 1992).

In **the immunosorbent dilution-plating technique (ISDP),** rod-shaped bacterial pathogens are trapped on antibody-coated Petri dishes or inoculation rods so that they can be incubated on selective media (Van Vurdee, 1987). Any unbound bacteria are removed by washing. The colonies that form are detected by drying down the agar and staining by immunofluorescence. After rehydration, bacteria can be reisolated from them as they are still viable.

Immunodiffusion is a development of the precipitin test in which the antiserum that diffuses through agar precipitates along the line of contact with the antigen diffusing in the other direction. Often, a central well cut into an agar Petri plate containing antigen is surrounded by similar wells filled with serial dilutions of antiserum. After incubation in high humidity, lines of precipitation arc away from the wells. The highest dilution of antiserum that displays an arc represents the titre of the antiserum. The **double diffusion exoantigen** technique is a variation of this method in which the samples of different antigens surround a well filled with antiserum. The antigens are deemed identical if the resultant arcs link; but if they cross, the antigens are considered to be different. In other tests, if the antigens to be compared are positioned parallel to the antiserum, the antigen-antibody zones of precipitation also link if they match.

Gel electrophoresis can be used to separate mixtures of antigens prior to immunodiffusion. A narrow trough is cut in a thin gel parallel to an electric current which passes close to the antigens along its length. Depending on its distinctive charge, each antigen moves in a separate wave at a characteristic rate so the proteins separate into bands. The current is then switched off and antiserum is added to the trough cut in the gel so that precipitin arcs composed of complexes of antibodies and antigens form at the individual bands of antigens in characteristic shapes and patterns. Cross-reactions with other antigens can be reduced by initially absorbing the antibody with the interfering antigen. The gel can also be split up after electrophoresis so that and one half can be treated with antiserum and the other stained to reveal the bands of antigen. Bands of pure antigen can be cut out from such gels.

1.4.3 ELISA

Since the introduction of enzyme-linked immunosorbent assays (ELISA) for plant virus detection (Clark and Adams, 1977), they have increasingly become the means to rogue out infected plants from plant certification schemes. Only polyclonal antiserum was available at first but problems arose as carbohydrates from the host plants can induce non-specific antibodies. These may impede ELISA tests, even when double gel diffusion and other serological tests are largely unaffected. With viruses, this can be partially overcome by cross-absorbing with healthy sap beforehand, but there is no necessity for this extra stage if a specific monoclonal antibody is substituted for the mixture of polyclonal antibodies. However, many tests still rely on polyclonal antibodies because they give a clearer result as they can bind to several characteristic epitopes. In conventional or 'direct' ELISA tests using 96-well immunoplates, the end point is when the colourless substrate is converted to a coloured product by an appropriate enzyme in proportion to the concentration of antigen in the original sample. This obvious colour change can either be scanned electronically with a spectrophotometer or assessed with the naked eye. The specific antibody/antigen complex is detected by using a conjugated second antibody that recognizes the non-specific antibody in 'indirect' ELISA. Morley and Jones (1980) describe an interesting modification of the ELISA technique in which a poorly antigenic microorganism can be more easily detected by incubating the substrate to increase its fluorescence.

In addition to the relatively inconvenient conventional 96-well immunoplates, cheaper and more portable assay systems have been developed using flexible nitrocellulose dip-sticks surface-coated with a capture antibody. Antigens of the pathogen present in sap squeezed from the host plant permeate through the plastic, where it is detected after incubation with the antibody-enzyme conjugate, followed by washing, then dipping in substrate, further washing, and finally adding the stopping solution. Hence ELISA reactions can be carried out *in situ* quickly and easily in the field as the dip-stick becomes coloured if the specified pathogen is present. If the test specimen is healthy the dip-stick remains colourless. Both positive and negative controls should be included (Mitchell *et al.,* 1988). In other variations, an ELISA reaction can be carried out in a similar way on membrane enzyme/substrate systems. A drop of the test sample is added to a 'dot' containing the specific MCA which has already been absorbed on the membrane and then blotted. Plant disease epidemiologists can benefit from the numerous advantages of portability and adaptability shared by dot-blot and dip-stick tests (Hill, 1984; Cooper and Edwards, 1986), as well as the extra bonus that both the protein and carbohydrate constituents of fungi can be detected.

The detection of prokaryotes and viruses in diseased plants is not prone to a number of serious difficulties that have delayed the widespread application of ELISA to fungi. Most antisera against mycelial fragments, extracts from lyophilized mycelium, surface washings of solid cultures or culture filtrates, cross-react widely with host tissues or extracts as well as related and unrelated fungi when tested by ELISA or similar techniques. The same antisera may appear species-specific when tested by immunodiffusion. Non-specific antigens may be common to both the insoluble and soluble fractions of fungal material (Chard *et al.,* 1985a,b). Efforts have been made to improve specificity by diluting out non-specific antibodies or cross-absorbing antisera with related fungi (Kitagawa *et al.,* 1989). Non-specific immunodominant carbohydrates or glycoproteins induce non-T cell stimulated responses. Better results have been obtained with antisera raised against protein precipitates from either culture filtrates or mycelial extracts (Gerik *et al.,* 1987; Gleason *et al.,* 1987; Barker and Pitt, 1988; Mohan, 1989) or specific fungal fractions such as enzymes, toxins or soluble carbohydrates (Johnson *et al.,* 1982; Notermans *et al.,* 1987). Where the fungus is present on or near the surface of the infected tissue, overnight soaking enables detection by polyclonal antisera to be effected at very low levels (Gleason *et al.,* 1987). This is important as specific antigens at the species- and subspecies-specific level may often be present in the walls and cross-walls of the hyphae but not the spores (Dewey *et al.,* 1989b).

While it is conceivable that there is an almost infinite supply of both epitopes on fungal protein and polysaccharide antigens and hence a virtually inexhaustible potential for the specific antibodies to them, this also implies that there would be an increased chance of cross-reactivity with fungi and other eukaryotes than with the simpler viruses and prokaryotes which consist of far fewer antigenic sites. Nonetheless, many popular serological techniques are still based on reactions between as yet unidentified antigens with relatively crude immune sera (antisera).

Although these and other simple methods involving purified antibody mixtures (polyclonal antibodies) are still widely used, pure single (monoclonal) antibodies can be used to attain greater levels of specificity. However, with increased selectivity there is a concomitant loss of potential binding sites and hence often a fainter reaction. For this reason, polyclonal antibodies are often preferred. These are purified from crude immune serum or antiserum humanely collected from the blood of living animals that have been injected with doses of antigenic material from the pathogen; usually this must be done only in registered premises.

In this way, a variety of antigenic substances have been used as immunogens, including peptides (e.g. Robard, 1987; Fox *et al.*, 1989; Martin *et al.*, 1992a) and proteins (e.g. Benhamou *et al.*, 1984; Groves *et al.*, 1988). The sources of these immunogens have included purified material, electrophoresis bands, crude mycelial culture extracts (Ouellette and Benhamou, 1987) and freeze dried mycelium (Fox and Hahne, 1988). Since many antigens in plants are shared by microorganisms, polyclonal antibodies to fungi are often ineffective for diagnosing them in host tissues, even when unwanted cross-reactions have been reduced by using pure antigen. Nonetheless, antisera to some fungi have been used successfully (e.g. Mohan, 1989).

One possible way around the problems caused by the mixed population of antibodies present in antisera is to produce a uniform population of a single monoclonal antibody (MCAs, mAbs or MAbs) that only reacts with a particular individual antigenic determinant (epitope). Such a homogeneous population can be produced because it is possible to select a particular stable hybridoma created by the hybridization between the B lymphocyte spleen cell responsible for the production of that particular antibody and a myeloma cell (Kohler and Milstein, 1975). However, only a few per cent (at most) of these hybridomas usually produce a useful antibody (Lin and Chen, 1985; Ouellette and Benhamou, 1987). Once a hybridoma has been selected by monoclonal antibody enzyme-linked immunosorbent assay (MCA, mAb or MAb ELISA) of its particular antibody, it can be cloned *in vitro*.

Hybridomas are cheap to maintain and can be used to produce monoclonal antibodies when required, so an ample and theoretically unending supply of a highly specific monoclonal antibody can continue to be provided routinely by resuscitating and growing subcultures of clones of a hybridoma stored at low temperature. This technique has allowed adequate supplies of antibodies specific to immunological markers to a particular strain, species, genus or any other taxon to be provided sufficiently economically for commercial kits as well as research. Nonetheless, it should also be understood that coupled with the low probability of obtaining an appropriate antibody, it might not form precipitin bands in double gel diffusion tests or readily conjugate to enzymes. The probability of success can be anticipated by obtaining small amounts of polyclonal antiserum from some drops of blood squeezed from a small nick in the tail of the immunized animal to confirm the presence of general activity by carrying out polyclonal antibody ELISA. The veracity of this prediction must be treated with caution since an intense colour reaction in indirect polyclonal antibody ELISA may be largely due to non-specific binding. A hybridoma that produces an antibody unique to any qualitative or quantitative immunogenic difference can be selected by screening the latter for specificity, thus obviating the need for preliminary purification of the immunogen (Goding, 1983).

Many ELISA protocols have been devised for particular purposes (Fox, 1993a). In practice, polyclonal antibody ELISA is still predominant in these. Most MCA ELISA have been used in academic research, for example, for taxonomy (Benhamou *et al.,* 1984; Hardham *et al.,* 1985; Clarke *et al.,* 1986; Fox and Hahne, 1988; Dewey *et al.,* 1989a; Estrada-Garcia *et al.,* 1989), research on fine structure of pathogens (Hardham *et al.,* 1985; Day *et al.,* 1986) or the detection of their toxins (Benhamou *et al.,* 1985a,b).

The monoclonal antibodies that have been raised against fungi (Hardham *et al.*, 1986; Dewey, 1988; Wong *et al.*, 1988; Dewey *et al.*, 1989a,b; Estradia-Garcia *et al.*, 1989), including seedborne fungi (Mitchell and Sutherland, 1986), form the basis of many effective diagnostics. **Several double antibody sandwich assays (DAS-ELISA)** based on monoclonal antibodies and polyclonal antisera have been developed as commercial kits for the detection of turf grass diseases by Miller *et al.* (1988, 1990) for Agri-Diagnostics.

1.4.4 Immunohistology

A sensitive indirect 'sandwich' method of immunolabelling a distinctive protein *in situ* can improve the light or electron microscopic examination of a pathogen in its host (Polak and Vardnell, 1984). In this technique, a suspension of MCA binds to the specific antigen epitope on the pathogen, then a second antibody against the first is added which is conjugated with a fluorochrome, enzyme substrate or heavy metal. If the **fluorochrome fluorescein isothiocyanate (FITC)** is used, a microscope with an expensive adaptation for fluorescence using ultraviolet light is required and only sections cut from fresh or frozen material can be used. When **peroxidase-antiperoxidase (PAP)** from horseradish is conjugated to the second antibody in the presence of hydrogen peroxide and **diaminobenzene (DAB),** it liberates a permanent brown polymeric pigment which stains the sections so that they can be counterstained. However, DAB is a suspected carcinogen, and safer techniques have now been developed by conjugating heavy metals (such as gold) to the specific monoclonal antibody, which produces a reddish appearance under the light microscope (e.g. Benhamou *et al.,* 1985a,b). The subsequent addition of silver enlarges the particles and at the same time makes them appear blacker and hence more distinct.

As the cell wall and tissues of plants prevent antibody probes from reaching the antigens of their pathogens, immunocytochemical techniques are less popular in plant pathology than in medical and animal pathology (Finan, 1984; Bullock and Petrusz, 1985-1989). Despite this, immunolabelling is now a routine procedure in the light and electron microscopy of plants (Pertot-Rechenmann and Gadal, 1986) as a variety of techniques for tissue preparation have been devised to overcome these constraints. It is far easier and more practical to squeeze out some sap from a plant to inspect for virus particles directly under the transmission electron microscope (TEM) than to go through the process of fixing, embedding and cutting sections. However, when virus particles are present in low numbers, this effective technique can be greatly enhanced (Barton, 1985) by immunosorbent electron microscopy (ISEM or IEM). In this procedure, an electron microscope grid coated with carbon is floated on a drop containing labelled antibodies which become strongly bound to it before it is placed on a drop of the sap of the host plant to be tested. Virus particles adsorbed to the antibodies can be clearly seen under a TEM as this method is at least 5000 times more sensitive than conventional transmission electron microscopy of thin sections of the same material.

1.4.5 Future potential for immunoassay

There is clearly considerable potential for the use of antibodies in plant disease epidemiology now that so many applications to detect plant pathogens are well established and will probably continue to expand considerably (Barnes, 1986; Klausner, 1987). Most of the present assays detect viruses but many more commercial assays are likely for diseases caused by both fungi and bacteria (including phytoplasmas). **Polyacrylamide gel electrophoresis (PAGE)** can be used to separate some different *Armillaria* species (Lung-Escarmant and Dunez, 1979). PAGE and Western blotting can be used to identify the proteins of many more different species of pathogens to specific antibodies. Any unique protein bands that are detected using PAGE may be cut out and used as immunogens to raise monoclonal antibodies. A greater selection of antibodies should help to clarify the differences between the epidemiology of closely related pathogens or monitor important subspecific variations (Fox *et al.,* 1989; Martinet *et al.,* 1992a). As a result, we may learn more about the evolution of pathogenic strains and interspecies competition.

In human medicine, toxins and other metabolites produced by pathogens and parasites are often detected in blood samples, particularly if they are antigenic. In infected plants, it is usually not possible to monitor such toxins as easily. Mycotoxins in food or feed are important, as tainted produce can cause illness or even death when consumed by humans or their animals. Many mycotoxins are simple non-antigenic chemicals, so a branch of diagnostics based on hapten technology has to be employed, in which the mycotoxin is bound to a known antigen (Klausner, 1987; Candlish *et al.,* 1989). This has resulted in routine tests that are simple to use *in situ.* Other antibodies produced by such techniques have also become important in crop protection (Klausner, 1987), where minute levels of pesticides may be similarly confirmed, also without the necessity for expensive laboratory equipment (Niewola *et al.,* 1983, 1985, 1986; Van Emmon *et al.,* 1986, 1987; Coxon *et al.,* 1988; Tomita *et al.,* 1988). These products may also have a role in plant disease epidemiology. It is possible that a number of diseases with deep-seated or otherwise difficult to expose pathogens, such as a number of the wilts could be detected by the presence of one of their characteristic metabolites.

In future, many more portable user-friendly ELISA test kits are likely to be developed in which the specific antibodies are bound to either inert granules or small dip-sticks so they can easily be used in the field (see also Chapter 2). Sample collection will probably be simplified, enabling even laboratory-based assays to be faster (Lange *et al.,* 1989). If monoclonal antibodies are used, initial purification of the antigen does not appear to be essential for diagnosis since it has been shown that pure specific monoclonal antibodies can be produced even when a complex of impure hyphal antigens is used as the immunogen. Synthetic peptides can also be used as vaccines (Robard, 1987). A monoclonal antibody raised to one synthetic peptide by Martin *et al.* (1992a) is able to determine whether tubulin in crude preparations from fungi is from strains resistant or susceptible to the fungicide carbendazim, based on only one different amino acid.

It may no longer be essential to inject laboratory animals with antigens each time a new antibody is required. Splenocytes from non-immunized mice can be stimulated to produce antibodies *in vitro.* It is also possible to substitute complete antibodies by fragments of antibodies (Fab or $F(ab')_2$ fragments) comprising only the heavy and light chain from one arm of a Y-shaped antibody molecule. The genes for the light and the heavy chains may be extracted from antibody-producing cells and multiplied by the **polymerase chain reaction (PCR)** to copy them many times over. These are stored in bacterial viruses to be expressed in bacteria. As a result, many more antibodies can be grown much more cheaply, quickly and easily in bacterial cultures and screened for specificity than if mouse hybridomas were used. If the antibodies were grown in crop plants in a similar way, antibodies could be produced at 0.0001 of the cost, which should allow antibodies to be mass-produced cheaply enough to be used very widely in plant disease epidemiology, diagnosis and control as well as other far-reaching applications.

Serological tests are restricted by the relative accessibility of the epitopes on the proteins and polysaccharides. However, although many are inaccessible, there are sufficient epitopes for the development of a wide range of immunological diagnostic tests. Others can be created by hapten technology. Since many of these tests are considered satisfactory, they are currently more widely used for routine practical diagnosis than methods based on recognizing the similarities between nucleic acids.

According to Robinson (1988), these established immunological tests are not going to be readily supplanted.

1.5 METHODS BASED ON THE NUCLEIC ACIDS OF PATHOGENS

Many diagnostic methods based on the detection of similarities between nucleic acids have been successfully adapted for identification of pathogens. Most of these techniques depend on the use of restriction enzymes to cleave DNA into fragments at or near a defined recognition sequence. This allows the enzymes to be used in the differentiation of non-clonal organisms, as each inherits a unique distribution of restriction sites.

In DNA **restriction fragment polymorphisms (RFLPs**), a combination of hybridization, Southern blot and restriction mapping forms the basis for the genetic analysis and characterization of pathogens (Botstein *et al.,* 1980). After the DNA has been digested into fragments, these are separated by agarose gel electrophoresis and then stained with ethidium bromide, which can be seen under UV light. The separated bands that result represent a genomic 'fingerprint' which can be analysed in a way that is somewhat analagous to 'reading' a 'bar chart code'. Although their chief advantage over antibodies is that they can detect any part of the genome of the pathogen, however large, nucleic acid techniques are more common for viral

(Robinson, 1988) rather than fungal or bacterial plant pathogens as the banding pattern becomes more complex as the genome becomes larger.

These techniques have been particularly popular for detecting unidentified viruses or viroids in plants. This is important where plants are propagated vegetatively or by seed (e.g. Robinson, 1988); unexpected pathogens cannot be detected by specific antibodies as these must be raised in advance to familiar pathogens. Methods based on nucleic acid differences are valuable as only these techniques can be used to detect if any alien nucleic acid is present in plants that are being held in quarantine. Discovering the presence of a foreign nucleic acid in addition to that of the host plant can give warning of infection by any pathogen, albeit of some unknown or symptomless disease. This diagnosis includes viroids that lack a protein coat and so cannot be detected by antibodies. In this case, the identification of the pathogen is of secondary importance since it is essential that the diagnostic test should not be specific to any particular disease agent but be capable of detecting all types of invading organisms. By constructing nucleic acid probes with various breadths of specificity to pathogens, it is therefore possible to make probes that are able to sort out the genetic relationships between them. Although a well chosen collection of antibodies may also be capable of this, it is typically achieved only by the chance selection of an appropriate antibody – hence success is generally more elusive and laborious. For this reason, while it is often not feasible to select an antibody to detect a strain, a suitable individual nucleic acid probe is often capable of detecting a range of strains (Robinson *et al.*, 1987; Robinson and Legorburu, 1988). Although methods based on differences in nucleic acid have so far been less widely used in epidemiological studies than immunological techniques, they have proved their value for determining the extent to which genets of *Armillaria* spp. have spread and vegetatively mutated (Smith *et al.*, 1992).

A multitude of methods to compare the genetic codes that were mostly pioneered in other areas of biology and medicine have now been investigated for the diagnosis of plant pathogens. Despite procedures as varied as melting nucleic acid to determine its mean content (Motta *et al.*, 1986) and restriction site mapping (Anderson *et al.*, 1987), the most widely used techniques adopted involve the detection of RFLP in the nucleic acid of pathogens by hybridization, versions of which have become widely used to 'fingerprint' minor differences in genetic polymorphisms in human populations.

Nucleic acid hybridization, or reassociation, is a process by which complementary single-stranded (ss) nucleic acid anneals to double-stranded (ds) nucleic acid. This process occurs because of the hydrogen bonds between the two strands of the DNA:DNA, DNA:RNA or RNA:RNA duplex and base stacking within the strands. The subsequent avidity and the energy released during hybridization ensure the stability of the hybrids (Tijssen, 1993). The homology of complementary base-pairs ensures that pairing occurs with great specificity as adenine (A) pairs only with thymine (T) – or uracil (U) in RNA – and guanine (G) with cytosine (C). For most conventional hybridization tests, sensitivity of detection is comparable with ELISA (Maule *et al.*, 1983). A 17 base-pair (bp) oligonucleotide probe has been shown to detect a single gene in a genome of 3×10^9 (Berent *et al.*, 1985). Even gene mutations of a single base-pair can be detected in pathogenic fungi (Martin *et al.*,

1992b). Hybridization is invaluable as a diagnostic tool since a selective probe may detect less of the pathogen's nucleic acid than is present in the inoculum necessary to infect the host. Universally conserved nucleic acid molecules such as ribosomal RNA (rRNA) may be adopted because of their wide recognition spectra.

The potential of nucleic acid hybridization to determine the genetic and hence the taxonomic relatedness of organisms was recognized by Britten and Kohne (1968). The conditions that favour the reassociation of single-stranded DNA (ssDNA) include an adequate concentration of cations, a sufficiently high temperature to weaken the intrastrand secondary structure, sufficiently long incubation time and high concentration of DNA to permit an adequate number of collisions and hence the opportunities to reassociate. The size of the fragments is also important as DNA sheared into small fragments favours reassociation. If suitable fragments are used, hybridization is now regarded as the conventional method for determining subgeneric taxonomic relationships to the species level in bacteria (Schleifer and Stackerbrandt, 1983).

Hybridization now generally takes place in a solid rather than liquid phase. The target nucleic acid from the pathogen to be detected is immobilized on a nitrocellulose or nylon membrane. DNA is first cut into fragments by a restriction enzyme, often following gel electrophoresis and denaturation (Southern, 1975). RNA is usually transferred by a similar technique known as Northern blotting (Alwine *et al.,* 1977). After it has been blotted onto the support, the target nucleic acid is fixed by baking or cross-linking under UV light. The labelled probe hybridizes to the target DNA. After the excess unbound probe has been removed by washing, the hybrid (target:probe) is detected by a suitable assay.

Most nucleic acid probes for detecting plant pathogens, particularly viruses, are usually used in routine spot hybridization or 'dot-blot' tests. In many dot-blot diagnoses, the DNA is neither cut by restriction enzymes nor fractioned but merely applied as a small drop of sap extracted from the infected plant directly onto a nitrocellulose sheet, dried, then hybridized with the probe in a sealed plastic bag using a buffer extract (Maule *et al.,* 1983). The nucleic acid is loaded into a multisample vacuum manifold which is used to spot it onto the membrane. This boosts sensitivity by increasing the volume of sample that is able to pass through the area of membrane under the 'dot' until the pores of the membrane become clogged with material from the plant sap (Robinson, 1988). However, this dot-blot method may be modified further by lysing cells of the target microorganism *in situ* on the membrane with alkalis or enzymes, using the colony hybridization method (Grunstein and Hogness, 1975). It is also possible to hybridize probes directly to infected plant material, soil or water in order to avoid the need to isolate and thus include obligate pathogens, and to preserve a clearer idea of their initial abundance and genetic integrity, as well as saving time. These detection techniques have had to be modified in order to allow the extraction of sufficient pure DNA to be freed from inhibition by organic matter, clay and other materials (Steffan *et al.,* 1988; Rasmussen and Reeves, 1992).

At present, routine hybridizations are not as easy as an immunological method even though prior extraction of nucleic acid from the test sample may be avoided in some cases. In the laboratory, molecular hybridization can be used after the separation of RNA and restriction fragments of DNA by gel electrophoresis in order to detect specific sequences among the bands by Northern and Southern blotting in an analogous way in which antibodies are used in Western blotting.

For a long time, since most hybridization was done with radioactive probes and filter-bound nucleic acids, it was reserved for specialists. Because of the development of non-radioactive methods it is now considered safer, less troublesome but still rather time-consuming to perform. Many current commercial nucleic-based tests employ the sandwich assay (Sylvanen *et al.,* 1986), in which adjacent sections of the target nucleic acid are hybridized by two probes. One is the capture probe which, by hybridization, links the target to the solid support to which this probe is already bound. The other probe is labelled with a reporter group, which may be a fluorescent molecule, an enzyme or a radioactive atom that hybridizes with an adjacent section of the target nucleic acid. As the reporter group must be readily distinguished even in minute amounts, fluorescent molecules have many advantages. However, there are some disadvantages since many other molecules in biological samples, including several components of healthy plants, can fluoresce without being labelled. Serious background fluorescence can often compete with the weaker signals from the probe, thus limiting sensitivity to unsatisfactory levels.

Enzyme-labelled probes can be used to generate coloured molecules which are then readily detected. In spite of this, thorough washing is imperative to remove impurities derived from the hybridization reaction as they inhibit the enzyme. Probes attached to biotin (a vitamin) bind to avidin (a protein) extremely tightly. As a result, a more effective complex of many enzyme molecules can be produced by binding enzymes first to biotin and then to avidin, around which they cluster. Digoxigenin may be used in a similar way with anti-digoxigenin enzymes. There are several other ways of enhancing detection systems, employing complexes of enzymes or luminescent reactions. It is also possible to fasten more than one reporter group to the labelled probe by making the probe very long or branched. In these ways, non-radioactive methods are starting to attain a sensitivity equivalent to that expected from radioactive labels. Improvement in the automation of hybridization is another beneficial development. The procedures for hybridization developed for these 'cold' non-radioactive probes are similar to those used for the 'hot' radioactive probes.

Probes of DNA are more widely used than those of RNA. In addition to DNA viruses, DNA probes can be used to detect RNA viruses by reverse transcription to produce a complementary DNA (cDNA) copy (Hull, 1986). Generally, probes consist of a few kilobases (kb) of DNA but may be as short as 15-20 bp or as large as 30 kb. Most often, the probes for a specific purpose are screened from randomly cloned fragments of DNA. Even a genomic library covering the whole pathogen can eventually be built up from such fragments. However, in most cases only a few specific probes need to be selected as diagnostics. This selection process can be simplified by subtractive hybridization or genomic subtraction. The non-specific DNA that is common between the pathogen and another organism, such as its host, is depleted while that specific to the pathogen only is enriched. This is done by hybridizing labelled DNA from the target organism – the pathogen – with an excess (100-200 times the amount) of DNA from the driver, which can be a related organism or the host. The driver DNA mops up the target DNA common to it, leaving behind the specific sequences to be recovered (Avery *et al.,* 1980; Smith *et al.,* 1987; Cook and Sequeira, 1991; Seal *et al.,* 1992). Genomic subtraction has been used to detect the gene for the distinctive toxin, phaseolotoxin, and hence specifically diagnose the bacterium *Pseudomonas syringae* pv. *phaseolicola* itself (Schaad *et al.,* 1989).

Often the genetic information responsible for the characteristic pathogenicity of a bacterial pathovar resides on its indigenous extrachromosomal plasmids, which are stable enough to be used as natural genetic markers (Lazo and Gabriel, 1987). Probes may also be developed from chemically synthesized oligonucleotides. Usually, these are less than 40 nucleotide sequences and can be sufficiently specific to detect point mutations and single base-pair changes (Wallace *et al.,* 1979) (Fig. 1.1). Such oligonucleotide probes can easily be synthesized cheaply and labelled non-radioactively – but only when the relevant genetic code has been characterized. They have been constructed for use as primers for amplification by the polymerase chain reaction (e.g. Martin *et al.,* 1992b).

The polymerase chain reaction (PCR) is an *in vitro* method in which a DNA fragment with known end sequences can be amplified exponentially into billions of copies, making detection very much easier (Saiki *et al.,* 1985; Mullis and Faloona, 1987). Since an initially very low number of DNA molecules can be multiplied enormously, PCR has now been widely used in plant pathology as well as numerous other areas, including criminal forensic medicine. Only a few nanograms of the initial template DNA is necessary, either in the form of a discrete molecule or as part of a larger one. Both dsDNA and ssDNA can be amplified by PCR. It is possible to amplify RNA by reverse transcription into a cDNA copy by RT-PCR (Sambrook *et al.,* 1989). Synthetic oligonucleotides (primers) can be produced that are complementary to the end sequences and can hybridize with them. The three stages in amplification are not complicated. First, the target DNA is melted, then the two oligonucleotide primers are annealed to the denatured DNA strands and finally the primer is extended, using the thermostable enzyme *Taq* polymerase. Since the PCR process does not depend on the use of purified DNA, the pathogen does not have to be cultured before its DNA is amplified but any substances that could inhibit the PCR are easily removed by purification (Steffan and Atlas, 1991). As the choice of primers determines whether the selectivity of the PCR is either specific or broad, this flexibility has proved especially useful in disease detection (Henson and French, 1993). The major practical problem experienced with PCR is that contamination from extraneous sources can cause false positive results, so both positive and negative controls are essential. A 'hot start' is also often adopted in which the reagents are heated to avoid premature misreactions (Chou *et al.,* 1992). By introducing a second round of amplification, nested primers can be used to allow a set of outside primers to amplify a DNA fragment (Henson *et al.,* 1993), but this is an expensive option. In addition to PCR, the ligase chain reaction is another technique that is sensitive enough to detect single base-pair differences (Wiedmann *et al.,* 1992).

Inverse PCR is used to analyse unknown sequences that flank a region that is known (Triglia *et al.,* 1988). In competitive PCR, the target DNA is co-amplified in the presence of known quantities of a competitive DNA (Gilliland *et al.,* 1990a,b), thus enabling the amount of target DNA in the original sample to be determined by titration.

In random amplified polymorphic DNA (RAPD), short oligonucleotides of about 10 oligomers are used as primers which should be complementary to sequences in the target genome (Williams *et al.*, 1990). Although no prior knowledge of the actual sequences in the target genome is necessary, the pathogens must be purified – for example, by culturing – before fingerprinting (Crowhurst *et al.*, 1991; Guthrie *et al.*, 1992; Welsh *et al.*, 1992). Therefore, this method is unsuitable for the direct detection of pathogens *in situ* in plants or soil. This procedure has also been termed arbitarily primed PCR (AP-PCR) by Welsh and McClelland (1990) and produces relatively simple fingerprints. DNA amplification fingerprinting (DAF) uses DNA silver stains and polyacrylamide gel electrophoresis (PAGE) to produce more complex fingerprints (Cactano-Anolles *et al.*, 1991).

Inter Simple Sequence Repeats (ISSRs) are an anonymous, RAPDs-like approach that access variation around the numerous microsatellite regions dispersed throughout almost all genomes (Zietkiewicz *et al.*, 1994). Simple sequence repeats (SSRs) (Tautz, 1989), also known as microsatellites (Litt and Luty, 1989) are repeats of 1-6 nucleotides that are distributed ubiquitously in the genomes of higher organisms (Toth *et al.*, 2000). Almost synonymous terms are variable number of tandem repeats (VNTR) or short tandem repeats (STR) (Edwards *et al.*, 1991). They can be found anywhere in the genome including protein coding and non-coding regions and are characterized by mono-, di- or trinucleotide repeats, eg., GG..., or GC..., AGT..., repeat units side by side. ISSR is a PCR-based method that involves amplification of DNA present between such repeats (hence the name inter-SSR). Mononucleotide repeats are characteristic of the chloroplast genome whereas di- and trinucleotide repeats are characteristic of the nuclear genome and hence ISSRs specifically target the di- and trinucleotide repeats of microsatellites.

Typically a single primer consisting of SSRs plus a 1-3 base pair randomly selected anchor is used for the generation of a multiband pattern on gel. Di-nucleotide, tri-nucleotide, tetra-nucleotide and penta-nucleotide repeats have been used as primers. The primers can be anchored or non-anchored. Because the length of the primers used is long enough (16-25 bp), ISSRs have high reproducibility as compared with RAPD markers (Zietkiewicz *et al.*, 1994). In fact, these markers combine the universality of RAPD with benefits of SSRs and AFLP (Reddy *et al.*, 2002). When the primer successfully locates two microsatellite regions within an amplifiable distance away on the DNA sample, the PCR reaction will generate a band of particular molecular size for that locus. Usually several such paired microsatellite regions exist in the nuclear genome, thereby resulting in many bands of different molecular sizes. Studies of ISSR locus heritability have demonstrated an exceedingly close approximation to classic Mendelian ratios (Tsumura *et al.*, 1996). The existence of high variability and high mapping density when compared to RAPD and RFLP data make these new dominant molecular markers ideal for producing genetic maps of individual species (Nagaoka and Ogihara, 1997). Microsatellites and ISSR markers are useful in species identification (Niesters *et al.*, 1993; Graser *et al.*, 1998) and also in the characterization of fungal strains due to their high levels of polymorphism (Schonian *et al.*, 1993; Longato and Bonfante,

1997). While RAPD analyses can provide good discrimination within and between species, it is generally not considered a robust enough technique (e.g. for comparisons at different times and in different laboratories) and has been slowly abandoned due to its poor reproducibility. Since ISSR primers are longer than RAPD primers, higher annealing temperatures can be used, which result in stable and reproducible markers (Han *et al.,* 2002).

ISSR markers are generally detected on agarose gels with ethidium bromide staining. However, detection on acrylamide gels with silver-staining or radioactive detection increases sensitivity. Recently, the introduction of fluorescently labelled primers combined with fragment analysis on DNA sequencers has increased sensitivity and throughput of these markers tremendously (Yasodha *et al.,* 2004). Fragments detected are most commonly scored as biallelic dominant markers (presence/absence) (Gupta *et al.,* 1994; Tsumura *et al.,* 1996) but co-dominant segregation has been shown in some cases (Wang *et al.,* 1998b; Sankar and Moore, 2001).

ISSR markers have been used to address a variety of issues related to genomic fingerprinting (Charters and Wilkinson, 2000), assessment of genetic diversity and phylogeny analysis (Tsumura *et al.,* 1996; Assefa *et al.,* 2003; Bornet *et al.,* 2004), detection of somaclonal variation (Albani and Wilkinson, 1998), genome mapping (Arcade *et al.,* 2000), determining frequency of microsatellites in the genome (Yasodha *et al.,* 2004), gene tagging and marker-assisted selection (Akagi *et al.,* 1996) and evolutionary biology (Wolfe *et al.,* 1998). The ISSR-based fingerprinting technique has been used successfully to determine genetic variability within various fungal species (Hantula *et al.,* 1996; Zhou *et al.,* 2001; Grunig *et al.,* 2001, 2002; Rodrigues *et al.,* 2004).

Using ISSR analysis, considerable variability was observed among isolates from different geographical locations and hosts in different fungal populations (Han *et al.,* 2002; Mishra *et al.,* 2003). These markers will also enable a greater understanding of the movement of economically important pathogens around the world. ISSR fingerprinting was used to differentiate *Botryosphaeria* species and related anamorphic fungi indicating its powerfulness in examining species aggregates or complexes, justifying newly identified species, and differentiating closely related fungi with very similar morphology and ITS sequences (Zhou *et al.,* 2001). Successful discrimination between *Pisolithus albus* and *P. microcarpus* was achieved by deploying ISSR markers (Hitchcock *et al.,* 2003).

DNA sequencing gained its importance in pathogen diagnostics by the revolutionary discovery of the DNA structure by Watson and Crick (1953). Since then the DNA-based techniques paved their way into the discipline of fungal, bacterial and viral systematics and proved capable of resolving the problems related to classification, detection and discrimination between and within the closely related genera and species. DNA sequence techniques have an added advantage of sensitivity. This sometimes cannot be obtained and is far beyond the range of detection by traditional methods (Mills *et al.,* 1992; O'Donnell, 1992; Chen *et al.,* 1992; Levy *et al.,* 1991). A better understanding of disease diagnosis, host-pathogen interaction and disease management can only be achieved with precise definition of the pathogen. The first stage in development of a diagnostic assay is to select the nucleic acid sequences to be used to identify the organism. A method for identifying the presence of the target sequences in the sample also needs to be concentrated to achieve successful diagnosis of the pathogen. Identification of DNA sequences can either be achieved by developing a method that can exploit the variations present in the unknown conserved genes or by screening random parts of the fungal genome that exhibit the required specificity (Ward *et al.,* 2004).

The **ribosomal DNA (rDNA)** operon is mainly targeted for development of diagnostic markers because of their presence in all organisms at high copy number that allow very sensitive detection. The ribosomal DNA operon comprises three functionally and evolutionary conserved genes, the large subunit gene, the small subunit gene and the 5.8S gene interspaced with a variable spacer region called the internal transcribed spacer (ITS) in a unit, which is repeated many times in the nuclear genome (Schmidt, 1994). Some of the rDNA regions are well conserved throughout the evolutionary process and other parts are variable even within a species. Based on these variations, multiple strategies have been adopted to characterize pathogens (Louws *et al.,* 1999). The conserved regions allow probes or primers (often referred to as universal primers) from one species to be used to detect and amplify ribosomal gene fragments from a broad range of phylogenetically diverse genera, families or even kingdoms (McCartney *et al.,* 2003). However, the equivalent rDNA gene fragments detected in different organisms have sequence variations that can be exploited for the subsequent selection of pathogen-specific primer sequences allowing the identification of the pathogen in question. rDNA sequences of different organisms available in publicly-available databases are far greater than any other region of the genome (Ward *et al.,* 2004).

The **internal transcribed spacer (ITS) region** that separates repeat units within tandem arrays of the nuclear DNA (nrDNA) genes are ubiquitous in nature and found in all eukaryotes. The ITS region is a part of the transcriptional unit of nrDNA and in part it has been demonstrated to reflect some functional value (Morales *et al.,* 1993). The ITS region has been used extensively in fungal taxonomy (Seifert *et al.,* 1995) and is known to show variation between and within species (Nazar *et al.,* 1991; O'Donnell, 1992). Species-specific diversity within the ITS region has been demonstrated across a variety of fungi. Mugnier (1998) has used the ITS and 5.8S sequences to infer the relationships between the classes within the division *Euascomycotina*. Studies of ITS regions from powdery mildews (Mori *et al.,* 2000), *Aspergillus* (Henry *et al.,* 2000), *Alternaria* (Pryor and Gilbertson, 2000); *Pythium* (Paul, 2000); *Verticillium* (Zare, 2000); *Tilletia tritici* (Josefsen and Christiansen, 2002); *Fusarium culmorum* (Mishra *et al.,* 2002); *Phytophthora* (Blomquist and Kubisiak, 2003) and *Phellinus* (Fischer and Binder, 2004) have revealed significant sequence divergence at the interspecific level indicating the potential of using the ITS region for the identification of pathogens at species level.

Other genes are also becoming more widely used as targets for diagnostic development and pathogen characterization. The beta-tubulin genes are widely used as second common targets for diagnostic development in fungi (McCartney *et al.,* 2003) and have been extensively studied by Hirsch *et al.* (2000, 2002), Schroeder *et al.* (2002) and Drogemuller *et al.* (2004). Apart from the ITS and beta-tubulin region, regions pertaining to the endochitinase gene (Lieckfeldt *et al.*, 1999, 2000), transcription elongation factor 1 (Bissett *et al.,* 2003), mating-type genes (Dyer *et al.,* 2001; Foster *et al.,* 2002) and mitochondrial small subunit rDNA (Smith *et al.,* 1996; Sirven *et al.,* 2002; Martin *et al.,* 2004) have also been deployed for detection and differentiation of strains and or species.

In bacteria, genus-specific rDNA sequences have been well documented by Deparasis and Roth (1990) and primers have been developed for pseudomonads, xanthomonads and most notably for phytoplasmas (Louws *et al.,* 1999). Specific primer pairs were designed for the detection of *Pseudomonas* (Widmer *et al.,* 1998; Alvarez, 2004), *Xanthomonas* (Maes, 1993) and *Ralstonia solanacearum* (Seal *et al.,* 1993) based on sequence variations present in 16S and 23S rDNA regions. Genes associated with pathogenicity (Prosen *et al.,* 1993; Manulis *et al.,* 2002) and those present on plasmid DNA (Bereswill *et al.,* 1992; Audy *et al.,* 1994) have also been used to design specific PCR assays.

The introduction of **real-time PCR** in the last few years has improved and simplified methods for PCR-based quantification. Real-time PCR has many advantages over conventional PCR: it is faster and higher throughput is possible, post-reaction processing is not required as the amplified products are detected by a built-in fluorimeter thereby eliminating the risk of contamination, and it is more specific than conventional PCR if a specific probe is used in addition to the two specific primers. Real-time PCR involves the detection and measurement of amplification products at each cycle of amplification. Over last few years many systems have been developed for performing amplification and detection of nucleic acids in a single tube. These can be classified into non-specific detection systems and specific detection systems.

The standard method for non-specific detection is using a DNA binding dye that fluoresces after binding. The most commonly used dye is SYBR® green I (Morrison *et al.,* 1998); this is excited at 497 nm and emits at 520 nm. When in solution, the dye exhibits no fluorescence. However, as PCR products accumulate during the amplification process, increasing amounts of dye bind to the double stranded DNA. So in each PCR cycle, the dye fluoresces during the elongation step as it binds to the DNA and then during the denaturation step, it falls off leading to a decrease in fluorescence. In this way, when monitored in real time at the end of each cycle, measurement of SYBR green-borne fluorescence can indicate an accumulation of the PCR product as the reaction proceeds. Use of SYBR green eliminates the need for target-specific fluorescent probes as specificity is entirely determined by the primers. SYBR green will bind to any double stranded DNA present in the reaction including primer dimmers; this makes it very important for the primers to be very specific and to generate a single product (Ririe *et al.,* 1997). There are many other detection methods available that will detect only a specific fragment. These methods include TaqMan® probes, double dye oligonucleotide probes, molecular beacon probes, Scorpions primers, hybridization probes, TaqMan® MGB® probes, MGB Eclipse probes and many others. As the use of qPCR has grown since its development in 1996, better and more advanced instruments capable of more accurate quantification and handling more samples in a given time have been developed. Some of the more popular machines are ABI PRISM® 7000/7700, ABI PRISM® 7900HT, Roche LightCycler® and Bio-rad iCycler iQ®.

Quantification of a real-time PCR product revolves around the concept of Ct values (Higuchi *et al.,* 1993). Fluorescence values recorded during each cycle represent the amount of product amplified until that point in the PCR reaction. The template DNA present at the start of the reaction is directly proportional to the cycles for fluorescence and is first recorded as being statistically significant above the baseline (Gibson *et al.,* 1996). This fractional cycle number value is known as the Ct value and always occurs in the exponential phase of the reaction. The two most widely used methods for quantifying an unknown sample using Ct values obtained from an amplification plot (Giulietti *et al.,* 2001) are the standard curve method and the comparative threshold method. In the former, a standard curve is prepared by plotting Ct values against the log of template DNA. Depending on the template used, an absolute or relative standard curve can be prepared. Because of the complication associated with an absolute standard curve, a relative standard curve has often been the method of choice. In a relative standard curve, a dilution series of one of the samples being compared is usually used as the template. Quantities of target 'unknown' samples are extrapolated from the standard curve and expressed in terms of arbitrary units. In the comparative threshold method, only relative quantification can be performed. In this method, arithmetic formulae are used to calculate relative expression levels; these are then compared with a calibrator, which can be one of the samples being compared, for example, the untreated control.

By using probes with different fluorescent reporter dyes, amplification of two or more PCR products representing different organisms, polymorphisms or SNPs can be detected in a single PCR reaction (McCartney *et al.,* 2003). Real-time PCR has been deployed for the detection of human pathogens, scanning SNPs associated with human diseases and resistance to anti-microbial agents, and detecting genetically modified foods and other agricultural products. This technique is not used routinely in plant pathogen detection assays; however certain detection assays for research purposes have been developed for fungi (Bohm *et al.,* 1999; Fraaije *et al.,* 2001, 2002), bacteria (Schaad *et al.,* 1999; Schaad and Frederick, 2002), viroids (Mumford *et al.,* 2000) and viruses (Boonham *et al.,* 2000).

Different molecular marker systems have become available during the last two decades to deal with all aspects of molecular biology applications, such as crop improvement, genotyping and diagnostics of pathogens. In recent years, there has been an emphasis on the development of newer and more efficient molecular marker systems involving inexpensive non-gel-based assays with high throughput detection systems. Availability of **Single Nucleotide Polymorphisms (SNPs)** is one such development of new generation molecular markers used for individual genotyping. RFLPs, RAPDs and SSRs which were the markers of choice during the last two decades, need gel-based assays and are, therefore, time-consuming and expensive. SNPs, in contrast, represent sites where the DNA sequence differs by a single base and can be detected by different non-gel-based methods (Gupta *et al.,* 2001). Single nucleotide polymorphism has been shown to be the most abundant, so that at least one million SNPs should be available, only in the non-repetitive transcribed region of the human genome (Wang *et al.,* 1998a). According to recent estimates, one SNP is known to occur in every 100-300 bp in any genome, much higher in order of magnitude than that of SSRs, thus making SNPs the most abundant molecular markers known so far (Gupta *et al.,* 2001). SNPs, at a particular site in a DNA molecule should in principle involve four possible nucleotides, but in actual practice only two of these four possibilities have been observed at a specific site in the population (Brookes, 1999). Consequently, SNP is a biallelic marker system as against the polyallelic marker systems such as RFLPs and microsatellites.

Identification of SNPs within a genetic locus can be achieved by using different methods. Prior to the detection of polymorphism, the sequence of the locus for a reference genotype needs to be determined. Once determined, this sequence can be used in designing oligonucleotide primers for use in the PCR, which forms the cornerstone of all subsequent SNP-based technology (Erlich, 1989). **Direct sequencing** is the most common and direct way of identifying SNPs; however, it is also the most time-consuming and expensive method. Sequencing errors are also a restrain and if not detected in early stages, would result in a considerable waste of resources in both designing allele-specific nucleotides and carrying out SNP assays. **Single-strand conformation polymorphism (SSCP)** uses variation in the mobility of small polymorphic single-stranded DNA fragments in non-denaturing acrylamide gels (Jordan *et al.,* 1998). SSCP is based on the assumption that the mobility of any DNA fragment will vary based on the exact sequence, and even a single base change can modify the mobility. This technique works well with small fragments (100-400 bp) that can be generated by PCR. Any change in mobility across the genotype would indicate a sequence change that could be targeted by direct sequencing. Etscheid and Riesner (1998) proposed several modifications and improvements to the SSCP technique and at present temperature gradient gel electrophoresis and denaturing gradient gel electrophoresis are the most widely used for SNP detection. **Chemical cleavage of mismatches (CCM)** relies on the ability of certain chemicals like piperidine (Prosser, 1993), to cleave a chemically modified heteroduplex precisely at mismatched bases. This technique involves a reference sample that is labelled with radioactivity and a sample under question. Both samples can be generated by PCR and after denaturation, the two samples are brought together and allowed to anneal to form homoduplexes and heteroduplexes. When a heteroduplex with sequence difference has been formed, it is chemically modified using osmium tetraoxide and then subjected to peperdine treatment that cleaves the mismatch. The cleaved and uncleaved products are then run on a denaturing polyacrylamide gel and autoradiographed. CCM is capable of scanning regions of up to 3 kb with cent per cent accuracy despite the disadvantages of being time-consuming and the handling of hazardous chemicals. To overcome the latter, a technique called **Enzyme mismatch cleavage (EMC)** has been introduced for SNP identification. EMC is very similar in principle to CCM except osmium tetraoxide is replaced by the enzyme 'resolvase'. Similar to CCM, the amplified DNA from either an individual plant or two individual plants is heat-denatured and allowed to re-anneal to form homo- and heteroduplexes which are then labelled with radioactivity. Any heteroduplex structures present are then subjected to resolvase enzyme treatment that cleaves the mismatch; this is then examined on a denaturing polyacrylamide gel. EMC has been found to be more efficient in scanning for SNPs (several kilo bases of DNA) and is gaining importance in SNP identification (Edwards and Mogg, 2001). All the methods listed above are used for the fingerprinting of two or more uncharacterized genotypes, but for commercial purposes **DNA chips** are used in SNP identification. A DNA chip is suitable for scoring several SNPs in parallel from each sample in a multiplexed fashion. DNA chips have immobilized oligonucleotides of known sequences that differ at specific sites of individual nucleotides and are used for SNP detection. This makes use of the technique 'sequencing by hybridization' and involves tiling strategy. Four oligonucloetides in a column of an array will differ only at the SNP site and only one will be fully homologous. When such an array is hybridized with biotinylated PCR product, the perfect match will allow binding and mismatched products will be washed away. The perfect match in each case can be detected by using this detection system (Gupta *et al.,* 2001). SNPs can also be detected by **minisequencing** using Sanger's dideoxynucleotide method, where the oligonucleotide primer has a sequence one or more bases upstream of the SNP site. A mixture of all the four dNTPs and one of the four possible ddNTPs that corresponds to the SNP locus, is used for primer extension, so that the incorporation of a single ddNTP at the SNP site will terminate the reaction and will allow detection of the SNP. A new sequencing method called **Pyrosequencing** has been developed to obtain short DNA sequences for SNP detection. This method relies on the step-wise addition of individual dNTP (with simultaneous release of pyrophosphates) and monitoring their template guided incorporation into the growing DNA chain via chemiluminescent detection of the formation of pyrophosphates (Ronaghi and Elahi, 2002). The unincorporated dNTP will be degraded using the enzyme apyrase. The pyrophosphate released is utilized to convert 5' amino phosphosulfurate into ATP with the help ATP sulfurylase, and the ATP produced drives luciferase-mediated conversion of luciferin into oxyluciferin, generating light. The light produced is proportionate to the ATP released, which in turn is directly proportional to dNTP consumption. The emitted light is detected by a charge coupled device camera and seen in a programme as a peak; whole height will reveal the number of dNTP molecules consumed. Pyrosequencing is particularly suitable for SNP genotyping as genotyping of previously identified SNPs by this method requires sequencing of only a few nucleotides (Gupta *et al.,* 2001).

Kroon *et al.* (2004) distinguished isolates of *Phytophthora ramorum* from Europe and America based on a point mutation in the mitochondrial *Cox* gene by developing a single nucleotide polymorphism-based method in which a multiple copy marker was amplified in *P. ramorum* isolates; this was later digested with the restriction enzyme *Apo*I. SNP in the *cytochrome b* gene has been found to confer resistance to strobilurin fungicides in *Blumeria* (*Erysiphe*) *graminis* f.sp. *tritici.* Based on this point mutation, three different types of molecular markers have been developed by Baumler *et al.* (2003) to score resistant and sensitive isolates from specifically selected regional populations across Europe. The results of molecular tests were in complete concordance to those revealed by *in vivo* tests, indicating the potential of SNP applications on a large scale.

DNA microarrays offer the latest technological advancement for multi-gene detection and diagnostics (Call, 2001). DNA microarrays were first described by Schena *et al.* (1995) for the simultaneous analyses of large-scale gene expressions by a large number of genes (Schena *et al.,* 1996; Lochhart *et al.,* 1996; DeRisi *et al.,*

1996; Wang *et al.,* 2002), but have also been applied to DNA sequence analysis (Pease *et al.,* 1994), cancer biology (Khan *et al.,* 1998; Perou *et al.,* 1999; Alizadeh *et al.,* 2000), immunology (Heller *et al.,* 1997), genotyping and diagnostics (Yershov *et al.,* 1996; Drmanac *et al.,* 1998; Hadidi *et al.,* 2004). The flexibility and high throughput capabilities of DNA microarrays have tremendous potential for pathogen detection, identification, and genotyping in molecular diagnostic laboratories.

DNA microarrays are typically composed of DNA probes that are bound to a solid substrate such as glass. Probes used for genotyping microarrays are either PCR products from cloned genes or oligonucloetide DNA. Each spot in the array lattice is composed of many identical probes that are complementary to the gene of interest. During hybridization, DNA diffuses passively across the glass surface, when sequences complementary to the probe will anneal and form a DNA duplex. Hybridized targets are then detected using reporter molecule systems. Biotin, Cy-3 and Cy-5 are generally used as reporter molecules for genotyping arrays and are incorporated directly into probes via 5-prime primer modifications. Multiple reporter molecules can be incorporated into genomic DNA using nick translation, random priming, or chemical incorporation. Regardless of the system being used, detection of hybridized targets is usually done by fluorescent scanner, or by phosphoimager.

The ability to accurately detect and identify microorganisms that are capable of causing a severe impact on production is becoming increasingly important for successful control of the pathogen. DNA microarrays are being used in the detection of multiple pathogens based on the differences in 16S rDNA sequences (Call, 2001). Due to high throughput, this technology has a potential for successful diagnosis and is extensively applied in detecting several human pathogens (Wang *et al.,* 2002). *Escherichia coli* enterohemorrhagic bacteria were detected in laboratory and environmental samples with precision using DNA microarrays (Vora *et al.,* 2004). Microarray technology is now gaining entrance into plant pathogen diagnostics and has been deployed in detecting several plant viruses (DeYong *et al.,* 2005), viroids (Hadidi *et al.,* 2004), bacteria (Call *et al.,* 2003) and also to study host-microbe interactions (Cummings and Relman, 2000).

The new molecular biology techniques that have been developed for rapid, accurate and sensitive detection of plant pathogens need to be used extensively in routine practices. This can be achieved by generating less expensive and hand-driven portable equipment that can be directly used for pathogen detection at the field level. The application of novel diagnostic methods to detect pathogen inoculum will allow progress to be made in our understanding of the disease dynamics that will finally result in precise forecasting and management. Use of SNPs and microarray technologies can enhance the efficiency of sampling strategies and have great potential for developing new diagnostic tools. Long-term goals of functional genomics and microarray technology need to be focused on describing host-pathogen interactions and identifying critical pathways for the diagnostics of economically important pathogens. Novel approaches using biochips and sensor arrays to detect a quarantined pathogen in a large-volume sample, such as an entire seed lot, have interesting prospects for the near future.

1.6 FUTURE TRENDS IN DIAGNOSIS

Since there is no immediate prospect of any exclusive or reliably simple method of rapidly identifying plant pathogens or the diseases that they cause, most of the diagnostic methods currently used in plant disease epidemiology will probably continue to coexist and be used to some degree in various forms well into the foreseeable future. It is rather more difficult to decide how extensively methods based on pathogen biochemistry, microscopy, immunology and nucleic acid similarities will become established compared with what are now considered the more traditional methods such as identification by expert visual inspection or microscopic examination of pathogens *in situ* or *in vitro* as pure cultures (e.g. Cook and Fox, 1992; Fox and Hart, 1993). Many experienced plant disease epidemiologists can inspect plant specimens visually for symptoms at the later stages of an epidemic far quicker than they could carry out most other diagnostic tests. However, for many epidemiological investigations this is too late and so other more sensitive methods will have to be adopted either by the investigator or with support from trained experts, as in clinical diagnosis. These methods will be essential for the more rapid identification of many types of diseases where current methods are complicated – for example, in those cases such as the wilts and other diseases where the pathogen is deep-seated in the tissues, which therefore demands the destruction of the plant. Certain areas are more likely to benefit than others. The epidemiological studies of soilborne pathogens would be made easier if the methods were based on biochemical analysis rather than inspection. Diseased roots traditionally take far longer to inspect than foliage because the plants have to be dug out or pulled up first, and even then in nearly every examination there is a delay if a thick covering of soil masks the symptoms (Dusunceli and Fox, 1992).

Although methods based on nucleic acid can detect the presence of alien commensals within or on a host, none of the current methods that can be used to identify an unfamiliar pathogen is without problems. If the epidemiologist lacks appropriate training and experience, the traditional methods are ineffective without ready access to accurate disease descriptions that also identify the pathogens in some detail. At present, even with the major crops, published descriptions can be inadequate for foreign material. The difficulties are especially daunting on some of the economically less important and unusual crops – for example, many ornamentals. As it also takes time to visit libraries and search for reference books, it seems likely that this task will be made easier and quicker with the availability of electronic information systems such as the World Wide Web (www) and as rapid communications with expert consultants becomes cheaper and more feasible through the Internet. In addition to improved methods of communication for searching the literature and to satisfy current awareness, when investigating an unfamiliar disease this information can now be supplemented by data that has been electronically stored and retrieved from video disks. This system can provide a library of specialist information on even modest personal computers linked to a CD-ROM (compact-disk read-only memory) disk player to tap an extensive library of information on the literature, including illustrations. Eventually, it could become feasible to connect these with an intelligent scanning system analogous to those used by the police to scan fingerprints and portraits of suspects to provide a semi-automated system of identification. Similar technology has already been coupled to an intelligent computer programme to produce an interactive diagnosis 'key' for general practitioners in medicine. Although such developments may herald a renaissance in the classical visual methods of identification of the late stages of disease development, it is also likely that it will become even more imperative for plant disease epidemiologists as well as farmers to be able to detect the earliest stages of infection.

Even if electronic systems could be used for quick validation of a routine visual or microscopic diagnosis, without characteristic fruiting bodies, the isolation and growth of an unfamiliar microorganism would still be necessary on specialized media under controlled conditions. This obstacle may eventually encourage other techniques to be used to induce the formation of reproductive structures, but not every fungal pathogen is capable of producing fruiting structures. Currently, when no reproductive structure is formed and the mycelium is non-septate, records of Oomycetes or Zygomycetes should be examined. Although it is likely to remain considerably more tedious, it is frequently possible to distinguish at least the septate mycelium of an ascomycete from that of a basidiomycete by transmission electron microscopy. At the present time this sort of microscopic examination is restricted to those pathogens whose morphology has been sufficiently well defined to detect some distinct taxonomic traits. Electron microscopy can also be used to identify viruses and bacteria in this way. Despite this, it is not only relatively costly and tedious, but in the majority of cases at present, no immunological or specific stain is yet available and so many pathogens can prove rather troublesome to locate in a section or on a coated grid if present in low numbers. In the future, it is highly likely that a far wider selection of labelled antibodies will become accessible. Nonetheless, it would be sensible to seek easier ways of using hyphae for preliminary identification, exploiting whatever histological peculiarities that they possess. Unfortunately, hyphal characteristics alone are rarely adequate as the sole basis for diagnosis. Also, few diagnostic stains are available for the light microscopy of fungi to supplement those based on specific antibodies. It would be a valuable breakthrough if more simple fungal stains were developed analagous to the range used for bacteria.

It is a daunting task for a plant disease epidemiologist to confirm the identity of not only the main pathogen being studied but also other pathogens, commensals and saprophytes that may be present. However, in the case of the more common diseases of the staple crops, Cummins (1969) suggested that it would be reasonable to authenticate the identification of most common diseases routinely by merely using a relatively crude identification of the causal pathogen to check whether the symptoms on the host plant correspond to the description published for that pathogen in the host index, or simply the index present in the disease literature on that crop. This short-cut approach is limited at present as adequate detailed descriptions of some unusual organisms and their symptoms are rarely published or easily accessible. This omission may soon become increasingly serious for post-harvest disease epidemiology as the trend towards free trade may allow the import of a greater variety of previously exotic produce – and, with it, uncommon pathogens. There is also the possibility that lax or inadequate methods of examining familiar foreign crops may have already allowed previously unknown strains of indigenous pathogens to enter new territory, including some strains that are resistant to some widely used fungicides (Fox, 1993b). This could subsequently force plant disease epidemiologists to use more appropriate molecular diagnostic techniques (Martin *et al.*, 1992b).

Another handicap when using the mycological, bacteriological and viral literature at present is that the once familiar names of the pathogens of even some very common diseases are perpetually and unnecessarily being replaced. Hawksworth and Kirsop (1988) have protested that this is largely the consequence of our incomplete information on many genera and species, principally because of the scarcity of mycologists with appropriate taxonomic talents. If they are correct, ultimately there should be a time after the introduction of authoritative rules of nomenclature when plant pathologists, including epidemiologists, will no longer be frustrated by the apparently continual short-term instability of current taxonomy. Although it is against the current trends in taxonomy (and possibly also its basic ethos), a start has been made in the direction of a more permanent nomenclature for fungi since 1986 when the International Commission of the Taxonomy of Fungi (ICTF) of the International Union of Microbiological Societies (IUMS) started to publish current changes in the names of fungi of importance in the IUMS journal, *Microbiological Sciences* (Cannon, 1986). As well as requiring these publications to provide sound reasons for name changes and guidance on their adoption, the ICTF has prepared a Code of Practice for mycological taxonomists in order to encourage permanence by minimizing any changes due to bad practice (Sigler and Hawksworth, 1987). Well used names for fungi are also retained under a procedure known as 'conservation', designed to ensure the maintenance of well known generic names which now require a strict application of the ICBN by review and vote by the Special Committee before they can be changed.

Despite the benefits bestowed on visual methods by electronic information systems, a fixed taxonomic system and advances in microscopy, they will still suffer from a handicap shared with other laboratory-based tests. Like those methods based on the identification of pure cultures of pathogens, biochemistry, microscopy, immunology or molecular genetics, they do not allow a direct opportunity for Koch's postulates to be satisfied to provide proof of the pathogenicity of the suspected organism. When a newly identified microorganism is thought to be a pathogen, pathogenicity must be established before its confirmation as the causal agent of the disease can be accepted. This extra authentication stage should be introduced, unless the microorganism is already a familiar pathogen or its pathogenity is otherwise clearly evident. Conventional pathogenicity tests have the disadvantage of consuming time, space and materials, as well as being subject to environmental conditions that affect symptom expression or even the characteristics of the pathogen. For plant pathogenic bacteria, a hypersensitivity reaction with tobacco may be sufficient to confirm pathogenicity.

For an unfamiliar fungus, once its pathogenic nature has been corroborated, the currently available keys and descriptions used for classification are useful for diagnosis. However, there are often some problems with methods based on identification by the visual inspection of a pathogen *in situ* if it is not sufficiently well developed to bear its characteristic reproductive structures. The microscopic examination *in vitro* of pure cultures after isolation may also involve coaxing the pathogen into producing any identifiable sexual or asexual reproductive structures. Such manipulation is neither rapid nor completely dependable and requires some experience of mycological techniques. Isolation would also be made easier and more successful if a wider range of standard media were developed.

Frequently, isolating the pathogen responsible for a disease from the host, regardless of whether it is a fungus, bacterium, virus or another agent possibly as yet unknown, is not straightforward, particularly when the exact region of infection is not clearly defined. In this case, the whole plant must be thoroughly examined for the pathogen, including roots plus attached soil, as well as the aerial parts of the plant. When there are no localized symptoms but just a general malaise, where the pathogen could be present in the roots, stem or leaves but could result in similar host reactions (such as during wilting due to root rot, vascular blockage or leaf infection), it would be very helpful if there was a rapid non-specific scanning system to locate the presence of the pathogen. Similar problems in detecting breast cancer in humans have been tackled by screening patients with infrared detectors.

In general, traditional methods still require more experience on the part of an investigator than do tests based on differences in nucleic acids and immunology, where knowledge is increasingly being replaced by expensive equipment and reagents. At present there is often a hierarchy of methods. Visual inspection is unavoidable and can often give an instantaneous diagnosis when symptoms clearly conform to a well known syndrome or when signs of infection are revealed. When they are not, the next traditional route to identification in the laboratory is the incubation of the specimen to allow the pathogen to develop sufficiently to be identified and then, if necessary, isolated into culture. Biochemical and/or immunological tests may then follow. Immunoassays have been revolutionized by ELISA, which make them routinely completed within hours rather than the days or even weeks required if isolation proved necessary. The problem of non-specific binding has often meant that rapid diagnosis by monoclonal antibody ELISA has been substituted by PCR (Martin *et al.,* 1992a,b). Nonetheless, techniques based on PCR and other methods that involve matching nucleic acid from the pathogen with known sequences have been less suited for field use as they have not yet been adapted to be as portable as many techniques based on immunology.

Traditional methods of diagnosis by experts, probably with increasing help from electronic aids, will doubtless survive and even thrive but farmers and growers are increasingly choosing to detect and monitor low but treatable levels of disease on the spot under field conditions with a wide selection of relatively inexpensive diagnostic kits that are simple to use (Klausner, 1987; Miller and Martin 1988; Miller *et al.,* 1988, 1990; MacAskill, 1989). Immunology has already provided cheap kits for the pesticide industry that are sensitive enough for lower numbers of pathogens than previously to be detected, and hence treated with less fungicide. Each year, more immunological and nucleic acid hybridization techniques are being developed for the rapid detection of many of those plant pathogens that cannot be easily identified by other routine ways. Even pathogens that cause diseases with variable or latent symptoms on the host plant and those with an indistinct structure or an undistinguished morphology, including many viruses, bacteria and some fungi (particularly *fungi imperfecti* and *mycelia sterilia),* can now be detected and identified in host tissue and hence be eradicated at such early stages that losses can be minimized. At the same time, such techniques allow the complete progress of an epidemic to be observed. These methods are both sensitive and specific. An extreme example of this is the use of PCR to detect strains of fungi that are resistant to fungicides because of a single base-pair mutation (Martin *et al.,* 1992b). This technique also allows other minor changes in the genome of intraspecific strains, races and genets of a pathogen to be monitored quickly, and even the origin and period of mutation and its subsequent evolution may be identified (Smith *et al.,* 1992).

The majority of nucleic acid hybridization dot-blot tests may escape from the confines of the laboratory in the future if immunocapture allows the handicap of non-mobility to be overcome. Portability will increase the usefulness of this valuable technique to plant disease epidemiologists, who have generally rejected any method that is more restricted than immunological techniques and involves the services of professional operators. In time, both immunological and nucleic acid identification techniques are increasingly likely to be used in plant disease epidemiological studies as they become automated to make them more rapid and affordable.

REFERENCES

Akagi, H., Yokozeki, Y., Inagaki, A. *et al.* (1996) A codominant DNA marker closely linked to the rice nuclear restorer gene, Rf-1, identified with inter-SSR fingerprinting. *Genome,* **39**, 1205-1209.

Albani, M.C. and Wilkinson, M.J. (1998) Inter simple sequence repeat polymerase chain reaction for the detection of somaclonal variation. *Plant Breeding,* **117**, 573-575.

Alizadeh, A.A., Eisen, M.B., Davis, R.E. *et al.* (2000) Distinct types of diffuse large B-cell lymphoma identified by gene expression profiling. *Nature,* **286**, 503-511.

Alvarez, A.M. (2004) Integrated approaches for detection of plant pathogenic bacteria and diagnosis of bacterial diseases. *Annual Review of Phytopathology,* **42**, 339-366.

Alwine, J.C., Kemp, D.J. and Stark, G.R. (1977) Method for detection of specific RNAs in agarose gels by transfer to diazobenzyloxy-methyl paper and hybridization with DNA probes. *Proceedings of the National Academy of Sciences USA,* **74**, 5350-5354.

Anderson, J.B., Petsche, D.M. and Smith, M.L. (1987) Restriction fragment polymorphisms in biological species of *Armillaria mellea* in North America. *Mycologia,* **79**, 69-76.

Arcade, A., Anselin, F., Rampant, P.F. *et al.* (2000) Application of AFLP, RAPD and ISSR markers to genetic mapping of European and Japanese larch. *Theoretical and Applied Genetics,* **100**, 299-307.

Assefa, K., Merker, A. and Tefera, H. (2003) Inter simple sequence repeat (ISSR) analysis of genetic diversity in tef Eragrostis tef (Zucc.) Trotter. *Hereditas,* **139**, 174-183.

Audy, P., Laroche, A., Saindon, G. *et al.* (1994) Detection of the bean common blight bacteria, *Xanthomonas campestris* pv. *phaseoli* and *X. c. phaseoli* var. *fuscans,* using the polymerase chain reaction. *Phytopathology,* **84**, 1185-1192.

Avery, R.J., Norton, J.D., Jones, J.S. *et al.* (1980) Interferon inhibits transformation by murine sarcoma viruses before integration of provirus. *Nature,* **289**, 93-95.

Barker, I. and Pitt, D. (1988) Detection of the leaf curl pathogen of anemones in corms by enzyme-immunosorbent assay (ELISA). *Plant Pathology,* **37**, 417-422.

Barnes, L.W. (1986) The future of phytopathological diagnostics. *Plant Disease,* **70**, 180.

Barton, R.J. (1985) The development of sensitive mushroom virus detection methods. *Mushroom Journal,* **151**, 223-228.

Baumler, S., Sierotzki, H., Gisi, U. *et al.* (2003) Evaluation of *Erysiphe graminis* f.sp. *tritici* field isolates for resistance to strobilurin fungicides with different SNP detection systems. *Pest Management Science*, **59**, 310-314.

Benhamou, N., Ouellete, G.B., Lafontaine, J.G. and Joly, J.R. (1984) Monoclonal antibodies to a glycopeptide produced by *Ophiostoma ulmi. Canadian Journal of Plant Pathology*, **6**, 260.

Benhamou, N., Ouellette, G.B., Lafontaine, J.G. and Joly, J.R. (1985a) Use of monoclonal antibodies to detect a phytotoxic glyopeptide produced by *Ophiostoma ulmi,* the Dutch elm disease pathogen. *Canadian Journal of Botany,* **63**, 1177-1184.

Benhamou, N., Lafontaine, J.G., Joly, J.R. and Ouellette, G.B. (1985b) Ultrastructural localisation in host tissues of a toxic glyopeptide produced by *Ophiostoma ulmi,* using monoclonal antibodies *Canadian Journal of Botany,* **63**, 1185-1195.

Berent, S.L., Mahmoudi, M., Torcynski, R.M. *et al.* (1985) Comparison of oligonucleotide and long DNA fragments as probes in DNA and RNA dot, southern, northern, colony and plaque hybridization. *Biotechniques,* **3**, 208-220.

Bereswill, S., Pahl, A., Bellemann, P. *et al.* (1992) Sensitive and species-specific detection of *Erwinia amylovora* by polymerase chain reaction analysis. *Applied and Environmental Microbiology,* **58**, 3522-3526.

Bissett, J., Szakacs, G., Nolan, C.A. *et al.* (2003) New species of *Trichoderma* from Asia. *Canadian Journal of Botany*, **81**, 570-586.

Blomquist, C. and Kubisiak, T. (2003) Laboratory diagnosis of *Phytophthora ramorum* from field samples. Sudden Oak Death Online Symposium. www.apsnet.org/online/SOD.

Bohm, J., Hahn, A., Schubert, R. *et al.* (1999) Real-time quantitative PCR: DNA determination in isolated spores of the mycorrhizal fungus *Glomus mosseae* and monitoring of *Phytophthora infestans* and *Phytophthora citricola* in their respective host plants. *Journal of Phytopathology*, **147**, 409-416.

Boonham, N., Walsh, K., Mumford, R.A. and Baker, I. (2000) Use of multiplex real-time PCR (TaqMan) for the detection of potato viruses. *EPPO Bulletin*, **30**, 427-430.

Bornet, B. and Branchard, M. (2004). Use of ISSR fingerprints to detect microsatellites and genetic diversity in several related *Brassica* taxa and *Arabidopsis thaliana. Hereditas,* **140**, 245-248.

Bornet, B., Antoine, E., Bardouil, M. and Marcaillou-Le Baut, C. (2004) ISSR as new markers for genetic characterization and evaluation of relationships among phytoplankton. *Journal of Applied Phycology,* **16**, 285-290.

Botstein, D., White, R.L., Skolnick, M. and Davis, R.W. (1980) Construction of a genetic linkage in man using restriction fragment length polymorphism. *American Journal of Human Genetics,* **32**, 314-331.

Bove, IM. (1984) Wall-less prokaryotes of plants. *Annual Review of Phytopathololog*y, **22**, 361-396.

Britten, R.J. and Kohne, D.E. (1968) Repeated sequences in DNA. *Science,* **161**, 529-540.

Brookes, A. J. (1999) The essence of SNPs. *Gene,* **234**, 177-186.

Bullock, G.R. and Petrusz, P. (1985-1989) *Techniques in Immunocytochemistry,* Vols l-IV, Academic Press, London.

Caetano-Anolles, G., Bassam, B.J. and Gresshoff, P.M. (1991) DNA amplification fingerprinting using very short arbitary oligonucleotide primers. *Biotechnology*, **9**, 553-557.

Call, D. (2001) DNA microarrays-their mode of action and possible applications in molecular diagnostics. *Veterinary Sciences Tomorrow*, **3**, 1-9.

Call, D.R., Borucki, M.K. and Loge, F.J. (2003) Detection of bacterial pathogens in environmental samples using DNA microarrays. *Journal of Microbiological Methods*, **53**, 235-243.

Candlish, A.A.G., Stimson, W.H. and Smith, J.E. (1989) Assay methods for mycotoxins using monoclonal antibodies. *Abstracts of Papers, The British Society for Plant Pathology/British Crop Protection Council Conference on Techniques for the Rapid Diagnosis of Plant Disease,* University of East Anglia, Norwich, p. 11.

Candlish, A.A.G., Stimson, W.H. and Smith, J.E. (1992) Assay methods for mycotoxins using monoclonal antibodies, in *Techniques for the Rapid Diagnosis of Plant Pathogens,* (eds J.M. Duncan and L. Torrance) Blackwell, Oxford, pp. 63-75.

Cannon, P.F. (1986) Name changes in fungi of microbiological, industrial and medical importance, I-II. *Microbiological Sciences,* **3**, 168-171, 285-287.

Chard, J.M., Gray, T.R.G. and Frankland, J.C. (1985a) Purification of an antigen characteristic for *Mycena galopus. Transactions of the British Mycological Society,* **84**, 235-241.

Chard, J.M., Gray, T.R.G. and Frankland, I.C. (1985b) Use of an *anti-Mycena galopus* serum as an immunofluorescent reagent. *Transactions of the British Mycological Society,* **84**, 243-249.

Charters, Y.M. and Wilkinson, M.J. (2000). The use of self-pollinated progenies as 'in-groups' for the genetic characterization of cocoa germplasm. *Theoretical and Applied Genetics,* **100**, 160-166.
Chen, W., Hoy, W. and Schneider, R.W. (1992) Species-specific polymorphisms in transcribed ribosomal DNA of five *Pythium* species. *Experimental Mycology*, **16**, 22-34.
Chirkov, SN., Olovnikov, A.M., Surgachyova, H.A. and Atabekov, J.G. (1984) Immunodiagnosis of plant viruses by a virobacterial agglutination test. *Annals of Applied Biology,* **104**, 477-483.
Chou, Q., Russell, M., Birch, D.E. *et al.* (1992) Prevention of pre-PCR mis-priming and primer dimerization improves low-copy-number amplifications. *Nucleic Acid Research,* **20**, 1717-1723.
Clark, M.F. (1992) Immunodiagnostic techniques for plant mycoplasma-like organisms, in *Techniques for the Rapid Diagnosis of Plant Pathogens,* (eds J.M. Duncan and L. Torrance), Blackwell, Oxford, pp. 34-45.
Clark, M.F. and Adams, A.N. (1977) Characteristics of the microplate method of enzyme-linked immunosorbent assay for the detection of plant viruses. *Journal of General Virology,* **34**, 475-483.
Clark, ME., Morton, A. and Buss, S.L. (1989) Preparation of mycoplasma immunogens from plants and a comparison of polyclonal and monoclonal antibodies made against Primula Yellows mycoplasma-like organisms. *Annals of Applied Biology,* **114**, 111-124.
Clarke, J.H., MacNicoll, A.D. and Norman, l.A. (1986) Immunological detection of fungi in plants, including stored cereals, in *Spoilage and Mycotoxins of Cereals and other Stored Products,* (ed. B. Flannigan), *International Biodeterioration Supplement,* **22**, 123-130.
Cook, D. and Sequeira, L. (1991) The use of subtractive hybridization to obtain a DNA probe specific for *Pseudomonas solanacearum* race 3. *Molecular and General Genetics,* **227**, 401-410.
Cook, R.T.A. and Fox, R.T.V. (1992) Powdery mildew on faba beans and other legumes in Britain. *Plant Pathology,* **41**, 506-512.
Cooper, J.I. and Edwards, M.L. (1986) Variations and limitations of enzyme-amplified immunoassays, in *Development and Applications of Virus Testing,* (eds R.A.C. Jones and L. Torrance), Association of Applied Biologists, Wellesbourne, pp. 139-155.
Coxon, R.E., Rae, C., Gallacher, G. and Landon, J. (1988) Development of a simple fluoroimmunoassay for paraquat. *Clinica Chimica Acta,* **175**, 297-305.
Crowhurst, R.N., Hawthorne, B.T., Rikkerink, E.H.A. and Templeton, M.D. (1991) Differentiation of *Fusarium solani* f.sp. *cucurbitae* races 1 and 2 by random amplification of polymorphic DNA. *Current Genetics,* **20**, 391-396.
Cummings, C.A. and Relman, D.A. (2000) Using DNA microarrays to study host-microbe interactions. *Emerging Infectious Diseases*, **6**, 513-525.
Cummins, G.B. (1969) Identification of fungal pathogens, in *Plant Pathological Methods, Fungi and Bacteria,* (ed. J. Tuite), Burgess Publishing, Minneapolis, pp. 224-226.
Day, A.W., Gardinier, R.B., Smith, R. *et al.* (1986) Detection of fungal fimbrae by protein A-gold immunocytochemical labelling in host plants infected with *Ustilago heufleri* or *Peronospora hiyoscami* f.sp. *tabacina. Canadian Journal of Microbiology,* **32**, 577-584.
Deparasis, J. and Roth, D.A. (1990) Nucleic-acid probes for identification of genus-specific 16S ribosomal-RNA sequences. *Phytopathology*, **80**, 618-621.
DeRisi, J., Penland, L., Brown, P.O. *et al.* (1996) Use of a cDNA microarray to analyze gene expression patterns in human cancer. *Nature Genetics*, **14**, 457-460.
Dewey, F.M. (1988) Development of immunodiagnostic assays for fungal plant pathogens. *Proceedings Brighton Crop Protection Conference Pests and Diseases,* British Crop Protection Council, Thornton Heath, pp. 777-786.
Dewey, F.M. (1992) Detection of plant-invading fungi by monoclonal antibodies, in *Techniques for the Rapid Diagnosis of Plant Pathogens,* (eds J.M. Duncan and L. Torrance), Blackwell, Oxford, pp. 47-62.
Dewey, F.M., Munday, C.J. and Brasier, C.M. (1989a) Monoclonal antibodies to specific components of the Dutch Elm Disease pathogen *Ophiostoma ulmi. Plant Pathology,* **38**, 9-20.
Dewey, F.M., MacDonald, M.M. and Philips, S.I. (1989b) Development of monoclonal antibody -ELISA, -DOT-Blot and -DIP-Stick immunoassays for *Humicola lanuginosa* in rice. *Journal of General Microbiology,* **135**, 361-374.
DeYong, Z., Willingmann, P., Heinze, C. *et al.* (2005) Differentiation of Cucumber mosaic virus isolates by hybridization to oligonucleotides in a microarray format. *Journal of Virological Methods*, **123**, 101-108.

Drmanac, S., Kita, D., Labat, I. *et al.* (1998) Accurate sequencing by hybridization of DNA diagnostics and individual genomics. *Nature Biotechnology*, **16**, 54-58.

Drogemuller, M., Schnieder, T., Himmelstjerna, S. and Von, G. (2004) Beta-tubulin complementary DNA sequence variations observed between cyathostomins from benzimidazole-susceptible and -resistant populations. *Journal of Parasitology*, **90**, 868-870.

Dusunceli, F. and Fox, R.T.V. (1992) The accuracy of methods for estimating the size of *Thanatephorus cucumeris* populations in soil. *Soil Use and Management*, **8**, 21-26.

Dyer, P.S., Furneaux, P.A., Douhan, G. and Murray, T.D. (2001) A multiplex PCR test for determination of mating types applied to the plant pathogens *Tapesia yallundae* and *Tapesia acuformis*. *Fungal Genetics and Biology*, **33**, 173-180.

Edwards, A., Civitello, A., Hammond, H.A. and Caskey, C.T. (1991) DNA Typing and Genetic-Mapping with Trimeric and Tetrameric Tandem Repeats. *American Journal of Human Genetics*, **49**, 746-756.

Edwards, K.J. and Mogg, R. (2001) Plant genotyping by analysis of single nucleotide polymorphisms, in *Plant genotyping: the DNA fingerprinting of plants*. (ed. R.J. Henry), CABI Publishing, Wallingford, UK, pp. 1-13.

Erlich, H.A. (1989) *PCR technology*. (1st edn). Macmillan Publishers, Basingstoke, UK.

Estrada-Garcia, M.T., Green, J.R., Booth, I.M. *et al.* (1989) Monoclonal antibodies to cell surface components of zoospores and cysts of the fungus *Pythium aphanidermatum* reveal species-specific antigens. *Experimental Mycology*, **13**, 348-356.

Etscheid, M. and Riesner, D. (1998) TGGE and DGGE, in *Molecular tools for screening biodiversity* (eds A. Karp, P.G. Isaac, and D.S. Ingram), Chapman and Hall, London, pp. 133-156.

Finan, P. (1984) Histology, in *Monoclonal Antibodies*, (eds K. Sikora and H.M.Smedley), Blackwell Scientific, Oxford, pp. 45-52.

Fischer, M. and Binder, M. (2004) Species recognition, geographic distribution and host-pathogen relationships: a case study in a group of lignicolous basidiomycetes, *Phellinus s.l. Mycologia*, **96**, 799-811.

Foster, S.J., Ashby, A.M. and Fitt, B.D.L. (2002) Improved PCR-based assays for pre-symptomatic diagnosis of light leaf spot and determination of mating types of *Pyrenopeziza brassicae* on winter oilseed rape. *European Journal of Plant Pathology*, **108**, 374-383.

Fox, R.T.V. (1990a) Rapid methods for diagnosis of soil-borne plant pathogens, in *Soil-borne Diseases*, (ed. D. Hornby), Special Issue, *Soil Use and Management*, **6**, 179-184.

Fox, R.T.V. (1990b) Diagnosis and control of *Armillaria* honey fungus root rot of trees. *Professional Horticulture*, **4**, 121-127.

Fox, R.T.V. (1993a) *Principles of Diagnostic Techniques in Plant Pathology*, CAB International, Wallingford, 220 pp.

Fox, R.T.V. (1993b) Prospects for the diagnosis and control of soil-borne pathogens, in *International Symposium on Plant Health and the European Single Market*, (eds R.G. McKinlay and D. Atkinson), British Crop Protection Council, Thornton Heath, pp. 409-412.

Fox, R.T.V. and Hahne, K. (1988) Prospects for the rapid diagnosis of *Armillaria* by monoclonal antibody ELISA. *Proceedings of the 7th International Conference IUFRO Working Party S2.06.01 on Root and Butt Rots of Forest Trees*, Vernon & Victoria, British Columbia, Canada, August 9-16 1988, pp. 458-468.

Fox, R.T.V. and Hart, C.A. (1993) Microscopy in the study of plant disease. *Microscopy*, **39**, 64-72.

Fox, R.T.V., Martin, L.-A., Groves, J.G. and Baldwin, B.C. (1989) Screening phytopathogenic fungi for resistance to fungicides using monoclonal antibodies. *Abstracts of Papers, The British Society for Plant Pathology/British Crop Protection Council Conference on Techniques for the Rapid Diagnosis of Plant Disease*, University of East Anglia, Norwich, July 11-13 1989, p. 9.

Fox, R.T.V., Manley, H.M., Culham, A. *et al.* (1994) Methods for detecting *Armillaria mellea*, in *Ecology of Plant Pathogens*, (ed. J.P. Blakeman), Blackwell/BSPP, Oxford, pp. 119-133.

Fraaije, B.A., Lovell, D.J. and Baldwin, S. (2002) Septoria epidemics on wheat: Combined use of visual assessment and PCR-based diagnostics to identify mechanisms of disease escape. *Plant Protection Science*, **38**, 421-424.

Fraaije, B.A., Lovell, D., Coelho, J.M. *et al.* (2001) PCR-based assays to assess wheat varietal resistance to blotch (*Septoria tritici* and *Stagonosporum nodorum*) and rust (*Puccinia striiformis* and *Puccinia recondita*) diseases. *European Journal of Plant Pathology*, **107**, 905-917.

Gerik, J.S., Lommel, S.A. and Huisman, O.C. (1987) A specific serological staining procedure for *Verticillium dahliae* in cotton root tissue. *Phytopathology*, **77**, 261-266.

Gibson, U.E.M., Heid, C.A. and Williams, P.M. (1996) A novel method for real time quantitative RT-PCR. *Genome Research,* **6**, 995-1001.

Gilliland, G., Perrin, S., Blanchard, K. and Bunn, H.F. (1990a) Analysis of cytokine mRNA and DNA: Detection and quantitation by competitive polymerase chain reaction. *Proceedings of the National Academy of Sciences USA,* **87**, 2725-2729.

Gilliland, G., Perrin, S. and Bunn, H.F. (1990b) Competitive PCR for quantitation of mRNA, in *PCR Protocols: A Guide to Methods and Applications,* (eds M. Innis, D. Gelfand, D. Snisky and T. White), Academic Press, New York, pp. 60-69.

Giulietti, A., Overbergh, L., Valckx, D. *et al.* (2001) An overview of real-time quantitative PCR: Applications to quantify cytokine gene expression. *Methods,* **25**, 386-401.

Gleason, M.L., Ghabrial, S.A. and Ferriss, R.S. (1987) Serological detection of *Phomopsis longicolla* in soybean seeds. *Phytopathology,* 77, 371-375.

Goding, J.W. (1983) *Monoclonal Antibodies: Principles and Practice,* Academic Press, London and New York, 329 pp.

Graser, Y., El-Fari, M., Presber, W. *et al.* (1998) Identification of common dermatophytes (*Trichophyton, Microsporum, Epidermophyton*) using polymerase chain reactions. *British Journal of Dermatology,* **138**, 576-582.

Groves, J.G., Fox, R.T.V. and Baldwin, B.C. (1988) Tubulin from *Botrytis cinerea* and the potential for development of an immunodiagnostic for benzimidazole resistance. *Proceedings 1988 Brighton Crop Protection Conference – Pests and Diseases,* **1**, 415-420.

Grunig, C.R., Sieber, T.N. and Holdenrieder, O. (2001) Characterisation of dark septate endophytic fungi (DSE) using inter-simple-sequence-repeat-anchored polymerase chain reaction (ISSR-PCR) amplification. *Mycological Research,* **105**, 24-32.

Grunig, C.R., Sieber, T.N., Rogers, S.O. and Holdenrieder, O. (2002) Spatial distribution of dark septate endophytes in a confined forest plot. *Mycological Research,* **106**, 832-840.

Grunstein, M. and Hogness, D.S. (1975) Colony hybridization: a method for the isolation of cloned DNAs that contain a specific gene. *Proceedings of the National Academy of Sciences USA,* **72**, 3961-3965.

Gupta M, Chyi Y-S, Romero-Severson J and Owen JL (1994) Amplification of DNA markers from evolutionarily diverse genomes using single primers of simple-sequence repeats. *Theoretical and Applied Genetics,* **89**, 998-1006.

Gupta, P.K., Roy, J.K. and Prasad, M. (2001). Single nucleotide polymorphisms: A new paradigm for molecular marker technology and DNA polymorphism detection with emphasis on their use in plants. *Current Science,* **80**, 524-535.

Guthrie, P.A.I., Magill, W.W., Federicksen, R.A. and Odvody, G.N. (1992) Random amplified polymorphic DNA markers: a system for identifying and differentiating isolates of *Colletotrichum graminicola. Phytopathology,* **82**, 832-835.

Hadidi, A., Czosnek, H. and Barba, M. (2004) DNA microarrays and their potential applications for the detection of plant viruses, viroids, and phytoplasmas. *Journal of Plant Pathology,* **86**, 97-104.

Han, Q., Inglis, G.D. and Hausner, G. (2002) Phylogenetic relationships among strains of the entomopathogenic fungus, *Nomuraea rileyi*, as revealed by partial β-tubulin sequences and inter-simple sequence repeat (ISSR) analysis. Letters in Applied Microbiology, **34**, 376-383.

Hantula, J., Dusabenyagasani, M. and Hamelin, R.C. (1996) Random amplified microsatellites (RAMS) – a novel method for characterizing genetic variation within fungi. *European Journal of Forest Pathology,* **26**, 159-166.

Hardham, A.R., Suzaki E. and Perkin, J.L. (1985) The detection of monoclonal antibodies specific for surface components on zoospores and cysts of *Phytophthora cinnamomi. Experimental Mycology,* **9**, 254-268.

Hardham, A.R., Suzaki, F. and Perkin, J.L. (1986) Monoclonal antibodies to isolate-, species-, and genus-specific components on the surface of zoospores and cysts of the fungus *Phytophthora cinnamomi. Canadian Journal of Botany,* **64**, 311-321.

Hawksworth, D.L. and Kirsop, B.E. (1988) (eds) *Filamentous Fungi. Living Resources for Biotechnology,* Cambridge University Press, Cambridge, 209 pp.

Heller, R.A., Schena, M., Chai, A. *et al.* (1997) Discovery and analysis of inflammatory disease-related genes using cDNA microarrays. *Proceeding of National Academy of Sciences USA,* **94**, 2150-2155.

Henry, T., Iwen, P.C. and Hinrichs, S.H. (2000) Identification of *Aspergillus* species using internal transcribed spacer regions 1 and 2. *Journal of Clinical Microbiology,* **38**, 1510-1515.

Henson, J.M. and French, R. (1993) The polymerase chain reaction and plant disease diagnosis. *Annual Review of Phytopathology,* **31**, 81-109.

Henson, J.M., Goins, T., Grey, W. *et al.* (1993) Use of polymerase chain reaction to detect *Gaeumannomyces graminis* DNA in plants grown in artificially and naturally infested soil. *Phytopathology,* **83**, 283-287.

Higuchi, R., Fockler, C., Dollinger, G. and Watson, R. (1993) Kinetic PCR analysis-Real-time monitoring of DNA amplification reactions. *Biotechnology,* **11**, 1026-1030.

Hill, S.A. (ed.) (1984) *Methods in Plant Virology,* Blackwell Scientific, Oxford, 175 pp.

Higuchi, R., Fockler, C., Dollinger, G. and Watson, R. (1993) Kinetic PCR analysis-Real-time monitoring of DNA amplification reactions. *Biotechnology,* **11**, 1026-1030.

Hirsch, P.R., Atkins, S.D., Mauchline, T.H. *et al.* (2002) Methods for studying the nematophagous fungus *Verticillium chlamydosporium* in the root environment. *Interactions in the root environment: an integrated approach. Proceedings of the Millennium Conference on Rhizosphere Interactions,* IACR-Rothamsted, UK, pp. 21-30.

Hirsch, P.R., Mauchline, T.H., Mendum, T.A. and Kerry, B.R. (2000) Detection of the nematophagous fungus *Verticillium chlamydosporium* in nematode-infested plant roots using PCR. *Mycological Research,* **104**, 435-439.

Hitchcock, C.J., Chambers, S.M., Anderson, I.A. and Cairney, J.W.G. (2003) Development of markers for simple sequence repeat-rich regions that discriminate between *Pisolithus albus* and *P. microcarpus. Mycological Research,* **107**, 699-706.

Holliday, P. (1989) *A Dictionary of Plant Pathology,* Cambridge University Press, Cambridge, 369 pp.

Hull, R. (1986) The potential for using dot-blot hybridization in the detection of plant viruses 3-12 in *Developments and Applications in Virus Testing,* (eds R.A.C. Jones and L. Torrance), Association of Applied Biologists, Cambridge, 300 pp.

Johnson, M.C., Pirone, T.P., Siegel, M.R. and Varney, D.R. (1982) Detection of *Epichloe typhina* in tall fescue by means of enzyme-linked immunosorbent assay. *Phytopathology,* **72**, 647-650.

Jordan, W.C., Foley, K. and Bruford, M.W. (1998) Single-strand conformation polymorphism (SSCP) analysis, in *Molecular tools for screening biodiversity* (eds. A. Karp, P.G. Isaac, and D.S. Ingram), Chapman and Hall, London, pp. 152-156.

Josefsen, L. and Christiansen, S.K. (2002) PCR as a toll for early detection and diagnosis of common bunt in wheat, caused by *Tilletia tritici. Mycological Research,* **106**, 1287-1292.

Khan, J., Simon, R., Bittner, M. *et al.* (1998) Gene expression profiling of alveolar rhabdomyosarcoma with cDNA microarrays. *Cancer Research,* **58**, 5009-5013.

Kitagawa, T., Sakamoto, Y., Furumi, K. and Ogwra, H. (1989) Novel enzyme immunoassays for specific detection of various *Fusarium* species. *Phytopathology,* **79**, 162-165.

Klausner, A. (1987) Immunoassays flourish in new markets. *BioTechnology,* **5**, 551-556.

Kohler, G. and Milstein, C. (1975) Continuous culture of fused cells secreting antibody of predefined specificity. *Nature, London,* **256**, 495-497.

Kroon, L.P.N.M., Verstappen, E.C.P., Kox, L.F.F. *et al.* (2004) A rapid diagnostic test to distinguish between American and European populations of *Phytophthora ramorum. Phytopathology*, **94**, 613-620.

Lange, L., Heide, M., Holbolth, L. and Olson, L.W. (1989) Serological detection of *Plasmodiophora brassicae* by dot immunolabelling and visualization by scanning electron microscopy. *Phytopathology,* **79**, 1066-1071.

Lazo, G.R. and Gabriel, D.W. (1987) Conservation of plasmid DNA sequences and pathovar identification of strains of *Xanthomonas campestris. Phytopathology*, **77**, 448-453.

Lazo, G.R., Roffey, R. and Gabriel, D.W. (1987) Pathovars of *Xanthomonas campestris* are distinguishable by restriction length polymorphism. *International Journal of Systematic Bacteriology,* **37**, 214-221.

Levy, M., Romao, J., Marchetti, M.A. and Hamer, J.E. (1991) DNA fingerprinting with a dispersed repeated sequence resolves pathotype diversity in the rice blast fungus. *The Plant Cell,* **3**, 95-102.

Lieckfeldt, E., Cavignac, Y., Fekete, C. and Borner, T. (2000) Endochitinase gene-based phylogenetic analysis of *Trichoderma. Microbiological Research,* **155**, 7-15.

Lieckfeldt, E., Samuels, G.J., Nirenberg, H.I. and Petrini, O. (1999) A morphological and molecular perspective of *Trichoderma viride*: is it one or two species? *Applied and Environmental Microbiology*, **65**, 2418-2428.

Lin, C.P. and Chen, T.A. (1985) Monoclonal antibodies against the aster yellows agent. *Science,* **227**, 1233-1235.

Lin, C.P. and Chen, T.A. (1986) Comparison of monoclonal and polyclonal antibodies in detection of Aster Yellows mycoplasma-like organism. *Phytopathology,* **76**, 45-50.

Litt, M. and Luty, J.A. (1989) A hypervariable microsatellite revealed by *in vitro* amplification of a dinucleotide repeat within the cardiac-muscle actin gene. *American Journal of Human Genetics*, **44**, 397-401.

Lockhart, D.J., Dong, H., Byrne, M.C. *et al.* (1996) Expression monitoring by hybridization to high-density oligonucleotide arrays. *Nature Biotechnology*, **14**, 1675-1680.

Longato, S. and Bonfante, P. (1997) Molecular identification of mycorrhizal fungi by direct amplification of microsatellite regions. *Mycological Research,* **101**, 425-432.

Louws, F.J., Rademaker, J.L.W. and de Bruijn, F.J. (1999) The three Ds of PCR-based genomic analysis of phytobacteria: diversity, detection and disease diagnosis. *Annual Review of Phytopathology*, **37**, 81-125.

Lung-Escarmant, B. and Dunez, J. (1979) Differentiation of *Armillariella* and *Clitocybe* species by the use of the immunoenzymatic ELISA procedure. *Annals of Phytopathology,* **11**, 515-518.

MacAskill, J. (1989) Diagnosic kits for cereal diseases. *Abstracts of Papers, The British Society for Plant Pathology/British Crop Protection Council Conference on Techniques for the Rapid Diagnosis of Plant Disease,* University of East Anglia, Norwich, p. 10.

Maes, M. (1993) Fast classification of plant-associated bacteria in the Xanthomonas genus. *FEMS Microbiology Letters,* **113**, 161-166.

Manulis, S., Chalupowicz, L., Dror, O. and Kleitman, F. (2002) Molecular diagnostic procedures for production of pathogen-free propagation material. *Pest Management Science,* **58**, 1126-1131.

Martin, F.N., Tooley, P.W. and Blomquist, C. (2004) Molecular detection of *Phytophthora ramorum*, the causal agent of sudden oak death in California, and two additional species commonly recovered from diseased plant material. *Phytopathology,* **94**, 621-631.

Martin, L.-A., Fox, R.T.V. and Baldwin, B.C. (1992a) Rapid methods for the detection of MBC resistance in fungi: I. Immunological approaches. *Proceedings of 10th International Reinhardsbrunn Symposium,* pp. 209-218.

Martin, L.-A., Fox, R.T.V., Baldwin, B.C. and Connerton, I.F. (1992b) Rapid methods for the detection of MBC resistance in fungi: II. Use of the polymerase chain reaction as a diagnostic tool. *Proceedings 1992 Brighton Crop Protection Conference – Pests and Diseases,* **1**, 207-214.

Maule, A.J., Hull, R. and Donson, J. (1983) The application of spot hybridization to the detection of DNA and RNA viruses in plant tissues. *Journal of Virology Methods,* **6**, 183-191.

McCartney, H.A., Foster, S.J., Fraaije, B.A. and Ward, E. (2003) Molecular diagnostics for fungal plant pathogens. *Pest Management Science*, **59**, 129-142.

Miller, S.A. and Martin, R.R. (1988) Molecular diagnosis of plant disease. *Annual Review of Phytopathology,* **26**, 409-432.

Miller, S.A., Rittenburg, J.H., Petersen, F.P. and Grothaus, G.D. (1988) Application of rapid, field-useable immunoassays for the diagnosis and monitoring of fungal pathogens in plants. *Brighton Crop Protection Conference – Pests and Diseases,* pp. 795-803.

Miller, S.A., Rittenburg, J.H., Petersen, F.P. and Grothaus, G.D. (1990) Development of modern diagnostic tests and benefit to the farmer, in *Monoclonal Antibodies in Agriculture,* (ed. A. Schots), Pudoc, Wageningen, Netherlands, pp. 15-21.

Mills, P.R., Sreenivasaprasad, S. and Brown, A.E. (1992) Detection and differentiation of *Colletotrichum gloeosporides* isolates using PCR. *FEMS Microbiology Letters,* **98**, 137-144.

Mishra, P.K., Fox, R.T.V. and Culham, A. (2003) Inter-simple sequence repeat and aggressiveness analyses revealed high genetic diversity, recombination and long-range dispersal in *Fusarium culmorum. Annals of Applied Biology*, **143**, 291-301.

Mishra, P.K., Fox, R.T.V. and Culham, A. (2002) Restriction analysis of PCR amplified nrDNA regions revealed intraspecific variation within populations of *Fusarium culmorum. FEMS Microbiology Letters*, **215**, 291-296.

Mitchell, D.H., Rose, D.G. and Howell, P. (1988) European Patent Application No. 88300937.5, Squash Blot Device, European Patent Office.

Mitchell, L.A. and Sutherland, J.K. (1986) Detection of seed-borne *Sirococcus strobilinus* with monoclonal antibodies in an enzyme-linked immunosorbent assay. *Canadian Journal of Forestry Research,* **16**, 945-948.

Mohan, S.B. (1989) Cross-reactivity of antiserum raised against *Phytophthora* species and its evaluation of a genus detecting antiserum. *Plant Pathology,* **38**, 352-363.

Morales, V.M., Pelcher, L.E. and Taylor, J. (1993) Comparison of the 5.8S rDNA and internal transcribed spacer sequences of isolates of *Leptosphaeria maculans* from different pathogenicity groups. *Current Genetics*, **23**, 490-495.

Mori, Y., Sato, Y. and Takamatsu, S. (2000) Evolutionary analyses of the powdery mildew fungi using nucleotide sequences of the nuclear ribosomal DNA. *Mycologia*, **92**, 74-93.

Morley, S.I. and Jones, D.G. (1980) A note on a highly sensitive modified ELISA technique for *Rhizobium* strain identification. *Journal of Applied Bacteriology,* **49**, 103-109.

Morrison, T.B., Weis, J.J. and Wittwer, C.T. (1998) Quantification of low-copy transcripts by continuous SYBR (R) green I monitoring during amplification. *Biotechniques,* **24**, 954-962.

Motta, J.J., Peabody, D.C. and Peabody, R.B. (1986) Quantitative differences in nuclear DNA content between *Armillaria mellea* and *Armillaria bulbosa. Mycologia*, **78**, 963-965.

Mugnier, J. (1998) Molecular evolution and phylogenetic implications of ITS sequences in plants and in fungi, in *the Molecular variability of fungal pathogens* (eds P.D. Bridge, Y. Couteaudier and J. Clarkson) Wallingford, UK. CAB International. pp. 253-277.

Mullis, K.B. and Faloona, F.A. (1987) Specific synthesis of DNA *in vitro* via a polymerase-catalysed chain reaction. *Methods in Enzymology,* **155**, 335-351.

Mumford, R.A., Walsh, K. and Boonham, N. (2000) A comparison of molecular methods for the routine detection of viroids. *EPPO Bulletin*, **30**, 431-435.

Nagaoka, T. and Ogihara, Y. (1997) Applicability of inter-simple sequence repeat polymorphisms in wheat for use as DNA markers in comparison to RFLP and RAPD markers. *Theoretical and Applied Genetics,* **94**, 597-602.

Nazar, R.N., Hu, X., Schmidt, J. *et al.* (1991) Potential use of PCR-amplified ribosomal intergenic sequences in the detection and differentiation of *Verticillium* wilt pathogens. *Physiological and Molecular Plant Pathology*, **39**, 1-11.

Niesters, H.G.M., Goessens, W.H.F., Meis, J.F.M.G. and Quint, W.G.V. (1993) Rapid, polymerase chain reaction-based identification assays for Candida species. *Journal of Clinical Microbiology,* **31**, 904-910.

Niewola, Z., Walsh, S.T. and Davies, G.E. (1983) Enzyme-linked immunosorbent assay (ELISA) for paraquat. *International Journal of Immunopharmacology,* **5**, 211-218.

Niewola, Z., Hayward, C., Symington, B.A. and Robson, R.T. (1985) Quantitative estimation of paraquat by an enzyme-linked immunosorbent assay using a monoclonal antibody. *Clinica Chimica Acta,* **148**, 149-156.

Niewola, Z., Benner, J.R and Swaine, H. (1986) Determination of paraquat residues by an enzyme-linked immunosorbent assay. *Analyst,* **111**, 399-402.

Notermans, S., Wieten, G., Engel, H.W.B. *et al.* (1987) Purification and properties of extracellular polysaccharide (EPS) antigens produced by different mould species. *Journal of Applied Bacteriology,* **62**, 157-166.

O'Donnell, K. (1992) Ribosomal DNA internal transcribed spacers are highly divergent in the phytopathogenic ascomycete *Fusarium sambucinum* (*Gibberella pulicaris*). *Current Genetics,* **22**, 213-220.

Ouellette, G.B. and Benhamou, N. (1987) Use of monoclonal antibodies to detect molecules of fungal plant pathogens. *Canadian Journal of Plant Pathology,* **9**, 167-176.

Paul, B. (2000) ITS1 region of the rDNA of *Pythium megacarpum* sp. nov., its taxonomy, and its comparison with related species. *FEMS Microbiology Letters*, **186**, 229-233.

Pease, A.C., Solas, D., Sullivan, E.J. *et al.* (1994) Light-generated oligonucleotide arrays for rapid DNA sequence analysis. *Proceeding of National Academy of Sciences USA*, **91**, 5022-5026.

Perou, C.M., Jeffrey, S.S., van de Rijn, M. *et al.* (1999) Distinctive gene expression patterns in human mammary epithelial cells and breast cancers. *Proceedings of National Academy of Sciences USA*, **96**, 9212-9217.

Pertot-Rechenmann C. and Gadal, P. (1986) Enzyme immunocytochemistry, in *Immunology in Plant Science,* (ed. T.L. Wang), Cambridge University Press, Cambridge, pp. 59-88.

Polak, J.M. and Varndell, I.M. (1984) *Immunolabelling for Electron Microscopy,* Elsevier, Amsterdam.

Prosen, D., Hatziloukas, E., Schaad, N.W. and Panopoulos, N.J. (1993) Specific detection of *Pseudomonas syringae* pv. *phaseolicola* DNA in bean seed by polymerase chain reaction-based amplification of a phaseolotoxin gene region. *Phytopathology,* **83**, 965-970.

Prosser, J. (1993) Detecting single base mutations. *Trends in Biochemistry*, **11**, 238-246.

Pryor, B.M. and Gilbertson, R.L. (2000) Molecular phylogenetic relationships amongst *Alternaria* species and related fungi based upon analysis on nuclear ITS and mtSSU rDNA. *Mycological Research*, **104**, 1312-1321.

Rasmussen, O.F. and Reeves, J.C. (1992) DNA probes for the detection of plant pathogenic bacteria. *Journal of Biotechnology*, **25**, 203-220.

Reddy, M.P., Sarla, N. and Siddiq, E.A. (2002) Inter simple sequence repeat (ISSR) polymorphism and its application in plant breeding. *Euphytica*, **128**, 9-17.

Ririe, K.M., Rasmussen, R.P. and Wittwer, C.T. (1997) Product differentiation by analysis of DNA melting curves during the polymerase chain reaction. *Analytical Biochemistry*, **245**,154-160.

Robard, J. (1987) Synthetic peptides as vaccines. *Nature, London*, **330**, 106-107.

Robinson, D.J. and Legorburu, F.J. (1988) Detection of tobacco rattle tobravirus by spot hybridisation. *Annual Report of the Scottish Crop Research Institute for 1987*, pp. 195-196.

Robinson, D.J. (1988) Prospects for the application of nucleic acid probes in plant virus detection. *Proceedings 1988 Brighton Crop Protection Conference – Pests and Diseases*, **2**, 805-810.

Robinson, D.J., Hamilton, W.D.O., Harrison, B.D. and Baulcombe, D.C. (1987) Two anomalous tobraviruses: evidence for RNA recombination in nature. *Journal of General Virology*, **68**, 2551-2561.

Rodrigues, K.F., Sieber, T.N., Grunig, C.R. and Holdenrieder, O. (2004) Characterization of *Guignardia mangiferae* isolated from tropical plants based on morphology, ISSR-PCR amplifications and ITS1-5.8S-ITS2 sequences. *Mycological Research*, **108**, 45-52.

Ronaghi, M. and Elahi, E. (2002) Discovery of single nucleotide polymorphisms and mutations by Pyrosequencing. *Comparative and Functional Genomics*, **3**, 51-56.

Saiki, R.K., Scharf, S.J., Faloona, F. *et al.* (1985) Enzymatic amplification of β-globulin genomic sequences and restriction site analysis for diagnosis of sickle cell anemia. *Science*, **230**, 1350-1354.

Sambrook, J., Fritsch, E.F. and Maniatis, T. (1989) *Molecular Cloning. A Laboratory Manual*, 2nd edn, Cold Spring Harbor Laboratory Press.

Sankar, A.A. and Moore, G.A. (2001) Evaluation of inter-simple sequence repeat analysis for mapping in Citrus and extension of the genetic linkage map. *Theoretical and Applied Genetics*, **102**, 206-214.

Schaad, N.W., Azad, H., Peet, R.C. and Panapoulos, N.J. (1989) Identification of *Pseudomonas syringae* pv. *phaseolicola* by a DNA hybridization probe. *Phytopathology*, **79**, 903-907.

Schaad, N.W. and Frederick, R.D. (2002) Real-time PCR and its application for rapid plant disease diagnosis. *Canadian Journal of Plant Pathology*, **24**, 250-258.

Schaad, N.W., Schaad, B.Y., Sechler, A. and Knoor, D. (1999) Detection of *Clavibacter michiganensis* subsp. *sepedonicus* in potato tubers by BIO-PCR and an automated real-time fluorescence detection system. *Plant Disease*, **83**, 1095-1100.

Schena, M., Shalon, D., Davis, R.W. and Brown, P.O. (1995) Quantitative monitoring of gene expression patterns with a complementary DNA microarray. *Science*, **270**, 467-470.

Schena, M., Shalon, D., Heller, R. *et al.* (1996) Parallel human genome analysis: microarray-based expression monitoring of 1000 genes. *Proceedings of National Academy of Sciences USA*, **93**, 10614-10619.

Schleifer, K.H. and Stackerbrandt, E. (1983) Molecular systematics of prokaryotes. *Annual Review of Microbiology*, **37**, 143-147.

Schmidt, T. (1994) Fingerprinting bacterial genomes using ribosomal RNA genes and operons. *Methods in Molecular and Cell Biology*, **5**, 3-12.

Schonian, G., Meusel, O., Tietz, H.J. *et al.* (1993) Identification of clinical strains of *Candida albicans* by DNA fingerprinting with the polymerase chain reaction. *Mycoses*, **36**, 171-179.

Schroeder, S., SeongHwan, K., Sangwon, L. *et al.* (2002) The beta -tubulin gene is a useful target for PCR-based detection of an albino *Ophiostoma piliferum* used in biological control of sapstain. *European Journal of Plant Pathology*, **108**, 793-801.

Seal, S.E., Jackson, L.A. and Daniels, M.J. (1992) Isolation of a *Pseudomonas solanacearum*-specific DNA probe by subtraction hybridization and construction of species-specific oligonucleotide primers for sensitive detection by the polymerase chain reaction. *Applied and Environmental Microbiology*, **58**, 3751-3758.

Seal, S.E., Jackson, L.A., Young, J.P.W. and Daniels, M.J. (1993) Differentiation of *Pseudomonas solanacearum, Pseudomonas syzygii, Pseudomonas pickettii* and blood disease bacterium by partial

16S rRNA sequencing: construction of oligonucleotide primers for sensitive detection by polymerase chain reaction. *Journal of General Microbiology*, **139**, 1587-1594.

Seifert, K.A., Wingfield, B.D. and Wingfield, M.J. (1995) A critique of DNA sequence analysis in the taxonomy of filamentous ascomycetes and ascomycetous anamorphs. *Canadian Journal of Botany*, **73**, S760-S767.

Sigler, L. and Hawksworth, D.L. (1987). Code of practice for systematic mycologists. *Microbiological Sciences*, **4**, 83-86; *Mycopathologia*, **99**, 3-7; *Mycologist*, **21**, 101-105.

Sirven, C., Gonzalez, E., Bufflier, E. *et al.* (2002) PCR-based method for detecting mutation allele frequencies for QoI resistance in *Plasmopara viticola*. *The BCPC Conference: Pests and diseases, Volumes 1 and 2. Proceedings of an international conference held at the Brighton Hilton Metropole Hotel, Brighton, UK*, 823-828.

Smith, M.L., Bruhn, J.N. and Anderson, J.B. (1992). The fungus *Armillaria bulbosa* is among the largest and oldest living organisms. *Nature*, **356**, 428-431.

Smith, O.P., Peterson, G.L., Beck, R.J. *et al.* (1996) Development of a PCR-based method for identification of *Tilletia indica*, causal agent of Karnal bunt of wheat. *Phytopathology*, **86**, 115-122.

Smith, T.J., Wilson, L., Kenwrick, S.J. *et al.* (1987) Isolation of a conserved sequence deleted in Duchenne muscular dystrophy patients. *Nucleic Acids Research*, **15**, 2167-2174.

Southern, E. (1975) Detection of specific sequences among DNA fragments separated by gel electrophoresis. *Journal of Molecular Biology*, **98**, 503-517.

Steffan, R.J. and Atlas, R.M. (1991) Polymerase chain reaction: applications in environmental microbiology. *Annual Review of Microbiology*, **45**, 137-161.

Steffan, R.J., Goksoyr, J., Bej, A.K. and Atlas, R.M. (1988) recovery of DNA from soils and sediments. *Applied Environmental Microbiology*, **54**, 2908-2915.

Sylvanen, A.C., Laaksonen, M. and Soderlund, H. (1986) Fast quantification of nucleic acid hybrids by affinity-based hybrid collection. *Nucleic Acids Research*, **14**, 5037-5048.

Tautz, D. (1989) Hypervariability of simple sequences as a general source for polymorphic DNA markers. *Nucleic Acids Research*, **17**, 6463-6471.

Tijssen, P. (1993) *Hybridization with Nucleic Acid Probes. Part 1. Theory and Nucleic Acid Preparation*, Elsevier, Amsterdam, 268 pp.

Tomita, M., Suzuki, K., Shimosato, K. *et al.* (1988) Enzyme-linked immunosorbent assay (ELISA) for plasma paraquat levels of poisoned patients. *Forensic Science International*, **37**, 11-18.

Toth, G., Gaspari, Z. and Jurka, J. (2000) Microsatellites in different eukaryotic genomes: Survey and analysis. *Genome Research*, **10**, 967-981.

Triglia, T., Peterson, M.G. and Kemp, D.J. (1988) A procedure for the *in vitro* amplification of DNA segments that lie outside the boundaries of known sequences. *Nucleic Acids Research*, **16**, 8186.

Tsumura, Y., Ohba, K. and Strauss, S.H. (1996) Diversity and inheritance of inter-simple sequence repeat polymorphisms in Douglas-fir (*Pseudotsuga menziesii*) and sugi (*Cryptomeria japonica*). *Theoretical and Applied Genetics*, **92**, 40-45.

Van Emmon, J., Hammock, B. and Seiber, J.N. (1986) Enzyme-linked immunosorbent assay (ELISA) for paraquat and its application to exposure analysis. *Analytical Chemistry*, **58**, 1866-1873.

Van Emmon J., Seiber, J.N. and Hammock, B. (1987) Application of an enzyme-linked immunosorbent assay (ELISA) to determine paraquat residues in milk, beef and potatoes. *Bulletin of Environmental Contamination Toxicology*, **39**, 490-497.

Van Vurdee, J.W.L. (1987) New approach in detecting phytopathogenic bacteria by combined immunoisolation and immunoidentification assays. *OEPP/EPPO Bulletin*, **17**, 139-148.

Vora, G.J., Meador, C.E., Stenger, D.A. and Andreadis, J.D. (2004) Nucleic acid amplification strategies for DNA microarray-based pathogen detection. *Applied and Environmental Microbiology*, **70**, 3047-3054.

Wallace, R.B., Shaffer, J., Murphy, R.F. *et al.* (1979) Hybridization of synthetic oligonucleotides to phi chi 174 DNA: the effect of a single base mismatch. *Nucleic Acids Research*, **6**, 3543-3557.

Wang, D., Coscoy, L., Zylberberg, M. *et al.* (2002) Microarray-based detection and genotyping of viral pathogens. *Proceeding of National Academy of Sciences USA*, **12**, 15687-15692.

Wang, D.G., Fan, J.B., Siao, C.J. *et al.* (1998a) Large-scale identification, mapping, and genotyping of single-nucleotide polymorphisms in the human genome. *Science*, **280**, 10777-10782.

Wang, G., Mahalingam, R. and Knap, H.T. (1998b). (C-A) and (G-A) anchored simple sequence repeats (ASSRs) generated polymorphism in soybean, *Glycine max* (L) Merr. *Theoretical and Applied Genetics*, **96**, 1086-1096.

Ward, E., Foster, S.J., Fraaije, B.A. and McCartney, H.A. (2004) Plant pathogen diagnostics: immunological and nucleic acid-based approaches. *Annals of Applied Biology*, **145**, 1-16.

Watson, J.D. and Crick, F.H.C. (1953) A structure for deoxyribose nucleic acid. *Nature,* **171**, 737.

Welsh, J. and McClelland, M. (1990) Fingerprinting genomes with arbitary primers. *Nucleic Acids Research*, **18**, 7213-7218.

Welsh, J., Chada, K., Dalal, S.S. *et al.* (1992) Arbitary primed PCR fingerprinting of RNA. *Nucleic Acids Research*, **20**, 4965-4970.

Widmer, F., Seidler, R.J., Gillevet, P.M. *et al.* (1998) A highly selective PCR protocol for detecting 16S rRNA genes of the genus *Pseudomonas* (*sensu stricto*) in environmental samples. *Applied and Environmental Microbiology,* **64**, 2545-2553.

Wiedmann, M., Czajika, J., Barany, F. and Batt, C.A. (1992) Discrimination of *Listeria monocytogenes* from other *Listeria* species by ligase chain reaction. *Applied Environmental Microbiology,* **58**, 3443-3447.

Williams, J.G.K., Kubelik, A.R., Livak, K.J. *et al.* (1990) DNA polymorphisms amplified by arbitary primers are useful as genetic markers. *Nucleic Acids Research,* **18**, 6531-6535.

Wolfe, A.D. and Liston, A. (1998) Contributions of PCR-based methods to plant systematics and evolutionary biology, in *Plant Molecular Systematics II* (eds D.E. Soltis, P.S. Soltis and J.J. Doyle), pp. 43-86.

Wong, W.C., White, M. and Wright, I.G. (1988) Production of monoclonal antibodies to *Fusarium oxysporum* f.sp. *cubense* race 4. *Letters Applied Microbiology,* **6**, 39-42.

Yasodha, R., Kathirvel, M., Sumathi, R. *et al.* (2004) Genetic analyses of Casuarinas using ISSR and FISSR markers. *Genetica,* **122**, 161-172.

Yershov, G., Barsky, V., Belgovskiy, A. *et al.* (1996) DNA analysis and diagnostics on oligonucleotide microchips. *Proceedings of National Academy of Sciences USA*, **93**, 4913-4918.

Zare, R., Gams, W. and Culham, A. (2000) A revision of *Verticillium* sect. Prostrata. I. Phylogenetic studies using ITS sequences. *Nova Hedwigia,* **71**, 465-480.

Zhou, S., Smith, D.R. and Stanosz, G.R. (2001) Differentiation of *Botryosphaeria* species and related anamorphic fungi using Inter Simple or Short Sequence Repeat (ISSR) fingerprinting. *Mycological Research,* **105**, 919-926.

Zietkiewicz, E., Rafalski, A. and Labuda, D. (1994) Genome fingerprinting by simple sequence repeat (SSR)-anchored polymerase chain reaction amplification. *Genomics,* **20**, 176-183.

CHAPTER 2

DISEASE ASSESSMENT AND YIELD LOSS

B.M. COOKE

2.1 INTRODUCTION

Disease assessment, or phytopathometry, (Large, 1966), involves the measurement and quantification of plant disease and is therefore of fundamental importance in the study and analysis of plant disease epidemics. Nutter *et al.* (2006) distinguished between disease assessment and phytopathometry, the former being defined as the process of quantitatively measuring disease intensity and the latter as the theory and practice of quantitative disease assessment. The importance of accurate disease assessment methods was identified in early reviews on phytopathometry and crop loss assessment by Chester (1950) and Large (1966). Kranz (1988) stated that without quantification of disease no studies in epidemiology, no assessment of crop losses and no plant disease surveys and their applications would be possible. Lucas (1998) advanced the idea that disease assessment includes a number of interrelated activities, such as the future progress of the disease, disease diagnosis, forecasting and crop loss. Strange (2003) maintained that that the measurement of plant disease and its effects on crop yield, quality and value are crucial for control priorities. Traditional methods of disease assessment, such as the use of pictorial keys derived from standard area diagrams to evaluate disease severity on a 0-100% scale, have now been joined by several new approaches made possible by rapid advances in computer technology. In addition, modern assays using immunological and molecular techniques for the identification, detection and quantification of plant pathogenic organisms are used. Other new approaches to phytopathometry have evolved in which remote sensing, image analysis and the detection of crop stress caused by disease (using changes in chlorophyll fluorescence and foliage temperature) are involved.

The relationship between phytopathometric data and yield loss has always suffered from a number of confounding factors and, as a result, many authors including James (1983), have criticized the lack of reliable estimates of crop losses due to plant disease. Chiarappa (1981) defined a crop loss model as 'a mathematical method used to describe the relationship between yield reduction and the intensity of harmful organisms'. It is therefore not surprising that current papers on crop loss assessment deal with recent developments in modelling crop losses. Hughes (1996) dealt with incorporating spacial patterns of harmful organisms into crop loss models; Madden and Nutter (1995) discussed this and other developments that have changed our understanding of the disease-yield loss relationship, such as healthy leaf area

B. M. Cooke, D. Gareth Jones and B. Kaye (eds.), The Epidemiology of Plant Diseases, 2nd edition, 43–80.

duration, radiation interception and time of infection, and Gaunt (1995) reviewed technologies in disease measurement and yield loss appraisal.

Of paramount importance in disease assessment and yield loss appraisal is the standardization of concepts and terms in order to improve communication between plant pathologists and across scientific disciplines (Nutter *et al.*, 1991). Standardization would also permit industry to compare assessment data from several field trials for product evaluation (Watson and Morton, 1990). Consequently, the reader is referred here to glossaries of disease assessment terms and concepts (Nutter *et al.*, 1991) and terms and concepts for yield, crop loss and disease thresholds (Nutter *et al.*, 1993).

2.2 WHY ASSESS DISEASE AND YIELD LOSS IN PLANTS?

The assessment of the amount of disease on a plant or a crop of plants is essential in any quantitative epidemiological study. Jones and Clifford (1978) and James (1983) identified a number of important reasons for phytopathometric and crop loss measurements, the most important of which must surely be that if we are not in a position to estimate the losses from diseases, then how can we decide rationally on how much to spend on control? Other reasons identified include the importance of disease survey data to farmers, plant breeders, fungicide manufacturers, economists and government agencies in determining the priorities for allocating resources and timing control measures. Finally, researchers and extension workers require precise methods for evaluating their experiments, particularly plant breeders where potential resistant germplasm is being screened. Unfortunately, as pointed out by Campbell and Madden (1990c), there is often a lack of a perceived need for estimates of yield loss by some scientists and administrators; this translates into a lack of funding for the research needed to provide such estimates, particularly at regional and national levels. On a global basis, James (1983) pointed out the urgent need for effective crop loss assessment programmes in the context of a world food situation that continues to be precarious, especially in those developing countries which can least afford to lose crop yield. Oerke and Dehne (1997) calculated actual and potential crop losses of eight major food and cash crops by evaluating data from the literature and field experiments from 1965 onwards. Yield limiting factors identified were water (the most important), genetic yield potential and adaptation, and crop losses due to plant pathogens, pests and weeds. Disease assessment and crop loss appraisal will become especially important in sustainable systems of crop protection, where critical evaluation of disease levels is required in order to assess the effectiveness of proposed low-input, environmentally friendly strategies, such as the use of cereal cultivar mixtures.

2.3 METHODS USED IN SAMPLING PLANTS FOR DISEASE

Any sampling method used in disease assessment must be random, representative and objective and, depending on the disease involved, can be destructive or non-destructive (Jones and Clifford, 1978). Until recently, little research in plant pathology has been devoted to sampling methods appropriate for assessing disease incidence and severity (Hughes, 1999). Traditional sampling methods involve diagonal sampling in farmers' fields where at least 50 tillers are sampled at random along each diagonal; the pattern of disease spread, whether it is scattered or uniform, may influence the number of samples taken, which in turn is related to the standard deviation of disease incidence (Church, 1971). In small experimental plots, sampling may not be customary as replication produces the needed accuracy (Zadoks and Schein, 1979); however, a minimum of 10 samples is often used for small cereal plots. Depending on the disease, the usual emphasis in disease measurement is given to incidence or severity within the sampling unit. Chaube and Singh (1991) reviewed a number of terms used in sampling, including entity, sample size, sample point and sampling fraction, all of which need to be considered for the satisfactory measurement of disease, particularly over large field areas.

Disease incidence, severity and spatial pattern depend on data obtained from field samples. The accuracy of these data, as well as the time and effort required to obtain them, are affected by the sampling technique used. In a study of three naturally occurring epidemics of leek rust (caused by *Puccinia allii*), de Jong and de Bree (1995) concluded that in the development of a practical sampling method for detection of the disease, it was necessary to take into account a clustered distribution of diseased plants. In their study, the Black-White (BW) join-count statistic was used to detect non-randomness in the spatial distribution of rust-infected leek plants. Delp *et al.* (1986a) developed a computer software system called Field Runner to simplify the task of sampling fields. The system uses the stratified random sampling design (SRSD) with single-stage cluster sampling; this provides an unbiased sample and a lower error of disease incidence estimates than conventional diagonal, 'X' or 'W' sampling designs. A hand-held microcomputer is used to direct the operator to each sample site within a sector, each site being composed of a cluster or transect (see Fig. 2.1); fields can be assessed for severity of one disease or for incidence of one to several diseases.

In a further paper, Delp *et al.* (1986b) evaluated field sampling techniques for estimation of disease incidence using computer-simulated field tests. Disease incidence and aggregation were varied to determine their effects on sampling techniques. The authors concluded that SRSD required the least number of samples and the lowest sample intensity to estimate disease incidence within a 95% confidence interval for all field types.

In a three-year study of winter wheat diseases, Parker and Royle (1993) developed a novel large-scale sampling procedure using randomly positioned transects based on the theory of autocorrelation. The procedure allowed valid tests of significance to be made on the autocorrelation coefficients calculated; the sample data obtained were also suitable for use in mapping analysis and the production of semivariograms.

A series of papers (Fleischer *et al.,* 1999; Hughes, 1999; Madden and Hughes, 1999; Morrison, 1999 and Nyrop *et al.,* 1999) confirmed that data obtained by sampling are crucial for improved decision making by farmers and growers in crop loss assessment and disease management. However, decision making will be imperfect

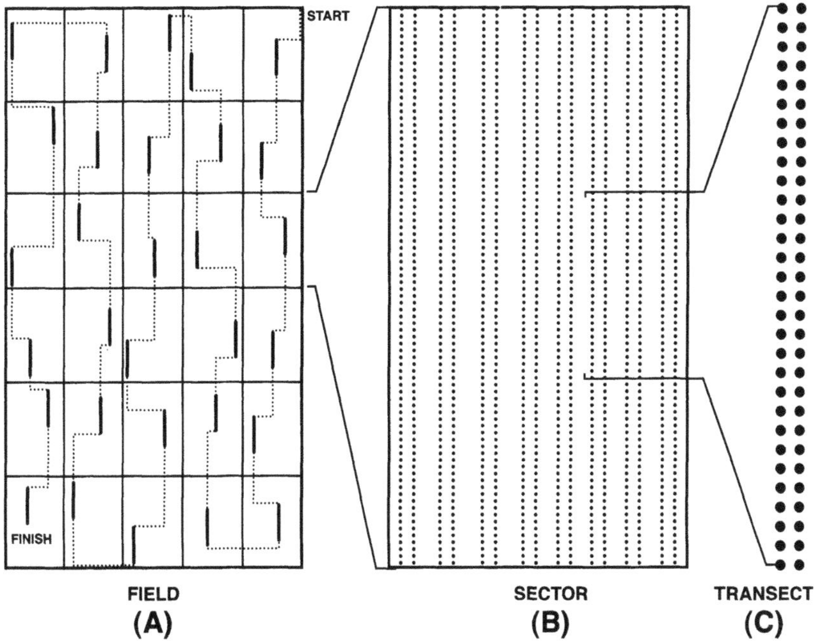

Figure 2.1. (A) Path (broken line) generated by the computer system (Field Runner) used to direct the operator to the sampling sites (solid lines) of a stratified random design. (B) One sector composed of two-row beds of individual plants (points) and a randomly located transect selected within the sector (bracketed area). (C) Two-row transect with 30 plants per row. (From Delp et al., *1986a).*

whatever the sampling regime and Hughes (1999) described operating characteristic (OC) curves as instrumental in establishing and evaluating the performance of sampling schemes; these were affected by the sampling distribution of pathogen intensity and by sampling size (Madden and Hughes, 1999b). In a separate paper, Madden and Hughes (1999a) explored the relationship between disease incidences at two levels (field and individual plants) in a spatial hierarchy in order to find a simple approximation (effective sample size) to interrelate incidence at the two levels; they concluded that it was possible to predict incidence at the lower level from incidence at the higher level in the presence of aggregation without using the complicated beta-binomial function, making it easier to use group sampling.

2.4 TIMING AND FREQUENCY OF DISEASE ASSESSMENT

Disease assessment data must be qualified by the growth stage of the crop or plant at the time of the assessment. This is because the effects of a given level of disease on plant growth and yield and the importance of that disease level in relation to the progress of an epidemic will vary at different plant growth stages. Consequently, it is important to be familiar with the keys currently available and other methods, for determining stages of plant growth. It is also pertinent here to briefly consider the frequency with which disease assessments should be carried out, as this will obviously relate to the type of disease being assessed.

Simple interest (Van der Plank, 1963) diseases (monocyclic or polyetic) may well require fewer assessments than compound interest or polycyclic diseases. As pointed out by Schumann (1991), with monocyclic pathogens the amount of disease at the end of the season should be proportional to the initial inoculum but, with polycyclic pathogens, the relationship is less direct. Factors such as temperature, moisture and crop plant resistance will influence the final disease level more than initial inoculum. Indeed, with some polycyclic pathogens (e.g. *Phytophthora infestans,* cause of late blight of potato), increase through secondary inoculum production is so rapid that different levels of initial inoculum can still result in the total destruction of a potato field. Clearly then, several disease assessments would be necessary to effectively monitor the progress of a potato blight epidemic in order to implement appropriate control methods using disease threshold values. It should be remembered, however, that some polycyclic pathogens might not always cause as much damage as those monocyclic pathogens that reach critical levels over a relatively short number of years.

Disease assessments should be related to a stage of plant development that determines an important physiological function - for example grain filling in cereals. For many years, growth stages in cereals were scored on the Feekes scale, illustrated by Large (1954) (Fig. 2.2). The Feekes scale was superseded by the decimalised key of Zadoks *et al.* (1974) that facilitates computerized data processing (Table 2.1); this key is essentially a further development of the Feekes scale and provides better descriptions of the earlier stages of cereal growth for all small-grain species growing in a wide range of climatic conditions. The decimalized key of Zadoks *et al.* (1974) was illustrated by Tottman *et al.* (1979) and Tottman and Broad (1987) (Fig. 2.3), and differs from the Feekes scale in describing individual plants rather than classifying crop growth stages.

Correlations between the growth stages defined by the decimal code and apical development in the cereal plant are often poor but certain stages give a useful indication of such development. Examination of the shoot apex requires plant dissections under a microscope or use of a cereal diagnostic system such as that produced by the Bayer Company based on an original design by Verreet and Hoffmann (Technical University of Munich, Germany). Such an approach is considered to be more precise than examination of growth stage by the naked eye and, therefore, permits more exact timing of plant protection measures such as disease measurement and fungicide application. The Cereal Development Guide by Kirby and Appleyard (1981) clearly illustrates important stages of early apical development in wheat and barley, such as the vegetative stage, double ridge stage, floret primordium stage and the terminal spikelet stage.

A uniform decimal code for growth stages of crops and weeds was produced by Lancashire *et al.* (1991) and is known as the BBCH (BASF, Bayer, Ciba-Geigy and Hoechst) scale. The scale and codes were based on those of Zadoks *et al.* (1974) for cereals but, in addition, deals with rice, maize, oilseed rape, field beans, peas,

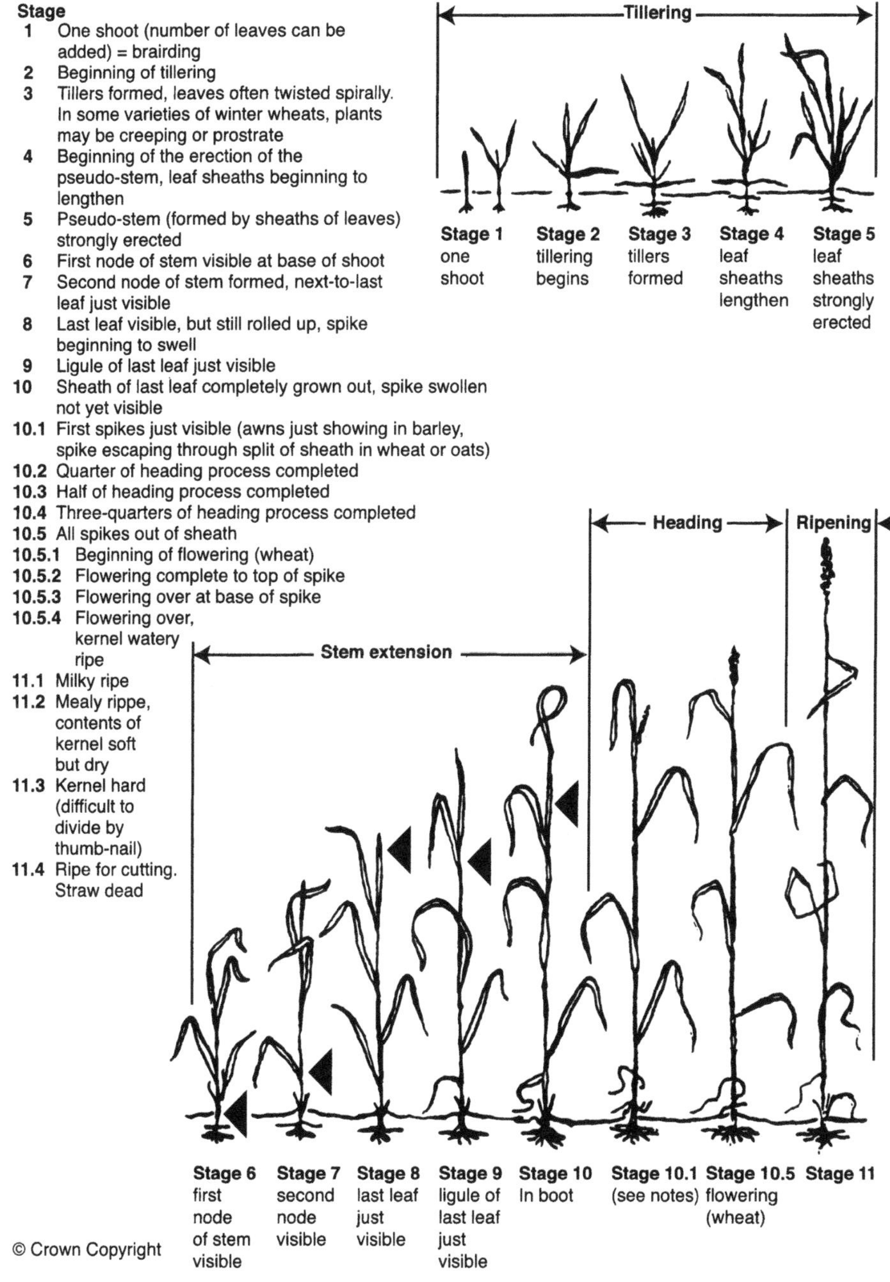

Figure 2.2. The Feekes growth stage scale for cereals, illustrated by Large (1954).

Table 2.1. Decimalised key of Zadoks et al. *(1974) for the growth stages of cereals*

code	
0	**Germination**
00	Dry seed
01	Start of imbibition
02	–
03	Imbibition complete
04	–
05	Radicle emerged from caryopsis
06	–
07	Coleoptile emerged from caryopsis
08	–
09	Leaf just at coleoptile tip
1	**Seedling growth**
10	First leaf through coleoptile
11	First leaf unfolded
12	2 leaves unfolded
13	3 leaves unfolded
14	4 leaves unfolded
15	5 leaves unfolded
16	6 leaves unfolded
17	7 leaves unfolded
18	8 leaves unfolded
19	9 or more leaves unfolded
2	**Tillering**
20	Main shoot only
21	Main shoot and 1 tiller
22	Main shoot and 2 tillers
23	Main shoot and 3 tillers
24	Main shoot and 4 tillers
25	Main shoot and 5 tillers
26	Main shoot and 6 tillers
27	Main shoot and 7 tillers
28	Main shoot and 8 tillers
29	Main shoot and 9 or more tillers
3	**Stem elongation**
30	Pseudo stem erection
31	1st node detectable
32	2nd node detectable
33	3rd node detectable
34	4th node detectable
35	5th node detectable
36	6th node detectable
37	Flag leaf just visible
38	–
39	Flag leaf ligule/collar just visible
4	**Booting**
40	–
41	Flag leaf sheath extending
42	–
43	Boots just visibly swollen
44	–
45	Boots swollen
46	–
47	Flag leaf sheath opening
48	–
49	First awns visible
5	**Inflorescence emergence**
50*, 51	First spikelet of inflorescence just visible
52*, 53	1/4 of inflorescence emerged
54*, 55	1/2 of inflorescence emerged
56*, 57	3/4 of inflorescence completed
58*, 59	Emergence of inflorescence completed
6	**Anthesis**
60, 61	anthesis (not easily detectable in barley)
62	–
63	–
64, 65	Anthesis half-way
66	–
67	–
68, 69	Anthesis complete
7	**Milk development**
70	–
71	Caryopsis water ripe
72	–
73	Early milk
74	–
75	Medium milk (Increase in solids of liquid endosperm notable when crushing the caryopsis between fingers)
76	–
77	Late milk
78	–
79	–
8	**Dough development**
80	–
81	–
82	–
83	Early dough
84	–
85	Soft dough (Finger-nail impression not held)
86	–
87	Hard Dough (Finger-nail impression held, infloresence losing chlorophyll)
88	–
89	–
9	**Ripening**
90	–
91	Caryopsis hard (difficult to divide by thumb-nail)
92	Caryopsis hard (can no longer be dented by thumb-nail)
93	Caryopsis loosening in daytime
94	Over-ripe, straw dead and collapsing
95	Seed dormant
96	Viable seed giving 50% germination
97	Seed not dormant
98	Secondary dormancy induced
99	Secondary dormancy lost

*Even code numbers refer to crops in which this stage is reached by all shoots simultaneously and odd numbers to unevenly developing crops when 50% of the shoots are at the stage given.

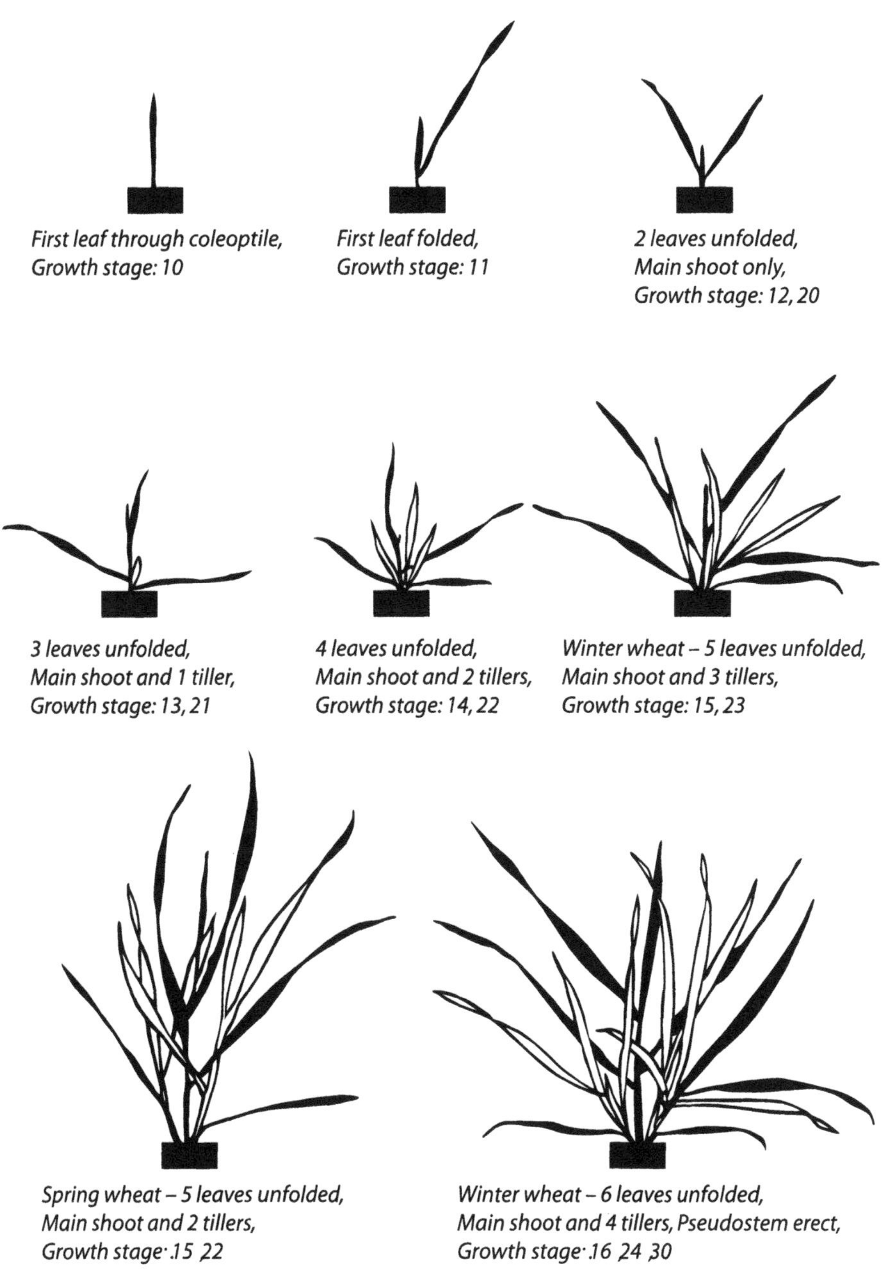

Figure 2.3. The decimalized key of Zadoks et al. *(1974) for cereal growth stages, illustrated by Tottman* et al. *(1979) and Tottman and Broad (1987).*

sunflower and weeds, the aim being to establish a universal scale using a consistent set of numeric codes that can readily be adapted to all crops (Table 2.2). Individual scales for maize, pea, faba bean, sunflower and oilseed rape growth stages were published respectively by Hanway (1963), Knott (1987, 1990), Schneiter and Miller (1981), and Sylvester-Bradley *et al.* (1984); Chiarappa (1971) published keys for numerous crops. The use of remote sensing with hand-held radiometers (section 2.5.4) offers possibilities for indirectly measuring crop growth stages based on spectral reflectance changes from a healthy crop during plant growth. However, accurate calibration of radiometer readings with existing decimalised codes for crop growth stages in order to aid and standardize disease assessment would be desirable.

2.5 METHODS OF DISEASE ASSESSMENT

In any disease assessment or phytopathometric method, two criteria must be satisfied; these were described by James (1983) as consistency between observers and simplicity for speed of operation. These criteria, therefore, dictate that all assessment methods should be well defined and standardized at the earliest possible stage of their development. Campbell and Madden (1990a) pointed out that a successful system for the assessment of disease gives results that are accurate, precise and reproducible and presented the analogy of the target used by an archer where the objective was to shoot all arrows into the centre circle of the target (Fig. 2.4): obviously, option A would be the most desirable for any assessment method. Strange (2003) pointed out that although there is generally little disagreement between observers at either end of a descriptive disease severity scoring scale (such as 1-9), wide variation can occur in the central (often critical) part of the scale especially if there are no visual prompts.

Disease can be measured using direct methods (i.e. assessing disease in or on plant material) or indirect methods (e.g. monitoring the spore population using spore traps). Obviously direct methods are likely to be more strongly correlated with yield losses in the crop and are therefore to be preferred. However, recent methods involving remote sensing and detection of crop stress due to disease are likely to increase the accuracy of indirect disease measurements. Direct methods are concerned with both the quantitative and qualitative estimations of disease.

2.5.1 Direct quantitative methods

Direct quantitative methods are largely concerned with measurements of incidence or severity, defined as follows.

$$\text{Disease incidence } (I) = (\text{number of infected plant units} / \text{total number of plant units assessed}) \times 100$$

$$\text{Disease severity } (S) = (\text{area of diseased tissue} / \text{total tissue area}) \times 100$$

Table 2.2. Extract from the BBCH growth stage scale for cereals, rice and maize (Lancashire et al., *1991)*

0 *Germination*

BBCH code	*Cereals*	*Rice*	*Maize*
00	Dry seed (caryopsis)	Dry seed (caryopsis)	Dry seed (caryopsis)
01	Beginning of imbibition	Beginning of imbibition	Beginning of imbibition
02	–	–	–
03	Imbibition complete	Imbibition complete (pigeon breast)	Imbibition complete
04	–	–	–
05	Radicle emerged from caryopsis	Radicle emerged from caryopsis	Radicle emerged from caryopsis
06	Radicle elongated, root hairs and/or side roots visible	Radicle elongated, root hairs and/or side roots visible	Radicle elongated, root hairs and/or side roots visible
07	Coleoptile emerged from caryopsis	Coleoptile emerged from caryopsis (in water-rice stage occurs before stage 05)	Coleoptile emerged from caryopsis
08	–	–	–
09	Emergence: coleoptile penetrates soil surface (cracking stage)	Imperfect leaf emerges (still rolled) at the tip of the coleoptile	Emergence: coleoptile penetrates soil surface (cracking stage)

1 *Leaf development*

BBCH code	*Cereals*	*Rice*	*Maize*
10	First leaf through coleoptile	Imperfect leaf unrolled, tip of first true leaf visible	First leaf through coleoptile
11	First leaf unfolded	First leaf unfolded	First leaf unfolded
12	2 leaves unfolded	2 leaves unfolded	2 leaves unfolded
13	3 leaves unfolded	3 leaves unfolded	3 leaves unfolded
14	4 leaves unfolded	4 leaves unfolded	4 leaves unfolded
15	5 leaves unfolded	5 leaves unfolded	5 leaves unfolded
16	6 leaves unfolded	6 leaves unfolded	6 leaves unfolded
17	7 leaves unfolded	7 leaves unfolded	7 leaves unfolded
18	8 leaves unfolded	8 leaves unfolded	8 leaves unfolded
19	9 or more leaves unfolded	9 or more leaves unfolded	9 or more leaves unfolded

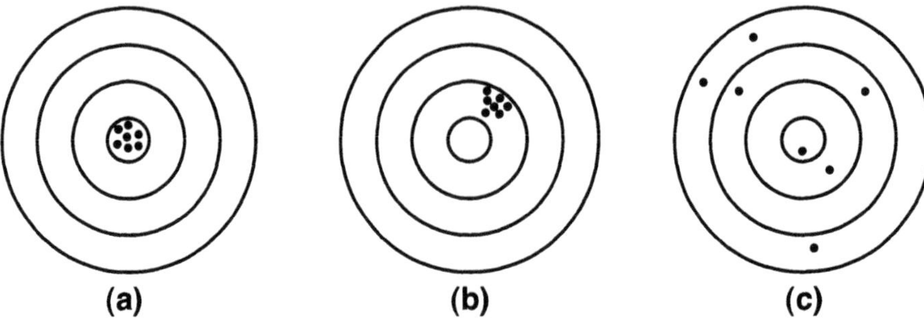

Figure 2.4. Accuracy and precision of an archer when the objective is to place all arrows in the central circle: (a) accurate and precise; (b) not accurate, but precise; (c) not accurate and not precise. (From Campbell and Madden, 1990a).

Although assessment of disease incidence is traditionally based on visual disease symptoms, the definition can easily accommodate other more modern methods such as the enzyme-linked immunosorbent assay (ELISA) and polymerase chain reaction (PCR) (section 2.5.3); disease incidence is a binary variable, that is, a plant unit is either (visibly) diseased or not (Madden and Hughes, 1999). Disease incidence would be suitable for assessing systemic infections which may result in total plant loss (e.g. viruses or cereal smuts) as well as many root diseases, or where a single lesion causes leaf death (e.g. axil lesions in barley caused by *Rhynchosporium secalis)* but may also be useful in the early stages of an epidemic caused by a cereal foliar pathogen when both incidence (number of tillers affected) and severity (leaf area affected) are related and increase simultaneously (James, 1983). Hughes and Madden (1995) and Madden and Hughes (1999) reviewed the methodology for the analysis of disease incidence data especially where aggregated patterns of disease occur. The authors pointed out that generalized linear models (GLMs) can be used for binomially distributed (i.e. random) data and overcome the problems of applying analysis of variance (ANOVA) to proportions; however if diseased plants or leaves are aggregated or in clusters (i.e. beta-binomially distributed), difficulties can arise in determining statistical significance between experimental treatments. Ridout and Xu (2000) explored the relationships between several quadrat-based statistical methods that have been applied to spatial aspects of disease incidence data. In contrast to incidence, disease severity is a continuous variable typically bound by 0 and 1, and a measure of the quality of plant tissue rather than the number of plant units affected (Madden and Hughes, 1999). In general, incidence is easier and quicker to assess than severity and is therefore more convenient to use in disease surveys where many observations are needed or when non-experts are used to collect data (Madden and Hughes, 1995); however, severity may be a more important and useful measurement for many diseases and is sometimes measured as the number of colonies (or lesions) per plant unit (disease density) (Xu and Madden, 2000).

Relationships between incidence and severity (*I-S* relationships) are examples of data comprising a spatial hierarchy and are an epidemiologically significant concept; any quantifiable relationship between the two parameters may permit more precise measurements of severity. Three types of analysis have been used to describe the *I-S* relationship: these are correlation and regression, multiple infection models and the measurement of aggregation (Seem, 1984). Drawing on Seem's approach, Hughes *et al.* (1997) formulated relationships between measurements of disease incidence made at two levels in a spatial hierarchy. Disease incidence at the higher scale was shown to be an asymptotic function of incidence at the lower scale, the degree of aggregation at that scale, and the size of the sampling unit. Turechek and Madden (2003) extended this work to a three-scaled hierarchial system. Daamen (1986b) studied the *I-S* relationship in powdery mildew of winter wheat. The model explained the effects of leaf, pustule and cluster size on the *I-S* relationship and could also be applicable to systemic foliar diseases when a minimum lesion size is defined. Hughes *et al.* (2004) revisited Daamen's *I-S* relationship with the aim of widening the application of the model. Fitt *et al.* (1998) working with light leaf spot (caused by *Pyrenopeziza brassicae*) on winter oilseed rape studied the *I-S* relationship by assessing the disease as % plants, % leaves or % leaf area (severity); regression analyses showed good relationships between % leaves (incidence at the leaf scale and severity at the plant scale) and % plants (incidence) until % plants approached 100%. Silva-Acuna *et al.* (1999) working with coffee rust found incidence could be used to estimate two measures of severity. Groth and Ozmon (1999) tested the repeatability and relationship of incidence and severity assessments in *Fusarium* ear blight of wheat over a three-year period and found them to be highly correlated. Xu *et al.* (2004) also investigated the *I-S* relationship in *Fusarium* ear blight of wheat in order to predict disease severity (number of infected spikelets) using infected ear incidence. This relationship was considered important for predicting the risk of mycotoxin contamination in the grain. The authors found a robust *I-S* relationship assuming a fixed variance-mean relationship and a negative binomial distribution for the number of infected spikelets and based on the complementary log-log or logit transformation of ear and spikelet infection incidence. However, incidence can be overestimated in an entire field and a more reliable method might be to visually estimate the incidence of *Fusarium* ear blight in several sub-samples and then to calculate an average for the field or plot (Shaner, 2003). Xu and Madden (2002) explored the relationships between disease incidence and colony density at the same scale and across scales (leaf and shoot) in apple powdery mildew with the aim of finding a robust relationship for predicting density using leaf or shoot incidence. The authors pointed out that with binomially distributed random data, incidence at the lower scale could be predicted from that at the higher scale, but with aggregated data, the beta-binomial distribution could not easily predict incidence of the lower scale from that at the higher scale.

Other direct methods of quantifying disease may involve estimations of disease intensity or prevalence. Intensity is often used to denote measures of the number of fungal colonies on leaves; it is also measured as both incidence and severity. Jeger (1981) found consistent relationships between incidence and intensity for apple scab caused by *Venturia inaequalis.* Daamen (1986a) working with wheat powdery mildew concluded that at low disease intensities (<5 pustules per leaf) and small sample sizes (<12 leaves) it was more efficient to sample the upper surface only than both surfaces. Prevalence is an ambiguous term and usually refers to disease incidence within a geographical area. For example, ten fields in an area are inspected for disease and six are found to be infected; the disease prevalence for that area is 60%.

Most assessment keys have been designed to measure disease severity using either descriptive or pictorial (picture) keys. With either type of key, it is essential that standardization is maintained and the use of arbitrary categories such as slight, moderate or severe should be avoided. Such broad categories take no account of the fact that the eye apparently assesses diseased areas in logarithmic steps, as stated by the Weber-Fechner law for visual acuity (for appropriate stimuli, visual response is proportional to the logarithm of the stimulus). Thus up to 50% disease severity, the eye reads diseased tissue but above this value healthy tissue is judged. Horsfall and Barratt (1945) therefore proposed a logarithmic scale for the measurement of plant disease severity, in which grades were allotted according to the leaf area diseased: 1 = nil, 2 = 0-3%, 3 = 3-6% and so on to 11 = 97-100% and 12 = 100%. This scale reads the diseased tissues in logarithmic units below 50% and healthy tissue in the same units above 50%. Thus, if the Horsfall-Barratt hypothesis is correct, the least reliable estimates of severity should occur at the 50% level. Forbes and Jeger (1987) found the greatest overestimation of severity occurred at levels below 25%, suggesting that the Horsfall-Barratt hypothesis over-simplifies the stimulus response relationship of visual disease severity assessment. Hebert (1982) pointed out that some visual estimates might not obey the Weber-Fechner law. Nutter and Esker (2006) revisited the Weber-Fechner law using a classical method developed in the field of psychophysics (the method of comparison of stimuli) and concluded that although Weber's law appeared to hold true, Fechner's law did not. Furthermore, the relationship between actual disease severity and estimated severity was found to be linear rather than logarithmic as proposed by Horsfall and Barratt (1945). There is, therefore, no single accepted method of making visual estimates of disease severity, and a linear percentage scale is often used.

Chaube and Singh (1991) and James (1983) identified the advantages of the percentage scale as: the upper and lower limits are always uniquely defined; the scale is flexible and can be divided and subdivided; it is universally known and can be used to measure incidence and severity by a foliar or root pathogen; and it can easily be transformed for epidemiological analysis, e.g. transformation to logits for calculation of *r,* the apparent infection rate. The best-known descriptive key to utilize the percentage scale was that published by the British Mycological Society (Anon., 1947) for measuring potato late blight (Table 2.3).

The pictorial disease assessment key uses standard area diagrams that illustrate the developmental stages of a disease on small simple units (leaves, fruits) or on large composite units such as branches or whole plants. Such standard diagrams are derived from a series of disease symptom pictures that may be in the form of line drawings, photographs or even preserved specimens. The assessment scale of Cobb (1892) for wheat rust was among the first to use standard area diagrams; this was joined by numerous others for disease assessment on a wide range of crops (e.g. Dixon and Doodson, 1971; James, 1971) (Fig. 2.5). Campbell and Madden (1990a) provided a useful tabular summary of pictorial disease assessment keys available for measuring disease severity on a range of hosts using the principle of standard area

Table 2.3. Descriptive key for assessment of late blight of potatoes caused by Phytophthora infestans *(Anon., 1947)*

Blight (%)	*Disease severity description*
0	Not seen on field
0.1	Only a few plants affected here and there; up to 1 or 2 spots in 12 yards radius
1	Up to 10 spots per plant, or general light spotting
5	About 50 spots per plant or up to 1 leaflet in 10 attacked
25	Nearly every leaflet with lesions, plants still retaining normal form: field may smell of blight, but looks green although every plant is affected
50	Every plant affected and about half of leaf area destroyed by blight; field looks green flecked with brown
75	About $^{3}/_{4}$ of leaf area destroyed by blight: field looks neither predominantly brown nor green. In some varieties the youngest leaves escape infection so that green is more conspicuous than in varieties like King Edward, which commonly shows severe shoot infection
95	Only a few leaves left green, but stems green
100	All leaves dead, stems dead or dying

Notes on assessment

In the earlier stages of a blight epidemic parts of the field sometimes show more advanced decay than the rest and this is often associated with the primary foci of the disease. Records may then be made as, say 1 + pf 25, where pf 25 means 25% in the area of the primary foci.

Make successive assessments at intervals of 7-14 days to record progress of blight. Begin in good time as both nil and starting date records (0.1%) are important. Difficulties in judging allowance to be made for stem blight on a particular date are overcome by making another assessment later.

diagrams. Since the ultimate aim is to relate disease to yield loss, the plant units assessed should ideally be important contributors to yield, for example, the top two leaves of a cereal plant. Standard area diagrams were traditionally and painstakingly prepared using graph paper outlines but the use of planimeters, electronic scanners and image analyzers have improved and quickened their production.

Despite the above measures to standardize assessment keys and to eliminate as far as possible operator error (subjectivity), the visual assessment of disease severity suffers from fundamental errors. Standard area diagrams do not display the variegated patterns of disease so commonly caused by a plant pathogen, especially on a leaf. Thus an observer is compelled to visualize the total area that the various lesion shapes would cover if they could be combined and then expressed as a percentage of the total area of the leaf. A second problem relates to variation in leaf

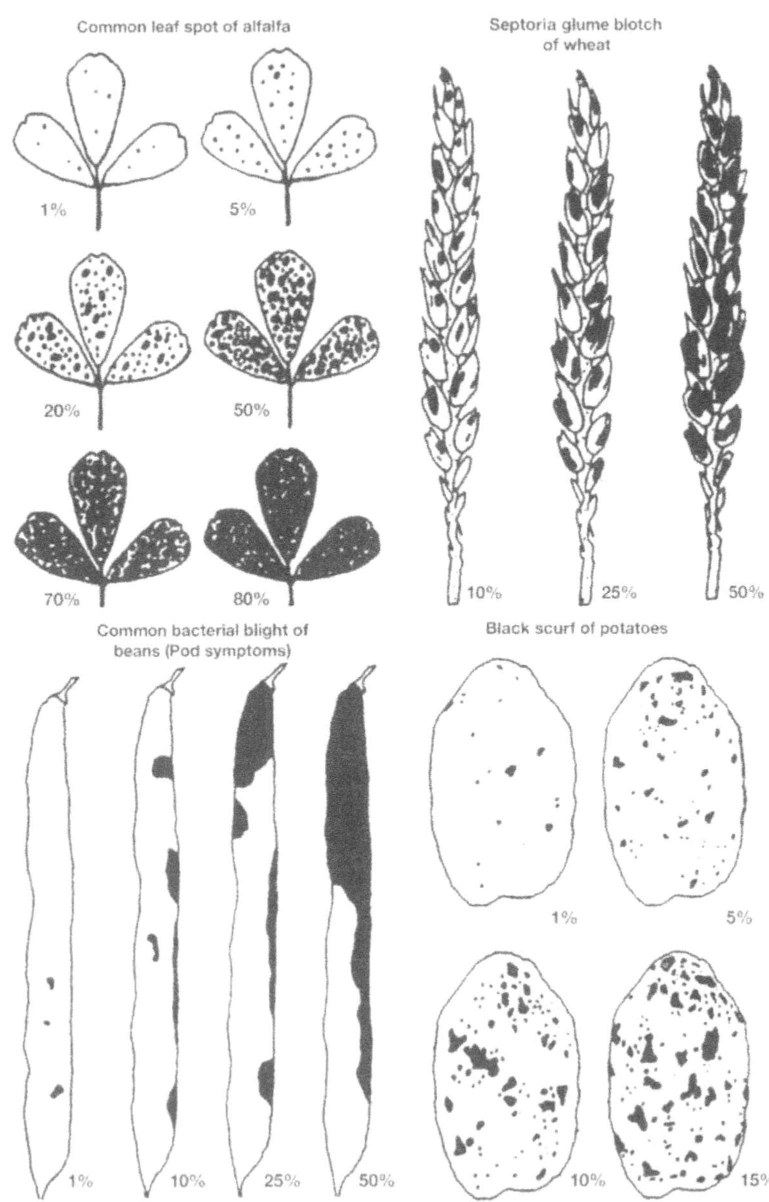

Figure 2.5. Examples of pictorial assessment keys for estimating disease severity (after James, 1971).

size and how this affects the observer's assessment of severity. The key for the assessment of barley leaf blotch disease caused by *Rhynchosporium secalis,* devised by James *et al.* (1968) (Fig. 2.6), usefully attempted to relate comparable percentage areas of disease on four standard area leaf diagrams of barley of differing size classes divided into 10% divisions.

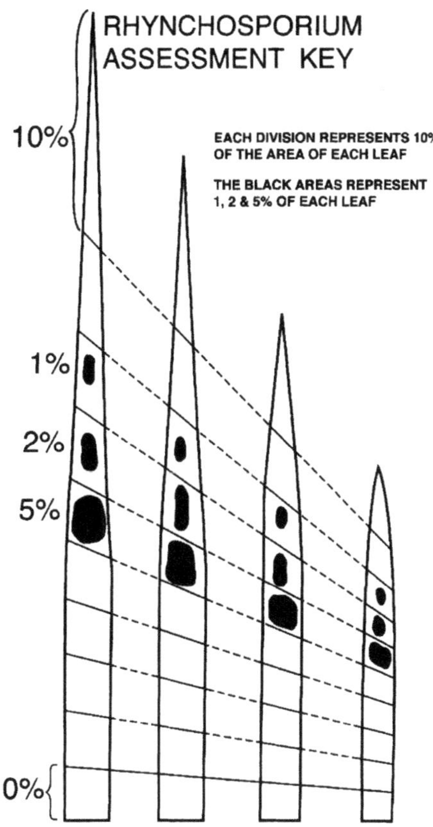

Assessment key for *Rhynchosporium* leaf blotch or scald of barley. Match the leaf to one of the diagrams and use the black areas (representing 1%, 2% and 5% of each leaf) as a guide in assessing the percentage leaf (lamina) area covered by small isolated lesions, and the 10% sections for the larger lesions that have coalesced.

Figure 2.6. Pictorial assessment key for leaf blotch of barley caused by Rhynchosporium secalis *(from James* et al., *1968).*

Parker *et al.* (1995) compared visual estimates of wheat disease severity with actual severities using image analyses of tracings of diseased leaves infected by *Septoria tritici* and *Blumeria (Erysiphe) graminis;* results showed that observer estimates were imprecise, inaccurate and varied considerably over short time-scales, but that relative bias decreased with increasing disease severity, so that overestimations occurred at low (<10%) disease severity, or 30-40% leaf senescence. The work also showed that such visual assessment errors could alter experimental conclusions. Parker and Royle (1993) and Lovell *et al.* (1997) therefore developed and used a new key for assessing foliar disease severity in wheat, in which pictorial representation of disease levels was presented on a generic cereal leaf divided into a grid of 1% sectors of leaf lamina (Fig. 2.7) with the following claimed advantages: a reduced need for interpolation; no reliance on high contrast diagrams; and avoidance of the need for different keys for assessments of more than one disease. Use of the new key revealed more comparable assessments were possible between observers, although precision of assessments did not improve.

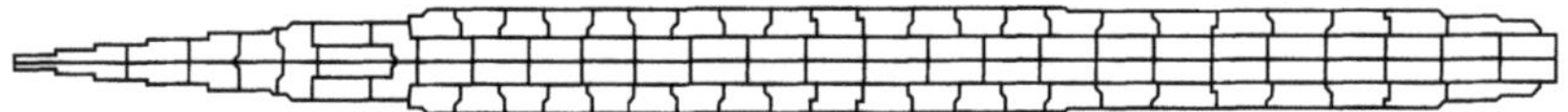

Figure 2.7. Prototype of a new wheat leaf disease assessment aid to avoid some of the disadvantages associated with the use of conventional disease assessment keys. Each sector of the grid is equal to 1%. (Parker and Royle, 1993).

Nutter and Schultz (1995) concluded that the accuracy and precision of disease assessments was improved simply by selecting the most appropriate methods and by training observers to assess disease severity using computerized disease assessment training programmes such as AREAGRAM, DISTRAIN and Disease.Pro. Although AREAGRAM (Shane *et al.*, 1985) graded user's performance, it generated only standard area diagrams with fixed disease patterns. DISTRAIN (Tomerlin and Howell, 1988) was developed as a training programme for disease assessment using variegated patterns of disease severity for eight common foliar diseases of cereals; the programme also allowed a comparison of estimated severity with actual severity. Nutter and Worawitlikit (1989) expanded the computer training concept in their advanced programme for peanut diseases, Disease.Pro and later in 1998, developed a more generic disease assessment programme, Severity.Pro, that allowed the user to select from a menu of leaf shapes (e.g. alfalfa, apple, barley, cucumber, grape, tomato) and lesion types (e.g. anthracnose, blotch, downy mildew, target spot, powdery mildew) so mimicking almost any foliar pathosystem (Nutter *et al.*, 2006).

There are many variations and modifications of the standardized pictorial disease assessment key described so far in this chapter. One of the more useful of these is the Saari-Prescott 0-9 scale (Saari and Prescott, 1975) incorporating a double digit 00-99 scale (Fig. 2.8) for evaluating the intensity (severity and vertical disease progress) of foliar diseases (except rusts) in wheat, triticale and barley. In this system, the first digit gives the relative height of the disease using the original 0-9 Saari-Prescott scale as a measure and the second digit shows disease severity but in terms of 0-9 (0%-90% coverage in equal divisions of 10%). So in a plant with a disease height of 5 and an average disease coverage on the upper four leaves of 10%, the numerical disease description is 51. A further variation is the Eyal-Brown diagrammatic scale for estimating pycnidial density of *Septoria tritici* per unit wheat leaf area (Eyal and Brown, 1976); the scale uses actual observed pycnidial coverage and a scaled percentage possible pycnidial coverage.

Figure 2.8. Saari-Prescott (0-9) scale (Saari and Prescott, 1975) for appraising the intensity of foliar diseases in wheat and barley. (See text for explanation).

Other direct quantitative methods of measuring disease involve computing coefficients and indices, and measuring components of partial disease resistance (PDR). Eyal and Ziv (1974) used a Septoria Progress Coefficient (SPC) in which plant and disease height were determined, where SPC = disease height (cm)/plant height (cm). SPC indicates the position of pycnidia relative to plant height regardless of pycnidial coverage and allows a comparison of infection placement on cultivars with different plant stature. A disease index for measuring eyespot infection on wheat caused by *Pseudocercosporella herpotrichoides* was produced by Scott and Hollins (1974). Here, tillers taken at random from the field are assigned to one of the infection categories and an index calculated from the formula (Table 2.4). Oyarzun (1993) computed a root disease index (RDI) - a measure of soil inoculum potential - for root rot of pea in which disease of the epicotyl and root was weighted more heavily than that of the cotyledons and xylem. Chaube and Singh (1991) defined an index number as a 'specialized type of average' used to measure the changes in some quantity that we cannot observe directly. The authors concluded that weighted indices may be preferable to unweighted indices and reviewed the following four methods for computing a disease index: as a simple arithmetic average; as a weighted arithmetic average; as a logarithm of the geometric mean of the percentage assessment; and as severity estimates for larger areas. Diamond and Cooke (1999) developed a novel *in vitro* detached leaf assay for early screening of *Fusarium* ear blight resistance in wheat using *Microdochium nivale*, a member of the *Fusarium* ear blight complex capable of forming discrete leaf lesions. PDR components measured on the leaf segments were incubation period (time to first appearance of symptoms), latent period (time to sporulation) and lesion size; correlations were obtained between these components and the development of *Fusarium* ear blight on whole wheat plants. The bioassay was further developed and evaluated by Browne and Cooke (2004), and later extended to barley and oats (Browne and Cooke, 2005).

Table 2.4. Calculation of a disease index for eyespot of wheat caused by Pseudocercosporella herpotrichoides *(Scott and Hollins, 1974)*

Infection category	*Disease severity description*
0	**Uninfected**
1	**Slight eyespot** (one or more small lesions occupying less than half the circumference of the stem)
2	**Moderate eyespot** (one or more lesions occupying at least half the circumference of the stem)
3	**Severe eyespot** (stem completely girdled with lesions; tissue softened so that lodging would readily occur)

Notes on assessment

1. Examine 20 tillers per 20 m^2 plot.
2. Assign each tiller to one of the infection categories above.
3. Write the number of tillers in each category on the record sheet.
4. An index will be calculated from the data as follows:

$$\text{Disease index} = \frac{(0\times a)+(1\times b)+(2\times c)+(3\times d)}{(a+b+c+d)}\times\frac{100}{3}$$

where *a, b, c* and *d* are the number of tillers examined which fall into the categories 0, 1, 2, and 3, respectively.

2.5.2 Direct qualitative methods

Direct qualitative assessments of disease are used to differentiate host responses or interactions, ideally under controlled conditions, where resistance or susceptibility is determined by genetic systems in the host and pathogen. Thus, responses to individual virulences (physiologic races), as required in breeding programmes or race surveys, are measured using a qualitative method of assessment as shown in Table 2.5 for cereal rusts and *Pyrenophora teres* (cause of barley net blotch disease). Such qualitative keys clearly differentiate resistant from susceptible responses; in the case of cereal rusts 0-, 1- or 2-type responses are resistant and 3- and 4-types are susceptible, whereas for net blotch disease 0, 1 and 2 are resistant (no chlorosis), and 3 and 4 are susceptible (chlorosis present) (Khan and Boyd, 1969). Rosielle (1972) developed a six-point qualitative assessment scale for *Septoria tritici* in which 0 = an immune response - no pycnidia or leaf symptoms; 1 = highly resistant (HR) - occasional isolated pycnidia with hypersensitive flecking; 2 = resistant (R) - very light pycnidial formation with some lesion coalescence; 3 = intermediate (I) - light pycnidial formation with lesion coalescence; 4 = susceptible (S) - moderate pycnidial formation with considerable lesion coalescence; and 5 = very susceptible (VS) - large abundant pycnidia with extensive lesion coalescence. Reddy *et al.* (1981) provided a qualitative assessment key for infection type in ascochyta blight

Table 2.5. Examples of qualitative keys for disease assessment

An assessment key for cereal rust virulence response

Symbol	*Host: parasite interaction*
Oi	Immune; no visible signs of infection
Oc	Highly resistant; minute chlorotic flecks
On	Highly resistant; minute necrotic flecks
1	Resistant; small pustules with necrotic surrounding tissue
2	Moderately resistant; medium-sized pustules with necrotic surrounding tissue
3	Moderately susceptible; medium-sized pustules with chlorotic surrounding tissue
4	Susceptible; large pustules with little or no chlorosis
X	Mesothetic reaction; mixed reaction types on one leaf

Reaction-type classes for *Pyrenophora teres* on barley (Khan and Boyd, 1969)

Class	*Reaction*
0	No observable infection.
1	Pin-point brown lesions, no chlorosis.
2	Small dark brown lesions, no chlorosis.
3	Restricted long brown streaks, slight associated chlorosis.
4	Brown elongated lesions with net-like cross variations, marked chlorosis.

of chickpea, caused by *Ascochyta rabiei,* and Subba Rao *et al.* (1990) for infection type in groundnuts *(Arachis hypogaea)* infected by rust *(Puccinia arachidis).* Edwards *et al.* (1997) developed an assessment key for lesion responses of celery and related Umbelliferae to *Septoria apiicola* infection, where 0 = an extremely resistant host response (no lesions) and 3 = very susceptible (grey lesions, leaf chlorosis and numerous pycnidia). In the field, the assessment of reaction types such as those described here is often more difficult than under controlled conditions, as host-pathogen interactions can be modified by environmental variables such as temperature and leaf surface wetness.

2.5.3 Indirect methods

Indirect methods of disease assessment have increased in number with the development of new technologies. Traditional methods rely on monitoring pathogen spore populations over infected crops or trapping insect vectors of a virus to estimate the level of crop infection. Fox (1993a) identified two basic methods for air-borne fungal spores: measuring the concentration of spores in a given volume of air (concentration methods); and counting the number of spores deposited on a surface (deposition methods). The correlation between the two methods is poorly understood and will obviously depend on meteorological factors. Concentration methods involve sophisticated spore traps with a power source (e.g. the original Hirst Volumetric Spore Trap), whereas deposition methods often comprise simple sticky, horizontal or vertical surfaces exposed to the air under a rain shelter (Wheeler, 1976). Rain-dispersed spores can be effectively caught in funnel traps positioned within the infected crop (Deadman and Cooke, 1989); these are then emptied after rainfall and the spores counted on a haemacytometer slide. Methods used to trap spores in this way therefore involve estimates of spores using microscopy, or colony counts in culture or on living plants used themselves as spore traps. An extension of the latter is the use of trap nurseries and mobile nurseries (Eyal *et al.,* 1973), in which sets of genotypes are assembled that carry specific resistances to the target pathogen in different geographic locations. Standardized methods of sowing and disease assessment are used and samples sent to a testing centre for virulence identification, usually as part of a race survey. Other indirect methods of assessing disease include measuring the effect of the pathogen on host parameters such as (for cereals) stunting, increased or decreased tillering, root growth, premature or delayed ripening and reductions in ear number, grain number, size and quality. Deadman and Cooke (1987) used such methods to assess the effects of net blotch disease on the growth and yield of spring barley.

It is often the case that data from the visual assessment of plant disease severity do not correlate with the amount of fungal biomass colonizing host tissue; this lack of correlation inevitably leads to inaccurate disease-yield loss relationships (section 2.6.1). Whereas diseases such as powdery mildew, which has a superficial ectotrophic growth habit on the host, may well show a close correlation between visual assessment and tissue colonization, most other diseases, where the pathogen is more invasive of the host tissue, are unlikely to show such a relationship. In order to test these assumptions, several workers developed more precise techniques of quantifying fungal biomass within host tissues, either by measuring fungal chitin (Ride and Drysdale, 1972; Parker and Royle, 1993) or ergosterol (Griffiths *et al.,* 1985; Gunnarsson *et al.*, 1996). Chitin is not found in plant tissue but is a principal component of fungal cell walls and, similarly, ergosterol is a fungal membrane-specific component. Thus, the chemical assays used for these biomarkers provide sophisticated quantitative techniques for the indirect assessment of disease severity in plant tissue. Strange (2003) provides a good overview of the methodology used in these techniques. Other indirect methods use spore production as a measure of severity; Gough (1978) described a method for evaluating wheat cultivar response to *Septoria tritici* based on pycnidiospore production from soaked leaf segments using haemacytometer counts.

Indirect methods of assessing disease incidence traditionally rely on the use of selective agar media on which infected plant materials are plated out. Methods for quantifying seed infection in this way are described by de Tempe (1964) and Neergaard (1979), whilst Pettitt *et al.* (1993) described a technique using potato dextrose agar supplemented by benomyl (Benlate fungicide, Du Pont U.K. Ltd.) for the improved estimation of the incidence of *Microdochium nivale* in winter wheat stems; the technique exploits the widespread incidence of benomyl resistance in

Microdochium nivale populations in the U.K. The dilution plate method is a way of visualizing and quantifying soil pathogen propagules using a serially diluted soil suspension in sterile distilled water plated out on selective media; results are expressed as colony forming units (cfu) per gram of soil.

Traditional methods, although still widely used, are rapidly being replaced by immunological and nucleic acid-based techniques. Of particular interest in the quantitative assessment of plant disease are user-friendly enzyme-linked immunosorbent assay (ELISA) kits for use in the field and the use of the polymerase chain reaction (PCR), particularly quantitative PCR (qPCR), for determining infection in plant material (see also Chapter 1). Fluorescent *in situ* hybridization (FISH) is a recent technique that is used to identify and quantify soil bacteria and fungi using complementary probes to DNA or RNA sequences of the organism of interest labelled with a fluorochrome. Further development of these techniques for use by the farmer or grower as dip-sticks or dot-blots will provide more precise methods of indirectly assessing plant diseases on site. Fox (1993b), Oliver (1993) and Schots *et al.* (1994) provide reviews of assays available for use by the plant pathologist.

2.5.4 Remote sensing

The use of aerial photography and photogrammetry using infrared film or colour filter combinations to enhance the differentiation between healthy and diseased tissue, represent a separate approach to disease assessment and were first used by Neblette (1927) and Taubenhaus *et al.* (1929) for surveying infection by cotton root rot (caused by *Phymatotrichum omnivorum)* in Texas and by Bawden (1933) in studies of virus diseases of potato and tobacco. Aerial photography was an example of remote sensing, defined by Nilsson (1995) as 'the measurement of an object without physical contact between the measuring device and the object'. Quality of results possible depends on the properties of the photographic film used, such as grain size and spectral sensitivity. Infrared film is usually used because near-infrared and infrared light are reflected deeper in leaf tissue than visible light (Campbell and Madden, 1990a). Early films were mainly analyzed using densitometry but, in later years, advanced image processing and spectral analysis were employed. Remote sensing now relies on digital image processing and image analysis, including advanced nuclear magnetic resonance imaging (NMRI), for the interpretation and quantification of non-destructive disease measurements in crops.

Remote sensing uses the properties of the electromagnetic spectrum and is based on the principle that any body reflects or absorbs radiant energy as electromagnetic waves with specific properties. Such properties of plant vegetation, such as whether it is healthy or diseased, influence the amount and quality of radiation reflected or emitted from the canopy. As such, this technology provides a useful tool in phytopathometry. A distinction should be made between the more commonly used passive remote sensing which measures (via films or electronic instruments) the electromagnetic solar energy reflected from vegetation, and the newer active remote sensing, where intensive energy pulses of specific wavelengths are directed against the vegetation and the interaction is exploited and analyzed, such as in LIDAR (light detection and ranging).

Remote sensing for detecting and estimating severity of plant diseases is used at three altitudes or levels above the crop canopy. At the lowest altitude, within 1.5-2.0 m above crop height, hand-held multispectral radiometers or multiple waveband video cameras are used; at 75-1500 m, aerial photography is used, whereas at the highest altitude, satellite imagery is employed utilizing satellites orbiting at 650-850 km above the earth's surface. In addition, video image analysis systems, such as that described by Lindow and Webb (1983), (Fig. 2.9), which uses a video camera interfaced through a digitizer to a microcomputer and display monitor, can be used under laboratory conditions for measuring diseased or damaged tissue at close quarters; systems such as the Delta-T Devices WinDIAS true-colour Windows-based system are able to differentiate the primary colours of diseased and healthy tissue (brown, yellow and green) in order to analyze percentage diseased leaf area automatically. In 2002, image analysis software called ASSESS was made available by The American Phytopathological Society for plant disease quantification. The software was optimized for the measurement of leaf area, percent area infected, lesion/pustule count, root length and ground cover. ASSESS relies on the Hue-Saturation-Intensity colour model enabling the user to effectively extract the leaf from the background and then the lesions from the leaf. Bannon and Cooke (1998) used NIH Image 1.60/fat (National Institute of Mental Health, USA), a public domain image analysis programme for the Apple Macintosh, to assess splashed areas on white card in simulated studies on the dispersal of *Septoria tritici* pycnidiospores; the NIH programme can be used for a wide variety of image analyses including the measurement of diseased areas on plant organs.

Whereas hand-held multispectral radiometers or multiple waveband video cameras are most appropriate for disease measurements on plants or pots within fields, aerial infrared photography is most useful at field level, and satellite imagery has been used since 1972 for large areas or regions of the earth's surface devoted to agriculture and forestry. Images are transmitted to earth stations by satellites such as the American National Oceanic and Atmospheric Administration (NOAA) and LANDSAT series (1, 4 and 5), and the French SPOT satellite (which uses 10 metre resolution imagery), that feature advanced very high resolution radiometer (AVHRR) optical and thermal sensors; these have been joined by IRS, Ikonos and EROS satellites. However, the importance of ground truth, that is actual visits to the target crop to verify remote sensing data, is an important part of the process. The persistence of cloud cover in countries such as the UK and Brazil has been a serious impediment to the progress of this technology; however synthetic aperture radar (SAR) high-resolution technology can overcome this problem and was used in 1991 on board the European Remote Sensing Satellite ERS-1. An excellent overview of digital imaging was published by Graham in 1998.

The application of remote sensing to plant pathology lies mainly in the detection of crop stress. A plant or plant population becomes stressed when a biotic or abiotic factor adversely affects growth and development (Nilsson, 1995). Stress or disease can be expressed in various ways, such as imbalance in water supply leading to stomatal closure, decreased photosynthesis with associated changes in leaf fluorescence

(Daley, 1995) and evapotranspiration, and increased leaf surface temperature. Other symptoms may include leaf curling, wilting, stunting, chlorosis and necrosis of plant parts. Remote sensing provides a method for detecting and assessing such changes. However, it is likely that remote sensing will remain an indirect method of assessing plant disease through the interpretation of deviations from the norm, such as leaf temperature, rather than directly measuring reductions in leaf area due to disease.

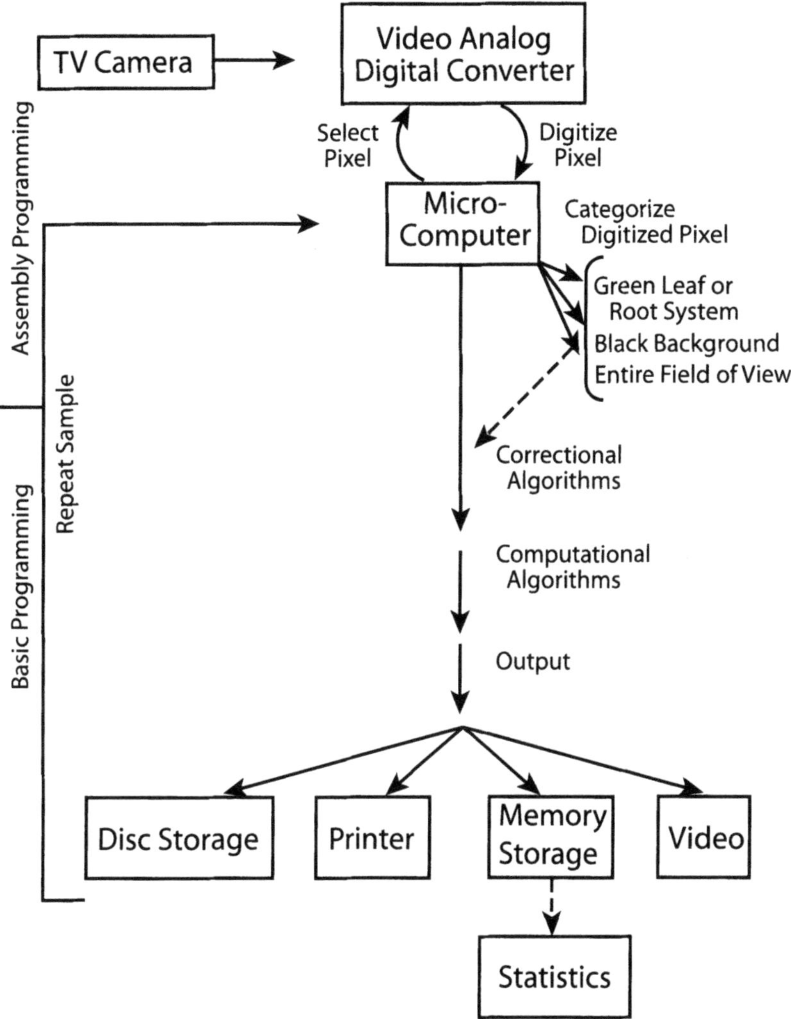

Figure 2.9. Video image analysis system for measuring diseased or damaged plant tissue (Lindow and Wenn, 1983).

2.6 ASSESSMENT OF YIELD LOSS

2.6.1 Confounding factors

Relating the yield of crops to varying levels of plant disease has remained a complex task both in theory and practice. The difficulty of interpreting disease-yield loss relationships was recognized by James (1983), one of the pioneers in this field of research. Madden and Nutter (1995) argued that the failure to measure disease intensity (severity and incidence) accurately was a major factor contributing to this difficulty; they concluded, however, that recent research in crop loss assessment would lead to a better understanding and to predictions of crop losses through sophisticated modelling, progress in sampling theory and application, and advances in instrumentation.

There may be several confounding factors that weaken the statistical relationship between the independent variable (assessment) and the corresponding dependent variable (yield loss). Such factors have been identified as: interactions between diseases and between the pathogen and an environmental factor; the self-limiting effect of local lesions; overcapacity of the host plant (within-plant compensation); and between-plant compensation and lesion position effects. In relation to the latter, Zadoks and Schein (1979) cautioned against the conclusion that lesions of equal size always had equal effects on yield and crop loss, and cited the example of *Phytophthora infestans* (cause of potato late blight) in which a stem lesion can kill a haulm with 10 leaves and 50 leaflets whereas the same-sized lesion on a leaflet kills only one leaflet at most. The same caution would apply to the effects of axil and leaf blade lesions of equal size caused by *Rhynchosporium secalis* on barley.

Another identifiable confounding factor is the often poor correlation between visible symptoms and amount of tissue colonization. Precise techniques can now measure fungal biomass using chitin or ergosterol as biomarkers (section 2.5.3). In *Fusarium* ear blight of wheat, kernels in asymptomatic spikelets may be infected and mycotoxin content may not be correlated with visible symptoms. Mycotoxin produced in grain can have serious consequences for the food chain; assays for mycotoxins may be more important for the milling and baking industries than estimates of disease symptom incidence and severity (Shaner, 2003). Madden and Nutter (1995) reviewed the approaches for modelling crop losses in relation to disease intensity and identified additional factors that might change our understanding of the disease-yield loss relationship, such as the relevance of healthy leaf area duration, radiation interception, spatial pattern of disease intensity and time of infection.

Bryson *et al.* (1995) suggested that a simple model relating loss of green area within a winter wheat crop canopy to changes in light interception might be useful in predicting disease-induced yield losses by yellow rust (caused by *Puccinia striiformis)*; such a yield loss model is thus based on crop function rather than measurements of disease severity and related area under the disease progress curve (AUDPC) values which have no mechanistic link to the productivity of the host plant. Bryson *et al.* (1995) obtained a significantly better relationship between area under leaf green area index progress curve (AULGAIPC-synonymous with healthy area duration, HAD) and yield compared with disease severity assessments. Waggoner and Berger (1987) suggested that a logical adaptation of HAD (or AULGAIPC) would be to integrate radiation by green tissue to give healthy area absorption (HAA). Here, yield loss was related not just to disease intensity but also to crop physiological variables. Traditional single-point, multiple point and integral models such as AUDPC based on disease intensity do not give a complete description of the disease-yield loss relationship, as crop yield is determined by the magnitude of photosynthesis, a function of HAD (or AULGAIPC) and HAA. Finally, it should be remembered that a complete disease-yield loss relationship should also take account of economic thresholds for crop loss due to disease and the assessment of any loss in crop quality.

2.6.2 Reference points, terms and concepts

In describing crop-yield loss relationships, it is important to establish reference points, terms and concepts in order to standardize communication between workers. Zadoks and Schein (1979), Campbell and Madden (1990c) and Nutter *et al.* (1993) reviewed concepts and terminology for crop losses and differentiated between potential losses (in the absence of control measures) and actual losses in crops, the latter being sub-divided into direct (loss in quantity or quality of yield) and indirect (the economic or social impact of losses). Similarly, yield was divided into attainable yield (when crops were grown under optimum conditions), primitive yield (when no disease control was applied), economic yield (highest net return on expenditure), actual yield (obtained using disease management programmes) and theoretical yield (obtained using calculations based on crop physiology or crop growth simulation models). The difference between actual and attainable yield was the method used by the Food and Agriculture Organization (FAO) to report crop losses; most disease management programmes aim to close the gap between these two yield concepts (Fig. 2.10).

2.6.3 Statistical and experimental methods

The assessment of yield loss is carried out using statistical and experimental methods. Statistical methods were reviewed by Chaube and Singh (1991) and Parry (1990); such methods utilize reports of disease incidence, estimated crop losses and yield for analyzing yield losses. The accuracy of such methods is obviously inferior to that obtainable from scientifically designed experiments. Briefly, statistical methods involve the following: analysis of yields in relation to estimated disease incidence over many seasons (but other factors such as weather, pests, farming practice and plant varieties must be taken into account); comparisons of expected and actual yields when it is known that a pathogen is the major cause of yield loss; yield analysis before and after control measures are applied; and the use of holistic synoptic methodology which involves grower questionnaires or national disease surveys by agricultural officers to gather information on particular diseases (e.g. severity, date of appearance, weather conditions, varietal susceptibility, estimated crop loss).

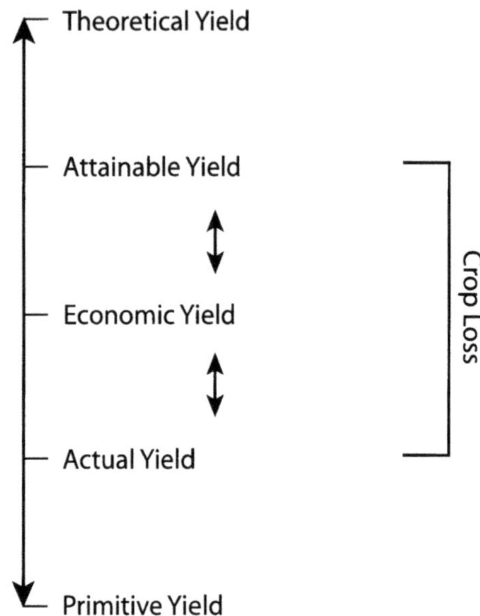

Figure 2.10. Yield levels and crop loss (Campbell and Madden, 1990b).

Experimental methods are mainly based on yield comparisons between infected and healthy plants or between plants with different disease severities using field plots, microplots (hill plots), single plants or tillers; between resistant and susceptible varieties; between infected plants and plants treated with fungicides; or between healthy plants and plants where disease damage has been simulated by the removal of essential plant organs, such as the flag leaf on a cereal plant. In all of these methods, it is important that experiments are properly designed so that results can be analyzed statistically and, if possible, they should be repeated over a number of seasons and in different areas. Gaunt (1995) pointed out that in the general area of yield loss experimentation, two methodologies have gained less acceptance than might have been anticipated: the use of disease gradients in large field plots and the use of non-replicated designs. In both cases, difficulties have arisen with experimental design and interpretation of results.

2.6.4 Empirical yield loss models

Each experimental method, such as those described above, will always have advantages and disadvantages but conventional field experiments will probably continue to provide most information for the mathematical modelling of disease-yield loss relationships. Most models describe losses at the field level; the type and complexity depend on the pathosystem and the disease descriptor employed as the independent variable using least-squares regression analysis. Such models do not directly use information on crop physiology and are, therefore, empirical or descriptive rather than mechanistic in nature. However, they can be consistent with known physiological processes or can be expanded to accommodate such processes. Teng (1985) classified empirical loss models into six categories: (1) single-point (critical-point) models, (2) multiple-point (multiple regression) models, (3) response-surface models, (4) integral models, (5) generalized or non-linear models, and (6) synoptic models. The first five models describe losses in yield due to one disease, whereas synoptic models include variables for several diseases and non-disease factors. Another approach to classifying such models is to consider whether the model uses one or more independent variables or how many estimates of disease are made over time.

In single-point or critical-point models, yield loss is related to disease measurement at one specific time during the growing season or at a specific growth stage. Models using time to a certain disease severity are also considered as critical-point models. It should be remembered that a critical-point model does not imply that a host plant responds to a disease at only one time or growth stage, but rather that a good statistical relationship is found at one specific time. This type of model is probably the most commonly used because of the small amount of data required and has been heavily employed for grain crops where epidemics with a reasonably stable infection rate occur near to grain-filling. Single-point models may be linear or non-linear in their parameters and can be written in the form:

$$\% \text{ loss (L)} = a + bX \qquad (2.1)$$

in which a and b are parameters and X is the disease measurement or a transformation of disease measurement at a given time. Examples of critical-point models are those developed for cereal diseases (Table 2.6) and that of Large (1952) for late blight of potato (Fig. 2.11). The models shown for cereal foliar diseases were developed to estimate yield loss from corresponding disease severity estimates at particular growth stages, whereas those for cereal stem-base diseases (eyespot and sharp eyespot) were developed for use with disease incidence values (Cook and King, 1984). Large's critical point model for estimating yield losses from late blight of potato uses time to a critical disease severity: the model assumes bulking up of potato tubers ceases when 75% blight severity on the haulm is reached. A major problem with critical-point models is that they fail to accommodate variables in infection rates and shape of the disease progress curve.

Multiple-point models can be used for diseases with high variability in infection rates and where the disease progress curves can be markedly different. These models can be used for epidemics that develop over a long time period relative to the life of the crop and where yield accumulation is a prolonged process (e.g. potatoes). Multiple-point models relate yield loss to assessments of disease made at several times during the growth season. Assessments can be made at specific times or at specific host plant growth stages. Loss is then related to disease measured at each of these points during the epidemic or to the change in disease between assessments using a multiple regression equation, with the general form:

$$\% \text{ loss (L)} = b_1X_1 + b_2X_2 + b_3X_3 \ldots b_nX_n \qquad (2.2)$$

Table 2.6. Examples of critical-point yield loss models for cereal diseases (Cook and King, 1984)

Disease	*Relationship*
Spring barley	
Mildew	$y = 2.5\sqrt{x_i}$
Brown rust	$y = 0.4x_{ii}$
Rhynchosporium	$y = 0.5x_{ii}$
Winter wheat	
Mildew	$y = 2.0\sqrt{x_i}$
Septoria	$y = x_{iii}$
Yellow rust	$y = 0.4x_{iii}$
Eyespot	$y = 0.1x_m + 0.36x_s$
Sharp eyespot	$y = 0.05x_m + 0.26x_s$

y = % loss in grain yield,
x_i = % disease on leaf 3 at GS 58,
x_{ii} = % disease on leaf 2 at GS 75,
x_{iii} = % disease on flag leaf at GS 75,
x_m = % tillers with moderate symptoms at GS 75,
x_s = % tillers with severe symptoms at GS 75.

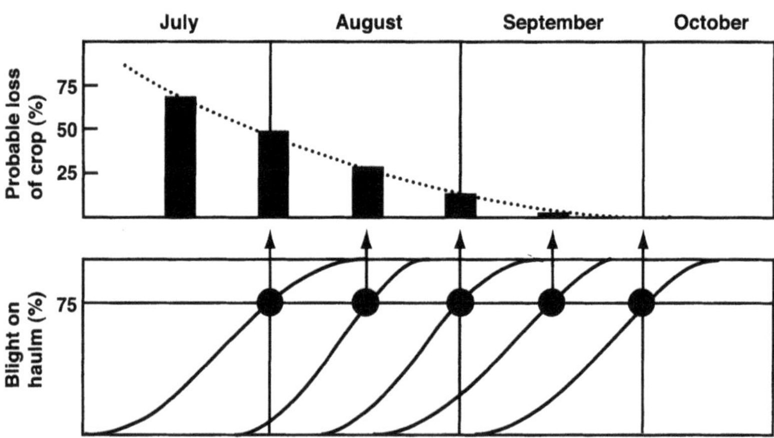

Figure 2.11. Estimated yield loss from epidemics of potato late blight in which the 75% disease level (the critical disease severity) is reached at differing times during the season. (After Large, 1952; Crown copyright, 1986).

where b_1 ... b_n are partial regression coefficients for the first and nth week respectively, and X_1 ... X_n are the corresponding weekly disease increments for the first and the nth week, respectively. In a classic study, James *et al.* (1972) developed a multiple-point model for estimating loss in tuber yield due to late blight of potato using disease increments during weekly intervals as the independent variable. Using the equation, the estimated loss was within 5% of the actual loss in nine cases out of 10. However, there is a need to examine whether a multiple-point model is biologically meaningful, as stepwise-selected regression models often contain both positive and negative estimated coefficients. Careful evaluation of regression results, especially intercorrelation of the Xs, is essential to avoid conclusions that imply an increase in disease severity at a particular time or growth stage produces an increase in yield, rather than a decrease.

Calpouzos *et al.* (1976) developed another form of multiple-point model for estimating losses due to wheat stem rust. Yield loss was plotted as a response-surface (a three-dimensional graph) and was a function of the slope of the epidemic and the growth stage at the time of epidemic onset using the equation:

$$\%\ \text{loss (L)} = f(X_1 X_2) \tag{2.3}$$

where X_1 = slope of the epidemic (infection rate) and X_2 = growth stage at epidemic onset.

Van der Plank (1963) proposed a modification of the multiple-point model in which the area under the disease progress curve (AUDPC) is used as a descriptor for the epidemic to measure crop loss. The AUDPC, an integral model, relates loss to a summing of disease measurements over a specific period of crop growth. AUDPC can be estimated using the following equation of Shaner and Finney (1977):

$$\text{AUDPC} = \sum_{i}^{n-1} \left(\frac{y_i + y_{i+1}}{2} \right) (t_{i+1} - t_i) \tag{2.4}$$

in which n is the number of assessment times, y is the disease measurement and t is time (usually in days or degree days). AUDPC is simply y integrated between two times and can be approximated using the midpoint rule or trapezoidal integration method. As shown in Fig. 2.12 (Campbell and Madden, 1990b), the disease progress is divided into a series of rectangles, the areas of which are summed to approximate the total area under the curve. The narrower the intervals between assessments, the more accurate is the determination of AUDPC, which can be standardized by dividing its value with the total time duration (t_n – t_1) of the epidemic. This allows for comparisons between epidemics of differing durations and allows two epidemics to be distinguished which have different progress curves but the same disease severity at a critical date.

AUDPC models make two assumptions: injury to the host is proportional to the amount of tissue infected; and injury is proportional to the duration of the disease. Most AUDPC models have been used for epidemics of relatively short duration,

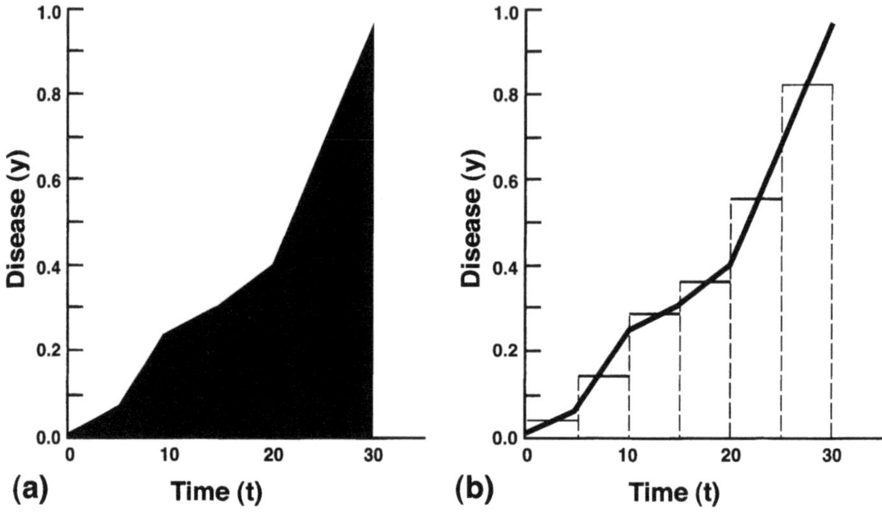

Fig. 2.12. (a) Area under the disease progress curve (AUDPC) for a hypothetical epidemic. (b) Illustration of the midpoint rule (or trapezoidal integration method) for calculating AUDPC. (Campbell and Madden, 1990b).

which are late in the crop growth cycle and where yield is accumulated over a short period of time. However, integral models that use AUDPC cannot distinguish between early and late-occurring epidemics without applying weighting factors to assessments at different growth stages (Hills *et al.*, 1980).

Generalized or non-linear models are sometimes more appropriate where the shape of the loss-disease curve dictates that this approach should be used; many such models can have variability in the shape of the curve relating yield to the disease descriptor. Synoptic or multivariate statistical models are used where multiple diseases and other constraints may be determining the yield-loss relationship, a situation often encountered in actual crop production systems; data for such models often derive from surveys, in which no manipulation has been carried out to obtain specific disease levels. Complex multivariate techniques for analysis of the data, such as principal components and correspondence analyses, may be required. Models can be expanded to account for control costs and resulting economic yield, both in quantity and quality. Expert systems and geographic information systems (GIS) can also be used to provide regional estimates of losses in agricultural production.

2.7 CONCLUSIONS AND FUTURE DEVELOPMENTS

The future of agricultural research will depend on understanding higher levels of organization, and these are the levels addressed by the science of applied ecology; thus, the crop is a plant population and pathogens are populations of organisms with which the plant population interacts (Weiner, 2003). This concept will demand that agriculture is considered as ecosystem manipulation rather than production, and this approach could influence such activities as the assessment of disease in crop plants. There exists little doubt that the need for disease assessment and yield loss appraisal in the world's food crops is of paramount importance. Oerke and Dehne (1997) concluded that although worldwide production of most crops has increased considerably since 1965, losses in wheat, potatoes, barley and rice from pathogens, pests and weeds simultaneously increased by 4% to 10% whereas in maize, soybean, cotton and coffee losses remained unchanged or slightly decreased. These authors also concluded that the efficacy of crop protection varied for different crops, being highest (55%) in cotton but reaching only 34% to 38% in rice, wheat and maize. However variability between world cropping areas was high in terms of actual prevention of crop losses with a worldwide crop protection efficiency of only 40%. Johnson *et al.* (2003) quantified the economic impact of *Fusarium* ear blight in wheat in the USA. During 1991-1997 production losses and price effects resulted in direct losses totalling more than $1.3 billion whereas the total economic impact of the disease on affected rural communities and agribusinesses related to grain production and marketing was three to four times this amount. Not since potato late blight has a plant disease caused as much social havoc as *Fusarium* ear blight (McMullen, 2003).

Although traditional methods of disease assessment will continue to find a place in modern agriculture, technological advances in the application of computer programmes for estimating disease severity with variegated disease patterns have improved the accuracy of observers in the field and have removed, at least in part, the subjective element of phytopathometry and the effect of several confounding factors. Detailed data analyses such as those undertaken by Hughes *et al.* (1997) have revealed the complexities and consequences of aggregated data in spacial hierarchies when estimating disease incidence. Furthermore, the now widespread use of immunological and molecular nucleic acid-based techniques for the identification of disease organisms has increased our capability to accurately measure disease incidence in infected plant tissue; in the last few years the development of robust diagnostic kits for use by the farmer or grower (as dip-sticks or dot-blots) has simplified such technology in the field for the detection and quantification of fungal and bacterial plant pathogens (Strange, 2003).

The increasing use of remote-sensing technology for estimating disease damage to crops is having a continuing impact on the science of phytopathometry, from the use of sophisticated hand-held radiometers to satellite imagery. The development of high resolution technology, such as synthetic aperture radar (SAR) has greatly improved the effectiveness of satellite imagery in countries where cloud cover has been a serious impediment to the progress of this methodology. Monitoring agriculture with remote sensing (MARS) has become an increasingly common form of GIS technology. However, it is likely that remote sensing at this magnitude will remain an indirect method of assessing plant disease through the detection of crop stress due to biotic or abiotic factors, rather than directly measuring reductions in leaf area due to disease. In the laboratory, the availability of relatively inexpensive true-colour image analysis systems for analyzing diseased leaf area percentages has increased the accuracy of disease assessments at this level.

The relationship between disease assessment and crop loss has always presented problems to the plant pathologist, who has traditionally relied on empirical yield loss models. These have no mechanistic link to the productivity of the host plant. Bryson *et al.* (1995) suggested a yield loss model for yellow rust of wheat based on loss of green leaf area and changes in canopy light interception. Such a model is therefore based on crop function and utilizes crop physiological variables rather than measurements of disease; this and other developments such as incorporating spatial patterns into crop loss models (Hughes, 1996) have changed our understanding of the disease-yield loss relationship. It will be important in the future to validate new individual loss models and to compare alternative models for describing disease-yield loss relationships.

REFERENCES

Anon. (1947) The measurement of potato blight. *Transactions of the British Mycological Society,* **31**, 140-141.

Bannon, F.J. and Cooke, B.M. (1998) Studies on dispersal of *Septoria tritici* pycnidiospores in wheat-clover intercrops. *Plant Pathology*, **47**, 49-56.

Bawden, F.C. (1933) Infra-red photography and plant virus diseases. *Nature,* **132**, 168.

Browne, R.A. and Cooke, B.M. (2004) Development and evaluation of an *in vitro* detached leaf assay for pre-screening resistance to *Fusarium* head blight in wheat. *European Journal of Plant Pathology*, **110**, 91-102.

Browne, R.A. and Cooke, B.M. (2005) A comparative assessment of potential components of partial disease resistance to *Fusarium* head blight using a detached leaf assay of wheat, barley and oats. *European Journal of Plant Pathology*, **112**, 247-258.

Bryson, R.J., Sylvester-Bradley, R., Scott, R.K. and Paveley, N.D. (1995) Reconciling the effects of yellow rust on yield of winter wheat through measurements of green leaf area and radiation interception. *Aspects of Applied Biology,* **42**, 9-18.

Calpouzos, L., Roelfs, A.P., Madson, M.E. *et al.* (1976) A new model to measure yield losses caused by stem rust in spring wheat. *Technical Bulletin Agriculture Experimental Station No. 397,* University of Minnesota, St Paul, Minnesota.

Campbell, C.L. and Madden, L.V. (1990a) Monitoring epidemics: disease, in *Introduction to Plant Disease Epidemiology,* John Wiley, New York, pp. 107-128.

Campbell, C.L. and Madden, L.V. (1990b) Temporal analysis of epidemics 1: description and comparison of disease progress curves, in *Introduction to Plant Disease Epidemiology,* John Wiley, New York, pp. 161-202.

Campbell, C.L. and Madden, L.V. (1990c) Crop loss assessment and modelling, in *Introduction to Plant Disease Epidemiology,* John Wiley, New York, pp. 393-422.

Chaube, H.S. and Singh, V.S. (1991) Pathometry - assessment of disease incidence and loss, in *Plant Disease Management, Principles and Practices,* CRC Press, Florida, pp. 119-131.

Chester, K.S. (1950) Plant disease losses: their appraisal and interpretation. *Plant Disease Reporter Supplement* No. 193, 189-362.

Chiarappa, L. (1971) (ed.) *Crop Loss Assessment Methods,* Commonwealth Agricultural Bureaux, Slough, UK, 200 pp.

Chiarappa, L. (1981) (ed.) *Crop Loss Assessment Methods - Supplement 3,* Commonwealth Agricultural Bureaux, Slough, UK, 123 pp.

Church, B.M. (1971) The place of sample survey in crop loss estimation, in *Crop Loss Assessment Methods,* (ed. L. Chiarappa), Commonwealth Agricultural Bureaux, Slough, UK, pp. 2.2/1-2.2/8.

Cobb, N.A. (1892) Contribution to an economic knowledge of the Australian rusts (Uredineae). *Agricultural Gazette N. S. Wales,* **3**, 60.

Cook, R.J. and King, J.E. (1984) Losses caused by cereal diseases and the economics of fungicidal control, in *Plant Diseases, Infection, Damage and Loss* (eds R.K.S. Wood and G.J. Jellis), Blackwell, Oxford, 238 pp.

Daley, P.F. (1995) Chlorophyll fluorescence analysis and imaging in plant stress and disease. *Canadian Journal of Plant Pathology*, **17**, 167-173.

Daamen, R.A. (1986a) Measures of disease intensity in powdery mildew (*Erysiphe graminis*) of winter wheat. 1. Errors in estimating pustule number. *Netherlands Journal of Plant Pathology*, **92**, 197-206.

Daamen, R.A. (1986b) Measures of disease intensity in powdery mildew (*Erysiphe graminis*) of winter wheat. 2. Relationships and errors of estimation of pustule number, incidence and severity. *Netherlands Journal of Plant Pathology*, **92**, 207-222.

Deadman, M.L. and Cooke, B.M. (1987) Effects of net blotch on growth and yield of spring barley. *Annals of Applied Biology*, **110**, 33-42.

Deadman, M.L. and Cooke, B.M. (1989) An analysis of rain-mediated dispersal of *Drechslera teres* conidia in field plots of spring barley. *Annals of Applied Biology*, **115**, 209-214.

de Jong, P.D. and de Bree, J. (1995) Analysis of the spatial distribution of rust-infected leek plants with the Black-White join-count statistic. *European Journal of Plant Pathology*, **101**, 133-137.

Delp, B.R., Stowell, L.J. and Marois, J.J. (1986a) Field Runner: a disease incidence, severity and spatial pattern assessment system. *Plant Disease*, **70**, 954-957.

Delp, B.R., Stowell, L.J. and Marois, J.J. (1986b) Evaluation of field sampling techniques for estimation of disease incidence. *Phytopathology*, **76**, 1299-1305.

de Tempe, J. (1964) Recent developments in seed health testing. *Proceedings of the International Seed Testing Association*, **29**, 479-486.

Diamond, H. and Cooke, B.M. (1999) Towards the development of a novel *in vitro* strategy for early screening of *Fusarium* ear blight resistance in adult winter wheat plants. *European Journal of Plant Pathology*, **105**, 363-372.

Dixon, G.R. and Doodson, J.K. (1971) Assessment keys for some diseases of vegetable, fodder, and forage crops. *Journal of the National Institute of Agricultural Botany*, **23**, 299-307.

Edwards, S.J., Cohn, H.A. and Isaac, S. (1997) The response of different celery genotypes to infection by *Septoria apiicola*. *Plant Pathology*, **46**, 264-270.

Eyal, Z. and Brown, M.B. (1976) A quantitative method for estimating density of *Septoria tritici* pycnidia on wheat leaves. *Phytopathology*, **66**, 11-14.

Eyal, Z. and Ziv, O. (1974) The relationship between epidemics of septoria leaf blotch and yield losses in spring wheat. *Phytopathology*, **64**, 1385-1389.

Eyal, Z., Yurman, R., Moseman, J.G. and Wahl, I. (1973) Use of mobile nurseries in pathogenicity studies of *Erysiphe graminis hordei* on *Hordeum spontaneum*. *Phytopathology*, **63**, 1330-1334.

Fitt, B.D.L., Doughty, K.J., Gilles, T. *et al.* (1998) Methods for assessment of light leaf spot (*Pyrenopeziza brassicae*) on winter oilseed rape (*Brassica napus*) in the UK. *Annals of Applied Biology*, **133**, 329-341.

Fleischer, S.J., Blom, P.E. and Weisz, R. (1999) Sampling in precision IPM: when the objective is a map. *Phytopathology*, **89**, 1112-1118.

Forbes, G.A. and Jeger, M.J. (1987) Factors affecting the estimation of disease intensity in simulated plant structures. *Zeitschrift fuer Pflanzen krankheiten und Pflanzenschutz*, **94**, 113-120.

Fox, R.T.V. (1993a) Detection of disease outbreaks in the field, in *Principles of Diagnostic Techniques in Plant Pathology*, CAB International, Wallingford. UK, pp. 1-7.

Fox, R.T.V. (1993b) Immunological techniques for identification, in *Principles of Diagnostic Techniques in Plant Pathology*, CAB International, Wallingford, UK, pp. 129-151.

Gaunt, R.E. (1995) New technologies in disease measurement and yield loss appraisal. *Canadian Journal of Plant Pathology*, **17**, 185-189.

Gough, F.J. (1978) Effect of wheat host cultivars on pycnidiospore production by *Septoria tritici*. *Phytopathology*, **68**, 1343-1345.

Graham, R. (1998) Digital Imaging. Whittles Publishing, UK, 212 pp.

Griffiths, H.M., Jones, D.G. and Akers, A. (1985) A bioassay for predicting the resistance of wheat leaves to *Septoria nodorum*. *Annals of Applied Biology*, **107**, 293-300.

Groth, J.V. and Ozmon, E.A. (1999) Repeatability and relationship of incidence and severity measures of scab of wheat caused by *Fusarium graminearum* in inoculated nurseries. *Plant Disease*, **83**, 1033-1038.

Gunnarsson, T., Almgren, I., Lydén, P. *et al.* (1996) The use of ergosterol in the pathogenic fungus *Bipolaris sorokiniana* for resistance rating of barley cultivars. *European Journal of Plant Pathology*, **20**, 883-889.

Hanway, J.J. (1963) Growth stages of corn (*Zea mays* L.). *Agronomy Journal*, **55**, 487-492.

Hebert, T.T. (1982) The rationale for the Horsfall-Barratt plant disease assessment scale. *Phytopathology,* **72**, 1269.
Hills, F.J., Chiarappa, L. and Geng, S. (1980) Powdery mildew of sugarbeet: disease and crop loss assessment. *Phytopathology,* **70**, 680-682.
Horsfall, J.G. and Barratt, R.W. (1945) An improved grading system for measuring plant disease. *Phytopathology,* **35**, 655 (Abstr.).
Hughes, G. (1996) Incorporating spatial patterns of harmful organisms into crop loss models. *Crop Protection,* **15**, 407-421.
Hughes, G. (1999) Sampling for decision making in crop loss assessment and pest management: introduction. *Phytopathology*, **89**, 1080-1083.
Hughes, G. and Madden, L.V. (1995) Some methods allowing for aggregated patterns of disease incidence in the analysis of data from designed experiments. *Plant Pathology*, **44**, 927-943.
Hughes, G., McRoberts, N., Madden, L.V. and Gottwald, T.R. (1997) Relationship between disease incidence at two levels in a spatial hierarchy. *Phytopathology,* **87**, 542-550.
Hughes, G., McRoberts, N. and Madden, L.V. (2004) Daamen's incidence-severity relationship revisited. *European Journal of Plant Pathology*, **110**, 759-761.
James, W.C. (1971) An illustrated series of assessment keys for plant diseases, their preparation and usage. *Canadian Plant Disease Survey,* **51**, 39-65.
James, W.C. (1983) Crop loss assessment, in *Plant Pathologist's Pocketbook,* 2nd edn, (eds A. Johnston and C. Booth), Commonwealth Mycological Institute, Kew, UK, pp. 130-143.
James, W.C., Jenkins, J.E.E. and Jemmett, J.L. (1968) The relationship between leaf blotch caused by *Rhynchosporium secalis* and losses in grain yield of spring barley. *Annals of Applied Biology,* **62**, 273-288.
James, W.C., Shih, C.S., Hodgson, W.A. and Callbeck, L.C. (1972) The quantitative relationship between late blight of potato and loss in tuber yield. *Phytopathology,* **62**, 92-96.
Jeger, M.J. (1981) Disease measurement in a study of apple scab epidemics. *Annals of Applied Biology,* **99**, 43-51.
Johnson, D.D., Flaskerud, G.K., Taylor, R.D. and Satyanarayana, V. (2003) Quantifying economic impacts of Fusarium head blight in wheat, in *Fusarium Head Blight of Wheat and Barley* (eds K.J. Leonard and W.R. Bushnell), APS Press, St. Paul, Minnesota, 512 pp.
Jones, D.G. and Clifford, B.C. (1978) Pathological techniques, in *Cereal Diseases, their Pathology and Control,* BASF, Ipswich, UK, pp. 52-94.
Khan, T.N. and Boyd, W.J.R. (1969) Physiologic specialisation in *Drechslera teres. Australian Journal of Biological Science,* **22**, 1229-1235.
Kirby, E.J.M. and Appleyard, M. (1981) *Cereal Development Guide.* National Agricultural Centre, Stoneleigh, UK, 80 pp.
Knott, C.M. (1987) A key for the stages of the development of the pea. *Annals of Applied Biology,* **111**, 233-244.
Knott, C.M. (1990) A key for stages of development of the faba bean *(Vicia faba). Annals of Applied Biology,* **116**, 391-404.
Kranz, J. (1988) Measuring plant disease, in *Experimental Techniques in Plant Disease Epidemiology*, (eds. J. Kranz and J. Rotem), Springer-Verlag, Berlin, pp. 35-50.
Lancashire, P.D., Bleiholder, H., Van den Boom, T. *et al.* (1991) A uniform decimal code for growth stages of crops and weeds. *Annals of Applied Biology,* **119**, 561-601.
Large, E.C. (1952) The interpretation of progress curves for potato blight and other plant diseases. *Plant Pathology,* **1**, 109-117.
Large, E.C. (1954) Growth stages in cereals. Illustration of the Feekes' scale. *Plant Pathology,* **3**, 128-129.
Large, E.C. (1966) Measuring plant disease. *Annual Review of Phytopathology,* **4**, 9-28.
Lindow, S.E. and Webb, R.R. (1983) Quantification of foliar plant disease symptoms by micro-computer-digitized video image analysis. *Phytopathology,* **73**, 520-524.
Lovell, D.J., Parker, S.R., Hunter, T. *et al.* (1997) Influence of crop growth and structure on the risk of epidemics by *Mycosphaerella graminicola (Septoria tritici)* in winter wheat. *Plant Pathology,* **46**, 126-138.
Lucas, J.A. (1998) Plant pathology and plant pathogens (3rd edn). Blackwell Science, UK, 274 pp.
Madden, L.V. and Hughes, G. (1995) Plant disease incidence: distributions, heterogeneity, and temporal analysis. *Annual Review of Phytopathology*, **33**, 529-564.

Madden, L.V. and Hughes, G. (1999a) An effective sample size for predicting plant disease incidence in a spatial hierarchy. *Phytopathology*, **89**, 770-781.
Madden, L.V. and Hughes, G. (1999b) Sampling for plant disease incidence. *Phytopathology*, **89**, 1088-1103.
Madden, L.V. and Nutter, F.W. (1995) Modelling crop losses at the field scale. *Canadian Journal of Plant Pathology*, **17**, 124-137.
McMullen, M. (2003) Impacts of Fusarium head blight on the North American agricultural community: the power of one disease to catapult change, in *Fusarium Head Blight of Wheat and Barley* (eds K.J. Leonard and W.R. Bushnell), APS Press, St. Paul, Minnesota, 512 pp.
Morrison, R.H. (1999) Sampling in seed health testing. *Phytopathology*, **89**, 1084-1087.
Neblette, C.B. (1927) Aerial photography for the study of plant diseases. *Photo-Era Magazine*, **58**, 346.
Neergaard, P. (1979) Seed health testing methods, in *Seed Pathology*, Vol. I, Macmillan, London, pp. 715-802.
Nilsson, H.-E. (1995) Remote sensing and image analysis in plant pathology. *Canadian Journal of Plant Pathology*, **17**, 154-166.
Nutter, F.W. and Esker, P.D. (2006) The role of psychophysics in phytopathometry: the Weber-Fechner law revisited. *European Journal of Plant Pathology*, (in press).
Nutter, F.W., Esker, P.D. and Netto, R.A.C. (2006) Disease assessment concepts and the role of psychophysics in phytopathology. *European Journal of Plant Pathology*, (in press).
Nutter, F.W. and Schultz, P.M. (1995) Improving the accuracy and precision of disease assessments: selection of methods and use of computer-aided training programs. *Canadian Journal of Plant Pathology*, **17**, 174-184.
Nutter, F.W. and Worawitlikit, O. (1989) Disease.Pro: A computer program for evaluating and improving a person's ability to assess disease proportion. *Phytopathology*, **79**, 1135 (Abstr.).
Nutter, F.W., Teng, P.S. and Shokes, F.M. (1991) Disease assessment terms and concepts. *Plant Disease*, **75**, 1187-1188.
Nutter, F.W., Teng, P.S. and Royer, M.H. (1993) Terms and concepts for yield, crop loss, and disease thresholds. *Plant Disease*, **77**, 211-215.
Nyrop, J.P., Binns M.R. and van der Werf, W. (1999) Sampling for IPM decision making: where should we invest time and resources? *Phytopathology*, **89**, 1104-1111.
Oerke E.-C. and Dehne, H.-W. (1997) Global crop production and the efficacy of crop protection - current situation and future trends. *European Journal of Plant Pathology*, **103**, 203-215.
Oliver, R.P. (1993) Nucleic-acid based methods for detection and identification, in *Principles of Diagnostic Techniques in Plant Pathology* (ed. R.T.V. Fox), CAB International, Wallingford, UK, pp. 153-169.
Oyarzun, P.J. (1993) Bioassay to assess root-rot in pea and effect of root-rot on yield. *Netherlands Journal of Plant Pathology*, **99**, 61-75.
Parker, S.R. and Royle, D.J. (1993) Sampling and monitoring disease in winter wheat. *HGCA Project Report No. 71*, 66 pp.
Parker, S.R., Shaw, M.W. and Royle, D.J. (1995) The reliability of disease severity on cereal leaves. *Plant Pathology*, **44**, 856-864.
Parry, D.W. (1990) What effect does disease have on the crop? in *Plant Pathology in Agriculture*, Cambridge University Press, pp. 69-85.
Pettitt, T.R., Parry, D.W. and Polley, R.W. (1993) Improved estimation of the incidence of *Microdochium nivale* in winter wheat stems in England and Wales, during 1992, by use of benomyl agar. *Mycological Research*, **97**, 1172-1174.
Reddy, M.V., Singh, K.B. and Nene, Y.L. (1981) Screening techniques *for Ascochyta* blight of chickpea, in *Proceedings of the Workshop on Ascochyta Blight and Winter Sowing of Chickpea, Aleppo, Syria, May 4-7, 1981*, (eds M.C. Saxena and K.B. Singh), Martinus Nijhoff and W. Junk, Aleppo, Syria. [Cited in Strange, R.N. (1993) *Plant Disease Control, Towards Environmentally Acceptable Methods*, Chapman & Hall, London, p. 93.]
Ride, J.P. and Drysdale, R.B. (1972) A rapid method for the chemical estimation of filamentous fungi in plant tissue. *Physiological Plant Pathology*, **2**, 7-15.
Ridout, M.S. and Xu, X.-M. (2000) Relationships between several quadrat-based statistical measures used to characterize spacial aspects of disease incidence data. *Phytopathology*, **90**, 568-575.
Rosielle, A.A. (1972) Sources of resistance in wheat to speckled leaf blotch caused by *Septoria tritici*. *Euphytica*, **21**, 152-161.

Saari, E.E. and Prescott, J.M. (1975) A scale for appraising the foliar intensity of wheat diseases. *Plant Disease Reporter,* **59**, 377-380.

Schneiter, A.A. and Miller, J.F. (1981) Description of sunflower growth stages. *Crop Science,* **21**, 901-903.

Schots, A., Dewey, F.M. and Oliver, R.P. (eds) (1994) *Modern Assays for Plant Pathogenic Fungi: Identification, Detection and Quantification,* CAB International, Wallingford, U.K., 267 pp.

Schumann, G.L. (1991) Plant disease epidemics and their management, in *Plant Diseases: Their Biology and Social Impact,* APS Press, Minnesota, pp. 123-148.

Scott, P.R. and Hollins, T.W. (1974) Effects of eyespot on the yield of winter wheat. *Annals of Applied Biology,* **78**, 269-279.

Seem, R.C. (1984) Disease incidence and severity relationships. *Annual Review of Phytopathology,* **22**, 133-150.

Shane, W.W., Thompson, C.E. and Teng, P.S. (1985) AREAGRAM - A standard area diagram program for the Apple computer. *Phytopathology,* **75**, 1363 (Abstr.).

Shaner, G.E. (2003) Epidemiology of Fusarium head blight of small grain cereals in North America, in *Fusarium Head Blight of Wheat and Barley* (eds K.J. Leonard and W.R. Bushnell), APS Press, St. Paul, Minnesota, 512 pp.

Shaner, E. and Finney, R.E. (1977) The effect of nitrogen fertilization on the expression of slow-mildewing resistance in Knox wheat. *Phytopathology,* **67**, 1051-1056.

Silva-Acuna, R., Maffia, L.A., Zambolim, L. and Berger, R.D. (1999) Incidence-severity relationships in the pathosystem *Coffea arabica-Hemileia vastatrix. Plant Disease,* **83**, 186-188.

Strange, R.N. (2003) Introduction to plant pathology. John Wiley & Sons Ltd., UK, 464 pp.

Subba Rao, P.V., Subrahmanyam, P. and Reddy, P.M. (1990) A modified 9-point disease scale for assessment of rust and late leaf spot of groundnut. *Proceedings of a Meeting of the French Phytopathological Society, Montpellier, France.* [Cited in R.N. Strange (1993) *Plant Disease Control, Towards Environmentally Acceptable Methods,* Chapman & Hall, London, pp. 94-95.]

Sylvester-Bradley, R., Makepeace, R.J. and Broad, H. (1984) A code for stages of development in oilseed rape *(Brassica napus* L.). *Aspects of Applied Biology* **6**, *Agronomy, physiology, plant breeding and crop protection of oilseed rape,* 399-419.

Taubenhaus, J.J., Ezekiel, W.N. and Neblette, C.B. (1929) Airplane photography in the study of cotton root rot. *Phytopathology,* **19**, 1025-1029.

Teng, P.S. (1985) Construction of predictive models. II. Forecasting crop losses. *Advances in Plant Pathology, Vol. 3: Mathematical Modelling of Crop Disease* (ed. C.A. Gilligan), Academic Press, UK, pp. 179-206.

Tomerlin, J.R. and Howell, T.A. (1988) DISTRAIN: A computer program for training people to estimate disease severity on cereal leaves. *Plant Disease,* **72**, 455-459.

Tottman, D.R. and Broad, H. (1987) The decimal code for the growth stages of cereals, with illustrations. *Annals of Applied Biology,* **110**, 441-454.

Tottman, D.R., Makepeace, R.J. and Broad, H. (1979) An explanation of the decimal code for the growth stages of cereals, with illustrations. *Annals of Applied Biology,* **93**, 221-234.

Turechek, W.W. and Madden L.V. (2003) A generalized linear modeling approach for characterizing disease incidence in a spatial hierarchy. *Phytopathology,* **93**, 458-466.

Van der Plank, J.E. (1963) *Plant Diseases: Epidemics and Control.* Academic Press, London, 349 pp.

Waggoner, P.E. and Berger, R.D. (1987) Defoliation, disease and growth. *Phytopathology,* **77**, 393-398.

Watson, G. and Morton, V. (1990) Standardisation of disease assessment and product performance reporting: an industry perspective. *Plant Disease,* **74**, 401-402.

Weiner, J. (2003) Ecology – the science of agriculture in the 21st century. *Journal of Agricultural Science* **141**, 371-377.

Wheeler, B.E.J. (1976) Development of diseases in crops, in *Diseases in Crops,* Edward Arnold, London, pp. 22-36.

Xu, X.-M. and Madden, L.V. (2002) Incidence and density relationships of powdery mildew on apple. *Phytopathology,* **92**, 1005-1014.

Xu, X.-M., Parry, D.W., Edwards, S.G. *et al.* (2004) Relationship between the incidences of ear and spikelet infection of *Fusarium* ear blight in wheat. *European Journal of Plant Pathology,* **110**, 959-971.

Zadoks, J.C., Chang, T.T. and Konzak, C.F. (1974) A decimal code for the growth stages of cereals. *Weed Research,* **14**, 415-421.

Zadoks, J.C. and Schein, R.D. (1979) Disease and crop loss assessment, in *Epidemiology and Plant Disease Management,* Oxford University Press, New York, pp. 237-281.

CHAPTER 3

SURVEYS OF VARIATION IN VIRULENCE AND FUNGICIDE RESISTANCE AND THEIR APPLICATION TO DISEASE CONTROL

JAMES K.M. BROWN

3.1 INTRODUCTION

Genetic variation has been discovered in almost all plant pathogens which have been studied, both in characters which affect their ability to infect host plants and in traits which are not related to pathogenicity. Knowledge of variation in pathogenicity characters, gained through surveys of pathogen populations, is valuable in planning effective measures for disease control, such as the use of resistance genes in crop varieties or the application of fungicides. Conversely, knowledge about such epidemiological factors as the dispersal of inoculum helps in the planning of pathogen surveys and in predicting how pathogens may adapt to control measures. This chapter reviews methods of surveying variation in traits related to pathogenicity in fungi and ways of applying the results of surveys to disease control. In particular, it focuses on traits important in agriculture, especially virulences and responses to fungicides. It also reviews two sets of recent developments: the application of molecular genetics to surveys of pathogenic variation and the dissemination of the results of surveys and recommendations arising from them *via* the internet.

3.2 CHARACTERISING INDIVIDUAL PATHOGENS

The two kinds of phenotypic variation that are most often studied in pathogen surveys are adaptation to different host cultivars and responses to fungicides.

3.2.1 Host cultivar range

The most intensively studied type of adaptation of pathogens to different host cultivars is that which follows the gene-for-gene relationship, first defined for flax rust by Flor (1956). Many aspects of the genetics and molecular biology of gene-for-gene interactions were reviewed by Crute *et al.* (1997) and Dangl and Jones (2001). A gene-for-gene interaction operates between a specific avirulence gene in a pathogen and a specific resistance gene in a plant. The resistance protein detects the presence of a pathogen with the corresponding functional avirulence gene, most likely by detecting the interaction of the avirulence protein with a

B. M. Cooke, D. Gareth Jones and B. Kaye (eds.), The Epidemiology of Plant Diseases, 2nd edition, 81–115.

pathogenicity target in the plant (Mackey *et al.,* 2002). Detection of the avirulence protein results in induction of effective host defences and therefore causes the host-parasite interaction to be incompatible. If the pathogen lacks a functional avirulence gene, its presence cannot be detected by the corresponding resistance gene, while a plant which lacks a functional resistance gene cannot detect pathogens with the corresponding avirulence gene. In many diseases, host-parasite interactions are controlled by several avirulence-resistance gene pairs but there need only be one matching pair of resistance and avirulence genes for an interaction to be incompatible.

(a) Host differentials

In order to identify the virulences a pathogen isolate has (or more accurately, the avirulence functions it lacks), it is inoculated onto a set of host varieties with different specific resistances. A universally susceptible control variety should always be included to check that the inoculation has been successful. Many diseases, such as powdery mildews and rusts, have qualitative symptoms such as necrosis or chlorosis in an incompatible interaction and conversely, growth and sporulation of the parasite in a compatible interaction. A pathogen isolate may therefore be classified as virulent on a differential variety if it is compatible with a high infection type (IT), or as avirulent if the interaction is incompatible with a low IT. An ordinal IT scale of this kind is widely used for research on cereal powdery mildew (Moseman *et al.*, 1965).

In other diseases, the distinction between an avirulent pathogen and a virulent one is quantitative. Here, the ratio between the amount of disease produced on a differential variety and that on the susceptible control is much lower for an isolate which is avirulent on the differential than it is for a virulent isolate. In such cases, classification of isolates as avirulent or virulent requires a quantitative statistical technique, such as median tetrad analysis, used by Brown *et al.* (2001) to investigate septoria tritici blotch of wheat. The fact that this method (and any other method used for the same purpose) relies on analysis of quantitative data gives rise to the usual statistical problems: Type I error, when an avirulent isolate is identified incorrectly as virulent, or Type II error, when a weakly virulent isolate is identified as avirulent.

Two types of differential sets of host lines are generally used. One consists of near-isogenic lines, in which different resistance genes have been bred into a common genetic background. Some widely used near-isogenic sets are those in Pallas barley, with different genes for resistance to powdery mildew (*Blumeria graminis* [formerly *Erysiphe graminis*] f.sp. *hordei*) (Kølster *et al.*, 1986), Chancellor wheat, also for resistance to powdery mildew (*B. graminis* f.sp. *tritici*) (Briggle, 1969) and Thatcher wheat for resistance to brown (leaf) rust (*Puccinia triticina* [formerly *P. recondita*]) (Samborski and Dyck, 1976).

The second important type of differential set includes cultivars, landraces, breeding lines and other material, each of which has a different resistance specificity but also differs from the other lines in its genetic 'background' – that is to say, genes other than those controlling the specific resistance. Such a set offers the flexibility to choose material suited to the purpose of the survey. In the annual study of barley powdery mildew undertaken by the UK Cereal Pathogen Virulence Survey (UKCPVS), for example, the differential set includes not only cultivars with known resistance genes but also two breeding lines and one landrace which express certain resistance genes clearly, as well as several cultivars with unknown resistances and others which are of interest because they are resistant to the current pathogen population (Slater, 2005b).

(b) Resistance genes in differential varieties

If a differential variety is known to have one particular resistance gene, it may then be postulated that the pathogen has the corresponding avirulence or virulence gene. It is often assumed that near-isogenic lines are most suitable for identifying pathogen virulences because each line should have a single specific resistance gene. In fact, this is not always so. Some of the Pallas near-isogenic lines of barley, for instance, were known to have more than one mildew resistance gene when they were released (Kølster *et al.*, 1986) and other specificities have since been discovered. The Pallas line P17, for example, was thought to have only *Mlk1* (formerly *Mlk*) but was later found to have an additional specificity, called *Ml(P17)* (Brown *et al.*, 1996). Some isolates which are virulent on *Mlk1* and on the UKCPVS differential Hordeum 1063 are therefore avirulent on P17.

Inevitably, near-isogenic sets cannot include all recently discovered resistance genes. Many modern, European spring barley varieties have *Ml(Ab)* mildew resistance, first used in the variety Triumph (Brown and Jørgensen, 1991; Slater, 2005b). As there is no Pallas near-isogenic line with *Ml(Ab)*, barley mildew workers may include either Triumph, which has *Mla7* as well as *Ml(Ab)*, or Lotta (formerly Sv.83380), with *Ml(Ab)* only, in their differential sets (e.g. Wolfe *et al.*, 1992).

Nevertheless, it is fair to say that there is usually greater uncertainty about the resistance genes in a differential set of varieties than in a near-isogenic set, especially when varieties are included in the set because they show differential interactions with some pathogen isolates but little or nothing is known about the genetics of their resistance. One of the best characterised differential sets is that used in the UK barley mildew survey; not only are the resistance genes in the core set of varieties known – in most cases, they can be traced back to the introduction of the genes into plant breeding programmes – but most lines have only one gene (Slater, 2005b). In the UK wheat mildew differential set, by contrast, several varieties have poorly characterised resistances (Slater, 2005a). Attempts to resolve ambiguities in some sets have been made. For example, the international differential set of wheat varieties for resistance to yellow (stripe) rust (caused by *Puccinia striiformis* f.sp. *tritici*) includes Heines VII, which has the resistance gene *Yr2*. Subsequently, Heines VII was found to have another resistance specificity, which is effective against isolate WYR 85-22 from Ecuador but does not differentiate European isolates. Isolates may therefore be avirulent on Heines VII even if they are virulent towards *Yr2* (Johnson, 1992; Calonnec *et al.*, 1997a).

The reliability of conclusions about pathogen virulences therefore depends on the extent of knowledge about the genetics of resistance in the differential set. One cannot assume that simply because two host lines are known to share one or more genes, they can be treated as alternative differentials; they may in fact differ in some other gene. This must be borne in mind when comparing results of experiments in which different differential sets have been used.

Finally, a good differential variety should discriminate virulent and avirulent isolates clearly, but the expression of a resistance gene's incompatible IT may vary between varieties. This implies that that gene interacts with other, unknown genes. For example, the international wheat yellow rust differential for *Yr6*, Heines Kolben, also has *Yr2*, but the latter gene is expressed more weakly in that variety than in Heines VII or Heines Peko, which also has *Yr2+Yr6* (Calonnec *et al.*, 1997b).

(c) Genetics of avirulence

The classic model of gene-for-gene interactions in which, as the name implies, one pathogen avirulence gene matches one plant resistance gene does indeed apply quite widely. As with most rules, however, there are exceptions – see reviews of plant pathogenic fungi in general by Christ *et al.* (1987) and of powdery mildews in particular by Brown (2002). In a common type of exception, two genes confer avirulence to one resistance gene. For example, avirulence of *B. graminis* f.sp. *hordei* to *Mla13* resistance was shown by genetic analysis to be controlled by two genes (Caffier *et al.*, 1996). Without comprehensive molecular genetic evidence, however, one cannot determine if the two avirulence genes do indeed match a single resistance gene, as has been demonstrated for *Pseudomonas syringae* infecting *Arabidopsis thaliana* (Bisgrove *et al.*, 1994), or actually match two closely linked resistance genes with different specificities. *Mla13* has now been shown to be an allele of the *Mla* gene (Halterman *et al.*, 2003) but there remains the possibility that one of the avirulence genes discovered by Caffier *et al.* (1996) matches a different gene, closely linked to the *Mla* locus. The complexity of avirulence genetics may be still deeper; to continue with the example of *Mla13*, not only do two avirulence genes match this resistance, but there may be other genes which inhibit these avirulences (unpublished data). Genes that inhibit or suppress avirulence have been identified in *Melampsora lini*, the flax rust fungus (Lawrence *et al.*, 1981; Ellis *et al.*, 1997).

At one level, these complexities do not matter: if one is simply interested in the frequency of a virulence phenotype in the pathogen population, perhaps to predict the level of risk to current cultivars, one does not need to know whether the genetics of avirulence are simple or complex. Knowledge about the genetics of avirulence may be desirable, however, if one wishes to use survey data to make quantitative predictions about pathogen evolution (Hovmøller *et al.*, 1993), which is likely to be affected by the details of the genetic control of avirulence.

A common misconception is that resistances which are expressed in a quantitative manner are under polygenic control and are not race-specific. There are very many exceptions to this supposed rule. One such is *Ml(Ab)* mildew resistance in Triumph barley, which is a single gene (unpublished data) that causes a 95% reduction in the number of colonies formed by an isolate with the avirulence allele, Avr_{Ab}, compared to virulent isolates (Brown and Jessop, 1995).

(d) Environmental effects

The expression of some resistances is affected by the environmental conditions in which tests are done. In the UK, plants used in the virulence survey of wheat brown rust are grown, before inoculation, in standard glasshouse conditions. After inoculation, they are transferred to either a low or a high temperature regime (10°C or 25°C respectively), because some varieties are resistant to certain *P. triticina* isolates at the higher temperature but not the lower, and *vice-versa* (Jones and Clifford, 1997). These interactions therefore depend on the environmental conditions after inoculation.

One can therefore only make rigorous comparisons between the results of different surveys if the resistance genes studied are expressed similarly not only in different differential sets of varieties but also in the environmental conditions in which each investigation is done. A standard set of isolates may be used to test whether or not the conditions do indeed allow similar expression of resistance genes.

(e) Other variety-isolate interactions

There are many ways in which plants and pathogens may interact, other than by the particular type of molecular interaction that characterises gene-for-gene relationships. A particularly important and well-understood resistance which does not have a gene-for-gene relationship with the pathogen is *mlo* in spring barley. Non-functional alleles of *mlo*, a gene required for susceptibility of barley to mildew (Büschges, 1997), especially *mlo11* (Piffanelli *et al.*, 2004), have provided durable resistance in cultivars over 30 years. Variation in the ability of *B. graminis* f.sp. *hordei* isolates to infect *mlo* barley varieties has been detected in the UK since 1998 (Slater and Clarkson, 2001). However, this has not led to *mlo* resistance becoming noticeably less effective in the field. It is not known why this should be so; one possibility is that increased aggressiveness to *mlo* barley reduces the pathogen's fitness in some other respect.

(f) New differential sets

The UKCPVS differential sets for barley mildew and wheat yellow rust illustrate the value of long-term research on a disease, in which plant pathologists collaborate closely with plant breeders. Consequently, these sets of varieties are relevant to the resistance genes used by breeders and are also informative in research on the pathogen. Quite often, however, it may be necessary to construct a differential set to study a disease despite little being known about the genetics of resistance. Yu (2000) wished to construct a different set of wheat varieties to study the dispersal of *B. graminis* f.sp. *tritici* in central China, where little was previously known about the resistance genes present in local varieties. He found that all varieties tested had at least one mildew resistance gene that was not present in European wheat cultivars. He therefore chose a differential set of Chinese wheats that maximised the ability to detect diversity in the local pathogen population while including, so far as possible, varieties with different known resistance genes.

Septoria tritici blotch has become an important disease of wheat in Europe in the last two decades (Bearchell *et al.*, 2005) and research on this disease illustrates the challenges involved in constructing a new differential set. In an unpublished study, J.C. Makepeace *et al.*, investigated variation in virulence of the pathogen, *Mycosphaerella graminicola*, in the UK. Their differential set included the cultivars Longbow and Riband as susceptible controls, the differentially resistant varieties Hereward, Tonic, Frontana, Equinox and Chaucer and the breeding line Kavkaz-K4500, an important source of resistance in wheat breeding. Hereward and Tonic were subsequently found to have a single resistance gene each, *Stb6* (Brading *et al.*, 2002) and *Stb9* (Chartrain, 2004) respectively, but both, especially Tonic, could be replaced with varieties which express those resistances more clearly. Frontana appears to have *Stb10* resistance (Chartrain *et al.*, 2004) but this is not present in any northern European wheat studied to date. The resistance of Equinox is weakly expressed and depends on environmental conditions, so this is not a good differential variety. Chaucer has resistance which is clearly expressed but appears to be cytogenetically unstable. Kavkaz-K4500 has at least five, possibly six resistance genes (Chartrain *et al.*, 2005b) so an isolate avirulent to this line may have any one or more of the corresponding avirulence genes. Finally, both of the susceptible controls, Longbow and Riband (as well as many other European cultivars), were found to be resistant to an African isolate of *M. graminicola* though not to any European isolate tested to date. It would be desirable to use a control susceptible variety which does not have this resistance. The survey by Makepeace *et al.*, produced valuable new information about the distribution of virulences in the UK population of *M. graminicola*. Given that septoria workers do not have the benefit of a long history of research on the genetics of resistance, it is not surprising that the choice of varieties included in the differential set could be improved for use in future research.

3.2.2 Responses to fungicides

Fungicides are one of the major elements of modern crop protection strategies but resistance to most modern, systemic fungicides has evolved in at least some pathogen populations (Russell, 1995).

(a) All-or-nothing responses

In some cases, there is a very great difference between the responses of resistant and sensitive isolates; the resistant isolates are not killed by the maximum dose of the fungicide recommended for field use. A simple test – essentially an all-or-nothing test of the ability to grow in the presence of a critical dose of the fungicide – is sufficient to classify an isolate's response one way or the other. A currently important example is the response of many fungi to Q_o inhibitor (QoI, also known as strobilurin) fungicides (Sierotzki *et al.*, 2000a, 2000b; Chin *et al.*, 2001; Robinson *et al.*, 2002).

(b) Minimum inhibitory concentrations

Resistance to many other fungicides is quantitative in nature, so isolate responses need to be measured by scoring these responses to several doses of the fungicide. The most widely used measure of resistance is the median effective dose (ED_{50}), that which kills half the sample being tested. Another is the minimum inhibitory concentration (MIC), the lowest dose sufficient to kill the pathogen.

The MIC might appear at first sight to be the easier of these two quantities to estimate, as it is simply an empirical observation of the lowest dose that kills all individuals. However, the estimated MIC is positively correlated with the inoculation density. Suppose that a certain MIC is observed at a low inoculation density: at a higher density of inoculation, there is a greater probability that the inoculum will contain at least some individuals capable of surviving at the MIC observed at the lower density. The MIC observed at the higher dose is therefore likely to be higher than that at the lower dose.

(c) Effective doses

The ED_{50} is estimated by fitting a curve of pathogen response (or survival) to the dose series. An appropriate curve is generally that of probits or logits of pathogen response against the logarithm of the dose. Such models can only be fitted accurately if there are several doses covering the steep part of the dose-response curve, some doses (including the zero dose – i.e. untreated material) to which the pathogen does not respond at all and at least one dose which is completely effective or nearly so (Brown, 1998).

The aim of applying a fungicide to a crop is to achieve near-complete disease control. It is therefore often useful to estimate effective doses (EDs) higher than the ED_{50} such as the ED_{95}, which kills 95% of the target pathogen. The value of estimating higher EDs depends on the material being tested: they may be useful for samples which are mixtures of genotypes, representing the whole target population, but are all but meaningless for isolates which are genetically uniform. In the former case, the difference between the ED_{95} and the ED_{50} reflects genetic variation in responses because the greater the difference between the two EDs, the greater the variance of response in the sample and therefore the greater the variation in the population as a whole, if the sample is a good representation of the population. In the latter case, however, all individuals of the fungus have the same genotype (barring mutation) so any difference between the ED_{95} and the ED_{50} is merely the result of environmental variation in the experiment.

If a sample contains many genotypes, the difference between the ED_{50} and a higher ED, such as ED_{95}, may be of value in planning fungicide application regimes because the rate of evolution is proportional to the additive genetic variance (Hedrick, 2004). However, interpretation of effective doses other than the ED_{50} is not necessarily straightforward. The widely-used probit model assumes that the logarithm of *tolerances* is normally distributed in the pathogen sample (Finney, 1971), an individual's tolerance being the dose of fungicide which is just sufficient to kill it. If tolerance is not normally distributed – if, for example, there are several discrete levels of resistance in a fungal population, as is the case for resistance of *B. graminis* f.sp. *hordei* in the UK to triazoles (Brown and Wolfe, 1991; Brown *et al.*, 1991a; Blatter *et al.*, 1998) and to morpholines (Brown *et al.*, 1991b; Brown and Evans, 1992) –EDs calculated by probit analysis may be inaccurate (Brown, 1998).

3.3 POPULATIONS AND SAMPLES

The aim of most surveys is to understand the composition of a pathogen's population and so to assess the value of different ways of controlling the disease. Some surveys, such as the UKCPVS (Bayles *et al.*, 1997), surveys of cereal rusts in the USA and Canada (Kolmer, 2005), and those of barley mildew and wheat yellow rust in Denmark (Hovmøller, 2001, 2004; Justesen *et al.*, 2002), collect data which are applied quite directly to disease control by indicating which resistant varieties and fungicides are most likely to be effective. Many other pathogens have been surveyed less frequently for similar practical purposes. Other surveys are done to answer more general questions about the structure or evolution of a population, often to support the medium to long-term development of disease control strategies. When planning and conducting a survey, one should aim to be clear about what population is being sampled and should understand how different schemes for sampling from a population might affect the interpretation of survey data.

3.3.1 What is a population?

(a) Defining a population

There are two reasons why a pathogen population must be defined carefully before carrying out a survey. The first is to ensure that the population sampled is representative of the population of interest. This is especially important if a survey is intended to support disease control measures. Secondly, one should be sure that the samples collected are representative of the population from which they are drawn. There may be no single definition of a population which applies to a pathogen species in all circumstances. Indeed, it may not be easy to define a population at all. Milgroom and Fry (1997) suggest that a useful definition of a pathogen population is an operational one:

> A fundamental concept ... is that the samples ... are so structured that the inferences made can be applied to more general phenomena beyond the objects being observed. The key question is to what populations will inferences be made?

In other words, the pathogen population to be sampled in a survey should be typical of the population to which disease control is to be applied.

From the point of view of population genetics, a useful definition of a population is 'a group of interbreeding individuals that exist together in time and space' (Hedrick, 2004). This definition comprises both gene flow, as interbreeding causes allele frequencies to be more or less uniform throughout a population, and natural selection, in that any one area at any one time will have particular features which affect the relative fitnesses of different genotypes. For sexual species, definitions of a population based on interbreeding and on natural selection overlap to a large extent, because individuals must be present in the same area at the same time in order to interbreed, and are therefore likely to be subject to similar selection pressures. However, many plant pathogenic fungi have a prolonged asexual phase. For instance, *P. striiformis* has no known sexual stage (Hovmøller *et al.*, 2002; Enjalbert *et al.*, 2005) so while individual fungi which coexist in time and space – perhaps even in a single field – cannot interbreed with one another, they are nonetheless subject to the same selective forces. In practical terms, therefore, they are members of a single population, though not one in which interbreeding occurs.

(b) Changes in populations

More than in most other organisms, the nature of a population of a plant pathogen may change radically through a year. There are two especially striking ways in which this can happen. In the first case, there may be distinct populations of a pathogen on different varieties of a crop, but these may merge to form a single population on another crop. In the 1970s and 1980s, many spring barley varieties in the UK had gene-for-gene resistances to mildew whereas winter barley varieties generally had either no gene-for-gene resistance at all or only ineffective resistances. The mildew fungus, *B. graminis* f.sp. *hordei*, therefore existed as several distinct populations on spring cultivars, in that colonies on one variety produced conidia which were not necessarily able to infect other varieties. However, almost all clones could successfully infect winter barley cultivars, on which there was therefore essentially a single population (Wolfe, 1984). Such a process of successive division and amalgamation of populations has two consequences for the pathogen's population genetics. Firstly, negative linkage disequilibrium between virulence genes required to overcome the resistances of different spring barley varieties may arise, because different pathogen clones carry those virulences (Wolfe and Knott, 1982; Østergård and Hovmøller, 1991). Secondly, pathogen virulence genes may become recombined through sexual reproduction on winter barley plants, giving rise to clones able to overcome a previously effective combination of resistances (Brown *et al.*, 1993).

In the second case, there may be a great difference between the mobility of the asexual and sexual phases of a pathogen, so that there are two different types of population. In its asexual phase, the septoria tritici blotch fungus, *M. graminicola* (anamorph *Septoria tritici*), is dispersed over very short distances, as splash-borne pycnidiospores. Consequently, there is no correlation between the genotypes of single-pycnidium isolates of *M. graminicola* even between sampling sites just 10 m apart (McDonald *et al.*, 1995). Hence, as a result of low mobility within a cropping period, the *M. graminicola* population in a wheat field is a meta-population, composed of very many sub-populations, each confined to a very small area. By contrast, the ascospores of *M. graminicola* are wind-dispersed and are formed throughout much of the growing season (Hunter *et al.*, 1999; Eriksen and Munk, 2003) Since ascospores can be dispersed over long distances, the selection pressures which act on them are quite different from those acting on the small sub-populations in which the ascospores are formed. Furthermore, the same selection pressures act on ascospores over a large area, so those spores form a single, geographically extensive population.

3.3.2 Sampling populations

(a) Single genotype and bulk isolates

Samples from pathogen populations may be collected as pure genotypes, such as single colony isolates of powdery mildew fungi, single pustule isolates of rust fungi, single pycnidium isolates of *Mycosphaerella* spp. and so on. Alternatively, bulk isolates consist of many individuals, usually of many genotypes, and are usually intended to be representative of the pathogen population on a plant, in a field or in some larger area.

More precise information can be obtained from single colony isolates than from bulk isolates, but if it is more important to collect information from many sampling sites than to have very precise data about the population at any individual site, it may be more cost-effective to use bulk isolates. However, if a bulk isolate has to be maintained for several generations before being tested, there is a danger that some genotypes will out-compete others, with the result that the bulk isolate, when it is tested, is not actually representative of the population from which it was sampled.

(b) Sampling schemes

For population genetic analysis, the ideal sample is a random one, obtained from a defined population. This presents two problems: how is the target population to be defined and how can a random sample be collected from it.

Wind-dispersed spores of a few fungi can be sampled randomly and such a sample can be treated as a random sample from the source population if it is assumed that this population of spores will form the next generation of inoculum. This is done by many laboratories for *B. graminis*, which is sampled by allowing airborne conidiospores to infect susceptible trap plants. Such a sample should be randomly drawn from the local conidial population from fields upwind of the sampling site. However, a sample should not only be random, but also be representative of the population in which one is interested. Hovmøller *et al.* (1995) investigated stationary and mobile traps of *B. graminis* f.sp. *hordei*, the former being plants of the susceptible variety Pallas placed more than 1 km from the nearest barley field, the latter being leaves of Pallas placed in a spore trap mounted on the roof of a car, driven around the area close to the stationary site. There was less variation in virulence frequencies over the three dates on which samples were collected in the mobile sample than in the stationary sample. This was probably because the latter was strongly influenced by the wind direction at the time of sampling, and therefore consisted of isolates from different, nearby fields on the three occasions, whereas the mobile sample included isolates from many fields, so differences due to wind direction were smoothed out.

If it is not possible to sample airborne spores, a sample from a variety which is susceptible to all pathotypes may be considered to be a random sample from the local population. Replicate fields should be sampled to ensure that the sample is indeed representative. Samples from varieties with resistance genes that differentiate the current pathogen population are clearly not random. This point needs to be emphasised, because statistical analyses of such samples are sometimes presented as if they were actually random, as discussed by Barrett (1987).

3.4 MOLECULAR DETECTION OF VIRULENCE AND FUNGICIDE RESISTANCE

Cloned genes now offer alternatives to pathology testing to detect virulence and fungicide resistance. At present, the number of avirulence genes and fungicide resistance genes in important pathogens that can be screened in this way is limited but this is sure to be a growth area in plant pathology in the next decade or so. For tests based on molecular information to be effective, the relevance of sequence variation to pathogen phenotypes must be considered carefully. In addition, molecular markers other than cloned genes may provide useful information about the evolution of virulence and fungicide resistance.

3.4.1 Avirulence genes

Avirulence genes have been isolated from bacteria, fungi and oomycetes, including several economically important pathogens (Skamnioti and Ridout, 2005). DNA sequence information offers the opportunity not only to estimate frequencies of avirulence and virulence phenotypes but also to investigate their distribution, dispersal and evolution by examining variation between functional (avirulence) and non-functional (virulence) alleles. The only example of this to date (Schürch *et al.*, 2004) illustrates both the potential of this approach and also indicates some potential pitfalls which may need to be considered.

(a) Alleles of avirulence genes

In *Rhynchosporium secalis*, (cause of barley leaf blotch or scald), an avirulence gene, *NIP1* or *AvrRrs1*, produces a necrosis-inducing peptide, NIP1. The *NIP1* gene is involved in a gene-for-gene interaction because *R. secalis* isolates with a functional *NIP1* allele are unable to overcome resistance encoded by the *Rrs1* gene in barley. Eighteen *NIP1* alleles of full or nearly full length have been detected (Rohe *et al.*, 1995; Schürch *et al.*, 2004). Of these, five were isolated from virulent isolates and are non-functional alleles of the avirulence gene, producing variants of NIP1 which are not detected by *Rrs1* resistant plants. No single feature of the *NIP1* sequence distinguishes these alleles from those in avirulent isolates and the proposed causes of their non-functionality are different: alleles 3 and 4 produce ineffective NIP1 elicitor proteins (Fiegen and Knogge, 2002) and allele 8 may also be ineffective, whereas NIP1 may not be expressed in isolates with alleles 7 and 17. In addition to mutations in the *NIP1* sequence, deletions of *NIP1* were found in other virulent isolates of *R. secalis* and in fact were much more frequent than the mutant, non-functional alleles.

(b) Use of gene sequences in population studies

Sequences of alleles of an avirulence gene may be used in population genetic studies of a pathogen. For example, Schürch *et al.* (2004) estimated the gene diversity of *NIP1* in populations from four continents. DNA sequences have the potential to be used as a replacement or a supplement to pathology tests because well-designed PCR tests of sequence variation are now relatively quick and use techniques which are standard in any molecular genetics laboratory. Pathology tests of virulence, by contrast, may take several weeks and require highly specialised skills.

However, if sequences are to be used widely in pathogen surveys, several potential obstacles need to be avoided, the first three of which may be illustrated by reference to the work of Schürch *et al.* (2004). The source of most of them is the large effective population size of most plant pathogen species, which allows many alleles of any one gene to be maintained in the population. The first is that no single nucleotide base may distinguish precisely between avirulence and virulence. A second issue, which arises from the first, is that when a new allele is detected, pathology tests are essential to determine whether it encodes a functional avirulence allele or not. Thirdly, the testing procedure must be sufficiently reliable that the experimenter can be sure that null PCR amplifications result from the absence of the gene (or at least the part of the gene encoding the primer hybridisation sites) and not merely from a poor sample. Fourthly, an efficient PCR diagnostic test requires amplification of a DNA fragment by a single pair of primers. Any sequence variation outside the amplified region that contributes to phenotypic variation will not be detected. Lastly, virulence may result from non-expression of the avirulence gene, which might be caused by variation in a different gene that interacts with the avirulence gene.

Despite these scientific and technical challenges, it is likely that cloned avirulence genes will begin to be used in pathogen surveys and that their use in population genetic research will increase. If the aim is to estimate frequencies of virulent and avirulent phenotypes of the pathogen in target populations, misidentification of occasional isolates may not matter very much. If, however, accurate identification of individual isolates is essential, the 'gold standard' remains the pathology test but sequence data may provide useful supporting information.

3.4.2 Fungicide response genes

DNA sequences of genes offer the same potential and problems for research on fungicide resistance as they do for avirulence and have already been used in population research. There is the substantial additional complication, however, that in most cases, several genes are involved in control of the response to a fungicide. All site-specific fungicides have a target enzyme and in some cases, mutation of the target site has been shown to be involved in resistance, including mutation of β-tubulin in resistance to benzimidazole fungicides (Koenraadt *et al.*, 1992), cytochrome b in resistance to QoIs (Sierotzki *et al.,* 2000a, 2000b; Zheng *et al.,* 2000) and the cytochrome P450 14α-demethylase (P45014DM) in resistance to triazoles (Délye *et al.* 1997, 1998; Wyand and Brown, 2005). In other cases, however, resistance is conferred by mutations in non-target proteins, such as those involved in active transport of the fungicide across the plasma membrane (Hayashi *et al.*, 2001; Schoonbeek *et al.*, 2001; Vermeulen *et al.*, 2001).

(a) A single gene encoding the target protein

QoI resistance is associated with mutation of amino acid residue 143 in the wild-type cytochrome b sequence from glycine to alanine (G143A). Although this is not the only mutation of cytochrome involved in resistance to QoI fungicides (Avila-Adame and Koller, 2003; Kim *et al.*, 2003; Pasche *et al.*, 2005), it is much the most common and gives rise to complete resistance. Fraaije *et al.* (2002) used PCR primers to devise a real-time PCR system to estimate the frequencies of G143 and A143 alleles in *B. graminis* f.sp. *tritici* and showed that the frequency of A143 increases on exposure of powdery mildew to QoI fungicides applied to wheat crops. Fraiije *et al.* (2005) used a similar system to investigate frequencies of the A143 allele in field populations of *M. graminicola* and the dispersal of QoI-resistant ascospores.

Another resistance caused by a mutation of a single nucleotide is that of single fungi to benzimidazoles. Wheeler *et al.* (1995) devised a PCR test based on sequence variation in the β-tubulin gene to diagnose resistance in *R. secalis*.

(b) More complex genetic systems

Resistances to most fungicides are genetically more complex, and for these, it will therefore be correspondingly more challenging to devise diagnostic tests for fungicide resistance based on DNA sequences. Délye found that mutation from tyrosine to phenylalanine at amino acid residue 136 (Y136F) of P45014DM correlated with resistance to triadimenol, a triazole fungicide, in *Erysiphe necator* (formerly *Uncinula necator*), the grapevine powdery mildew fungus (Délye *et al.*, 1997) and in *B. graminis* f.sp. *hordei* (Délye *et al.*, 1998).

Research by Wyand and Brown (2005) on *B. graminis* supported this conclusion but also showed that the situation is considerably more complex, such that a diagnostic test for the Y136F mutation is not sufficient to characterise resistance. In *B. graminis* f.sp. *hordei*, three of the four known phenotypes of triazole resistance were associated with sequence variation in the *CYP51* gene which encodes P45014DM. All of these had F136 in P45014DM but isolates with two of the resistance levels, known as low and high, had identical sequences of *CYP51*. It was inferred that a second gene, very closely linked to *CYP51*, must be involved in differentiation between low and high resistance. Isolates with a third phenotype, very high resistance, also had F136 but in addition had a second mutation, from lysine to glutamine at residue 147 (K147Q). Very high resistance co-segregated with the F136+Q147 allele of P45014DM but it is not yet known if the K147Q mutation is directly involved in determining the very high resistance phenotype.

Three factors complicate the picture further (Wyand and Brown, 2005). Firstly, isolates of the wheat powdery mildew fungus, *B. graminis* f.sp. *tritici* with very high resistance did not have the K147Q mutation, nor did they all share any other mutation in P45014DM. Secondly, in a cross between triazole-resistant and sensitive isolates of *B. graminis* f.sp. *tritici*, resistance co-segregated with the *CYP51* allele of the resistant parent, supporting the hypothesis that alleles of this gene confer triazole resistance. The distribution of resistances, however, indicated that several genes were involved in resistance, not just *CYP51*. This suggests that *B. graminis* f.sp. *tritici* has a polygenic or oligenic system controlling resistance to triazoles, in addition to a major gene at the *CYP51* locus. Finally, the fourth known phenotype of triazole resistance in *B. graminis* f.sp. *hordei*, the medium level of resistance, had the same sequence of *CYP51* as triazole-sensitive isolates. Medium resistance is controlled by a single gene unlinked to the high resistance phenotype (Brown *et al.*, 1996). Together, the genetic and molecular data indicate that medium resistance must be controlled by a gene other than *CYP51*.

Resistance to triazole fungicides and probably to sterol demethylation inhibitors (DMIs) in general, of which triazoles are one type, therefore presents a complex picture. Even in such closely-related fungi as *B. graminis* f.sp. *hordei* and *tritici*, there are rather distinct genetic systems of triazole resistance. Both involve *CYP51*, but also different, uncharacterised genes, a single gene unlinked to *CYP51* in *B. graminis* f.sp. *hordei* and several genes, the segregation of which gives rise to a continuous distribution of ED_{50}s, in *B. graminis* f.sp. *tritici*. In *Botrytis cinerea*, by contrast, resistance to DMIs is associated not with the target site of the fungicides but with increased transport of the fungicide out of the fungal cell (Hayashi *et al.*, 2001).

(c) Applications of fungicide resistance gene sequences

When they become widely-used, PCR-based tests will offer major advantages to farmers as they cut the time needed to characterise the fungicide resistance of isolates from several weeks to just a day or two. This will make it possible to give rapid advice to farmers on whether or not a control failure was due to the pathogen population being resistant to the fungicide or to some other cause. They will also facilitate research on the population genetics of resistance, as it will be much quicker and easier to distinguish resistant and sensitive isolates than it generally is nowadays, given that pathology tests of fungicide resistance are often laborious and expensive. The same warnings apply, however, as with the use of DNA sequences to identify virulences, with the substantial additional complication that several genes are involved in resistance to most groups of fungicides, notably the triazoles, the most important group of all.

3.4.3 Indirect uses of molecular markers

DNA markers and sequence data have been widely used in population genetics research on pathogens. In the context of surveys of variation in phenotypes related to pathogenicity, an important use is tracing the spread of new races or clones of a pathogen. This may lead to strategic advice of medium to long-term value, on how strategies of disease control could be improved.

(a) Clones of Blumeria graminis

The earliest research of this kind concerned the spread of clones of *B. graminis* f.sp. *hordei* bearing virulences which enabled the fungus to overcome important resistance genes. This showed that on several occasions when a race-specific resistance gene was first introduced into popular varieties, the matching virulence first appeared in one or a very few clones of the pathogen with characteristic patterns of DNA markers. The virulent clones then spread rapidly because of the availability of large areas of varieties with the corresponding resistance, and the virulence gene became recombined into diverse genotypes, with diverse sets of DNA markers (Brown *et al.*, 1990, 1991a; Wolfe *et al.*, 1992; Brown, 1994b; Wolfe and McDermott, 1994; Brändle *et al.*, 1997; Brown *et al.*, 1997).

Some practical conclusions of this research offered by the authors of these reviews are that breeding should be based on durable resistance, not on the ephemeral effectiveness of gene-for-gene resistance, and that more diverse resistances should be used by breeders, especially those in different countries. Much of the research cited here concerned the spread of clones of *B. graminis* f.sp. *hordei* virulent to *Mla13* resistance, which was used simultaneously by many barley breeders during the 1980s. The fact that there has been no more recent known instance of the widespread dissemination of clones with virulence to a new resistance gene may be significant. Cereal breeders in the UK, at least, have based their selection programmes on durable resistance and there has been little enthusiasm for the deliberate introduction of new genes conferring gene-for-gene resistance.

(b) Pathogenic and molecular variation in Puccinia striiformis

Molecular markers are especially useful in surveys of pathogens that lack a sexual phase in their life cycle. Populations of these organisms consequently consist of a set of clonal types, each of which has a distinct genotype and may have a distinct phenotype, such as that controlling virulence. In order to track the movement of these clonal types and to estimate their frequencies in field crops, it is not necessary to have sequence data of the avirulence genes: information about the genotypes, as determined by molecular markers is sufficient for this purpose.

One of the most extensive applications of indirect marker systems to investigate the evolution of virulence has been in *P. striiformis* f.sp. *tritici*. Justesen *et al.* (2002) developed a set of amplified fragment length polymorphisms (AFLP) for this fungus and showed that groups of related AFLP genotypes were often associated with characteristic pathotypes, as determined by virulences to a differential set of wheat varieties. Hovmøller *et al.* (2002), using information from the distribution of AFLP variants, showed that long-distance dispersal of *P. striiformis* f.sp. *tritici* isolates from the UK to Denmark occurred on at least five occasions, each of which was associated with an epidemic of yellow rust on wheat varieties possessing certain resistance genes. The role of long-distance dispersal in the evolution of plant pathogens in general was reviewed by Brown and Hovmøller (2002).

AFLP markers were also used by Enjalbert *et al.* (2005) to investigate the evolution of *P. striiformis* f.sp. *tritici* in France. As in the UK and Denmark, the pathogen has a population structure consistent with clonal reproduction but populations in the north and south of France were quite distinct in their distribution of AFLP types. It was proposed that this has resulted partly from selection by local wheat varieties but also perhaps from adaptation to local climatic conditions.

Another significant conclusion from the work of Justesen *et al.* (2002) and Hovmøller *et al.* (2002) is that the rate of evolution of virulence and avirulence, under pressure of selection by resistance genes in crop varieties, is an order of magnitude faster than that of the DNA markers used. This implies that molecular markers may be appropriate for delimiting the likely range of pathotypes of an isolate but cannot be treated as a substitute for pathology tests of virulence.

3.5 CHARACTERISING PATHOGEN POPULATIONS

For practical use, data on a set of isolates must be reduced to a summary form so that the composition of the sample, and by inference that of the population, as well as changes in response to natural selection and other factors, can be described concisely. Here, the analysis of virulence and fungicide response data is discussed.

3.5.1 Virulences

The frequency of virulence to a resistance gene or a differential variety is straightforward to estimate from a random sample of a pathogen population (assuming that a reasonably random sample can be obtained). Two other important aspects of the characterisation of virulence frequencies which have concerned pathologists involve associations between virulences and frequencies of pathotypes defined by several virulences.

(a) Associations between virulences

Associations between alleles of pairs of genes may be summarised by the linkage disequilibrium, D (also known as the gametic disequilibrium or the haploid phase disequilibrium), which in statistical terms is the covariance between allele frequencies. *D* and other statistics derived from it can only be estimated accurately from a random sample from a defined population. However, other ways of describing associations between virulences can be used if a random sample cannot be obtained, especially if the emphasis is on practical utility rather than statistical purity. If samples are collected directly from varieties, one can determine if the population virulent on one variety has a particularly high or low frequency of virulence on another variety or another resistance gene of interest. This is especially useful in indicating if a new variety with a combination of resistance genes is at risk from disease by new pathotypes with the matching combination of virulence. In 2000, for instance, wheat varieties in the UK with a combination of resistances to yellow rust including the genes *Yr9* and *Yr17* and the then poorly characterised Carstens V (CV) resistance, notably Oxbow, became susceptible to a group of *P. striiformis* f.sp. *tritici* clones which were virulent on all three resistances (Bayles and Stigwood, 2001). Until then, varieties with *Yr9* + *Yr17* but not CV resistance or *vice-versa* were susceptible to yellow rust but not those with the combined resistance. CV resistance is genetically complex but the most important gene in the Carstens V source is probably *Yr32* (Eriksen *et al.,* 2004). Isolates with virulence to the combination *Yr9* + *Yr17* + CV are now common, as are those with virulence to the combination *Yr6* + *Yr9* + *Yr17* but isolates with virulence to all four genes have not yet been detected in the UK (Bayles and Hubbard, 2005). When they do appear, another group of wheat varieties will become susceptible to yellow rust.

(b) Pathotypes

If a group of isolates are identical in their avirulence or virulence to all varieties in a differential set, they are said to be of the same pathotype or race (although they may differ in avirulence to varieties which were not included in the differential set). Pathotype frequencies are often useful summaries of the results of pathogen surveys.

Some pathologists, notably those who work on cereal rusts, have agreed standard systems for describing pathotypes (McIntosh *et al.,* 1995). These involve testing isolates on a standard set of differential varieties, and using a numerical or alphabetical code as a brief description of the pathotype. Such codes help pathologists to exchange information about surveys in different countries or to compare populations between years. They are most useful for asexual pathogens, such as *P. striiformis*, or *P. triticina* in much of North America, in which distinct pathotypes exist over wide areas or for long periods of time (Hovmøller *et al.,* 2002; Justesen *et al.,* 2002; Kolmer, 2005). They are less useful for describing pathogens with a life-cycle which is at least partly sexual, because pathotypes lose their distinct identity through recombination.

Although standardised pathotype codes can help the sharing of information, they have the disadvantage that, to describe a pathotype, one may have to test isolates on several differential varieties that have no relevance to the local situation. Indeed, one has the feeling on reading the reports of some surveys that those conducting them have been more concerned with characterising virulence on a set of differential varieties for the sake of giving isolates names that are familiar to international researchers, than with investigating the threat that the pathogen may pose to local crops. The latter is a much more important goal and should be emphasised. When a pathogen is studied by several research groups, different authors may use a variety of codes, either because they have used different differential sets or because they wish to emphasise different features of their data. So long as a paper explains the race codes clearly, there is no problem with several systems existing side-by-side. Standardisation at the expense of scientific insight and practical application is not desirable.

(c) Bulk isolates

A bulk isolate can be used to obtain approximate estimates of a virulence frequency in the population from which it was sampled, by inoculating the isolate on a differentially resistant host and a near-isogenic susceptible variety. The number of successful infections on the former, divided by the number on the latter, is an estimate of the frequency of virulent isolates in the bulk. It is not possible to estimate associations between virulences in individual bulk isolates, but Wolfe *et al.* (1983) showed how bulk isolates could be used to estimate associations between virulences in a population.

3.5.2 Responses to fungicides

(a) Frequencies of resistance

How the responses of a pathogen population to a fungicide should be summarised depends on the nature of resistance to the chemical, the kind of samples tested, the quantity estimated (i.e. ED_{50} or MIC) and the use to which the data will be put. We can distinguish three kinds of distribution of responses to fungicides: the 'all-or-nothing' type, a continuous distribution of responses and the intermediate case of several distinct levels of resistance. These three cases are all conveniently illustrated by *B. graminis*: the all-or-nothing type by responses to QoIs (Sierotzki *et al.*, 2000a; Chin *et al.*, 2001; Robinson *et al.*, 2002), the continuous distribution by responses of *B. graminis* f.sp. *tritici* to triadimenol (Robinson *et al.*, 2002; Wyand and Brown, 2005) and the several levels by the resistance of *B. graminis* f.sp. *hordei* to triazoles and to fenpropimorph (Brown and Wolfe, 1990; Brown *et al.*, 1991a, 1991b, 1992, 1996; Brown and Evans, 1992; Brown, 1994a, 1996a; Blatter *et al.*, 1998; Wyand and Brown, 2005). The response of *B. graminis* f.sp. *hordei* to ethirimol is either sensitive or partially resistant (Brown and Wolfe, 1990; Brown *et al.*, 1991a) and is controlled by a single gene (Brown *et al.*, 1992) but the quantitative difference between the two responses is much smaller than that for the QoIs.

Samples may be single genotypes or bulk isolates, and the purpose of the tests may either be to characterise the population itself or to develop advisory recommendations. If single genotype isolates are tested, the severity of resistance of an all-or-nothing kind is easily estimated as the frequency of resistant isolates, as for virulences. In a bulk isolate, the frequency of resistance can be measured in a similar way as for that of virulence, as the amount of disease on plants treated with a differential dose of the fungicide compared to that on untreated plants.

Analysis of continuously distributed resistance generally requires the assumption that log-tolerances are normally distributed. If this is the case, the mean and standard deviation of the log-ED_{50}s of a random sample of isolates should be the same as the log-ED_{50} and the standard deviation of log-ED_{50} in the population. For a bulk isolate, log-ED values are estimates of the corresponding log-ED values in the population as a whole (e.g. the log-ED_{95} of a bulk isolate is an estimate of the log-ED_{95} of the population), provided that the bulk isolate is indeed a random sample of genotypes in the population.

Matters are a little trickier with a fungicide to which there several distinct levels of resistance. If individual isolates are tested, a population may be described by a table of frequencies of the various levels of resistance, as estimated by a measure such as the ED_{50} or MIC. Changes in resistance, for example in response to selection by the use or withdrawal of a fungicide, may be summarised as shifts to increased frequencies of higher or lower levels of resistance. There is no fully satisfactory way of summarising the response of a bulk isolate when there are two or more discrete levels of resistance to a fungicide. Probit models are often used but the same ED_{50} may be estimated if, on the one hand, there is a high frequency of genotypes with an intermediate level of resistance, or on the other hand, if there are high frequencies of genotypes with either high or low levels of resistance.

The ED_{50} is not always the most useful measure of resistance from the point of view of providing guidance on fungicide applications to crops, see section 3.2.2(c). Higher EDs can be estimated from tests on bulk isolates, provided that an appropriate model can be fitted to the data; i.e. a probit model will produce an unbiased estimate of ED_{95} if log-tolerances are normally distributed but an inaccurate estimate if there are several discrete levels of resistance (Brown, 1998). They can also be estimated from the distribution of ED_{50}s in a sample of single-genotype isolates, even though estimates of higher EDs of the individual isolates themselves have no value.

MICs are often estimated for bulk isolates and it is well known that in doing so, one is essentially estimating the MIC of the most resistant genotype in the bulk, even if that genotype is only a small fraction of the total. This makes data of this kind difficult to use in describing a population. However, if the aim of an experiment is to predict the dose of a fungicide needed to control a particular pathogen population or to judge the risk of a control failure should a particular dose be applied, estimating the MIC of a bulk isolate may give a cheap, quick, useful answer to the question.

(b) Baseline data

When one wishes to understand how the use of a fungicide has affected a pathogen's resistance, one needs to know the level of resistance both before and after the fungicide was used. In order to obtain baseline data on the responses, pathogen samples should ideally be collected before the fungicide is in widespread use, so that the response of a population which has not been exposed to the fungicide and selected by it can be estimated. Alternatively, if a pathogen has a low dispersal rate, as do many soil-borne pathogens, populations in fields which have been treated with the fungicide may be compared to those which have not; e.g. King and Griffin (1985), who studied benzimidazole resistance in *Tapesia* spp. Lastly, if sampling was not started before the fungicide was widely used, baseline responses may be estimated from historical collections which include isolates dating from before the introduction of the fungicide.

Appropriate baseline data are needed if changes in resistance are to be properly understood. In a study of resistance to cymoxanil in grapevine downy mildew (caused by *Plasmopara viticola*), only one vineyard which had never been treated with the fungicide could be sampled (Gullino *et al.*, 1997). There, *P. viticola* was much more sensitive to cymoxanil than in the other vineyards, where resistance was common and often severe. The lack of replication of untreated vineyards limits the extent to which one can conclude that the use of cymoxanil caused resistance to develop. The sensitivity of *P. viticola* in the untreated vineyard may have been representative of the baseline sensitivity before cymoxanil was introduced, but it might conceivably have been a fluke result.

(c) Cross-resistance

Most fungicides used commercially belong to just a few groups of compounds. Resistance to one fungicide of a group is very often correlated with resistance to other members of that group. In *B. graminis* f.sp. *hordei*, for instance, a gene identified by its effect in causing resistance to a triazole fungicide, triadimenol, also reduced sensitivity to four other triazoles (Blatter *et al.*, 1998). However, cross-resistance is not always so predictable. In the same fungus, genes which cause resistance to a morpholine, fenpropimorph, also cause resistance to a closely-related fungicide, fenpropidin, but have little or no effect on responses to another morpholine, tridemorph (Brown *et al.*, 1996). This reflects the pattern of cross-resistance in the *B. graminis* f.sp. *hordei* population (Brown *et al.*, 1991b; Brown and Evans, 1992; Brown, 1994a). In the barley net blotch pathogen, *Pyrenophora teres*, resistances to pairs of sterol biosynthesis-inhibiting fungicides were correlated in some populations but not in others, implying genetic variation in the cross-resistance relationships in different populations of *P. teres* (Peever and Milgroom, 1993). These results imply that data on the development of the resistance of a pathogen to a fungicide may be extrapolated to other fungicides in the same group and to other pathogens, but only tentatively. Any such extrapolations are hypothetical and must be tested.

3.6 APPLICATIONS OF PATHOGEN SURVEY DATA

Many pathogen surveys have immediate practical value, in guiding the choice of disease-resistant varieties and fungicides. They can also help in planning longer-term strategies for using these methods of controlling disease. The results of well-designed pathogen surveys should ultimately be of benefit to farmers but the primary users, whose work supports that of farmers, may be plant breeders and fungicide manufacturing companies. Agronomists often mediate between those two groups and farmers, helping to translate technical information into practical recommendations.

3.6.1 Breeding resistant varieties

Pathogen surveys are used in many countries to support the breeding and use of resistant varieties. Results of surveys enable breeders to assess the risk to existing sources of resistance, validate potential new sources of resistance and identify sources of durable resistance. In addition, several proposals have been made to manage the deployment of resistance genes in time or space, generally on a scale of several years and tens of thousands of square kilometres. Such deployment schemes would make use of data on frequencies of virulences matching resistance genes which are already present in commercial varieties and have been overcome by part of the pathogen population. With one exception, see section 3.6.1(d), none of these schemes are achievable without regulation by government and are therefore not considered further here (see review by Brown, 1995).

The examples below are drawn from obligate pathogens of cereals in the UK. They could equally well have been taken from other countries, crops and diseases in which long-term surveys have been carried out.

(a) Existing sources of resistance

A major application of survey data is in identifying current sources of resistance that are threatened by the emergence of new pathotypes. If a differential set includes varieties which are being used currently by breeders as sources of resistance, new pathogen strains, virulent towards that resistance, can be detected while they are still at a low frequency in the population. Perhaps the most important aim of the UKCPVS is to provide such an early warning system. The current differential sets for powdery mildew of wheat and barley and for yellow rust of wheat include several such varieties (Bayles and Hubbard, 2005; Slater, 2005a, 2005b). When survey data indicate that a hitherto effective resistance gene or combination of genes is at risk from new pathotypes, breeders may wish to cease using a threatened variety in their breeding programmes. They may, however, wish to continue using such a variety if it has other desirable properties (Brown, 1995). Farmers may use such information either to choose not to grow a variety or – more usually in Western Europe – to budget for additional fungicide applications.

An isolate with a new pathotype, virulent towards a previously effective resistance, may be useful to plant breeders. If the source of the resistance which is now at risk has been used in a breeding programme, tests with the new, virulent isolate may give a better prediction of a line's future disease performance than field trials in which lines are exposed to the current, natural population of the pathogen, in which the new pathotype is still at a low frequency. Trials of varieties with selected isolates, including those which overcome new resistances and new combinations of resistance genes, may be included in surveys as a means of detecting varieties which are susceptible to new pathotypes (Bayles and Hubbard, 2005).

(b) Identification of varietal resistances

The integration of information about host resistances with data on pathogen virulences is crucial to the effective use of survey data. An integrated approach makes it possible to know when several varieties are all at risk from the same pathogen races and therefore require similar additional control measures. The process of identifying varietal resistances begins with information about the interaction of a set of isolates with a differential set of varieties. Each differential variety should ideally have just one resistance gene, so the gene-for-gene relationship can be applied: if an isolate has an incompatible interaction with a differential, it can be presumed to have the avirulence gene matching the variety's resistance gene. Consequently, when a variety with unknown resistance is susceptible to an isolate, it can be presumed to have none of the resistance genes which match that isolate's avirulences. The resistances that a test variety might have can thus be determined by a process of elimination of those resistances that it does not have.

Knowledge about the genetics of resistance can also be used, because a variety of a self-pollinating species cannot have two different alleles of a single locus, such as the homologous, multi-allelic, mildew resistance genes, *Mla* in barley (Jørgensen, 1994; Halterman *et al.*, 2003; Shen *et al.*, 2003) and *Pm3* in wheat (Huang *et al.*, 2004; Yahiaoui *et al.*, 2004). Note, however, that some breeding programmes produce named varieties which have a considerable degree of genetic diversity. Different inbred lines of such a variety may have different resistance genes, even different alleles of a single multi-allelic resistance locus. Two further kinds of information may be helpful but should be treated with caution. A variety's pedigree may help to delimit the genes that it might have, if the resistances of its parents are known. However, errors in published pedigrees are quite common. Secondly, the IT might help to indicate which of several candidate resistance genes a variety might have. However, one should not rely on this information too heavily because ITs may be modified by other genes. Many features of the system of identifying resistance genes are exemplified by a series of papers about mildew resistances of barley varieties grown in Denmark (Torp *et al.*, 1978; Jensen and Jørgensen, 1981; Jensen *et al.*, 1992).

The logic of the process of inferring resistance genes relies on the principle of Ockham's Razor, that 'no more things should be presumed to exist than are absolutely necessary'. Although one can exclude the presence of certain resistance genes, it is not possible to confirm absolutely the presence of a gene by pathology tests alone. If a test variety is resistant to a set of isolates and those isolates are all avirulent on a differential variety, the test variety may have the same resistance gene as the differential but alternatively, it may have a previously unknown resistance, to which the isolates in question all happen to be avirulent. Here, pedigree information is especially useful. The determination of a variety's gene-for-gene resistances by pathology tests is therefore ultimately hypothetical, and, if it is crucial to know exactly which resistances it has, the hypothesis should be tested by appropriate crosses. However, it is not at all practicable to do test-crosses to determine the resistances of all commercially significant varieties to all important diseases.

Inferences about varietal resistances are most reliable if a well-characterised set of isolates is used. Such collections exist for many important pathogens and have usually been built up over several decades, reflecting the successive introduction of resistance genes to breeding programmes and the development of knowledge about the pathogen itself. They are therefore themselves important outputs of surveys of variation in pathogen populations.

It is now possible to identify some commercially important resistance genes by the DNA sequences of the genes. Inference of resistance genes by molecular data is simpler than for avirulence or fungicide resistance genes. Since most commercial crop species have a very small effective population size (Barrett, 1981), the allelic variation at any one resistance locus is usually limited. In particular, when a resistance gene has been introgressed into commercial breeding programmes from a landrace or a wild plant related to the crop species, there may only be one resistance allele in current use. Continuing with the example of the *Mla* mildew resistance genes of barley, this is probably the case with most commercially important alleles, including *Mla1*, *Mla6*, *Mla9*, *Mla12* and *Mla13* (Jørgensen, 1992b, 1994). *Mla7* represents a more complex situation, however, as it appears to have been introgressed into commercial breeding programmes on four separate occasions (Brown and Jessop, 1995). In a still more complex case, certain important resistance genes have not been introduced deliberately into breeding programmes so the number of functional resistance alleles is not known. For instance, *Stb6* in wheat for resistance to septoria tritici blotch has entered modern commercial breeding programmes from at least ten independent landrace sources (Chartrain *et al.*, 2005a) and that number may be an underestimate, as Chartrain *et al.* studied only one wheat variety from China and none from North Africa or South Asia, which are important secondary centres of diversity for wheat. Even when *Stb6* has been isolated, it will be a challenging task to enumerate every independent functional allele of this gene used in world wheat breeding.

Molecular sequences do not give complete information, however, as they describe the genotype at a particular locus, not the resistance phenotype. Some resistances are suppressed by modifier genes in certain varieties, for example the *Pm8* gene of wheat for resistance to mildew (Ren *et al.*, 1997). In such a case, use of the DNA sequence alone would incorrectly identify some varieties as resistant although they had a susceptible phenotype. However, the widespread use of DNA sequence data, using standard molecular genetic methods, will at least help to delimit the range of resistances that should be studied by pathology tests, which require more specialised skills.

(c) Identification of durable resistances

An especially useful output from a long-term survey programme is knowledge about durable resistance. For example, the *mlo* gene has provided resistance to powdery mildew in spring barley varieties in northern Europe since 1979 (Jørgensen, 1992a; Brown, 2002b) and, despite the evolution of *B. graminis* f.sp. *hordei* genotypes partially aggressive to *mlo* varieties (Slater and Clarkson, 2001), continues to be fully effective in the field. It is worth reflecting on the research that makes it possible to write that last sentence. On the plant side, the resistances carried by barley varieties have been determined, in the UK and several other countries, since the mid-1960s so that information may be combined from several sources – the fact that they are resistant to all isolates, their pedigrees and the distinctive defence response of *mlo* plants – to identify varieties which carry *mlo*. The recent publication of the sequence of *mlo11* (Piffanelli *et al.*, 2004), which is much the most commercially significant *mlo* resistance allele, will in future allow the use of DNA sequences to identify *mlo11* varieties. On the pathogen side, *mlo* varieties such as Alexis, Atem and Apex have been used in differential sets since the resistance was introduced (Wolfe *et al.*, 1992; Slater 2005b) and no isolate highly virulent to *mlo* has been collected among the many thousands tested since then – the level of disease caused by *mlo*-aggressive isolates on *mlo* varieties is much lower than on non-*mlo* varieties (Slater and Clarkson, 2001; Slater, 2005b). If both these aspects of the barley mildew survey had not been carried out, breeders would simply be left with the observation that the resistance of some varieties – those which we in fact know do not have *mlo* resistance – become ineffective, while the resistance of others – those which have *mlo* – is durable. Long-term pathogen survey data, collected over nearly 30 years, has made it possible to ascribe the durability of the mildew resistance of *mlo* varieties to the presence of that gene, and has therefore enabled breeders to use *mlo* systematically as a source of durable resistance.

Survey data can be used indirectly to identify varieties with partial resistance, quantitatively effective against all pathogen isolates. Quantitative resistance is often durable, in the sense of being effective over a wide area for a long time (Johnson, 1984), though is sometimes race-specific and non-durable – see the discussion of *Ml(Ab)* in section 3.2.1(c). Even when virulence towards a resistance gene is very common, there is usually variation in the quantitative level of resistance of varieties with that gene. For example, the virulence Vra, corresponding to the *Mlra* resistance gene, has been fixed (i.e. avirulent isolates have not been detected) in the *B. graminis* f.sp. *hordei* population since at least 1985 (Brown and Wolfe, 1990; Slater, 2005b) and probably for even longer (Wolfe, 1984). Since then, winter barley varieties with *Mlra* resistance have had NIAB Recommended List ratings ranging from 2 to 8 on a scale which ranges from 1 (extremely susceptible) to 9 (nearly immune). Those at the resistant end of the scale need much less fungicide to be applied to control mildew and are also useful in breeding for durable, partial resistance. Many other winter barley varieties with different specific mildew resistances also have good quantitative resistance.

Quantitative resistance can be detected, though more tentatively, in varieties with a specific resistance which is still partially effective by means of tests with newly detected isolates which overcome the specific resistance (Bayles and Hubbard, 2005). For instance, when the *Yr17* wheat yellow rust resistance became ineffective in the mid-1990s through the evolution of virulence to *Yr17* (V17) in the pathogen population, the variety Brigadier became very susceptible to V17 isolates, with a

NIAB rating of 1, whereas Hussar appears to have some additional, so-called background resistance, with a rating of 5 (Anon., 1998). The latter variety therefore needed much less expenditure on fungicide treatments than the former and in fact, the use of Brigadier in commercial wheat production in the UK ceased because of its susceptibility to yellow rust.

(d) Use of resistance genes

Priestley and Bayles (1980) proposed the use of variety diversification schemes as a means of reducing the risk of disease. They suggested that when a farmer grows several fields of a crop species, they should be of varieties with different resistances, particularly in neighbouring fields, to restrict the ability of a pathogen to spread between fields on a single farm. NIAB publishes diversification schemes for wheat yellow rust and barley powdery mildew each year, to assist the choice of varieties for this purpose. The schemes are based on the frequencies of virulences matching the specific resistances of commercially important varieties and on associations between those virulences.

The value of diversification schemes in reducing the spread of disease on the scale of a farm depends on the mobility of the pathogen population. O'Hara and Brown (1998) showed that very few *B. graminis* f.sp. *hordei* spores move more than a very few metres in a field of spring barley, while O'Hara and Brown (1996) showed that almost all immigration of *B. graminis* f.sp. *hordei* spores into a plot of a susceptible variety occurs early in the season. It is therefore unlikely that diversification of varieties among fields which are typically a few hundred metres across would restrict the development of disease within each field, because relatively little inoculum moves between fields.

However, diversification schemes have considerable value at a national or regional level, because, if farmers do indeed choose several varieties with different genes for resistance to a disease, the acreage covered by varieties with any one resistance gene will be limited, which reduces the opportunity for pathogen clones virulent to that resistance to spread.

3.6.2 Effective use of fungicides

Information about fungicide resistance is important and valuable to enable farmers to implement cost-effective disease control. It also enables the agrochemical industry to adapt the marketing of fungicides to the altered response of the pathogen population.

(a) Choice of a fungicide

If resistance has evolved to a fungicide, the simplest decision for a farmer to make is to choose to spray a different fungicide to which the pathogen population is still sensitive, if one is available. The emergence of resistance to QoI fungicides in several important plant diseases during the late 1990s and early 2000s was a major problem both for farmers and for the agrochemical industry (Chin *et al.*, 2001; Ishii *et al.*, 2001; Gisi *et al.*, 2002). However, information about the appearance and spread of resistance allowed farmers to choose alternative fungicides. An example of a difficulty that arose from QoI resistance is that the effect of QoIs in controlling septoria tritici blotch of wheat was rather robust whereas the timing of application of triazole fungicides to control this disease must be much more accurate, hence the efficacy of triazoles depends to a significant extent on weather conditions around the time of spraying (Cook, 1999).

The appearance of resistance to many widely-used fungicides has caused the agricultural industry to consider strategies to delay the emergence of resistance in a pathogen population or to reduce its rate of increase once it has appeared. Most resistance management schemes rely on the use of mixtures of fungicides from different chemical groups, or alternating treatments with such chemicals (Russell, 1995). In designing practical disease control strategies, it must be remembered that almost all crops are at risk from more than one disease. Consequently, a fungicide to which resistance has evolved in one pathogen may still be useful if it protects the crop against other diseases. For instance, resistance to triazoles developed in both wheat and barley mildew shortly after the introduction of these chemicals in the UK in 1977 (Fletcher and Wolfe, 1981; Fletcher *et al.*, 1987) and severe resistance had appeared by the late 1980s (Brown *et al.*, 1991a; Blatter *et al.*, 1998). However, triazoles still protect cereals against other foliar diseases and are therefore important components of strategies for broad-spectrum disease control.

(b) Combined use of fungicides and resistant varieties

Since fungicides are expensive, advice to farmers should help them to use cost-effective strategies to minimise the number of sprays needed. This is also perceived by the general public as desirable from the point of view of environmental protection. Survey data may occasionally reveal opportunities for adapting fungicide recommendations to different varieties. Such a possibility arose between 1988 and 1990, in yellow rust of wheat, when isolates virulent on several important varieties with *Yr6*+*Yr9*, such as Hornet, were more sensitive to both triadimenol and fenpropimorph than were other isolates (Bayles *et al.*, 1992, 1994).

More generally, durable and effective use of fungicides should be based on the use of these chemicals to complement varietal resistance. Too often, fungicides have been seen as a 'quick fix' to control disease on a susceptible crop variety (this may have been the case when QoIs were used to control septoria tritici blotch on such susceptible wheat varieties as Riband, Consort and others). It must be recognised that, valuable though systemic fungicides have been in controlling crop diseases for the past 30 years, genetic disease resistance has been useful to farmers for thousands of years and will still be important once the era of systemic fungicides has passed. *Blumeria graminis*, on both wheat and barley, has evolved complete or partial resistance to almost all important fungicides used to control powdery mildew, including ethirimol (Brown and Wolfe, 1990), the triazole group (Brown and Wolfe, 1991; Blatter *et al.*, 1998; Robinson *et al.*, 2002; Wyand and Brown, 2005), the morpholine and piperidine group (Brown and Evans, 1992) and the QoI group (Chin *et al.*, 2001), while there have been recent reports of resistance to an important preventative, quinoxyfen (Walker *et al.*, 2004). Has this spelt disaster for cereal farming in Northern Europe? No, because in the meantime, breeding for resistance to mildew has been effective in both barley and wheat. Almost all varieties of wheat released in the UK in the last 20 years and many varieties of winter barley have had high partial resistance to mildew, while many spring barley varieties have had the *mlo* resistance, which has been durably effective.

3.7 DISSEMINATION OF SURVEY RESULTS

Until recently, the only practical option for disseminating the results of pathogen surveys was in printed form, which limited the audience for the research. Research papers reporting the results of surveys remain important, as they provide the technical and scientific information required by the community of researchers interested in a disease. The internet, however, is beginning to make a radical difference to the dissemination of information about surveys to farmers and agricultural advisors, just as it has in so many other walks of life.

3.7.1 Presentation of survey results

In the UK and several other countries, the only significant route for the dissemination of the results of pathogen surveys used to be technical bulletins, available as books known as Annual Reports of the UKCPVS. These were primarily aimed at specialists, including plant breeders and agricultural advisors, who could then transmit messages of practical importance to farmers. This still continues, although dissemination is now more often via the internet. The UKCPVS is supported by the Home-Grown Cereals Authority, which in turn raises funds from a levy on the sales of cereals. The main route for publication of the technical bulletins of the UKCPVS is now *via* the website of the HGCA (http://www.hgca.com/content.output/56/56/Crop-Research/Crop-Research/UK-Cereal-Pathogen Virulence-Survey.mspx). For those interested in survey information at the technical level, this has the very great advantage of ready access to summaries of the data.

3.7.2 Incorporation of survey results with advice to farmers

The internet also offers much more immediate opportunities to incorporate the results of surveys with advice to farmers on the choice of crop varieties and fungicides. To date, this has been done most extensively in Denmark, where relevant results of surveys of barley powdery mildew and wheat yellow rust are incorporated into the variety recommendation system available on the internet (http://www.sortinfo.dk). This includes, for example, information on the responses of leading wheat varieties to new races of yellow rust, tested in protected conditions by the Danish Institute of Agricultural Science, and the presence of the durable *mlo* resistance to mildew in varieties of spring barley.

3.7.3 Early warning systems

Occasionally, the results of survey indicate the need for action by farmers. The appearance of new pathotypes of an important fungus may suggest that they should choose alternative varieties for the next cropping year or else budget for additional fungicide applications. In the UK, such information of immediate value is announced in the farming press. The fact that this is done rarely, however, is seen as a mark of success for the UKCPVS, the aims of which are to:

- protect cereal growers from unexpected epidemics
- support plant breeding for improved and more durable resistance
- underpin cultivar evaluation and recommendation schemes

(see http://www.hgca.com/content.output/56/56/Crop-Research/Crop-Research/ UK Cereal-Pathogen-Virulence-Survey.mspx). Frequent press releases announcing, for example, the unexpected appearance of new races of fungi would indicate failure to achieve those aims.

In Denmark, use of the internet for this purpose has gone further than in any other country, just as it has with variety recommendations. Early warnings of important problems are disseminated to farmers through the national crop advisory system, PlanteInfo (http://www.planteinfo.dk), which is now used by the majority of Danish farmers.

3.7.4 Decision support

Computerised decision support systems are becoming important in agricultural advisory services. In particular, they can help farmers to manage variable costs such as that of pesticide applications. Systems such as PlanteInfo in Denmark offer the potential to provide advice relevant to different varieties, rather than to a crop species as a whole. One of the criteria on which the decision support should be based is disease resistance, so accurate, up-to-date knowledge about the susceptibility of varieties to current pathogen populations and new pathotypes is essential.

An important influence on the susceptibility of a variety to a disease is the frequency of matching virulence. A simple model has been devised to predict the frequencies of virulences in the barley mildew pathogen population in a six-month period, from information about the host and pathogen populations in the previous six months. Hovmøller *et al.* (1993) applied this model to data from an area of 1 km radius. The model was largely successful in predicting virulence frequencies but there was one striking exception, in that a much higher frequency of the virulence Va9, corresponding to *Mla9* resistance, was observed in one sample than the model predicted. This was because there was a particularly large number of volunteer seedlings of Ida, which has the resistances *Mla9+Mlk1+Mlg*.

If such a model is to be incorporated into a decision support system, predicted virulence frequencies must be converted into predictions of disease severity. This will need two additional kinds of information: meteorological data, to predict mean disease severities over all varieties, and data on varietal partial resistances, to predict relative disease severities on different varieties with the same specific resistance. Furthermore, the ability of this model to provide quantitative risk predictions may be limited by such unexpected events as the heavy infection of volunteeers of Ida, mentioned above.

Another issue to be considered is how closely spaced sampling sites need to be in order to provide risk predictions which are robust over a reasonably wide area. In effect, we return to the question of what is meant by 'a population'. The spacing should be related to the pathogen's mobility; in the case of cereal mildews, the large population of wind-dispersed spores should allow sampling sites to be spaced much further apart than would be the case for splash-borne spores, dispersed over a very small area.

3.8 PATHOGEN SURVEYS AND DISEASE MANAGEMENT

Several examples discussed in this chapter have illustrated the value of epidemiological knowledge in the execution of pathogen surveys, notably in the effective design of sampling schemes and the application of the resulting data. In some of the better-studied pathogens, it is becoming possible to begin to integrate the pathogen's epidemiology and population genetics, in order to develop a deeper understanding of how populations adapt to crops (Brown and Hovmøller, 2002; Milgroom and Peever, 2003).

Conversely, surveys of pathogenic variation clearly have the potential to continue to contribute to improving the efficiency of disease management. They should be particularly valued in the present climate of increased concern about the environmental impact of farming, as they can help to optimise the use of resistant varieties and reduce unnecessary fungicide applications. For this desirable situation to continue – indeed, for the value of pathogen surveys to be increased further – management of pathogen surveys must be seen as a long-term exercise. Many examples cited in this review illustrate the benefit of being able to place new results, whether on virulences or on responses to fungicides, in a historical perspective of years or even decades of continuous research on a disease (Bayles *et al.*, 1997; Kolmer, 2005).

Unfortunately, pathogen survey work has something of the image of a Cinderella, compared with the much more fashionable subject of molecular genetics. However, while molecular genetics research is helping to provide medium to long-term guidance on strategies for crop management and is also beginning to increase the efficiency of pathogenicity testing, it is no substitute, in terms of its value for disease control, for the kind of direct knowledge about variation in characters relevant to pathogenicity that comes from well-managed, long-term surveys.

ACKNOWLEDGEMENT

The writing of this chapter was supported by the Biotechnology and Biological Sciences Research Council.

REFERENCES

Anon. (1998) *NIAB 1998 Cereals Variety Handbook*, National Institute of Agricultural Botany, Cambridge.

Avila-Adame, C. and Koller, W, (2003) Characterization of spontaneous mutants of *Magnaporthe grisea* expressing stable resistance to the Q_o-inhibiting fungicide azoxystrobin. *Current Genetics*, **42**, 332-338.

Barrett, J.A. (1981) The evolutionary consequences of monoculture, in *Genetic Consequences of Man Made Change* (eds J.A. Bishop and L.M. Cook), Academic Press, London pp. 209-248.

Barrett, J.A. (1987) The dynamics of genes in populations, in *Populations of Plant Pathogens: their Dynamics and Genetics* (eds M.S. Wolfe and C.E. Caten), Blackwell Scientific Publications, Oxford, pp. 39-53.

Bayles, R.A. and Hubbard, A.J. (2005) Yellow rust of wheat. *U.K. Cereal Pathogen Virulence Survey: 2004 Annual Report*, 13-22.

Bayles, R.A. and Stigwood, P.L. (2001) Yellow rust of wheat. *UK Cereal Pathogen Virulence Survey: 2000 Annual Report*, 28-36.

Bayles, R.A., Barnard, E.G. and Stigwood, P.L. (1992) Fungicide sensitivity in yellow rust of wheat (*Puccinia striiformis*). *Brighton Crop Protection Conference - Proceedings - 1992*, **1**, 165-170.

Bayles, R.A., Stigwood, P.L. and Barnard, E.G. (1994) Sensitivity to morpholine fungicides in yellow rust of wheat (*Puccinia striiformis*), in *Fungicide Resistance* (eds S. Heaney, D. Slawson, D.W. Hollomon, M. Smith, P.E. Russell and D.W. Parry), British Crop Protection Council, Farnham, pp. 309-314.

Bayles, R.A., Clarkson, J.D.S. and Slater, S.E. (1997) The UK Cereal Pathogen Virulence Survey, in *The Gene-for-Gene Relationship in Plant-Parasite Interactions* (eds I.R. Crute, E.B. Holub and J.J. Burdon), CAB International, Wallingford, pp. 103-117.

Bearchell, S.J., Fraaije, B.A., Shaw, M.W. and Fitt, B.D.L. (2005) Wheat archive links long-term fungal pathogen population dynamics to air pollution. *Proceedings of the National Academy of Sciences of the USA*, **102**, 5438-5442.

Bisgrove, S.R., Simonich, M.T., Smith, N.M. *et al.* (1994) A disease resistance gene in *Arabidopsis* with specificity for two different pathogen avirulence genes. *Plant Cell*, **6**, 927-933.

Blatter, R.H.E., Brown, J.K.M. and Wolfe, M.S. (1998) Genetic control of the resistance of *Erysiphe graminis* f.sp. *hordei* to five triazole fungicides. *Plant Pathology*, **47**, 570-579.

Brading, P.A., Verstappen, E.C.P., Kema, G.H.J. and J.K.M. Brown (2002) A gene-for-gene relationship between wheat and *Mycosphaerella graminicola*, the septoria tritici blotch pathogen. *Phytopathology*, **92**, 439-445.

Brändle, U.E., Haemmerli, U.A., McDermott, J.M. and Wolfe, M.S. (1997) Interpreting population genetic data with the help of genetic linkage maps, in *The Gene-for-Gene Relationship in Plant-Parasite Interactions* (eds I.R. Crute, E.B. Holub and J.J. Burdon), CAB International, Wallingford, pp. 157-171.

Briggle, L.W. (1969) Near-isogenic lines of wheat with genes for resistance to *Erysiphe graminis* f.sp. *tritici*. *Crop Science*, **9**, 70-72.

Brown, J.K.M. (1994a) The genetics of the response of barley mildew to morpholine and piperidine fungicides , in *Fungicide Resistance* (eds S. Heaney, D. Slawson, D.W. Hollomon, M. Smith, P.E. Russell and D.W. Parry), British Crop Protection Council, Farnham, pp. 291-296.

Brown, J.K.M. (1994b) Chance and selection in the evolution of barley mildew. *Trends in Microbiology*, **2**, 470-475.

Brown J.K.M. (1995) Pathogens' responses to the management of disease resistance genes. *Advances in Plant Pathology*, **11**, 75-102.

Brown, J.K.M. (1996) Fungicide resistance in barley powdery mildew: from genetics to crop protection, in *Proceedings of the 9th European and Mediterranean Cereal Rusts and Powdery Mildews Conference, 2-6 September 1996, Lunteren, The Netherlands*, (eds G.H.J. Kema, R.E. Niks and R.A. Daamen), European and Mediterranean Cereal Rust Foundation, Wageningen, pp. 259-267.

Brown, J.K.M. (1998) Surveys of variation in pathogen populations and their application to disease control, in *The Epidemiology of Plant Diseases* (ed D. Gareth Jones), Kluwer Academic Publishers, Dordrecht, pp. 73-102.

Brown, J.K.M. (2002a) Comparative genetics of avirulence and fungicide resistance in the powdery mildew fungi, in *The Powdery Mildews: a Comprehensive Treatise* (eds R.R. Belanger, A.J. Dik and W.R. Bushnell), APS Press, St. Paul, pp. 56-65.

Brown, J.K.M. (2002b) Yield penalties of disease resistance in crops. *Current Opinion in Plant Biology*, **5**, 339-344.

Brown, J.K.M. and Evans, N. (1992) Selection on responses of barley powdery mildew to morpholines and piperidine fungicides. *Crop Protection*, **11**, 449-457.

Brown, J.K.M. and Hovmøller, M.S. (2002) Aerial dispersal of fungi on the global and continental scales and its consequences for plant disease. *Science*, **297**, 537-541.

Brown, J.K.M. and Jessop, A.C. (1995) Genetics of avirulence in *Erysiphe graminis* f.sp. *hordei*. *Plant Pathology*, **44**, 1039-1049.

Brown, J.K.M. and Jørgensen, J.H. (1991) A catalogue of mildew resistance genes in European barley varieties, in *Integrated Control of Cereal Mildews: Virulence Patterns and their Change* (ed. J.H. Jørgensen), Risø National Laboratory, Roskilde, pp. 263-286.

Brown, J.K.M. and Wolfe, M.S. (1990) Structure and evolution of a population of *Erysiphe graminis* f.sp. *hordei*. *Plant Pathology*, **39**, 391-401.

Brown, J.K.M. and Wolfe, M.S. (1991) Levels of resistance of *Erysiphe graminis* f.sp. *hordei* to the systemic fungicide triadimenol. *Netherlands Journal of Plant Pathology*, **97**, 251-263.

Brown, J.K.M., O'Dell, M., Simpson, C.G. and Wolfe, M.S. (1990) The use of DNA polymorphisms to test hypotheses about a population of *Erysiphe graminis* f.sp. *hordei*. *Plant Pathology*, **39**, 391-401.

Brown, J.K.M., Jessop, A.C. and Rezanoor, H.N. (1991a) Genetic uniformity in barley and its powdery mildew pathogen. *Proceedings of the Royal Society of London, Series B*, **246**, 83-90.

Brown, J.K.M., Slater, S.E. and See, K.A. (1991b) Sensitivity of *Erysiphe graminis* f.sp. *hordei* to morpholine and piperidine fungicides. *Crop Protection*, **10**, 445-454.

Brown, J.K.M., Jessop, A.C., Thomas, S. and Rezanoor, H.N. (1992) Genetic control of the response of *Erysiphe graminis* f.sp. *hordei* to ethirimol and triadimenol. *Plant Pathology*, **41**, 126-135.

Brown, J.K.M., Simpson, C.G. and Wolfe, M.S. (1993) Adaptation of barley powdery mildew populations in England to varieties with two resistance genes. *Plant Pathology*, **42**, 108-115.

Brown, J.K.M., Le Boulaire, S. and Evans, N. (1996) Genetics of responses to morpholine-type fungicides and of avirulences in *Erysiphe graminis* f.sp. *hordei*. *European Journal of Plant Pathology*, **102**, 479-490.

Brown, J.K.M., Foster, E.M. and O'Hara, R.B. (1997) Adaptation of powdery mildew populations to cereal varieties in relation to durable and non-durable resistance, in *The Gene-for-Gene Relationship in Plant-Parasite Interactions* (eds I.R. Crute, E.B. Holub and J.J. Burdon), CAB International, Wallingford, pp. 119-138.

Brown, J.K.M., Kema, G.H.J., Forrer, H.R. *et al.* (2001) Resistance of wheat cultivars and breeding lines to septoria tritici blotch caused by isolates of *Mycosphaerella graminicola* in field trials. *Plant Pathology*, **50**, 325-338.

Büschges, R., Hollricher, K., Panstruga, R. *et al.* (1997) The barley *mlo* gene: a novel control element of plant pathogen resistance. *Cell*, **88**, 695-705.

Caffier, V., de Vallavieille-Pope, C. and Brown, J.K.M. (1996) Segregation of avirulences and genetic basis of infection types in *Erysiphe graminis* f.sp. *hordei*. *Phytopathology*, **86**, 1112-1121.

Calonnec, A., Johnson, R. and de Vallavieille-Pope, C. (1997a) Genetic analysis of resistance to *Puccinis striiformis* in the wheat differential cultivars Heines VII, Heines Peko and Strubes Dickkopf. *Plant Pathology*, **46**, 373-386.

Calonnec, A., Johnson, R. and de Vallavieille-Pope, C. (1997b) Identification and expression of the gene *Yr2* for resistance to *Puccinia striiformis* in the wheat differential cultivars Heines Kolben, Heines Peko and Heines VII. *Plant Pathology*, **46**, 387-396.

Chartrain, L. (2004) Genes for Isolate-Specific and Partial Resistance to septoria tritici blotch (*Mycosphaerella graminicola*) in Wheat. Ph.D. thesis, University of East Anglia.

Chartrain, L., Brading, P.A., Makepeace, J.C. and Brown, J.K.M. (2004) Sources of resistance to septoria tritici blotch and implications for wheat breeding. *Plant Pathology*, **53**, 454-460.

Chartrain, L., Brading, P.A., Makepeace, J.C. and Brown, J.K.M. (2005a) Presence of the *Stb6* gene for resistance to septoria tritici blotch (*Mycosphaerella graminicola*) in cultivars used in wheat-breeding programmes worldwide. *Plant Pathology*, **54**, 134-143.

Chartrain, L., Berry S.T. and Brown, J.K.M. (2005b) Resistance of the wheat line Kavkaz-K4500 L.6.A.4 to septoria tritici blotch controlled by isolate-specific resistance genes. *Phytopathology*, **95**, 664-671.

Chin, K.M., Chavaillaz, D., Kaesbohrer, M. *et al.* (2001) Characterizing resistance risk of *Erysiphe graminis* f.sp *tritici* to strobilurins. *Crop Protection*, **20**, 87-96.

Christ, B.J., Person, C.O. and Pope, D.D. (1987) The genetic determination of variation in pathogenicity, in *Populations of Plant Pathogens: their Dynamics and Genetics* (eds M.S. Wolfe and C.E. Caten), Blackwell Scientific Publications, Oxford, pp. 7-19.

Cook, R.J. (1999) Management by chemicals, in *Septoria of cereals: a Study of Pathosystems* (eds J.A. Lucas, P. Bowyer and H.M. Anderson), CABI Publishing, Wallingford, pp. 286-298.

Crute, I.R., Holub, E.B. and Burdon, J.J. (eds) (1997) *The Gene-for-Gene Relationship in Plant-Parasite Interactions*, CAB International, Wallingford, UK.

Dangl, J.L. and Jones, J.D.G. (2001) Plant pathogens and integrated defence responses to infection. *Nature*, **411**, 826-833.

Délye, C., Laigret, F. and Corio-Costet, M.F. (1997) A mutation in the 14α-demethylase gene of *Uncinula necator* that correlates with resistance to a sterol biosynthesis inhibitor. *Applied and Environmental Microbiology*, **63**, 2966-2970.

Délye, C., Bousset, L. and Corio-Costet, M.F. (1998) PCR cloning and detection of point mutations in the eburicol 14 α-demethylase (CYP51) gene from *Erysiphe graminis* f. sp. *hordei*, a "recalcitrant" fungus. *Current Genetics*, **34**, 399-403.

Ellis, J., Lawrence, G., Ayliffe, M. *et al.* (1998) Advances in the molecular genetic analysis of the flax-flax rust interaction. *Annual Review of Phytopathology*, **35**, 271-291.

Enjalbert, J., Duan, X., Leconte, M. *et al.* (2005) Genetic evidence of local adaptation of wheat yellow rust (*Puccinia striiformis* f. sp *tritici*) within France. *Molecular Ecology*, **14**, 2065-2073.

Eriksen, L. and Munk, L. (2003) The occurrence of *Mycosphaerella graminicola* and its anamorph *Septoria tritici* in winter wheat during the growing season. *European Journal of Plant Pathology*, **109**, 253-259.

Fiegen, M. and Knogge, W. (2002) Amino acids alterations in isoforms of the effector protein NIP1 from *Rhynchosporium secalis* have similar effects on in avirulence and virulence associated activities on barley. *Physiological and Molecular Plant Pathology*, **61**, 299-302.

Finney, D.J. (1971) *Probit Analysis* (3rd edn), Cambridge University Press, Cambridge.

Fletcher, J.T. and Wolfe, M.S. (1981) Insensitivity of *Erysiphe graminis* f.sp. *hordei* to triadimefon, triadimenol and other fungicides. *Proceedings of the 11th British Crop Protection Conference*, **2**, 633-640.

Fletcher, J.T., Cooper, S.T and Prestidge, A.L.H. (1987) An investigation of the sensitivity of *Erysiphe graminis* f.sp. *tritici* to various ergosterol inhibiting fungicides, in *Integrated Control of Cereal Mildews: Monitoring the Pathogen* (eds M.S. Wolfe and E. Limpert), Martinus Nijhoff Publishers, Dordrecht, pp. 129-135.

Flor, H.H. (1956) Current status of the gene-for-gene concept. *Advances in Genetics*, **8**, 29-54.

Fraaije, B.A., Butters, J.A., Coelho, J.M. *et al.* (2002) Following the dynamics of strobilurin resistance in *Blumeria graminis* f.sp *tritici* using quantitative allele-specific real-time PCR measurements with the fluorescent dye SYBR Green I. *Plant Pathology*, **51**, 45-54.

Fraaije, B.A., Cools, H.J., Fountaine, J. *et al.* (2005) Role of ascospores in further spread of QoI-resistant cytochrome b alleles (G143A) in field populations of *Mycosphaerella graminicola*. *Phytopathology*, **95**, 933-941.

Gisi, U., Sierotzki, H., Cook, A. and McCaffery, A. (2002) Mechanisms influencing the evolution of resistance to Qo inhibitor fungicides *Pest Management Science*, **58**, 859-867.

Gullino, M.L., Mescalchin, E. and Mezzalama, M. (1997) Sensitivity to cymoxanil in populations of *Plasmopara viticola* in northern Italy. *Plant Pathology*, **46**, 729-736.

Halterman, D.A., Wei, F.S. and Wise, R.P. (2003) Powdery mildew-induced *Mla* mRNAs are alternatively spliced and contain multiple upstream open reading frames. *Plant Physiology*, **131**, 558-567.

Hayashi, K., Schoonbeek, H.J., Sugiura, H. and de Waard, M.A. (2001) Multidrug resistance in *Botrytis cinerea* associated with decreased accumulation of the azole fungicide oxpoconazole and increased transcription of the ABC transporter gene *BcatrD*. *Pesticide Biochemistry and Physiology*, **70**, 168-179.

Hedrick, P.W. (2004) *Genetics of Populations* (3rd edn), Jones and Bartlett Publishers, Boston, U.S.A.

Hovmøller, M.S. (2001) Disease severity and pathotype dynamics of *Puccinia striiformis* f.sp. *tritici* in Denmark. *Plant Pathology*, **50**, 181-189.

Hovmøller, M.S. (2004) Den danske virulensovervågning af meldug og rust 1985-2003 - hvad har vi lært? [The Danish virulence survey for cereal powdery mildews and rusts, 1985-2003; what have we learnt?] *Danmarks Jordbrugsforskning Rapport*, **98**, 83-91.

Hovmøller, M.S., Munk, L and Østergård, H. (1993) Observed and predicted changes in virulence gene frequencies at 11 loci in a local barley powdery mildew population. *Phytopathology*, **83**, 253-260.

Hovmøller, M.S., Munk, L and Østergård, H. (1995) Comparison of mobile and stationary spore-sampling techniques for estimating virulence frequencies in aerial barley powdery mildew populations. *Plant Pathology*, **44**, 829-837.

Hovmøller, M.S., Justesen, A.F. and Brown, J.K.M. (2002) Clonality and long-distance migration of *Puccinia striiformis* f.sp. *tritici* in north-west Europe. *Plant Pathology*, **51**, 24-32.

Huang, X.Q., Hsam, S.L.K., Mohler, V. *et al.* (2004) Genetic mapping of three alleles at the *Pm3* locus conferring powdery mildew resistance in common wheat (*Triticum aestivum* L.). *Genome*, **47**, 1130-1136.

Hunter, T., Coker, R.R. and Royle, D.J. (1999) The teleomorph stage, *Mycosphaerella graminicola*, in epidemics of septoria tritici blotch on winter wheat in the UK. *Plant Pathology*, **48**, 51-57.

Ishii, H., Fraaije, B.A., Sugiyama, T. *et al.* (2001) Occurrence and molecular characterization of strobilurin resistance in cucumber powdery mildew and downy mildew. *Phytopathology*, **91**, 1166-1171.

Jensen, H.P. and Jørgensen, J.H. (1981) Powdery mildew resistance genes in north-west European winter barley varieties. *Danish Journal of Plant and Soil Science*, **85**, 303-319.

Jensen, H.P., Christensen, E. and Jørgensen, J.H. (1992) Powdery mildew resistance genes in 127 northwest European spring barley varieties. *Plant Breeding*, **108**, 210-228.

Johnson, R. (1984) A critical analysis of durable resistance. *Annual Review of Phytopathology*, **22**, 309-330.

Johnson, R. (1992) Reflections of a plant pathologist on breeding for disease resistance, with emphasis on yellow rust and eyespot of wheat. *Plant Pathology*, **41**, 239-254.

Jones, E.R.L., and Clifford, B.C. (1997) Brown rust of wheat. *U.K. Cereal Pathogen Virulence Survey: 1996 Annual Report*, 18-28.

Jørgensen, J.H. (1992a) Discovery, characterization and exploitation of *mlo* powdery mildew resistance in barley. *Euphytica*, **63**, 141-152.

Jørgensen, J.H. (1992b) Multigene families of powdery mildew resistance genes in locus *Mla* on barley chromosome-5. *Plant Breeding*, **108**, 53-59.

Jørgensen, J.H. (1994) Genetics of powdery mildew resistance in barley. *Critical Reviews in Plant Sciences*, **13**, 97-119.

Justesen, A.F., Ridout, C.J. and Hovmoller M.S. (2002) The recent history of *Puccinia striiformis* f.sp *tritici* in Denmark as revealed by disease incidence and AFLP markers. *Plant Pathology*, **51**, 13-23.

Kim, Y.S., Dixon, E.W., Vincelli, P. and Farman, M.L. (2003) Field resistance to strobilurin (Q_oI) fungicides in *Pyricularia grisea* caused by mutations in the mitochondrial cytochrome b gene. *Phytopathology*, **93**, 891-900.

King, J.E. and Griffin, M.J. (1985) Survey of benomyl resistance in *Pseudocercosporella herpotrichoides* on winter wheat and barley in England and Wales in 1983. *Plant Pathology*, **34**, 272-283.

Koenraadt, H., Somerville, S.C. and Jones, A.L. (2002) Characterization of mutations in the β-tubulin gene of benomyl-resistant field strains of *Venturia inaequalis* and other plant pathogenic fungi. *Phytopathology*, **82**, 1348-1354.

Kolmer, J.A. (1997) Virulence dynamics and genetics of cereal rust populations in North America, in *The Gene-for-Gene Relationship in Plant-Parasite Interactions* (eds I.R. Crute, E.B. Holub and J.J. Burdon), CAB International, Wallingford, pp. 139-156.

Kolmer, J.A. (2005) Tracking wheat rust on a continental scale. *Current Opinion in Plant Biology*, **8**, 441-449.

Kølster, P., Munk, L., Sløen, O. and Løhde, J. (1986) Near-isogenic barley lines with genes for resistance to powdery mildew. *Crop Science*, **26**, 903-907.

Lawrence, G.J., Mayo, G.M.E. and Shepherd, K.W. (1981) Interactions between genes controlling pathogenicity in the flax rust fungus. *Phytopathology*, **71**, 12-19.

Mackey, D., Holt, B.F., Wiig, A. and Dangl, J.L. (2002) RIN4 interacts with *Pseudomonas syringae* type III effector molecules and is required for RPM1-mediated resistance in *Arabidopsis*. *Cell*, **108**, 743-754.

McDonald, B.A., Pettway, R.E., Chen, R.S. *et al.* (1995) The population genetics of *Septoria tritici* (teleomorph *Mycosphaerella graminicola*). *Canadian Journal of Botany*, **73** (supplement 1), S292-301.
McIntosh, R.A., Wellings, C.R. and Park, R.F. (1995) *Wheat Rusts: an Atlas of Resistance Genes*, CSIRO, East Melbourne, Australia.
Milgroom, M.G. and Fry, W.E. (1997) Contributions of population genetics to plant disease epidemiology and management. *Advances in Botanical Research Incorporating Advances in Plant Pathology*, **24**, 1-30.
Milgroom, M.G. and Peever, T.L. (2003) Population biology of plant pathogens. *Plant Disease*, **87**, 608-617.
Moseman, J.G., Macer, R.C.F. and Greeley, L.W. (1965) Genetic studies with cultures of *Erysiphe graminis* f.sp. *hordei* virulent on *Hordeum spontaneum*. *Transactions of the British Mycological Society*, **48**, 479-498.
O'Hara, R.B. and Brown, J.K.M. (1996) Immigration of the barley mildew pathogen into field plots of barley. *Plant Pathology*, **45**, 1071-1076.
O'Hara, R.B. and Brown, J.K.M. (1998) Movement of barley powdery mildew within field plots. *Plant Pathology*, **47**, 394-400.
Østergård, H. and Hovmøller, M.S. (1991) Gametic disequilibria between virulence genes in barley powdery mildew populations in relation to selection and recombination. I. Models. *Plant Pathology*, **40**, 166-177.
Pasche, J.S., Piche, L.M. and Gudmestad, N.C. (2005) Effect of the F129L mutation in *Alternaria solani* on fungicides affecting mitochondrial respiration. *Plant Disease*, **89**, 269-278.
Peever, T.L. and Milgroom, M.G. (1993) Genetic correlations in resistance to sterol biosynthesis-inhibiting fungicides in *Pyrenophora teres*. *Phytopathology*, **83**, 1076-1082.
Piffanelli, P., Ramsay, L., Waugh, R. *et al.* (2004) A barley cultivation-associated polymorphism conveys resistance to powdery mildew. *Nature*, **430**, 887-891.
Priestley, R.H. and Bayles, R.A. (1980) Variety diversification as a means of reducing the spread of cereal diseases in the United Kingdom. *Journal of the National Institute of Agricultural Botany*, **15**, 205-214.
Ren, S.X., McIntosh, R.A. and Lu, Z.J. (1997) Genetic suppression of the cereal rye-derived gene *Pm8* in wheat. *Euphytica*, **93**, 353-360.
Robinson, H.L., Ridout, C.J., Sierotzki, H. *et al.* (2002) Isogamous, hermaphroditic inheritance of mitochondrion-encoded resistance to Qo inhibitor fungicides in *Blumeria graminis* f. sp. *tritici*. *Fungal Genetics and Biology*, **36**, 98-106.
Rohe, M., Gierlich, A., Hermann, H. *et al.* (1995) The race-specific elicitor, NIP1, from the barley pathogen, *Rhynchosporium secalis*, determines avirulence on host plants of the *Rrs1* resistance genotype. *EMBO Journal*, **14**, 4168-4177.
Russell, P.E. (1995) Fungicide resistance: occurrence and management. *Journal of Agricultural Science, Cambridge*, **124**, 317-323.
Samborski, D.J. and Dyck, P.L. (1976) Inheritance of virulence in *Puccinia recondita* on six backcross lines of wheat with single genes for resistance to leaf rust. *Canadian Journal of Botany*, **54**, 1666-1671.
Schoonbeek, H., del Sorbo, G. and de Waard, M.A. (2001) The ABC transporter BcatrB affects the sensitivity of *Botrytis cinerea* to the phytoalexin resveratrol and the fungicide fenpiclonil. *Molecular Plant-Microbe Interactions*, **14**, 562-571.
Schürch, S., Linde, C.C., Knogge, W. *et al.* (2004) Molecular population genetic analysis differentiates two virulence mechanisms of the fungal avirulence gene *NIP1*. *Molecular Plant-Microbe Interactions*, **17**, 1114-1125.
Shen, Q.H., Zhou, F.S., Bieri, S. *et al.* (2003) Recognition specificity and RAR1/SGT1 dependence in barley *Mla* disease resistance genes to the powdery mildew fungus. *Plant Cell*, **15**, 732-744.
Sierotzki, H., Wullschleger, J. and Gisi, U. (2000a) Point mutation in cytochrome b gene conferring resistance to strobilurin fungicides in *Erysiphe graminis* f. sp. *tritici* field isolates. *Pesticide Biochemistry and Physiology*, **68**, 107-112.
Sierotzki, H., Parisi, S., Steinfeld, U. *et al.* (2000b) Mode of resistance to respiration inhibitors at the cytochrome bc(1) enzyme complex of *Mycosphaerella fijiensis* field isolates. *Pest Management Science*, **56**, 833-841.

Skamnioti, P. and Ridout, C.J. (2005) Microbial avirulence determinants: guided missiles or random antigenic flak? *Molecular Plant Pathology*, **6**, 551-559.

Slater, S.E. (2005a) Mildew of wheat. *U.K. Cereal Pathogen Virulence Survey: 2004 Annual Report*, 1-11.

Slater, S.E. (2005b) Mildew of barley. *U.K. Cereal Pathogen Virulence Survey: 2004 Annual Report*, 33-44.

Slater, S.E. and Clarkson, J.D.S. (2001) Reaction of spring barley cultivars carrying the *mlo* resistance to infection by powdery mildew isolates in the UK. *Cereal Rusts and Powdery Mildews Bulletin*, (online) http://www.crpmb.org/2001/0724slater.

Torp, J., Jensen, H.P. and Jørgensen, J.H. (1978) Powdery mildew resistance genes in 106 northwest European spring barley varieties. *Den Kgl. Veterinær- og Landbohøjskoles Årkskrift 1978*, 75-102.

Vermeulen, T., Schoonbeek, H. and de Waard, M.A. (2001) The ABC transporter BcatrB from *Botrytis cinerea* is a determinant of the activity of the phenylpyrrole fungicide fludioxonil. *Pest Management Science*, **57**, 393-402.

Walker, A.S., Leroux, P., Bill, L. *et al.* (2004) Wheat powdery mildew: what about fungicide resistance in France? *Phytoma*, **577**, 49-54.

Wheeler, J.E., Kendall, S.J., Butters, J. *et al.* (1995) Using allele-specific oligonucleotide probes to characterize benzimidazole resistance in *Rhynchosporium secalis*. *Pesticide Science*, **43**, 201-209.

Wolfe, M.S. (1984) Trying to understand and control powdery mildew. *Plant Pathology*, **33**, 451-466.

Wolfe, M.S. and Knott, D.R. (1982) Populations of plant pathogens: some constraints on analysis of variation in pathogenicity. *Plant Pathology*, **31**, 79-90.

Wolfe, M.S. and Mc Dermott, J.M. (1994) Population genetics of plant-pathogen interactions: the example of the *Erysiphe graminis* – *Hordeum vulgare* pathosystem. *Annual Review of Phytopathology*, **32**, 89-113.

Wolfe, M.S., Barrett, J.A. and Slater, S.E. (1983) Pathogen fitness in cereal mildews, in *Durable Resistance in Crops* (eds F. Lamberti, J.M. Waller and N.A. van der Graaff), Plenum Press, New York, pp. 81-100.

Wolfe, M.S., Brändle, U., Koller, B. *et al.* (1992) Barley mildew in Europe: population biology and host resistance. *Euphytica*, **63**, 125-139.

Wyand, R.A. and Brown, J.K.M. (2005) Sequence variation in the *CYP51* gene of *Blumeria graminis* associated with resistance to sterol demethylase inhibiting (DMI) fungicides. *Fungal Genetics and Biology*, **42**, 726-735.

Yahiaoui, N., Srichumpa, P., Dudler, R. and Keller, B. (2004) Genome analysis at different ploidy levels allows cloning of the powdery mildew resistance gene *Pm3b* from hexaploid wheat. *Plant Journal*, **37**, 528-538.

Yarden, O. and Katan, T. (1993) Mutations leading to substitutions at amino-acids 198 and 200 of β-tubulin that correlate with benomyl resistance phenotypes of field strains of *Botrytis cinerea*. *Phytopathology*, **83**, 1478-1483.

Yu, D.Z. (2000) Wheat Powdery Mildew in Central China: Pathogen Population Structure and Host Resistance. Ph.D. thesis, University of Wageningen.

Zheng, D.S., Olaya, G. and Koller, W. (2000) Characterization of laboratory mutants of *Venturia inaequalis* resistant to the strobilurin-related fungicide kresoxim-methyl. *Current Genetics*, **38**, 148-155.

CHAPTER 4

INFECTION STRATEGIES OF PLANT PARASITIC FUNGI

C. STRUCK

4.1 INTRODUCTION

How plant pathogens attack their host plants is one of the most interesting questions in plant pathology. Numerous fungi and fungus-like organisms cause some 10,000 different diseases in plants (Pennisi, 2001). Fungal diseases of plants cause economic losses by reducing seed germination, destroying plants or the harvesting products and by secreting secondary metabolites that are toxic for man and animals. Successful colonization of the habitat plant, including uptake of nutrients and reproduction, greatly depends on an efficient mode of infection. Plant parasitic fungi have developed various strategies to enter their hosts, and to establish direct contact with them. Successful interactions result in devastating plant diseases. In many cases this means the production of very large amounts of spores that are wind dispersed from one susceptible host to another. Furthermore, there are several examples described of long-distance dispersal of fungal spores by wind that can spread plant diseases across and even between continents and allow invasion into new territories (Brown and Hovmøller, 2002).

The infection mechanisms of pathogenic fungi are highly variable. During the early phases of the infection process before invading the plant tissue, the development of fungi greatly depends on favourable environmental conditions such as surface moisture, relative humidity, temperature and light. In some cases, the supply of nutrients on the surface may also have an influence on germination.

Morphogenetic and physiological differentiation of the invading fungus depends on the mode of penetration. The infection process consists of a number of morphologically more or less distinguishable stages: spore germination, formation of an appressorium and a penetration hypha that penetrates the cuticle and the cell wall (Mendgen *et al.*, 1996). Within the host tissue, pathogenic fungi develop infection hyphae and some biotrophs form haustoria. A review of this subject with more details on cell biology is given by Hardham (2001).

The diverse strategies that fungal pathogens use to infect their host plants are becoming much better understood by means of molecular genetics techniques that have allowed the identification of fungal genes which are crucial to disease development.

B. M. Cooke, D. Gareth Jones and B. Kaye (eds.), The Epidemiology of Plant Diseases, 2nd edition, 117–137.

In this chapter examples of fungi and fungi-like parasites with widely differing pathogenic lifestyles are introduced and phases of their infection processes are described.

4.2 THE PRE-PENETRATION PHASE

4.2.1 Attachment of fungal spores

The initial step of establishing infection is the adhesion of fungal propagules to the plant surface. A binding process is essential to resist displacement by wind or water. Although adhesion to plant surfaces is common among all fungal species, there are differences in the factors that induce the attachment process and in the composition of adhesive material. Reviews concerning these earliest events of fungal infection are reviewed by Nicholsen (1996) and Tucker and Talbot (2001).

Of special importance is the attachment of zoospores of soilborne oomycetes living in water-saturated soil to make sure that the pathogen is not washed away by water before invasion of the host tissue can occur. Zoospores of *Phytophthora cinnamoni* that adhere to a surface during the first 3-4 min of encystment remain strongly attached. It was found that, although the adhesive material was still present, cell adhesiveness had declined rapidly 5 min later (Gubler *et al.*, 1989).

Considering the differences in surface composition and properties (e.g. hydrophobicity) of aerial plant organs and roots, it seems obvious that fungi use different mechanisms to bind to the host surface. One important factor is the amount of water available during infection. In some cases, the hydration of fungal propagules leads to a rapid release of mucilage that is involved in a passive, non-specific adhesion to a variety of substrates. For example, the spores of the rice blast fungus *Magnaporthe grisea* release a carbohydrate-containing adhesive material from the tip region of the germ tube as a result of hydration expansion (Howard *et al.*, 1991). About 20 min after hydration, conidia of *Cochliobolus heterostrophus* secrete a material at the spore tips that serves as non-specific adhesive (Braun and Howard, 1994).

Ungerminated conidia of *Colletotrichum graminicola*, the causal agent of anthracnose (leaf blight of corn), begin to adhere within minutes of their contact with the leaf surface. Adhesion is shown to be essential to the success of disease establishment (Mercure *et al.*, 1994). The spores produce a water-soluble glyco-protein-rich material with properties to protect spores against unfavourable conditions such as dry periods (Mercure *et al.*, 1995).

Weak adhesion of germlings of *Botrytis cinerea* to hydrophobic substrates occurs immediately upon hydration. In a second stage, viable conidia strongly attach to either hydrophobic or hydrophilic substrates by means of secretion of a sheath of material (Doss *et al.*, 1995).

Contact and adhesion to the substratum has been found as a prerequisite for the germination of several phytopathogenic fungi, for example *Bipolaris sorokiniana* (Apoga *et al.*, 2001). The uredospores of the rust fungus *Uromyces fabae* form an adhesion pad and release a cutinase and two specific esterases after contacting the host cuticle. Apparently, adhesion of the pads is improved by these enzymes. The spores have reduced ability to attach to the leaf surface when these enzymes are inactivated (Deising *et al.*, 1992).

In contrast to propagules of most fungi, conidia of barley powdery mildew (*Blumeria graminis,* syn. *Erysiphe graminis*) begin the process of adhesion in the absence of free water in a wide range of high relative humidities. Carver *et al.* (1995) showed that the extracellular material released by germlings of *B. graminis* f.sp. *avenae* sticks the fungus firmly to the leaf surface. The conidia of *B. graminis* f.sp. *hordei* produce an esterase containing liquid in response to a non-specific contact stimulus with the barley leaf surface or to a moistened cellophane surface (Nicholson *et al.*, 1988). Furthermore, this exudate contains cutinase activity (Pascholati *et al.*, 1992) that alters the cuticle surface and may help the fungus to penetrate the leaf surface more efficiently.

4.2.2 Host surface perception

During the early phase of the infection process before invading the plant tissue, the developing fungus greatly depends on favourable environmental conditions such as surface moisture, relative humidity, temperature and light. In addition, nutrient availability is an important stimulus especially for necrotrophic pathogens. The precise mechanism by which germination occurs is an interplay of several factors. Spore germination, growth direction of germ tubes and subsequent induction of appressorium formation all have been shown to be triggered by chemical or physical features of the substratum. To place the appressorium in the optimal site for penetration, a recognition event is required which includes topographical, chemical or environmental signals (for review, see Staples and Hoch, 1997).

The germ tubes of several pathogenic fungi have been described as growing along the anticlinal cell walls of their host plants. *In vitro* studies of germinating spores of *Cochliobolus sativus* indicate that both chemical and topographic signals given by anticlinal cell wall junctures over epidermal cells are involved in the appressorium induction (Clay *et al.*, 1994).

Appressorium formation by urediospore germ tubes of the bean rust *Uromyces appendiculatus* is induced by physical differences in the topography of the leaf surface, such as stomatal lips of guard cells, or by defined ridges of 0.5 μm height formed on an artificial surface (Hoch *et al.*, 1987; Kwon and Hoch, 1991). In addition, it has been shown that many rust fungi exhibit species-specific responses on membranes with defined topographies (Allen *et al.*, 1991).

Surface contact on host leaves or artificial substrates was found to be essential for the formation of appressoria by germ tubes of *Magnaporthe grisea* (Xiao *et al.*, 1994). Furthermore, a high surface hydrophobicity and light favoured the formation of appressoria but these factors were not essential (Jelitto *et al.*, 1994). Gilbert *et al.* (1996) reported evidence for the regulation of appressorium formation of *M. grisea* by chemical signals: they found that plant cutin monomers of host plants and non-host plants induced infection structure formation. From *Colletotrichum trifolii*, the causal agent of alfalfa anthracnose, a lipid-induced protein kinase (LIPK) required for appressorium formation was induced specifically by plant cutin as well as cutin monomers, but not by structurally-related synthetic analogues (Dickman *et al.*, 2003).

4.3 ENTERING THE PLANT TISSUE

Fungal pathogens may penetrate through wounds, through natural openings such as stomata and lenticels, through stigmas, or by breaching the plant surface and entering by active penetration.

4.3.1 Entry through wounds or natural openings

Entering a host plant requires the recognition of possible openings. Post-harvest diseases caused by *Botrytis* or *Monilinia* are often the result of infections through wounds that result from handling injuries during or after harvest. In addition, *B. cinerea* (syn. *Botryotinia fuckeliana*), often invades senescent or damaged plant tissue. Although this and other wound-infecting fungi are able to penetrate the cuticle and cell wall directly, some factors of the wound, such as humidity or nutrients, may stimulate spore germination. However, conidia from *B. cinerea* do not strictly need a specific signal in order to germinate, as they usually germinate under humid conditions. Germination is strongly stimulated by the presence of low concentrations of glucose and organic phosphate (Rijkenberg *et al.*, 1980). It is assumed that nutrients released from wounds of various host plants lead to greater susceptibility to infection (Harrison, 1988). Apparently the fungus senses conditions that are well suited for growth, resulting in the stimulation of germination. Furthermore, some pathogens show preferential growth toward stomata. For example, zoospores of pathogenic oomycetes enter their host plants through both stomata and lenticels. Penetration through these natural openings requires that the fungus can locate them. It is assumed that fungal zoospores have receptors able to detect signals that influence the direction of motility; an excellent review of the molecular basis of recognition between *Phytophthora* pathogens and their hosts was published by Tyler in 2002.

Studies of *Phialophora malorum*, the causal agent of side rot (a post-harvest disease of pears), demonstrated that infection depends on the relationship between wound size and hydrostatic pressure in immersion tanks. Infection of wounds less than 1 mm in diameter generally depended on immersion depth, whereas with wound diameters more than 1 mm infection took place at all immersion depths. Furthermore, it was shown that wound exudates stimulate spore germination (Sugar and Spotts, 1993).

Wounds caused by physical damage from insects, hail and wind stress plants and encourage numerous pathogens by creating entry points. For example, in maize injuries due to the feeding of the European corn borer (*Ostrinia nubilalis*) encourage the development of both *Fusarium* ear rot and maize stalk rot (Christensen and Schneider, 1950). Recently, it was shown that the reduced *Ostrinia* injury in Bt-maize hybrids lead to a shift in species composition. The species diversity of the stalk rot complex was lower in Bt-maize hybrids than in non-Bt-maize (Gatch and Munkvold, 2001). Also the *Fusarium* ear rot incidence was reduced (Munkvold *et al.*, 1997). The type of interaction between insect and pathogen remains unclear. Besides the formation of entry wounds for the fungi the corn borer larvae carry spores of *Fusarium* species to the maize plant surface (Gatch and Munkvold, 2001). Disease incidence was positively correlated with damage caused by the European corn borer.

4.3.2 Entry by active penetration

Fungi that are able to penetrate the plant surface directly do so by enzymes that can digest components of both the cuticle and the cell wall or by mechanical force. A combination of both mechanisms is most likely but the relative contribution of both are unclear. Recently, two species of the fungus-like *Oomycota* group were compared regarding the penetration process: the phytopathogen *Pythium graminicola*, a widespread pathogen of *Graminaceae*, and *P. insidiosum*, a mammalian pathogen (MacDonald *et al.*, 2002). Hyphal apices of both species exerted minor pressures (0.19 MPa for *P. graminicola* and 0.14 MPa for *P. insidiosum*). Measurements of the mechanical resistance of the epidermis of the grass roots performed by micropenetration with glass microprobes showed values of 1-12 MPa. That of mammalian skin reached 10-47 MPa (Ravishankar *et al.*, 2001). These results show clearly that the tissue strength exeeds the pressures exerted by hyphae of these pathogens. Thus, the force of the hyphae of these species is not sufficient to penetrate the tissue surfaces of their hosts without the concerted activity of secreted tissue degrading enzymes (MacDonald *et al.*, 2002).

(a) Formation of appressoria

In many fungi, wall penetration, whether by enzyme action or by pressure, is initiated by the differentiation of appressoria (for review, see Deising *et al.*, 2000). Appressoria attach themselves firmly to their substrate and increase the area of contact between the fungus and the host. Depending on the species, appressoria are positioned over stomata or they develop in random distribution over the leaf surface. From a pore in the appressorial base a penetration hypha (or infection peg) grows directly through the cuticle and cell wall into the host tissue or, in the case of dikaryotic stages of rust fungi, through the stomatal aperture into the substomal chamber.

The production of appressoria is under both environmental and genetic control. For several plant pathogenic fungi, such as *Magnaporthe grisea* (Xu and Hamer, 1996), *Colletotrichum lagenarium* (Lev *et al.*, 1999), *Cochliobolus heterostrophus* (Takano *et al.*, 2000) and *Pyrenophora teres* (Ruiz-Roldan *et al.*, 2001) production was found to be under the control of an homologous mitogen-activated protein (MAP) kinase gene. In each case these genes are not necessary for vegetative growth but for successful infection of the host plant. As found by fusion with the green fluorescent protein (GFP) in *M. grisea* this gene (*PMK1*) was mainly expressed in appressoria and developing conidia (Bruno *et al.*, 2004).

Furthermore, fungi without clearly differentiated appressoria show *PMK1* homologue genes which are essential for fungal pathogenicity, such as the causal agent of grey mould, *Botrytis cinerea* (Zheng *et al.*, 2000), the vascular wilt pathogen *Fusarium oxysporum* f.sp. *lycopersici* (Di Pietro *et al.*, 2001), *Gaeumannomyces graminis* (Dufresne and Osbourn, 2001) and *Claviceps purpurea*, the causal agent of ergot of grasses (Mey *et al.*, 2002). Obviously, the signal transduction via the MAP kinase *PMK1* homologues play an important role in the pathogenicity of several plant pathogenic fungi.

A further question arising is what kind of genes are regulated by this MAP kinase pathway. Two genes (*GAS1* and *GAS2*) that were regulated by *PMK1* and specifically expressed in appressoria of *M. grisea* could be identified (Xue *et al.*, 2002). Mutants deleted in one or both genes were reduced in appressorial penetration. Interestingly, homologues of these genes were found in several filamentous fungi but not in yeasts (Xue *et al.*, 2002). However, these two genes may function as real virulence factors in fungal pathogens.

(b) Penetration by mechanical force

Appressoria create a high turgor pressure that allows the penetration peg to penetrate the plant cell wall (for review, see Bastmeyer *et al.*, 2002). Well studied examples of penetration by mechanical force are that of the rice blast fungus *M. grisea* (Howard and Valent, 1996) and of *Colletotrichum graminicola*, the causal agent of anthracnose of numerous grasses (Bechinger *et al.*, 1999). In both cases cell walls of mature appressoria are melanized and melanin plays a crucial role in pathogenicity. Experiments with melanin synthesis inhibitors and with melanin-deficient mutants (Chumley and Valent, 1990) show that non-melanized appressoria have lost the ability to penetrate the plant surface as well as artificial membranes. Melanin lowers the porosity of appressoria and thus helps to increase the osmotic pressure – up to 8 MPa in appressoria of *M. grisea* (Howard *et al.*, 1991). How such high pressure can be generated within living cells remained unclear until de Jong *et al.* (1997) showed that glycerol accumulates in the appressoria to very high concentrations (more than 3.0 M). The melanized appressorium wall is impermeable to glycerol and thus leads to the generation of turgor pressure. Both glycogen and lipid mobilization during conidial germination and subsequent degradation contribute to glycerol biosynthesis (Thines *et al.*, 2000).

(c) Penetration by enzymatic digestion

The destruction of the plant cell wall by numerous phytopathogenic fungi is believed to be an important aspect of the infection process. Although there is serious debate whether microbial cuticle degrading and cell wall degrading enzymes (CWDEs) like cutinases, cellulases and pectinases in general have a major impact on pathogenesis, they are ubiquitously produced in many plant-pathogen interactions. Penetration of plant surfaces might be supported by these enzymes.

Essential to an understanding of cuticle and CWDEs is the knowledge of the complex structure of these surface barriers. The first barrier to be breached is the plant cuticle which covers the epidermal cells. It consists of two lipid polymers, which may occur in any ratio, embedded into wax. The dominant structural component is the lipid polyester cutin, a polar cross-linked polymer containing predominantly C16 and C18 fatty acids. It is readily solubilized by alkaline hydrolysis. A very insoluble residue remains after saponification – the polymethylene polymer cutan. For detailed reviews, see Jeffree (1996) and Riederer and Schreiber (2001).

After breaching the host cuticle, the next barrier to the invading fungus is the cell wall. In mature plant cells, the cell wall mainly consists of two layers: the primary and secondary cell wall. The structural complexity of the primary cell wall is described in Carpita and Gibeaut (1993) and Carpita *et al.* (1996).

The major polysaccharides of the walls of many flowering plants have been described. In most flowering plants the cell wall consists of chains of ß-1,4-linked glucose interwoven with a xyloglucan polymer embedded in a matrix of pectin. In contrast, the cellulose microfibrils of primary cell walls of the *Poaceae* contain chains of ß-1,4-xylose, instead of xyloglucan, which are connected by arabinose and less frequently by glucoronic acid. Both types of primary cell wall are associated with protein components. The major structural protein is extensin, a hydroxyprolin-rich glycoprotein (Showalter, 1993).

Cuticle degrading enzymes. The diversity of the physical structure and chemical composition varies not only in different plant species. Even within the same species, it depends on the type of tissue, environmental conditions and the age of the plant.

Cuticle penetration could be achieved by degradation of the cutin polymer by cutinases which are serine esterases belonging to the ∝-ß hydrolase fold class of lipases (Longhi and Cambillau, 1999). Numerous plant pathogenic fungi produce cutinases but the role of these enzymes in pathogenicity is disputed. In some cases the fungi only penetrate the cuticle and grow between the cuticle and cell wall. *Rhynchosporium secalis* (the causal agent of barley leaf scald) and *Venturia inaequalis* (causal agent of apple scab) develop such subcuticular mycelia.

The first cutinase studied in plant pathogens was that of the pea pathogen *Fusarium solani* f.sp. *pisi* (*Nectria haematococca*). Kolattukudy and co-workers demonstrated the secretion of this enzyme during penetration of the host cuticle (Shaykh *et al.*, 1977). Inhibitors of cutinase as well as antibodies to it prevented fungal infection on intact host surfaces but no effect has been found on wounded cuticle (Maiti and Kollatukudy, 1979; Köller *et al.*, 1982). Furthermore, insertion of a cutinase gene derived from *F. solani* f.sp. *pisi* to the cutinase-deficient wound parasite *Mycosphaerella* spp. enabled the transformants to infect intact surfaces of papaya fruits (Dickman *et al.*, 1989).

However, gene disruption studies performed with *F. solani* (Stahl and Schäfer, 1994; Stahl *et al.*, 1994), *Magnaporthe grisea* (Sweigard *et al.*, 1992) and *Botrytis cinerea* (Van Kan *et al.*, 1997) have questioned the importance of cutinase for fungal pathogenicity. These results suggest a saprophytic role for the cutinases and imply that these enzymes are not important in plant infection. Thus, there may be additional cutinases that function during pathogen ingress in a localized manner.

Only recently Li *et al.* (2003) provided molecular evidence for a requirement of cutinase for the pathogenicity of *Pyrenopeziza brassicae,* cause of light leaf spot on oilseed rape. A cutinase deficient mutant failed to penetrate the cuticle and was unable to develop disease symptoms. The complementation of this mutant with the single-copy *P. brassicae* cutinase gene *Pbc1* restored both cutinase activity and pathogenicity.

However, the general importance of fungal cutinases for the direct penetration of plant surfaces remains unclear. The lifestyle of the pathogen – saprophytic or parasitic – may play a role and/or the site of pathogen ingress. *Pyrenopeziza brassicae* does not enter the leaf through stomata, but directly penetrates the cuticle (Li *et al.*, 2003). Thus, enzymatic degradation of the cuticle might be important.

Besides the role of fungal cutinolytic enzymes for the penetration process, the regulation, especially the induction of these enzymes has been questioned. Host factors such as cutin monomers could serve as signals that activate the pathogen enzyme synthesis (Kolattukudy *et al.*, 1995). Furthermore, for some pathogens cutin monomers have been described as being important for the induction of appressorium formation.

Another possible aspect of cutinase function was reported for *Monilia fructicola.* The cutinase production was inhibited by antioxidants such as caffeic acid at the transcription level (Wang *et al.*, 2002). Thus, it could be assumed that the host microenvironment, for example the ripening state, is an important factor for cutinase activity.

Cell wall degrading enzymes (CWDEs). In general, the degradation of the plant cell wall involves a concerted and/or synergistic action of several enzyme families including cellulases, xylanases, pectic enzymes and proteases adapted to the different cell wall polymers (Walton, 1994; Annis and Goodwin, 1997). Matching to the complexity of the components that make up the plant cell wall, fungal plant pathogens are able to produce a broad range of extracellular enzymes capable of degrading the plant cell wall. These enzymes may be essential for pathogenicity. However, similar to the cutinolytic enzymes, the role of CWDEs is not clear. The redundancy – most enzymes are encoded by multigene families – makes it difficult to prove their importance for the infection process.

Nevertheless, penetration by obligately biotrophic parasites, such as rust fungi and powdery mildews or some hemibiotrophs, requires only minor damage of the cell wall. Degradation of the cell wall is limited to the site of penetration as shown by Xu and Mendgen (1997). Secretion of cellulytic enzymes of these pathogens is either developmentally regulated or triggered by environmental signals (Mendgen *et al.*, 1996).

For example, cellulase activity of *Uromyces fabae* germlings has been shown to be strictly regulated by differentiation. It increases during appressorium formation and reaches a maximum during development of infection hyphae and haustorial mother cells (Heiler *et al.*, 1993). Also, the production of the pectic enzymes pectin methylesterase and polyglacturonate lyase (Deising *et al.*, 1995) and extracellular proteases (Rauscher *et al.*, 1995) of this rust fungus depends on the differentiation of infection structures. Apparently, the concerted action of CWDEs enables the hyphal growth through the leaf tissue but prevents extensive cell wall maceration and cell death which would interfere with the biotrophic lifestyle of the fungus.

For the biotrophic fungus *Blumeria graminis* f.sp. *hordei* there are clear indications of the function of CWDEs. The cellobiohydrolase I is present at the tip of the appressorial germ tube whereas isoform II is present at the tip of the primary germ tube (Pryce-Jones *et al.*, 1999).

During its biotrophic phase, the mycelium of *Venturia inaequalis* does not macerate the host tissue and is restricted to the area between the cuticle and the outer epidermal cell wall. Low amounts of CWDEs are necessary to remove physical barriers and to release nutrients. Kollar (1994) detected a pattern of twelve cellulase isoenzymes that are produced constitutively in very low amounts *in situ* as well as in *in vitro* cultures of different isolates of *V. inaequalis*. This complex cellulytic system, with low variability, appears to be correlated with virulence of the fungus or may give flexibility with properties that contribute to the performance of the enzyme.

Xylan is the predominant hemicelluose in the cell walls of plants containing 1, 4-ß linked xylose residues, which can be substituted by different side groups. The biodegradation of the xylan backbone depends mainly on two classes of enzymes, endo-1,4-ß -xylanases (EC 3.2.1.8), which hydrolyse the 1,4-ß -linked xylose backbone, and ß-xylosidases (EC 3.2.1.37), which hydrolyse xylobiose and other short xylooligosaccharides resulting from the action of endoxylanases.

Because of their potential role in fungal pathogenicity, xylanases have been purified and characterised from an increasing number of phytopathogenic fungi and the genes encoding xylanases have been cloned and characterised. Wu *et al.* (1997) reported as many as five xylanases from the rice blast fungus *Magnaportha grisea,* and at least four different xylanases have been identified from the maize leaf spot fungus, *Cochliobolus carbonum* (Apel *et al.,* 1993; Apel-Birkhold and Walton, 1996), each differing in molecular weight and pI values. Mutants of the maize pathogen *C. carbonum* that specifically lacked a functional gene for a xylan-degrading enzyme showed 85-94% reduced activity but growth of this strain was indistinguishable from the wild-type in media containing corn cell walls or xylan as the sole carbon source (Apel *et al.*, 1993). Some of the xylanases are induced only during infection (Apel-Birkhold and Walton, 1996) suggesting different sets of endoxylanases function in saprophytic and pathogenic growth of fungi.

Vast amounts of CWDEs are secreted by saprophytes and necrotrophic pathogens. In contrast to biotrophic fungi, the involvement of these enzymes in the penetration process of necrotrophic fungi is unclear. These fungi produce a high level of CWDEs during infection. Thus, it is difficult to make a distinction between the penetration of the plant surface and the maceration of plant tissue for nutrient supply. Due to the importance of CWDEs, especially the pectic enzymes, for colonising plant tissue, this subject will be discussed later.

In conclusion, CWDEs are important determinants of the life style of numerous plant pathogenic fungi and may be involved in the virulence of certain plant pathogens. However, evidence of a role for these enzymes in any aspect of pathogenesis is difficult to obtain. Nevertheless, an alternative approach to the disruption of individual genes encoding CWDEs was shown by Tonukari *et al.* (2000), who analysed a gene encoding a protein kinase (*SNF1*) in *C. carbonum* known as a regulatory element of catabolite repressed genes including the CWDEs; their results indicated that this gene was important for pathogenesis especially for the penetration process.

4.4 STRATEGIES FOR COLONIZING THE HOST TISSUE

Plant pathogens can be broadly divided into those that kill the host and feed on the contents, the so-called necrotrophs, and those that require a living host to complete their life cycle, the biotrophs.

4.4.1 Necrotrophs

Necrotrophic parasites obtain their nutrients from dead plant tissues. These fungi establish themselves inside the host tissue by releasing macerating enzymes and/or detoxifying enzymes and phytotoxins, which disrupt cell integrity and cause cell death immediately.

(a) Colonization supported by enzymes

Different groups of fungal enzymes can play important roles during any or all stages of infection: initial penetration as described above, supporting the spread through the host tissue and in addition, especially in the case of the necrotrophic lifestyle, providing a food source. Cuticle degrading enzymes and CWDEs are purified from a wide range of fungi. One of the most important groups of enzymes of necrotrophs are the pectic enzymes causing maceration of plant tissue.

Other types of enzymes that might contribute to colonising plant tissue are those that degrade antifungal substances of host plants – the so-called detoxifying enzymes.

Pectic enzymyes. For a long time research on cell wall degrading enzymes has focused on pectinolytic enzymes (endo- and exo-pectin lyase, endo- and exo-polygalacturonases and pectin methyl esterases) from pathogens of dicotyledonous plants because pectin is the main compound of the middle lamella and primary cell wall of dicotyledonous plants; thus these enzymes are able to cause maceration.

For example, during all stages of infection *B. cinerea* produces a broad set of pectinases, including pectin methylesterase (Reignault *et al.*, 1994), pectin lyase (Movahedi and Heale, 1990), and exo- and endopolygalacturonases (Johnston and Williamson, 1992). All are believed to degrade the cell wall in concerted action but their involvement in pathogenicity is still unknown. As shown by Ten Have *et al.*

(1998) inactivation of the endopolygalacturonase gene *Bcpg1* has no effect on penetration efficiency, but results in a reduction of the spread of the pathogen into the host tissue. Similar results have been obtained for *Fusarium oxysporum* f.sp. *lycopersici*. Neither endopolygalacturonases nor exopolygalacturonases are essential for pathogenicity on tomato plants (Di Pietro and Roncero, 1998; Garcia-Maceira *et al.*, 2000).

Gene disruption experiments constructing specific mutants lacking one or more CWDEs have been used in order to better understand the role of these enzymes in pathogenicity. The first targeted mutation of a CWDE gene in a fungus was reported for an endopolygalacturonase from *C. carbonum* (Scott-Craig *et al.*, 1990). The mutant displayed no reduction in virulence on maize, its natural host.

Detoxifying enzymes. Pathogenic fungi are confronted with toxic secondary metabolites of their host plants. These substances may be preformed, or as in the case of phytoalexins, induced by pathogen attack. Some of these metabolites have been shown to induce the synthesis of fungal enzymes required for their detoxification (for review, see Morrissey and Osbourn, 1999).

In *Nectria haematococca*, the causal agent of stem and root rot disease of the common pea, pisatin demethylase – a cytochrome P450 monooxygenase – is induced to break down the pea antimicrobial isoflavonoid compound pisatin (Matthews and VanEtten, 1983; Hirschi and VanEtten, 1996). Disruption of a pisatin demethylase gene showed that lack of the encoded enzyme reduces but does not eliminate the virulence of *N. haematococca* (Wasmann and VanEtten, 1996). Consequently, this enzyme is not essential for pathogenicity.

Another detoxifying enzyme is avenacinase produced by the oat root-infecting pathogen *Gaeumannomyces graminis* var. *avenae* which detoxifies the triterpenoid avenacin. Fungal defect mutants that have lost the ability to produce avenacinase have been shown to be unable to infect oats, indicating that this enzyme is an essential determinant of host range for *G. graminis* var. *avenae* (Bowyer *et al.*, 1995).

Different pathogens of tomato, including *Septoria lycopersici*, *Botrytis cinerea*, *Fusarium oxysporum* f. sp. *lycopersici*, produce tomatinases that detoxify the steroidal glycoalkaloid tomatine (Sandrock and VanEtten, 1998; Osbourn, 1996). As in the case of pisatin demethylase, the tomatinase of *S. lycopersici* was not essential for pathogenicity (Osbourn *et al.*, 1995). Nevertheless, recently it was shown that the degradation product of the *S. lycopersici* tomatinase suppresses the host plant defense responses, indicating a dual function for this enzyme: the degradation of the tomatine and suppression of disease resistance (Bouarab *et al.*, 2002).

(b) Colonization supported by toxins

One successful strategy adopted by many necrotrophic pathogens is the production of low-molecular weight secondary metabolites with phytotoxic activity. There is great diversity of fungal toxins in structure as well as in their modes of action. The non-host selective toxins typically affecting fundamental processes potentially have activity on the host as well as on non-host plants. Consequently, evidence for their role in disease establishment is not as clear and their significance in pathogenesis is difficult to establish (Panaccione *et al.*, 2002). Nevertheless, as shown by disruption experiments in which toxin synthesis has been negated, non-specific toxins may contribute to virulence.

A very well studied example of a non-host toxin is cercosporin produced by several species of the genus *Cercospora* causing leaf spot diseases on a diversity of crop species world-wide. Cercosporin is a photosensitising compound that has been shown to be toxic not only to plants but also to mice, bacteria and many species of fungi (Daub and Ehrenshaft, 2000). Nevertheless, the role of cercosporin in pathogenicity is not clarified. First evidence came from *C. kikuchii* mutants deficient in cercosporin production. Inoculation of soybean with these mutants resulted in reduced virulence (Upchurch, 1991). In addition, only recently, a functional and molecular characterisation of a polyketide synthase was presented that is involved in cercosporin biosynthesis. Virulence of these mutants is also markedly reduced (Choquer *et al.*, 2005).

In contrast, the host-selective toxins, most of which are produced by species of the genera *Alternaria* and *Cochliobolus*, are determinants of specificity (for review, see Markham and Hille, 2001; Wolpert *et al.*, 2002). The function of some of these compounds is regulated by single host plant genes (Walton, 1996), making them ideal subjects for examining host-parasite specificity.

Examples of *Cochliobolus* toxins are HC-toxin (Walton *et al.*, 1997), synthesized by races of *C. carbonum* that cause leaf spot disease of maize; T-toxin, synthesised by *C. heterostrophus* causing Southern Corn Leaf Blight disease (Dewey *et al.*, 1988; Rose *et al.*, 2002), and victorin synthesised by *C. victoriae* responsible for Victoria blight disease in oats (Meehan and Murphy, 1946; Navarre and Wolpert, 1999; Curtis and Wolpert, 2004). The cellular targets and possible mechanisms are divers. However, in most cases these toxins as well as the *Alternaria* toxins kill the plant cells of their specific host plants and enable the fungus to spread throughout the host tissue, with one exception: the HC-toxin does not kill the host cells but is able to prevent the plant defence reaction. HC-toxin is a cyclic peptide functioning as an inhibitor of histone deacetylase (HDAC) from many organisms. As a result, the hyperacetylated form of histon is accumulated. It is supposed that this leads to the inhibition of the synthesis of plant defense proteins (Brosch *et al.*, 1995; Ransom and Walton, 1997).

However, the activities and roles of phytotoxins in different pathosystems are highly variable. Most of them are clearly involved in virulence. In contrast, the so-called NIP2 protein from *Rhynchosporium secalis* that was originally detected as a non-specific phytotoxin inducing necrosis of barley leaves (Wevelsiep *et al.*, 1993) has been shown to function as an elicitor in the recognition process of plants carrying the Rrs1 resistance gene. As a result plants react with the onset of the defence response. Consequently, NIP1 functions as an avirulence determinant (Rohe *et al.*, 1995).

4.4.2 Biotrophs

Plant-infecting biotrophic fungi have developed an intimate relationship with their host plants. In contrast to necrotrophs, these fungi colonize and draw nutrients only from living tissue. This group represents economically the most important plant parasitic fungi including the downy mildews, powdery mildews and rusts. Besides the loss of photosynthetically active leaf area, the enormous spore production of these fungi results in a significant loss of biomass by the diseased crop.

Hemibiotrophic pathogens are typified by a transient phase of the biotrophic life style. These fungi have initial biotrophic growth phases before switching to killing the host. Examples of hemibiotrophs are *Phytophthora infestans*, *Cladosporium fulvum*, *Colletotrichum* spp. and *Magnaporthe grisea*. A predominantly biotrophic phase until reproduction occurs has been observed in parasites such as *Septoria* spp., *Claviceps* spp. and *Venturia inaequalis* (Parbery, 1996).

In contrast to hemibiotrophs, obligate biotrophic fungi require living host plants to complete their life cycle. A particular requirement for infection success is the ability to keep host cells alive. Obligate pathogens form stable intimate associations with their hosts with which they may co-exist for a period of time. Thus, they cause very little systemic damage to host plants. A comparison of plant infection of biotrophs and hemibiotrophs is given by Mendgen and Hahn (2002).

(a) Nutrient uptake at biotrophic plant-fungus interfaces

The biotrophic way of life requires a high degree of adaptation to the metabolism of the living host to ensure the supply of nutrients from the living plant cell. A prerequisite for nutrient movement between host and biotrophic pathogen is the development of a functional interface which promotes the nutrient transfer. During infection some of these pathogens, such as downy mildews, powdery mildews and rust fungi, form specialized physiological and morphological adaptations – the haustoria – that represent the host–parasite interface specialized in nutrient uptake (Harder and Chong, 1984; Staples, 2001).

A typical example is the powdery mildew fungus, *Blumeria graminis*, because the haustoria of this species provide the only interface with the host cell, while the mycelium develops on the leaf surface (for review, see Aist and Bushnell, 1991).

The rust fungus *Uromyces fabae* is one of the best studied examples of functional aspects of haustoria (Mendgen *et al.*, 2000). Differential screening of a haustorium-specific cDNA library of the rust fungus *Uromyces fabae* resulted in the isolation of a large number of rust genes, showing preferential expression in parasitically growing hyphae and haustoria (Hahn and Mendgen, 1997). Several of these genes were further analysed to obtain more insight into the functional aspects of haustoria. The role of nutrient transfer across the haustorial interface was confirmed by several proton-symport-driven transporters. Both the genes *Hxt1*, encoding a hexose transporter (Voegele *et al.*, 2001) and *AAT2*, encoding an amino acid transporter (Hahn *et al.*, 1997) are preferentially expressed in the haustorial membrane. Furthermore, using immuno-localisation it was shown that both of the transporters were restricted to the haustoria; the signals were absent in other fungal structures. Functional characterisation revealed that HXT1p has a transport preference for D-glucose and D-fructose, suggesting that these might be the main carbohydrates assimilated by haustoria of *U. fabae*. Two further amino acid transporters (AATp1 and AATp3) showed strongest expression in haustoria and weak expression in intercellular hyphae. Interestingly, both amino acid transporters showed broad substrate affinities with preferences for *in planta* apoplast-scarce amino acids such as histidine and lysine in the case of AAT1p and leucine and the sulphur-containing amino acids cysteine and methionine in the case of AAT3p (Struck *et al.*, 2002; Struck *et al.*, 2004a). Furthermore, AAT1p showed no transport of cysteine and only intermediate transport of methionine and leucine. On the other hand, AAT3p showed no transport of lysine and had only weak affinity for histidine. Thus, it appears that these amino acid transporters complement each other. This might reflect specific adaptations of the fungus to its environment, the host plant.

However, there is little information concerning both the nutrient supply of biotrophic pathogens and the composition of nutrients (Solomon *et al.*, 2003). Only recently, a comparison of expression profiles of genes involved in the primary metabolic pathways of *Blumeria graminis* f.sp. *hordei* performed by microarray analysis was published (Both *et al.*, 2005). The results showed a dynamic regulation of genes encoding key metabolic steps during the different developmental stages of germination, penetration, infection establishment and conidial production. Studies of genes encoding glycolytic enzymes confirmed that glucose is the predominant source of carbon (Sutton *et al.*, 1999) taken up by haustoria; they also showed that *B. graminis* had not lost metabolic capacity nor had it lost the ability to modulate its metabolism. Thus, the question remains: Why is *B. graminis* an obligate pathogen?

(b) Suppression of plant defense reactions

To establish a successful infection, biotrophic pathogenic fungi are dependent on living host cells. Thus, the fatal defense reaction of plants is the so-called hypersensitive reaction – a localized programmed cell death at infection sites. For a long-term biotrophic interaction the suppression or avoidance of the plant defense is an important pre-requisite (Panstruga, 2003). Nevertheless, several fungal avirulent genes have been described that may prevent a compatible interaction in host plants containing the corresponding resistance gene. The best studied examples of this group are the avirulence (Avr) genes of *Cladosporium fulvum* (Laugé and de Wit, 1998). Only recently, a direct interaction between *C. fulvum* avirulence proteins and tomato disease resistance proteins leading to the activation of the hypersensitive reaction was demonstrated (Rooney *et al.*, 2005).

Do the obligate biotrophic fungi have the competence to influence processes in their hosts? Interestingly, the first example of a possible fungal effector protein of the rust fungus *Uromyces fabae* was recently reported: the protein (RTP1p: rust transferred proteins) with a potential signal sequence was shown to be translocated from the extrahaustorial matrix to the cytoplasm of the infected host cell (Struck *et al.*, 2004b). This protein might be involved in signalling phenomena between host and parasite.

4.5 CONCLUDING REMARKS

Pathogenic fungi vary greatly in the extent to which they colonise the host plant. Advances in the techniques of molecular genetics have made significant contributions to our understanding of plant-pathogen interactions (Yoder and Turgeon, 2001). In combination with optical imaging techniques, for example, expression of green fluorescent protein (GFP) examined by confocal laser scanning microscopy (CLSM), new structural information has been provided (Howard, 2001). The knowledge of cellular and genetical aspects of fungal pathogenicity and virulence could contribute greatly to different approaches for controlling disease epidemics in agricultural crops.

However, unsolved questions are: Why are fungi pathogens and not just saprophytes? What are the regulation factors that are essential for the ability to infect the host plant, or in the case of biotrophs, that are involved in retaining the biotrophic interaction?

Pathogens have evolved strategies to overcome the plant barriers, either the aerial plant structures or the roots. Recently, it has been shown that the rice leaf pathogen *Magnaporthe grisea* is able to infect rice roots in a different way compared to leaf surfaces (Sesma and Osbourn, 2004). On roots it forms infection pads typical for root-infecting pathogens. Furthermore, it invades the vascular system of the plant leading to systemic invasion. Interestingly, defect mutants unable to infect rice leaves were fully pathogenic on roots, indicating that there are clear differences between the factors required for penetration of leaves and roots. These findings show that the fungus might be able to switch its infection strategy and to change niches. This could be of great epidemiological significance and has important implications for the development of new strategies for disease control.

REFERENCES

Aist, J.R. and Bushnell, W.R. (1991) Invasion of plants by powdery mildew fungi, and cellular mechanisms of resistance, in *The Fungal Spore and Disease Initiation in Plants and Animals*, (eds G.T. Cole and H.C. Hoch), Plenum Press, New York, London, pp. 321-345.

Allen, E.A., Hazen, B.E., Hoch, H.C. *et al.* (1991). Appressorium formation in response to topographical signals by 27 rust species. *Phytopathology*, **81**, 323-331.

Annis, S.L. and Goodwin, P.H. (1997). Recent advances in the molecular genetics of plant cell wall-degrading enzymes produced by plant pathogenic fungi. *European Journal of Plant Pathology*, **103**, 1-14.

Apel, P.C., Panaccione, D.G., Holden, F.T. and Walton, J.D. (1993). Cloning and targeted gene disruption of *XYL1*, a ß1,4-xylanase gene from the maize pathogen *Cochliobolus carbonum*. *Molecular Plant-Microbe Interactions*, **6**, 467-473.

Apel-Birkhold, P.C. and Walton, J.D. (1996). Cloning, disruption, and expression of two endo-ß 1, 4-xylanase genes, *XYL2* and *XYL3*, from *Cochliobolus carbonum*. *Applied Environmental Microbiology*, **62**, 4129-4135.

Apoga, D., Jansson, H-B. and Tunlid, A. (2001). Adhesion of conidia and germlings of the plant pathogenic fungus *Bipolaris sorokiniana* to solid surfaces. *Mycological Research*, **105**, 1251-1260.

Bastmeyer, M., Deising, H.B. and Bechinger, C. (2002). Force Exertion in Fungal Infection. *Annual Reviews Biophysical Biomolecular Structure*, **31**, 321-341.

Bechinger, C., Giebel, K-F., Schnell, M. *et al.* (1999). Optical measurements of invasive forces exerted by appressoria of a plant pathogenic fungus. *Science*, **285**, 1896-1899.

Both, M., Csukai, M., Stumpf, M.P.H. and Spanu, P. (2005). Gene expression profiles of *Blumeria graminis* indicate dynamic changes to primary metabolism during development of an obligate biotrophic pathogen. *Plant Cell*, **17**, 2107-2122.

Bouarab, K., Melton, R.J.P., Baulcombe, D. and Osbourn, A. (2002). A saponin-detoxifying enzyme mediates suppression of plant defences. *Nature*, **418**, 889-892.

Bowyer, P., Clarke, B.R., Lunness, P. *et al.* (1995) Host range of a plant pathogenic fungus determined by a saponin detoxifying enzyme. *Science*, **267**, 371-374.

Braun, E.J. and Howard, R.J. (1994). Adhesion of *Cochliobolus heterostophus* conidia and germlings to leaves and artificial surfaces. *Experimental Mycology*, **18**, 211-220.

Brosch, G., Ramsom, R., Lechner, T. *et al.* (1995). Inhibition of maize histone deacetylases by HC toxin, the host-selective toxin of *Cochliobolus carbonum*. *The Plant Cell*, **7**, 1941-1950.

Brown, J.K.M. and Hovmøller, M.S. (2002). Aerial dispersal of pathogens on the global and continental scales and its impact on plant disease. *Science*, **297**, 537-541.

Bruno, K.S., Tenjo, F., Li, L. *et al.* (2004). Cellular localization and role of kinase activity of *PMK1* in *Magnaporthe grisea*. *Eukaryotic Cell*, **3**, 1525-1532.

Carpita, N., Mccann, M. and Griffing, L.R. (1996). The plant extracellular matrix: News from the cell's frontier. *The Plant Cell*, **8**, 1451-1463.

Carpita, N.C. and Gibeaut, D.M. (1993). Structural models of primary cell walls in flowering plants: consistency of molecular structure with the physical properties of the walls during growth. *The Plant Journal*, **3**, 1-30.

Carver, T.L.W., Thomas, B.J. and Ingerson- Morris, S.M. (1995). The surface of *Erysiphe graminis* and the production of extracellular material at the fungus-host interface during germling and colony development. *Canadian Journal of Botany*, **73**, 272-287.

Choquer, M., Dekkers, K.L., Chen, H.Q. *et al.* (2005). The *CTB1* gene encoding a fungal polyketide synthase is required for cercosporin biosynthesis and fungal virulence of *Cercospora nicotianae*. *Molecular Plant-Microbe Interactions*, **18**, 468-476.

Christensen, J.J. and Schneider, C.L. (1950). European corn borer (*Pyrausta nubilalis* Hbn.) in relation to shank, stalk, and ear rots of corn. *Phytopathology*, **40**, 284-291.

Chumley, F.G. and Valent, B. (1990). Genetic analysis of melanin-deficient, nonpathogenic mutants of *Magnaporthe grisea*. *Molecular Plant-Microbe Interactions*, **3**, 135-143.

Clay, R.P., Enkerli, J. and Fuller, M.S. (1994). Induction and formation of *Cochliobolus sativus* appressoria. *Protoplasma*, **178**, 34-47.

Curtis, M.J. and Wolpert, T.J. (2004). The victorin-induced mitochondrial permeability transition precedes cell shrinkage and biochemical markers of cell death, and shrinkage occurs without loss of membrane integrity. *The Plant Journal*, **38**, 244-259.

Daub, M.E. and Ehrenshaft, M. (2000). The photoactivated *Cercospora* toxin cercosporin: Contributions to plant disease and fundamental biology. *Annual Reviews of Phytopathology*, **38**, 461-490.

De Jong, J.C., McCormack, B.J., Smirnoff, N. and Talbot, N.J. (1997). Glycerol generates turgor in rice blast. *Nature*, **389**, 244-245.

Deising, H., Frittrang, A.K., Kunz, S. and Mendgen, K. (1995). Regulation of pectin methylesterase and polygalacturonate lyase activity during differentiation of infection structures in *Uromyces viciae-fabae*. *Microbiology*, **141**, 561-571.

Deising, H., Nicholson, R.L., Haug, M. *et al.* (1992). Adhesion pad formation and the involvement of cutinase and esterases in the attachment of uredospores to the host cuticle. *The Plant Cell*, **4**, 1101-1111.

Deising, H.B., Werner, S. and Wernitz, M. (2000). The role of fungal appressoria in plant infection. *Microbes and Infection*, **2**, 1631-1641.

Dewey, R.E., Siedow, J.N., Timothy, D.H. and Levings, C.S., 3rd. (1988). A 13-kilodalton maize mitochondrial protein in *E. coli* confers sensitivity to *Bipolaris maydis* toxin. *Science*, **239**, 293-295.

Di Pietro, A., Garcia-MacEira, F.I., Meglecz, E. and Roncero, M.I.G. (2001). A MAP kinase of a vascular wilt fungus *Fusarium oxysporum* is essential for root penetration and pathogenesis. *Molecular Microbiology*, **39**, 1140-1152.

Di Pietro, A. and Roncero, M.I.G. (1998). Cloning, expression, and role in pathogenicity of *pg1* encoding the major extracellular endopolygalacturonase of the vascular wilt pathogen *Fusarium oxysporum*. *Molecular Plant-Microbe Interactions*, **11**, 91-98.

Dickman, M.B. (2003). A protein kinase from *Colletotrichum trifolii* is induced by plant cutin and is required for appressorium formation. *Molecular Plant-Microbe Interactions*, **16**, 411-421.

Dickman, M.B., Patil, S.S. and Kolattukudy, P.E. (1989). Insertion of cutinase gene into a wound pathogen enables it to infect intact host. *Nature*, **342**, 446-448.

Doss, R.P., Potter, S.W., Soeldner, A.H. *et al.* (1995). Adhesion of germlings of *Botrytis cinerea. Applied and Environmental Microbiology,* **61**, 2260-2265.

Dufresne, M. and Osbourn, A. (2001). Definition of tissue specific and general requirements for plant infection in a phytopathogenic fungus. *Molecular Plant-Microbe Interactions*, **14**, 300-307.

Gatch, E.W. and Munkvold, G.P. (2002). Fungal species composition in maize stalks in relation to European corn borer injury and transgenic insect protection. *Plant Disease*, **86**, 1156-1162.

Garcia-Maceira, F.I., Di Pietro, A. and Roncero, M.I.G. (2000). Cloning and disruption of *pgx4* encoding an *in planta* expressed exopolygalacturonase from *Fusarium oxysporum*. *Molecular Plant-Microbe Interactions*, **13**, 359-365.

Gilbert, R.D., Johnson, A.M. and Dean, R.A. (1996). Chemical signals responsible for appressorium formation in the rice blast fungus *Magnaporthe grisea*. *Physiological and Molecular Plant Pathology*, **48**, 335-346.

Gubler, F., Hardham, A.R. and Duniec, J. (1989). Characterising adhesiveness of *Phytophthora cinnamomi* zoospores during encystment. *Protoplasma*, **149**, 24-30.

Hahn, M. and Mendgen, K. (1997). Characterization of *in planta* induced rust genes isolated from a haustorium-specific cDNA library. *Molecular Plant-Microbe Interactions*, **10**, 427-437.

Hahn, M., Neef, U., Struck, C. *et al.* (1997). A putative amino acid transporter is specifically expressed in haustoria of the rust fungus *Uromyces fabae*. *Molecular Plant-Microbe Interactions,* **10**, 438-445.

Harder, D.E. and Chong, J. (1984). Structure and physiology of haustoria, in *The Cereal Rusts: Origins, Specificity, Structure, and Physiology* (eds W.R. Bushnell and A.J. Roelfs) (Vol. 1, pp. 431-476), Academic Press.

Hardham, A.J. (2001). Cell biology of fungal infection of plants, in *The Mycota: Biology of the Fungal Cell* (eds R.J. Howard and N.A.R. Gow) (Vol. VIII, pp. 91-124), Springer.

Harrison, J.G. (1988). The biology of *Botrytis* spp. on *Vicia* beans and chocolate spot disease – a review. *Plant Pathology,* **37**, 168-201.

Heiler, S., Mendgen, K. and Deising, H. (1993). Cellulotic enzymes of the obligately biotrophic rust fungus *Uromyces viciae-fabae* are regulated differentiation-specifically. *Mycological Research,* **97**, 77-85.

Hirschi, K. and VanEtten, H.D. (1996). Expression of the Pisatin Detoxifying Genes (*PDA*) of *Nectria haematococca in vitro* and *in planta*. *Molecular Plant-Microbe Interactions*, **9**, 483-491.

Hoch, H.C., Staples, R.C., Whitehead, B. *et al.* (1987). Signaling for growth orientation and cell differentiation by surface topography in *Uromyces*. *Science*, **235**, 1659-1662.

Howard, R.J. (2001). Cytology of fungal pathogens and plant-host interactions. *Current Opinion in Microbiology*, **4**, 365-373.

Howard, R.J., Ferrari, M.A., Roach, D.H. and Money, N.P. (1991). Penetration of hard substrates by a fungus employing enormous turgor pressures. *Proceedings of the National Academy of Sciences USA,* **88**, 11281-11284.

Howard, R.J. and Valent, B. (1996). Breaking and entering: Host penetration by the fungal rice blast pathogen *Magnaporthe grisea*. *Annual Review of Microbiology*, **50**, 491-512.

Jeffree, C.E. (1996). Structure and ontogeny of plant cuticles, in *Plant Cuticles* (ed. G. Kerstiens), Oxford: BIOS Scientific Publishers Ltd., pp. 33-82.

Jelitto, T.C., Page, H.A. and Read, N.D. (1994). Role of external signals in regulating the pre-penetration phase of infection by the rice blast fungus, *Magnaporthe grisea*. *Planta,* **194**, 471-477.

Johnston, D.J. and Williamson, B. (1992). Purification and characterization of four polygalacturonases from *Botrytis cinerea*. *Mycological Research*, **96**, 343-349.

Kolattukudy, P.E., Rogers, L.M., Li, D. *et al.* (1995). Surface signaling in pathogenesis. *Proceedings of the National Academy of Sciences USA*, **92**, 4080-4087.

Kollar, A. (1994). Characterization of specific induction, activity, and isoenzyme polymorphism of extracellular cellulases from *Venturia inaequalis* detected *in vitro* and on the host plant. *Molecular Plant-Microbe Interactions*, **7**, 603-611.

Köller, W., Allan, C.R. and Kolattukudy, P.E. (1982). Role of cutinase and cell wall degrading enzymes in infection of *Pisum sativum* by *Fusarium solani* f.sp. *pisi*. *Physiological Plant Pathology,* **20**, 47-60.

Kwon, Y.H. and Hoch, H.C. (1991). Temporal and spatial dynamics of appressorium formation in *Uromyces appendiculatus*. *Experimental Mycology,* **15**, 116-131.

Laugé, R. and DeWit, P.J.G.M. (1998). Fungal avirulence genes: Structure and possible functions. *Fungal Genetics and Biology,* **24**, 285-297.
Lev, S., Sharon, A., Hadar, R. *et al.* (1999). A mitogen-activated protein kinase of the corn leaf pathogen *Cochliobolus heterostrophus* is involved in conidiation, appressorium formation and pathogenicity: diverse roles for mitogen-activated protein kinase homologs in foliar pathogens. *Proceedings of the National Academy of Sciences USA,* **96**, 13542-13547.
Li, D., Ashby, A.M. and Johnstone, K. (2003). Molecular evidence that the extracellular cutinase Pbc1 is required for pathogenicity of *Pyrenopeziza brassicae* on oilseed rape. *Molecular Plant-Microbe Interactions,* **16**, 545-552.
Longhi, S. and Cambillau, C. (1999). Structure-activity of a cutinase, a small lipolytic enzyme. *Biochimical Biophysical Acta*, **1441**, 185-196.
MacDonald, E., Millward, L., Ravishankar, J.P. and Money, N.P. (2002). Biomechanical interaction between hyphae of two *Pythium* species (Oomycota) and host tissues. *Fungal Genetics and Biology,* **37**, 245-249.
Maiti, I.B. and Kolattukudy, P.E. (1979). Prevention of fungal infection of plants by specific inhibition of cutinase. *Science*, **205**, 507-508.
Markham, J.E. and Hille, J. (2001). Host-selective toxins as agents of cell death in plant-fungus interactions. *Molecular Plant Pathology,* **2**, 229-239.
Matthews, D.E. and VanEtten, H.D. (1983). Detoxification of the phytoalexin pisatin by a fungal cytochrome P-450. *Archives of Biochemistry and Biophysics,* **224**, 494-505.
Meehan, F. and Murphy, H.C. (1946). A new *Helminthosporium* blight in oats. *Science,* **104**, 13-14.
Mendgen, K., Struck, C., Voegele, R.T. and Hahn, M. (2000). Biotrophy and rust haustoria. *Physiological and Molecular Plant Pathology,* **56**, 141-145.
Mendgen, K. and Hahn, M. (2002). Plant infection and the establishment of fungal biotrophy. *Trends in Plant Science*, **7**, 352-356.
Mendgen, K., Hahn, M. and Deising, H. (1996). Morphogenesis and mechanisms of penetration by plant pathogenic fungi. *Annual Review of Phytopathology,* **34**, 367-386.
Mercure, E.W., Kunoh, H. and Nicholson, R.L. (1994). Adhesion of *Colletotrichum graminicola* conidia to corn leaves: a requirement for disease development. *Physiological and Molecular Plant Pathology*, **45**, 407-420.
Mercure, E.W., Kunoh, H. and Nicholson, R.L. (1995). Visualization of materials released from adhered, ungerminated conidia of *Colletotrichum graminicola. Physiological and Molecular Plant Pathology,* **46**, 121-135.
Mey, G., Oeser, B., Lebrun, M.H. and Tudzynski, P. (2002). The biotrophic non-appressorium-forming grass pathogen *Claviceps purpurea* needs a Fus3/*PMK1* homologous mitogen activated protein kinase for colonization of rye ovarian tissue. *Molecular Plant-Microbe Interactions,* **15**, 303-312.
Morrissey, J.P. and Osbourn, A.E. (1999). Fungal resistance to plant antibiotics as a mechanism of pathogenesis. *Microbiology and Molecular Biology Reviews,* **63**, 708-724.
Movahedi, S. and Heale, J.B. (1990). The roles of aspartic proteinase and endo-pectinlyase enzymes in the primary stages of infection and pathogenesis of various host tissues by different isolates of *Botrytis cinerea* Pers ex. Pers. *Physiological and Molecular Plant Pathology,* **36**, 303-324.
Munkvold, G.P., Hellmich, R.L. and Showers, W.B. (1997). Reduced *Fusarium* ear rot and symptomless infection in kernels of maize genetically engineered for European corn borer resistance. *Phytopathology,* **87**, 1071-1077.
Navarre, D.A. and Wolpert, T.J. (1999). Victorin induction of an apoptotic/senescence-like response in oats. *The Plant Cell,* **11**, 237-249.
Nicholson, R., Yoshioka, H., Yamaoka, N. and Kunoh, H. (1988). Preparation of the infection court by *Erysiphe graminis.* II. Release of esterase enzyme from conidia in response to a contact stimulus. *Experimental Mycology,* **12**, 336-349.
Nicholson, R.L. (1996). Adhesion of Fungal Propagules, in *Histology, Ultrastructure and Molecular Cytology of Plant-Microorganism Interactions* (eds M. Nicole and V. Gianinazzi-Pearson). Dordrecht: Kluwer Academic Publishers, pp. 117-134.
Osbourn, A.E. (1996). Saponins and plant defense - a soap story. *Trends in Plant Science,* **1**, 4-9.
Osbourn, A.E., Bowyer, P., Lunness, P. *et al.* (1995). Fungal pathogens of oat roots and tomato leaves employ closely related enzymes to detoxify host plant saponins. *Molecular Plant-Microbe Interactions*, **8**, 971-978.

Panaccione, D.G., Johnson, R.D., Rasmussen, J.B. and Friesen, T.L. (2002). Fungal Phytotoxins, in *The Mycota* (ed. Kempken) (Vol. XI). Berlin, Heidelberg: Springer, pp. 311-340.

Panstruga, R. (2003). Establishing compatibility between plants and obligate biotrophic pathogens. *Current Opinion in Plant Biology*, **6**, 320-326.

Parbery, D.G. (1996). Trophism and the ecology of fungi associated with plants. *Biological Review of Cambridge Philosophical Society*, **71**, 473-527.

Pascholati, S.F., Yoshioka, H., Kunoh, H. and Nicholson, R.L. (1992). Preparation of the infection court by *Erysiphe graminis* f.sp. *hordei*: cutinase is a component of the conidial exudate. *Physiological and Molecular Plant Pathology*, **41**, 53-59.

Pennisi, E. (2001). The push to pit genomics against fungal pathogens. *Science*, **292**, 2273-2274.

Pryce-Jones, E., Carver, T. and Gurr, S.J. (1999). The roles of cellulase enzymes and mechanical force in host penetration by *Erysiphe graminis* f.sp. *hordei*. *Physiological and Molecular Plant Pathology*, **55**, 175-182.

Ransom, R.F. and Walton, J.D. (1997). Histone hyperacetylation in maize in response to treatment with HC-Toxin or infection by the filamentous fungus *Cochliobolus carbonum*. *Plant Physiology*, **115**, 1021-1027.

Rauscher, M., Mendgen, K. and Deising, H. (1995). Extracellular proteases of the rust fungus *Uromyces viciae-fabae*. *Experimental Mycology*, **19**, 26-34.

Ravishankar, J.P., Davis, C.M., Davis, D.J. *et al.* (2001). Mechanics of solid tissue invasion by the mammalian pathogen *Pythium insidiosum*. *Fungal Genetics and Biology*, **34**, 161-175.

Reignault, P., Mercier, M., Bompeix, G. and Boccara, M. (1994). Pectin methylesterase from *Botrytis cinerea*: Physiological, biochemical and immunochemical studies. *Microbiology*, **140**, 3249-3255.

Riederer, M. and Schreiber, L. (2001). Protecting against water loss: analysis of the barrier properties of plant cuticles. *Journal of Experimental Botany*, **52**, 1023-1032.

Rijkenberg, F.H.J., De Leeuw, G.T.N. and Verhoeff, K. (1980). Light and electron microscopy studies on the infection of tomato fruits by *Botrytis cinerea*. *Canadian Journal of Botany*, **58**, 1394-1404.

Rohe, M., Gierlich, A., Hermann, H. *et al.* (1995). The race-specific elicitor, NIP1, from the barley pathogen, *Rhynchosporium secalis*, determines avirulence on host plants of the Rrs1 resistance genotype. *EMBO Journal*, **14**, 4168-4177.

Rooney, H.C., Van't Klooster, J.W., van der Hoorn, R.A. *et al.* (2005). *Cladosporium* Avr2 inhibits tomato Rcr3 protease required for Cf-2-dependent disease resistance. *Science*, **308**, 1783-1786.

Rose, M.S., Yun, S.H., Asvarak, T. *et al.* (2002). A decarboxylase encoded at the *Cochliobolus heterostrophus* translocation-associated Tox1B locus is required for polyketide (T-toxin) biosynthesis and high virulence on T-cytoplasm maize. *Molecular Plant-Microbe Interactions*, **15**, 883-893.

Ruiz-Roldan, M.C., Maier, F.J. and Schäfer, W. (2001). *PTK1*, a mitogen-activated protein kinase gene is required for conidiation, appressorium formation, and pathogenicity of *Pyrenophora teres* on barley. *Molecular Plant-Microbe Interactions*, **14**, 116-125.

Sandrock, R.W. and VanEtten, H.D. (1998). Fungal sensitivity to and enzymatic degradation of the phytoanticipin α-tomatine. *Phytopathology*, **88**, 137-143.

Scott-Craig, J.S., Panaccione, D.G., Cervone, F. and Walton, J. (1990). Endopolygalacturonase is not required for pathogenicity of *Cochliobolus carbonum* on maize. *The Plant Cell*, **2**, 1191-1200.

Sesma, A. and Osbourn, A.E. (2004). The rice leaf blast pathogen undergoes developmental processes typical of root-infeccting fungi. *Nature*, **431**, 582-586.

Shaykh, M., Soliday, C.I. and Kolattukudy, P.E. (1977). Proof for the production of cutinase by *Fusarium solani* f.sp. *pisi* during penetration into its host, *Pisum sativum*. *Plant Physiology*, **60**, 170-172.

Showalter, A.M. (1993). Structure and function of plant cell wall proteins. *The Plant Cell*, **5**, 9-23.

Solomon, P.S., Tan, K-C. and Oliver, R.P. (2003). The nutrient supply of pathogenic fungi; a fertile field for study. *Molecular Plant Pathology*, **4**, 203-210.

Stahl, D.J. and Schäfer, W. (1994). Cutinase is not required for fungal pathogenicity on pea. *The Plant Cell*, **4**, 621-629.

Stahl, D.J., Theuerkauf, A., Heitefuss, R. and Schäfer, W. (1994). Cutinase of *Nectria haematococca* (*Fusarium solani* f.sp. *pisi*) is not required for fungal virulence or organ specificity on pea. *Molecular Plant-Microbe Interactions*, **7**, 713-725.

Staples, R.C. (2001). Nutrients for a rust fungus: the role of haustoria. *Trends in Plant Science*, **6**, 496-498.

Staples, R.C. and Hoch, H.C. (1997). Physical and chemical cues for spore germination and appressorium formation by fungal pathogens, in *The Mycota* (eds G. Caroll and P. Tudzynski) (Vol. V, part A). Berlin, Heidelberg, New York: Springer, pp. 27-40.

Struck, C., Ernst, M. and Hahn, M. (2002). Characterization of a developmentally regulated amino acid transporter (AAT1p) of the rust fungus *Uromyces fabae*. *Molecular Plant Pathology*, **3**, 23-30.

Struck, C., Mueller, E., Martin, H. and Lohaus, G. (2004a). The *Uromyces fabae UfAAT3* gene encodes a general amino acid permease that prefers uptake of *in planta* scarce amino acids. *Molecular Plant Pathology*, **5**, 183-189.

Struck, C., Voegele, R.T., Hahn, M. and Mendgen, K. (2004b). Rust haustoria as sink in plant tissues or - how to survive in leaves, in *Biology of Plant-Microbe Interactions* (eds I. Tikhonovich, B. Lugtenberg and N. Provorov) (Vol. 4). St. Paul: International Society for Molecular Plant-Microbe Interactions, pp. 177-179.

Sugar, D. and Spotts, R.A. (1993). The importance of wounds in infection of pear fruit by *Phialophora malorum* and the role of hydrostatic pressure in spore penetration of wounds. *Phytopathology*, **83**, 1083-1086.

Sutton, P.N., Henry, M.J. and Hall, J.L. (1999). Glucose, and not sucrose, is transported from wheat to wheat powdery mildew. *Planta*, **208**, 426-430.

Sweigard, J.A., Chumley, F.G. and Valent, B. (1992). Disruption of a *Magnaporthe grisea* cutinase gene. *Molecular General Genetics*, **232**, 183-190.

Takano, Y., Kikuchi, T., Kubo, Y. *et al.* (2000). The *Colletotrichum lagenarium* MAP kinase gene *CMK1* regulates diverse aspects of fungal pathogenesis. *Molecular Plant-Microbe Interactions*, **13**, 374-383.

Ten Have, A., Mulder, W., Visser, J. and van Kan, J.A. (1998). The endopolygalacturonase gene *Bcpg1* is required for full virulence of *Botrytis cinerea*. *Molecular Plant-Microbe Interactions*, **11**, 1009-1016.

Thines, E., Weber, R.W. and Talbot, N.J. (2000). MAP kinase and protein kinase A-dependent mobilization of triacylglycerol and glycogen during appressorium turgor generation by *Magnaporthe grisea*. *The Plant Cell*, **12**, 1703-1718.

Tonukari, N.J., Scott-Craig, J.S. and Walton, J.D. (2000). The *Cochliobolus carbonum SNF1* gene is required for cell wall-degrading enzyme expression and virulence on maize. *Plant Cell*, **12**, 237-247.

Tucker, S.L. and Talbot, N.J. (2001). Surface attachment and pre-penetration stage development by plant pathogenic fungi. *Annual Review of Phytopathology*, **39**, 385-417.

Tyler, B.M. (2002). Molecular basis of recognition between *Phytophthora* pathogens and their hosts. *Annual Review of Phytopathology*, **40**, 137-167.

Upchurch, R.G., Walker, D.C., Rollins, J.A. *et al.* (1991). Mutants of *Cercospora kikuchii* altered in cercosporin synthesis and pathogenicity. *Applied Environmental Microbiology*, **57**, 2940-2945.

Van Kan, J.A.L., VantKlooster, J.W., Wagemakers, C.A.M. *et al.* (1997). Cutinase A of *Botrytis cinerea* is expressed, but not essential, during penetration of gerbera and tomato. *Molecular Plant-Microbe Interactions*, **10**, 30-38.

Voegele, R.T., Struck, C., Hahn, M. and Mendgen, K. (2001). The role of haustoria in sugar supply during infection of broad bean by the rust fungus *Uromyces fabae*. *Proceedings of the National Academy of Sciences USA*, **98**, 8133-8138.

Walton, J.D. (1994). Deconstructing the cell wall. *Plant Physiology*, **104**, 1113-1118.

Walton, J.D. (1996). Host-selective toxins: Agents of compatibility. *The Plant Cell*, **8**, 1723-1733.

Walton, J.D., Ranson, R. and Pitkin, J.W. (1997). Northern corn leaf spot: chemistry, enzymology, and molecular genetics of a host-selective phytotoxin, in *Plant-Microbe-Interactions III* (eds G. Stacey and N.T. Keen). New York: Chapman & Hall, pp. 94-123.

Wang, G.Y., Michailides, T.J., Hammock, B.D. *et al.* (2002). Molecular cloning, characterization, and expression of a redox-responsive cutinase from *Monilinia fructicola* (Wint.) Honey. *Fungal Genetics and Biology*, **35**, 261-276.

Wasmann, C.C. and VanEtten, H.D. (1996). Transformation-mediated chromosome loss and disruption of a gene for pisatin demethylase decrease the virulence of *Nectria haematococca* on pea. *Molecular Plant-Microbe Interactions*, **9**, 793-803.

Wevelsiep, L., Rüpping, E. and Knogge, W. (1993). Stimulation of barley plasmalemma H^+-ATPase by phytotoxic peptides from the fungal pathogen *Rhynchosporium secalis*. *Plant Physiology*, **101**, 297-301.

Wolpert, T.J., Dunkle, L.D. and Ciuffetti, L.M. (2002). Host-selective toxins and avirulence determinants: what's in a name? *Annual Review of Phytopathology*, **40**, 251-285.

Wu, S-C., Ham, K-S., Darvill, A.G. and Albersheim, P. (1997). Deletion of two endo- ß-1,4-xylanase genes reveals additional isozymes secreted by rice blast fungus. *Molecular Plant-Microbe Interactions,* **10**, 700-708.

Xiao, J.Z., Ohshima, A., Kamakura, T. *et al.* (1994). Extracellular glycoprotein(s) associated with cellular differentiation in *Magnaporthe grisea. Molecular Plant-Microbe Interactions*, **7**, 639-644.

Xu, H.X. and Mendgen, K. (1997). Targeted cell wall degradation at the penetration site of cowpea rust basidiosporelings. *Molecular Plant-Microbe Interactions,* **10**, 87-94.

Xu, J.R. and Hamer, J.E. (1996). MAP kinase and cAMP signalling regulate infection structure formation and pathogenic growth in the rice blast fungus *Magnaporthe grisea. Genes and Development,* **10**, 2696-2706.

Xue, C., Park, G., Choi, W. *et al.* (2002). Two novel fungal virulence genes specifically expressed in appressoria of the rice blast fungus. *The Plant Cell*, **14**, 2107-2119.

Yoder, J.I. and Turgeon, B.G. (2001). Fungal genomics and pathogenicity. *Current Opinion in Plant Biology*, **4**, 315-321.

Zheng, L., Campbell, M., Murphy, J. *et al.* (2000). The *BMP1* gene is essential for pathogenicity in the gray mold fungus *Botrytis cinerea. Molecular Plant-Microbe Interactions*, **13**, 724-732.

CHAPTER 5

EPIDEMIOLOGICAL CONSEQUENCES OF PLANT DISEASE RESISTANCE

M.L. DEADMAN

5.1 INTRODUCTION

The concept of resistance is, or should be, central to any plant disease management programme. Other disease control practices, including the use of chemical intervention techniques, cultural control methods and biological control, can all be used to minimize crop damage. Each of them, however, can be seen essentially as a complement to plant disease resistance. For farmers and growers, plant resistance offers the most cost-effective front line method for disease control, since its adoption requires little alteration to existing practices. This is especially important for the so-called resource-poor producers - those without the financial or technological wherewithal to adopt chemical or other high cost or high input strategies.

Resistance refers to the ability of the host plant to overcome, either completely or in part, the effect of a pathogen. The scale of this ability may vary from small, where there may only be a slight suppression of disease development, to large, where pathogenesis is incomplete. Incomplete pathogenesis can adequately suppress disease. If the effects are large enough, pathogen reproduction rates are slowed to the extent that the pathogen population merely replaces lost individuals but fails to increase in size. If the resistance effects are small, disease increases more or less rapidly and, as a consequence, other control techniques will be required. However, the use of resistance is not without its own problems. Genetic uniformity within a plant population can lead to pressure for new virulences to develop within the pathogen population, or minor pathogens to become increasingly important in some agricultural production systems. When identical plants are sown over large areas there are no spatial or temporal obstacles to epidemic progress.

Resistance in plants can take two possible, though not mutually exclusive, forms; Van der Plank (1963) termed these horizontal and vertical resistance. Horizontal resistance is expressed against all races of the pathogen, i.e. it is race, pathotype or biotype non-specific. Vertical, race-specific or specific resistance can be described as resistance that is race or biotype specific. In addition, consideration needs to be given to the concepts of induced resistance, non-host immunity and tolerance. Each of these will be considered in turn in this chapter.

B. M. Cooke, D. Gareth Jones and B. Kaye (eds.), The Epidemiology of Plant Diseases, 2nd edition, 139–157.

5.2 HORIZONTAL RESISTANCE

Other terms have been used in some cases more or less synonymously, to describe horizontal resistance including race non-specific resistance, general resistance, durable resistance, polygenic resistance and partial resistance. However, Van der Plank (1963,1968) provides compelling reasons why the term horizontal is the most appropriate. Horizontal resistance slows down the rate at which disease increases within a single plant or population of identical plants. It can, therefore, usefully be described as rate-reducing resistance. The mechanisms by which horizontal resistance slows the rate of epidemic development may be active or passive and may be expressed through reduced infectivity, lower levels of sporulation (in both cases resulting in a reduction in the basic infection rate), lengthened latent periods, or increased rate of removal of infectious tissue (reducing the infectious period). Frequently, experimenters have examined the nature of individual components of horizontal resistance to specified pathogens within selected varieties and have correlated these with the established reactions of the same varieties when infected by that same pathogen under field conditions. Under such circumstances, it is relatively easy to establish a relationship between, for example, lesion development and field response. Guzman-N (1964) showed that the numbers of late blight lesions established per unit of inoculum by *Phytophthora infestans* on a known susceptible potato variety were 50% higher than on a known resistant variety (Table 5.1). Similarly, the numbers of sporangia per unit area were seven times greater on a susceptible compared with a resistant variety (Guzman-N, 1964) (Table 5.1).

In the same way, Hakiza (1997) reported that the latent period for robusta coffee leaves infected by rust (*Hemileia vastatrix*) was over three times longer in an established resistant variety compared with a known susceptible variety (Table 5.1). In a study of epidemics of *Diaporthe adunca* on experimental and natural populations of *Plantago lanceolata,* Linders *et al.* (1996) used non-linear models to examine the rate of disease increase and the final disease levels in susceptible and partially resistant cloned genotypes. For both of the models used, the estimate of final disease level was significantly lower in the partially resistant genotype (69% and 66% for logistic and Gompertz models) compared with the susceptible genotype (100% for both models, Table 5.1). Similarly the rate of disease growth was lower in the partially resistant genotype (0.048 and 0.083 for the logistic and Gompertz models) than in the susceptible type (0.078 and 0.111 for the logistic and Gompertz models, Table 5.1).

When novel plant selections are generated from a breeding programme, initial resistance screening is often conducted under controlled conditions. Although individual components of horizontal resistance are relatively easy to measure under controlled conditions, they are difficult to assess in the field. Furthermore, advanced knowledge about which component of resistance may relate to field response is lacking. There is no guarantee that values obtained for each of the individual components measured under controlled conditions will be mirrored in the field response. Some of the reasons for this lack of correlation arise from the differences in environmental conditions and their individual effects on the expression of resistance.

Table 5.1. Some examples of reduction in disease parameters brought about by horizontal resistance in various crop-pathogen combinations

Resistance component	*Host*	*Pathogen*	*Resistant response*	*Susceptible response*	*Reference*
Latent period	Potato	*P. infestans*	5 days	3 days	Guzman-N, 1964
	Coffee	*H. vestatrix*	64 days	20 days	Hakiza, 1997
	Barley	*B. graminis*	13 days	9 days	Asher and Thomas, 1984
Relative lesion number (per unit leaf area)	Rice	*P. oryzae*	8	100	Roumen, 1993
	Coffee	*H. vestatrix*	13	100	Hakiza, 1997
	Barley	*B. graminis*	68	100	Asher and Thomas, 1984
Spore production per unit leaf area	Potato	*P. infestans*	700	4900	Guzman-N, 1964
	Barley	*B. graminis*	10934	22787	Asher and Thomas, 1984
Final disease level	Plantago	*D. adunca*	69% (logistic model) 66% (Gompertz model)	100% (logistic model) 100% (Gompertz model)	Linders *et al.*, 1996
Rate of growth	Plantago	*D. adunca*	0.048 (logistic model) 0.083 (Gompertz model)	0.078 (logistic model) 0.111 (Gompertz model)	Linders *et al.*, 1996

Frequently then, new breeding lines are assessed for their reaction to a particular pathogen or group of pathogens using a wide range of potential indicators. Kong *et al.* (1997) assessed the response of 17 sunflower accessions to *Alternaria helianthi* under field conditions. At the same time, each line was also assessed under controlled conditions to examine variation between them in terms of incubation period (defined here as the time between inoculation and 25% of lesions becoming visible), infection frequency (expressed as a proportion of the mean number of lesions on a susceptible check), mean lesion size and spore production. The resulting correlation matrix, derived following regression of means of components with field severity ratings, showed mean lesion size to be the best indicator of field response (Table 5.2).

Kong *et al.* (1997) found little difference in incubation periods between the accessions used. Consequently, a poor correlation with field results was obtained. As they point out, it may be that small differences in incubation period have little epidemiological consequence compared with other components. They did, however, find a high degree of correlation between infection frequency and mean lesion size which, they suggest, could be due to these components being controlled either by the same gene(s) or by having at least some gene(s) in common.

Table 5.2. Matrix of correlation coefficients (r) *for components of horizontal resistance to* Alternaria helianthi *in 17 sunflower accessions (from Kong* et al., *1997)*

	Infection frequency	*Mean lesion size*	*Incubation period*	*Spore production*	*IF x MLS index*
Infection frequency	1.00				
Mean lesion size	0.75	1.00			
Incubation period	0.12	-0.03	1.00		
Spore production	0.26	0.22	-0.26	1.00	
Field 1994	0.55	0.80	-0.03	-0.03	0.73
Field 1995	0.73	0.76	-0.01	-0.18	0.83
Field (both years)	0.58	0.74	-0.02	-0.07	0.72

Components were regressed with one another, with severity ratings obtained from field trials conducted in 1994 and 1995, the pooled ratings for both years and with a severity index calculated from infection frequency (IF) and mean lesion size (MLS).

Parlevliet (1992) pointed out that infection frequency is the most likely of all components to be independent of the others, even though the apparent association of two or more components of resistance does make potential exploitation through breeding more feasible. In reality, it may be desirable for components not to be controlled by linked genes as durability is likely to be reduced if this were to be the case.

Plant breeders in the past have mainly avoided horizontal resistance due to its more complex inheritance. Indeed, according to Van der Plank (1968), no identification of horizontal resistance is possible since it can only be acquired indirectly through the selection of linked genes. For the successful identification of horizontal resistance, therefore, breeders are faced with difficulties in the demand for adequate and meaningful assessments of disease progress, the influence of environmental interactions, inoculum potential, changes to the physiological status of the host plant with age and the manner in which it may be obscured by vertical resistance. Furthermore, the disparity shown by breeding lines when components of resistance are compared indicates the severe limitations raised when initial screening is conducted under controlled conditions, especially given the sensitivity of components to environmental fluctuations. However, a controlled-environment screening system may have considerable advantages over a full-scale field test in terms of time and manpower cost effectiveness.

Although horizontal resistance may be inherited either polygenically or oligogenically, the former is far more common (Robinson, 1969). In a number of cases, a small degree of race specificity has been demonstrated - for example, in potatoes without resistance to potato blight (R0 varieties) when challenged by individual isolates of *Phytophthora infestans* (Caten, 1974). However, Caten found no evidence for increasing performance by *P. infestans* isolates on individual potato varieties and, in general, reductions in the level of expression of horizontal resistance occur rarely and are quantitative and partial (Bennett, 1984). Van der Plank (1971) illustrated the stability of R0 potato varieties in their resistance to late blight by referring to varieties that had appeared on an annual list of reactions to blight continuously between 1938 and 1968 (Table 5.3). In terms of foliar resistance, the variety that changed most was Record, which fell from equal top in 1938 to sixth in 1968. Over the same time period, the ranking for the tuber resistance of Record showed a change from second to first place (Van der Plank, 1971) and it is unlikely that a new race of *P. infestans* arose which was capable of causing greater damage to foliage whilst at the same time being less well adapted to attack the tubers.

Table 5.3. Stability of horizontal resistance as illustrated by the field response of 10 potato varieties to infection by Phytophthora infestans *in 1938 and 1968 (from Van der Plank, 1971)*

Cultivar	*Resistance level*	
	1938	1968
Alpha	8	7
Bevelander	8	7
Furore	8	7
Record	8	6
Noordeling	7	7
Voran	7	7
Ultimus	6	5
Eigenheimer	4	5
Bintje	3	3
Eersteling	3	3

The effect of lengthening the latent period, reducing the rate of lesion expansion and lowering the numbers of spores produced on each lesion is to elongate the disease progress curve horizontally, as classically shown by Van der Plank (1968) using the example of potato varieties without R genes for resistance to late blight. Disease levels on Bintje, Eigenheimer and Voran were surveyed over time in potato fields in The Netherlands. The progress of disease, which had first been noticeable on all varieties at the beginning of July, was much slower in Varan than in Bintje, with the result that all of the foliage of the latter variety was destroyed by the start of August whilst that of Voran survived, at least in part, into September.

Sorghum, a tropical cereal grown as a subsistence crop in many drier parts of the world, is affected by a wide range of foliar diseases. Among them, anthracnose (*Colletotrichum graminearum*) is of major importance, especially in the tropics. Rate-reducing resistance in sorghum to anthracnose has been identified (Casela *et al.*, 1993) using isolated plots in the United States and Brazil and calculating the area under disease progress curves (AUDPC). More recently, Peacocke (1995) made an extensive experimental study of the reactions of sorghum varieties to anthracnose infection in southern Africa. Using three cultivars over two growing seasons at two locations, clear differences in the rate of epidemic progress were determined; mean estimates of the rate parameter were significantly greater for the varieties MMSH 413 and Kuyuma than for Sima (Table 5.4). Estimates ranged from 0.09 for Sima to 0.15 for Kuyuma; in all cases, estimates of the intercept were low, corresponding to moderate levels of initial infection, and final disease severities were high (Table 5.4).

Table 5.4. Logistic parameter estimates for the progress of anthracnose on three varieties of sorghum with different levels of horizontal resistance (from Peacocke, 1995)

Year	*Site*	*Parameter*	*Sorghum variety*		
			Kuyuma	*MMSH 413*	*Sima*
1992/93	Mansa	Intercept	-13.10	-14.26	-12.12
		Point of inflection	89.67	100.10	143.27
		Rate	0.15	0.14	0.09
	Golden Valley	Intercept	-10.51	-10.62	-10.73
		Point of inflection	87.73	87.55	113.77
		Rate	0.12	0.12	0.09
1993/94	Mansa	Intercept	-10.04	-11.46	-10.28
		Point of inflection	112.82	115.25	110.80
		Rate	0.09	0.10	0.09
Mean		Intercept	-11.21	-12.11	-11.04
		Point of inflection	96.74	100.97	122.61
		Rate	0.12	0.12	0.09

Such rapid rates of epidemic progress as those reported by Peacocke (1995) have important ramifications for the determination of effective control strategies. Van der Plank (1963) considered that ‘in the long run, a high rate of interest is more important than a large balance in the bank today’. In other words, whereas a modest decrease in the rate of epidemic development may prevent significant yield losses, considerable efforts expended in sanitation practices which reduce the initial amount of inoculum may have little effect on disease control. Empirically, this is confirmed by the high levels of disease found in each of Peacocke's (1995) trials where, in certain cases, trials were planted on land following lengthy fallow periods of up to 25 years.

Although in Peacocke's trials the observed rates of disease progress differed significantly among cultivars, none showed complete immunity to infection. It is possible that genotypes carrying high levels of monogenic, race-specific resistance identified in earlier pathogenicity studies were absent from the breeding programme, or overlooked in the selection process. In the case of both Sima and Kuyuma, initial crosses had identified several genotypes showing immunity to anthracnose infection. As a result, it was apparent that despite high disease pressure in Peacocke's trials, in which cryptic error may have caused difficulties in the identification of resistance genotypes (*sensu* Van der Plank, 1963), durable rate-reducing resistance was observed during the trials.

In certain circumstances, horizontal resistance can delay the start of an epidemic. Since horizontal resistance reduces the percentage of spores that successfully infect, the first cycle of a polycyclic epidemic could be delayed in the same way as vertical resistance delays the start of an epidemic. Alternatively, by reducing the general level of disease, horizontal resistance may lower the level of initial inoculum surviving the winter to start a new disease cycle. Horizontal resistance may also reduce the progeny/parent ratio to 1 or less, this being below the level required for disease to increase.

5.3 VERTICAL RESISTANCE

Vertical resistance operates against a specific genetic component of the pathogen species but not against all and it is equated with the gene-for-gene concept of Flor (1955). Vertical resistance usually involves resistance mechanisms inherited monogenically or oligogenically. Such genes are matched, at least potentially, by corresponding genes for pathogenicity within the parasite population. Vertical resistance may be applied to both complete resistance and the components of incomplete resistance that interact differently with components of the pathogen population (Johnson, 1984). It has generally been considered that vertical resistance serves to reduce the infection frequency of inoculum, thereby blocking or delaying the potential epidemic (Van der Plank, 1963).

The advantages of vertical resistance to the breeder are that it is often easily identifiable, manipulated and incorporated into cultivars as a result of its simple inheritance and high levels of resistance. However, in selecting for such complete resistance, the genetic basis of resistance is narrowed, since other forms of resistance are hidden and may be lost in the selection process. Van der Plank (1963) termed this the 'vertifolia' effect. Typically, selection in the pathogen population has eventually led to the emergence of a pathotype virulent with respect to the genetic, vertical resistance of the host. This cycle of resistance identification, incorporation and breakdown has been termed the 'boom and bust' cycle (Priestley, 1978). Although examples of durable vertical resistance forms have been reported, durability may be influenced by the size of the pathogen population, the environment in which the host is grown, the mechanisms by which resistance is conferred and its genetic background (Parlevliet, 1979). Such factors are generally beyond the control of plant breeders.

The epidemiological effects of vertical resistance in its simplest form are clear. When the initial inoculum of a pathogen population is heterogeneous for pathogenic compatibilities, the plant's resistance will be effective against some of the races but not against others. In other words, the amount of inoculum capable of causing disease will have been vertically reduced. This results in a delay in the onset of disease. Classic examples of vertical resistance are frequently taken from crop-pathogen interactions where there is a well established gene-for-gene relationship. The potato variety Pentland Dell has the *Rl, R2* and *R3* genes for resistance to *P. infestans.* In the UK, the variety was first released in the early 1960s at a time when there were no compatible races of *P. infestans.* As a consequence, no late blight developed on the variety. During 1966, late blight was first recorded on this variety but only in a limited number of fields. This suggests that the initial selection of biotypes compatible with Pentland Dell occurred during or before the 1966 season. During the 1967 season race (1,2,3) was prevalent and the resistance within Pentland Dell was ineffective over large parts of the country. In 1968, this race became even more common and disease levels on Pentland Dell were similar to those on other potato varieties. During 1966 and 1967, when race (1,2,3) was rare in some potato fields, disease onset on Pentland Dell was delayed relative to that on other varieties and also delayed relative to the time of first appearance of symptoms on the same variety in 1968.

An illustration of the rate at which vertical resistance can be overcome is provided by Parry (1990) using the example of the spring barley cultivar, Triumph. A race of mildew with a virulence gene capable of overcoming the vertical resistance present in Triumph was identified in 1977, yet it was not until 1983 that the frequency of the gene had increased to the extent that major outbreaks of disease occurred. The rating for resistance of Triumph fell from 8 in 1983, when the popularity of the variety was at its peak, to 2 in 1985 (Table 5.5).

The resistance gene, *Mlg,* effective against powdery mildew, was the first major gene deliberately introduced into barley. The gene remained effective against mildew in Germany in the 1930s and 1940s while the area under cultivation to the variety containing the gene remained small. Beginning in the late 1940s the area under cultivation increased rapidly. This was followed by an increase in the frequency of the corresponding virulence gene, *Vg.* Eventually the *Mlg* varieties lost their mildew resistance as the pathogenicity gene became increasingly common (Wolfe, 1984).

The *Mlg* gene was widely used in European breeding programmes and consequently the matching pathogenicity, *Vg,* also became widespread. This increase in the frequency of *Vg* was partly through increased selection and partly through the movement of spores from areas where it was already common. It was only during the mid 1980s that there was a decline in the frequency of *Vg;* for over 50 years its frequency remained high because of the large proportion of spring barley crops possessing *Mlg* (Wolfe, 1984).

Table 5.5. Popularity and resistance of cv. Triumph spring barley to powdery mildew (from Parry, 1990)

Harvest year	*NIAB Recommended List resistance rating for mildew (1-9)*[a]	*Certified seed (t)*[b]	*% total barley market*[b]
1980	8	4609	2
1981	8	41329	13
1982	8	58370	20
1983	8	65480	22
1984	7	49049	19
1985	2	45164	16
1986	2	36091	14
1987	2	30204	12
1988	2	28232	10

[a] A high figure indicates a high degree of resistance

[b] From MAFF Seed Certification Scheme statistics

The history of the *Mlg* gene in barley is mirrored in most of the major gene introductions in barley and other crops. For example, the wheat cultivar Eureka, released as a rust-resistant variety in 1938, began to show signs of susceptibility in 1941; by 1942 it was becoming heavily rusted (Johnson, 1961). Arabica coffee is known to have five resistance genes to leaf rust (*SH1, SH2, SH3, SH4, SH5*); however, these have not provided durable resistance. Eskes (1983) observed that within three years of rust presence in Brazil, the resistance genes had lost their effectiveness.

Not every plant-pathogen interaction that is characterized by single dominant genes is liable to rapid breakdown of resistance. Examples of major gene resistance that have a proven durability include the single gene for resistance in cabbage which has not been overcome by the known races of *Fusarium oxysporum* f.sp. *conglutinans* (Fry, 1982) (Table 5.6).

Table 5.6. Examples of crop plants with single gene resistances against which pathogen races have evolved slowly or not at all (from Fry, 1982)

Crop	*Pathogen*
Cabbage	*Fusarium oxysporum*
Cucumber	*Cladosporium cucumerinum*
	Corynespora melonis
Maize	*Helminthosporium carbonum*
Oats	*Helminthosporium victoria*

More recently, van den Bosch and Gilligan (2003) have argued the need to redefine the criteria by which the durability of resistance is measured, since conventional models for this durability focus on the dynamics of the frequency of resistance genes. Consequently, the durability of resistance was defined as the time from the introduction of the cultivar to the time when the frequency of the virulence gene reaches a preset threshold (the loosely defined 'bust' of the boom and bust cycle). Van den Bosch and Gilligan (2003) argued for a more sophisticated approach by comparing three potential measures of durability: the time taken until the virulent genotype invades by mutation or immigration and then establishes itself within the population; the time taken for the virulent genotype to take over the pathogen population as measured by virulence frequencies, and the additional yield that might be expected based on the benefit accruing from uninfected host growth days. Based on computer model outputs, van den Bosch and Gilligan (2003) showed how these additional measures of durability actually depend on the interaction between population dynamics and population genetics and they suggested that these interactions can have major effects on the outcomes of resistance deployments.

5.4 CULTIVAR MIXTURES

Models describing the simplest mixtures of one susceptible and one immune plant genotype, responding to a single pathogen genotype (Leonard, 1969) have been termed classic models (Garrett and Mundt, 1999). In models of this type, one can envisage mixtures of two species, only one being a host to the same pathogen, or mixtures of two genotypes of the same species but with different race-specific resistance, with one component being immune to all local races. In this model Leonard (1969) predicted that the reduction in disease would follow as:

$$x'/x_0 = m^n \, x/x_0 \tag{5.1}$$

where x is the proportion of infected host tissue in a population composed only of the susceptible genotype, x' is the proportion of infected host tissue in the mixture, x_0 is the proportion of host tissue initially infected, m is the proportion of susceptible plants in the host mixture and n is number of generations of disease increase. As Garrett and Mundt (1999) illustrated, this means that the proportion of infected tissue for the susceptible genotype will be m^n times the proportion in a population composed solely of the susceptible genotype. For example, there should be approximately 12% of the disease on susceptible plants in a 50% susceptible mixture after three generations of pathogen increase and disease severity should decrease logarithmically as the proportion of resistant plants in the mixture is increased.

Much of the early work using cultivar mixtures concentrated on the resultant effects in mixtures of small-grain cereals infected by pathogens with resistance and pathogenicity both varying qualitatively. For example, Wolfe (1985) showed that in successful mixtures of spring barley cultivars there was an 80% reduction in powdery mildew compared with mean disease levels of the components of the mixture when they were grown as pure stands (see also Chapter 10). Less work has been reported on the effects of mixtures where resistance and pathogenicity vary quantitatively, or for crops other than small-grain cereals.

The use of cultivar mixtures against non-specialized pathogens has also been studied, though to a far lesser extent than with specialized pathogens such as rusts and powdery mildews. Against *Stagonospora (Septoria) nodorum*, Jeger *et al.* (1981) found that mixtures of winter wheat cultivars reduced the severity of disease almost to levels equivalent to those found in pure stands of a resistant cultivar (Table 5.7). Linders *et al.* (1996) found no reducing effect when mixtures of susceptible and partially resistant genotypes of *Plantago lanceolata* were exposed to epidemics of *Diaporthe adunca*. They suggested the reason for this discrepancy could be either that in their experiment disease incidence was assessed, whilst Jeger *et al.* (1981) measured disease severity, and/or that in the former study disease measurements were made when overall disease levels were still low.

Table 5.7. Observed percentage severity of Stagonospora nodorum *on component cultivars in mixed and pure stands (from Jeger* et al., *1981)*

Mixture	*Susceptible*	*Resistant*	*Mixture mean*
Pure susceptible			3.57
3 susceptible : 1 resistant	1.36	0.48	1.14
1 susceptible : 1 resistant	1.44	0.52	0.98
1 susceptible : 3 resistant	1.36	0.60	0.79
Pure resistant			0.52

As Garrett and Mundt (1999) have pointed out, mixture efficiency tends to be greater for certain epidemiological conditions than others. The effects of host diversity tend to be larger when the unit area of the host genotype is small (smaller plant types), when there is strong host specialization, when the dispersal gradient of the pathogen is shallow (air dispersed rather than splash-dispersed), when the pathogen lesion size is small (as in rusts and powdery mildews, for example) and when the pathogen has a high reproductive capacity with many generations per season. Overall then, cultivar mixtures should be expected to have less effect on diseases that are monocyclic, splash-dispersed or soil-borne. However, recent studies (Cox *et al.*, 2004) have shown that mixtures can function in the simultaneous control of such contrasting diseases as tan spot (residue/soil-borne, few disease cycles, splash-dispersed, steep dispersal gradient, limited cultivar specificity) and leaf rust (polycyclic, windborne, host specific) of wheat even though the degree of mixture efficiency was greater for leaf rust than for tan spot. The mechanisms by which disease reduction occurs for soil- and residue-borne pathogens are poorly understood. It is possible that resistance induced by avirulent spores may contribute to disease reductions in mixtures (Lannou *et al.*, 1995).

Van der Plank (1968) predicted a small effect of host diversity on epidemics in larger plants. This is because inoculum on small plants may be more effectively mixed throughout the host population rather than landing principally on the source host individual. Nonetheless, Garrett and Mundt (2000) showed a significant effect of host diversity on the severity of foliar symptoms of late blight in potato cultivar mixtures consisting of cv. Red LaSoda (susceptible) and a resistant breeding selection. The AUDPC was reduced by an average of 36 and 37% in consecutive years although the yield and the level of tuber infection increased and decreased respectively only in the first year. Similar trends were observed in experiments using inoculum sites throughout the field and sites restricted to one corner of the field. The mechanisms postulated for the observed effects included a reduction in the proportion of susceptible tissue, physical barriers to inoculum spread, or compensation or competition between host genotypes. It is probable that the resistant plants produced little inoculum as the epidemic proceeded on the susceptible plants resulting, overall, in a greatly reduced inoculum level in the mixtures (Garrett and Mundt, 2000). The size of potato plants apparently failed to produce significant autoinfection effects as inoculum was probably being spread far enough from source individuals. The plant size effect may have been further reduced by the intertwined nature of foliage growth (Garrett and Mundt, 2000).

Predictions of the likely extent of host diversity effects in specific host–pathogen systems can be made on the basis of the five major characteristics described by Garrett and Mundt (1999) (Table 5.8). Of course, the extent of the host diversity effect on disease progress will also be affected by additional factors including growth compensation, plant competition and competitive interactions between pathogen genotypes. Furthermore, environmental conditions and crop management activities will also influence the host diversity effect by imposing changes on the host density, epidemic length and epidemic intensity (Garrett and Mundt, 1999).

5.5. INDUCED RESISTANCE

At present, reference to induced resistance is fraught with the problems of semantics. Clarifications have been attempted (see for example, Kloepper *et al.*, 1992), but confusion remains. Induced systemic resistance (ISR) can best be described as: activation, by biotic or abiotic agents of a resistance dependent on the host plant's physical or chemical barriers. Systemic acquired resistance (SAR) is regulated through a distinct signal transduction pathway. Salicylic acid is implicated in SAR regulation, whereas jasmonic acid is implicated in ISR regulation (Graham and Leite, 2004).

The effectiveness of induced resistance under controlled environmental conditions now appears not to be under any doubt. Although successful induced resistance under field conditions has been demonstrated (see for example Reglinski *et al.*, 1994), as Lyon and Newton (1997) point out, the question is one of how well the treatment works under field conditions and what are the epidemiological consequences when the plants come under attack from a variety of pathogens and other stresses.

Table 5.8. Inherent characteristics that can be used to predict the host-diversity effect for reduced disease, showing whether selected host-pathogen systems possess these attributes (from Garrett and Mundt, 1999)

		Characteristic[a]				
Host	*Pathogen*	Small host genotype unit area	Shallow dispersal gradient	Small lesion size	Short pathogen generation time	Strong host specialization[b]
Coffee	*Hemileila vestatrix*	-	+	+	-	+
Pepper	*Xanthomonas campestris* pv. *vesicatoria*	-	-	-	+	+
Potato	*Phytophthora infestans*	-	+	-	+	+
Rice	*Magnaporthe grisea*	+	+	+	+	+
Wheat	*Blumeria graminis* f.sp. *tritici*	+	+	+	+	+
	Puccinia recondita	+	+	+	-	+
	Puccinia striiformis	+	+	-	-	+
	Mycosphaerella graminicola	+	-	-	-	-
	Rhizoctonia cerealis	+	-	-	-	-

[a] -, Host-pathogen system does not have the characteristic, so a host-diversity effect for reduced disease is less likely; +, host-pathogen system has the characteristic, so a host-diversity effect for reduced disease is more likely.
[b] High degree of host specialization in local pathogen populations.

To date there is little evidence of the effect of induced resistance on the natural spread of disease. Hervas *et al.* (1997) have reported the use of non-pathogenic *Fusarium oxysporum* as a seed treatment to control *Fusarium oxysporum* f.sp. *ciceris* on chickpea. They describe an increase in the incubation period from 35.4 days to 46.8 days and from 21.7 days to 46.5 days for the chickpea varieties ICCV 4 and PV 61, respectively, when the inoculum level of *F. oxysporum* f.sp. *ciceris* in the soil was 500 chlamydospores per g of soil. At higher inoculum concentrations, the treatments were less effective. Similarly, decreases in the final disease level and overall area under the disease progress curve were observed (Table 5.9). These results indicate a quantitative effect of induced resistance against soil-borne pathogens.

Table 5.9. Effect of seed treatment with non-pathogenic Fusarium oxysporum *on fusarium wilt in chickpea sown in soil infested with* F. oxysporum *f.sp.* ciceris *(from Hervas* et al., *1997)*

Cultivar	F. oxysporum *concentration (chamydospores g^{-1} soil)*	*Seed treatment*	*Disease assessment*		
			Incubation period	*Disease intensity*	*Area under the disease progress curve*
ICCV 4	500	Non-pathogenic *Fusarium*	46.8	20.5	0.07
	500	Control	35.4	73.7	0.32
	1000	Non-pathogenic *Fusarium*	29.7	100.0	0.61
	1000	Control	28.3	97.9	0.39
PV 61	500	Non-pathogenic *Fusarium*	46.5	4.3	0.01
	500	Control	21.7	75.2	0.36
	1000	Non-pathogenic *Fusarium*	30.3	90.8	0.45
	1000	Control	20.0	89.6	0.50

More detailed analyses of dose-response relationships in the amounts of pathogenic and non-pathogenic soil microorganisms and disease incidence levels, suggest that it is possible to differentiate between biological control agents that act via induced resistance and those that are effective as competitors for nutrients even though the end result, lower disease, may be the same. Such a relationship, involving non-pathogenic *Fusarium* species, against *Fusarium* wilt of tomato (Larkin and Fravel, 1999) has shown that the resistance-inducing microorganisms are effective at much lower doses than the competitors. This was reflected in the consequent impact on overall disease incidence, with inducing agents causing a greater reduction in the disease level compared to the competitors.

Jørgensen, *et al.* (1998) elegantly described the quantitative ability of the non-barley pathogens, *Bipolaris maydis* and *Stagonospora (Septoria) nodorum* to induce resistance in barley against *Drechslera teres*. Relative to a control pre-treatment with water, pre-treatment with both non-pathogens reduced the percent conidia forming appressoria, the percent conidia causing penetrations, the percent conidia forming intracellular hyphae and the percent conidia causing fluorescent epidermal cells. The percent conidia causing fluorescent papillae was increased. The enhancement of resistance was associated with an early defense reaction characterized by the increase in fluorescent papillae formation, and a late defense reaction linked to multicellular hypersensitive responses. The authors provided evidence in these induced responses for the expression of defense response genes that are involved in barley attacked by powdery mildew.

In one of the relatively few reports of the use of induced resistance under field epidemic conditions, Calonnec *et al.* (1996) showed that resistance in wheat induced by inducer races of *Puccinia striiformis*, applied two days prior to infection, reduced tiller disease severity due to virulent *P. striiformis* by 44-57%. However, the durability of this resistance needs to be established. Although it is possible that resistance elicitors act through the induction of horizontal resistance mechanisms, it is possible that the widespread use of elicitors could lead to some erosion of their effectiveness; however this is unlikely to lead to a sudden loss of effectiveness or to the production of a new pathogen 'race' (Lyon and Newton, 1997). Variety-elicitor interactions in the field have been demonstrated. Such differences in varietal response could be due to the uptake of the elicitor, the pathogen recognition mechanism, the control of resistance induction, the resistance mechanism available, or to the delivery of the resistance response (Lyon and Newton, 1997). The variability in varietal response does suggest that reaction to resistance elicitors could be selected for in a plant breeding programme.

5.6 NON-HOST IMMUNITY

At present it is not known whether there are any differences between the mechanisms or basis of resistance expressed by non-host plants and by resistant cultivars of the host (Heath, 1981). Immunity is definable only as resistance, of whatever kind, which renders most plants resistant to most pathogens and most pathogens avirulent to most plants (Browning, 1974). The fact that host specificity exists among parasites, and in most cases has not changed over recorded history, suggests that this form of resistance is not easily overcome (Heath, 1981). It is, therefore, not only absolute but also of great durability. For example, Bhat and Subbarao (2001) demonstrated that the immunity shown by broccoli (*Brassica oleracia* var. *botrytis*) against *Verticillium dahliae* is stable both with respect to host age and continued exposure of *V. dahliae* isolates to brassica hosts. These authors proposed that the mechanism of immunity was active since neither autoclaved broccoli leaves nor a broccoli extract had any effect on the growth of the pathogen in soil, or its ability to produce microsclerotia.

In theory, therefore, the exploitation of non-host immunity could provide an effective and lasting control of crop diseases (Peacocke, 1995). Non-host immunity may provide a significant contribution to resistance breeding as our knowledge of the mechanisms involved increases. There is only limited knowledge of the possibly complex additive and complementary nature of this form of resistance and of the possible influence of environmental criteria for its expression (Peacocke, 1995).

5.7 TOLERANCE

Although tolerance has been expressed in a number of ways (Posnette, 1969; Robinson, 1969; Schafer, 1971), by definition tolerant plants are susceptible to pathogen attack but show lower levels of yield loss than those expressed by other susceptible lines supporting the same levels of infestation (Browning, 1974;

Buddenhagen and de Ponti, 1983). Nelson (1973) suggested that tolerance negates infection by desensitization, whereas hypersensitivity negates infection through the localization of infection sites. Recently, more refined definitions of tolerance have been proposed. For ecologists, plant fitness, measured for example by flowering frequency and plant survival under epidemic conditions, have been used as indicators of tolerance. In natural plant communities, Roy *et al.* (2000) cited examples of small fitness consequences of infection despite levels of disease incidence ranging as high as 100%, suggesting low resistance and either high host tolerance or low pathogen virulence. Evidence for tolerance would be indicated if host genotypes that were exposed to the same pathogen isolate, and showed identical amounts of infection (and thus showing similar resistances), had different fitness consequences of infection (Roy *et al.*, 2000) (Fig. 5.1). Thus when disease pressure is high tolerance is likely to have physiological costs due to shifting allocation away from growth to re-growth (Simms and Triplett, 1994).

For agronomists and plant pathologists, an examination of the physiological manifestations of tolerance and the consequent impact on disease progress are usually more desirable. Thus, comparisons of citrus rootstocks have demonstrated that infection of young roots occurs very rapidly and at the same rate on susceptible and tolerant hosts, but that root rot damage, due to *Phytophthora nicotianae* and fungal reproduction are subsequently limited on tolerant hosts (Graham, 1995). A further refinement, for quantitative epidemiological studies was made by Dan *et al.* (2001) reporting the use of PCR diagnostics to identify tolerance in potato clones to *Verticillium dahliae*, where a clone was designated as tolerant if the amount of fungus present in the host tissue was equal to or greater than the collective average amount for all clones in the symptom category above.

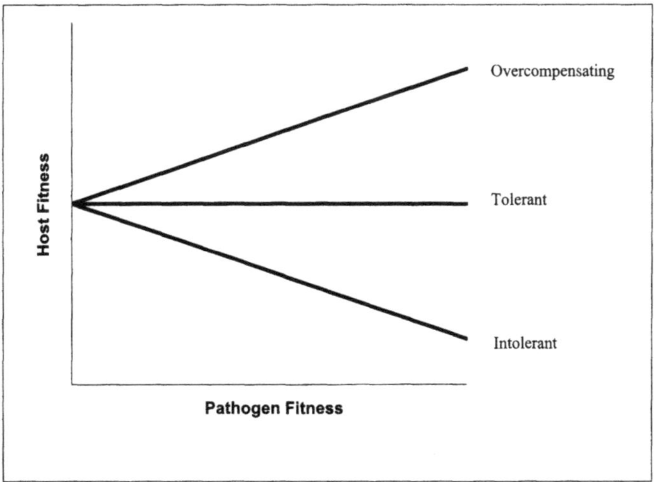

Figure 5.1. Norms of reaction diagram illustrating differences among three host genotypes in their ability to tolerate one pathogen strain. Each line represents an idealized regression line, the slope of which measures tolerance (adapted from Roy et al.*, 2000).*

In some legume crops it is possible to identify the mechanisms by which tolerance operates. Allen (1983) provided several examples of legume crops that have the ability to compensate for damage, to both vegetative and reproductive structures. Beans can compensate for infection, at or below ground level, by *Fusarium solani* f.sp. *phaseoli* through the production of adventitious roots. In other crops physiologically surplus foliage is an effective compensation mechanism. Pigeon pea varieties that have an indeterminate growth pattern can tolerate some infection damage caused by rust (*Uredo cajani*); varieties with a determinate growth pattern can suffer much more extreme yield losses. Similarly, alternately branched varieties of groundnut can compensate for defoliation caused by cercospora leaf spot but sequentially branched varieties cannot (Allen, 1983). Compensation for damage caused by disease is likely to have a physiological mechanism which is independent from 'true' resistance and the extent to which this compensation can be made will certainly be affected by the stage at which the damage occurs and the extent of that damage. Furthermore, it is likely to be a function of the partitioning and distribution of assimilates within the plant (Allen, 1983).

The physiological mechanisms underlying tolerance and the ability of certain crop varieties to produce high yields under disease severity levels that might be expected to constrain these yields have been examined in cereal cultivars. Zuckerman *et al.* (1997) found that under equivalent disease progress curve and disease severity levels the tolerant cultivar 'Miriam' showed significantly smaller losses in thousand kernel weight than the non-tolerant cultivar Barkai. In this host-pathogen system, it appears that the rate of carbon fixation per unit area of chlorophyll and per residual green leaf area of the tolerant cultivar was higher than in healthy plants. This enhancement of photosynthesis in residual green tissue compensated for the loss of photosynthesizing tissue due to septoria tritici blotch.

As is the case for horizontal resistance, arguments against the use of tolerance in crops stem from difficulties in breeding. In a practical sense, whether a cultivar yields well in the presence of disease due to tolerance or resistance makes little difference. However, where tolerance is operative and crop losses are disproportionately less than those commensurate with observed levels of infection, it is only possible to measure tolerance effectively by comparing symptom expression at the same stage of host development with the final yield of plants (Parlevliet, 1979). Since such assessments would prove both time-consuming and difficult, selection would not be possible among early segregating generations (Buddenhagen and de Ponti, 1983). As a result, breeding for disease resistance has, in the past, generally been limited to the selection of host resistance and little attention has been paid to the selection of host genotypes showing evidence of tolerance to infection. More recently however, with the introduction of molecular marker technologies, it has been possible to identify, for example, quantitative loci for tolerance to virus diseases (Jin *et al.*, 1998). This suggests that the process of selection, identification and transfer of genes controlling complex traits such as tolerance to disease into new lines, could be accelerated.

REFERENCES

Allen, D.J. (1983) *The Pathology of Tropical Food Legumes,* John Wiley, Chichester, 413 pp.

Bennett, F.G.A. (1984) Resistance to powdery mildews in wheat: a review of its use in agriculture and plant breeding programmes. *Plant Pathology,* **33**, 279-300.

Bhat, R.G. and Subbarao, K.V. (2001) Reaction of broccoli to isolates of *Verticillium dahliae* from various hosts. *Plant Disease*, **85**, 141-146.

Browning, J.A. (1974) Relevance of knowledge about natural ecosystems to development of pest management programmes for agro-ecosystems. *Proceedings of the American Phytopathological Society,* **1**,191-199.

Buddenhagen, I.W. and de Ponti, O.M.B. (1983) Crop improvement to minimise future losses to diseases and pests in the tropics. *FAO Plant Protection Bulletin,* **39**, 385-409.

Calonnec, A., Goyeau, H. and de Vallavielle-Pope, C. (1996) Effects of induced resistance on infection efficiency and sporulation of *Puccinia striiformis* on seedlings in varietal mixtures and on field epidemics in pure stands. *European Journal of Plant Pathology,* **102**, 733-741.

Casela, C.R., Frederiksen, R.A. and Ferreira, A.S. (1993) Evidence for dilatory resistance to anthracnose in sorghum. *Plant Disease*, **77**, 908-911.

Caten, C.E. (1974) Intra-racial variation in *Phytophthora infestans* and adaptation to field resistance for potato blight. *Annals of Applied Biology,* **77**, 259-270.

Cox, C.M., Garrett, K.A., Bowden, R.L. *et al.* (2004) Cultivar mixtures for the simultaneous management of multiple diseases: tan spot and leaf rust of wheat. *Phytopathology*, **94**, 961-969.

Dan, H., Ali-Khan, S.T. and Robb, J. (2001) Use of quantitative PCR diagnostics to identify tolerances and resistance to *Verticillium dahliae* in potato. *Plant Disease*, **85**, 700-705.

Eskes, A.B. (1983) Qualitative and quantitative variation in pathogenicity of races of coffee leaf rust (*Hemileia vastatrix*) detected in the state of Sao Paulo, Brazil. *Netherlands Journal of Plant Pathology,* **89**, 31-35.

Flor, H.H. (1955) Host-parasite interaction in flax rust - its genetics and other implications. *Phytopathology,* **45**, 680-685.

Fry, W.E. (1982) *Principles of Plant Disease Management,* Academic Press, New York, 378 pp.

Garrett, K.A. and Mundt, C.C. (1999) Epidemiology in mixed host populations. *Phytopathology*, **89**, 984-990.

Garrett, K.A. and Mundt, C.C. (2000) Host diversity can reduce potato late blight severity for focal and general patterns of primary inoculum. *Phytopathology*, **90**, 1307-1312.

Graham, J.H. (1995) Root regeneration and tolerance of citrus rootstocks to root rot caused by *Phytophthora nicotianae. Phytopathology*, **85**, 111-117.

Graham, J.H. and Leite, R.P. (2004) Lack of control of citrus canker by induced systemic resistance compounds. *Plant Disease*, **88**, 745-750.

Guzman-N, J. (1964) Nature of partial resistance of certain clones of three *Solanum* species to *Phytophthora infestans. Phytopathology,* **54**, 1398-1404.

Hakiza, G.J. (1997) Characterisation of the epidemiology of coffee leaf rust caused by *Hemileia vastatrix* on robusta coffee *(Coffea canephora)* in Uganda. PhD Thesis, University of Reading, UK.

Heath, M.C. (1981) Non-host resistance, in *Plant Disease Control* (ed. R.C. Staples), Wiley, New York, pp. 201-220.

Hervas, A., Landa, B. and Jimenez-Diaz, M. (1997) Influence of chickpea genotype and *Bacillus* sp. on protection from Fusarium wilt by seed treatment with non- pathogenic *Fusarium oxysporum. European Journal of Plant Pathology,* **103**, 631-642.

Jeger, M.J., Jones, D.G. and Griffiths, E. (1981) Disease progress of non-specialised fungal pathogens in intraspecific mixed stands of cereal cultivars. II. Field experiments. *Annals of Applied Biology,* **8**,199-210.

Jin, H., Domier, L.L., Kolb, F.L. and Brown, C.M. (1998) Identification of quantitative loci for tolerance to barley yellow dwarf virus in oat. *Phytopathology*, **88**, 410-415.

Johnson, R. (1984) A critical analysis of durable resistance. *Annual Review of Phytopathology,* **22**, 309-330.

Johnson, T. (1961) Man-guided evolution in plant rusts. *Science,* **133**, 357-362.

Jørgensen, H.J.L., Lubeck, P.S., Thordal-Christensen, H. *et al.* (1998) Mechanisms of induced resistance in barley against *Drechslera teres. Phytopathology*, **88**, 698-707.

Kloepper, J.W., Tuzun, S. and Kuc, J.A. (1992) Proposed definitions related to induced resistance. *Biocontrol Science and Technology*, **2**, 349-351.

Kong, G.A., Simpson, G.B., Kochman, J.K. and Brown, J.F. (1997) Components of quantitative resistance in sunflower to *Alternaria helianthi. Annals of Applied Biology,* **130**, 439-451.

Lannou, C., de Vallavieille-Pope, C. and Goyeau, H. (1995) Induced resistance in host mixtures and its effect on disease control in computer-simulated epidemics. *Plant Pathology*, **44**, 478-489.

Larkin, R.P. and Fravel, D.R. (1999) Mechanisms of action and dose-response relationships governing biological control of *Fusarium* wilt of tomato by non-pathogenic *Fusarium* spp. *Phytopathology*, **89**, 1152-1161.

Leonard, K.J. (1969) Factors affecting rates of stem rust increase in mixed plantings of susceptible and resistant oat varieties. *Phytopathology*, **59**, 1845-1850.

Linders, E.G. A., Van Damme, J.M.M. and Zadoks, J.C. (1996) Epidemics of *Diaporthe adunca* in experimental and in natural populations of *Plantago lanceolata* and the effect of partial resistance on disease development. *Plant Pathology,* **45**, 70-83.

Lyon, G.D. and Newton, A.C. (1997) Do resistance elicitors offer new opportunities in integrated disease control strategies? *Plant Pathology,* **46**, 636-641.

Nelson, R.R. (ed.) (1973) *Breeding Plants for Disease Resistance: Concepts and Applications,* Pennsylvania University Press, University Park, 401 pp.

Parlevliet, J.E. (1979) Components of resistance that reduce the rate of epidemic development. *Annual Review of Phytopathology,* **17**, 203-222.

Parlevliet, J.E. (1992) Selecting components of partial resistance, in *Plant Breeding in the 1990s,* (eds H.T. Stalker and J.P. Murphy), CAB International, Wallingford, UK, pp. 281-302.

Parry, D.W. (1990) *Plant Pathology in Agriculture,* Cambridge University Press, Cambridge, 385 pp.

Peacocke, B.J. (1995) Epidemiology and management of anthracnose from sorghum. PhD Thesis, University of Reading, UK.

Posnette, A.F. (1969) Tolerance of virus infection in crop plants. *Review of Applied Mycology,* **48**, 113-118.

Priestley, R.H. (1978) Detection of increased virulence in populations of wheat yellow rust, in *Plant Disease Epidemiology,* (eds P.R. Scott and A. Bainbridge), Blackwell, Oxford, pp. 63-70.

Reglinksi, T., Newton, A.C. and Lyon, G.D. (1994) Assessment of the ability of yeast-derived resistance elicitors to control barley powdery mildew in the field. *Journal of Plant Diseases and Protection,* **101**, 1-10.

Robinson, R.A. (1969) Disease resistance terminology. *Review of Applied Mycology,* **48**, 593-606.

Roy, B.A., Kirchner, J.W., Christian, C.E. and Rose, L.E. (2000) High disease incidence and apparent disease tolerance in a North American Great Basin plant community. *Evolutionary Ecology*, **14**, 421-438.

Schafer, J.F. (1971) Tolerance to plant disease. *Annual Review of Phytopathology,* **2**, 235-252.

Simms, E.L. and Triplett, J. (1994) Costs and benefits of plant response to disease: resistance and tolerance. *Evolution*, **48**, 1933-1945.

van den Bosch, F. and Gilligan, C.A. (2003) Measures of durability of resistance. *Phytopathology*, **93**, 616-625.

Van der Plank, J.E. (1963) *Plant Diseases: Epidemics and Control,* Academic Press, New York, 349 pp.

Van der Plank, J.E. (1968) *Disease Resistance in Plants,* Academic Press, New York, 206 pp.

Van der Plank, J.E. (1971) Stability of resistance to *Phytophthora infestans* in cultivars without R genes. *Potato Research,* **14**, 263-270.

Wolfe, M.S. (1984) Trying to understand and control powdery mildew. *Plant Pathology,* **33**, 451-466.

Wolfe, M.S. (1985) The current status and prospects of multiline cultivars and variety mixtures for disease resistance. *Annual Review of Phytopathology,* **23**, 251-273.

Zuckerman, E., Eshel, A. and Eyal, Z. (1997) Physiological aspects related to tolerance of spring wheat cultivars to Septoria tritici blotch. *Phytopathology*, **87**, 60-65.

CHAPTER 6

DISPERSAL OF FOLIAR PLANT PATHOGENS: MECHANISMS, GRADIENTS AND SPATIAL PATTERNS

H.A. McCARTNEY, B.D.L. FITT AND J.S. WEST

6.1 INTRODUCTION

Dispersal has long been recognised as fundamental to the development of plant disease epidemics, for without dispersal many epidemics would fail to progress. In recent years, agriculture, especially in the developed world, has come under increasing pressure to produce crops in sustainable and environmentally friendly ways. Consequently, there is an urgent need for more efficient disease management systems. Understanding the temporal and spatial dynamics of disease epidemics is crucial to the development of such systems. For example, when developments in precision agriculture lead to spatially targeted crop spraying, there will be a need to understand and predict disease patch dynamics. The role of dispersal in gene-flow within plant pathogen populations is little understood, but may be crucial to understanding fungicide resistance breakdown or the distribution of alleles conferring virulence within populations. Knowledge of dispersal processes is also needed to understand movement of new pathogens into a landscape, for example the introduction of exotic pathogens into a country or the movement of pathogens due to climate change. Thus, knowledge of dispersal will increasingly be needed by policy makers devising plant health protection strategies. It is evident that now, perhaps more than ever, there is a need to understand the nature and scope of the dispersal of plant pathogen propagules.

Plant pathogen propagules include fungal spores, virus particles and bacteria (cells and spores) and there are many different mechanisms by which each can be dispersed from infected host plants. For example, many plant viruses are dispersed by insect vectors; insects, birds and farming activities can spread both bacteria and fungal spores. Many soil-borne pathogens can be spread in ground water or by agricultural operations (see also Chapter 14). However, many economically important crop diseases are caused by foliar fungal pathogens, for which the main routes of dispersal are wind-borne (Chapter 15) or splash-borne (Chapter 16) spores. The scale of dispersal by these processes ranges from a few centimetres for some spores spread by rain-splash up to hundreds of kilometres for some spores carried by the wind. This chapter will concentrate on the spread of such pathogens.

For foliar pathogens, disease spread is the direct consequence of spore dispersal, although spatial patterns of disease may be quite different from the spore dispersal

B. M. Cooke, D. Gareth Jones and B. Kaye (eds.), The Epidemiology of Plant Diseases, 2nd edition, 159–192.

patterns which cause them. This is partly because spore dispersal is a short term phenomenon compared to most other stages of disease development. For example, conidia of *Pyrenopeziza brassicae*, the cause of light leaf spot on oilseed rape (*Brassica napus* ssp. *oleifera*), take about 18 hours to germinate under optimum conditions, while splash dispersal of conidia over typical distances of 20-30 cm takes less than one second and wind dispersal of ascospores of *P. brassicae* over 100 m takes 1-2 minutes. Even for long distance dispersal, such as for tobacco blue mould (www.ces.ncsu.edu/depts/pp/bluemold/) that can spread from Cuba to the southern USA (Aylor, 1999) or cereal rusts in the USA or India (Hamilton and Stakman, 1967; Nagarajan and Singh, 1990), dispersal events (hours or days) may be short compared with infection processes. Disease patterns are often the result of many individual dispersal events from many sources over periods of days or even weeks. Environmental and biological factors that affect infection and disease development can add further complications. The development of real epidemics is, therefore, a complex process; as Waggoner wrote in 1962 "we shall find the real epidemic muddy and uncomfortable" (quoted by Gregory, 1973). This chapter aims to make the water a little clearer by considering the underlying processes which govern spore dispersal, examining the relationship between disease and spore dispersal gradients, presenting some examples of field studies and considering how disease spread can be modelled.

6.2 UNDERLYING MECHANISMS: SPORE DISPERSAL

This section will discuss dispersal from the point of view of the spore; from the source (lesion, pustule, fruiting body) to the new host (or loss to the ground or non-host surface). It is important to appreciate these 'primary' physical mechanisms of spore dispersal as a foundation for understanding the spread of disease epidemics.

6.2.1 Dispersal by wind

Winds are highly variable in both time and space (McCartney and Fitt, 1985; Aylor, 1990). This variability or turbulence causes individual spores, released from the same source under the same wind conditions, to follow different paths and travel different distances. Therefore, as spore plumes disperse downwind from sources their concentrations in the air decrease. The decreases in concentration are frequently referred to as 'concentration gradients' (Gregory, 1973). Mean wind speed characteristics above crops are fairly well understood; wind speeds increase with height depending on the nature of the crop (height, architecture, density) and the stability of the atmosphere (temperature profile) (McCartney and Fitt, 1985). For example, in neutrally stratified atmospheres when buoyancy effects can be neglected, over open terrain with uniform vegetation, wind speed $u(z)$ increases logarithmically with height z (Grace, 1977; Monteith and Unsworth, 1990):

$$u(z) = 2.5u_* \ln\left(\frac{z-d}{z_0}\right) \quad (6.1)$$

The constant u_*, the friction velocity, scales the wind speed and defines the amount of turbulence; z_0, the roughness length, scales the height and d is a datum level less than the crop height called the zero plane displacement. Equation 6.1 predicts that $u = 0$ at height d but the equation is not valid within a crop. For most crops, z_0 is an order of magnitude smaller than the height of the crop h, and d is between 0.6 and $0.8h$ (Thom, 1975). On sunny days when there is convective activity (unstable temperature lapse rate) or in the evening when atmospheric mixing is suppressed (stable temperature lapse rate), the wind profile deviates from equation 6.1 (Thom, 1975; Grace, 1977; McCartney and Fitt, 1985). The wind profile is logarithmic only with well formed surface boundary layers over large uniform areas. Wind profiles near obstructions such as hedges or near changes in terrain, for example woodland boundaries, may be more complex than suggested by equation 6.1. However, the general form of the profile is usually similar and the equation can be used as an approximation to estimate wind speeds at crop height from local synoptic measurements (usually made at heights of 10 m). The vertical turbulence mixing scales (the dispersive ability) above crops also increase with height (Thom, 1975). This can be parameterised as a diffusion coefficient K that increases with height. For a neutral atmosphere, K is proportional to the friction velocity u_*, but the relationships are more complex for stable and unstable conditions (Thom, 1975).

Mean wind profiles within crops depend greatly on crop architecture, particularly the vertical distribution of foliage and the size, shape and density of leaves. For crops where the leaves are relatively uniformly distributed with height, such as cereals, wind speed profiles can often be estimated using the equation (Cionco *et al.*, 1963):

$$u(z) = u(h)\exp\left[a\left(\frac{z}{h} - 1\right)\right] \quad (6.2)$$

where h is the crop height and the attenuation coefficient a has a value between 0.3 and 3, depending upon crop type and leaf area density. In crops with a 'canopy and stem' structure (e.g. orchards), the wind speed profile may be S-shaped, with wind speeds greater in the 'stem' layer than the 'canopy' layer (Grace, 1977). It has been suggested that turbulent diffusivity is more or less constant within the upper two-thirds of crop canopies (Thom, 1975) and decreases towards the ground level. However, wind speeds within crops are highly intermittent (Aylor, 1990), with air flow at relatively low speeds interspersed with sporadic bursts at high speeds with increased turbulence. This produces highly skewed distributions of wind speed, with a long 'tail' of high wind speeds (gusts) occurring at low frequencies (Shaw *et al.*, 1979; Shaw and McCartney, 1985). These high wind speeds are caused by relatively large eddies which penetrate the canopy (Aylor, 1990). The corollary of gust penetration is the gradual ejection of air from the canopy after gusts, which can be responsible for the upward transport of spores. In another phenomenon, called outward interactions, air moves upwards at speeds greater than the local average wind speed (Aylor, 1990). These complex flow patterns have consequences for spore removal and transport within and out of the canopy. Unfortunately, current models of air flow within crop canopies do not describe these phenomena well and new models are needed to improve understanding of spore dispersal close to sources (Aylor, 1990). However, recent advances in modelling in fluid dynamics may begin to address these questions.

Wind not only transports spores but also removes them from infected plants. Although many fungi have evolved active spore release mechanisms to eject spores directly into the air (Lacey, 1986), spores of a large number of foliar pathogens are simply passively blown or shaken off their hosts. To remove spores, the aerodynamic or mechanical forces generated by wind must overcome the forces holding the spore to the host surface (Aylor and Parlange, 1975). The wind speeds needed to remove spores are not known for many fungi but can be relatively large (Grace, 1977). Conidia of *Blumeria* (*Erysiphe*) *graminis* f.sp. *hordei* (cause of barley powdery mildew), which form in chains above the leaf surface, were released by wind speeds greater than 0.5 m s^{-1} (Hammett and Manners, 1974) and conidia of *Drechslera maydis* (cause of southern leaf blight of maize) were removed only by wind speeds of more than 5 m s^{-1} (Aylor, 1975). The wind intermittency observed in crop canopies probably plays an important role in spore removal because it is only in gusts that wind speeds are large enough to remove spores (Aylor, 1978; Aylor *et al.*, 1981). The importance of gusts in the removal of conidia of *Passalora personata* (cause of late leaf spot of groundnut) has been demonstrated in wind tunnel experiments (Wadia *et al.*, 1998). Wind gusts can also remove bacterial cells (*Pseudomonas syringae*) and spores (*Bacillus subtilis*) from leaves on which they had previously been sprayed (Lighthart *et al.*, 1993). Spores can also be dislodged by shaking (Bainbridge and Legg, 1976); thus wind gusts may indirectly remove spores by moving the crop canopy. The removal of spores by gusts of wind has important implications for dispersal, particularly within crop canopies.

The turbulent nature of wind causes a dilution in the concentration of a spore plume as it moves down-wind. Within crops, concentrations are also depleted by deposition of spores on the crop and the ground. Spores can be deposited by gravitational settling and inertial impaction (Legg and Powell, 1979). The rate at which spores settle onto surfaces, S, is proportional to the spore fall speed, V_s, and the spore concentration above the surface, C ($S = CV_s$). V_s is generally in the range 0.1 to 3 cm s^{-1} for most fungal spores (Gregory, 1973). Deposition by impaction I is dependent on C and wind speed u ($I = CuE$); the constant of proportionality, E (the impaction efficiency), increases with increasing spore size and wind speed but decreases with increasing width of the impaction surface (Chamberlain, 1975; Aylor, 1982). When spores are released in wind gusts their inertial impaction efficiency is enhanced because they are travelling at relatively high wind speeds (Aylor, 1978; Aylor *et al.*, 1981), especially in crop canopies where mean wind speeds are small (McCartney and Bainbridge, 1987). Simulations using random walk models (see below) suggest that this enhanced inertial impaction can steepen gradients when plume depletion is predominantly by deposition (Legg, 1983).

However, Aylor (1987) suggests that deposition will determine the shape of concentration gradients only when wind speeds in the canopy are low and when turbulence is slight. Thus, for spores released in gusts the effects of enhanced turbulence on diffusion may be much greater than the enhanced deposition by inertial impaction. This conclusion was supported by an analysis of deposition gradients of wind-dispersed urediniospores of *Puccinia recondita* (cause of brown rust of wheat) (Aylor, 1987). The escape of spores from a canopy into the atmosphere also depends on the relative effects of turbulent transport and deposition and may depend on how they are released (Aylor, 1990). Spores that are removed passively from leaves are usually released only when turbulence and vertical mixing are large; such conditions also favour escape from the canopy. Vertical transport may be affected by spore size; concentrations of ascospores of *P. brassicae* decreased less quickly with height above an oilseed rape crop than concentrations of larger oilseed rape pollen grains (McCartney, 1990a). Once released into the atmosphere, spores have the potential to disperse over large distances and heights. Fungal spores have been found in the atmosphere at heights of 500-1000 m above the north sea many kilometres from any potential sources (Hirst *et al.*, 1967) and spores and pollen from South America have been found in air samples taken in Antarctica (Marshall, 1996). Long distance aerial transport of inoculum has been cited as the probable mechanism for the invasion of disease into new territories, although such events are rare (Brown and Hovmøller, 2002). Long-distance transport of particles may be enhanced by natural events such as bush fires, which were implicated in spread of viable bacteria and fungal spores over 1450 km from Yucatan to Texas and from south east Asia to Hawaii (Mims and Mims, 2004).

Three dimensional time-averaged spore concentration or deposition patterns round a point source (infected plant) are complex. However, average concentrations measured in one direction away from the source decrease monotonically with distance. These dispersion gradients have been described by a number of different equations (McCartney and Fitt, 1985; Fitt and McCartney, 1986; Minogue, 1986; Fitt *et al.*, 1987; Aylor 1990). Two of the most commonly used are a negative exponential equation:

$$C = C_0 \exp(-\alpha x) \tag{6.3}$$

And an inverse power law equation:

$$C = Ax^{-\beta} \tag{6.4}$$

where C is the concentration or deposition rate, x is the distance from the source and C_0, α, A and β are constants. The coefficients α and β determine the rate of decrease in spore concentration (or deposition) with distance.

Although the two functions have similar shapes, there is a fundamental difference between them. The exponential equation implies a constant length scale (C decreases by the same proportion over equal distances), while the inverse power law implies a length scale which changes proportionally with distance. The fixed length scale for the exponential function leads to the concept of a half-distance $d^1/_2$ (= 0.693/α), the distance in which C decreases by half, as an easy method for visualising gradients (McCartney and Fitt, 1985; Fitt and McCartney, 1986). The length scale predicted for an exponential model usually reflects the distance over which it is measured (Aylor, 1990) and should not be used to infer concentrations beyond this distance. Ferrandino (1996) showed that dispersal length scales for a larger number of data sets can be estimated from only the geometrical extent of the measurements. The proportional increase in length scales with distance characteristic of a power law is similar to one of the characteristics of turbulent diffusion, where larger and larger eddies act on a cloud of spores as they diffuse down-wind (Aylor, 1990). Thus the power law equation may be more appropriate for describing dispersal over longer distances when the effects of atmospheric turbulence are dominant (Brown and Hovmøller, 2002). However, the power law equation predicts an infinite concentration at $x = 0$, a physically impossible result that can cause problems when applying the equation at distances close to the source. Mundt and Leonard (1985) modified the equation by effectively introducing a virtual source upwind of the actual source:

$$C = A(c + x)^{-\beta} \tag{6.5}$$

They interpreted c as related to the size of the source (lesion, leaf, plant). Equations 6.4 and 6.5 converge when x is much larger than c.

Studies comparing the fit of power law and exponential models to spore deposition and disease gradients (Gregory, 1968; Fitt *et al.*, 1987; Ferrandino, 1996) have shown that in many cases both models fitted observed gradients equally well but sometimes one or other fitted better. For physical reasons, a negative exponential model is appropriate when plume depletion is predominantly by deposition (low turbulence within crop canopies) but an inverse power law is appropriate when turbulent diffusion is dominant. Bullock and Clarke (2000) have suggested a combined exponential and power law equation for wind-borne seed dispersal:

$$C = A(ae^{-\alpha x} + bx^{-\beta}) \tag{6.6}$$

that attempts to allow for two different components of the dispersal process: a steep short distance gradient and a flatter long distance gradient. The mixed model fitted the tail of the seed deposition gradients of *Calluna vulgaris* (heather) and *Erica cinerea* (bell heather).

Little systematic work has been done to relate dispersal gradient parameters to wind characteristics, other weather factors, crop structure or spore characteristics. Aylor and Ferrandino (1990) proposed a simple equation to describe deposition in crop canopies close to the source of spores, based on the assumption that, close to the source, spores travel in straight lines radially away from the source:

$$D = \gamma\left(1 + \frac{x^2}{L^2}\right)^{-1} \tag{6.7}$$

Where γ is related to the strength of the source of spores and L is a length scale characteristic of the plant. The model adequately predicted the spread of phaseolus bean rust (causal agent *Uromyces appendiculatus*) within 3 m of the source but relied on averaging times long enough to give fairly uniform dispersal round the source.

Two or three dimensional empirical or semi-empirical models have rarely been used to describe plant pathogen dispersal. Tufto *et al.* (1997) compared exponential and Weibull distribution models to two semi-empirical models using data for maize pollen dispersal. One of the semi-empirical models was based on Brownian diffusion and the other included a wind threshold for pollen release and a wind direction. The Brownian diffusion model can be summarised as:

$$C(x,y) = \frac{1}{2\pi r\sqrt{\gamma}}.\exp\left\{1/\gamma(\tau_x x + \tau_y y) - \sqrt{1/\gamma + \tau_x^2/\gamma^2 + \tau_y^2/\gamma^2 r}\right\} \tag{6.8}$$

Where $C(x, y)$ is the concentration or deposition rate at position x, y from the source and r is the distance from the source ($r^2 = x^2+y^2$). The coefficient γ is related to the variance of the distribution and τ_x and τ_y are related to the mean wind speeds in the x and y directions. The other model, also based on the idea of Brownian diffusion, can be written as:

$$C(x,y) = \frac{\sqrt{\lambda}}{\sqrt{2}(\pi r)^{2/3} I_0(\kappa)}.\exp\{-\lambda r + \kappa\cos(\theta - \theta_0)\} \tag{6.9}$$

Where r and θ are the polar co-ordinates of x and y ($r^2 = x^2+y^2$ and $\theta = \tan^{-1}(y/x)$. I_0 is a modified Bessel function of the first kind. The coefficient λ is related to the threshold wind speed, κ is related to the variance of the distribution and θ_0 to the mean wind direction. Both semi-empirical models fitted the maize pollen deposition data better than either exponential or Weibull models, partly because these two models could not describe the effects of wind direction. Although both semi-empirical models could describe experimental data better, they are relatively difficult to fit to measured data. As with other empirical approaches, it is difficult to estimate the values of model parameters *a priori*.

Three dimensional patterns of mean pollutant concentrations emitted from point and line sources can be estimated using Gaussian Plume models (Pasquill and Smith, 1983), which assume that concentration profile distributions are Gaussian in both cross-wind and vertical directions. The standard deviations of the cross-wind (σ_y) and vertical (σ_z) distributions are functions of distance and determine the shape of down-wind gradients (McCartney and Fitt, 1985; Fitt and McCartney, 1986). For a ground level point source, concentrations at the ground are given by:

$$C = \frac{Q}{u} \cdot \frac{\exp(-y^2 / \sigma_y^2)}{\pi \sigma_z \sigma_y} \tag{6.10}$$

where Q is the rate of release of material and y is the distance from the centre-line of the plume. As σ_y and σ_z are functions of x which depend on atmospheric turbulence (Pasquill and Smith, 1983; McCartney and Fitt, 1985; Fitt and McCartney, 1986), this equation operates like an inverse power law. Equation 6.10 does not account for the loss of spores by deposition. However, Gregory (1973) and Chamberlain (1953) accounted for spore deposition in their Gaussian plume models by decreasing the source term, Q, exponentially with distance downwind of the source of spore dispersal. These 'source depletion' models assume that the spores are well mixed within the plume, so that the cross-wind and vertical profiles remain Gaussian, and ignore effects of deposition to the ground on vertical concentration profiles. This can overestimate surface concentrations and deposition rates at large distances (Horst, 1977).

Within crop canopies, deposition occurs within the body of the plume and the assumptions of source depletion models may be more appropriate, especially while the plume is confined within the canopy. McCartney and Fitt (1985) suggested a method for calculating the source depletion term within crops from estimates of deposition rates and showed that, for a homogeneous crop with a uniform wind profile, $Q(x)$ can be described by an exponential equation. Aylor (1989) developed a source depletion model with two equations: one describing dispersal wholly within the crop and one describing dispersal when part of the plume had escaped from the crop. Both models contain terms which behave in an exponential manner to describe deposition and terms which behave in an inverse power manner to describe dispersal. Gaussian plume models cannot accurately predict spore dispersal when wind speeds and deposition rates vary with height (as in most crops). There is also little information on how σ_y and σ_z should be formulated within crops. Nevertheless, Gaussian plume models are useful for understanding the spatial development of epidemics because they are easy to formulate and incorporate into disease epidemic simulations.

Gaussian plume models were the basis for early atmospheric pollutant dispersal models (Pasquill and Smith, 1983) and much effort has been put into deriving appropriate parameters for their use. A number of current atmospheric dispersal models that are based on the Gaussian plume approach are used in air pollution regulation and in emergency planning (Caputo *et al.*, 2003). Examples of such models are AEROMOD, developed for the USA Environmental Protection Agency for regulatory purposes (USEPA, 1999) and PCCOSYMA, developed by the Forschungszentrum Karlsruhe for the National Radiological Protection Board to help assess the environmental impact of radiological accidents (Brown and Ehrhardt, 1999). This type of model can be used to predict atmospheric dispersal over mesoscale distances, but, as far as we are aware, has not yet been used for plant pathogen inoculum dispersal. However, in the early 1980s Gaussian plume models were developed in the UK to assess risk of aerial transmission of cattle Foot and

Mouth Disease (FMD) (Blackall and Gloster, 1981; Gloster, 1983a) and Newcastle Disease in poultry (Gloster, 1983b). This model was also used to investigate the 2001 UK FMD outbreak (Mikkelsen *et al.*, 2003).

Several studies of potential long distance aerial transport of plant pathogens have used air parcel trajectory analysis to establish links between source and receptor regions (Aylor, 1986; Davis, 1987). Trajectory analysis is a standard tool in the study of air pollutant movement and it tracks the movement of air parcels using information on wind fields and atmospheric temperature structures (Davis, 1987; Stohl, 1998). It is widely used in air pollution studies and computational methods and applications have been reviewed by Stohl (1998). Back-trajectory analysis of wind contributed to evidence for long distance dispersal of exotic *Bacillus* bacteria 1800 km from the black sea to Sweden, where the species was isolated from red-pigmented snow (Bovallius *et al.*, 1978). Web-based trajectory models are available from the USA National Ocean and Atmosphere Administration (HYSPLIT model http://www.arl.noaa.gov/) and the British Atmospheric Data Centre (NERC Centre for Atmospheric Sciences, http://badc.nerc.ac.uk/community/). Trajectory modelling can account for large scale movement of air parcels due to wind direction changes and track air movements over large distances, although errors propagated during the calculations must be carefully considered (Kottmeier and Fay, 1998). Trajectory models can also be adapted to take into account particle dispersal within the air parcel (Aylor, 1986; Davis, 1987; Aylor, 1999). In this approach, the dispersing spores are treated as an expanding 'puff' travelling along the path of the trajectory. The expanding 'puff' can be treated in a similar manner to the Gaussian plume models, where the vertical (z), cross-wind (y) and along-wind (x) concentrations in the puff are assumed to follow Gaussian distributions, unless constrained by an atmospheric boundary, such as an inversion (Aylor, 1986, 1999). The spore concentration (C) in the puff is:

$$C(x,y,z,t) = \frac{2Q(t)}{(2\pi)^{3/2}\sigma_x\sigma_y\sigma_z}.\exp\left[\frac{-(x-U_c t)^2}{2\sigma_x^2}\right].\exp\left[\frac{-y^2}{2\sigma_y^2}\right].\exp\left[\frac{-z^2}{2\sigma_z^2}\right] \quad (6.11)$$

where $Q(t)$ is the proportion of released spores that are viable and still air-borne at time after release t; σ_x, σ_y and σ_z are the standard deviations of the distributions in C in the along-wind, cross-wind and vertical directions. U_c is a constant transport speed representative of the atmospheric layer in which the spores are travelling. The form of $Q(t)$ can take account of loss of spores by deposition by sedimentation (dry deposition), rainfall (wet deposition) and loss in viability (Aylor, 1999). Davis and Main (1986) used a similar approach to investigate the spread of tobacco blue mold (causal agent *Peronospora tabacina*) in south-eastern USA. This led to the development of the North American Plant Disease Forecast Centre, North Carolina State University, Raleigh, NC, that provides an internet-based disease risk forecasting system for tobacco and cucurbit growers (Main *et al.*, 2001). The scheme has been operating since 1996 and uses the NOAA HYSPLIT trajectory model to calculate potential inoculum dispersal from areas where the disease is known to be present and estimates the risk that disease will develop in other areas. A Gaussian puff model (RIMPUFF) was one of the dispersal models used to investigate the UK 2001 FMD outbreak (Mikkelsen *et al.*, 2003). Four atmospheric dispersion models (AEROMOD, HPDM, PCCOSYMA and HYSPLIT), based on Gaussian plume or Gaussian puff principles, have recently been reviewed by Caputo *et al.* (2003).

Other physical models developed to describe spore and pollen dispersal (McCartney and Fitt, 1985; Fitt and McCartney, 1986; Aylor, 1990; Aylor *et al.*, 2003; Di Giovanni and Kevan, 1991) have not been widely used in spore dispersal or disease progress studies. This is perhaps because they are perceived to be complex and need detailed micro-meteorological information to operate them. However, recent advances in computer technology and in mathematical computing make such models more accessible for general use. Two modelling approaches based on the physical description of the dispersal process have been used to model spore dispersal: Eulerian Advection-Diffusion models (EAD) (gradient transfer or K-theory models) and Lagrangian Stochastic models (LS) (random walk models), (McCartney and Fitt, 1985; Fitt and McCartney, 1986; Aylor, 1990, 1999; McCartney, 1997).

EAD models assume that atmospheric diffusion is analogous to molecular diffusion and obeys Fick's Law (i.e. the rate of diffusion is proportional to the concentration gradient of the diffusing material). The approach can be illustrated by considering dispersal from an infinite line source orientated at a right angle to the prevailing wind direction and releasing particles continuously. For an infinite line source, cross-wind diffusion (y-direction) can be ignored, and along-wind diffusion (x-direction) is assumed to be negligible relative to down-wind transport. The model is based on the number balance of particles entering or leaving small volumes of air (i.e. dispersed spores are neither created nor destroyed). The difference between the rates at which particles enter or leave the volume horizontally (by wind, LHS in Equation 6.12) is balanced by the difference between the rates at which particles enter or leave the volume vertically (by diffusion and sedimentation, terms 1 and 2, RHS in Equation 6.12) and the rate of loss of particles by deposition onto surfaces within the volume (term 3, RHS in Equation 6.12). This can be expressed by the differential equation:

$$u\frac{\partial C(x,z)}{\partial x} = \frac{\partial}{\partial z}(K_z\frac{\partial C(x,z)}{\partial z}) + v_s\frac{\partial C(x,z)}{\partial z} + S(x,z) \tag{6.12}$$

where $C(x, z)$ is the particle concentration at height z and distance x down-wind of the source. The wind speed u determines the rate of horizontal advection of spores into and out of the volume. The rate of vertical diffusion is determined by K_z, the vertical diffusion coefficient. The rate of settling of particles depends on the fall speed, v_s, and $S(x, z)$ defines the rate of removal of particles by deposition. Additional terms can be added to account for loss of viable spores by spore death or wash-out by rain (Aylor 1999). EAD models can be formulated for point and area sources (Yao *et al.*, 1997; Aylor, 1999) and include terms describing horizontal diffusion and advection.

To solve EAD equations, it is necessary to define diffusion coefficients (K), wind speeds and the spore removal term (S) at all points in the model volume and to define appropriate boundary conditions at the edges of the volume (i.e. at the ground and some upper limit in the atmosphere). K and wind field values may be derived from micro-meteorological measurements, or theoretically estimated for different atmospheric flow conditions (McCartney and Fitt, 1985; McCartney, 1997; Yao *et al.*, 1997; D'Amours, 1998). $S(x, z)$ has to account for deposition by both sedimentation and inertial impaction (McCartney and Fitt, 1985; Aylor, 1986, 1990; Yao *et al.*, 1997). In validations against observed measurements of the dispersal of fungal spores (Legg and Powell, 1979; Aylor and Ferrandino, 1989), the model over-estimated concentrations close to the source, through either underestimation of diffusion or enhancement of deposition by gusts. EAD models make the assumption that the length scales of vertical movement are small compared with the size of the plume, which is generally invalid within crops, and cannot easily accommodate the effects of gustiness (Aylor, 1990). However, this type of model can give useful results for spore dispersal over distances far enough from the source such that the dominant turbulent eddies are small compared with the vertical width of the plume (Aylor, 1999). For example, EAD models are suitable for calculating dispersal from the downwind edge of a field (Yao *et al.*, 1997), providing the initial spore profile can be defined. EAD models have recently been combined with models of atmospheric thermo-hydrodynamics to calculate pollutant dispersal over regional scales taking into consideration orographic and thermal inhomogeneities and terrain effects (Aloyan, 2004; D'Amours, 1998).

Lagrangian stochastic (LS) atmospheric dispersal models employ a different approach to Gaussian plume or EAD models. LS models simulate the paths of individual air parcels or particles as a pseudo-random walk using the turbulence statistics of the air flow. Particle trajectories are represented by a series of discrete steps determined partly by a correlation between successive velocities (representing velocity 'memory') and partly by a random component to simulate turbulent fluctuations. The formulation of LS models (reviewed by Wilson and Sawford, 1996) can be illustrated by considering dispersal from an infinite line source where cross-wind diffusion can be ignored (McCartney and Fitt, 1985; Aylor, 1990, 1999; Aylor *et al.*, 2003; Jarosz *et al.*, 2004). Particle trajectories are simulated as a series of short straight-line segments, each representing the motion of the air parcel over a short time step dt. If the horizontal and vertical speeds of the air parcel at the start of the step are u and w, the horizontal (dx) and vertical (dz) displacements of a spore contained in the air parcel are (Aylor *et al.*, 2003; Jarosz *et al.*, 2004):

$$dx = (u + du)dt \quad \text{and} \quad dz = (w + dw - v_s)dt \tag{6.13}$$

where

$$du = a_u dt + b_u d\xi_u \quad \text{and} \quad dw = a_w dt + b_w d\xi_w \tag{6.14}$$

where coefficients a_u, b_u, a_w and b_w are functions of velocity and position, and $d\xi_u$ and $d\xi_w$ are random numbers selected from a Gaussian distribution with mean 0 and variance *dt*. The variable v_s, the fall speed of the spore, is included to account for gravitational settling. The first term in the expressions for *du* and *dw* contains information on the velocity 'memory' of the air parcel and the second term represents fluctuations caused by turbulence. The formulation of the coefficients depends on the nature of the flow (Legg 1983; Wilson and Sawford, 1996; Wilson and Flesch, 1997).

LS models are particularly useful for estimating the dispersal of spores close to sources within plant canopies or close to the ground (Aylor, 1989, 1999). Consequently they have been used to estimate the escape of *Venturia inaequalis* (cause of apple scab) ascospores from ground cover (Aylor and Flesch, 2001) and *Phytophthora infestans* (cause of potato late blight) sporangia from a potato crop (Aylor *et al.*, 2001). More recently LS models have been used to investigate the dispersal of pollen from maize crops (Aylor *et al.*, 2003; Jarosz *et al.*, 2004) to assess the risk of gene flow from transgenic to conventional crops. Because LS models simulate the flight of 'individual' spores, they have the potential to account for the effects of wind gusts on spore dispersal. LS model predictions suggest that, within crops, deposition by impaction could be enhanced by gust release (Legg, 1983) and above crops the concentration and deposition curves will be displaced down-wind by an amount proportional to the speed of the gusts (McCartney, 1990b). The success of LS models depends on the accurate parameterisation of air flow and turbulence, which can be difficult in complex air flows such as at crop boundaries. Jarosz *et al.* (2004) found that an LS model tended to underestimate maize pollen deposition within 10 m of the boundary between the crop and bare soil. They attributed this to an incorrect parameterisation of turbulence in the transition between the crop and the surroundings. Since LS models used for spores and pollen ignore effects of particle inertia, they are applicable only to particles less than c. 300 μm in diameter (Walklate, 1987; Wilson, 2000). However, they have been adapted to describe the application of agricultural sprays in orchards (Walklate, 1992; Xu *et al.*, 1998). There has recently been interest in use of LS models for pollutant dispersal in the convective boundary layer (Raza *et al.*, 2001; Oettl *et al.*, 2001; Franzes, 2003) and as understanding of atmospheric flow improves the accuracy and applicability of LS based models should increase.

Atmospheric dispersal models are becoming more sophisticated as understanding of mechanisms of atmospheric flow increases. Atmospheric dispersal models are being developed that can account for not only dispersal processes but also changes in topography and surface characteristics (e.g. Aloyan, 2004; Wang and Ostoja-Starzewski, 2004). Such models, although complex, have the potential to enhance understanding of spore and pollen dispersal within realistic landscapes. This information is needed to understand gene flow in fungal pathogen communities, for example, the movement of fungicide resistance or virulence genes. Computational fluid dynamic systems, developed to calculate air flows in complex terrains such as around buildings, are now being coupled with dispersal models to investigate pollutant dispersal in urban or industrial landscapes (Riddle *et al.*, 2004). Such techniques can potentially help in the understanding of inoculum dispersal within plant communities, particularly at crop boundaries.

6.2.2. Dispersal by rain

Rain or spray irrigation can remove spores from infected leaves in run-off water or in splash droplets (Fitt *et al.*, 1989; Madden, 1992, 1997). The spores of many plant pathogens can be dispersed only by water because they are contained in mucilage which prevents dispersal by wind (Gregory, 1973; Fitt *et al.*, 1989). However, raindrop impacts can also dislodge 'dry' spores from leaves to allow them to be dispersed by wind (Wadia *et al.,* 1998; Geagea *et al.,* 2000). For example, wind tunnel and field observations on late leaf spot of groundnut suggest that rain may play an important role in releasing conidia of the causal agent, *P. personata*, into the air.

Water splash directly removes spores from leaf surfaces by incorporating them into splash droplets. Such droplets can travel more than a metre from the point of impact but most travel only a few centimetres (Fitt *et al.*, 1989; Madden, 1992, 1997). Consequently, dispersal gradients for splash-dispersed spores are generally much shorter than those for wind-dispersed spores. Splash can also transport inoculum vertically and can play an important role in vertical disease movement, for example in cereals (Fitt *et al.*, 1989; Shaw and Royle, 1993) and oilseed rape (Pielaat *et al.*, 2002). The effectiveness of splash in removing spores depends on the size and velocity of the incident drop and on the orientation and mechanical properties of the surface, but the physical mechanisms involved are not well understood.

Raindrop size influences both the removal of spores (Gregory, 1973; Fitt and McCartney, 1986; Fitt *et al.*, 1989; Madden, 1992) and the distances of dispersal (Yang *et al.*, 1991; Yang *et al.*, 1992; Butterworth and McCartney, 1992). Large raindrops are more effective than small ones; they remove more spores and splash them further (Fitt *et al.*, 1988). Raindrops less than 0.5 mm in diameter contribute little to direct dispersal by splash but can contribute to the wetting of leaf surfaces (Madden, 1992). Spore removal and dispersal are dependent on the force of impact (Walklate, 1989) or the kinetic energy (Yang *et al.*, 1991) of the incident water drops. Therefore, large slow-moving drops dripping from leaves may remove spores as effectively as small raindrops falling at their terminal velocity. Sensors that respond to the kinetic energy of impacting drops have been developed to estimate the 'splash dispersal potential' of rainfall (Madden *et al.,* 1998; Lovell *et al.*, 2002).

The potential for dispersal by rain-splash depends partly on the size distribution of the raindrops (Fitt *et al.*, 1989; Walklate, 1989; Walklate *et al.*, 1989), which depends on the type of rainfall (Ulbricht, 1983). For example, spores of *Septoria tritici* (anamorph of *Mycosphaerella graminicola,* cause of septoria tritici blotch of wheat) were splashed from the base of a wheat canopy to the upper leaves only during heavy summer showers (Shaw and Royle, 1993). The texture, angle and flexibility of leaves and other surfaces in the canopy all influence the splash process and the amount of water splashed (Huber *et al.*, 1997; Madden, 1997; Ntahimpera *et al.*, 1999). This may then affect the energy imparted to the droplets and dispersal distances. Distances to which spores are dispersed also depend on the effects of canopy structure and density and mulching material, if present, on splash drop trajectories (Fitt *et al.*, 1992; Madden, 1997). The efficiency with which spores are incorporated into splash droplets also affects spore dispersal gradients and the number of spores carried per droplet is influenced by spore size (Fatemi and Fitt, 1983; Fitt and Lysandrou, 1984; Brennan *et al.*, 1985). The incorporation of spores into splash drops can be modelled as the product of three functions of the droplet diameter, which may take similar forms for a range of pathogens: the diameter frequency distribution, the proportion of droplets carrying spores and the mean number of spores in each number category (Huber *et al.*, 1996).

Most splash droplets that carry spores are very much larger than wind-dispersed spores and are therefore affected less by turbulence. When effects of turbulence are small, splash droplet trajectories can be computed using conventional Newtonian dynamics (Macdonald and McCartney, 1987). The trajectory of a splash droplet is determined by solving the equation of motion which defines its velocity v:

$$m\frac{dv}{dt} = F_g + F_A + F_D + F_a \tag{6.15}$$

where m is its mass, F_g is the force of gravity, F_A is a buoyancy force, F_D is the drag force and F_a is a force due to acceleration of the droplet. Functional representations are available for each of these forces and only the initial velocity (speed and direction) and the local wind speed are needed to calculate the droplet trajectory using iterative procedures. However, few studies of splash dispersal from leaves (Reynolds *et al.*, 1987; Macdonald and McCartney, 1988; Yang *et al.*, 1991; Yang and Madden, 1993; Ntahimpera *et al.*, 1999) have measured initial velocities of splash droplets. Initial speeds of up to 10 m s^{-1}, with median values of 2 m s^{-1}, have been measured for splash droplets dispersed from field bean (*Vicia faba*) or barley leaves (Macdonald and McCartney, 1988). The greatest speeds were associated with small droplets (less than 250 μm diameter) while the largest droplets (greater than 900 μm) rarely reached speeds greater than 3 m s^{-1}. Average initial velocities of droplets less than 1.2 mm in diameter measured from splashes on a number of different surfaces were consistent (~1 m s^{-1}) but there was greater variability for larger droplets (Ntahimpera *et al.*, 1999). The angle of ejection, which is important in determining the droplet trajectory, depends on the flexibility, angle and texture of the leaf. For flexible leaves, many droplets are ejected downwards or parallel to the leaf surface (Macdonald and McCartney, 1988).

Physics-based modelling of dispersal of inoculum in splash droplets in plant canopies is complex due to the absence of detailed information on the splash process. Pietravalle *et al.* (2001) used a mechanistic trajectory model to predict maximum splash height from impacting drop kinetic energy for droplets splashed from a horizontal water surface. The model assumed that droplet trajectories could be approximated as parabolas. Model parameters were estimated from controlled indoor experiments and validated against natural rain events. Maximum splash heights were well predicted. If the structure of the model is assumed to be unaffected by the nature of the splash surface then, as a first approximation, the model could be used to predict droplets splashing from leaves. In a more complex approach, Saint-Jean *et al.* (2004) developed a framework for modelling splash in a three-dimensional plant canopy. The model used a combination of stochastic and mechanistic methods and included initial velocity and droplet diameter distributions and a Newtonian approach to droplet trajectory modelling. Splash dispersal patterns simulated by the model compared well with measured patterns for simple artificial canopies consisting of vertical cylinders. Such complex models may give insights into the important factors governing splash dispersal, but are probably too complex for routine use in disease risk assessment.

The distances travelled by primary splash droplets splashing directly from lesions are affected by crop canopy structure, position of the lesion in the crop, the nature of the water source (rain type, irrigation) and the leaf surface, and the wind speed. Crop canopy structure affects the deposition of splashed droplets and the potential for spread by secondary splash (Madden, 1992, 1997). Dispersal gradients of droplets splashed from strawberries were steeper with a straw surface on the ground than with a polyethylene surface (Yang and Madden, 1993). Dispersal gradients of *Oculimacula yallundae* (teleomorph of *Pseudocercosporella herpotrichoides,* cause of eyespot disease of cereals) conidia were steeper in a uniform wheat seedling canopy than in wheat intercropped with clover, suggesting that the clover was acting as a source of secondary splash (i.e. conidia previously splashed from the wheat) (Soleimani *et al.*, 1996). Thus, duration of exposure to rain and rain intensity may modify 'primary splash' gradients. The spread of the bacterial disease, citrus canker (*Xanthomonas axonopodis* pv. *citri*), in Florida is enhanced by strong winds simultaneous with rain in storms or hurricanes. Disease spread from a source to the nearest newly diseased tree within a 30-day period was estimated at up to 3.5 km (Gottwald *et al.*, 2002).

Primary dispersal is dominant at the beginning of a rain shower while the canopy is being wetted and the initial source of inoculum is being dispersed. Most of the bacteria that cause citrus canker were released within the first 10 minutes in simulated wind-driven rain experiments (Bock *et al.,* 2005). However, as rain duration continues, secondary spread may begin to be important as previously splashed inoculum is transported by further splashes. If the rain persists for sufficient time to deplete the source, inoculum deposited may be lost by wash-off. This conclusion is supported by the results of experiments on the effects of rain intensity (increasing intensity for fixed time durations is equivalent to increasing duration for fixed intensities) on the dispersal of spores of *Colletotrichum acutatum* (cause of black spot or anthracnose) from strawberry fruit (Madden *et al.*, 1996). After an initial increase, disease incidence declined with increasing intensity and the intensity at which maximum infection occurred decreased with longer periods of exposure.

Splash dispersal is a complex process that is difficult to model, particularly if the effects of secondary splash and wash-off are to be incorporated. Since dispersal may involve more than one event, Yang *et al.* (1991) developed a diffusion model to describe effects of multiple splashing on dispersal gradients. The effective diffusion coefficient was derived from knowledge of the primary splash dispersal patterns and the spore loss defined by a rate parameter. The approach was able to describe both the increase and decrease in deposition with time at a fixed point and the decrease in deposition with distance at a fixed time. Pielaat and van den Bosch (1998) used a 'random jump' modelling approach to describe the effects of secondary splash. The model was based on three parameters: the probability per unit time that a spore is splashed; a spore dispersal distance probability function (primary gradient function); and the probability that the spore is not lost from the system. Compared to the diffusion approximation, this approach appears to be more realistic for short and medium time scales, although the two models converge at long time scales.

6.3 SPORE DEPOSITION AND DISEASE GRADIENTS

6.3.1 Relationship between spore deposition and disease gradients

For both wind and splash-dispersed plant pathogen inoculum, deposition rates decrease with distance away from the inoculum source. Under environmental conditions favourable to infection, dispersed inoculum will produce further infections on susceptible hosts. The disease pattern that develops will also show a decrease in disease with increasing distance away from the source, i.e. a disease gradient. Disease gradients can also result from gradients in host or environmental factors but these will not be considered in this chapter. The observation of a gradient, therefore, implies the existence of a local source of inoculum, since background inoculum from a large number of distant sources produces a uniform distribution of disease with distance across a crop (Gregory, 1968, 1973). Vertical disease gradients (i.e. disease decreasing with height) can also be observed when inoculum sources are at ground level, for example with black pod of cocoa (caused by *Phytophthora megakarya*) (McCartney and Fitt, 1998). Disease gradients produced by splash-dispersed inoculum are usually steeper than those produced by wind-dispersed inoculum, reflecting the differences in dispersal length scales between the two mechanisms (Fitt and McCartney, 1986).

Gradients of monocyclic or polycyclic diseases in crops can provide much information about the role of the wind-dispersed or splash-dispersed pathogen spores in the development of epidemics. Monocyclic diseases produce only primary disease gradients, in which all the lesions arise from the same inoculum source. For example, gradients of the phoma leaf spot stage of stem canker (causal agent *Leptosphaeria maculans*) can be produced by the wind-borne ascospores in winter oilseed rape crops in the autumn (Gladders and Musa, 1980). However, spores of pathogens causing monocyclic diseases may be released over long periods of time so that the disease gradients gradually become less steep as the growing season progresses. This may explain why gradients of wheat eyespot in inoculated winter wheat plots became less steep with successive observations, although removal of inoculum suggested that there was no secondary disease spread (Rowe and Powelson, 1973).

Many studies on disease gradients have been done with polycyclic diseases spread by wind-dispersed spores, such as potato late blight (Minogue, 1986), yellow rust of wheat (Zadoks and Schein, 1979) or powdery mildew of barley (Welham *et al.*, 1995). Typically, such diseases are first observed in a crop as primary disease foci resulting from a single lesion; by the time a yellow rust focus 1 m^2 in diameter is observed, four pathogen generations of infection, latent period and sporulation have occurred (Zadoks and Schein, 1979). Initially disease gradients away from these foci are steep but spores which escape from the crop canopy soon establish secondary foci; primary disease gradients become more shallow as foci expand and, with the expansion of secondary foci disease, disease is soon distributed uniformly across the crop (Gregory, 1968, 1973).

Less work has been done with polycyclic diseases spread by splash-dispersed spores, such as wheat glume blotch (caused by *Phaeosphaeria nodorum,* teleomorph, *Stagonospora nodorum*, anamorph; Jeger *et al.*, 1983) or barley leaf blotch (caused by *Rhynchosporium secalis*; McCartney *et al.*, unpublished). The concept of isopaths, to describe the spread of these diseases in time and space, was introduced by Berger and Luke (1979) and reviewed by Minogue (1986). Often the rate of spread of isopaths was faster for pathogens spread by wind-dispersed spores (e.g. *P. infestans*, 3-4 m day^{-1}) than for pathogens spread by splash-dispersed spores (e.g. *S. nodorum*, 0.3 m day^{-1}).

Spread of other polycyclic diseases involves both wind-dispersed ascospores, which initiate epidemics at the beginning of the growing season, and splash-dispersed conidia, responsible for subsequent cycles of disease spread. For example, initial horizontal gradients of white leaf spot (causal agent *Mycosphaerella capsellae*) in winter oilseed rape (McCartney and Fitt, 1998) are caused by wind-dispersed ascospores but subsequent horizontal spread and vertical spread up the crop canopy is achieved by splash-dispersed conidia, with an estimated 9-13 pathogen generations per season (Inman, 1993). A similar pattern of disease spread is observed for septoria tritici blotch in winter wheat crops (Shaw, 1987). Other pathogens with both ascospores and conidia are apparently monocyclic because either the ascospores (e.g. *O. yallundae*) or the conidia (e.g. *L. maculans*; anamorph *Phoma lingam*) seem to play little part in epidemics in practice.

6.3.2 Measurement of gradients

To measure a spore dispersal or disease gradient in a natural or experimental situation, measurements of spore numbers per m^3 (spore concentration gradient) or per m^2 (spore deposition gradient) (C) or disease incidence or severity (Y) at different distances (x) from the source are needed. Spore numbers can be estimated with artificial samplers but the choice of sampler and timing of sampling depend on the size of the spores, their mode of dispersal and concentration and the objective of the investigation (Fitt and McCartney, 1986; McCartney *et al.*, 1997). Generally, samplers need to be simple and easy to use because the measurement of gradients requires the use of at least 10-20 identical samplers simultaneously. To measure spore deposition gradients, passive samplers such as horizontal slides under rain-shields for wind-dispersed spores or beakers for splash-dispersed spores can be used. Concentration gradients can be measured with volumetric samplers such as rotorods; vertical sticky cylinders can be used effectively for large spores, such as those of *B. graminis*. Rain-activated switches can be used to confine sampling to periods of rainfall for spores released by rain (Fitt *et al.*, 1989) and sampling can be confined to specific times for spores with known diurnal or seasonal periodicities (Fitt and McCartney, 1986). Conventional spore samplers usually use microscopy to quantify concentration or deposition. This can be time-consuming and often restricts the number of samples that can be collected. Spore sampling methods based on the use of serological or molecular pathogen diagnostics are being developed (McCartney *et al.*, 2003; Ward *et al.*, 2004). A spore sampler that collects spores directly into microtitre wells, developed by the Burkard Manufacturing Company (Rickmansworth, U.K.) (Wakeham *et al.*, 2004), has been used to quantify *Mycosphaerella brassicicola* (cause of ringspot of cabbage) inoculum using monoclonal antibodies (Kennedy *et al.*, 2000). PCR-based diagnostics have also been used to detect air-borne inoculum of oilseed rape pathogens (Calderon *et al.*, 2002; Freeman *et al.*, 2002).

The disease component of disease gradients has been measured as numbers of lesions, numbers of infected leaves, numbers of infected plants, the percentage leaf area affected or the percentage of the population of plants which is affected (Fitt *et al.*, 1987). For most diseases, only some of these measurements are appropriate; for example sorghum downy mildew (causal agent *Peronosclerospora sorghi*) infects plants systemically so cannot be assessed on individual leaves and barley leaf blotch lesions merge so that they cannot be assessed individually (Bock *et al.*, 1997). When there is a choice of methods available, the selection of the method of assessment may be influenced by the objectives of the work. Furthermore, it should be appreciated that whether disease is measured as incidence or severity affects the form of the gradient since incidence and severity gradients measured simultaneously can have different slopes (Minogue, 1986). The advent of molecular diagnostic methods (McCartney *et al.*, 2003; Ward *et al.*, 2004) could change our view of crop disease assessment. For example, such methods have the potential to quantify the amount of pathogen DNA, and perhaps the biomass, in leaf tissue (see also Chapter 2).

The choice of distances (x) at which to assess spore numbers or disease is influenced by the geometry of the source, the scale of the gradient and the objectives of the investigation. Sources can be classified as point, line or area sources (Gregory, 1968, 1973). Point sources may be individuals or small groups of infected plants; ideally a point source should have a diameter of less than 1% of the length of the gradient although the diameter is often 5-10% of the length in practice (Zadoks and Schein, 1979). Line sources may be hedges containing infected alternative host plants or strips of a susceptible cultivar. Area sources may be infected fields, although such area sources may become point sources if the distance over which the gradient is measured is kilometres rather than metres. Gradients from sources above ground level are generally less steep than those from ground level sources.

For measuring gradients a minimum of five distances should be sampled. Ideally there should be at least 10 distances, selected on a logarithmic scale with more samples near to the source for gradients within crops, and more distances for gradients between crops or over longer distances. Generally, it is easiest to measure gradients over short distances within crops, although gradients have been measured over km distances, for example down river valleys (Gregory, 1968; Fitt *et al.*, 1987). It may also be possible to observe gradients within crops from above using remote sensing techniques, such as infra-red aerial photography (Lacey *et al.*, 1997) and optical techniques from satellite or tractor-mounted platforms (West *et al.,* 2003; see also Chapter 2). Long distance transport of spores over 100s of kilometres is important in the spread of some epidemics, such as black stem rust (caused by *Puccinia graminis*) in North America; such spores can be sampled at heights of 1000 m with samplers on aircraft (Gregory, 1973) but are generally dispersed by wind and deposited by rain in relatively uniform clouds so that gradients are not observed.

When plotted on a linear scale, spore dispersal and disease gradients are generally hollow curves, which are difficult to compare. Therefore, to compare gradients, the empirical negative exponential (Equation 6.3) or inverse power law (Equation 6.4) models are generally log-transformed to give the forms for disease (Y):

$$\ln(Y) = \ln(Y_0) - \alpha x \qquad (6.16)$$

and

$$\ln(Y) = \ln(A) - \beta \ln(x) \qquad (6.17)$$

When the models are fitted in these forms, linear regression can be used to estimate parameters to describe and compare gradients. When disease gradients are fitted by an exponential model they can also be expressed as a half-distance (McCartney and Fitt, 1985; Fitt and McCartney, 1986). If disease incidence is expressed as the proportion of individual plants affected, then a multiple infection transformation must generally be used to account for multiple infections of the same plant by different spores (Gregory, 1973):

$$N_i = N_t(1 - e^{-Y_i / N_t}) \qquad (6.18)$$

This allows calculation of the probable number of infections Y_i that occurred when N_i leaves are diseased out of a total of N_t plants or leaves. Although only one new infection among 100 plants is required to increase the percentage of plants diseased from 1 to 2%, 69 new infections are required to increase it from 98 to 99%. A problem which arises in using these transformed models is that log-transformation cannot be used if the value of Y or C is zero. This can be overcome by adding a small quantity to each value (Gregory, 1968), which can distort the gradient, or by more complex procedures, including non-linear transformations (Minogue, 1986). Ideally, there should be sufficient measurements so that the gradient can be truncated at a distance before zero values occur.

Field experiments to study horizontal or vertical spore dispersal or disease gradients are difficult to design and to do. The experimenter has to contend with the unpredictability in the occurrence, direction and strength of the wind or rain. Consequently, some experiments on dispersal have been done in controlled wind tunnel or rain tower conditions (Fitt *et al.*, 1986); results obtained with these model systems have then been compared with those obtained in field crops. In controlled conditions, it is possible to replicate treatments in time but this is rarely feasible under natural conditions. Decisions about replication, the size and shape of experimental plots and of the inoculum source, the isolation of plots, the sampling positions and the frequency and methods of assessment, are often difficult to reconcile, especially if the objective is to compare treatments (e.g. cultivars or fungicides) or to study development of gradients with time. However, knowledge about the mode of spore dispersal (by wind or rain-splash) and thus the scale of the gradient to be measured helps in making these decisions.

Experiments with wind-dispersed pathogens, such as *Puccinia polysora* (cause of maize rust; Cammack, 1958), *P. infestans* (Minogue, 1986) or *B. graminis* (Welham *et al.*, 1995) have used point sources of inoculum to study horizontal dispersal and disease gradients. When inoculum has been placed in the centre of plots, disease assessments have been made at points on concentric circles around the source. By contrast, Minogue (1986) placed inoculum at the end of long thin plots and measured the potato blight gradients along the plots; this method does not provide any information about two-dimensional disease spread and is only practical if the prevailing wind direction is reasonably constant. An alternative method is to use a line source of inoculum, perpendicular to the prevailing wind direction, and measure the gradients along transects parallel with the wind direction (Vloutoglou *et al.*, 1995). If several transects through the same plot are used as replicates, they are not true replicates because values are more highly correlated than they would be in separate plots and experimental errors are underestimated. For splash-dispersed pathogens, such as *S. tritici* or *R. secalis*, it is not necessary to use such large plots and the problems with point sources of inoculum are fewer than for wind-dispersed pathogens, because the scale of dispersal is smaller. The isolation distance between plots or area of non-susceptible crop between plots required to prevent cross-contamination is also smaller. However, it is still advisable to include uninoculated control plots to assess levels of background contamination on the site.

The physical process of disease assessment may influence the dispersal of the disease, particularly for diseases spread by readily detached wind-borne spores. There can therefore be a conflict between the need to sample frequently enough to obtain useful disease progress information and the need not to disturb the crop. The use of long thin plots (Minogue, 1986), paths along transects (Vloutoglou *et al.*, 1995) or a ladder attached to a central pivot which can be rotated above plots (Welham *et al.*, 1995) can minimise the effects of disturbance to the crop on disease spread. For crops with relatively open canopies, such as maize, the risks of disturbing plants during the assessment are less than for crops with closed canopies (e.g. potatoes, barley) or with plants at high densities (e.g. linseed). For splash-dispersed pathogens, assessments can be done when leaves are dry to minimise the spread of pathogen spores.

Studies on vertical disease gradients for splash-dispersed pathogens have sometimes involved assessment of disease on successive leaf layers up plants in the crop (white leaf spot on oilseed rape; Inman, 1993) or sampled from the crop (septoria tritici blotch on wheat; Shaw, 1987). Vertical spore dispersal gradients for wind-borne spores have generally been studied with volumetric spore samplers, such as rotorods, placed at different heights above the crop (McCartney, 1990a; McCartney and Lacey, 1990; Vloutoglou *et al.*, 1995). The work on *Alternaria linicola* (seedling blight; Vloutoglou *et al.*, 1995) in linseed is one of the few studies which have combined successive measurements of both spore concentration gradients and disease gradients in the same crop. It clearly demonstrates the effects of background inoculum and the way in which secondary spore dispersal can flatten primary spore dispersal and disease gradients with time.

6.3.3 Uses of spore dispersal and disease gradients

The uses of spore dispersal and disease gradients have been discussed by Fitt and McCartney (1986), Minogue (1986) and Jeger (1999), so will only be summarised. Measurement of disease or spore gradients can be extremely important for identifying sources of disease, for identifying inoculum dispersal mechanisms, for assessing the effectiveness of some disease control strategies and for interpreting the results of field experiments.

Since the observation of a disease gradient implies the existence of a local source of inoculum, gradient measurements can be used to identify inoculum sources. In the Netherlands, disease gradient observations were combined with DNA fingerprinting to identify sources of *P. infestans* in regional potato late blight epidemics (Zwankhuizen *et al.*, 1998). The study identified infested refuse piles as primary inoculum sources for establishing early infection foci in fields and infected organic crops as secondary inoculum sources, resulting in the spread of at least two genotypes within the region. From their observations the authors were able to make recommendations on measures to reduce the risk of future epidemics. A vertical gradient of cocoa black pod, with decreasing numbers of lesions with increasing height above ground up to 3 m, suggested that the source of inoculum was at ground level (McCartney and Fitt, 1998). Inoculum of *P. megakarya* was found to be present in the soil and the gradient suggested that zoospores were being released into ground water during rainfall, carried up onto the lower cocoa fruit in the splash zone in large ballistic splash droplets and onto higher fruit in smaller air-borne splash droplets.

Dispersal gradients can be used to infer inoculum dispersal mechanisms; shallow gradients suggest wind dispersal and steep gradients imply splash dispersal. Primary gradients of mummy berry disease (causal agent *Monilinia vaccinii-corymbosi*) in blueberry crops were shallower downwind than upwind, suggesting that the ascospores causing the infections were wind dispersed (Cox and Scherm, 2001). In contrast, secondary disease gradients, caused by conidial infections, were generally shallower upwind than downwind, suggesting that conidia might be dispersed by insect pollinators. Gradients have also been used to assess the relative importance of primary and secondary inoculum in disease development. Gradients of pod rot (caused by *Botrytis cinerea*) at harvest of beans (*Phaseolus vulgaris*) were similar to spore dispersal gradients observed in inoculated plots during flowering, although the spore dispersal gradients had become much flatter by harvest (Johnson and Powelson, 1983). This suggested that the primary inoculum during flowering was the main cause of the pod rot. Gradients have helped to overcome the problem of inter-plot interference in small plot experiments with wind-borne pathogens. Paysour and Fry (1983) were able to use an exponential spore dispersal gradient model to identify plot sizes and spacings which would decrease interference to acceptable levels in their experiments with potato late blight. Knowledge of gradients can also be used to formulate recommendations for growers about distances away from sources of inoculum (e.g. infected stubble) at which a new susceptible crop may safely be planted.

The effectiveness of mixtures of susceptible and resistant cultivars in decreasing the rate of spread of epidemics can be influenced by disease gradients. Simulation models suggested that such mixtures would be most effective against pathogens with shallow spore dispersal gradients (Fitt and McCartney, 1986); these predictions were confirmed in experiments with the wind-borne barley powdery mildew pathogen. By contrast, such mixtures were relatively ineffective in decreasing spread of the splash-dispersed wheat glume blotch pathogen (*S. nodorum*), which has steep spore (conidia) dispersal gradients (Jeger, 1983; see also Chapter 10).

As many crop diseases occur as patches, especially early in the epidemic, it has been suggested that spraying patches to control disease would be less environmentally damaging than spraying whole fields (West *et al.*, 2003). Recent technological advances in optical sensors may make the detection of disease patches feasible (Bravo *et al.*, 2003). However, knowledge of disease gradients is required to make the best use of such measurements as areas of latent (symptomless) infection at the edges of the patches would also need to be sprayed for disease control to be effective (West *et al.*, 2003).

6.4 DISEASE SPREAD: MODELLING DEVELOPMENT OF FOCI

Epidemics may be considered as the consequence of the development of many individual disease foci. Therefore, understanding disease spread from an initial focus of infection is fundamental to understanding the spatial and temporal dynamics of epidemics. It is thus not surprising that much effort has been expended to develop models that describe disease focus development. This work has largely concentrated on the development of mathematically-based models that simulate focal expansion in time and space (Minogue, 1986; Jeger, 1990; Ferrandino, 1993; Shaw, 1994, 1995; Zadoks and van den Bosch, 1994; Mundt, 1995; Maddison *et al.*, 1996; Xu and Ridout, 1996, 1998)). Simulation models of disease spread vary in their complexity, formulation and methods of solution, but often have a similar conceptual basis. The concepts behind disease spread simulation can be illustrated by considering a simple one-dimensional model of disease spread within a row of identical plants (Minogue and Fry, 1983a; Fitt and McCartney, 1986). Assumptions are that the disease occurs as discrete identical lesions which produce spores for a fixed length of time (T_p, the infectious period) and after a latent period (T_l), each plant can sustain a fixed number of lesions and the spore production rate depends on the number of infectious lesions. Spores are dispersed from an infected plant and deposited on other plants according to a 'primary deposition function', *PDF*, which is a function of the distance between the source plant and the receptor plant. The model can be formulated in terms of the rate of production of new lesions and the rate of death of old lesions for all the plants. For a given plant, the rate of production of active lesions depends on the existing number of lesions (active and passive) and on the previous rate of spore deposition from all other plants at time T_l (effect of latent period). The rate of production of passive lesions depends on the rate of production of active lesions at time T_p. The model then calculates the development of lesions with time on each individual plant from some initial infection conditions, for example a single infected plant (Fig. 6.1). This type of model can be formulated as continuous differential equations, as discrete time difference equations or as stochastic functions (Shaw, 1994, 1995, 1996; Xu and Ridout 1996, 1998). For stochastic solutions, the functions defining lesion production, spore production and spore dispersal are formulated as probabilities. Stochastic simulation models are equivalent to Lagrangian stochastic models used in particle dispersal modelling in that they use many simulations of individual dispersal events to accumulate information on the spatial and temporal dynamics of the epidemic. Disease simulation models differ in the method by which the various steps in the disease cycle are formulated but each model includes the equivalent of a 'primary deposition' function.

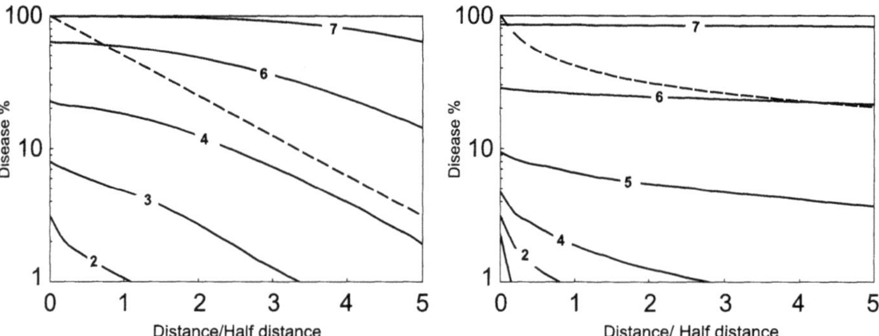

Figure 6.1. Simulated disease progress curves. Output of a simple one dimensional disease progress model illustrating the difference between exponential primary dispersal function (PDF) *(left hand graph) and a power law* PDF *(long tailed distribution) (right hand graph). Both models had the same infection and sporulation characteristics and latent and sporulation periods. The* PDFs *are shown as dashed lines. The exponent of the power law* PDF *was chosen so that it had a similar slope to the exponential* PDF *when plotted over three exponential half distances. Initial disease intensity at the focus was set at 1% area infected. Distance is plotted in units of exponential half distance (distance/ half distance). The numbers associated with each curve are latent periods from the start of the epidemic. Exponential* PDFs *result in a 'travelling disease wave'. Long-tailed* PDFs *give a dispersive wave front (the gradient flattens with time).*

Focus expansion simulation models have highlighted the variability in disease gradient development but all suggest that the shape of primary deposition function plays an important role in both the shape of disease gradients and the expansion rate of the disease focus. The shapes of disease gradients are affected by the shape of the *PDF*, although they are not necessarily similar (Minogue, 1986) (Fig. 6.1). Once the focus begins to expand, disease gradients are often sigmoid in shape (i.e. they start shallow, steepen and then flatten out). Shallow *PDFs* tend to produce shallow disease gradients but the shape of the gradient is also affected by factors such as the sporulation rate and infectious period (Minogue, 1986). Models in which the primary dispersal function has an exponential tail produce foci which develop as a constant velocity travelling wave with a fixed disease pattern at the leading edge of the wave (Minogue and Fry, 1983a; van den Bosch *et al.*, 1988a, 1988b; Zawolek and Zadoks, 1992; Maddison *et al.*, 1996) (Fig. 6.1). The velocity of the travelling wave depends on the length scale of the *PDF* and on factors such as the reproduction rate and the latent period (Zadoks and van den Bosch, 1994). When turbulent diffusion is dominant, functions which behave more like an inverse power law (Equation 6.4) may be more appropriate to describe spore dispersal gradients; when the *PDF* has these properties (i.e. infinite variance), disease focus simulations produce a continuously accelerating and flattening wave (a dispersive wave) which never reaches a steady state (Minogue, 1986). Ferrandino (1993) demonstrated this with a simulation model in which the *PDF* was derived to account for loss of spores both by escape from the canopy and by deposition. The formulation, based on gradient transfer theory (Equation 6.12), was rather complex and did not behave exponentially at large distances from the source. This model produced disease gradients which became shallower as the epidemic progressed, with the leading edge of the wave spreading more quickly than the trailing edge. If the model was formulated with a log-transformed distance it produced travelling waves in 'log-transformed space'. Epidemics in which disease gradients become shallower as the epidemic progresses have been observed in crops, for example with potato late blight (Minogue and Fry, 1983b), *A. linicola* on linseed (Vloutoglou *et al.*, 1995) and *P. striiformis* on wheat (West and McCartney, 2002) (Fig. 6.2).

The influence of primary dispersal on focus development has also been demonstrated by Scherm (1996) using Fickian diffusion to describe disease spread and the logistic equation of Van der Plank (1963) to describe disease intensification:

$$\frac{\partial Y}{\partial t} = \frac{1}{x}\frac{\partial}{\partial x}\left(K_d x \frac{\partial Y}{\partial x}\right) + rY(1-Y) \qquad (6.19)$$

where Y is disease density, x is the distance from the initial infection source, r is the apparent infection rate and K_d is a diffusion coefficient (not to be confused with K_z in Equation 6.12) which defines the dispersal of inoculum from the source (equivalent to a *PDF*). A constant value of K_d (independent of x) corresponds to a *PDF* with an exponential tail and the model produces a travelling wave with velocity $2(rK_d)^{1/2}$. However, when K_d is an increasing function of x (i.e. the rate of diffusion increases with distance from the source) the solution is a dispersive wave.

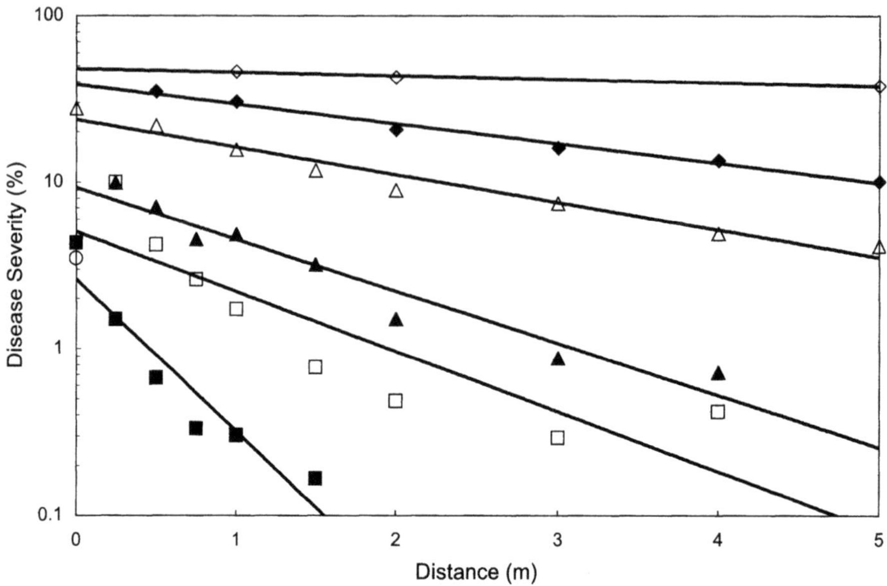

Figure 6.2. Disease progress curve for yellow rust (Puccinia striiformis) *on wheat. The plot was inoculated at its centre in mid-March and disease severity assessed periodically at different radial distances from the centre of the plot in four directions. Disease severity on the single leaf layer with maximal sporulation was measured on the 9 April* (○), *15 April* (■), *23 April* (□), *29 April* (▲) *15 May* (△), *29 May* (♦) *and 11 June* (◇) *at Rothamsted Research in 2003. Values plotted are severity averaged over all four directions for each radial distance.*

Deterministic differential (or difference) equation-based models can predict only mean focus expansion patterns and always produce a monotonic disease front. Stochastic models, however, by their nature, have the potential to yield information on the variability in disease patterns. Such models simulate individual dispersal/infection events, and the basic structure of the approach can be illustrated by considering a very simple model. Assume that a field can be represented by a two-dimensional grid of identical plants (Fig. 6.3) that can be either infected or healthy. Disease spread is simulated as a series of discrete time steps. At each time step, each infected individual releases spores (in the example, we assume that three spores are released). Each spore travels a different distance, chosen randomly from a distribution with the same shape as the primary dispersal function, and in a different random direction. The spore is assumed to infect the grid square on which it lands if that square is not already infected. In the next time step, newly-infected squares also produce spores that travel in random directions over distances estimated by the *PDF* (Fig. 6.3). The disease pattern is constructed from simulations of many time steps. Stochastic simulation models can be much more sophisticated than this example, and can include latency, stochastic spore production functions and infection rates (Xu and Ridout, 1998), variability in wind speed and direction (Xu and Ridout, 2001) and aspects of crop growth (Fink and Kofoet, 2005). *PDFs* used with stochastic simulations include exponential, power law (Shaw, 1994, 1995, 1996; Gibson and Austin, 1996) and half Cauchy distributions (a 'long-tailed' distribution) (Xu and Ridout, 1996, 1998, 2000, 2001; Diggle *et al.*, 2002; Fink and Kofoet, 2005). A simple 'nearest neighbour' dispersal function was used to model barley yellow dwarf virus transmission by aphids (Chaussalet *et al.*, 2000).

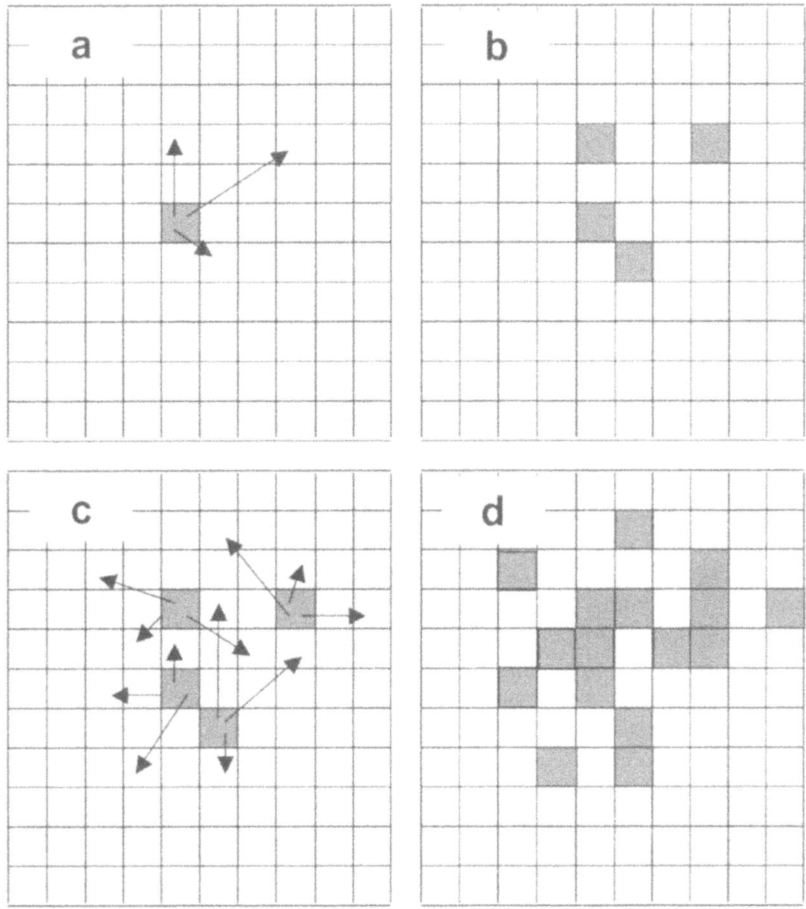

Figure 6.3. A simplified stochastic disease spread model. Grey squares are infected plants. Plants release spores that travel in straight lines (a) to infect other plants (b) which in turn release spores (c) to infect other plants (d).

Stochastic disease modelling suggests that *PDFs* with 'long tails' are more likely to produce discrete 'daughter' foci separate from the main focus than exponential *PDFs* (Shaw 1994, 1995). Stochastic models can be used not only to model disease foci expansion but also to examine effects of development of multiple foci on epidemics (Xu and Ridout, 1996, 1998, 2000, 2001; Diggle *et al.*, 2002; Fink and

Kofoet, 2005). Studies of the effects of *PDFs* on multiple focus epidemics suggest that disease patterns and epidemic size are strongly influenced by dispersal patterns. Long range dispersal (shallow *PDFs*) favours more widespread epidemics and increases the likelihood of disease persistence (Filipe and Maul, 2004). Stochastic models have been used to study effects of sample quadrat size on measurement of spatial variability of disease (Xu and Ridout, 2000), effects of sporulation rate, dispersal distance (Xu and Ridout, 1998) and wind (Xu and Ridout (2001) on epidemic development. Simulation models are beginning to be used to investigate real epidemics, such as citrus tristeza virus (Gibson and Austin, 1996), citrus variegated chlorosis (Martins *et al.,* 2000), anthracnose in lupins (Diggle *et al.*, 2002) and downy mildew of radish (Fink and Kofoet, 2005).

Although sophisticated disease simulation models may give insight into factors affecting epidemic development, they may be too complex for routine disease risk forecasting. For example, West *et al.* (2003) suggest using a simple exponential disease decline model to estimate areas of latent infection to guide application of sprays to disease patches. Shaw (2003) concluded that mathematical representations of epidemics should be simple and robust as they are often based on limited information available about the complex, heterogeneous environment of the agricultural ecosystem.

6.5 CONCLUSIONS

We hope that in this chapter we have shown that inoculum dispersal is of fundamental importance in the spread of plant disease and that understanding the processes involved greatly enhances our knowledge of disease epidemic development. To fully understand the spread of disease epidemics, detailed knowledge of the patterns of spore dispersal by wind and rain-splash is needed. This knowledge can be obtained by a judicious combination of experimental measurements in field and controlled conditions with modelling to improve the theoretical understanding of the fundamental mechanisms which govern dispersal processes. The combination of recent advances in molecular diagnostics and fluid dynamics offer new opportunities for the in-depth study of pathogen dispersal over complex landscapes and should lead to a better understanding of the role dispersal plays in agricultural and natural ecosystems. However, we need now to begin to translate improvements in the understanding of the dispersal phase of epidemics into recommendations for improving strategies for plant disease control in real agricultural systems. This could involve changes in agronomic practices resulting from a better understanding of the effects of crop management on disease spread, or the development of more accurate disease forecasting schemes as aids to disease management decision making.

ACKNOWLEDGEMENTS

Rothamsted Research receives grant aided support from the Biotechnology and Biological Sciences Research Council.

REFERENCES

Aloyan, A.E. (2004) Numerical modelling of minor gas constituents and aerosols in the atmosphere. *Ecological Modelling*, **179**, 163-175.

Aylor, D.E. (1975) Force required to detach conidia of *Helminthosporium maydis*. *Plant Physiology*, **55**, 99-101.

Aylor, D.E. (1978) Dispersal in time and space: aerial pathogens, in *Plant Disease; an Advanced Treatise* (eds J.G. Horsfall and E.B. Cowling), Academic Press, New York, pp. 159-180.

Aylor, D.E. (1982) Modeling spore dispersal in a barley crop. *Agricultural Meteorology*, **26**, 215-219.

Aylor, D.E. (1986) A framework for examining inter-regional aerial transport of fungal spores. *Agricultural and Forest Meteorology*, **38**, 263-288.

Aylor, D.E. (1987) Deposition gradients of urediniospores of *Puccinia recondita* near a source. *Phytopathology*, **77**, 1442-1448.

Aylor, D.E. (1989) Aerial spore dispersal close to a focus of disease. *Agriculture and Forest Meteorology*, **47**, 109-122.

Aylor, D.E. (1990) The role of intermittent wind in the dispersal of fungal pathogens. *Annual Review of Phytopathology*, **28**, 73-92.

Aylor, D.E. (1999) Biophysical scaling and passive dispersal of fungus spores: relationship to integrated pest management strategies. *Agricultural and Forest Meteorology*, **97**, 275-292.

Aylor, D.E. and Ferrandino, F.J. (1989) Dispersion of spores released from an elevated line source within a wheat canopy. *Boundary-Layer Meteorology*, **46**, 251-273.

Aylor, D.E and Ferrandino, F.J. (1990) Initial spread of bean rust close to an inoculated bean leaf. *Phytopathology*, **80**, 1469-1476.

Aylor, D.E. and Flesch, T.K. (2001) Estimating spore release rates using a Lagrangian stochastic simulation model. *Journal of Applied Meteorology*, **40**, 1196-1208.

Aylor, D.E., Fry, W.E., Mayton, H. and Andrade-Piedra, J.L. (2001) Quantifying the rate of release and escape of *Phytophthora infestans* sporangia from a potato canopy. *Phytopathology*, **91**, 1189-1196.

Aylor, D.E., McCartney, H.A. and Bainbridge, A. (1981) Deposition of particles liberated in gusts of wind. *Journal of Applied Meteorology*, **20**, 1212-1221.

Aylor, D.E. and Parlange, J-Y. (1975) Ventilation required to entrain small particles from leaves. *Plant Physiology*, **56**, 97-99.

Aylor, D.E. Schultes, N.P. and Shields, E.J. (2003) An aerobiological framework for assessing cross-pollination in maize. *Agricultural and Forest Meteorology*, **119**, 111-129.

Bainbridge, A. and Legg, B.J. (1976) Release of barley-mildew conidia from shaken leaves. *Transactions of the British Mycological Society*, **66**, 495-498.

Berger, R.D. and Luke, H.H. (1979) Spatial and temporal spread of oat crown rust. *Phytopathology*, **69**, 1199-1201.

Blackall, R.M. and Gloster, J. (1981) Forecasting the airborne spread of foot and mouth disease. *Weather*, **36**, 162-167.

Bock, C.H., Jeger, M.J., Fitt, B.D.L. and Sherington, J. (1997) Effect of wind on the dispersal of oospores of *Peronosclerospora sorghi* from sorghum. *Plant Pathology*, **46**, 439-449.

Bock, C.H., Parker, P.E. and Gottwald, T.R. (2005) Effect of simulated wind-driven rain on duration and distance of dispersal of *Xanthomonas axonopodis* pv. *citri* from canker-infected citrus trees. *Plant Disease*, **89**, 71-80.

Bovallius, A., Bucht, B., Roffey, R. and Anäs P. (1978) Long-range air transmission of bacteria. *Applied and Environmental Microbiology*, **35**, 1231-1232.

Brown, J. and Ehrhardt, J. (1995) EUR16240 EN-PCCOSYMA version 2.0 User Guide, NRPB-SR280. National Radiological Protection Board, Forschungszentrum Karlsruhe GmbH, 443pp.

Brown, J.K.M. and Hovmøller, M.S. (2002) Aerial dispersal of pathogens on the global and continental scales and its impact on plant disease. *Science*, **297**, 537-541.

Brennan, R.M., Fitt, B.D.L., Taylor, G.S. and Colhoun, J. (1985) Dispersal of *Septoria nodorum* pycnidiospores by simulated rain and wind. *Phytopathologische Zeitschrift*, **112**, 291-297.

Bullock, J.M. and Clarke, R.T. (2000) Long distance seed dispersal by wind: measuring and modelling the tail of the curve. *Oecologia*, **124**, 506-521.

Butterworth, J. and McCartney, H.A. (1992) Effect of drop size on the removal of *Bacillus subtilis* from foliar surfaces by water splash. *Microbial Releases*, **1**, 177-185.

Bravo C, Moshou D, West J. *et al.* (2003) Early disease detection in wheat fields using spectral reflection. *Biosystems Engineering,* **84**, 137-145.

Calderon, C., Ward, E., Freeman, J. *et al.* (2002) Detection of airborne inoculum of *Leptosphaeria maculans* and *Pyrenopeziza brassicae* in oilseed rape crops by polymerase chain reaction (PCR) assays. *Plant Pathology,* **51**, 303-310.

Cammack, R.H. (1958) Factors affecting infection gradients from a point source of *Puccinia polysora* in plots of *Zea mays*. *Annals of Applied Biology,* **46**, 186-197.

Caputo, M., Giménez, M. and Schlamp, M. (2003) Intercomparison of atmospheric dispersion models. *Atmospheric Environment*, **37**, 2435-2449.

Chamberlain, A.C. (1953) Aspects of travel and deposition of aerosol and vapour clouds. *Report AERE-R, UK*. Atomic Energy Research Establishment, Harwell, UK.

Chamberlain, A.C. (1975) The movement of particles in plant communities, in *Vegetation and the Atmosphere*, Vol 1 (ed. J.L. Monteith), Academic Press, London, pp. 155-203.

Chaussalet, T.J., Mann, J.A., Perry, J.N. and Francos-Rodriguez, J.C. (2000) A nearest neighbour approach to the simulation of spread of barley yellow dwarf virus. *Computers and Electronics in Agriculture*, **28**, 51-65.

Cionco, R.M., Ohmstede, W.D. and Appleby, J.F. (1963) *Meteorological Research Notes Number 5*, USAERDAA, Fort Huachuca, Arizona.

Cox K.D. and Scherm H. (2001) Gradients of primary and secondary infection by *Monilinia vaccinii-corymbosi* from point sources of ascospores and conidia. *Plant Disease*, **85**, 955-959.

D'Amours, R. (1998) Modeling the ETEX Plume dispersion with the Canadian emergency response model. *Atmospheric Environment*, **32**, 4335-4341.

Davis, J.M. (1987) Modeling the long-range transport of plant pathogens in the atmosphere. *Annual Review of Phytopathology*, **25**, 169-188.

Davis, J.M. and Main, C.E. (1986) Applying atmospheric trajectory analysis to problems in epidemiology. *Plant Disease*, **70**, 490-497.

Diggle, A. J., Salam, M. U., Thomas, G. J. *et al.* (2002) AnthracnoseTracer: A spatiotemporal model for simulating the spread of anthracnose in a lupin field, *Phytopathology*, **92**, 1110-1121.

Di-Giovanni F. and Kevan P.G. (1991) Factors affecting pollen dynamics and its importance to pollen contamination: a review. *Canadian Journal of Forest Research*, **21**, 1155-1170.

Fatemi, F. and Fitt B.D.L. (1983) Dispersal of *Pseudocercosporella herpotrichoides* and *Pyrenopeziza brassicae* spores in splash droplets. *Plant Pathology*, **32**, 401-404.

Ferrandino, F.J (1993) Dispersive epidemic waves; I: focus expansion in a linear planting. *Phytopathology*, **83**, 795-802.

Ferrandino, F.J. (1996) Length scale of disease spread: fact or artefact of experimental geometry. *Phytopathology*, **86**, 807-811.

Filipe, J.A.N. and Maule, M.M. (2004) Effects of dispersal mechanisms on spatio-temporal development of epidemics. *Journal of Theoretical Biology,* **226**, 125-141.

Fink, M. and Kofoet, A. (2005) A two-dimensional stochastic model of downy mildew of radish. *Ecological Modelling*, **181**, 139-148.

Fitt, B.D.L. and Lysandrou, M. (1984) Studies on mechanisms of splash dispersal of spores, using *Pseudocercosporella herpotrichoides* spores. *Phytopathologische Zeitschrift*, **111**, 323-331.

Fitt, B.D.L. and McCartney, H.A. (1986) Spore dispersal in relation to epidemic models, in *Plant Disease Epidemiology, Vol.1: Population Dynamics and Management* (eds K.J. Leonard and W.E. Fry), Macmillan, New York, pp. 311-345.

Fitt, B.D.L., McCartney, H.A., Creighton, N.F. *et al.* (1988). Dispersal of *Rhynchosporium secalis* conidia from infected barley leaves or straw by simulated rain. *Annals of Applied Biology*, **112,** 49-59.

Fitt, B.D.L., McCartney, H.A. and Walklate, P.J. (1989) Role of rain in the dispersal of pathogen inoculum. *Annual Review of Phytopathology*, **27**, 241-270.

Fitt, B.D.L, Inman, A.J., Lacey, M.E. and McCartney, H.A. (1992) Splash dispersal of spores of *Pseudocercosporella capsellae* (white leaf spot) from oilseed rape leaves of different inclination, flexibility and age. *Zeitschrift für Pflanzenkrankheiten und Pflanzenschutz*, **99**, 234-244.

Fitt, B.D.L., Gregory, P.H., Todd, A.D. *et al.* (1987) Spore dispersal and plant disease gradients: a comparison between two empirical models. *Journal of Phytopathology*, **118**, 227-242.

Fitt, B.D.L., Walklate, P.J., McCartney, H.A. *et al.* (1986) A rain tower and wind tunnel for studying the dispersal of plant pathogens by rain and wind. *Annals of Applied Biology*, **109**, 661-671.

Franzes, P. (2003) Lagrangian stochastic modeling of a fluctuating plume in the convective boundary layer. *Atmospheric Environment*, **37**, 1691-1701.

Freeman, J., Ward, E., Calderon, C. and McCartney, A. (2002) A polymerase chain reaction (PCR) assay for the detection of inoculum of *Sclerotinia sclerotiorum*. *European Journal of Plant Pathology*, **108**, 877-886.

Geagea, L., Huber, L., Sache, I. *et al.* (2000) Influence of simulated rain on dispersal of rust spores from infected wheat seedlings. *Agricultural and Forest Meteorology*, **101**, 53-66.

Gibson, G.J. and Austin, E.J. (1996) Fitting and testing spatio-temporal stochastic models with application in plant epidemiology. *Plant Pathology*, **45**, 172-184.

Gladders, P. and Musa, T.M. (1980) Observations on the epidemiology of *Leptosphaeria maculans* stem canker in winter oilseed rape. *Plant Pathology*, **29**, 28-37.

Gloster, J. (1983a) Forecasting the airborne spread of foot-and-mouth disease and Newcastle disease. *Philosophical Transactions of the Royal Society, London*, **B302**, 535-541.

Gloster, J. (1983b) Analysis of two outbreaks of Newcastle disease. *Agricultural Meteorology*, **28**, 177-189.

Gottwald, T.R., Sun, X., Riley, T. *et al.* (2002) Geo-referenced spatiotemporal analysis of the urban citrus canker epidemic in Florida. *Phytopathology* **92**, 361-377.

Grace, J. (1977) *Plant Response to Wind.* Academic Press, London. 204 pp.

Gregory, P.H. (1968) Interpreting plant disease dispersal gradients. *Annual Review of Phytopathology*, **6**, 189-212.

Gregory, P.H. (1973) *The Microbiology of the Atmosphere* (2nd edn), Leonard Hill, London, 377 pp.

Hamilton, L.M. and Stakman, E.C. (1967) Time of stem rust appearance on wheat in western Mississippi basin in relation to development of epidemics from 1921 to 1962. *Phytopathology*, **57**, 609-616.

Hammett, K.R.W. and Manners, J.G. (1974) Conidium liberation in *Erysiphe graminis*. III: wind tunnel studies. *Transactions of the British Mycological Society*, **62**, 267-282.

Hirst, J.M., Stedman, O.J. and Hurst, G.W. (1967) Long distance spore transport: vertical sections of spore clouds over the sea. *Journal of General Microbiology*, **48**, 357-377.

Horst, T.W. (1977) A surface depletion model for deposition from a Gaussian plume. *Atmospheric Environment*, **11**, 41-46.

Huber, L., Fitt, B.D.L. and McCartney, H.A. (1996) The incorporation of pathogen spores into rain-splash droplets: a modelling approach. *Plant Pathology*, **45**, 506-517.

Huber, L., McCartney, H.A. and Fitt, B.D.L. (1997) Influence of target characteristics on the amount of water splashed by impacting drops. *Agricultural and Forest Meteorology*, **87**, 201-211.

Inman, A.J. (1993) The biology and epidemiology of white leaf spot (*Pseudocercosporella capsellae*) on oilseed rape. Ph.D. thesis, University of London.

Jarosz, N, Loubet, B and Huber, L. (2004) Modelling airborne concentration and deposition rate of maize pollen. *Atmospheric Environment*, **38**, 5555-5566.

Jeger, M.J. (1983) Analysing epidemics in time and space. *Plant Pathology*, **32**, 5-11.

Jeger, M.J. (1990) Mathematical analysis and modeling of spatial aspects of plant disease epidemics, in *Epidemics of Plant Diseases* (ed J. Kranz), Springer Verlag, New York, pp. 53-95.

Jeger, M.J. (1999) Improved understanding of dispersal in crop pest and disease management: current status and future directions. *Agricultural and Forest Meteorology*, **97**, 331-349.

Johnson, K.B. and Powelson, M.L. (1983) Analysis of spore dispersal gradients of *Botrytis cinerea* and grey mold disease gradients in snap beans. *Phytopathology*, **73**, 741-746.

Kennedy, R., Wakeham, A.J., Byrne, K.G. *et al.* (2000) A new method to monitor airborne inoculum of the fungal plant pathogens *Mycosphaerella brassicicola* and *Botrytis cinerea*. *Applied and Environmental Microbiology*, **66**, 2996-3003.

Kottmeier, C. and Fay, B. (1998) Trajectories in the Antarctic lower atmosphere. *Journal of Geophysics Research*, **103**, 10947-10959.

Lacey, J. (1986) Water availability and fungal reproduction: patterns of spore production, liberation and dispersal, in *Water Fungi and Plants* (eds P.G. Ayres and L. Boddy), Cambridge University Press, Cambridge, pp. 65-86.

Lacey, J., Lacey, M.E. and Fitt, B.D.L. (1997) Philip Herries Gregory 1907-1986: Pioneer aerobiologist, versatile mycologist. *Annual Review of Phytopathology*, **35**, 1-14.

Legg, B.J. (1983) Movement of plant pathogens in the crop canopy. *Philosophical Transactions of the Royal Society*, London, **B302**, 559-574.

Legg, B.J. and Powell, F.A. (1979) Spore dispersal in a barley crop: a mathematical model. *Agricultural Meteorology*, **20**, 47-67.
Lighthart B., Shaffer, B.T., Marthi, B. and Giano, L.M. (1993) Artificial wind-gust liberation of microbial bioaerosols previously deposited on plants. *Aerobiologia*, **9**, 189-196.
Lovell, D.J., Parer, S.R, Van Peteghem, P. *et al.* (2002) Quantification of raindrop kinetic energy for improved prediction of splash-dispersed pathogens. *Phytopathology*, **92**, 497-503.
Macdonald, O.C. and McCartney, H.A. (1987) Calculation of splash droplet trajectories. *Agricultural and Forest Meteorology*, **39**, 95-110.
Macdonald O.C. and McCartney H.A. (1988) A photographic technique for investigating splashing of water drops on leaves. *Annals of Applied Biology*, **113**, 627-638.
Madden, L.V. (1992) Rainfall and dispersal of fungal spores. *Advances in Plant Pathology*, **8**, 39-79.
Madden, L.V. (1997) Effects of rainfall on splash dispersal of fungal pathogens. *Canadian Journal of Plant Pathology,* **19**, 225-230.
Madden, L.V., Yang, X. and Wilson, L.L. (1996) Effects of rain on splash dispersal of *Colletotrichum acutatum*. *Phytopathology*, **86**, 864-874.
Madden, L.V, Wilson, L.L. and Ntahimpera, N. (1998) Calibration and evaluation of an electronic sensor for rainfall kinetic energy. *Phytopathology*, **88**, 950-959.
Maddison, A.C., Holt, J. and Jeger, M.J. (1996) Spatial dynamics of a monocyclic disease in a perennial crop. *Ecological Modelling*, **88**, 45-52.
Main, C.E., Keever, T., Holmes, G.J. and Davis, J.M. (2001) Forecasting long-range transport of downy mildew spores and plant disease epidemics. APS*net,* May 2001, http://www.apsnet.org/online/feature/.
Marshall, W.A. (1996) Biological particles over Antarctica. *Nature*, **383**, 680.
Martins, M.L., Ceotto, G., Alves, S.G. *et al.* (2000) Cellular automata model for citrus variegated chlorosis. *Physical Review E,* **62**, 7024-7030.
McCartney, H.A. (1990a) The dispersal of plant pathogen spores and pollen from oilseed rape crops. *Aerobiologia*, **6**, 147-152.
McCartney, H.A. (1990b) Dispersal mechanisms through the air, in *Dispersal in Agricultural Habitats* (eds R.G.H. Bunce and D.C. Howard), Belhaven Press, London, pp. 133-158.
McCartney, H.A. (1997) Modelling the dispersal of fungal spores and pollens by wind, in *Aerobiology* (ed. S.N. Agashe), Oxford and IBH Publishing Co, New Delhi, pp. 327-332.
McCartney, H.A. and Bainbridge, A. (1987) Deposition of *Erysiphe graminis* conidia on a barley crop. I: sedimentation and impaction. *Journal of Phytopathology*, **118**, 243-257.
McCartney, H.A. and Fitt, B.D.L. (1985) Construction of dispersal models, in *Advances in Plant Pathology, Vol.3: Mathematical Modelling of Crop Disease* (ed. C.A. Gilligan), Academic Press, London, pp. 107-143.
McCartney, H.A. and Fitt, B.D.L. (1998) Dispersal of foliar plant pathogens: mechanisms, gradients and spatial patterns, in *The Epidemiology of Plant Diseases* (ed. D. Gareth Jones), Kluwer Academic Publishers, Dordrecht, pp. 138-160.
McCartney, H.A. and Lacey, M.E. (1990) The production and release of ascospores of *Pyrenopeziza brassicae* on oilseed rape. *Plant Pathology*, **39**, 17-32.
McCartney, H.A., Fitt, B.D.L. and Schmechel, D. (1997) Sampling bioaerosols in plant pathology. *Journal of Aerosol Science*, **28**, 349-364.
McCartney, H.A, Foster S.J., Fraaije B.A. and Ward, E. (2003) Molecular diagnostics for fungal plant pathogens. *Pest Management Science*, **59**, 129-142.
Mikkelsen, T., Alexandersen, S., Astrup, P. *et al.* (2003) Investigation of airborne foot-and–mouth disease virus transmission during low-wind conditions in the early phase of the UK 2001 epidemic. *Atmospheric Chemistry and Physics,* **3**, 2101-2110.
Mims, S.A. and Mims, F.M. (2004) Fungal spores are transported long distances in smoke from biomass fires. *Atmospheric Environment*, **38**, 651-655.
Minogue, K.P. (1986) Disease gradients and the spread of disease, in*: Plant Disease Epidemiology, Vol. 1: Population Dynamics and Management* (eds K.J. Leonard and W.E. Fry), Macmillan, New York, pp. 285-310.
Minogue, K.P. and Fry, W.E. (1983a) Models for the spread of disease: model description. *Phytopathology*, **73**, 1168-1173.
Minogue, K.P. and Fry, W.E. (1983b) Models for the spread of disease: some experimental results. *Phytopathology*, **73**, 1173-1176.

Monteith, J.L. and Unsworth, M.H. (1990) *Principles of Environmental Physics* (2nd edn.), Edward Arnold, London. 291pp.
Mundt, C.C. (1995) Models from plant pathology on the movement and fate of new genotypes of microorganisms in the environment. *Annual Review of Phytopathology*, **33**, 467-488.
Mundt, C.C. and Leonard, K.J. (1985) A modification of Gregory's model for describing plant disease gradients. *Phytopathology*, **75**, 930-905.
Nagarajan, S. and Singh, D.V. (1990) Long-Distance Dispersion of Rust Pathogens. *Annual Review of Phytopathology* **28**, 139-153.
Ntahimpera, N., Hacker, J.K., Wilson, L.L. *et al.* (1999) Characterization of splash droplets from different surfaces with a phase Doppler particle analyzer. *Agricultural and Forest Meteorology*, **97**, 9-19.
Oettl, D. Kukkonen, J., Almbauer, R.A. *et al.* (2001) Evaluation of a Gaussian and Lagrangian model against a roadside data set, with emphasis on low wind speed conditions. *Atmospheric Environment*, **35**, 2123-2132.
Pasquill, F. and Smith, F.B. (1983) *Atmospheric Diffusion* (3rd edn), Ellis Horwood, Chichester. 437 pp.
Paysour, R.E. and Fry, W.E. (1983) Interplot interference: a model for planning field experiments with aerially disseminated pathogens. *Phytopathology*, **73**, 1014-1020.
Pielaat, A. and van den Bosch, F. (1998) A model for dispersal of plant pathogens by rainsplash. *IMA Journal of Mathematics Applied in Medicine and Biology*, **15**, 117-134.
Pielaat, A., van den Bosch, F., Fitt, B.D.L., and Jeger, M.J. (2002) Simulation of vertical spread of plant diseases in a canopy by stem extension and splash dispersal. *Ecological Modelling*, **151**, 195-212.
Pietravalle, S., van den Bosch, F., Welham, S.J. *et al.* (2001) Modelling of rain splash trajectories and prediction of rain splash height. *Agricultural and Forest Meteorology*, **109**, 171-185.
Raza, S.S, Avila, R, Cervantes, J. (2001) A 3-D Lagrangian stochastic model for the meso-scale atmospheric dispersion applications. *Nuclear Engineering and Design*, **208**, 15-28.
Reynolds, K.M., Madden, L.V., Reichard, D.L. and Ellis, M.A. (1987) Method for study of raindrop impaction on plant surfaces with application for predicting inoculum dispersal by rain. *Phytopathology*, **77**, 226-232.
Riddle, A, Carruthers, D. Sharpe, A. *et al.* (2004) Comparison between FLUENT and ADMS for atmospheric dispersion modelling. *Atmospheric Environment*, **38**, 1029-1038.
Rowe, R.C. and Powelson, R.L. (1973) Epidemiology of *Cercosporella* foot rot of wheat-disease spread. *Phytopathology*, **63**, 984-988.
Saint-Jean, S., Chelle, M., Huber, L. (2004) Modelling water transfer by rain-splash in a 3D canopy using Monte Carlo integration. *Agricultural and Forest Meteorology*, **121**, 183-196.
Scherm, H. (1996) On the velocity of epidemic waves in model plant disease epidemics. *Ecological Modelling* , **87**, 217-222.
Shaw, M.W. (1987) Assessment of upward movement of rain splash using a fluorescent tracer method and its application to the epidemiology of cereal pathogens. *Plant Pathology*, **36**, 201-213.
Shaw, M.W. (1994) Modeling stochastic processes in plant pathology. *Annual Review of Phytopathology*, **32**, 523-544.
Shaw, M.W. (1995) Simulation of population expansion and spatial pattern when individual dispersal distributions do not decline exponentially with distance. *Proceedings of the Royal Society B*, **259**, 243-248.
Shaw, M.W. (1996) Simulating dispersal of fugal spores by wind, and the resulting patterns. *Aspects of Applied Biology*, **46**, 165-172.
Shaw, M.W. (2003) Mathematical representation of epidemiological complexity. Paper C20.1 of the *8th International Congress of Plant Pathology, 2-7 February 2003*, Christchurch, New Zealand.
Shaw, M.W. and Royle, D.J. (1993) Factors determining the severity of epidemics of *Mycosphaerella graminicola* (*Septoria tritici*) on winter wheat in the UK. *Plant Pathology*, **42**, 882-899.
Shaw, R.H. and McCartney, H.A. (1985) Gust penetration into plant canopies. *Atmospheric Environment*, **19**, 827-830.
Shaw, R.H., Ward, D.P and Aylor, D.E. (1979) Frequency of occurrence of fast gusts of wind inside a corn canopy. *Journal of Applied Meteorology*, **18**, 167-171.
Soleimani, M.J., Deadman, M.L. and McCartney, H.A. (1996) Splash dispersal of *Pseudocercosporella herpotrichoides* spores in wheat monocrop and wheat-clover bicrop canopies from simulated rain. *Plant Pathology*, **45**, 1065-1070.

Stohl, A. (1998) Computation, accuracy and applications of trajectories – a review and bibliography. *Atmospheric Environment*, **32**, 947-966.

Thom, A.S. (1975) Momentum, mass and heat exchange of plant communities, in *Vegetation and the Atmosphere, Vol 1* (ed. J.L. Monteith), Academic Press, London, pp. 57-109.

Tufto, J., Engen, S. and Hindar, K. (1997) Stochastic dispersal processes in plant populations. *Theoretical Population Biology*, **52**, 16-62.

USEPA (1999) Revised draft user's guide for the AEROMOD meteorological processor (aermet). EPA, 273 pp.

Ulbrich, C.W. (1983) Natural variations in the analytical form of the raindrop size distribution. *Journal of Climate and Applied Meteorology*, **22**, 1764-1775.

van den Bosch, F., Zadoks, J.C. and Metz, J.A.J. (1988a) Focus expansion in plant disease. I: the constant rate focus expansion. *Phytopathology*, **78**, 54-58.

van den Bosch, F., Zadoks, J.C. and Metz, J.A.J. (1988b) Focus expansion in plant disease. II: realistic parameter-sparse models. *Phytopathology*, **78**, 59-64.

Van der Plank, J.E. (1963) *Plant Disease: Epidemics and Control*. Academic Press, New York. 349 pp.

Vloutoglou, I., Fitt, B.D.L. and Lucas, J.A. (1995) Periodicity and gradients in dispersal of *Alternaria linicola* in linseed crops. *European Journal of Plant Pathology*, **101**, 639-653.

Wadia, K.D.R., McCartney, H.A. and Butler, D.R. (1998) Dispersal *of Passalora personata* conidia from groundnut by wind and rain. *Mycological Research*, **102**, 355-360.

Wakeham, A., Kennedy, R and McCartney, A. (2004) The collection and retention of a range of common airborne spore types trapped directly into microtiter wells for enzyme-linked immunosorbent analysis. *Journal of Aerosol Science*, **35**, 835-850.

Walklate, P.J. (1987) A random-walk model for dispersion of heavy particles in turbulent air flow. *Boundary-Layer Meteorology*, **39**, 175-190.

Walklate, P.J. (1989) Vertical dispersal of plant pathogens by splashing. Part I: the theoretical relationship between rainfall and upward rain-splash. *Plant Pathology*, **38**, 56-63.

Walklate, P.J. (1992) A simulation study of pesticide drift from an air-assisted orchard sprayer. *Journal of Agricultural Engineering Research*, **51**, 263-283.

Walklate, P.J., McCartney, H.A. and Fitt, B.D.L. (1989) Vertical dispersal of plant pathogens by splashing. II: experimental study of the relationship between rain drop size and maximum splash height. *Plant Pathology*, **38**, 64-70.

Wang, G. and Ostoja-Starzewski (2004) A numerical study of plume dispersion motivated by a mesoscale atmospheric flow over a complex terrain. *Applied Mathematical Modelling*, **28**, 957-981.

Ward, E., Foster, S.J., Fraaije B.A. and McCartney, H.A. (2004) Plant Pathogen diagnostics: immunological and nucleic acid-based approaches. *Annals of Applied Biology*, **145**, 1-16.

Welham, S.J., McCartney, H.A. and Fitt, B.D.L. (1995) A case study in measurement and analysis of disease gradients. *Aspects of Applied Biology*, **43**, 77-85.

West, J. and McCartney, A. (2002) Optical disease detection and estimation of latent infections around disease foci for targeted pesticide application. *Aspects of Applied Biology*, **66**, 463-468.

West, J.S., Bravo, C., Oberti, R. *et al.* (2003) The potential of optical canopy measurement for targeted control of field crop diseases. *Annual Review of Phytopathology*, **41**, 593-614.

Wilson, J.D. (2000) Trajectory models for heavy particles in atmospheric turbulence: comparison with observations. *Journal of Applied Meteorology*, **39**, 1894-1912.

Wilson, J.D. and Flesch, T.K. (1997) Trajectory curvature as a selection criterion for valid Lagrangian stochastic dispersion models. *Boundary-Layer Meteorology*, **84**, 411-425.

Wilson, J.D. and Sawford, B.L. (1996) Review of Lagrangian stochastic models for trajectories in the turbulent atmosphere. *Boundary-Layer Meteorology*, **78**, 191-210.

Xu, X.-M. and Ridout, M.S. (1996) Analysis of disease incidence data using a stochastic spatial-temporal simulation model. *Aspects of Applied Biology*, **46**, 155-158.

Xu, X.-M. and Ridout, M.S. (1998) Effects of initial conditions, sporulation rate, and spore dispersal gradient on the spatio-temporal dynamics of plant disease epidemics: *Phytopathology*, **88**, 1000-1012.

Xu, X.-M. and Ridout, M.S. (2000) Effects of quadrat size and shape, initial epidemic conditions, and spore dispersal gradient on spatial statistics of plant disease epidemics: *Phytopathology*, **90**, 738-750.

Xu, X.-M. and Ridout, M.S. (2001) Effects of prevailing wind direction on spatial statistics of plant disease epidemics: *Journal of Phytopathology*, **149**, 155-166.

Xu, Z.G., Walklate, P.J., Rigby, S.G. and Richardson, G.M. (1998) Stochastic modelling of turbulent spray dispersion in the near-field of orchard sprayers. *Journal of Wind Engineering and Industrial Aerodynamics*, **74-76**, 295-304.

Yang, X. and Madden, L.V. (1993) Effects of ground cover, rain intensity, and strawberry plants on splash of simulated raindrops. *Agricultural and Forest Meteorology*, **65**, 1-20.

Yang, X., Madden, L.V., Reichard, D.L. *et al.* (1991) Motion analysis of drop impaction on a strawberry surface. *Agriculture and Forest Meteorology*, **56**, 67-92.

Yang, X., Madden, L.V., Reichard, D.L. *et al.*, (1992) Splash dispersal of *Colletotrichum acutatum* and *Phytophthora cactorum* from strawberry fruit by single drop impactions. *Phytopathology*, **82**, 332-340.

Yao, C., Arya, S.P., Davis, J., Main, C.E. (1997) A numerical model of the transport and diffusion of *Peronospora tabacina* spores in the evolving atmospheric boundary layer. *Atmospheric Environment*, **31**, 1709-1714.

Zadoks, J.C. and Schein, R.D. (1979) *Epidemiology and Plant Disease Management*. Oxford University Press, New York. 427 pp.

Zadoks, J.C. and van den Bosch, F. (1994) On the spread of plant disease: a theory on foci. *Annual Review of Phytopathology*, **32**, 503-521.

Zawolek, M.W. and Zadoks, J.C. (1992) Studies in focus development: an optimum for dual dispersal of plant pathogens. *Phytopathology*, **82**, 1288-1297.

Zwankhuizen, M.J., Govers, F. and Zadoks, J.C. (1998) Development of potato late blight epidemics: disease foci, disease gradients, and infection sources. *Phytopathology*, **88**, 754-763.

CHAPTER 7

PATHOGEN POPULATION DYNAMICS

M.W. SHAW

7.1 INTRODUCTION

Population dynamics is the study of how the size and structure of populations respond to the forces that act on them. This chapter begins with attempts to clarify concepts of time-scale, of population structure, of how populations can be measured and of regulation of a population. The chapter continues by considering how pathogen populations change in fixed host populations over short time-scales, then how populations change when the host population changes on a time-scale comparable with the pathogen and, finally, how populations change over many generations of both host and pathogen. The argument uses both reasoning about drastically simplified settings and generalisations from experimental data.

7.2 THE MEASUREMENT OF POPULATIONS

Before it is possible to study changes in populations, it is first necessary to decide what the individuals are that are being counted. A population is composed of individuals at various stages in their life-cycles. It seems sensible to start by discussing a relatively easy example, *Blumeria graminis* (cause of cereal powdery mildew) on a cereal crop. Three components of the population can be immediately distinguished. First, there is a population of suspended or recently deposited and ungerminated conidia, each of which is potentially capable of infection. Second, there is a population of physiologically independent mycelia growing on leaf surfaces. Some of these are too young to sporulate, others are sporulating, still others have exhausted the food available to them and have ceased to sporulate. Finally, there may be a population of cleistothecia. A measure of the population will have to include several numbers, describing the number of each type of individual per square metre (or per square metre of host). The conidia could be measured by various techniques (Gregory, 1973; Kennedy *et al.*, 2000) and the sporulating mycelia counted visually. More sophisticated methods using staining and microscopy would be necessary to count the immature mycelia directly. There is still the problem that a count of visually distinct colonies is not necessarily a count of physiologically distinct pathogen individuals; one colony may contain several individuals. A final problem would be sampling cleistothecia, since dead host tissue will be difficult to survey adequately. An independent categorization could divide

B. M. Cooke, D. Gareth Jones and B. Kaye (eds.), The Epidemiology of Plant Diseases, 2nd edition, 193–214.

the population according to genotype or virulence, or by infection with a hyperparasite. Similar sampling and estimation problems would arise.

Regardless of which stages are used, the 'stage structure' of the population can be described by specifying the proportion of it that exists in each stage. If the stages are simply age groups, then the stage structure corresponds to the age structure, or the proportion of the population in each of a series of age groups.

A more difficult problem is posed by a leaf blotching or spotting infection, *Phaeosphaeria nodorum* (cause of wheat glume blotch) or *Mycosphaerella graminicola* (cause of septoria tritici blotch), for example. The extent of visible damage to the leaf does not necessarily correspond in any simple way to the number of physiologically independent infections that have occurred, and non-sporulating mycelia are hidden within the leaf. It may be possible to measure the amount of mycelium, or of mycelium of a particular type immunologically or by DNA sequence-based assays (see also Chapter 1) but this still does not correspond in a simple way to a notion of population. Estimates of spore density would be possible based on washing leaves to remove spores released or available for release but not yet bound to a leaf surface; this has proved useful in practice (Shaw and Royle, 1993).

It is still more difficult to define the population of a root-infecting fungus like *Gaeumannomyces graminis* (cause of take-all of cereals). Transmission from host to host is only very rarely by spores and substantial amounts of mycelium can exist in the soil as well as on the root surface of susceptible hosts. Single physiological individuals are hard to define, let alone count. In some soil-borne pathogens transitions between stages may be very fast, leading to very rapid changes in age-structure; if some stages are not easy to observe, because they do not infect test plants or cannot be cultured, it may be very hard to obtain sensible estimates of population sizes (Hardman *et al.*, 1989).

In practice, it is usually necessary to fall back on measuring the pathogen population indirectly, by correlation with a measure of disease or a biochemical or DNA-based assay. Biochemical assays have the advantage of precision, but by their nature do not distinguish changes in numbers of individuals from changes in size of individuals; empirical relations between the assay and population need to be constructed for each pathogen, and supplementary information on the stage-structure may be needed. Measurement of disease is discussed comprehensively in Chapter 2. Here, it is only necessary to note that most crops can be subdivided into units which can be categorised as diseased or not. The proportion of diseased units in an area will often be a useful proxy for the true population density of the life-stages which are reproducing actively on the host tissue under study.

The size of the unit used will depend on both the time-scale and on the purpose of the measurement. For example, with a systemic virus disease such as cocoa swollen shoot virus, the obvious unit is the infected plant; for a foliar fungal disease, such as septoria tritici blotch of wheat, it may be more appropriate to consider infected leaves or portions of leaves. However, progression into or from dormant forms independent of a host may require a different definition of the sampled units. The relationship between incidence, pathogen biomass and number of physiological or genetic units of pathogen in an area will vary for almost every pathosystem. It will usually be neither simple nor independent of the environment and host.

Finally, it should be noted that the study of pathogen population dynamics will rarely be possible in isolation from the host population dynamics, since most pathogens affect growth and reproduction of their hosts.

7.3 TIME-SCALES

A natural time-scale for changes in a pathogen population is set by the generation time of the pathogen, the time taken for one propagule to infect and give rise to another. However, for pathogens with complex life-cycles, there may be more than one natural time-scale, depending on which life stage is being considered. For example, in a heterocyclic rust, the generation time of asexually produced, physiologically independent, individuals in a uredinial cycle may be much shorter than the generation time of sexually produced genetically distinct individuals, say from telium to telium.

Often, this notion of generation time will broadly correspond to the latent period for a typical fungal pathogen. It differs in that it refers to the time when, on average, new infections are actually formed under the prevailing conditions, rather than to when infectious pathogen stages are produced. A detailed discussion of definitions of generation time is quite complicated (Charlesworth, 1994) but the average age of an individual when its offspring infect is adequate for the purposes of this chapter. As an example, Lovell *et al.* (2004) measured the latent period of *Mycosphaerella graminicola* on the flag leaf of wheat cv. Riband as about 300°C days above a base temperature of 2.4°C: this is, for example, 30 days at an average temperature of 8.6°C. If rainfall were rare, the generation time might be substantially longer than this. However, in a rainy period, the generation time would correspond closely to the latent period. For systemic virus diseases, the generation time may also be related to vector biology since, with a slow-moving vector, it may on average take a considerable extra time after a plant becomes infectious before it actually causes further infections.

'Short-term' may be used to refer to a few pathogen generations. For a rapidly reproducing pathogen such as *Phytophthora infestans* (cause of late blight of potato), this may mean a few weeks; for a forest pathogen like Cocoa swollen shoot badnavirus (CSSV), it may mean a few years. However, patterns of short-term change may differ greatly depending on whether the host population is also changing. Beresford and Royle (1988) introduced the term 'pathocron' defined as the ratio of the latent period to the 'phyllocron' or leaf emergence interval. They stressed its usefulness in clarifying how host growth affected the expression of disease. The case of an annually reproducing pathogen like *Sclerotinia sclerotiorum* (cause of pink rot of celery and other diseases of vegetables, and stem rot of sunflower) suggests that, in general, it will be useful to consider the *relative* time-scales over which pathogen and host populations change. For example, although CSSV has a long 'generation' time, its host the cocoa tree (*Theobroma cacao)* has an even longer generation time and the host population will normally change little over one or a few pathogen generations. Similarly, a potato crop stops producing new leaves a few months before harvest, while the life cycle of *P. infestans* can be completed in a week or so. So, once the crop has stopped producing new leaves, it is possible to ask how the potato blight epidemic develops for the remainder of the season in an approximately *fixed* host population; after harvest, it is possible to ask how the population in the tubers changes during the off-season, without considering changes in the tuber population.

Many very interesting processes require the simultaneous study of host and pathogen populations because both are changing at comparable rates. In the example of *P. infestans*, it is not possible to study the transfer of fungus from a predominantly tuber-borne population in the off-season to a predominantly foliar population in the growing season without considering how both populations are changing. Similarly, a fungus like *Sclerotinia sclerotiorum* infects annual crops like oilseed rape and sunflower once a year, so the generation times of host and pathogen are similar and neither population can be considered fixed.

7.4 CHANGES IN POPULATIONS

Changes in population size may come from birth, death, immigration or emigration (Begon *et al.*, 1996). Changes in the stage-structure of a population may arise in the same way: individuals may progress from one stage to the next (for example, cleistothecia → immature mycelia) or may enter the stage from a previous one in the life-cycle (for example, immature mycelium → sporulating mycelium). Immigration and emigration will usually only occur at certain stages.

If a population is described by specifying the number of individuals in each of a series of stages, matrices can be used to model the population and the ways in which it will change (Caswell, 2000). Such models describe the change of the population in jumps, specifying the numbers and structure at one time and then attempting to specify how these will have changed at a definite time, one time-step, later. The time-step appropriate depends on the biology of the organism and the time-scale of interest; for some pathogens, changes from hour to hour may be appropriate, for others, changes from year to year may be useful to study.

The data describing the population can be set out in a vector or list of numbers in each stage. The changes to this list from one time-period to the next can be set out in a matrix or table showing two things: the proportion of each stage which will have advanced to the next stage (for example, from latent infection to sporulating infection, or from sporulation to over-seasoning stage), and the number of new-born individuals. The number of new-borns will be proportional, among other things, to the total number of propagules produced by all the infectious stages of the pathogen. Emigration will usually be implicit in such a representation, causing a reduction in the number of births, since a live pathogen individual is usually associated with a plant host that cannot move far.

Immigration can be included by adding to each stage in the population vector the number of individuals in the stage that immigrate in one time-step. Often, it is useful to consider a large area of host, so that almost all new infections come from parent infections or over-seasoning stages within the crop and emigration and immigration can be ignored. How large an area this is depends on how the pathogen propagules move around. Fungal spores moved by rain splash (see also Chapter 16) move only a few metres (Madden, 1992), so a field 50 m on each side would be a reasonably self-contained population. It is the minimum dimension of the area that matters in this context. A hectare of ground consisting of a 1 m wide strip 10 km long is unlikely to be a closed system. On the other hand, for a virus moved by whitefly, such as African Cassava Mosaic geminivirus (Colvin *et al.*, 2004; see also Chapter 20) or a wind-blown pathogen such as powdery mildew of cereals, an important, if small, fraction of infections may occur hundreds, thousands or tens of thousands of metres from the source plant (Brown and Hovmøller, 2002). In such cases, areas measuring many kilometres in both dimensions are necessary in order to have an isolated system. For many purposes, this is not appropriate and immigration (especially) needs to be considered explicitly.

7.5 DENSITY-DEPENDENT AND DENSITY INDEPENDENT FACTORS

The factors controlling the growth rate of a pathogen population may be completely uninfluenced by the population: temperature, rainfall, or developmental changes in the susceptibility of hosts are good examples. These factors are as likely to increase a population when it is small as when it is large. If controlled entirely by such factors, a pathogen population would change by a random walk, with no predictable long-term trend and the certainty of eventual extinction when a long run of bad seasons occurred, or when it exterminated its host. However, there are also many factors that influence population growth rates in ways which tend to *regulate* the population, that is, to return it to some central value. These operate over both short and long time-scales. Over short time-scales, rare pathogens will not trigger secondary defences in their hosts, will not support large populations of hyperparasites or mycoviruses and will not compete with each other for infection sites or host tissue; common pathogens will do all three. Over longer time-scales, rare pathogens will not cause rapid evolution of resistance in their hosts and are not liable to reduce their host population density, while common pathogens will select strongly for resistance and may tend to reduce their host population, hence intensifying competition within the pathogen population. Examples of all types of factors will be given in subsequent sections.

7.6 SHORT-TERM CHANGE IN A STATIC HOST POPULATION

As discussed in section 7.3, the notion of short-term change encompasses a wide range of actual time-scales. A lettuce crop may be in the ground for only 12 weeks; a tree crop 150 years. Whether the host population can be considered constant over this time-scale depends on the nature of the pathogen: for a systemic virus, a single lettuce crop may represent a fixed population but for a pathogen whose host was leaf or root tissue, there would be considerable host turnover during the season.

It is useful to begin by considering a very simple setting in which there is negligible immigration and the host population does not change over the time to be considered. In this setting, a pathogen population may change in two ways. First, numbers overall may grow or decline. Second, the stage-distribution may alter. This in turn involves two types of change: the relative balance between active infectious phases of the pathogen, as when pathogen individuals advance from a latent to a sporulating relationship with their host, and alternate or resting stages that may increase or disappear as when over-seasoning or sexual stages develop. For example, consider an apple orchard in which there is a population of *Venturia inaequalis* (cause of apple scab). At the start of the host-growing season, the population is small and exists as perithecia and inactive mycelium on twigs. Then as the season develops, these stages decline and disappear, while an asexual population increases. At the end of the season, the asexual population generates over-wintering forms once again (see also Chapter 18).

It is useful to imagine a thought experiment in which a single young pathogen individual is placed in an otherwise healthy plant population. In the case of the example in the previous paragraph, this would be a newly infected apple leaf. Each subsequent unit infected by this initial one is immediately removed and replaced by a new healthy one, until no new infections are produced. The total count of new units infected is known as R_0. Real variants of this experiment involve complications but can be done: for example, van den Bosch (1988) measured R_0 for *Puccinia striiformis* (cause of yellow rust) growing on wheat in the Netherlands, between the start of stem extension and flowering, as 55 ± 16; for *Peronospora farinosa* (downy mildew) on spinach (*Spinacia oleracea*) in the Netherlands in the autumn the figure was 3 ± 2. It is obvious but important that a disease cannot increase in a crop unless R_0 is greater than 1: the 'threshold theorem'. This quantity is therefore of central importance in managing invasions of new disease, in predicting ranges, or in eliminating disease (Anderson and May, 1986; Gilligan, 2002).

An individual will not be infectious until some time after it has infected a host. This interval is the latent period (Butt and Royle, 1980). Then typically, infectiousness (for example, spore production) will increase before declining as the pathogen ages or runs out of food. The exact timing of the different phases will vary from individual to individual and will depend on the environment. However, it is helpful to do some thought experiments to see what would happen over some time, if such outside factors stayed the same.

To begin with, useful results can be derived by assuming that diseased hosts remain uncommon, so that the proportion of propagules which infect does not change. In this case, the conclusions important for pathology are as follows (Caswell, 2000). First, as time passes, the proportions of the pathogen population lying in each age class gradually stop changing. Second, these proportions are the same regardless of the initial age structure of the population. Third, the total population grows by a constant factor each day. Since the age structure is constant, the number of pathogen individuals at any given stage also grows exponentially at the same rate as the whole population. Because the age structure does not depend on the initial composition of the population, this exponential growth rate is characteristic of the host, pathogen and environmental circumstances but does not depend on how the epidemic started.

An important conclusion is that measuring a single stage of the population, which is often all that can be done in practice, is usually a reasonable relative measure of the entire population involved in the multiplication, although such a measure can give no information about life-stages which are not involved in the multiplication process. Thus, to return to the *Venturia inaequalis* example, the relative proportions of latent individuals, sporulating individuals and conidia will tend to become constant during a period of consistent weather. By contrast, the proportion of the population existing as perithecia – either from the previous or the current season – will bear no necessary relation to the other stages.

The initial *per capita* rate of multiplication of the population is usually given the symbol r. It can be estimated from a graph of log(disease) against time before disease becomes common (say, before about 10% of tissue is infected) (Zadoks and Schein, 1979). However, it is useful to know how it is related to the basic reproductive rate R_0 and to the latent period. The exact relationship is not expressible in a single simple formula but a crude approximation is:

$$r = \frac{\log_e(R_0)}{p} \qquad (7.1)$$

(Caswell, 2000; Charlesworth, 1994; Segarra *et al.*, 2001). This is a fairly good approximation provided r is not too large. The definition of the generation time to be used needs to be something like 'the average age of the lesion at which spores are produced' (Charlesworth, 1994). This means that the rate at which disease increases in the host ('the rate of the epidemic', or the 'apparent infection rate', or the 'malthusian parameter' or the 'intrinsic rate of increase') increases only slowly with R_0 unless R_0 is close to 1, but decreases inversely with the latent period (Fig.7.1).

The growth rate of a pathogen population will be affected by the environment and by crowding; the stage-structure of the population will alter at the same time. For example, the latent period may be decreased, or infection permitted or prevented for a short time; or when host condition, or light, or temperature provides appropriate signals, individuals may change physiologically to over-seasoning or migratory life-stages.

This model can be used to explore, in a preliminary way, the way in which environmental and management factors affect populations. For example, if fungicide is to be used to control an epidemic, it must be applied long before disease becomes sufficiently severe to cause economic loss, because a large proportion of disease may be latent. The effect of fungicide on an epidemic depends on how the fungicide acts, on the latent period and on R_0. Fungicide action can conceptually be split into protectant and eradicant components. Some fungicides act only as protectants: examples are Bordeaux mixture, chlorothalanil and the dithiocarbamates; others have both protectant and eradicant activity to varying extents. Protectant fungicides act only at the point of infection and therefore alter only the infection rate of the pathogen. Eradicant activity, residing inside the plant and therefore associate with systemic fungicides only, may affect R_0 by killing entire individual mycelia, by reducing the spore production or the length of time for which spores are produced, or by lengthening the latent period. Protectant activity, no matter how effective, will allow latent infections to grow. For disease with high r but fairly long latency, with an incubation period (Butt and Royle, 1980) similar to the latent period, this implies considerable increase in disease severity after the application of fungicide.

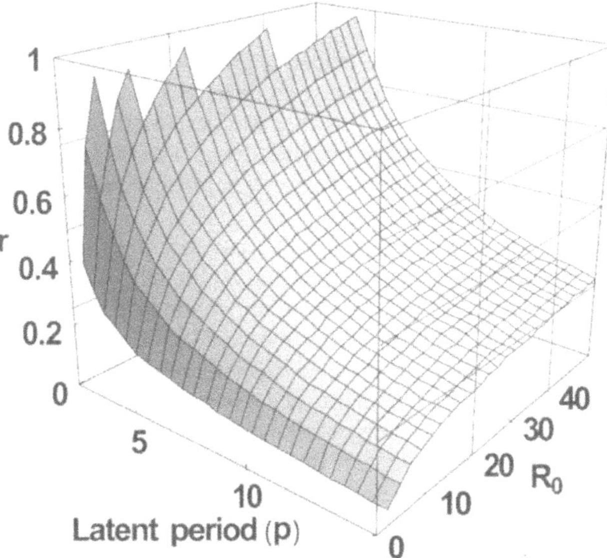

Figure 7.1. The relationship between the intrinsic rate of increase of a disease, r, *the basic reproductive rate* R_0, *and the generation time* p. *The time-units for* r *and* p *are the same. For a disease that reproduced slowly, they might be in years* $^{-1}$ *and years; for a faster disease, they might be in days*$^{-1}$ *and days. Control actions which reduce* R_0 *by a fixed proportion will work best if* R_0 *is already small; those which increase* p *by a fixed proportion will work best if* p *is small.*

Systemic fungicides, such as the azole group or the phenylamides, may have both eradicant and protectant effects. They may kill a proportion of established and growing pathogen individuals, they may prevent a proportion of new infections by interfering with infection processes or germination and they may lengthen latent periods, in some cases very dramatically. This means that they may be able to work better on diseases with high rates but long latent periods, because the latent disease will actually be eliminated to some extent. It will still be true that, if R_0 is large, changes in R_0 will have only slight effect on r. However, if latent periods of surviving pathogen individuals are lengthened, the control at any given actual time will be better than with the protectant, because fewer latent periods will have passed.

7.7 AFFECTED HOST TISSUE AND PATHOGEN MULTIPLY AT COMPARABLE RATES

This is the common situation with foliar and root diseases in arable agricultural and annual horticultural crops. After harvest, the pathogen population usually decreases over the off-season. Then the crop starts to grow again, typically producing leaves and roots in a roughly exponential pattern until competition slows growth to a steady rate, before ceasing at maturity or flowering. This 'expolinear' pattern (Goudriaan, 1973) has been shown to apply to many crops, from *Vicia* beans to wheat, although the rate of growth will vary, particularly because of temperature variations. Nonetheless many crops, such as cereals, produce new leaves at the same time as older leaves are dying off. Net crop growth arises because the new leaves appearing are more numerous or bigger than the old ones dying. Typical crop doubling times at the start of the growing season are a week or less, corresponding to leaf area index increasing by around 10% per day.

The pathogen population dynamics in this case are determined by the relative rates of multiplication of host and pathogen and by the fate of the pathogen on dead host tissue. Because both host and pathogen are continually being born and dying, disease severity may be a very poor measure of the pathogen population. Furthermore, the deductions of the previous section no longer apply and many more patterns of growth or decline are possible. If the pathogen can exist and sporulate on dead tissue, it may be necessary only to allow for host growth (Johnson and Teng, 1990; Lalancette and Hickey, 1986). If the pathogen is multiplying faster than the host tissue, disease severity and the pathogen population will increase; if the reverse occurs, the population will decline.

Host tissue born recently, within one latent period of the present, cannot be visibly diseased. If the pathogen did not affect the death rate of host tissue, the pathogen population would tend to increase so that all host tissue becomes infected, even if invisibly. However, most pathogens tend to accelerate the death of host tissue on which they are growing and more heavily infected tissue will tend to die faster than lightly infected tissue. If the pathogen also dies when the host dies, then an equilibrium may exist, with a steady but dynamic population density substantially less than the maximum possible (Shaw and Peters, 1994). Furthermore, the effect of the host tissue dying is to reduce the standing population of pathogen substantially *more* than the population of the host. In effect, therefore, the host is regulating the pathogen population; this can actually be an evolutionary force leading to more rapid death of infected tissue (van den Berg and van den Bosch, 2004).

There is evidence for temporary equilibria like this existing in some UK grasslands, where disease levels in perennial plants, principally the grass *Holcus lanatus*, were steady at about 6% of green tissue infected throughout the summer in several years (Peters, 1994). This may also be an informative way to look at cereal disease levels during the vegetative growth phase, although it is unlikely that an equilibrium would be reached because there is rapid net growth of the crop throughout the season.

Once the crop reaches maturity and tissue turnover ceases, disease will tend to increase according to the patterns discussed in the previous section. However, the lifetime of host tissue is often only one or a few latent periods long, so this epidemic phase will usually be short and will often start from a high initial inoculum level represented by the standing population during vegetative growth.

7.8 CHANGES OVER TIME-SCALES LONGER THAN EITHER CROP OR PATHOGEN LIFETIME

7.8.1 Possible dynamical patterns

It is helpful to distinguish internal and external forces acting on a population. This is largely an artificial distinction but makes it possible to think about patterns of change in a structured way. If mathematical models of how populations change over time are used, internal forces are those that change only because of processes described by the model. External forces are those that change in particular ways over time or at specific times, in ways for which the model offers no explanation.

The simplest type of model includes only the pathogen population. In a constant host population and a steady environment, if time lags are not considered, this kind of model will always show the pathogen population tending to return to an equilibrium level from arbitrary starting conditions. If environmental disturbances alter the position of the equilibrium, the population will tend to track these disturbances.

However, the state of the pathogen population at some past times may influence how it now grows: this influence may arise from the age structure, or from changes induced in the host. In this case, other, more complicated dynamical patterns can arise. Even in a steady environment, a pathogen population may tend to oscillate regularly between two or more distinct levels, or to oscillate irregularly in a quasi-periodic or chaotic pattern. A quasi-periodic pattern is one in which there are two fundamental frequencies of oscillation in population numbers present but the ratio of the two frequencies is an irrational number, so no matter how long a time is considered, no multiple of one frequency is ever equal to a multiple of the other. A chaotic pattern is one in which very small changes in population numbers cause very large changes in the population at later times and in which, even in a steady environment, no regularly repeating sequence of population numbers ever appears.

In general, populations are more likely to show complicated behaviour if they both (a) change very fast (on their natural time-scale as discussed in section 7.3) and (b) are strongly influenced by their state or size a considerable time before. The main ways in which the past state of a population is likely to influence its present growing conditions are through interactions with its host, through interactions with other populations which compete with it or consume it by parasitism or predation, or through strong natural selection operating differently on different sizes or types of population.

A model including the host population must have at least two variables, one each to represent the host and the pathogen populations. If the pathogen has a free-living stage, or the host can become immune, more complex models still may be appropriate. Any model with more than two variables considered, or which includes time lags, *may* show complicated dynamics. Simple models of two interacting populations in continuous time, without time lags, are often simple to construct and may capture the essentials of host-pathogen interaction. Mathematically, however, they can *only* converge to an equilibrium or to steady repeating oscillations. This can be a trap, since apparently similar descriptions of a pathosystem may be capable of showing very different behaviour and it may be difficult to distinguish the models by comparison with the very short runs of data that are available.

7.8.2 Host-pathogen population linkages

In agricultural settings, the host population may be only indirectly linked to the pathogen population. In the most extreme case the same crop cultivar (or one with equivalent disease resistance) may be planted, regardless of the disease level. This may occur because there is no economic alternative, as with black Sigatoka and Sigatoka diseases of banana caused by *Mycosphaerella fijiensis* and *M. musicola* (Marin *et al.*, 2003), where both marketing and breeding constraints restrict the available cultivars, or it may simply be that agronomic and market imperatives override consideration of disease resistance in choice of a cultivar, as with the growth of wheat susceptible to *M. graminicola* in the UK during the 1980s and 1990s (Bayles, 1991). In other cases, planting areas of cultivars that suffer severe disease may decrease, as in the classic boom-bust cycles of gene deployment in cereals (Wolfe, 1984).

In some natural habitats, by contrast, the success of reproduction by the host is directly linked to that of the pathogen. For example, *Linum marginale* is a wild relative of flax native to Australia. Plants infected with *Melampsora lini* (flax rust) do not perennate, whereas healthy plants may survive many years (Burdon and Thompson, 1995). Likewise, Carlsson and Elmqvist (1992) showed that populations of *Silene alba* infected with *Microbotryum violacearum* smut were smaller than those which were uninfected. They also showed that this was at least in part because the populations were seed limited and seed output was reduced by infection with the smut.

To analyze long-term dynamics, it is necessary to study population changes over the entire pathogen life-cycle, including if they exist, seasons where a new host is unavailable for infection: often in the dry season(s) in tropical areas, typically winter in temperate countries. The simplest example to examine is an annual crop plant, affected by a disease that infects entire plants annually from an over-seasoning source, then grows within plants and produces new propagules which will carry the disease forward to the next season. A well-known example would be *Sclerotinia sclerotiorum* infecting oilseed rape (canola) and a range of other crops. Wind-blown ascospores initially colonise fallen petals, producing a mycelium which is then capable of penetrating a leaf on which the petal may be lying. These leaf infections become systemic and the pathogen eventually produces cankers in the stem containing sclerotia which will survive in the soil until the following spring when sclerotia near the soil surface produce apothecia from which ascospores are discharged to infect the new season's crop. Thus, to understand population changes it is only necessary to consider the host and pathogen population at the time of flowering in each year. If the same host area is sown each year and conditions are similar on average each year, the pathogen population will tend to change by a constant factor each year. If only one individual existed in the first year, producing R_0 individuals in the following year, we would expect exponential growth, with R_0^2 in the following year and so on. Clearly, for the population to persist, R_0 must be more than 1. As stated earlier, an aim of management is to reduce R_0 as much as possible.

Oilseed rape is generally grown as part of a rotation. Infection will come from old sclerotia within the same field (since some sclerotia survive for several years) and from spores from more distant sources. The distant sources affect R_0 in a rather simple way since the ascospores are wind-blown and may be regarded as distributed more or less at random over a wide area. If the crop area is doubled, twice as many of these spores will land on susceptible tissue. The number of locally generated spores depends on the time since the last susceptible crop was grown. Since this depends on the particular rotation used on a farm, it may depend rather little on the area of crop grown regionally.

This also provides a way to think about the long-term dynamics of diseases that have a phase of rapid population growth during the crop season, like those discussed in sections 7.6 and 7.7. In those sections, the population at the start of the season was simply assumed to be a known quantity. However, the start of active host growth is simply the low point of the pathogen population cycle, while the end of that season is usually the high point. A pathogen population at a single time in the year can therefore be thought of as being in balance between an increase during the host's growing season and a decrease during the off-season. This is usually combined with a change from one phase to another, for example from the conidial stage to the sexual stage, or from mycelium to sclerotia. The chances that the increase would balance the decrease exactly over millennia are very small but, if the factor of decrease were greater than the factor of increase, the pathogen would become extinct. In fact, this would happen if the decrease outweighed the increase for more than a few years.

There are two possible ways to explain why pathogens do not become extinct. In the first place, it is unlikely that conditions over large areas would be exactly synchronised. Extinction would require simultaneous coincidences of poor conditions over the whole of the pathogen range. The time-scale for this may be so long that it very rarely occurs even over millions of years. An alternative explanation is that the potential increase during the cropping season is on average greater than the decrease in the off-season but that the population size must be limited during the growing season by some factor or factors that are density-dependent. The next two sub-sections discuss such factors.

7.8.3 Parasitism and predation

In very many cases, as already mentioned, it appears that the growth rate of an epidemic within an annual crop slows long before crowding for space is plausible, even if it is assumed that the area visibly damaged by a disease is much smaller than the area from which nutrients are withdrawn and direct competition with another pathogen individual can occur. This suggests the operation of a density-dependent factor early on. Perhaps the most obvious factor is that attack by a pathogen will stimulate various types of acquired resistance: hence the success rate of fungal spores on a new, juvenile host may be much greater than on a host already attacked (successfully or unsuccessfully) by other spores. However, Fleming (1980) has argued forcibly that the role of predation and hyperparasitism in reducing the growth rate of pathogen populations might be much greater than commonly recognised.

Off-season survival may also be density-dependent because of natural enemies. For example, sclerotia of *Sclerotinia* spp. are attacked by various hyperparasites, including both fungi and viruses (Boland, 2004; Budge and Whipps, 2001; Burgess and Hepworth, 1996). As *Sclerotinia* becomes commoner, so its hyperparasites will become more common and overwinter survival of the sclerotia will decrease. This means that natural regulation by such agents will tend to occur only when the pathogen is common.

Such density-dependent processes can lead to equilibria in which both pathogen and hyperparasite have an annual cycle in abundance, the pathogen peak populations being lower than they would otherwise be. If population growth rates are very rapid, the result could be massive fluctuations between years without simple repeating patterns: chaotic fluctuations (Shaw, 1994b). Chance effects would be likely to lead to extinction for the hyperparasite or for both hyperparasite and pathogen, unless the pathogen is sufficiently widespread (or its dispersal sufficiently restricted) that cycles in different parts of its range do not coincide.

An exception to the general case that hyperparasites cannot maintain pathogens at low density has been described in *Salix* species grown for fuelwood by Morris and Royle (1993). Two pathovars of *Melampsora epitaea* rust infect plantations in the south-west of England; one infects leaves, the other stems (Pei and Ruiz, 2000). The stem infecting form overwinters as cankers which are susceptible to the hyperparasite *Sphaerellopsis filum* and support a standing population of this hyperparasite. Where this pathovar of rust is present, the rate of increase of the leaf-infecting form is much reduced because the hyperparasite:host ratio is quite high even at the start of the summer, when leaves appear. This results in much lower peak populations of rust and there is obvious interest in attempting to manipulate this situation to provide biological control of the rust. The complexity of the biology in this case is substantial: an adequate model might assume that there was a fixed population of trees but that the lifetime of leaves, the host of the leaf-inhabiting form of the rust, is short. This means that a model would have to include, as distinct variables, the population of leaves, both forms of healthy rust, and both forms of rust infected by the hyperparasite.

7.8.4 Competition

Pathogens may compete simply by depleting the space or nutrients available in a host (exploitative competition), by some more active interaction such as the excretion of antibiotics (interference competition) or, indirectly, by triggering changes in their hosts, or supporting larger populations of hyperparasites than would otherwise exist; (the example at the end of section 7.8.3 can be seen as a competitive one: in the presence of a population of the stem-infecting rust, the leaf-infecting rust has a smaller population).

The degree to which two species compete is determined by the similarity with which they use available resources. Two species that use resources identically will not be able to co-exist indefinitely in a stable environment. One will inevitably be very slightly more efficient and will grow more rapidly than the other, supplanting it slowly or rapidly. This is the competitive exclusion principle (Begon *et al.*, 1996).

At first sight, this conclusion is incompatible with the regular occurrence, on the same plant organ, of many similar pathogens: for example, in the UK on wheat, it is easy to find leaves simultaneously infected with *Mycosphaerella graminicola, Didymella exitialis* and *Phaeosphaeria nodorum*. In fact, several of the assumptions underlying the competitive exclusion principle are violated: hosts vary dramatically in abundance and quality during the year, so that the environment for a plant pathogen is never stable; pathogens and hosts are spatially aggregated, so that the strength of competition between species is reduced relative to that between related individuals within a dense patch of disease and there may well be subtle differences in resource use by apparently similar pathogen individuals, related to temperature and wetness requirements for infection, preferred age of host tissue, etc. On the other hand, the apparent replacement on banana of *Mycosphaerella musicola* by the more pathogenic *Mycosphaerella fijiensis* (cause of sigatoka disease) in parts of the world where both have occurred (Mourichon and Fullerton, 1990) is in accord with the theory.

It seems common for competition between plant pathogens to be very asymmetric. For example, *Phaeosphaeria nodorum* reduces the growth rate of *Blumeria graminis* populations on the same leaves but preferentially infects leaves infected with *B. graminis* (Weber *et al.*, 1994). Similar predisposing effects are known in a number of pathogen combinations but evidence on epidemic progress in both pathogens is usually lacking (Brokenshire, 1974; Jones and Jenkins, 1978; Madariaga and Scharen, 1986; Stahle and Kranz, 1984).

An interesting case where competition has been revealed by agricultural change is the balance among the eyespot pathogens *Oculimacula yallundae* and *O. acuformis* and the sharp eyespot pathogen *Rhizoctonia cerealis*. Treatment of north European wheat crops with demethylation-inhibiting fungicides has had two effects. First, an increase in population density of *R. cerealis* correlates well in individual fields with a decrease in *Oculimacula* spp. and this is due to antagonism (Kapoor and Hoffmann, 1984). Second, *O. acuformis* increases relative to *O. yallundae* (Bateman *et al.*, 1995). This is more likely to be because of differential sensitivity to fungicide treatment than competitive release because the absolute numbers of *O. acuformis* do not necessarily increase and *O. acuformis* appears to multiply later in the season than *O. yallundae* (Goulds and Fitt, 1988), so it may partly escape the effects of spring fungicides.

It also appears that populations of *Oculimacula* spp. are partly controlled by competition with true saprophytes (Jalaluddin and Jenkyn, 1996). The evidence is that in fields in which the straw from preceding wheat crops was regularly incorporated, *Oculimacula* infection of subsequent crops was less than in fields in which the straw was burnt. The burning reduced the number of *Oculimacula* individuals surviving from the crop. However, it also reduced the amount of straw incorporated and so reduced the population of lignin- and cellulose-decomposing saprophytic fungi. In fields with straw incorporated, the straw on which the overseasoning *Oculimacula* survived was rapidly colonised by specialist decomposers and the *Oculimacula* was deprived of nutrients, so that fewer spores were produced than in the burnt fields.

7.8.5 Co-evolution of host and pathogen

The linkage between host and pathogen populations sets up a co-evolutionary race between pathogen and host, the host evolving towards resistance and the pathogen to virulence. In agriculture this can produce a boom and bust cycle of repeated release and failure of cultivars (Barrett, 1988). In natural settings also, the consequences of this evolution may be far from a smooth progression (Gilbert, 2002). For example, in the *Linum* case (Burdon, 1993), the rust can multiply dramatically within a season and therefore the genetic structure of the population may adapt very quickly to that of the host. The host, however, has a seed-bank. This will be well-stocked with the survivors of past years of severe disease, to which the pathogen may now be ill-adapted. This is likely to produce extremely complicated population dynamics but no models have yet been published.

In agricultural settings, the host population is linked to the pathogen population through farmers' and breeders' responses to disease. Much of the detail in the evolutionary and dynamical response of a pathogen population to the host composition depends on chance events (Shaw, 1994a). For example, in the mid 1980s barley cv. Triumph was widely grown in the UK, carrying effective resistance alleles to *Blumeria graminis*. After virulence against these alleles became common, almost half the population of *B. graminis* was represented by a single phenotype carrying at least three virulence alleles not needed on any variety then grown widely (Brown and Wolfe, 1990). This is almost certainly because the necessary mutations for virulence on cv. Triumph occurred in a single individual already carrying the extra virulences. The result was an abundant pathogen population, composed of a few very common clones and many much rarer ones (Brown *et al.*, 1990).

In populations of aerially dispersed pathogens, once a virulence allele is present at a moderate frequency its fate depends on how common the matching resistance is, and on the strength of selection against the virulence over the whole life-cycle of the pathogen. Many virulence alleles appear to carry moderately small selective disadvantages in the absence of the corresponding resistance, and will persist at moderate frequencies for many years. If the matching host resistance is reintroduced, or introduced into a new area, the population of the pathogen carrying the matching virulence allele is likely to become common faster than when the cultivar was first introduced, because the allele is already present in the population.

Wolfe and McDermott (1994) conclude that 're-cycling' of alleles in *B. graminis* has 'little prospect except in the short-term in limited locations'.

Under exceptional circumstances, the long-distance movement characteristic of aerially dispersed pathogens may make selection for virulence ineffective. For example, in the Canadian great plains wheat cv. Selkirk remained resistant to *Puccinia graminis* (cause of black stem rust) for many years. It contained a suite of at least six resistance alleles, to all of which individually virulence was present within North America. However, the rust did not overwinter in the area where Selkirk was grown, but migrated in each season from the south. The pathogen population probably did not evolve resistance because the precise combination of resistance alleles in Selkirk was never used in the hosts of the overwintering population (Wolfe and Knott, 1982).

With soil-borne pathogens of restricted mobility, the genetic structure of a population is likely to be much more fragmented, and rapid replacement of the whole genetic structure of the population is less likely. Use of single resistance genes against some soil-borne pathogens has been successful in some cases, and even where the pathogen has evolved to overcome the resistance in one place this is followed by sporadic rather than rapid and universal spatial spread.

There is increasing evidence that populations of many trees are limited by the build-up of root pathogens in the soil around a mature tree (Augspurger, 1983; Hood *et al.*, 2004; Packer and Clay, 2004). Most seedlings fall close to the parent tree and are killed by root damage from soil pathogens which have multiplied on the parent roots, without killing it because of its size and reserves. Only rare individual seedlings germinating far from the parent survive. This seems likely to produce a particularly fine-grained mosaic of host and pathogen co-adaptation, since selection on the pathogen around one parent tree will be long-continued, but surviving seedlings nearby are likely to select for different strains.

7.9 SPATIAL POPULATION STRUCTURE

As section 7.8.5 showed, the spatial structure of populations can have a profound influence on population dynamics. It can also make the study of dynamics much harder, because the population studied may, in fact, be composed of several independent populations which are being unknowingly averaged, or the population may be part of a much larger system and be controlled by factors which cannot be observed within the area studied.

In some pathosystems, vectors or propagules can move long distances (see also Chapter 6) (Brown and Hovmøller, 2002). Examples include rusts and powdery mildews of cereals or many whitefly- and aphid-borne viruses (Irwin and Thresh, 1988). In such cases it is not sensible to consider a crop area isolated unless it is a long distance from other such crops; often, this may mean many kilometres. The dynamics of the pathogen within a defined crop area will, therefore, be determined both by processes within the crop and by immigration and emigration of propagules from the surrounding crops. This can make experiments on control very difficult, because the small crop areas desirable to make it feasible to include many treatments may not only not be independent but will have population dynamics very different from a larger, effectively closed, system treated in the same way. For example, spore transport by wind may be best modelled by an inverse power-law. One way to understand this is to visualise spores travelling in wind, dying or settling out only slowly, so that their trajectories spread out in proportion to distance. In this case the chance of a spore passing through a point would decrease inversely to distance. In practice, of course, many other factors reduce long-distance transport, but more realistic models of turbulent atmospheric dispersal do predict a power-law relationship between distance and probability of a spore reaching a given place. The relationship has been observed for both pollen and spores (Aylor, 2003; Fitt *et al.*, 1987) and has been used in several models (Minogue, 1989; Paradis *et al.*, 2002; Xu and Madden, 2004).

The consequences of such a relationship can be quite surprising. If spores disperse approximately according to an inverse square law at intermediate scales (for example, falling to 1/41 m from the source, 1/92 m from the source and so on) and an experiment were conducted in a large uniform crop area, then a disc-shaped plot 10 m across would receive the overwhelming majority of its spores from outside (see Appendix 7A). Even with dispersal according to an inverse 2.5 power, such a plot would receive almost equal contributions from inside and outside. In practice, most spores do not escape a canopy to take part in long-distance air movements, so the position is slightly less extreme than this; but plots of the order of 100 m^2 should not be assumed to be independent units.

For example, Bierman *et al.* (2002) reported experiments on fungicide resistance selection of the eyespot pathogens, *Oculimacula* spp. They found that in the absence of selection by carbendazim, (1) the frequency of resistance to the fungicide in plots starting with a high frequency declined with time but that (2) the frequency stabilised at an intermediate level. This indicates either frequency-dependent selection for the trait - in other words, the resistance is valuable to an individual possessing it, even in the absence of fungicide - or that immigration from treated plots is common enough to counterbalance selection against the resistance. It would not be easy to distinguish the two hypotheses by field observations unless the experiment was designed with the problem in mind and the populations were carefully genetically marked (see also Chapter 3).

A slightly different example is provided by a study of the likely effect of a certification scheme for leek (*Allium porri*) seedlings carried out by de Jong (1996). This estimated the equilibrium fraction of fields infected, assuming that fields might be infected by airborne spores, or by planting infected seedlings. De Jong then calculated how the proportion of infected fields depended on the proportion planted with infected leeks. Certification could only be effective if the likelihood of infection by airborne spores was low; if leek cultivation became too intensive, so that fields were very close together, this was not possible.

A further point is that such calculations refer to averages. Because small packets of air may be caught up in very large-scale movements, stochastic effects may be very large, with many spores landing a long way from their source but relatively close together. This will make the local dynamics subject to large chance fluctuations.

Foliar pathogens and (non-waterborne) root pathogens have very different mobility. Splash-borne or airborne pathogens move on scales of metres or much more, as explained in the previous paragraph. Although infections are most likely on the immediate neighbours of an infected host, pathogen propagules may reach some distance and the approximation of random contact between healthy and infected individuals is useful, although it will overestimate the rate of disease progress.

In a soilborne disease, however, propagule movement may be very limited and infection rates will be very rapidly limited because most inoculum will reinfect an already infected host (this is an extreme form of the phenomenon discussed above). Thus, within a season, soil-borne disease typically has lower rates of increase than aerially dispersed disease. Furthermore, the movement is usually more by the host than the pathogen, putting a premium on the ability of the pathogen propagules to survive until a host root comes sufficiently close. The total population size of a pathogen like this will evidently have much smaller annual fluctuations than that of a typical foliar pathogen.

Because of the limited movement of soilborne pathogens, it is quite common for an area of study to include many effectively separate populations. In this case, it may be possible to model some effects of the averaging over populations by specifying that infection rates in the model should scale as some power of the population densities of host and pathogen. For example, if this power is more than 1 for the host population, then the infection rate of hosts by the pathogen is more efficient at high density than would be predicted from the performance at low density. An example of the application of this model is an analysis of data on the lettuce pathogen *Sclerotinia minor*, attacked by the hyperparasite *Sporodesmium sclerotivorum* (Adams and Fravel, 1990). Gubbins and Gilligan (1997) showed that it was necessary to include an assumption of heterogeneity, in this implicit form, in order to adequately fit these data. Without the assumption, the hyperparasite had to be assumed to be so efficient in order to fit the overall data, that sclerotia of the pathogen would have been eliminated from the soil; in fact it survived for a long time at a low density.

Soil-inhabiting pathogens are notoriously patchy on very small scales. This is in part a function of the limited opportunity for smoothing out fluctuations in population density because of the difficulty of moving through soil and partly as a result of heterogeneity in the soil itself. However, it is also possible that variability may be generated by population processes within the soil itself. Kleczkowski *et al.* (1996), in experiments with damping-off of radish (*Raphanus)* seedlings by *Rhizoctonia solani* in the presence of the antagonistic and non-pathogenic fungus *Trichoderma viride*, showed how the development of host resistance amplified very small initial variations in rates of attack and parasitism into very large final variations in pathogen population density. In the field, such patchiness could then persist for long periods, influencing both the vegetation and the location of pathogen populations. The dynamics of soil-borne pathogens are further discussed in Chapter 14.

If hosts exist in patches, it may be sensible to regard the patch as an individual in a population on a larger scale, the metapopulation. Then the host patches can be regarded as reproducing, becoming infected, dying and so forth. The metapopulation can be studied as an entity in its own right (Hanski and Gaggiotti, 2004). Although the time-scales involved are long, processes at the metapopulation level are probably as important in determining the genetic structure and size of pathogen populations as process at the population level (Burdon and Thrall, 1999). For example, Damgaard (1999) showed that variability in the spectrum of virulence alleles present in a population can be maintained without assuming a cost of virulence, in distinction to all models based in single well-mixed populations, but making it easy to understand why it has often been hard to show any cost associated with virulence alleles.

APPENDIX 7A

It is assumed that dispersal obeys an inverse power-law with an exponent *n*, on intermediate scales. On very long scales, typical of the distance travelled by wind in a day or so, there is an exponential decline, with a scale parameter *l*. To simplify the calculations, it is also supposed that the wind is equally likely to blow from any direction; this assumption can be relaxed without changing the important conclusions from the argument. The density of spores moving a distance *r*, from a source with density ρ per unit area is then described by:

$$k(r,\theta) \; \alpha \; e^{-\frac{r}{l}} \frac{\rho}{(1 + r/a)^n} \tag{7.2}$$

Here *a* sets the short-distance scale of dispersal. It has units of distance and the larger it is, the further spores travel. Now the density of spores arriving from within a distance r_0 can be compared with the density of spores arriving from outside that distance. Each overall density can be obtained by the mathematical technique of integrating over all possible directions and all possible distances less than or greater than r_0. The result of this can be expressed analytically, but is very complex. The numbers in the text are obtained by substituting $r_0 = 10$; $a = 1$; $l = 100$ and $n = 2$ or 2.5; the qualitative results are very little affected by the value of *l* chosen, provided it is much larger than r_0.

REFERENCES

Adams, P.B. and Fravel, D.R. (1990) Economical biological control of *Sclerotinia* lettuce drop by *Sporodesmium sclerotivorum*. *Phytopathology*, **80**, 1120-1124.

Anderson, R.M. and May, R.M. (1986) The invasion, persistence and spread of infectious diseases within animal and plant communities. *Philosophical Transactions of the Royal Society, B*, **314**, 533-570.

Augspurger, C. (1983) Seed dispersal of the tropical tree *Platypodium elegans* and the escape of its seedlings from fungal pathogens. *Journal of Ecology*, **71**, 759-771.

Aylor, D.E. (2003) An aerobiological framework for assessing cross-pollination in maize. *Agricultural and Forest Meteorology*, **119**, 111-129.

Barrett, J.A. (1988) Frequency-dependent selection in plant fungal interactions. *Philosophical Transactions of the Royal Society of London B*, **319**, 473-483.

Bateman, G.L., Dyer, P.S. and Manzhula, L. (1995) Development of apothecia of *Tapesia yallundae* in contrasting populations selected by fungicides. *European Journal of Plant Pathology*, **101**, 695-699.

Bayles, R.A. (1991) Varietal resistance as a factor contributing to the increased importance of *Septoria tritici* Rob. and Desm. in the UK wheat crop. *Plant Varieties and Seeds*, **4**, 177-183.

Begon, M., Mortimer, M. and Thompson, D.J. (1996) *Population Ecology: a unified study of animals and plants*. (3rd edn), Blackwell, Oxford.

Beresford, R.M. and Royle, D.J. (1988) Relationships between leaf emergence and latent period for leaf rust (*Puccinia hordei*) on spring barley, and their significance for disease monitoring. *Zeitschrift für Pflanzenkrankheiten und Pflanzenschutz*, **95**, 361-371.

Bierman, S.M., Fitt, B.D.L., van den Bosch, F. *et al*., (2002) Changes in populations of the eyespot fungi *Tapesia yallundae* and *T. acuformis* under different fungicide regimes in successive crops of winter wheat, 1984-2000. *Plant Pathology*, **51**, 191-201.

Boland, G.J. (2004) Fungal viruses, hypovirulence, and biological control of *Sclerotinia* species. *Canadian Journal of Plant Pathology-Revue Canadienne De Phytopathologie*, **26**, 6-18.

Brokenshire, T. (1974) Predisposition of wheat to *Septoria* infection following attack by *Erysiphe*. *Transactions of the British Mycological Society*, **63**, 393-397.

Brown, J.K.M. and Hovmøller, M.S. (2002) Aerial dispersal of pathogens on the global and continental scales and its impact on plant disease. *Science*, **297**, 537-541.

Brown, J.K.M., O'Dell, M., Simpson, C.G. and Wolfe, M.S. (1990) The use of DNA polymorphisms to test hypotheses about a population of *Erysiphe graminis* f.sp. *hordei*. *Plant Pathology*, **39**, 391-401.

Brown, J.K.M. and Wolfe, M.S. (1990) Structure and evolution of a population of *Erysiphe graminis* f.sp. *hordei*. *Plant Pathology*, **39**, 376-390.

Budge, S.P. and Whipps, J.M. (2001) Potential for integrated control of *Sclerotinia sclerotiorum* in glasshouse lettuce using *Coniothyrium minitans* and reduced fungicide application. *Phytopathology*, **91**, 221-227.

Burdon, J.J. (1993) The structure of pathogen populations in natural plant communities. *Annual Review of Phytopathology*, **31**, 305-323.

Burdon, J.J. and Thompson, J.N. (1995) Changed patterns of resistance in a population of *Linum marginale* attacked by the rust pathogen *Melampsora lini*. *Journal of Ecology*, **83**, 199-206.

Burdon, J.J. and Thrall, P.H. (1999) Spatial and temporal patterns in coevolving plant and pathogen associations. *American Naturalist*, **153**, Supplement, S15-33.

Burgess, D.R. and Hepworth, G. (1996) Biocontrol of sclerotinia stem rot (*Sclerotinia minor*) in sunflower by seed treatment with *Gliocladium virens*. *Plant Pathology*, **45**, 583-592.

Butt, D. and Royle, D. (1980) The importance of terms and definitions for a conceptually unified epidemiology, in *Comparative epidemiology: a tool for better disease management* (eds J. Palti and J. Kranz), pp. 29-45. PUDOC, Wageningen, The Netherlands.

Carlsson, U. and Elmqvist, T. (1992) Epidemiology of anther smut disease (*Microbotryum violaceum*) and numeric regulation of populations of *Silene dioica*. *Oecologia*, **90**, 509-517.

Caswell, H. (2000) *Matrix Population Models: Construction, Analysis and Interpretation* (2nd edn), Sinauer, Sunderland, Massachusetts.

Charlesworth, B. (1994) *Evolution in age-structured populations, vol. 1* (2nd edn), Cambridge University Press, Cambridge.

Colvin, J., Omongo, C.A., Maruthi, M.N. *et al*. (2004) Dual begomovirus infections and high *Bemisia tabaci* populations: two factors driving the spread of a cassava mosaic disease pandemic. *Plant Pathology*, **53**, 577-584.

Damgaard, C. (1999) Coevolution of a Plant Host+Pathogen Gene-for-gene System in a Metapopulation Model without Cost of Resistance or Cost of Virulence. *Journal of Theoretical Biology*, **201**, 1-12.

de Jong, P.D. (1996) A model to study the effect of certification of planting material on the occurrence of leek rust. *European Journal of Plant Pathology*, **102**, 293-295.

Fitt, B.D.L., Gregory, P.H., Todd, A.D. *et al*., (1987) Spore dispersal and plant disease gradients: a comparison between two empirical models. *Journal of Phytopathology*, **118**, 227-242.

Fleming, R.A. (1980) The potential for control of cereal rust by natural enemies. *Theoretical Population Biology*, **18**, 374-395.

Gilbert, G.S. (2002) Evolutionary ecology of plant diseases in natural ecosystems. *Annual Review of Phytopathology*, **40**, 13-43.

Gilligan, C.A. (2002) An epidemiological framework for disease management. *Advances in Botanical Research*, **38**, 1-64.

Goudriaan, J. (1973) Dispersion in simulation models of population growth and salt movement in soil. *Netherlands Journal of Agricultural Science*, **21**, 269-281.

Goulds, A. and Fitt, B.D.L. (1988) The comparative epidemiology of eyespot *(Pseudocercosporella herpotrichoides)* types in winter cereal crops. Paper presented at the Brighton Crop Protection Conference, Brighton, UK.

Gregory, P.H. (1973) *The microbiology of the atmosphere* (2nd edn). Leonard Hill, London.

Gubbins, S. and Gilligan, C.A. (1997) A test of heterogeneous mixing as a mechanism for ecological persistence in a disturbed environment. *Proceedings of the Royal Society B*, **264**, 227-232.

Hanski, I.A. and Gaggiotti, O.E. (eds) (2004) *Ecology, genetics, and evolution of metapopulations*. Academic Press, London.

Hardman, J.M., Pike, D.J. and Dick, M.W. (1989) Short-term retrievability of *Pythium* propagules in simulated soil environments. *Mycological Research*, **93**, 199-207.

Hood, L.A., Swaine, M.D. and Mason, P.A. (2004) The influence of spatial patterns of damping-off disease and arbuscular mycorrhizal colonization on tree seedling establishment in Ghanaian tropical forest soil. *Journal of Ecology*, **92**, 816-823.

Irwin, M.E. and Thresh, J.M. (1988) Long-range aerial dispersal of cereal aphids as virus vectors in North America. *Philosophical Transactions of the Royal Society B*, **321**, 421-446.

Jalaluddin, M. and Jenkyn, J.F. (1996) Effects of wheat crop debris on the sporulation and survival of *Pseudocercosporella herpotrichoides*. *Plant Pathology*, **45**, 1052-1063.

Johnson, K.B. and Teng, P.S. (1990) Coupling a disease progress model for early blight to a model of potato growth. *Phytopathology*, **80**, 416-425.

Jones, D.G. and Jenkins, P.D. (1978) Predisposing effects of eyespot (*Pseudocercosphorella herpotrichoides*) on *Septoria nodorum* infection of winter wheat. *Annals of Applied Biology*, **90**, 45-49.

Kapoor, I.J. and Hoffmann, G.M. (1984) Antagonistic action of *Pseudocercosporella herpotrichoides* on *Ceratobasidium* sp. and *Rhizoctonia cerealis* associated with foot rot of cereals. *Zeitschrift für Pflanzenkrankheiten und Pflanzenschutz*, **91**, 250-257.

Kennedy, R., Wakeham, A.J., Byrne, K.G. *et al.*, (2000) A new method to monitor airborne inoculum of the fungal plant pathogens *Mycosphaerella brassicicola* and *Botrytis cinerea*. *Applied and Environmental Microbiology*, **66**, 2996-3003.

Kleczkowski, A., Bailey, D.J. and Gilligan, C.A. (1996) Dynamically generated variability in plant-pathogen systems with biological control. *Proceedings of the Royal Society B*, **263**, 777-783.

Lalancette, N. and Hickey, K.D. (1986) Disease progression as a function of plant growth. *Phytopathology*, **76**, 1171-1175.

Lovell, D.J., Hunter, T., Powers, S.J. *et al.*, (2004) Effect of temperature on latent period of septoria leaf blotch on winter wheat under outdoor conditions. *Plant Pathology*, **53**, 170-181.

Madariaga, B.R. and Scharen, A.L. (1986) Interactions of *Puccinia striiformis* and *Mycosphaerella graminicola* on wheat. *Plant Disease*, **70**, 651-654.

Madden, L.V. (1992) Rainfall and the dispersal of fungal spores. *Advances in Plant Pathology*, **8**, 40-79.

Marin, D.H., Romero, R.A., Guzman, M. and Sutton, T.B. (2003) Black Sigatoka: an increasing threat to banana cultivation. *Plant Disease*, **87**, 208-222.

Minogue, K.P. (1989) Diffusion and spatial probability models for disease spread, in *Spatial components of plant disease epidemics* (ed. M.J. Jeger), pp. 127-143.

Morris, R.A.C. and Royle, D.J. (1993) Willow rust *(Melampsora spp.)* biocontrol by *Sphaerellopsis filum*, in *6th International Congress of Plant Pathology*, Montreal, Canada, p. 64.

Mourichon, X. and Fullerton, R.A. (1990) Geographical distribution of the two species of *Mycosphaerella musicola* Leach. (*Cercospora musae*), and *M. fijiensis* Morelet (*C. fijiensis*), respectively agents of Sigatoka and black leaf streak disease in bananas and plantains. *Fruits*, **45**, 213-218.

Packer, A. and Clay, K. (2004) Development of negative feedback during successive growth cycles of black cherry. *Proceedings of the Royal Society of London Series B-Biological Sciences*, **271**, 317-324.

Paradis, E., Baillie, S.R. and Sutherland, W.J. (2002) Modelling large-scale dispersal distances. *Ecological Modelling*, **151**, 279-292.

Pei, M.H. and Ruiz, C. (2000) AFLP evidence of the distinctive patterns of life-cycle in two forms of *Melampsora* rust on *Salix viminalis*. *Mycological Research*, **104**, 937-942.

Peters, J.C. (1994) Pattern and impact of disease in natural plant communities of different age. PhD thesis, University of Reading, UK.

Segarra, J., Jeger, M.J. and van den Bosch, F. (2001) Epidemic dynamics and patterns of plant diseases. *Phytopathology*, **91**, 1001-1010.

Shaw, M.W. (1994a) Modelling stochastic processes in plant pathology. *Annual Review of Phytopathology*, **34**, 523-544.

Shaw, M.W. (1994b) Seasonally induced chaotic dynamics and their implications in models of plant disease. *Plant Pathology*, **43**, 790-801.

Shaw, M.W. and Peters, J.C. (1994) The biological environment and pathogen population dynamics: uncertainty, coexistence and competition, in *The Ecology of Plant Pathogens* (eds P. Blakeman and B. Williamson), pp. 17-37. CABI, Wallingford, UK.

Shaw, M.W. and Royle, D.J. (1993) Factors determining the severity of *Mycosphaerella graminicola* (*Septoria tritici*) on winter wheat in the UK. *Plant Pathology*, **42**, 882-899.

Stahle, H. and Kranz, J. (1984) Interactions between *Puccinia recondita* and *Eudarluca caricis* during germination. *Transactions of the British Mycological Society*, **82**, 562-563.

van den Berg, F. and van den Bosch, F. (2004) A model for the evolution of pathogen-induced leaf shedding. *Oikos*, **107**, 36-49.

van den Bosch, F., Frinking, H.D., Metz, J.A.J. and Zadoks, J.C. (1988) Focus expansion in plant disease. 3: Two experimental examples. *Phytopathology*, **78**, 919-925.

Weber, G.E., Gulec, S. and Kranz, J. (1994) Interactions between *Erysiphe graminis* and *Septoria nodorum* on wheat. *Plant Pathology*, **43**, 158-163.

Wolfe, M.S. (1984) Trying to understand and control powdery mildew. *Plant Pathology*, **33**, 451-466.

Wolfe, M.S. and Knott, D.R. (1982) Populations of plant pathogens: some constraints on analysis of variation in pathogenicity. *Plant Pathology*, **31**, 79-90.

Wolfe, M.S. and McDermott, J.M. (1994) Population genetics of plant pathogen interactions: the example of the *Erysiphe graminis-Hordeum vulgare* pathosystem. *Annual Review of Phytopathology*, **32**, 89-113.

Xu, X.-M. and Madden, L.V. (2004) Use of SADIE statistics to study spatial dynamics of plant disease epidemics. *Plant Pathology*, **53**, 38-49.

Zadoks, J.C. and Schein, R.D. (1979) *Epidemiology and Plant Disease Management*. Oxford University Press, New York.

CHAPTER 8

MODELLING AND INTERPRETING DISEASE PROGRESS IN TIME

XIANGMING XU

8.1 INTRODUCTION

The primary objective of epidemiological research is to increase our understanding of how diseases develop in host crop populations and how other factors influence their development in order to develop sustainable and effective strategies for managing diseases. One of main epidemiological activities in achieving this objective is to understand the relationship of temporal disease dynamics with external factors. With improved knowledge, more efficient, sustainable and effective management strategies may be developed to reduce the impact of diseases on crop yield. In order to measure the temporal disease development, the amount of disease present in a population of plants is usually assessed several times. These data can be presented collectively as a disease progress curve, essentially depicting the dynamics of disease development with time. This simple temporal progress curve represents outcomes of complex interactions between host, pathogen, environments and crop husbandry. Depending on the nature of the pathosystem and how the disease is assessed, the observed disease progress curve does not necessarily describe monotonic increase.

Quantitative descriptions and analysis of temporal disease progress was recognised and applied in many occasions prior to the 1960s. For example, Ware *et al.* (1932) and Ware and Young (1934) presented curves illustrating the effects of cultivar resistance and fertilizer treatment on the dynamics of cotton wilt. Large (1945, 1952) proposed the use of disease progress curves and rate of disease progress as a means of demonstrating the benefits of fungicide applications on development of potato late blight. He also used transformation based on the normal distribution to linearise the observed disease progress curve and noticed the value of using the half-decay point for comparing epidemics. However, quantitative analysis of epidemiological data entered the mainstream of plant pathology only after the contribution by Van der Plank (1963) and the availability of modern computers. He not only introduced a new way of conceptualising disease increase in populations but also introduced important models into the mainstream of plant disease epidemiology. He introduced several quantities, such as apparent infection rate, infectious and latent periods, demonstrated how differential equations could form a basis for quantifying disease progress curves and

B. M. Cooke, D. Gareth Jones and B. Kaye (eds.), The Epidemiology of Plant Diseases, 2nd edition, 215–238.

finally, showed the use of model parameter estimates in comparing treatment effects on disease development.

This chapter aims to provide brief introductions to the various quantitative tools used to describe, analyse and compare temporal dynamics of epidemics. First, main methods for summarising or modelling each individual observed epidemic are described. Then, several methods aiming to describe observed temporal epidemic patterns in fewer dimensions (often 2-3) are briefly introduced; these methods are based on similarity or dissimilarity between observed/derived data describing temporal epidemic development over many epidemics or treatments. Finally, a brief introduction is given to several methods for comparing epidemic development between treatments using either original data and/or those derived from the original data, such as estimates of model parameters and newly transformed data in new dimensions. In this chapter, treatments are defined in a very broad sense. This may include epidemics at different sites, or on different cultivars, or subjected to different management programmes; it may also mean different pathosystems altogether.

8.2 GENERAL CONSIDERATIONS

The degree of precision and complexity of the required modelling and analysis is determined by the question(s) researchers are trying to answer. Each analytical method is best suited to answer specific questions. Even for the same data set, different methods may be needed to answer different questions. For example, if we have collated disease development on several occasions for one cultivar that received different fungicide spray regimes, we may simply apply analysis of variance to the final disease incidence/severity or to the areas under disease progress curves to determine whether overall disease development is significantly affected by the spray regimes. To answer the question on how temporal dynamics are affected by the fungicide spray programme, we may need to examine and summarise the pattern of disease progress over time for each individual spray programme and compare the patterns. We could fit growth curve models to each epidemic data set and then use analysis of variance to determine which and how each growth curve model parameter is affected by the spray treatment. Alternatively, we could apply canonical variate analysis to the original or derived (such as growth model parameters) data from which we can interpret the differences in observed temporal patterns between treatment groups in relation to each individual disease assessment, provided there are several replicate epidemic data sets for each treatment. There are several other methods that could be used to answer the same question. Precisely which one to use depends on many factors, including the temporal pattern of the observed epidemics, the nature of data and their collation, and availability of statistical software packages. In essence, there are no fixed prescriptions as to which method is the best for every possible scenario. Researchers need to consider many factors before applying one or more particular methods to the observed data in order to answer the predefined questions.

In all the methods of modelling and analysing data, certain assumptions are made in order to correctly conduct statistical analysis and interpret results. Some assumptions are specific to each analytical method and others are needed for most methods. The seriousness of the resulting bias in data analysis and interpretation due to the violation of assumptions varies with individual assumptions, analytical methods and the research question concerned. For example, one assumption is that there is no spatial aggregation of diseases since spatial aggregation may also influence temporal epidemic development. Analytical modelling has shown that aggregation will slow the rate of disease development (Waggoner and Rich, 1981; Yang and TeBeest, 1992; McRoberts *et al.*, 1996) and that assuming a constant degree of aggregation over time, the effect of aggregation on the rate of increase is less with low aggregation. In experimental studies with Zucchini yellow mosaic potyvirus, an aggregated pattern of initial disease resulted in lower disease incidence than a uniform pattern (Nelson, 1996). Unfortunately, the existence of varying degrees of spatial aggregation of plant diseases is a rule rather than an exception. Furthermore, often in field epidemiological investigations we do not know precisely the density and spatial configuration of initial inoculum. Both theoretical and experimental studies have shown that the rates of disease increase were considerably influenced by the number of initial disease foci (Smith *et al.*, 1988; Xu and Ridout, 1998; Jeger *et al.*, 2004b). In an experiment with southern blight of processing carrot, the rate of disease increase generally increased as the number of initial disease foci increased (Smith *et al.*, 1988).

Therefore certain preliminary data checking is needed to identify whether some assumptions are not met, and if so, the severity of the violations and whether there are means to reduce the severity of the violations, such as transforming data onto a different scale. However, often violations of some assumptions might be inevitable and researchers need to be aware of such violations and take them into consideration when interpreting results and make them clear in subsequent publications.

8.3 ANALYSING INDIVIDUAL EPIDEMICS

8.3.1 Simple growth models

One of the most common methods used to describe temporal disease progress is the use of simple growth models. Three common models used to describe disease progress curves are the monomolecular, Gompertz and logistic models (Table 8.1) and their application and interpretations are extensively reviewed (Campbell and Madden, 1990) and these models can provide a range of curve shapes (Fig. 8.1). More complicated models have also been used to describe disease progress curves, like the generalised Richards growth functions (Hunt, 1982). These models have been used almost exclusively as statistical means to describe the observed patterns and then use the estimated model parameters for comparing epidemics.

Table 8.1. Summary of differential and integrated equations and their linearised forms for common growth curve models used in plant disease epidemiology

Model	*dy/dt*	*y*	*Linearising transformation*
Monomolecular	$r_M(K-y)$	$K[1-B\exp(-r_M t)]$	$\ln\left(\frac{K}{K-y}\right)$
Logistic	$r_L y\left(\frac{K-y}{K}\right)$	$\frac{K}{1+B\exp(-r_L t)}$	$\ln\left(\frac{y}{K-y}\right)$
Gompertz	$r_G y[\ln(K)-\ln(y)]$	$K\exp[-B\exp(-r_G t)]$	$-\ln\left[-\ln\left(\frac{y}{K}\right)\right]$

K = maximum level of disease or asymptote of disease progress curve, y = disease at time of observation, B = a parameter related to the level of initial disease or point of inflection, r = rate of disease increase and t = time.

The biological basis of these common growth models results from the fact that these growth models share a common form of the differential equation:

$$\frac{dy}{dt} = r \bullet f(y) \bullet g(1-y) \tag{8.1}$$

where *dy/dt* is the temporal change in disease and r is a rate parameter. This equation states that the rate of disease (y) change at time (t) over time is related to the amount of diseased (y) and healthy susceptible area ($1-y$). In the above equation, the maximum disease is assumed to be 1 (i.e. all healthy tissues will be eventually diseased); otherwise we can use $K - y$ to represent healthy susceptible tissues where K is a variable representing the maximum disease. The precise form of two functions (f, g) may depend on biological characteristics of the pathosystem concerned. For example, in the case of monocyclic diseases, since newly diseased tissue in the current season will not produce new inoculum that may lead to new infections in the same season, $f(y)$ is omitted from the differential equation. Furthermore, if we assume that $g(1-y) = 1-y$, the resulting model is the monomolecular model. On the other hand, for polycyclic diseases, as the rate of disease change is expected to be dependent on both diseased and healthy susceptible tissues, the simplest model is a logistic model where $f(y) = y$ and $g(1-y) = 1-y$. Similarly, in the Gompertz model $g(1-y) = \ln(1) - \ln(y) = -\ln(y)$. Van der Plank used the exponential, monomolecular and logistic models primarily as biological models in his analysis. However, it should be pointed out that the biological nature of the pathosystem could not be ascertained purely on the basis that one particular model was statistically more appropriate than the other. This is particularly so since statistical criteria used to determine the appropriateness of a model do not usually consider the model's relevance to the characteristics of the pathosystem concerned.

To compare the rate of disease increase (i.e. parameter r) between different models, a weighted mean absolute rate of disease increase can be estimated as $\rho = rK/(2m+2)$ (Campbell and Madden, 1990), where r is the rate parameter of the specific model, K is the estimated maximum disease and m is the shape parameter (m = 0, 1 and 2 for the monomolecular, Gompertz and logistic models, respectively). This absolute rate of increase is extremely valuable since it allows comparisons between different types of models. Use of these models to describe observed epidemics is widespread and often becomes a routine first step analysis to evaluate treatment effects or establish relationships of disease development with crop growth or loss (Jeger, 2004).

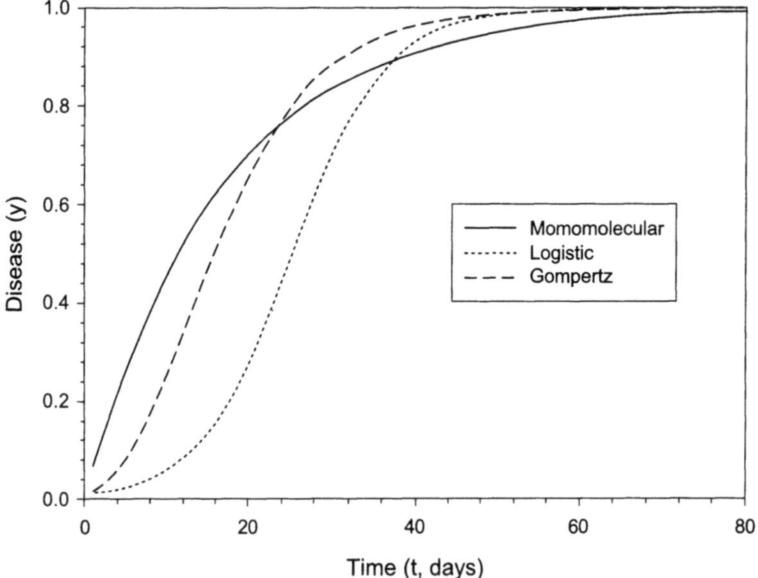

Figure 8.1. Examples of disease progress curves represented by monomolecular, logistic and Gompertz models with an equivalent weighted mean absolute rate (ρ = 0.03), y_0 = 0.01 and K = *1.0.*

In the past, linearised models (Table 8.1) were usually used to fit temporal disease data mainly because of the difficulties in performing non-linear regression analysis using the standard statistical software packages at the time. Nowadays, advances in computing power and widely available powerful statistical computing software have made use of model fitting based on linearised forms far less appealing and indeed appropriate than using non-linear regression analysis. In general, fewer assumptions are made with non-linear analysis than with linearised regression. Furthermore, the non-linear analysis of disease progress models allows additional parameters to be estimated, like the maximum disease parameter K.

Usually, calendar time (days, weeks, etc.) is used as the independent variable when fitting growth models to observed epidemic data. This is appropriate when the research objective is not to compare epidemics between treatments. Otherwise, use of time as an independent variable may result in incorrect conclusions if the host growth patterns differ significantly between epidemics (treatments). For example, suppose that all other data are the same except that the length of the host effective growing season (i.e. period when host tissues are susceptible) at one site is 80% of another site. It can be shown that the estimated apparent infection rate will be 25% greater for the epidemic at the site with the shorter growing season than at the other site. Although this analysis using time as an independent variable is technically correct, careful interpretation is needed as the rate difference is not attributed to the difference in disease development but rather to that in host growth pattern. We might overcome this problem by expressing the time as a proportion of the growing season and then using this 'biological time' as an independent variable when fitting curves. This is particularly important when we are comparing different diseases since the length of time for which host tissues are susceptible to infections may differ significantly between pathogens. Lalancette and Hickey (1986) developed a model that expressed the rate of disease change directly in terms of the change of host size.

A potential limitation of using physical time as a measure of time is illustrated by Lovell *et al.* (2004), who recommended instead the use of thermal time. Degree-days have been used to model disease progress of *Verticillium* wilt in cotton (Gutierrez and DeVay, 1986), tomato powdery mildew (Correll *et al.*, 1988), wheat take-all (Brasset and Gilligan, 1989; Colbach *et al.*, 1997; Schoeny and Lucas, 1999) and potato early blight (Johnson and Teng, 1990). There are several implicit assumptions associated with using degree-days. For example, it assumes that temperature is the most important factor driving growth rates of host, pathogen and disease and growth will stop if temperature is below the minimum or above the maximum temperature threshold values. Use of degree-days is very common in predicting spore maturity and subsequent discharge; for example, ascospores of *Venturia inaequalis* (cause of apple scab) (Gadoury *et al.*, 2004; see also Chapter 18) and apothecium production in *Sclerotinia sclerotiorum* (Sun and Yang, 2000).

8.3.2 Extensions to simple growth models

A growth model of two or three parameters often satisfactorily describes the temporal disease dynamics. However, there are situations where such a simple model may fail to capture the essential characteristics of the observed temporal pattern. This is not surprising given the fact that the simple models ignore several important factors that affect disease development, including host growth, fluctuating environmental conditions (and their effect on the rate of disease increase) and the length of latent and infectious periods. Significant advances have been made to incorporate these factors into disease progress models.

(a) Host growth

Host growth becomes important if there is large change in the size of susceptible host tissues relative to the duration of active epidemic development. Disease incidence or intensity may apparently decline over time even though the absolute disease level increases. For example, the proportion of leaves diseased may decline if the rate of new leaves produced is greater than the rate of leaf infection. Ignoring host growth may lead to underestimation or even negative values of the rate of disease increase (Kushalappa and Ludwig, 1982). Several methods have been proposed to incorporate the host growth into growth models.

Van der Plank (1963) proposed a correction for the rate of disease increase based on exponential change in mass of susceptible host tissues such that

$$r_L = \frac{1}{t_2 - t_1}\left[\ln\left(\frac{y_2}{1-y_2}\right) - \ln\left(\frac{y_1}{1-y_1}\right) + \ln(H_2) - \ln(H_1)\right] \tag{8.2}$$

where H_1 and H_2 represent the host size at time 1 and 2, respectively. This empirical method was judged to be satisfactory in a large comparative study (Kranz, 1975). However, even this corrected rate may have a negative value as demonstrated for coffee rust by Kushalappa and Ludwig (1982). To overcome this, these authors proposed a new method to correct host growth in calculating the apparent infection rate. Assuming a logistic disease increase, the corrected rate is

$$r_L = \frac{1}{t_2 - t_1}\left[\ln\left(\frac{y_2 h_2}{1-y_2 h_2}\right) - \ln\left(\frac{y_1 h_1}{1-y_1 h_1}\right)\right] \tag{8.3}$$

where $h = H/H_{max}$. In this model, no particular models were assumed for host growth.

A more general approach to accommodate host growth is to use linked differential/difference equations (LDE) (Jeger, 2000). This approach has been widely adopted and is regarded as a standard approach in medical epidemiology and modelling. In this approach, host growth and disease dynamics are described by separate equations but linked together via the fact that the change in disease at time t is a function of the total susceptible host area at that time. For example, if we assume that a logistic model can describe both host growth and disease development, then

$$\begin{cases} \dfrac{dY}{dt} = r_D Y\left(\dfrac{H-Y}{H}\right) \\ \dfrac{dH}{dt} = r_H H\left(\dfrac{K_H - H}{K_H}\right) \end{cases} \tag{8.4}$$

where r_H and K_H are parameters representing the rate of host growth and maximum host area, respectively. An analytical solution of disease incidence ($y = Y/H$) is possible for this set of two equations (Waggoner, 1986). The LDE approach is very flexible. For example, one could assume that absolute change in host size depends on the amount of disease-free tissues (Jeger, 1987b), then

$$\frac{dH}{dt} = r_H(H - Y)\left(\frac{K_H - H}{K_H}\right) \tag{8.5}$$

Sometimes, it is possible that dH/dt may be negative due to root loss or defoliation resulting from the disease. This was taken into consideration for the potato-*Verticillium dahliae* system (Johnson, 1988). In cases where linked equations cannot be readily integrated, iterative numerical methods may be used to estimate parameter values (Buwalda *et al.*, 1982; Gilligan and Kleczkowski, 1997).

Lalancette and Hickey (1986) took a different approach for incorporating host growth into disease progress models; instead of expressing disease and host growth as a function of time, they expressed disease as a function of host growth. One key area concerned with the effects of the host on disease development that still needs further research is ontogenic resistance, where host tissues become gradually resistant to diseases as they age. This ontogenic resistance has been observed in many diseases, for example apple scab (MacHardy, 1996; Li and Xu, 2002) and powdery mildew (Rogers, 1959; Mence and Hildebrandt, 1966). Equally, host tissues may become more susceptible with increasing age.

The effect of root growth on soil-borne disease development, particularly in the contact rate between inoculum (primary and secondary) and root, has received considerable interests. Jeger (1987a) showed that including root growth and inoculum density into a simple monomolecular model can result in a sigmoid, a monomolecular, a Gompertz-type or even an asymptotically exponential disease progress curve with the shape depending on the product of root growth and inoculum density. Gilligan and his co-workers developed various forms of models to investigate the effects of root growth, primary and secondary inoculum (infections), and biocontrol on soil-borne disease development (Gilligan and Bailey, 1997; Gilligan *et al.*, 1997; Gilligan and Kleczkowski, 1997). One of the important conclusions was the importance of host density and the rate of root production on disease development.

(b) Variable rate parameter

The rate of disease change depends on many factors, such as cultivar, environmental conditions and crop husbandry practices. Of these factors, environmental conditions are the most variable in terms of the amplitude and frequency of their variation within a season. In addition, environmental conditions not only directly but also indirectly affect disease development. For example, the apple scab fungus needs rainfall to discharge ascospores and free water or moisture to infect while the rate of infection depends critically on temperature (MacHardy, 1996; see Chapter 18). The length of latent period generally depends on temperature for most pathogens (Shaw, 1990; Davis and Fitt, 1994; Wadia and Butler, 1994; Webb and Nutter, 1997; Viljanen-Rollinson *et al.*, 1998; Xu, 1999b; Xu and Robinson, 2000, 2001). At the same time, weather conditions may affect the rate of fungicide decay, and in turn affect disease development. Further research is needed to investigate the severity of potential errors in conclusions drawn from models that assume a constant rate.

Campbell and Madden (1990) proposed a simple method to model a variable rate; they treated the rate parameter (r) as a specific function of time. This rate function can then be incorporated into any growth model. The rate parameter has also been expressed directly as a sinusoidal function of temperature in a logistic model (Waggoner, 1986). Another approach was taken to model a variable rate in order to understand the temporal spread of African cassava mosaic virus (Fargette and Vie, 1994; see also Chapter 20); they treated the rate as a product of two functions: one described the effects of the age and the other the effect of season on disease development.

(c) Disease components

Simple growth models when used to describe temporal disease progress have an inherent limitation in that they assume that the rate of absolute disease increase at a particular time is a function of the total disease at that time. However, not all lesions produce spores that may cause new infections at a given time; the absolute rate of disease increase does not depend on the total disease already present but on the proportion of the total diseased tissues that are producing viable spores. In order to overcome this limitation, we need to define age classes for diseases. Commonly, three important classes are defined: latent, infectious and removed. The period of time between initial infection and when new infections produce infective propagules (i.e. become infectious) is called the latent period; the period of time a single lesion can produce infectious propagules is called the infectious period; finally post-infectious disease is called removed.

A corrected basic infection rate (R) was proposed to take into account different classes of diseases (Van der Plank, 1963). A modified logistic model can be written as

$$\frac{dy}{dt} = R(y_{t-p} - y_{t-i-p})(1 - y_t) \tag{8.6}$$

where i and p are the length of infectious and latent periods, respectively. Thus, the absolute rate of change in disease depends on the amount of infectious disease (y_{t-p} - y_{t-i-p}) as well as total healthy susceptible area. The R parameter is related to the apparent infection rate (r):

$$R = \frac{r y_t}{y_{t-p} - y_{t-i-p}} \tag{8.7}$$

The explicit relationship of R with r for an entire epidemic in terms of y_0, i, p and t was discovered by Jeger (1984). R is an important parameter in determining when an epidemic can occur: in particular $iR > 1$ is necessary for an epidemic to occur (Jeger, 1986).

Another flexible and biologically intuitive approach to incorporate disease components is the use of linked differential equations. Often this approach may suffer from the difficulty in obtaining an analytical solution and hence numerical integration may be necessary. However, steady-state analysis of these equations may generate important results, for example, on criteria for persistence and invasion. Fitting a set of linked differential equations to a set of observed data is not a trivial matter though significant progress has been made in this area. For example, three linked differential equations can be used to describe the dynamics of disease components (latent - y_l, infectious – y_i, removed – y_r):

$$\begin{cases} \frac{dy_l}{dt} = bHy_i(1 - y_i - y_r - y_l) - gy_l \\ \frac{dy_i}{dt} = gy_l - hy_i \\ \frac{dy_r}{dt} = hy_i \end{cases} \quad (8.8)$$

where b, g and h are the infection, latent and infectious rate, respectively. This approach is very flexible, for example, it may allow rate parameters (b, g, h etc.) to be of any form of functions of other variables. As discussed before, host growth can also be included. Using this approach, Jeger (1982, 1984, 1986) successfully established the criteria for the establishment and subsequent spread of an epidemic, and the relationships among various components of the disease. This approach has been widely used in theoretical modelling of disease development for both soil- and air-borne pathogens, and virus diseases (Chan and Jeger, 1994; Gilligan *et al.*, 1997; Jeger *et al.*, 1998; Jeger, 2000; Madden *et al.*, 2000; Paulitz, 2000; Gilligan, 2002; Bailey and Gilligan, 2004; Bailey *et al.*, 2004; Jeger *et al.*, 2004a). A forecasting system for apple powdery mildew has been developed on the basis of this approach (Xu, 1999a).

8.3.3 Area under the disease progress curve (AUDPC)

AUDPC is the amount of disease integrated between two times of interest and is calculated without regard to curve shape (Shaner and Finney, 1977). This approach of summarising disease progress data into one value is appropriate when damages to host are proportional to the total amount and duration of the disease. However, if crop loss is related to a particular phase of the epidemic, for instance, during the blossom period, then AUDPC summarised over the period other than just over the particular period may lead to misleading conclusions. AUDPC can be a very useful alternative to fitting growth models, particularly if observed disease progress patterns cannot be fitted into simple growth models. Such situations can arise frequently in field epidemics. For example, incidence of diseased leaves may decrease in time due to rapid production of new leaves and/or defoliation of diseased leaves, commonly observed for rose downy mildew (Xu and Pettitt, 2003), or incidence of diseased leaves may stagnate for a long period of time and then increase rapidly, common for wet-loving pathogens in dry years. When observed disease patterns can be fitted satisfactorily to a model, then AUDPC can be directly obtained from the model integrated over time (Jeger and Viljanen-Rollinson, 2001).

AUDPC is usually estimated using the mid-point rule (so-called trapezoidal integration method), although other more complicated integration methods can be used as well:

$$AUDPC = \sum_{i=1}^{n-1}\left(\frac{(y_i + y_{i+1})}{2} \times (t_i - t_{i-1})\right) \tag{8.9}$$

When we compare different epidemics, it may be necessary to standardise AUDPC values in order to take into account the fact that epidemics may differ in their lengths of duration. A standardised AUDPC value is obtained by dividing the AUDPC by the total duration time and sometimes by the integration interval (Δt).

AUDPC is a simple statistic and yet is found to be very useful in many investigations in evaluating disease management practices (Jeger, 2004). For example, AUDPC calculated from two data points provides an equivalent amount of information as from repeated assessments in assessing wheat cultivar resistance to stripe rust (Jeger and Viljanen-Rollinson, 2001). In addition to several other variables describing various aspects of epidemic development, AUDPC is a useful predictor for crop loss in chickpea due to *Fusarium* wilt (Navas-Cortes *et al.*, 2000; see also Chapter 2).

8.3.4 Time series analysis

Successive disease assessments are usually correlated with each other. Analysis of such a variable with temporal autocorrelation can be better approached by time series methods (Wei, 1990). Time series analyses may reveal the nature of the system generating the series and can be used to study the dynamic relationship between two or more related series. As time series needs many temporal assessments, realistically only spore trapping data are suitable for time series analysis. There are two main methods of time series analyses: AutoRegressive Integrated Moving Average (ARIMA) models and transfer function analysis (TF). ARIMA models are used to study the nature of the series itself and are built from several simpler components. In an autoregressive (AR) model, the current value is linearly related to previous values. A random variable, a_t, is added to the AR component to account for any residual error. In an autoregressive moving average (ARMA) model, the current value is linearly related not only to previous values, the AR component, but also to the current and previous values of the random variable $\{a_t\}$. Terms involving the random variable are the moving average component. Trends in a series can often be removed by differencing the series. An ARIMA model is one in which a differenced series can be well described by the model. An

ARMA model can be further extended to describe periodic (seasonal, cyclic) patterns. Analysis of spore trapping data of apple powdery mildew over five years revealed that the temporal pattern of the number of airborne conidia was similar in all five years (Xu *et al.*, 1995).

TF models are an extension of ordinary regression models; they recognise temporal dependence within both the dependent and independent variables. In a TF model, the current value of the dependent variable can be related to the current and previous values of an independent variable. There may be a time lag before an independent variable affects the dependent variable. An important difference between TF models and ordinary regression models is in the assumption concerning the random component $\{e_t\}$. In ordinary regression, successive values of $\{e_t\}$ are assumed to be independent, whereas in TF models these values are assumed to follow an ARIMA model. It is also straightforward to model the persistent effect of an independent variable, for example, the long-lasting effects of a rain event on sporulation and dispersal. Results from TF analysis can be very useful in developing ordinary regression models for predicting spores. For example, a regression model was developed on the basis of TF analysis and this model satisfactorily predicted the temporal pattern of conidium numbers for apple powdery mildew (Xu *et al.*, 1995).

8.4 REDUCING DATA DIMENSION

All the methods described so far can be applied to original temporal disease assessment data for each individual epidemic, first with a view to understanding the epidemic pattern by summarising the observed pattern in a few parameters or statistics. These derived variables can then be used in a second stage analysis if necessary, where the emphasis is on evaluating treatment effects and comparing characteristics of different epidemics. For this second stage, the fewer variables describing observed epidemic patterns will result in easier analysis and interpretation of the results. Since a single method is not likely to describe an observed epidemic pattern completely, the derived statistics from more than one method can be included in the second stage analysis. Furthermore, in addition to these derived statistics as described above, researchers may also include other variables related to disease development or crop loss due to disease injuries, such as the level of initial inoculum and disease incidence/severity at a particular host growth stage. Since these variables describe the same epidemics, high correlations among them are expected. Given the high correlation, we may be able to describe key features with fewer variables. Then one can use these newly derived variables for further analysis. There are several methods that are designed for the purpose of reducing data dimensions while maintaining the key features in the original data sets. Here, only principal component and factor analyses are briefly described. For a more thorough understanding of these and other related methods, the reader is referred to many excellent books, such as Hair *et al.* (1998) and Tabachnick and Fidell (2000).

8.4.1 Principal component analysis

The goal of principal component analysis (PCA) is to extract maximum variance from the original data set with a few components. Each component, expressed as a linear combination of original variables, is a unique mathematical solution. In practical terms, PCA identifies variables that duplicate information held in other variables, and hence might be considered redundant. It is based on the study of the pairwise product-moment correlation coefficients computed in a data set. In general, if there are p variables, the relation between the p measured (and/or derived) variables on disease development, $y_1, y_2, \ldots, y_p$ and p principal component ($c_1, c_2, \ldots, c_p$) is

$$\begin{aligned} c_1 &= a_{11}y_1 + a_{12}y_2 + \ldots + a_{1p}y_p \\ c_2 &= a_{21}y_1 + a_{22}y_2 + \ldots + a_{2p}y_p \\ &\vdots \\ c_p &= a_{p1}y_1 + a_{p2}y_2 + \ldots + a_{pp}y_p \end{aligned} \tag{8.10}$$

Collectively, the information in the complete set of principal components is equivalent to that held in the original data set. Where only a reduced number of principal components is retained, there is some loss of information. The percentage of information retained is expressed in the cumulative percentage of variation explained. The first component explained the most variation in the original data; the second explained the next highest and so on. Normally, if there are strong correlations among original variables, as often is the case for data describing various aspects of temporal epidemic patterns, no more than three PCAs are retained. However, there are no formal criteria to verify the resulting PCA structure.

It is essential to understand how the information that was held in the original data is now contained in the reduced dimensions. A further rotation of axis in the reduced dimension may be necessary to better highlight the relations among the measured variables. Several ways are possible to achieve the 'best' rotation. One method often used is the varimax rotation, which selects axes that minimise the number of axes on which each measured variable is highly loaded. This tends to result in each variable highly correlated on only one axis, which helps interpretations. Usually, orthogonal rotation techniques are used, leading to statistically independent PCAs; varimax rotation is one of the orthogonal techniques. From the correlation matrix between original variables and retained newly rotated PCAs, we can interpret each PCA. We may now use the retained PCAs for further analysis of comparing epidemic development. Clear biological/epidemiological interpretations of each main PCA are crucial to the PCA analysis. If each of the retained PCAs cannot be explained biologically or epidemiologically, it may not be worthwhile proceeding to further analysis with the PCA scores.

In general, variables are standardised first before PCA analysis. Lack of standardisation makes sense only if the variables are all measured on the same scale; this is the case if PCA is applied to original disease incidence/severity data recorded over time. Standardisation should be applied to a collection of variables measured on different scales such as the estimated logistic rate, maximum disease and AUDPC.

PCA was applied to various epidemic variables in order to select the best subset of variables with the highest explanatory capacity to describe observed 60 disease progress curves of papaya ringspot (papaya ringspot virus type P [PRSV-P]) incidence on papaya (Mora-Aguilera *et al.*, 1996). Standardised area under disease progress curve, shape parameter of the Weibull distribution function, and time between transplant date and first symptoms were selected as the most important variables and represented 83.5% of the overall variance. These selected variables were then used in cluster analysis. In analysing the effects of sowing date on *Fusarium* wilt of chickpea, PCA was applied to a set of epidemic variables, including final disease intensity, AUDPC and parameters from the fitted Richards model over 108 epidemics (Navas-Cortes *et al.*, 1998). Three components, accounting for c. 98% of the total variance, were retained and they provided plausible epidemiological interpretations: representing a temporal positional factor, the AUDPC/final disease and the uniqueness of the estimated value for the point of inflection, respectively. PCA scores for the retained components can also be used as independent variables in regression analysis.

8.4.2 Factor analysis

The goal of factor analysis (FA) is to reproduce the correlation matrix of the original data set with a few factors. In general, the PCA and FA share the specific goal of summarising observed correlation patterns to a small number of components or factors. PCA analyses variance whereas FA analyses covariance. PCA is a unique mathematical solution whereas most forms of FA are not unique. In factor analysis, the variation in the original data set is assumed to arise from a set of common unknown explanatory variables (factors); these unknown factors are often called latent variables. Each original variable is a linear combination of these unknown factors. In general, if there are p variables, the relations between the p measured (and/or derived) variables on disease development, $y_1, y_2, \ldots, y_p$ and the n factors ($f_1, f_2, \ldots, f_n$) are

$$\begin{aligned} y_1 &= w_{11}f_1 + w_{12}f_2 + \ldots + w_{1n}f_n + e_1 \\ y_2 &= w_{21}f_1 + w_{22}f_2 + \ldots + w_{2n}f_n + e_2 \\ &\vdots \\ y_p &= w_{p1}f_1 + w_{p2}f_2 + \ldots + w_{pn}f_n + e_p \end{aligned} \tag{8.11}$$

The term e_i is an unexplained component (sometimes called unique component). The underlying theory for FA is that the observed variables are correlated because they are constructed from common factors. The greater the contribution from the common factor, the higher will be the correlation between observed variables. It is the correlation matrix of the original data that provides the information from which the number of common factors and the weighting (w_{ij}) are determined. Unlike PCA, it is possible to define a factor analysis model in which the unexplained components are chance effects, which leads to a statistical test as to whether a proposed model fits the data satisfactorily. Often, PCA is conducted first to suggest number of factors for FA.

The basic presentation of results, the terms used, and the interpretation are the same as in PCA. Similarly, the initial solutions are usually rotated to improve interpretation. Factor loadings of original variables are critical to interpreting factors biologically or epidemiologically. One useful outcome of FA is the factor score, which is the estimate the score subjects would have received on each of the factors had they been measured directly. Because there are normally fewer factors than the original variables, there is not a unique solution for expressing each factor as a function of original variables. In general, several options are offered in common statistical packages on how to calculate the factor scores.

Kranz (1968) identified six factors from 13 variables describing disease progress curves of 40 pathosystems. In a study of bean hypocotyl rot, four factors were identified from eight variables describing disease progress curves of 100 epidemics (Campbell *et al.*, 1980b). These four factors were all given plausible epidemiological interpretations.

8.5 COMPARING EPIDEMICS

One of the main objectives in epidemiological research is to compare epidemics so that key differences between epidemics can be established and attributed to treatments such as a cultivar and fungicide programme, to uncontrollable natural climatic factors, and to inherent biological/epidemiological differences between pathosystems. Usually, new variables derived from the original temporal disease assessment data using the methods described above are used as input for comparing epidemics, although original data can also be included.

Understanding the relationship of statistics describing temporal epidemic characteristics with underlying physical and biological factors is the key to the success of comparative epidemiology. Several studies (Xu and Ridout, 1998, 2000a, 2001) have clearly shown that spatio-temporal statistics are influenced greatly by sampling schemes, such as sampling quadrat size/shape and orientation in relation to prevailing wind, initial epidemic conditions and strength of prevailing wind, as well as by biological parameters. Thus, caution may also be needed in interpreting the observed differences between statistics reported in different studies since these differences may reflect, at least in part, differences in non-biological factors such as sampling details, initial conditions and wind conditions.

8.5.1 Analysis of variance (ANOVA)

A simple and straightforward analysis to compare different epidemics is to conduct either a univariate or multivariate analysis of variance. For example, if AUDPC is used to summarise temporal disease data, appropriate univariate ANOVA should be sufficient to illustrate whether disease development as described by AUDPC has been significantly affected by treatments. This approach is commonly used in many experimental or theoretical research investigations. AUDPC values revealed marked differences between fungicides with different properties (protectant, eradicant and curative) against leaf diseases in wheat (Hims and Cook, 1992). In a simulation study on the use of cultivar mixtures to control race-specific and non race-specific pathogens, the logistic rate parameter and AUDPC values were found to be satisfactory in describing the differences in development of simulated epidemics (Xu and Ridout, 2000b; see also Chapter 10).

If more than one variable is used to describe disease progress data, such as principal components/factors and growth curve parameters, multivariate analysis of variance can be conducted. Various test statistics may be employed to test the significance of treatment effects on epidemic development. However, existence of correlation between variables may reduce the power of MANOVA (Hair *et al.*, 1998). Analysis of variance of multivariate but correlated data should be analysed by the residual maximum likelihood (REML) method. This is particularly the case if original temporal disease assessment data (i.e., not derived variables) are used.

8.5.2 Residual (restricted) maximum likelihood (REML)

REML was introduced by Patterson and Thompson (1971). REML is used for analysis of linear mixed models (i.e., a linear model with both fixed and random effects), estimation of variance components and modelling of covariance structures. It has many applications, particularly in obtaining information on sources and sizes of variability in data sets. In the context of analysing disease epidemiological data, in addition to its use in analysing designed epidemiological experiments with more than one source of variation (Piepho, 1999; Madden *et al.*, 2002), it can be used to analyse temporal disease data as a design of repeated measurements.

The repeated measurement design is similar to split-plot designs. One of the main objectives is to measure the within-subject effect in addition to the usual objective of evaluating differences between subjects. In the epidemiological context, the subject is the group of plants receiving the same treatment, while disease assessed over time is the repeated measurement. In repeated measurements, variables are no longer independent but correlated with each other, i.e. disease measured on the same treatment at time t is expected to correlate with the disease at time t-1. One option of analysing repeated measurement data is to use univariate repeated-measures analysis of variance using the split-plot design. However, high temporal correlation between disease measurements may lead to overestimating the effects involving time.

In addition to the expected correlation between successive disease assessments, we also expect correlation to decrease with the interval between the two assessments. Hence, correlation between assessments at time t and $t+1$ is likely to be greater than between t and $t+2$. This variable correlation structure in repeated measurements can be accommodated in REML. REML has another advantage of being able to provide efficient estimates of treatment effects in unbalanced designs with more than one source of error.

8.5.3 Cluster analysis

Cluster analysis aims to classify individual units or subjects into clusters of similar units on the basis of a matrix of similarities (or distance) between all pairs of subjects to be clustered (Everitt *et al.*, 2001). The similarity (distance) matrix is calculated from the observed or measured disease-related variables. A hierarchical clustering approach is often used where an object is permitted to be a member of only one cluster.

There are two key decisions to be made when applying cluster analysis. First, which matrix (similarity or dissimilarity) should be used for clustering and which method is to be used for calculating this matrix? The second relates to the choice of clustering methods. For example, if two objects are already considered to form a cluster, what is the criterion for deciding whether a third object should join the same cluster? Should the criterion be based on its closeness at least to one member, or closeness to all members, or to the centre of this existing cluster? Different clustering methods may lead to different cluster structures. For instance, nearest-neighbour clustering tends to keep adding new members to existing clusters whereas furthest-neighbour clustering tends to produce more small clusters. A useful strategy is to apply several methods to investigate the consistency of the resulting clusters and possible biological implications of the formed clusters.

The resulting cluster structure is generally presented as a dendrogram. However, the clusters formed do not necessarily have any practical or biological significance. Forming a dendrogram is a simple matter given the accessibility of powerful general statistical software but interpreting the cluster structures is a more challenging task. Further discriminant and canonical variate analysis using the resulting cluster groups might assist in the interpretation. We must be aware of the danger that we invent a rational explanation for a particular clustering and then to argue that the cluster analysis supports this explanation. This is particularly true when the final clustering varies considerably with the clustering method and strategy used. Rather we should argue that the cluster analysis supports our prior expectation of cluster formation, which comes from independent sources or reasoning.

Kranz (1968) first used the cluster analysis in plant epidemiological research. In a cluster analysis of 40 pathosystems using 13 variables describing epidemic patterns, twelve groups were identified. This research indicated the huge variability that may be observed among epidemics even within the same pathosystem. Cluster analysis led to the identification of two types of epidemics of bean hypocotyl rot based on the analysis of six variables representing the characteristics of observed disease patterns (Campbell *et al.*, 1980a). Results from cluster analysis of experimental data indicated that epidemic development was related mainly to the date of sowing for *Fusarium* wilt of chickpea (Navas-Cortes *et al.*, 1998). Similarly, cluster grouping of observed epidemics of papaya ringspot was primarily related to site and transplanting date (Mora-Aguilera *et al.*, 1996).

8.5.4 Canonical variate analysis

In a univariate ANOVA, an F-test establishes evidence of differences among groups, and pairwise comparisons can then be applied to identify which groups are different and the pattern of the differences. However, in a multivariate case, this process becomes more complex and the representation of the differences may require two or more dimensions. Canonical variate analysis is a useful approach to determine the extent and nature of differences among groups in the multivariate case (Hair *et al.*, 1998). Suppose there are p disease-related variables, y_1, y_2,..., y_p and consider a linear transformation of these variables to a new variable, $c_1 = a_{11}y_1 + a_{12}y_2 + \ldots + a_{1p}y_p$. The c_1 can be thought of as the first of a series of new axes that provide an alternative coordinate system for defining the position of the data points. The c_1 is called the first canonical variate if the set of a_{11}, a_{12},..., a_{1p} values maximise the between group (or treatments) F-statistics on the c_1-axis. This process is repeated to form a second canonical variate c_2 on the condition that c_2 must use only information in the data that has not been used in the formation of c_1. This process continues until all canonical variates are computed. The important aspect of canonical variate analysis is to identify and interpret the significant canonical variables. Identification of significant canonical variables is usually achieved by the between group F-statistics and the cumulative percentage of variance explained. Interpretation of canonical variables in terms of original variables is via a_{11}, a_{12},..., and a_{1p}. If the first two canonical variables explained most variation, differences between groups can be easily visualised by plotting c_1 against c_2.

8.5.5 Discriminant function/logistic regression

Often temporal disease data for several treatments are collated. These data could refer to the development of the same disease in different management systems (chemical, cultural etc.), or on different cultivars, or in different locations; these data may also be obtained from different diseases subjected to the same treatment (conditions). We are interested not only in whether there are significant differences in the temporal disease patterns but also in whether it is possible to assign epidemics correctly to the predefined treatment categories. There are two distinct approaches that have the same aims and provide the same type of information but come from different statistical viewpoints. These are the discriminant analysis approach, which has the same basis as that underlying multivariate analysis of variance, and the logistic regression approach.

The presumption underlying the discriminant analysis approach is that the set of measured or derived disease data define the subjects. The probability of membership of the defined group is estimated on the basis of canonical variables: it is calculated as the total Mahandian distance in all canonical variables of the subject with the group concerned relative to the total Mahandian distance in all canonical variables of the subject with all the groups. Discriminant function has been used in plant pathological research to describe the relationship of physical and biological variables to disease epidemic development. For example, Hennessy *et al.* (1990) used discriminant function analysis to determine the relationship of climatic factors with severity of sorghum leaf blight in South Africa.

Logistic regression analysis, which may be termed as logistic discriminant analysis when used for classification, can be used where there are only two groups. In contrast to the discriminant analysis approach, the logistic model treats the classification variables (i.e. disease-related data) as predictors rather than response variables. Hence there are no probabilistic assumptions made about the distribution of these values. Consequently, the normality requirement of the discriminant model does not apply to the logistic model. Under the logistic model, the random variable is presumed to be the variable that indicates group membership. In addition, ordinal and categorical variables can also be included as predictors in the logistic model.

8.5.6 Survival analysis

Often in plant disease epidemiology we are interested in when an event will occur and how long it will take before it occurs. Examples of such an event may include time to defoliation and time to ultimate plant death; these type of data usually contain censored observation. In this sort of question, we are particularly interested in whether the probability and the time-course for a specific event are related to the amount of disease present. Several recent books provide a more detailed statistical treatment on analysis of survival data (Le, 1997; Kalbfleisch and Prentice, 2002) and its application in plant disease epidemiology is provided by Scherm and Ojiamho (2004). Weibull functions were derived for survivorship studies.

Principally, there are three applications of survival analysis in modelling temporal epidemic data. First, it allows the estimation of survival time distributions (e.g. survival and hazard function) for a group of individuals. Second, analysis can be conducted to determine whether there are significant differences in survival time distributions between groups. Finally, the effects of independent variables on the survival variables can be quantified. Two main mathematical functions in survival analysis are survival and hazard functions. The former is a cumulative distribution function describing the probability that an individual will survive until time t; the latter is the conditional probability density function describing the instantaneous risk that an event will occur at time t. Scherm and Ojiamho (2004) provided an example on how the survival analysis can be applied to plant disease epidemiology. Specifically, they demonstrated how the analysis can be used to determine whether *Septoria* leaf spot on blueberry could result in greater defoliation of the diseased leaves.

8.5.7 Fitting models with common parameters

Gilligan (1990) described a method for comparing epidemics by fitting common parameters to growth models between treatments. He illustrated the power of the method by constraining one or more common parameters to all treatments while fitting the remaining parameters separately for identifying which parameter(s) are affected by treatments. Essentially, this method fits nested models with a ranging number of parameters to the data set and then uses ANOVA to determine whether a model with common parameters results in similar goodness-of-fit as that by a more complex model with separate parameters for each treatment. This analysis is powerful in a sense that it can statistically assess which parameter(s) are significantly affected by treatments while simultaneously considering other model parameters. For example, suppose that logistic models fit all the observed temporal data, then it is possible to establish which of the three parameters (rate, maximum disease, and the time to the inflection point) is affected by treatments and if so by which treatments. An important prior requirement for using this method is that the same curve type can fit all (or nearly all) observed epidemics.

8.5.8 Canonical correlation

All the methods described so far are mainly concerned with describing and understanding the characteristics of observed temporal patterns of epidemics in relation to treatments. These treatment factors are usually qualitative variables, e.g. cultivars, fungicide, management strategy or quantitative variables but set at pre-defined levels, e.g. temperature and humidity set at different levels of constant values. However, we are often interested in relating temporal patterns of disease development to other non-controllable qualitative or quantitative variables, such as variables related to climatic conditions and soil structure etc. If only one disease variable, e.g. rate parameter and AUDPC, is used, then a regression analysis may be sufficient enough to establish a quantitative relationship. However, ordinary regression analysis cannot be used to establish a relationship of a set of more than one disease-related variable (dependent variables – DV) with another set of more than one variable (independent variables – IV), which can be approached by canonical correlation analysis.

Canonical correlation analysis is based on the canonical correlation matrix, which is a product of four correlation matrices, between DVs (inverted), between IVs (inverted), and between DVs and IVs, i.e. $R = R_{yy}^{-1} R_{yx} R_{xx}^{-1} R_{xy}$. The objective is to redistribute the variance in the original variables into very few pairs of canonical variates, each pair capturing a large share of variance and defined by combinations of IVs on one side and DVs on the other. Linear combinations are chosen to maximise the canonical correlation for each pair of canonical variates. There are two potential problems in using canonical correlation analysis. First, a procedure that maximises correlation does not necessarily maximise interpretation of pairs of canonical variates. Therefore, canonical solutions are often mathematically elegant but difficult to interpret biologically. Second, the procedure only maximises the linear relationship between two sets of variables. Scholosser *et al.* (2000) applied the canonical correlation analysis to characterise the relationship between plant morphological characters and disease variables in rice blast.

8.6 CONCLUDING REMARKS

Many new analytical and modelling tools have been applied to plant disease epidemiology in order to understand the temporal development of diseases. Generally speaking, there are two contrasting approaches in investigating temporal dynamics of plant pathogens. On one hand, for evaluating practical disease management strategies, the two most commonly used tools are comparing disease progress curves or AUDPCs between treatments (Jeger, 2004). On the other hand, the approach based on linked differential equations is often used to deal with more theoretical questions related to epidemic development (e.g., Gilligan and Bailey, 1997; Jeger *et al.*, 2002; Jeger *et al.*, 2004a). This chapter focuses on application of modelling tools to temporal epidemic data and hence does not attempt to include other more theoretical-orientated research on spatio-temporal modelling of plant disease epidemics (e.g., Bosch *et al.*, 1988a,b; Ferrandino, 1993; Scherm, 1996; Jeger *et al.*, 1998; Filipe and Gibson, 2001; Jeger *et al.*, 2002).

REFERENCES

Bailey, D.J. and Gilligan, C.A. (2004) Modeling and analysis of disease-induced host growth in the epidemiology of take-all. *Phytopathology,* **94**, 535-540.

Bailey, D.J., Kleczkowski, A. and Gilligan, C.A. (2004) Epidemiological dynamics and the efficiency of biological control of soil-borne disease during consecutive epidemics in a controlled environment. *New Phytologist,* **161**, 569-575.

Bosch, Fvd., Zadoks, J.C. and Metz, J.A.J. (1988a) Focus expansion in plant disease. 1: The constant rate of focus expansion. *Phytopathology,* **78**, 54-58.

Bosch, Fvd., Zadoks, J.C. and Metz, J.A.J. (1988b) Focus expansion in plant disease. 2: Realistic parameter-sparse models. *Phytopathology,* **78**, 59-64.

Brasset, P.R. and Gilligan, C. (1989) Fitting of simple models for field disease progress data for the take-all fungus. *Plant Pathology,* **38**, 397-407.

Buwalda, J.G., Ross, G., Stribley, D.P. and Tinker, P.B. (1982) The development of endomycorrhizal root systems: III. The mathematical representation of the spread of vesicular arbuscular mycorrhizal infection in root systems. *New Phytologist,* **91**, 669-682.

Campbell, C.L. and Madden, L.V. (1990) *Introduction to Plant Disease Epidemiology.* New York: John Wiley & Sons.

Campbell, C.L., Pennypacker, S.P. and Madden, L.V. (1980a) Progression dynamics of hypocotyl rot on snapbean. *Phytopathology,* **70**, 152-155.

Campbell, C.L., Pennypacker, S.P. and Madden, L.V. (1980b) Structural characterization of bean root rot epidemics. *Phytopathology,* **70**, 152-155.

Chan, M.S. and Jeger, M.J. (1994) An analytical model of plant disease dynamics with roguing and replanting. *Journal of Applied Ecology,* **31**, 413-427.

Colbach, N., Lucas, P. and Meynard, J.M. (1997) Influence of crop management on take-all development and disease cycles on winter wheat. *Phytopathology,* **87**, 26-32.

Correll, J.C., Gordon, T.R. and Elliott, V.J. (1988) Powdery mildew on tomato: the effect of planting date and triadimefon on disease onset, progress, incidence, and severity. *Phytopathology,* **78**, 512-519.

Davis, H. and Fitt, B.D.L. (1994) Effects of temperature and leaf wetness on the latent period of *Rhynchosporium secalis* (leaf blotch) on leaves of winter barley. *Journal of Phytopathology,* **140**, 269-279.

Everitt, B., Landau, S. and Leese, M. (2001) *Clustering analysis.* Arnold.

Fargette, D. and Vie, K. (1994) Modelling the temporal primary spread of african cassa mosaic virus into plantings. *Phytopathology,* **84**, 378-382.

Ferrandino, F.J. (1993) Dispersive epidemic waves: I. Focus expansion within a linear planting. *Phytopathology,* **83**, 795- 802.

Filipe, J.A.N. and Gibson, G.J. (2001) Comparing approximations to spatio-temporal models for epidemics with local spread. *Bulletin of Mathematical Biology,* **63**, 603-624.

Gadoury, D.M., Seem, R.C., MacHardy, W.E. *et al.* (2004) A comparison of methods used to estimate the maturity and release of ascospores of *Venturia inaequalis*. *Plant Disease,* **88**, 869-874.

Gilligan, C.A. (1990) Comparison of disease progress curves. *New Phytologist,* **115**, 223-242.

Gilligan, C.A. (2002) An epidemiological framework for disease management, in *Advances in Botanical Research,* **38**, 1-64.

Gilligan, C.A. and Bailey, D.J. (1997) Components of pathozone behaviour. *New Phytologist,* **136**, 343-358.

Gilligan, C.A., Gubbins, S. and Simons, S.A. (1997) Analysis and fitting of an SIR model with host response to infection load for a plant disease. *Philosophical Transactions of the Royal Society of London Series B-Biological Sciences,* **352**, 353-364.

Gilligan, C.A. and Kleczkowski, A. (1997). Population dynamics of botanical epidemics involving primary and secondary infection. *Philosophical Transactions of the Royal Society of London Series B-Biological Sciences,* **352**, 591-608.

Gutierrez, A.P. and DeVay, J.E. (1986) Studies of plant-pathogen-weather interaction: cotton and *Verticillium* wilt, in *Plant Disease Epidemiology: Population Dynamics and Management* (eds Leonard K. and Fry W.), New York: Macmillan, 205-231.

Hair, J.F.J., Anderson, R.E., Tatham, R.L. and Black, W.C. (1998) *Multivariate data analysis.* Englewood Cliffs, NJ: Prentice Hall.

Hennessy, G.G., de Milliano, W.A.J. and McLaren, C.G. (1990) Influence of primary weather variables on sorghum leaf blight severity in southern Africa. *Phytopathology,* **80**, 943-945.

Hims, M.J. and Cook, R.J. (1992) Disease epidemiology and fungicide activity in winter wheat. *Brighton Crop Protection Conference: Pests & Diseases 1992,* **2**, 615-620.

Hunt, R. (1982) *Plant growth curves: the functional approach to plant growth analysis.* Baltimore, MD: University Park Press.

Jeger, M.J. (1982) The relation between total, infectious, and postinfectious diseased plant tissue. *Phytopathology,* **72**, 1185-1189.

Jeger, M.J. (1984) Relation between rate parameters and latent and infectious periods during a plant disease epidemic. *Phytopathology,* **74**, 1148-1152.

Jeger, M.J. (1986) Asymptotic behaviour and threshold criteria in model plant disease epidemics. *Plant Pathology,* **35**, 355-361.

Jeger, M.J. (1987a) The influence of root growth and inoculum density on the dynamics of root disease epidemics: Theoretical analysis. *New Phytologist,* **a107**, 459-478.

Jeger, M.J. (1987b) Modelling the dynamics of pathogen populations, in *Populations of Plant Pathogens: Their Dynamics and Genetics* (eds Wolfe M.S. and Caten C.E.), Oxford: Blackwell, 91-107.

Jeger, M.J. (2000). Theory and plant epidemiology. *Plant Pathology,* **49**, 651-658.

Jeger, M.J. (2004) Analysis of disease progress as a basis for evaluating disease management practices. *Annual Review of Phytopathology,* **42**, 61-82.

Jeger, M.J., Holt, J., van den Bosch, F. and Madden, L.V. (2004a) Epidemiology of insect-transmitted plant viruses: modelling disease dynamics and control interventions. *Physiological Entomology,* **29**, 291-304.

Jeger, M.J., Termorshuizen, A.J., Nagtzaam, M.P.M. and Van den Bosch, F. (2004b). The effect of spatial distributions of mycoparasites on biocontrol efficacy: a modelling approach. *Biocontrol Science and Technology,* **14**, 359-373.

Jeger, M.J., van den Bosch, F. and Dutmer, M.Y. (2002) Modelling plant virus epidemics in a plantation-nursery system. *Ima Journal of Mathematics Applied in Medicine and Biology,* **19**, 75-94.

Jeger, M.J., van den Bosch, F., Madden, L.V. and Holt, J. (1998) A model for analysing plant-virus transmission characteristics and epidemic development. *Ima Journal of Mathematics Applied in Medicine and Biology,* **15**, 1-18.

Jeger, M.J. and Viljanen-Rollinson, S.L.H. (2001) The use of the area under the disease-progress curve (AUDPC) to assess quantitative disease resistance in crop cultivars. *Theoretical and Applied Genetics,* **102**, 32-40.

Johnson, K.B. (1988) Modelling the influence of plant infection rate and temperature on potato foliage and yield losses caused by *Verticillium dahliae*. *Phytopathology,* **78**, 1198-1205.

Johnson, K.B. and Teng, P.S. (1990) Coupling a disease progress model for early blight to a model of potato growth. *Phytopathology,* **80**, 416-425.

Kalbfleisch, J.D. and Prentice, R.L. (2002) *The statistical analysis of failure time data*. Boboken, NJ: John Wiley & Sons.

Kranz, J. (1968). Eine Analyse von annuellen Epidemien pilzlicher Parasiten. III. Uber Korrelationen zwischen quantitativen Merkmalen von Befallskurven und Ahnlichkeit von Epidemien. *Phytopathologische Zeitschrift,* **61**, 205-217.

Kranz, J. (1975) Das Abklingen von belfallskurven. *Zeitschrift fur Pflanzenkrankheiten und Pflanzenschutz,* **82**, 655-664.

Kushalappa, A.C. and Ludwig, A. (1982) Calculation of apparent infection rate in plant diseases: development of a method to correct for host growth. *Phytopathology,* **72**, 1373-1377.

Lalancette, N. and Hickey, K.D. (1986) Disease progression as a function of plant growth. *Phytopathology,* **76**, 1171-1175.

Large, E.C. (1945) Field trials of copper fungicides for the control of potato blight I. Foliage protection and yield. *Annals of Applied Biology,* **32**, 319-329.

Large, E.C. (1952) The interpretation of progress curves for potato blight and other plant diseases. *Plant Pathology,* **1**, 109-117.

Le, C.T. (1997) *Applied survival analysis*. New York: John Wiley & Sons.

Li, B-H. and Xu, X.-M. (2002) Infection and development of apple scab (*Venturia inaequalis*) on old leaves. *Journal of Phytopathology,* **150**, 687-691.

Lovell, D.J., Powers, S.J., Welham, S.J. and Parker, S.R. (2004) A perspective on the measurement of time in plant disease epidemiology. *Plant Pathology,* **53**, 705-712.

MacHardy, W.E. (1996) *Apple scab: Biology, Epidemiology, and Management*. St. Paul, MN: American Phytopathological Society.

Madden, L.V., Jeger, M.J. and van den Bosch, F. (2000) A theoretical assessment of the effects of vector-virus transmission mechanism on plant virus disease epidemics. *Phytopathology,* **90**, 576-594.

Madden, L.V., Turechek, W.W. and Nita, M. (2002) Evaluation of generalised linear mixed models for analyzing disease incidence data obtained in designed experiments. *Plant Disease,* **86**, 316-325.

McRoberts, N., Hughes, G. and Madden, L.V. (1996) Incorporating spatial variability into simple disease progress models for crop pathogens. In *Aspects of Applied Biology,* **46**, 75-82.

Mence, M.J. and Hildebrandt, A.C. (1966) Resistance to powdery mildew in rose. *Annals of Applied Biology,* **61**, 387-397.

Mora-Aguilera, G., Nieto-Angel, D., Campbell, C.L. *et al.* (1996). Multivariate comparison of papaya ringspot epidemics. *Phytopathology,* **86**, 70-78.

Navas-Cortes, J.A., Hau, B. and Jimenez-Diaz, R.M. (1998) Effect of sowing date, host cultivar, and race of *Fusarium oxysporum* f. sp. *ciceris* on development of fusarium wilt of chickpea. *Phytopathology,* **88**, 1338-1346.

Navas-Cortes, J.A., Hau, B. and Jimenez-Diaz, R.M. (2000) Yield loss in chickpeas in relation to development of Fusarium wilt epidemics. *Phytopathology,* **90**, 1269-1278.

Nelson, S.C. (1996) A simple analysis of disease foci. *Phytopathology,* **86**, 332-339.

Patterson, H.D. and Thompson, R. (1971) Recovery of inter-block information when block sizes are unequal. *Biometrika,* **58**, 545-554.

Paulitz, T.C. (2000) Population dynamics of biocontrol agents and pathogens in soils and rhizospheres. *European Journal of Plant Pathology,* **106**, 401-413.

Piepho, H-P. (1999) Analysing disease incidence data from designed experiments by generalised linear mixed models. *Plant Pathology,* **48**, 668-674.

Rogers, M.N. (1959) Some effects of moisture and host plant susceptibility on the development of powdery mildew of roses, caused by *Sphaerotheca pannosa* var. *rosae*. *Memoir, Cornell University Agriculture Experimental Station,* **363**, 3-37.

Scherm, H. (1996) On the velocity of epidemic waves in model plant disease epidemics. *Ecological Modelling,* **87**, 217-222.

Scherm, H. and Ojiambo, P.S. (2004) Application of survival analysis in botanical epidemiology. *Phytopathology,* **94**, 1022-1026.

Schlosser, I., Kranz, J. and Bonman, J.M. (2000) Characterization of plant type and epidemiological development in the pathosystem "upland rice/rice blast" (*Pyricularia grisea*) by means of multivariate statistical methods. *Zeitschrift Fur Pflanzenkrankheiten Und Pflanzenschutz-Journal of Plant Diseases and Protection,* **107**, 12-32.

Schoeny, A. and Lucas, P. (1999) Modeling of take-all epidemics to evaluate the efficacy of a new seed-treatment fungicide on wheat. *Phytopathology,* **89**, 954-961.

Shaner, G. and Finney, R.E. (1977). The effect of nitrogen fertilization on the expression of slow-mildewing resistance in Knox wheat. *Phytopathology,* **70**, 1183-1186.
Shaw, M.W. (1990) Effects of temperature, leaf wetness and cultivar on the latent period of *Mycosphaerella graminicola* on winter wheat. *Plant Pathology,* **39**, 255-268.
Smith, V.L., Campbell, C.L., Jenkins, S.F. and Benson, D.M. (1988) Effects of host density and number of disease foci on epidemics of southern blight of processing carrot. *Phytopathology,* **78**, 595-600.
Sun, P. and Yang, X.B. (2000) Light, temperature, and moisture effects on apothecium production of *Sclerotinia sclerotiorum. Plant Disease,* **84**, 1287-1293.
Tabachnick, B.G. and Fidell, L.S. (2000) *Using Multivariate Statistics.* Allyn and Bacon, Boston, USA.
Van der Plank, J.E. (1963) *Plant diseases: epidemics and control.* New York: Academic Press.
Viljanen-Rollinson, S.L.H., Gaunt, R.E., Frampton, C.M.A. *et al.*, (1998) Components of quantitative resistance to powdery mildew (*Erysiphe pisi*) in pea (*Pisum sativum*). *Plant Pathology,* **47**, 137-147.
Wadia, K.D.R. and Butler, D.R. (1994) Relationship between temperature and latent periods of rust and leaf-spot diseases of groundnut. *Plant Pathology,* **43**, 121-129.
Waggoner, P.E. (1986) Progress curves of foliar diseases: their interpretation and use, in *Plant Disease Epidemiology, Vol 1, Population Dynamics and Management* (eds Leonard K.J. and Fry W.E.), New York: MacMillan Publishing Company, 3-37.
Waggoner, P.E. and Rich, S. (1981) Lesion distribution, multiple infection, and the logistic increase of plant disease. *Proceedings of the National Academy of Sciences of the United States of America,* **78**, 3292-3295.
Ware, J.O. and Young, V.H. (1934) Control of cotton and 'rust'. *University of Arkansas Agricultural Experimental Station Bulletin 308, Fayetteville, AR*, 23 pp.
Ware, J.O., Young, V.H. and Janssen, G. (1932) Cotton wilt studies. III. The behavior of certain cotton varieties grown on soil artificially infested with the cotton wilt organism. *University of Arkansas Agricultural Experimental Station Bulletin 269, Fayetteville, AR*, 51 pp.
Webb, D.H. and Nutter, F.W.J. (1997) Effects of leaf wetness duration and temperature on infection efficiency, latent period, and rate of pustule appearance of rust in alfalfa. *Phytopathology,* **87**, 946-950.
Wei, W. (1990) *Time series analysis: univariate and multivariate methods.* New York: Addison-Wesley.
Xu, X.-M. (1999a) Modelling and forecasting epidemics of apple powdery mildew (*Podosphaera leucotricha*). *Plant Pathology,* **48**, 462-471.
Xu, X.-M. (1999b) Effects of temperature on the latent period of the rose powdery mildew pathogen, *Sphaerotheca pannosa. Plant Pathology,* **48**, 662-667.
Xu, X.-M., Butt, D.J. and Ridout, M.S. (1995) Temporal patterns of airborne conidia of *Podosphaera leucotricha*, causal agent of apple powdery mildew. *Plant Pathology,* **44**, 944-955.
Xu, X.-M. and Pettitt, T. (2003) Rose downy mildew, in *Encyclopaedia of rose science* (eds Roberts, A., Debener, T. and Gudin, S.), London: Academic Press, 154-158.
Xu, X.-M. and Ridout, M.S. (1998) Effects of initial epidemic conditions, sporulation rate, and spore dispersal gradient on the spatio-temporal dynamics of plant disease epidemics. *Phytopathology,* **88**, 1000-1012.
Xu, X.-M. and Ridout, M.S. (2000a) Effects of quadrat size and shape, initial epidemic conditions, and spore dispersal gradient on the spatio-temporal statistics of plant disease. *Phytopathology,* **90**, 738-750.
Xu, X.-M. and Ridout, M.S. (2000b) Stochastic simulation of the spread of race specific and non-specific aerial fungal pathogens in cultivar mixtures. *Plant Pathology,* **49**, 207-218.
Xu, X.-M. and Ridout, M.S. (2001) Effects of prevailing wind direction on spatial statistics of plant disease epidemics. *Journal of Phytopathology,* **149**, 155-166.
Xu, X.-M. and Robinson, J.D. (2000) The effects of temperature on the incubation and latent periods of hawthorn powdery mildew (*Podosphaera clandestina*). *Plant Pathology,* **49**, 791-797.
Xu, X.-M. and Robinson, J.D. (2001) The effects of temperature on the incubation and the latent periods of the clematis powdery mildew (*Erysiphe polygoni*). *Journal of Phytopathology,* **149**, 565-568.
Yang, X.B. and TeBeest, D.O. (1992) Dynamic pathogen distribution and logistic increase of plant disease. *Phytopathology,* **82**, 380-383.

CHAPTER 9

DISEASE FORECASTING

N.V. HARDWICK

9.1 INTRODUCTION

Indigenous diseases are controlled by a number of methods: cultural, crop rotation and by the use of chemicals as seed treatments, soil sterilants, soil amendments and foliar applications. Chemical control methods have played an important role in the control of plant pathogens. Because fungicides, by and large, are effective and, in some cases, the only practical methods of control, increasing reliance is being placed on their use. Governments, conscious of environmental lobbyists and increasing perception by the public that pesticides are bad for the environment and customers wishing to consume pesticide-free produce, are looking at ways through voluntary codes and legislation to restrict their use. One way of achieving reductions in pesticide use is to forecast when a crop is likely to be at economic threat from disease. Disease identification, disease assessment and yield loss, pathogen dispersal, modelling and data analysis, which have been discussed in preceding chapters are essential components of forecasting systems. However, we must ask a fundamental question – why forecast disease at all?

Disease forecasting is required for three main reasons, the first is economic the second safety and the third justified use. The economic issue is one of reducing the cost of production by timely application of control measures, usually in the form of fungicides. This avoids unnecessary wastage of fungicides as they are only applied when required, usually at the start of the epidemic. Safety covers not only the crop – reducing possible phototoxic effects of pesticides but also the environment – reducing exposure to non-target species, operators and consumers. Some countries, particularly Scandinavia, have a target to reduce pesticide loading by 50% (Jørgenson and Secher, 1996). Forecasting can play an important role in achieving this target. To meet consumer demand for pesticide-free produce supermarkets are seeking justification from growers for their use. The use of disease forecasting schemes can provide that justification.

Disease forecasting is the proper province of the extension pathologist and it is from this perspective that this chapter is written. Not only must extension pathologists be good diagnosticians but they must be able to assess the crop/pathogen situation in order to be able to advise their clients on appropriate disease control strategies.

Forecasting is pivotal to disease control. Not only does the subject fit in the middle of the range of skills required by an extension pathologist, from identification of the

B. M. Cooke, D. Gareth Jones and B. Kaye (eds.), The Epidemiology of Plant Diseases, 2nd edition, 239–267.

problem through to advising on control measures but it requires an understanding of all of them for forecasting systems to work at all. This was highlighted in a review of the subject by Miller and O'Brien (1952) where they stated "All sciences reach a point where future real progress, as distinguished from mere accumulation of additional information, depends on the organisation of the separate blocks of knowledge and application of them to accomplish practical results. It will be seen that forecasting is one of the ways of achieving such a synthesis. It is this inherent constructive feature that makes the development of forecasting one of the most interesting and rewarding phases of plant pathological investigation". Miller and O'Brien (1952) also provided a useful historical review of the subject using numerous examples from throughout the world and covering a wide range of commodities.

9.2 WHAT IS FORECASTING?

Forecasting is partly about defining the conditions under which a pathogen, when in contact with a susceptible host, can infect and become established. However, forecasting schemes have to have as their components data on, and an understanding of, the epidemiology and aetiology of the diseases and pathogens. Forecasting provides an indication when the disease is likely to go 'critical' and therefore have an economic impact. With some diseases, it is important to be able to predict the first occurrence of the disease, where subsequent control would be difficult due the absence of, for example, effective chemicals to eradicate the disease once established. For other diseases, a certain level may be tolerated, particularly on parts of the plant that make little contribution to yield or quality and, therefore, forecasting is about predicting when that disease is likely to affect the vital parts of the plant or critical periods in its development.

Understanding the components of the 'disease tetrahedron' (Fig. 9.1) (Zadoks and Schein, 1979), the interaction between host, inoculum, environment and human activity is essential for devising suitable forecasting systems.

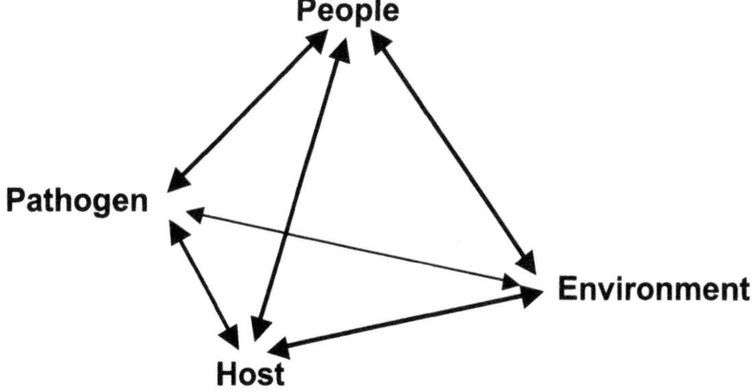

Figure 9.1. The 'Disease tetrahedron', illustrating the relationships between environment, host, pathogen and human activity (after Zadoks and Schein, 1979).

The importance of understanding the above relationships became apparent during the foot and mouth disease epidemic, which affected the United Kingdom (UK) in 2001. A highly virulent virus was infecting highly susceptible hosts (sheep and cattle) under favourable environmental conditions. It was only when the human factors were addressed by the strict policing of the movements of individuals and vehicles, coupled with thorough disinfection, that the disease was brought under control (Anon., 2002).

9.2.1 Environment

It is generally assumed that the environment is the driving force for diseases (Hardwick, 2002). Smith and Hugh-Jones (1969) analysed the foot and mouth disease outbreak of 1967 and concluded that weather played a greater part in the spread of the disease than was previously recognised.

Meteorological data can be collected on the macro-scale at synoptic weather stations (a network of weather stations measuring dry-bulb and wet-bulb temperature, cloud amount and type, cloud base, weather type, visibility, wind direction and speed and atmospheric pressure. The frequency of the observations vary, but the main synoptic hours are 0000, 0600, 1200 and 1600 GMT). Synoptic stations are generally sited at airfields which may be remote from the main arable areas or in atypical cropping situations and may therefore give readings less relevant to specific crops. The spatial scales are necessarily large. At the micro-scale, where data can be generated from in-field monitors by the minute, the data are spatially independent. The benefit of the micro-scale is that it can be identified with a specific crop and is, therefore, influenced more by the microclimate from within the crop or field.

The study of environmental factors play an important role in disease epidemiology and therefore forecasting. Much research is expended on capturing meteorological information and relating disease outbreaks to weather criteria. The basic systems consider only rainfall, its frequency and intensity; the more sophisticated include temperature, both maximum, minimum and mean, humidity, leaf wetness, wind speed and direction and hours of sunshine. The duration of these events and the period in which they fall, e.g. during night or day, are also important elements of data requirement.

9.2.2 Inoculum

In considering any forecasting scheme, inoculum is important and its source will greatly influence the type of forecasting scheme that is appropriate. These include inoculum that arises from within the crop, carried on the seed or on the perennating part of the plant, such as tubers and rhizomes. Seed certification schemes play an important role in restricting this source and, where strictly applied, can eliminate seed as a source of inoculum. Elimination of inoculum may involve micropropagation or treatment by physical and/or chemical means. Volunteers (plants that grow from seed or tubers left behind after harvest) carrying infection from the previous season can be an important source of inoculum for subsequent crops. Inoculum from infected crops in adjacent fields, or from longer distances – carried in air currents or rain generated aerosols, in vectors or on passive carriers – is a major source of in-season infection. It is the latter source that is generally the concern of the disease forecaster.

It is important to recognise that sole reliance on weather for forecasting must assume that inoculum is always present. The absence of inoculum, or delays in its initial arrival, will affect the reliability of forecasting systems dependent solely on weather. Van der Plank (1963) covers in detail the spread of disease, the influence of inoculum and cultivar resistance and fungicides in his seminal book on disease epidemics.

9.2.3 Host

The host, in terms of its resistance and the nature of that resistance, whether it is absent, partial or total, will affect the speed of development of the disease and therefore will interact with any forecasting scheme.

The habit of the host, for example the structure of the crop canopy, will influence the way disease moves about the plant and the architecture of the plant will influence the microclimate and also the amount of damage caused by the effects of wind – important where damage or wounding can aid pathogen entry.

9.3 POLYCYCLIC AND MONOCYCLIC DISEASES

Plant pathogens can be split into two groups, those where there are many generations in a season (polycyclic) and those where there is time to complete only one cycle from infection to spore production and re-infection of the host (monocyclic). For a disease that is capable of rapidly re-cycling, e.g. where the time from infection to spore production can be a matter of a few days under favourable conditions, such as potato blight (caused by *Phytophthora infestans*) and barley powdery mildew (caused by *Blumeria graminis* f.sp. *hordei*), information may also be required on the appropriate intervals between subsequent spray applications. With monocyclic diseases, where the disease only has one infection chance throughout the season, as the pathogen is slow to develop and will not produce a succession of spores capable of reinfecting and causing significant damage, a means of establishing the presence and amount of inoculum may be all that is required, e.g. *Sclerotinia* stem rot of oilseed rape (caused by *Sclerotinia sclerotiorum*) and eyespot of cereals (caused by *Oculimacula* spp.).

9.4 EQUIPMENT

Investigations into the aetiology and epidemiology of the pathogen usually require the definition of the optimum conditions required for the infection process, multiplication, survival and dispersal of the pathogen to occur. Much of this is to do with defining optimum temperature and water requirements for spore germination and hyphal growth. In order to devise a simple forecasting scheme for arable crops, there is generally a requirement for meteorological data; with air-borne pathogens, this usually requires data on rainfall – volume, duration and intensity, temperature, humidity and leaf wetness. Sometimes, solar radiation and wind speed and direction are also used. For soil-borne pathogens, soil temperatures and information on soil moisture deficit may be required.

It is not the purpose of this chapter to provide a detailed description of the type of equipment that is available for taking readings of the various meteorological parameters that are now built into sophisticated forecasting schemes. The standard Stevenson screen, usually found in a fenced area adjacent to laboratories, and a familiar sight to visitors to most research stations, is now superseded by finely tuned sensors linked to microprocessors. The data may be dispatched via satellite to computers at researchers' desks many kilometres away. Gone are the days when it is necessary to manually record at six hourly intervals. Equipment can now be placed in the field where the crop is grown and interrogated remotely.

Some of the systems available range from standard meteorological equipment, such as copper measuring cylinders and wet and dry bulb thermometers, to electronic sensors mounted on poles fixed in the ground and connected to a radio transmitter and receiver. They may be run on batteries or be solar powered. Many systems come already programmed with various warning systems and, at the press of a single button, provide an appropriate readout to advise farmers and consultants on the current risk for various diseases.

9.5 FORECASTING SCHEMES

There are numerous forecasting schemes published world-wide on a broad range of crops and pathogens. It is not possible to cover them all in a short chapter but only to highlight a few schemes that are illustrative of the development of forecasting, their uses, limitations and practical difficulties. As well as being written from an extension pathologist's perspective, it is also written largely from UK experience with arable crops. The three major arable crops in England are cereals (2.5 M ha), oilseed rape (0.42 M ha) and potatoes (0.11 M ha) in total occupying 80% of the total cropping area (figures taken from the Defra, June, 2003 Census). Potatoes are a particularly important crop as they are routinely sprayed against one particular disease, potato late blight, and receive per hectare nearly 3.5 times the weight of fungicide applied to winter wheat (Garthwaite *et al.*, 2002). Recent forecasting schemes rely heavily on computer modelling and processing, not generally the province of an extension pathologist. It is the output, reliability and robustness of such schemes that are of concern. Those researchers with modelling interests must look in the original papers for detailed information on the construction of specific forecasting models. However, the major difficulty for forecasters, as with all biological processes, is that they are attempting to establish markers or points on a continuum of infinite variability and combination of factors. The use of disease progress curves is an essential component of the forecasters' art. They can also be used to establish disease/yield loss relationships and justify the application of control measures (Large, 1952).

This chapter concentrates on potato blight forecasting, as it is here that the skills of the forecaster are tested to the full but comment is made also on cereals and oilseed rape. For those wishing to consider a wide range of schemes, Cambell and Madden (1990), in their book on plant disease epidemiology, list a number of forecasting systems covering a range of crops and diseases but they indicate that the schemes have not necessarily been implemented for individual plant diseases.

9.6 POTATOES

In the UK, all fungicides applied to the foliage of potatoes are for the control of late blight (caused by *Phytophthora infestans*). Other fungicides applied to the potato are to tubers for the control of blemish diseases and tuber rots. These are applied either as pre-storage or pre-planting treatments on the basis of experience and on a subjective assessment of risk.

9.6.1 Potato late blight

Potato late blight is a major disease world-wide and the key principles of forecasting can be best illustrated by the way blight forecasting has developed, particularly in Europe following its discovery on the Isle of Wight in the UK on 16 August 1845 (Nelson, 1995).

Potato late blight is a very important disease of the potato, for there are few diseases of major arable crops that can rapidly lead to the total destruction of the crop, both during the growing season and in store. Most cereal farmers will achieve some sort of a yield if the crop is left unsprayed. However, growers of potatoes, particularly in the UK, are so terrified of the consequence of blight that up to 15 spray applications in one season have been recorded (Hardwick and Turner, 1996). There is also evidence that the interval between subsequent sprays is reduced dramatically as the season progresses, indicating that farmers are attempting to halt the disease once established. This is a futile exercise, as fungicides for blight control are only effective as protectants and not eradicants. Large (1959a) states "The forecasts serve well to indicate the right time for the important spraying that is to protect the whole of the foliage when the real attack begins. After that - *and let it be said plainly* (my italics) – forecasting can give little help". There were hopes in the 1970s, with the development of the phenylamide group of fungicides, that eradication was a possibility (Urech *et al.*, 1977). This group of fungicides are systemic and highly effective against the pathogen and had eradicant properties. However, resistance soon developed in the pathogen population, so that in many regions they have ceased to be effective (Cooke, 1992).

The importance of the crop, and the destructive nature of the disease, has meant that a large number of forecasting schemes have been developed over the years. One of the reasons a scheme was desirable was to attempt to reduce the phytotoxic effect of the only fungicide available for control at the time – copper. Copper tended to be phytotoxic to young developing foliage and in dry years where blight was not a problem, spraying could reduce yield by 3% (Large, 1959a). Even spraying crops with the later-developed fungicides can also have a deleterious affect on yield in the absence of late blight (Taylor *et al.*, 2000). Delaying the first application until growth of the potato plant was nearly mature resulted in less damage from scorch. The development of a forecasting scheme was therefore important in delaying spraying until absolutely necessary to avoid too early an application and consequent damage to the potato foliage in drier and lower risk years.

The prime aim in devising a forecasting scheme was to define a set of conditions that, when satisfied, would lead to an outbreak of blight within the next 14 days (Large, 1959b). Under normal circumstances, this would allow for the timely application of a protectant fungicide. It is perhaps surprising to read in Large's (1959a) account of the major blight epidemic of 1958 that "Good commercial spraying, in most blight years, can be expected to give on average about two week's prolongation of useful growth of the haulm". The increase in yield from an attack that did not defoliate the crop until the end of August was estimated at 15%. Today this would seem marginal and the expectation is for a longer delay in defoliation.

It is important when considering a forecasting scheme to have an understanding of the aetiology of the pathogen. Most of the blight forecasting schemes have been based on the work of Crosier (1934). This showed that the fungus grows most rapidly at 18-21°C and slowly at below 9°C and above 24°C. Also, and importantly, in the initial phase of the disease, the growth of the fungus in an infected potato plant, either from a planted seed tuber, cull pile or from volunteers, is only governed by temperature. It is only when sporulation occurs that humidity, leaf wetness and their duration affects disease progress.

There is a major history to blight forecasting and this can be traced by looking at the increasing sophistication of systems devised over the years as technology has improved.

9.6.2 Dutch rules (van Everdingen, 1926)

The first national blight forecasting scheme was probably that devised by van Everdingen (1926) who analysed a number of weather parameters on the development of blight.

Four criteria (known as the Dutch rules), when satisfied, indicated the time to apply control measures:

- the occurrence of dew for at least four hours at night;
- a minimum temperature of 10°C;
- a mean cloudiness on the next day of 0.8 or more;
- at least 0.1 mm of rain in the following 24 h.

9.6.3 Beaumont period (Beaumont, 1947)

The Dutch rules were initially used in England but there was an unacceptable period between the predicted outbreak of blight and its actual occurrence. Beaumont (1947) recognised the importance of high humidity and its duration. The Dutch Rules relied upon dew. Beaumont reported that dew does not readily form on potato leaves and that there were also difficulties in measurement, as dew point is calculated rather than observed, and therefore the length of the actual period of dew present on the crop could not be determined with any degree of accuracy. Beaumont considered humidity to be more important as, during a period of high humidity, there is no wind and the nightly drop in temperature was most likely to result in dew formation and leaf wetness.

Beaumont combined the dew, cloud cover and rainfall requirements into a period of high humidity. His rules were that there should be at least two consecutive days with:

- a minimum temperature not less than 10°C;
- a relative humidity not below 75%.

When these conditions were met, growers in the vicinity were advised to begin their routine spray programmes. This became the standard blight warning scheme for the UK until the mid-1970s.

These rules were evaluated by Large (1956) over a period of five years (1950-1955). He concluded that "an 'operations-chart' method of interpreting the occurrence of Beaumont periods in screens at standard weather stations (about 40 synoptic weather stations were used) enables broad regional forecasts of the date of blight outbreaks to be made successfully in England and Wales, provided that the indications from the whole network of stations were taken into account and due regard paid to the seasonal and regional differences in the forwardness of crops". This is interesting as it illustrates precisely the problems with forecasting systems that will be discussed in more detail later. However, it is important to interpret the data intelligently. This statement basically means that the important element is a number of synoptic sites reporting Beaumont periods rather than relying on a single station for decision making for an individual locality.

It was also found that, with the exception of the UK's extreme south-west where crops were generally more advanced, Beaumont periods occurring before July could be ignored (Large, 1953; 1959b). This again demonstrates the importance of intelligent interpretation of a fixed set of rules.

9.6.4 Irish rules (Bourke, 1953)

The Irish rules (Bourke, 1953) differ from the Dutch rules and Beaumont period in that the relative importance of the duration of shorter spells of humid weather is also assessed. Also a forecast of blight-favourable weather is included in the forecast. The work of Crosier (1934) on the aetiology of the pathogen was used in defining the criteria.

The criteria to be satisfied are:

- a humid period of at least 12 h with the temperature at least 10°C and the RH 90% or above (conditions favourable for sporangial formation);
- free moisture on the leaves for a subsequent period of at least 4 h (conditions favourable for germination and reinfection). If there is no rainfall the alternative requirement is for a further 4 h beyond the initial 12 with RH at least 90%.

The duration of these periods was used to provide a means of weighting the importance of differing lengths of blight-favourable weather (Bourke, 1955).

Hourly records were made via the synoptic network and hours meeting the criteria were co-ordinated centrally and the duration of favourable weather recorded on a map. The data were used to plot the extent of the areas covered by blight favourable weather.

Bourke (1955) further reported that "The regular determination and analysis of the synoptic causes of weather spells favourable to potato blight also facilitates the forecasting of such spells, and indeed warnings were not normally issued in Ireland unless further periods of favourable weather within 3-8 days can be forecast". Bourke (1953) concluded that there were particular weather situations which favoured the development of blight in Ireland. He suggested that forecasts were possible for Ireland because the summer weather is generally settled and the precise timing of humid spells were of little importance, rather it was their possible duration. The warnings also included a forecast of periods of weather suitable for spraying. The current operation of the system is reported by Keane (1995).

9.6.5 Smith Period (Smith, 1956)

Smith (1956), an agricultural meteorologist, found that out of a total of 220 forecasts of Beaumont Periods, 43 from individual stations were not valid for their area. However, because of the way Beaumont should be interpreted, Large (1956) uses the term 'flushes' of periods from a number of stations, (i.e. periods reported from a single station should be ignored and only when a number of stations in a region are recording blight conditions should action be considered). Smith considered that periods reported from only a single station do not adversely affect the general decision. He found that an analysis of the weather data indicated that the failures were due to the humidity criterion. Smith reworked the 1950-1955 operations charts of Beaumont to test the validity of using a shorter period of higher humidity and found that 29 out of the 43 failures of Beaumont would have been valid using a 90% humidity criterion for 11 h in each of two days instead of the 75% for 48 h. However, nine of the Beaumont warnings would not have been trigged by the Smith modifications and, despite the reduction in the potential failure rate from 20% to 10%, Smith stated that specific failures were not important "as it is the occurrence of regional flushes of warning period that is important in our method of working and not individual station performance".

Smith also records that "The differences in effectiveness of the two systems, however, would appear to be small in practice and there would be little point in altering an established system unless the benefits are likely to be considerable". It was not until 1975 that the Smith Period came into full operation and formed the basis of blight forecasting in the UK. The Meteorological Office provided data fulfilling the Smith criteria from its synoptic network to help growers time the beginning of their routine spray programmes.

The Smith Period is still used widely in the UK and data from the Meteorological Office's synoptic network is now spatially interpreted and made available to the industry via the internet (Barrie and Bradshaw, 2001).

The Smith period, like the Beaumont, period consisted of two simple meteorological parameters – at least two days (ending at 0900) with:

- a minimum temperatures of 10°C or above;
- duration of humidity of 90% or above for at least 11 h each day.

9.6.6 Negative Prognosis Model (Schrödter and Ullrich, 1967)

Developed from laboratory studies on how temperature and humidity regulate the life cycle of the late blight fungus, Schrödter and Ullrich (1967) demonstrated that weather conditions exert differing influences on the likelihood of infection, sporulation and mycelial growth. These data were quantified in order to generate a daily severity figure. The daily figures were accumulated until a specific threshold was reached after which routine spraying should begin. The exact magnitude of the threshold could be varied depending on locality but once it passed a figure of 180, blight was most likely to be found. During highly favourable weather, the daily figure could be as high as 20 while, in hot dry conditions, there could be a negative value. This scheme had the advantage for the grower that he can have confidence that blight is unlikely to occur until the threshold is reached and sprays need not be applied.

9.6.7 Blitecast (Krause et al.*, 1975)*

Unlike the previous schemes, Blitecast (Krause *et al.*, 1975) attempts to advise not only on the date to begin spraying but also the timing of subsequent applications. Data required for the scheme were daily maximum and minimum temperatures, the number of hours when relative humidity was equal to or above 90%, the maximum and minimum temperature during the period when the RH was 90% and above and the daily rainfall figure to the nearest 1 mm. The RH and temperature sensors should be sited between the potato rows and within the potato canopy. The system is basically a combination of two programmes used in the USA, one (Hyre, 1954) based on rainfall and maximum and minimum temperatures, and the second (Wallin, 1962) based on relative humidity and temperature, and uses an accumulation of arbitrary severity values based on the relationships between the duration of relative humidity and temperature.

Blitecast is computer-driven programme, like the negative prognosis model, also based on severity values. An initial blight warning is given 7-10 days after either an accumulation of 10 rain-days (after Hyre, 1954) or 18 severity values (after Wallin, 1962) followed by a spray application programme based on the number of rain-favourable days and severity values accumulated during the previous seven days. Data are supplied to a central computer operator by the grower from in-crop thermohygrographs. Four recommendations are issued: no spray; a late blight warning; a 7-day spray schedule; and a 5-day spray schedule (recommended during severe blight weather). Recommendations are provided over the telephone and are completed within 3 minutes from the data being supplied. Fohner *et al.* (1984), using a computer simulation model found that Blitecast was no more effective in scheduling spray applications that a standard 7-day programme. However, they suggested that improvements could be made to the effectiveness of scheduling with better fungicides and weather forecasting.

9.6.8 Sparks's Risk Criteria (Sparks, 1984)

The Sparks model (Sparks, 1984) was developed as a direct result of the severe blight year of 1983, when major blight epidemics developed in the eastern counties of the UK despite there being no warnings under the Smith system. Early season temperatures were below the 10°C required to trigger a Smith period but there were exceptionally long periods of high humidity. Also, the plants were affected at a very early development stage, in many cases as they were appearing through the ridges. Potatoes are known to go through stages of varying degrees of susceptibility. This has been associated with the carbohydrate ratio (Grainger and Rutherford, 1963). Where the ratio between total carbohydrate of the plant and the non-carbohydrate of the shoot was above 1.0, the plant is more susceptible to blight. This occurs both early and late in the season.

In 1983, crops were at their first susceptible stage when conditions were extremely favourable for infection. Sparks tried to correct the errors he perceived in the Bourke and Smith schemes by taking account of lower temperatures and the development of separate generations of blight. The Sparks system was computerised and recommendations were issued as a single figure ranging from 0 (no risk) to 3 (high risk). Sparks also took account of canopy cover, open or closed, so that adjustment to risk depending on crop growth could be made at the start of the season. The scheme was never officially published and was considered to be too complex by the Agricultural Development and Advisory Service (ADAS) (formerly the Government's extension service for England and Wales). Keane (1995) compared the Sparks model with that of Bourke and concluded that the closed canopy model greatly over-estimated blight risk and was therefore not of practical use under Irish conditions.

9.6.9 NEGFRY (Hansen et al.*, 1995)*

Developed by the Danish Institute of Plant and Soil Science, NEGFRY is a combination of the negative prognosis scheme of Schrödter and Ullrich (1967) and

Blitecast (Krause *et al.*, 1975). The scheme takes account of cultivar susceptibility, emergence date and irrigation to advise on the date of the first fungicide application and the timing of subsequent sprays. It is distributed as an easy-to-use computer programme which growers can customise for their location and the level of infection they are willing to accept. In order to work effectively, the programme requires weather readings at three-hour intervals, or better, and preferably from sensors in or near the crop. In validation trials, the number of sprays was reduced by 50% in comparison with routine treatment. The reduction was based mainly on delays to the start of the spray programme and the longer interval between spraying as recommended by the system.

9.6.10 Developments

The introduction to Europe of new populations of the late blight fungus from Mexico in the late 1970s and particularly the A2 mating types in the early 1990s (Shattock, 2002) has cast doubt on the validity of current forecasting schemes. European blight forecasting systems were developed before oospore progeny were detected in field situations (Flier and Turkensteen, 1999). This has prompted moves to include more widespread monitoring of outbreaks and incorporating them within schemes and presenting them on maps using internet applications (Hansen *et al.*, 2000).

Some of the above schemes have been converted to computer programmes and are available for personal computers which can be fed data from 'in-crop' weather stations that provide read-outs giving an indication of when weather criteria have been satisfied and when sprays should be applied.

In England and Wales, blight forecasting has been revisited to determine whether, with the advent of inexpensive in-crop meteorological data capture equipment, it is possible to improve on blight forecasting from the current Smith periods. Detailed meteorological data, together with the record of first blight occurrence were obtained from a series of experiments conducted at five sites covering a range of infection pressures over four years (Taylor *et al.*, 2003) and tested against a number of schemes (Blitecast, NEGFRY, Schrödter and Ullrich, Smith and Sparks). The initial results were surprising, in that Smith was the only scheme to consistently predict the outbreak of blight in advance of it appearing in the crop. However, results from the 1997 season, a major blight year for the UK, indicated that the forecasting systems worked, in that they all triggered a spray application at least 14 days before the occurrence of blight. There would seem to be justification for confidence in forecasting in high risk years but less so where there was low risk.

The multiplicity of schemes for forecasting blight, even within single countries and the continued development of further schemes, indicates that forecasting for this disease is complex. It is highly unlikely that a standard scheme will be applicable to all countries, or even all regions within a country. The increasing sophistication of the data capture equipment may be providing the imperative to devise even more precise schemes. However, Royle and Shaw (1988) suggested that with an annual and substantial build-up of inoculum with a disease such a potato blight, then a "relatively complex tactical spray programme based on forecasting is unlikely to be worthwhile because it will so rarely differ from routine spraying". This is true in part but there are still savings to be made in defining the start of the programme in situations where inoculum is limiting. All blight forecasting schemes assume presence of inoculum. This is quite clearly not the case when a range of forecasting schemes trigger application based on weather weeks ahead of first occurrences.

9.7 CEREALS

The rapid and widespread introduction of fungicides to the cereal crop in the early 1970s has resulted in an increase in the number of schemes aimed at improving the efficacy of fungicide application. Unlike disease control in potatoes, failure to apply a fungicide, or to time an application with precision, will not result in total crop failure. The cereal crop has a lower profit margin than potatoes and therefore spraying at set and frequent intervals would not necessarily be economic, particularly at full doses. There are also stages in the growth of the cereal plant where the presence of disease is not damaging to yield and therefore control is not worthwhile. For winter wheat, control of disease in the autumn and generally up to the first node development stage (GS 31, Zadoks *et al.*, 1974) is unlikely to be necessary or economic.

A few schemes have been developed over the years in an attempt to define conditions where fungicide application is likely to be economic. Again, as with potato blight, it was the development of fungicides that were not phytotoxic to the crop which triggered an interest in disease forecasting and control.

9.7.1 Powdery mildew

Mildew of spring barley was probably the first disease targeted for fungicide control and the Polley Period was devised to indicate the optimum timing of sprays to control the disease (Polley and King, 1973; Polley and Smith, 1973). The barley mildew high-risk period was designed to detect days on which there was a high risk of a large number of spores being released in barley crops. It was intended that the criteria be used together with observing the presence of mildew in the crop. This was assessed as between 3 and 5% on the lower leaves before a spray was recommended.

A high-risk period occurred when:

- day maximum temperature >15.6°C;
- at 09.00 h the dew point deficit >5 °C;
- day rainfall <1 mm;
- day run of wind >246 km.

A high-risk period occurred on a day when all four criteria were satisfied, or the second day when three of the four factors occurred or the third day when at least two of the factors occurred but with one or two out of the three days having three factors satisfied. The high-risk period started with a high risk day and stopped when none, or only one of the factors had been satisfied. This scheme is not in current use, as it was recognised that the presence of mildew in the crop at the 3-5% level was itself an indication that conditions were generally suitable for the start of the epidemic. This advice then changed to when mildew was 'present' in the crop, as being able to detect mildew on plants at the tillering stage, without having search for it, or for it to be very obvious, equated to about 3-5%. This is a classic example of a complicated system being refined to simple practical advice.

9.7.2 Barley brown rust

A simple forecasting scheme for barley brown rust (caused by *Puccinia hordei*) was proposed by King and Polley (1976). Simkin and Wheeler (1974) reported that urediospores did not germinate at a relative humidity below 100%, germination was complete over a range of temperatures in 6 h and subsequent development was solely dependent on temperature. King and Polley (1976), suggested that a dew period of at least 5 h following a day when the maximum temperature was more than 15°C would indicate that susceptible crops were at risk.

9.7.3 Septoria

Septoria diseases of wheat (causal agents *Phaeosphaeria nodorum* and *Mycosphaerella graminicola*) are wet-weather diseases and ADAS developed a forecasting scheme based on a 'wet period' (Anon., 1984a) to assist in the timing of fungicide applications. A wet period was defined simply as 1 mm or more of rain occurring on any 4 days in the previous 14 days. However, under UK conditions, it was rare that these criteria were not satisfied, and there were risks of spraying crops prematurely or too frequently. Septoria spores are distributed largely in rain splash (Shaw, 1987), and Thomas *et al.* (1989) revised the scheme to take account of 'splashy' rain events so that: 'a total of 10 mm or more in up to 3 consecutive rain days, once the canopy has reach full height, although over 5 mm on any one day may be sufficient in shorter crops where stem elongation is incomplete'. Provision was made for a second spray if these conditions occurred after flag leaf emergence, or the protectant activity of the first spray had been exceeded.

Schöfl *et al.* (1994) proposed a scheme not only based on rainfall but also incorporating inoculum in the risk assessment:

Decision period: GS 33/37 to GS 55-71 (depending on cultivar)

Threshold value: 40-50% incidence per indication leaf layer infected (presence of disease)

Growth stage	Indication leaf layer
GS 33/37-43	leaf four
GS 45-65 (71)	leaf three

First application of fungicide: if the threshold is reached, spray timing depends on the occurrence of rain events in the past 14 days. If a rain event is recorded, a fungicide has to be applied within 10-17 days (dependent on cultivar). If no rain event occurs, treatment should follow the next infection event or when GS 51 is reached.

Rain event: - more than 10 mm within 2 or 3 consecutive days
- more than 5 mm in one day followed by 2 days leaf wetness.

Cultivar effects:	Susceptibility	max. interval	latest application trigger
	high	10	GS 71
	moderate	14	GS 65
	low	17	GS 59

Second application: necessary, if 10-20 days (dependent on protectant fungicide activity) after the first spray, there has been more that 10 mm rain within 2 or 3 consecutive days. If the first spray was at GS 47/51 or later, no second applications required.

The advantage of this scheme is that it does take account of inoculum, although in the UK inoculum is rarely limiting. Probably more importantly, the scheme attempts to take account of cultivar resistance, even if it is on a rather crude basis.

9.7.4 Eyespot

Eyespot forecasting was based on an assessment of the presence of the disease in the spring from the leaf sheath erect stage (GS 30) to first node (GS 31). Crops should be treated if the crop has obvious symptoms affecting more than 20% tillers and lesions penetrating at least two leaf sheaths (Anon., 1986). This could only be an insurance measure, because if subsequent weather turned out to be dry, the affected leaf sheaths would desiccate, become detached from the stem and remove the contact necessary to enable the disease to progress to the stem where it could cause damage.

Early drilling increases the risk from eyespot and a system of accumulated day-degrees was investigated by Croxall (1978). This was based on accumulation day-degrees above 6°C with a starting point on the first day after crop emergence with measurable rain and a mean temperature above 0°C. Croxall determined that visible eyespot lesions should be detectable at 100-125 day-degrees above 6°C and the optimum spray date between 200 and 250 days. However, he concluded, after four years of research, that it was impossible to predict when spraying against eyespot was certain to be beneficial. The most reliable advice, he thought, was a negative prognosis by eliminating crops that were unlikely to benefit from eyespot control. He listed a three-year break from cereals or following two years of grass, crops drilled too late to accumulate 250 day degrees above 6°C. Even after all this equivocation, he suggested taking into account cultivar susceptibility, crops with less than 20% tillers affected and previous experience with the disease on the farm. This well illustrates the dilemma faced by attempts to forecast disease by set rules.

Burnett *et al.* (2004), in a major series of experiments, have re-examined some of the criteria that predispose crops to infection and yield loss. Points are assigned to factors which contribute to disease risk so that the greater the number of points, the greater the risk. The factors identified as being major contributors to risk were sowing date, eyespot infection at GS 31-32, cumulative rainfall in March to May, soil preparation and type and previous crop. Crops at high risk would be those that were drilled before 6 October, had more that 7% tillers affected with eyespot at GS 31-32, had received more that 170 mm rain in the three month period beginning in March, where the field had been ploughed, the soil type was heavy and the crop followed another cereal (Table 9.1). The maximum number of accumulated risk points a crop could be assigned was 50. Two treatment thresholds were set – a risk-sensitive threshold and a risk-tolerant threshold. The advantage of a points system, when set against expected diseases levels, is that the farmers can decide on the level of risk they are prepared to accept before treating their crop.

Table 9.1. Factors affecting the severity of eyespot and risk scores

Factor	*Level*	*Risk points*
Sowing date	≤ 6 October	0
	> 6 October	5
Eyespot GS 31-32	≤ 7%	0
	> 7%	10
Rain (mm) in March / April / May	≤ 170 mm	0
	> 170 mm	5
Tillage	Minimum till	0
	Plough	10
Soil type *	Light	0
	Medium	1
	Heavy	5
Previous crop	Non-host	0
	Other cereal	10
	Wheat	15

1. Risk-sensitive user - treatment triggered at a Risk Score of 20
2. Risk-tolerant user - treatment triggered at a Risk Score of 29

*There was an increased risk of eyespot with brash or limestone soils; this would add a further 5 risk points and would reflect the increased risk observed.

9.7.5 Take-all

Forecasting the risk from take-all (caused by *Gaeumannomyces graminis*) is dependent on an understanding of cropping, soil type and drilling conditions. The take-all decline phenomenon (see also Chapter 14), makes it possible to predict those crops in cereal successions which are likely to be at risk and to take action by altering the rotation to avoid them. It is a classic example of a disease that integrates inoculum as judged by symptoms in the previous crop and agronomic factors, thus avoiding the need to use weather data. However, Clarkson and Polley (1981) suggested that there were general weather factors that do influence take-all but these are essentially short-term when, to be effective, information concerning any increased risk is needed at the time of drilling.

Unfortunately, plants are subjected to more than one disease. For this reason, a system of integrated disease control is required to take account of the development of, for example, rust, mildew and septoria that can attack cereals simultaneously. EPIPRE (Zadoks, 1981) was one such system developed in Holland with the specific aim of providing information specific to an individual crop. It was a centrally devised system that depended on farmers, or their consultants, returning disease observations. A central computerised system calculated the projected disease development and possible loss and generated a reply form - treat, no treatment necessary, and wait (Fig. 9.2).

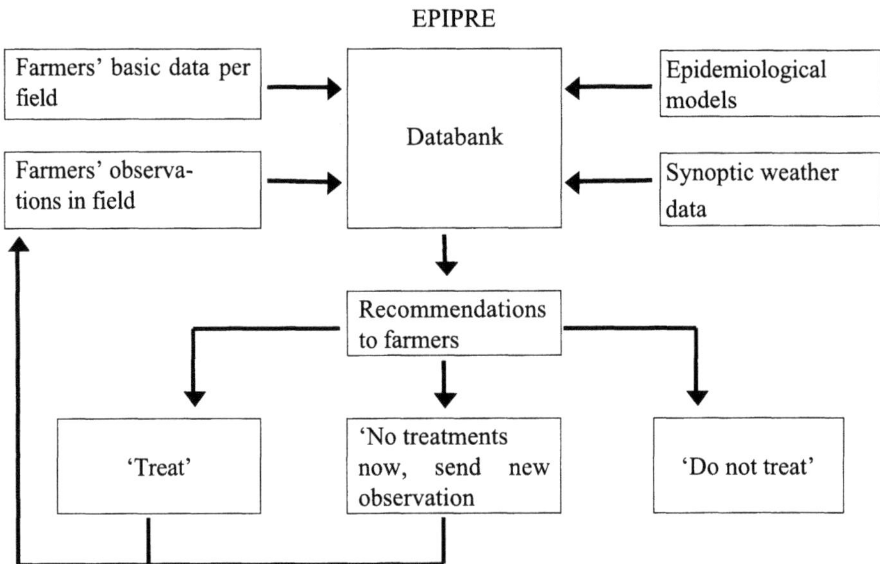

Figure 9.2. Block diagram of EPIPRE (redrawn from Zadoks, 1981).

In 1984, in England and Wales a system called 'Managed disease control' was introduced by ADAS (Anon., 1984a and b, 1985). This was a decision support system based on the result of numerous field experiments and experience. It was a series of decision algorithms and was published in the form of wall charts. This had the advantage that it involved no computing but, like EPIPRE, depended for its success on field observations made by farmers or their consultants who then went through a decision tree (Fig. 9.3). The one devised for barley yellow dwarf virus (BYDV) was modified for videotext and included region risk factors for each season. The videotext version was discontinued in the mid-1980s with the demise of one of the carriers dedicated to on-line farm information and remains to be developed into a PC programme (Foster *et al.*, 2004). However, with the increased use of computers on the farm some schemes have moved to CD-Rom and Web-based systems.

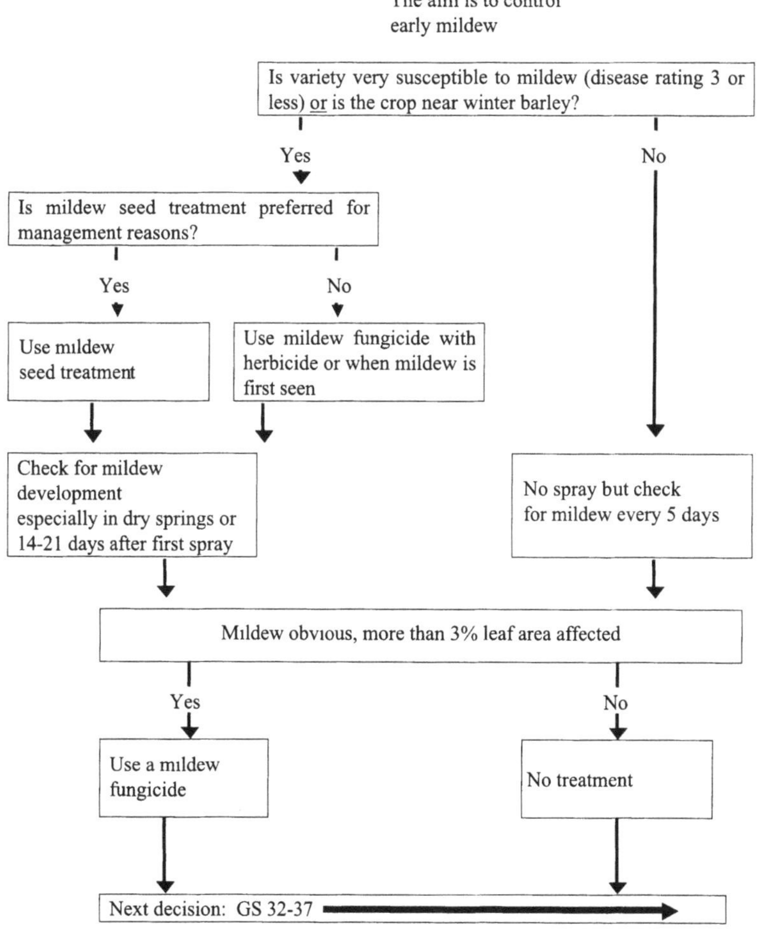

Figure 9.3. Managed disease control for spring barley mildew (Redrawn from Anon., 1986).

The mildew algorithm is fairly simple as it just covers one disease. The winter wheat chart is more complex (Fig. 9.4), as it becomes more demanding of the farmer, requiring him to make more accurate observations and diagnosis.

WINTER WHEAT GS 32 - 37 Second node - flag leaf visible

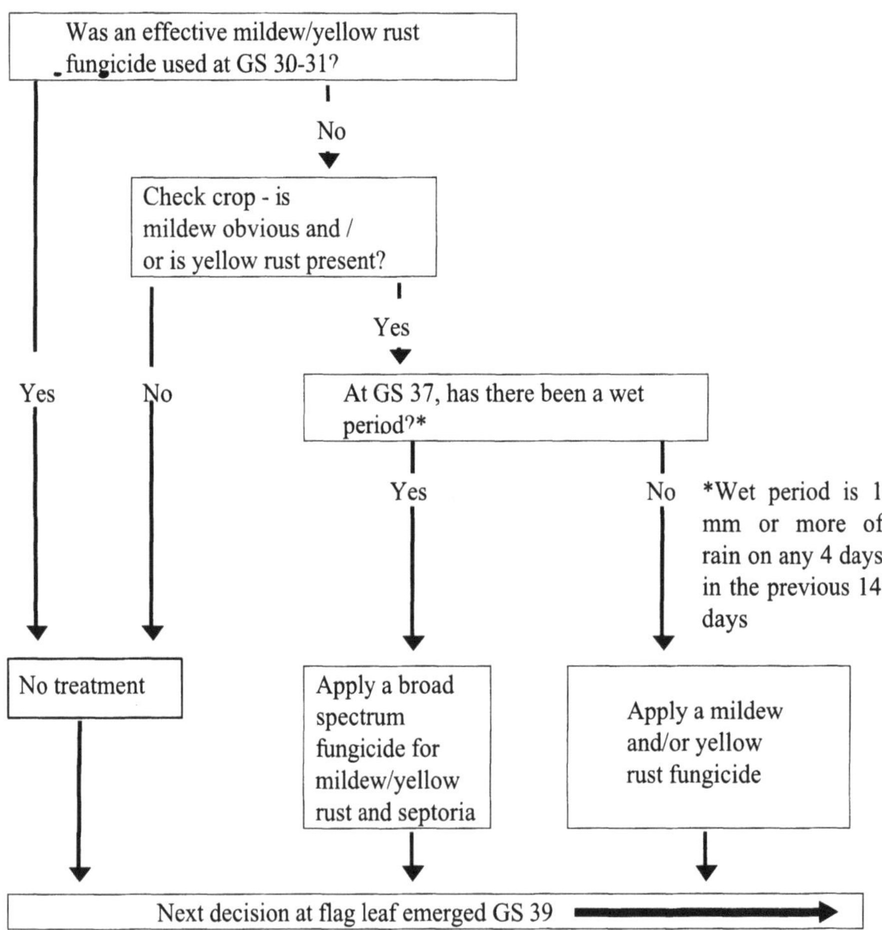

Figure 9.4. Managed disease control for winter wheat (Redrawn from Anon., 1986).

Forrer (1992) reported that, apart from EPIPRE, decision support systems have had little impact in Europe. One of the major deficiencies is observational weakness, for example in yellow rust where observations are required early in the season (Young *et al.*, 2003). The requirement is for accurate disease diagnosis – expertise not always present in farmers or indeed their consultants. Forrer (1992) proposed the integration of immunodiagnostics in the decision-making process to simplify the accuracy of disease identification.

Integrating observations, meteorological data and agronomic factors such as field history, cultivar performance and inputs into a computer-based system for general use is now possible with improved and readily available computer power at relatively reasonable costs. DESSAC (Decision Support System for Arable Crops) has been developed in the UK to provide the industry with sufficient detail to allow the user to make better informed decisions (Parsons *et al.*, 2004).

9.8 OILSEED RAPE

Compared to potatoes and cereals, oilseed rape is a relative new crop for forecasting disease. Disease/yield loss relationships have only relatively recently become understood (Sansford 1995a and b; Sutherland *et al.*, 1995).

9.8.1 Light leaf spot

Data on disease incidence and severity collected from crop surveys gathered between 1977 and 1993 (Fitt *et al.*, 1994, 1996; Gladders *et al.*, 1995) have been used to propose a prediction scheme for forecasting high-risk seasons for light leaf spot (caused by *Pyrenopeziza brassicae*). Gladders *et al.* (1995) found a good correlation between the incidence of light leaf spot on the pods in July and the incidence of light leaf spot on leaves in the following March. This has been proposed as part of the seasonal risk index for a more comprehensive forecasting scheme including an initial crop risk index – to identify crops at risk at the beginning of the season and an index to refine decisions as the season progresses (Fitt *et al.*, 1996). The seasonal risk element is the strongest of the three and the crop risk and progressive risk elements require further refinement. However, the use of the seasonal risk element on its own is a start and can give more confidence on spray recommendations required in the autumn in advance of disease symptoms being reported.

9.8.2 Canker

Canker (caused by *Leptosphaeria maculans*) of winter oilseed rape is a disease which affects the stem bases leading to premature ripening and, in severe cases, lodging. Early phoma leaf spot epidemics are a major cause of yield loss but can be managed economically with fungicides applied in autumn and winter. Conversely, phoma epidemics developing from December onwards may have little effect on yield. Gladders *et al.* (2001) found that canker incidence at harvest could be related to the development of phoma leaf spot in the previous autumn/winter.

Light leaf spot has been integrated into a canker warning system (Gladders *et al.*, 2004), which is available to farmers on the internet (http//phoma.csl.gov.uk).

9.8.3 Sclerotinia

Sclerotinia sclerotiorum, the cause of sclerotinia stem rot of winter oilseed rape, has a complex life cycle. Sclerotinia is a weak pathogen and generally requires a food base from which to infect undamaged tissue. With oilseed rape this food base is generally the senescing petals. In order to infect oilseed rape, sclerotia germinate in the soil to produce apothecia and ascospores are discharged into the atmosphere where they land on rape petals; when these senesce they are blown off the raceme and must adhere to the stems. They generally become trapped in the axils of petioles and racemes. The chance of all these necessary factors being coincident is low. However, such an occurrence took place in England and Wales in 1991 with the result that many crops became infected (Turner and Hardwick, 1995). This led to widespread spraying against the disease. Recent seasons have seen less severe attacks but spray applications have continued at a high level. It is a high profile disease, with only a few plants per field showing symptoms necessary for the disease to be alarmingly obvious. Farmers are not prepared to take the risk of crops being unprotected as action has to be taken long before symptoms appear in the crop.

There are basically three phases in the infection cycle:

- sclerotial germination and ascospore release;
- petal infection;
- petal retention on the canopy and infection.

A number of schemes to forecast the occurrence of sclerotinia stem rot have been devised and have concentrated on the first two elements in the infection cycle.

One of the simplest is the use of depots of buried sclerotia devised by the Danish extension services (Buchwaldt, 1986), where the carpogenic germination of buried sclerotia is monitored in established fields of oilseed rape. If germination is between 0 and 25%, there is a risk of a minor attack. If germination is from 26 to 30% a week before full flower, then the level of sclerotinia stem rot is likely to be above the economic threshold for spraying. It is also suggested that the system could be modified by a point system weighted by agronomic and edaphic factors as is the case in Sweden (Twengström and Sivald, 1995).

The use of depots of buried sclerotia was evaluated as part of a risk forecasting scheme to predict apothecial emergence dates (Sansford, 1995c). Sclerotial germination was immediately preceded by a dry/drier period of weather and/or a mean daily temperature (on a weekly basis) rising above 10°C. Sclerotial germination declined after temperatures reached a daily mean of >15.3°C.

Determination of the infectivity from petal infection is a further refinement and was commonly practised in Canada (Morrall and Thomson, 1995). Growers sampled petals from their crops and incubated them on agar plates. After 3-4 days incubation, any resultant fungal colonies were identified using high quality colour photographs of colonies and also compared with a pure culture of *Sclerotinia* grown at the same time. The percentage petal infection was used to assign a risk value to the crop. The results must be ready in 3-4 days so that spray decisions can be taken. This test was tried in England with limited success (Davies, 1995). The failures relate to the time taken to identify *Sclerotinia* from colony formations. Many of the petals are contaminated by *Botrytis cinerea* and it takes from 8-10 days for sclerotia to develop which can distinguish the colonies. This was considered to be an unacceptable delay. This highlights the requirement for immunodiagnostics/PCR tests specific to *Sclerotinia* to give more rapid identification.

9.9 CONCLUSIONS

Forecasting schemes, when successful, should provide the farmer with the information necessary to provide a cost-effective means of protecting crops from the ravages of disease. However, there are a number of factors which impinge on the success of forecasting schemes.

9.9.1 The models

The models are all trying to interpret the biology of the pathogen in the context of conditions which affect its development, survival and ability to infect and colonise the relevant host. There is an infinite variety of possible combinations. Fortunately, the advent of powerful computers has solved many of the technical constraints. However, biological processes are in a constant state of flux, particularly with the introduction of new strains and mating types, and therefore there is the potential for more rapid changes in aggressiveness and temperature adaptations; it is unlikely that all eventualities will be covered by even the most complex of models. This is an important constraint, for if the models are too complex they may be impractical, particularly if they are dependent on information being supplied by farmers or their consultants.

Most of the models impose a cut-off. This is not necessarily arbitrary but one that may be derived experimentally. However, it is likely to be a mean from a range of results. Fungi are not aware of means. What, for example, is special about 15.7°C but not 15.6°C (Polley and Smith, 1973), or 10°C and not 9.9°C, 90% RH and not 89.5%, 11 h and not 10 h 59 min (Smith, 1956)? This is clearly nonsense. It is now possible to collect data by the minute rather than every 6 h from the synoptic network. However, this gives a false sense of precision and can be likened to blowing up a photograph from a newspaper – the dots that make up the picture are seen in more detail but the subject loses focus and content. Using accumulated day degrees (Croxall, 1978) gives a feeling of greater confidence, as it takes account of seasonal variation, but why accumulate over 6°C and not 5°C? These points were eloquently put, in general terms, by Bourke (1953) – "Even the finely evolved model can scarcely avoid the defects of over-simplification and over-rigidity, over-simplification because the complexity of the phenomena involved cannot be precisely reflected in any easily handled formula; over-rigidity because, even if the criteria are to be in a form capable of objective application by a number of workers, they must introduce abrupt, and to a certain extent, arbitrary discontinuities which appear in nature only as gradual transitions. This does not mean that the evolution of a good working model would not be of considerable value...". The inclusion of husbandry factors, as proposed for the Danish system of sclerotinia risk forecasting, or of eyespot, by weighting each factor empirically, gives some confidence that the system might work as it does involve the crop and the field history, plus an assessment of potential inoculum pressure. Likewise, testing petals of oilseed rape for the presence of spores and the percentage that are affected gives an indication of potential infectivity (Morrall and Thomson, 1995). Computer models, such as NEGFRY (Hansen *et al.*, 1995), which can be modified to take account of local factors are an improvement but still include a level of subjectivity. A summary of some of the constraints are covered by Zadoks (1984).

9.9.2 Equipment

The move away from the synoptic network to the use of in-crop monitors to bring more precision to forecasting imposes its own problems. What is the accuracy of the data capture equipment, how regularly it is calibrated, what is the seasonal drift and where is it sited? All these have a bearing on the likely success of the model.

Sensors for determining relative humidity and leaf wetness are perhaps the two that are most likely to be subjected to inaccurate readings. Manufacturers generally only guarantee RH sensors to + or – 2% accuracy. What chance is there of triggering the 90% RH required for many of the potato late blight forecasts? Likewise, leaf wetness sensors are prone to false readings when exposed to the detritus blown onto them from the crop. Regular checking of the sensors is important but difficult if the equipment is remote from the user. The problem of continuous reading when failures occur due to battery life, solar panels becoming obscured and problems with the transmitters or cellphone networks all impose potential problems and poor delivery from single sites (Hims *et al.*, 1995).

Where should the equipment be sited in the field, near the farm building for ease of access or in an area thought to be more representative of the local climate? How valid will the forecast be with only one set of sensors? With a disease like potato blight, where precision is important, where in the field should the equipment be sited to give the most reliable reading if at one end there is a wood and the other a lake or river? The questions begs the answer that we may be looking for a precision that is not attainable. It is possible that Large (1956) had the answer, that it is the intelligent interpretation of a network of stations that is important. A network obviously has the advantage in that failure, or inaccurate readings, from one or two sites will not fail to trigger a general warning (Smith, 1956).

Some on-farm systems are linked on a network basis (e.g., PLANT-Plus, Raatjes *et al.*, 2004), supplementing the regional weather forecasts. It is the scale of the network required that is worthy of investigation, particularly in areas like the UK with its varied topography and weather systems when compared with countries like the Netherlands, with a more uniform weather system. Bourke (1953) was sceptical and commented "However, it is unfortunately true that even an expensive network of special weather observations from every potato field in the country would not fully cover the local variations in susceptibility to blight". It should, nevertheless, not prevent us from trying.

Whether an individual station or a network is appropriate will largely depend upon the pathogen. Forecasts of pathogens that build up rapidly and spread long distances, such as potato blight, may respond better to a network of stations while splash-borne pathogens, such as *Septoria*, can be covered from a single station within the field.

9.9.3 User

The second and potentially major constraint is the farmer. Will he use a forecasting system? Farmers are keen to produce their crops as economically as possible and any system that will enable them to save money will be favourably received. Also, farmers are coming under increasing pressure from their customers, particularly the large supermarket chains, to justify pesticide use. Forecasts provide this justification.

There are various degrees of providing the farmer with the necessary information on the actions that should be taken. Some systems are passive, in that data are obtained from a synoptic meteorological station, interrogated by the meteorological service for the appropriate criteria and the information fed to an extension service for interpretation and dissemination. Other systems may require farmer input, weather recording, monitoring crops for the presence of disease and identifying a threshold. The farmer may be prepared to do this himself or contract it out to a specialist consultant. Where computer-based systems are used, the latter is the most likely scenario. Where computer-based systems are required, it is obviously necessary not only to possess the hardware but to commit the time and effort to run the models.

A further constraint are farmers' concerns about the reliability of forecasting systems. Generally, a farmer's measurement of the success of advice is the appearance of his crop and whether the yield expectation is realised. For a commercial consultant, where repeat business is essential, it is likely that crop appearance is all important. This pushes consultants towards insurance and scheduled spraying rather than timely forecasts. There are also practical difficulties to overcome, scheduling being one, particularly when repeat spraying is required, as with potato blight. The ability to get round all fields at risk is a major logistical constraint and it is often easier, once started, to maintain a 7- or 10-day schedule rather than one varied according to risk. The increasing use of contractors for spray operation also means that they have to be reserved in advance and this imposes limitations on the ability to 'fine tune' the spray programme. Where spray intervals cannot be altered for practical reasons it may be possible to reduced dose according to forecast risk as suggested by Nielsen (2004). Spraying on a large scale also poses logistical problems in terms of sprayer capacity, adequate stock of fungicides and competing demands for the treatment of other crops (Hinds, 2000).

Where there is a one disease per crop situation, as with blight of potatoes and in some parts of the UK, mildew on spring barley, decisions can be relatively simple. Where several diseases are involved, timing and choice of product become more important. Dose and whether the product has protectant or eradicate activity are also important factors as these impinge on timing. The increased complexity of the decision-making process has lead to a more integrated approach to disease control through managed or decision support systems. EPIPRE (Zadoks, 1981), PRO_PLANT (Frahm *et al.*, 1996) and DESSAC (Parsons *et al.*, 2004) are such examples. However, Forrer (1992) suggests that such systems are short-lived after farmers have got used to them. They are also demanding of the farmer in terms of monitoring the crop for disease. Such systems also depend on the farmers' ability to identify disease, for example, being able to distinguish between eyespot and sharp eyespot is difficult even for experienced plant pathologists. Improved diagnostic techniques, using molecular methods (see Chapter 1), can provide assistance, but care in interpretation of the results is still required. However, decision support systems do provide a valuable training aid in crop management for farmers, consultants and students of crop protection.

With continuing reductions in financial support to agriculture, it is an inevitable consideration that forecasting systems become a substitute for the necessary number of experts required to provide an input into the daily decision-making process of disease management. However, what is forgotten is that these 'expert' systems very often require the expert to interpret the data generated from them. Accurate disease diagnosis and assessment (Chapter 2) are difficult for the non-specialist and they are not necessarily helped by the advent of immunodiagnostics and PCR methodology. Modern methods do not necessarily address the consequence of a specific diagnosis until there is a quantitative rather than qualitative basis for the result, or necessarily address growers' concerns in addressing their particular problems. Problem-solving is more complex as it is concerned with the practical possibilities, aspirations and even psychological security of the individual farmer. Decision support systems are what they say, they are a support to decisions and should not be viewed as an attempt to become a substitute for the absence of an effective extension service. It is important that the systems are over-ridden when the decisions being made are at variance with the 'sixth sense' of the experienced advisor. Douglas Bader (British World War II pilot) said that "Rules are for the obedience of fools and the guidance of wise men". Knowledge developed over a number of years of experience of local conditions together with results from appropriate experiments are sometimes more important than a computer read-out. However, forecasting systems will continue to be an essential part of the decision-making process on disease control, whether analytical or intuitive.

REFERENCES

Anon. (1984a) *Winter wheat managed disease control.* Alnwick, UK: Ministry of Agriculture, Fisheries and Food, Leaflet 831.

Anon. (1984b) *Winter barley managed disease control.* Alnwick, UK: Ministry of Agriculture, Fisheries and Food, Leaflet 843.

Anon. (1985) *Spring barley managed disease control.* Alnwick, UK: Ministry of Agriculture, Fisheries and Food, Leaflet 844.

Anon. (1986) *Use of fungicides and insecticides on cereals.* Alnwick, UK: Ministry of Agriculture, Fisheries and Food, Booklet 2257.

Anon. (2002). *Foot and Mouth Disease: Lessons to be learned.* Enquiry Report (HC888). London: The Stationery Office.

Barrie, I.A. and Bradshaw, N.J. (2001) Blight watch – a spatially interpolated system for the calculation of Smith-periods in the UK, in *Proceedings of the workshop on the European network for development of an integrated control strategy of potato late blight, 2001* (eds. Schepers H.T.A.M., Westerdijk, C.E.). Edinburgh, Scotland. PAV Report No. 8, 169-181.

Beaumont, A. (1947) The dependence on the weather of the dates of potato blight epidemics. *Transactions of the British Mycological Society,* **31**, 45-53.

Bourke, P.M.A. (1953) *Potato blight and the weather, a fresh approach.* Department of Industry and Commerce Meteorological Service, Dublin. Technical Note **No. 12**.

Bourke, P.M.A. (1955) *The forecasting from weather data of potato blight and other plant diseases and pests.* World Meteorological Organisation. Technical Note **No. 10**.

Buchwaldt, L. (1986) The Danish forecasting system for attack by stem rot (*Sclerotinia sclerotiorum*) in oilseed rape. *Nordisk Plantevaernskonference,* pp. 121-131.

Burnett, F.J. and Hughes, G. (2004) The development of a risk assessment method to identify wheat crops at risk from eyespot. *HGCA Project Report 347.* London: Home-Grown Cereals Authority.

Cambell, C.E. and Madden, L.V. (1990) *Introduction to plant disease epidemiology.* John Wiley & Sons Inc., New York.

Clarkson, J.D.S. and Polley, R.W. (1981) Diagnosis, assessment, crop-loss appraisal and forecasting, in *Biology and control of take-all* (eds M.J.C. Asher and P.J. Shipton). Academic Press, London pp. 251-269.

Cooke, L.R. (1992) The future of potato blight control: a more aggressive pathogen and fewer weapons? in *Disease Management in Relation to Changing Agricultural Practice 1992* (eds A.R. McCraken and P.C. Mercer). SIPP/BSPP, Belfast, pp. 65-73.

Crosier, W. (1934) Studies in the biology of *Phytophthora infestans* (Mont.) de Bary. Memoirs of the Cornell Agricultural Experimental Station No. 155.

Croxall, H.E. (1978) *Forecasting and the control of eyespot epidemics in the East Midlands.* Open conference of Advisory Plant Pathologists, Agricultural Development and Advisory Service, UK, PP/0/396.

Davies, J.M.Ll. (1995) Petal culturing to forecast *Sclerotinia* in oilseed rape. *Proceedings Ninth International Rapeseed Congress* 1995. GCIRC, Cambridge, pp. 1010-1012.

Everdingen, E. van. (1926) Het verband tusschen de weersgesteldheid en de aardappelziekte (*P. infestans*). *Tijdschr. Plantenziekten.* **32**, 129-140.

Fitt, B.D.L., Gladders, P., Figueroa, I. and Murray, G. (1994) Forecasting light leaf spot (*Pyrenopeziza brassicae*) on winter oilseed rape. *Proceedings Brighton Crop Protection Conference – Pest and Disease.* BCPC, Brighton, pp. 265-270.

Fitt, B.D.L., Gladders, P., Turner, J.A. *et al.*, (1996) Predicting risk of severe light leaf spot (*Pyrenopeziza brassicae*) on winter oilseed rape in the UK. *Proceedings Brighton Crop Protection Conference – Pest and Disease.* BCPC, Brighton, pp. 239-244.

Flier, W.G. and Turkensteen, L.J. (1999) Foliar aggressiveness of *Phytophthora infestans* in three potato growing regions in the Netherlands. *European Journal of Plant Pathology,* **15**, 381-388.

Fohner, G.R., Fry, W.E. and White, G.B. (1984) Computer simulation raises question about timing protectant fungicide application frequency according to a potato late blight forecast. *Phytopathology,* **74**, 1145-1147.

Forrer, H.R. (1992) Experiences with the cereal disease forecast system EPIPRE in Switzerland and prospects for the use of diagnostics to monitor disease state. *Proceedings Brighton Crop Protection Conference – Pest and Diseases.* BCPC, Brighton, pp. 711-770.

Foster, G.N., Blake, S., Tones, S.J. *et al.* (2004) Occurrence of barley yellow dwarf virus in autumn-sown cereal cops in the United Kingdom in relation to field characteristics. *Pesticide Management Science,* **60**, 112-125.

Frahm, J., Volk, T. and Johnen, A. (1996) Development of the PRO_PLANT decision-support system for plant protection in cereals, sugar beet and rape. *EPPO Bulletin,* **26**, 609-622.

Garthwaite, D.G., Thomas, M.R., Dawson, A. and Stoddart, H. (2002) Arable farm crops in Great Britain. *Pesticide usage survey report,* **187**. UK: Defra & SEERAD.

Gladders, P., Dyer, C., Fitt, B.D.L. *et al.* (2004) Development of a decision support system for phoma and light leaf spot in winter oilseed rape ("PASSWORD" Project). *HGCA Project Report* 357. London: Home-Grown Cereals Authority.

Gladders, P., Fitt, B.D.L. and Welham, S.J. (1995) Forecasting development of light leaf spot (*Pyrenopeziza brassicae*) epidemics on winter oilseed rape. *Proceedings Ninth International Rapeseed Congress* 1995. GCIRC, Cambridge, pp. 1001-1003.

Gladders, P., Green, M.R., Steed, J.M. *et al.* (2001) Improving stem canker control in winter oilseed rape by accurate timing of fungicide applications based on disease forecasts *HGCA Project Report* OS51. London: Home-Grown Cereals Authority.

Grainger, J. and Rutherford, A.A. (1963) Rapid determination of host receptivity in potato blight forecasting. *European Potato Journal,* **6**, 258-267.

Hansen, J.G., Andersson, B. and Hermansen, A. (1995) NEGFRY – A system for scheduling chemical control of late blight in potatoes, in *Phytophthora infestans 150* (eds L.J. Dowley, E. Bannon, L.R Cooke, T. Keane and E. O'Sullivan). EAPR, Dublin pp. 201-208.

Hansen, J.G., Lassen, P. and Röhrig, M. (2000) Monitoring of late blight based on a collaborative PC- and internet applications in *Proceedings of the workshop on the European network for development of an integrated control strategy of potato late blight, 2000.* (ed. H.T.A.M Schepers.) Munich, Germany. PAV Report No. 7, 39-54.

Hardwick, N.V. (2002) Weather and plant disease. *Weather,* **57**, 184-190.

Hardwick, N.V. and Turner, J.A. (1996) A survey of potato diseases in England and Wales, 1993-1996. *Abstracts 13th Triennial Conference, European Association of Potato Research 1996.* EAPR, Veldhoven, pp. 631-632.

Hims, M.J., Taylor, M.C., Leach, R.F. *et al.* (1995) Field testing of blight risk prediction models by remote data collection using cellphone analogue networks, in *Phytophthora infestans 150* (eds L.J. Dowley, E. Bannon, L.R Cooke, T. Keane and E. O'Sullivan). EAPR, Dublin, pp. 220-225.

Hinds, H. (2000) Can blight forecasting work on large potato farms? in *Proceedings of the workshop on the European network for development of an integrated control strategy of potato late blight, 2000.* (ed. H.T.A.M Schepers.) Munich, Germany. PAV Report No. 7, 99-106.

Hyre, R.A. (1954) Progress in forecasting late blight of potato and tomato. *Plant Disease Reporter,* **38**, 245-253.

Jørgenson, L.N. and Secher, B.J.M. (1996) The Danish action plan – ways of reducing inputs. *Proceedings Crop Protection in Northern Britain 1996*, 63-70.

Keane, T. (1995) Potato blight warning practice in Ireland, in *Phytophthora infestans 150* (eds L.J. Dowley, E. Bannon, L.R Cooke, T. Keane and E. O'Sullivan). EAPR, Dublin., pp. 191-200.

King, J.E. and Polley, R.W. (1976) Observations on the epidemiology and effect on grain yield of brown rust in spring barley. *Plant Pathology,* **25**, 63-73.

Krause, R.A., Massie, L.B. and Hyre, R.A. (1975). Blitecast: a computerized forecast of potato late blight. *Plant Disease Reporter,* **59**, 95-98.

Large, E.C. (1952) The interpretation of progress curves for potato blight and other plant diseases. *Plant Pathology,* **1**, 109-117.

Large, E.C. (1953) Potato blight forecasting investigation in England and Wales, 1950-1952. *Plant Pathology,* **2**, 1-15.

Large, E.C. (1956) Potato blight forecasting and survey work in England and Wales, 1950-1952. *Plant Pathology*, **5**, 39-52.

Large, E.C. (1959a) The battle against blight. *Agriculture,* **65**, 603-608.

Large, E.C. (1959b) Potato blight in England and Wales. *Proceedings IVth International Congress of Crop Protection.* Hamburg, pp. 215-220.

Miller, P.R. and O'Brien, M. (1952) Plant disease forecasting. *The Botanical Review,* **18**, 547-601.

Morrall, R.A.A. and Thomson, J.R. (1995) Four years' experience in western Canada with commercial petal testing to forecast sclerotinia. *Proceedings Ninth International Rapeseed Congress* 1995. GCIRC, Cambridge, pp. 1013-1015.

Nelson, E.C. (1995) The cause of the calamity: the discovery of the potato blight in Ireland, 1845-1847, and the role of the National Botanic Gardens, Glasnevin, Dublin, in *Phytophthora infestans 150* (eds L.J. Dowley, E. Bannon, L.R Cooke, T. Keane and E. O'Sullivan). EAPR, Dublin, pp. 1-11.

Nielsen, B.J. (2004) Control strategies against potato late blight using weekly model with fixed intervals but adjusted fungicide dose, in *Proceedings of the workshop on the European network for development of an integrated control strategy of potato late blight, 2004*. (eds C.E., Westerdijk and H.T.A.M Schepers). Jersey, Channel Islands. PAV Report No. 10, 233-235.

Parsons, D.J., Mayes, A., Meakin, P. *et al.* (2004) Taking DESSAC forward with the arable decision support community. *Aspects of Applied Biology,* **72**. Advances in applied biology: providing new opportunities for consumers and producers in the 21st century, Association of Applied Biologists, Warwick, UK, pp. 55-66.

Polley, R.W. and King, J.E. (1973) A preliminary proposal for the detection of barley mildew infection periods. *Plant Pathology,* **22**, 11-16.

Polley, R.W. and Smith, L.P. (1973) Barley mildew forecasting. *7th British Insecticide and Fungicide Conference.* BCPC, Brighton, pp. 373-378.

Raatjes, P., Hadders, J., Martin, D. and Hinds, H. (2004) PLANT-Plus: turn-key solutions for disease forecasting and irrigation management. *Decision support systems in potato management: bringing models into practice*. Wageningen Academic Publishers. Wageningen Netherlands, pp. 169-185.

Royle, D.J. and Shaw, M.S. (1988) The costs and benefits of disease forecasting in farming practice, in *Control of Plant Diseases: Costs and Benefits* (eds B.C. Clifford and E. Lester). Blackwell Scientific Publications, Oxford, pp. 231-246.

Sansford, C.E. (1995a) Oilseed rape: development of canker (*Leptosphaeria maculans*) and its effect on yield. *Proceedings Ninth International Rapeseed Congress* 1995 Cambridge: GCIRC, 577-579.

Sansford, C.E. (1995b) Oilseed rape: development of stem rot (*Sclerotinia sclerotiorum*) and its effect on yield. *Proceedings Ninth International Rapeseed Congress* 1995 Cambridge: GCIRC, 634-636.

Sansford, C.E. (1995c) Oilseed rape: an aid to Sclerotinia risk forecasting. *Proceedings Ninth International Rapeseed Congress* 1995 Cambridge: GCIRC, 1007-1009.

Schöfl, U.A., Morris, D.B. and Verreet, J.A. (1994) The development of an integrated decision model based on disease threshold to control *Septoria tritici* on winter wheat. *Proceedings Brighton Crop Protection Conference – Pest and Disease*, Brighton: BCPC 671-678.

Schrödter, H. and Ullrich, J. (1967) Eine mathematischstatistiche losung des problems der prognose von epidemien mit hilfe meteorologischer parameter, dargestellt am beispiel der kartoffelkrautfaule (*Phytophthora infestans*). *Agricultural Meteorology,* **4**, 119-135.

Shattock, R.C. (2002) *Phytophthora infestans*: populations, pathogenicity and phenylamides. *Pest Management Science,* **58**, 944-950.

Shaw, M.S. (1987) Assessment of upward movement of rain splash using a fluorescent tracer method and its implication to the epidemiology of cereal pathogens. *Plant Pathology,* **36**, 201-213.

Simkin, M.B. and Wheeler, B.E.J. (1974) The development of *Puccinia hordei* on barley cv. Zephyr. *Annals of Applied Biology,* **78**, 225-235.

Smith, L.P. (1956) Potato blight forecasting by 90% humidity criteria. *Plant Pathology,* **5**, 83-87.

Smith, L.P. and Hugh-Jones, M.E. (1969) The weather factor in foot and mouth disease epidemics. *Nature*, **223**, 712-715.

Sparks, W.R. (1984) The use of 'critical' weather periods in the prediction of potato blight outbreaks. *Agricultural Memorandum* **No. 1020**, Meteorological Office, Bracknell, UK.

Sutherland, K.G., Wale, S.J. and Sansford, C. (1995). Effect of different epidemics of *Pyrenopeziza brassicae* on yield loss in winter oilseed rape. *Proceedings Ninth International Rapeseed Congress* 1995. GCIRC, Cambridge, pp. 1004-1006.

Taylor, M.C., Hardwick, N.V., Bradshaw, N.J. and Hall, A.M. (2000). Are excessive blight sprays detrimental to potato yield? *Proceedings BCPC Conference, Pests & Diseases, 2000.* Brighton, UK: BCPC, 553-858.

Taylor, M.C., Hardwick, N.V., Bradshaw, N.J. and Hall, A.M. (2003) Relative performance of five forecasting schemes for potato late blight (*Phytophthora infestans*) I. Accuracy of infection warnings and reduction of unnecessary, theoretical, fungicide applications. *Crop Protection,* **22**, 275-283.

Thomas, M.R., Cook, R.J. and King, J.E. (1989) Factors affecting development of *Septoria tritici* in winter wheat and its effect on yield. *Plant Pathology,* **38**, 246-257.

Turner, J.A. and Hardwick, N.V. (1995) The rise and fall of *Sclerotinia sclerotiorum*, the cause of stem rot of oilseed rape in the UK. *Proceedings Ninth International Rapeseed Congress* 1995. GCIRC, Cambridge, pp. 640-642.

Twengström, E. and Sivald, R. (1995) A forecasting method for sclerotinia stem rot. *Danish Institute of Plant and Soil Science,* **10**, 75-76.

Urech, P.A., Schwinn, F. and Staub, F. (1977) CGA-48988, a novel fungicide for the control of late blight, downy mildews and related soil borne diseases. *Proceedings Brighton Crop Protection Conference - Pest and Disease.* BCPC, Brighton, pp. 671-678.

Van der Plank, J.E. (1963) *Plant diseases: epidemics and control.* Academic Press, New York.

Wallin, J.R. (1962) Summary of recent progress in predicting late blight epidemics in the United States and Canada. *American Potato Journal,* **39**, 306-312.

Young, C.S., Vaughan, T.B., Thomas, J.M. and Lockley, K.D. (2003). Predicting epidemics of yellow rust (*Puccinia striiformis*) on the upper canopy of wheat from disease observations on lower leaves. *Plant Pathology,* **52**, 338-349.

Zadoks, J.C. (1981) EPIPRE: a disease and pest management system for winter wheat developed in the Netherlands. *EPPO Bulletin*, **11**, 365-369.

Zadoks, J.C. (1984) A quarter of a century of disease warning, 1958-1983. *Plant Disease,* **68**, 352-355.

Zadoks, J.C., Chang, T.T. and Konzak, C.F. (1974) A decimal code for the growth stage of cereals. *Weed Research,* **14**, 415-421.

Zadoks, J.C. and Schein, R.D. (1979) *Epidemiology and Plant Disease Management*, Oxford University Press, New York, 427 pp.

CHAPTER 10

DIVERSIFICATION STRATEGIES

MARIA R. FINCKH AND MARTIN S. WOLFE

10.1 INTRODUCTION

From the early 1980s, black leg on oilseed rape (caused by *Leptosphaeria maculans*) became a serious disease problem in Canada. It was encouraged by a sudden export market opportunity which triggered a huge increase in the area grown to a single genotype of rape (Busch *et al.*, 1994; Juska *et al.*, 1997). This illustrates an important side-effect of industrialised agriculture (Clunies-Ross and Hildyard, 1992) which, over the past 50 years, has led to a sharp reduction in the number of different crops produced and in the number of crop varieties.

In contrast, the plant species and genotypes that survive in natural ecosystems do so because they are able to withstand, simultaneously, the depredations of many competitors, diseases and pests in addition to all other selective pressures. The defence mechanisms involved are still poorly understood but it seems clear that there are many and that they operate at different levels from the individual plant to the plant community. Nevertheless, infections or infestations may be sometimes severe, which can lead to local extinction of the host. Importantly, pure stands of species are rare in natural ecosystems; where they do occur, the individuals within the stand are thought to be genetically distinct (e.g. Wills, 1996). Indeed, genetic diversity within and among species appears to be a concomitant of survival and of stability in communities (see review by McCann, 2000).

The origins of monoculture are buried in our agricultural past but the concentration and isolation of crop plants for convenience of planting and harvesting was a development that defined and separated early agriculture from food-gathering. The emergence of species, variety and gene monocultures in agricultural ecosystems represented a further and even more radical departure from nature. Although the potential for yield gain is probably limited by the likelihood that inter-plant competition for resources is maximal in uniform stands, the convenience of monoculture outweighed the disadvantages.

Until recently, crop species were numerous and relatively undeveloped. Then, as the crop range decreased and human population increased, the disadvantages of monoculture in terms of greater and more consistent attacks by diseases and pests became more apparent because of the ease of spread of the organisms involved. For the major crops, there has also been a tendency to increase the harvest index for large-scale production, resulting usually in dwarf forms of the plant which may be less competitive towards weeds than their taller predecessors. Moreover, splash

B. M. Cooke, D. Gareth Jones and B. Kaye (eds.), The Epidemiology of Plant Diseases, 2nd edition, 269–307.

dispersed pathogens such as *Stagonospora nodorum* or *Fusarium* species that depend on spreading upwards on plants may also be more problematic in shorter wheat varieties (Scott *et al.*, 1982).

In addition, monoculture encourages the evolution, multiplication and spread of newly adapted weeds, pests and pathogens on massive and uniform crops. Often, our extensive, dense and continuous stands of single species can be supported only by frequent use of fungicides, insecticides and herbicides and, indeed, by a regular succession of novel biocides and of varieties with new resistances.

Different solutions to monoculture problems developed at different times. One of the first recorded followed the observation that stem rust (caused by *Puccinia graminis* f.sp. *tritici*) of wheat (*Triticum aestivum*) was always worse if the wheat was surrounded by barberry hedges; removal of the alternative host, barberry, allowed the monoculture of wheat to be maintained (Tozzetti, 1767). Fungicides were invented in the eighteenth century following Jethro Tull's observation that wheat seed accidentally wetted in sea water was freed from bunt. Biffen's application of Mendel's Laws to breeding for disease resistance in wheat (Biffen, 1905) led to the modern breeding industry. Such discoveries in epidemiology, fungicides and breeding made it possible or more profitable to grow monocultures, at least initially.

Because of the ability of pests, pathogens and weeds to adapt to resistances and pesticides, most of these solutions have been short-lived. It is fortunate that, by unconscious and indirect selection plant varieties with **durable disease resistance** (Johnson, 1984) emerged from the monoculture approach. Under the extreme selection imposed by monoculture, there may be a small proportion of resistant host genotypes that remains resistant. However, there are still problems with detecting and handling durable resistance (Johnson, 1993), not the least of which is that any resistant variety is only as durable as its least durable resistance to the range of diseases with which it is confronted. In addition to durably resistant host plants, some fungicides appear to be durable in that pathogen resistance to them is slow to develop. The full ecological impact of these toxic agents is still unknown (Colbourn *et al.*, 1996) but nevertheless they are used extensively to support monocultures. Because durability of resistance does not imply permanence, there is a risk in depending on a limited range of durably resistant genotypes and fungicides for long-term, large-scale food production. Moreover, such a limitation may also restrict the potential of the crop or the cropping system for further improvement.

The understanding of the interactions between host and pathogen populations has led to suggestions for alternative strategies for the use of resistance genes. The goals of such alternative strategies are (Finckh and Wolfe, 1997):

- to achieve an **acceptable level** of disease;
- to achieve **durable** disease control;
- to achieve **stable resistance,** effective in different environments;
- to control simultaneously all important diseases (and pests and weeds), i.e. the development of a **systems approach.**

Under the heading of diversification strategies, our objective here is to consider various options which can reverse the trend to monoculture back towards a degree of functional diversity which provides more sustainable forms of cropping. Such options differ greatly. At the simplest level, we consider cryptic diversification of resistance genes within monoculture systems. At the most complex, there are developments in new or improved systems of polyculture, for example in agroforestry (Nair, 1993) or perennial polyculture (Jackson, 1980; Soule and Piper, 1992). A major objective of these latter options is to develop systems that are able to deliver many ecological benefits and services simultaneously, for example, control of erosion, weeds, pests and diseases (Altieri, 1995; Altieri *et al.*, 1996), while maintaining or enhancing the scale and efficiency of production. Although we concentrate here on crop diseases, we comment also on pests, weeds and other problems since, in the real world, our concern is for simultaneous restraint of all major unwanted organisms and their side effects. The examples given show how system modifications can be introduced to reduce costs, increase production and reduce stress on the environment.

10.2 DEFINITIONS

Possibilities for host diversity at the three levels of species, variety and resistance genotype, both for monoculture and for polyculture, are listed in Table 10.1. Based on the ranges of diversity indicated in this table, we use the following definitions:

Monoculture refers usually to the continuous use of a single crop species over a large area. For the pathologist, however, the term monoculture alone is inadequate since it can be applied at the level of **species, variety** or **gene**. This additional specification is important; for example, within a species monoculture, attempts have been made to encourage farmers to diversify among different varieties with different disease resistances. But, if the varieties available within the species all possess the same gene for resistance to a particular pathogen, then the system is effectively a monoculture with respect to that resistance **(variety diversification with resistance gene monoculture)**. Conversely, the multiline approach is an attempt to combine the benefits of monoculture and diversity by widespread use, for example, of a population of plant genotypes that are uniform for desired quality and other characteristics but diversified for a range of resistance genes against a major pathogen of the crop **(variety monoculture with resistance gene diversification)**.

Polyculture refers usually to the use of a range of different species over time on a single management unit, which may be a farm or a field. Given that it may be difficult to define a field, we follow a modification of the scheme adopted by Vandermeer (1989).

Sole (single) cropping: different crops are used but are managed separately. This includes:

- **rotational cropping**, involving production of different crops among fields and over years or, **sequential cropping**, involving two or more crops in sequence in the same field in one year.

Inter-cropping: two or more crops or varieties are managed simultaneously in the same field. The degree of their interaction depends on their arrangement as follows:

- **mixed inter-cropping** - two or more crops are wholly interspersed, maximising their interaction. The crop is a species or variety mixture. It may also be a species *and* variety mixture;
- **row inter-cropping** - one or more of two or more crop species or varieties are planted in alternating rows, simplifying management but reducing species or varietal interactions;
- **strip inter-cropping** - two or more crop species or varieties are grown in strips; the strips are wide enough for independent management but narrow enough for some interaction among the crops;
- **plot intercropping** - the field is divided into plots or blocks, each of which contains a single species or variety; interactions will be relatively small.

Table 10.1. Diversification in monoculture and polyculture at the three production levels, species, variety and resistance character

Level	*Monoculture*	*Polyculture*
Species	Diversification still possible at the variety and/or resistance sub-levels, either in single or intercropping	Many possible arrangements among and within species, varieties and resistances using intercropping
Variety	Dangerous unless used as multilines (i.e. mixed crop diversification at the resistance level)	Many possible arrangements among and within varieties and resistances - includes variety mixtures, multilines and populations
Resistance	The highest risk for losing the effectiveness of the resistance	Many possible arrangements among resistances - multilines and populations

With mixed inter-cropping, the cycles of the crops tend to be similar except in some forms of intensive hand cultivation or where the crops are so different that their management is separate (e.g. cereal/clover). In the other forms of intercropping, the difference in the crop cycles may vary from a few days, in some forms of vegetable production, to many years, in all forms of agroforestry. A special form of deliberate mixed cycle cropping is **relay inter-cropping**, a form of polyculture intermediate between rotation and intercropping. It is often applied as a form of strip-intercropping; two or more crops are grown simultaneously during part of the life-cycle of each.

Different forms of intercropping can be combined, not only together but with sole cropping; for example, where an intercrop vegetable enterprise is considered as one element in a farm rotation. Such complex systems can have a major influence in limiting diseases, pests and weeds.

10.3 BENEFITS FROM SPATIAL DIVERSIFICATION: SMALL SCALE

The expected benefits from intercropping were summarised by Vandermeer (1989), although similar summaries have been published widely, for example in agroforestry (Nair, 1993):

- increased productivity and yield;
- better use of resources such as land, labour, time, water and nutrients;
- reduction of losses due to diseases, pests and weeds;
- socio-economic benefits such as stability of production, economic gain and nutrition.

To achieve such benefits requires great attention to management on the one hand, and to choice of crop partners on the other. Diversity simply as an end in itself may lead to losses in production and productivity. Partners need to be selected that have different useful characteristics, that occupy non-identical ecological niches and that are complementary (Tilman *et al.*, 2001).

10.3.1 Diversity and disease control

Mechanisms

Diversity can play an important role in the control of pests and pathogens (Wolfe, 1985; Wolfe and Finckh, 1997; Mundt, 2002) provided such diversity is functional. **Functional diversity** is diversity that limits pathogen and pest expansion and that is designed to make use of knowledge about host-pest/pathogen interactions to direct pathogen evolution (Schmidt, 1978; Mundt and Browning, 1985a).

Several mechanisms may contribute to changes in disease incidence or severity (usually a reduction) in host populations that are diverse for resistance (e.g., Van der Plank, 1968; Browning and Frey, 1969; Barrett, 1978; Burdon and Chilvers, 1982; Mundt and Browning, 1985a; Wolfe, 1985; Wolfe and Finckh, 1997; Mundt, 2002). Boudreau and Mundt (1997) argued that the most important mechanisms in variety mixtures are those that affect dispersal since competitive differences among the host components are relatively small (but see Finckh and Mundt, 1992b; Finckh *et al.*, 1999). On the other hand, they found that in species mixtures, competitive interactions among the components are relatively large and probably more important in determining disease levels.

The first four mechanisms apply to all variety mixtures, whether or not there is pathogen specialisation to the host in question:

1. Disease reduction due to **increased distance** between plants of the most susceptible component in the mixture. Burdon and Chilvers (1976) confirmed, experimentally, density dependence of the spread of powdery mildew (caused by *Blumeria graminis* f.sp. *hordei*) of barley (*Hordeum vulgare*). This space effect depends on the proportion of spores available for allo-infection, the density of planting and the number of components in the mixture. The first two of these factors

are fixed by the nature of the pathogen involved and the standard planting density used by farmers. The third factor, the number of components, can be changed but the law of diminishing returns may be limiting beyond three or four components (greater variation may be valuable for other characters).
2. Further restriction of pathogen spread can be caused by resistant plants acting as **barriers** ('fly-paper' effect; Trenbath, 1977). However, a particular space/barrier is variable in effectiveness depending on spore density in the canopy, gradient of spore dispersal, air movement and plant architecture.
3. **Selection in the host population** for the more competitive and/or more resistant genotypes can reduce (or sometimes increase) overall disease severity (Boudreau and Mundt, 1992; Finckh and Mundt, 1992a), together with possible effects of host competition on susceptibility (this mechanism may be important in species mixtures). For example, severity of yellow rust on wheat cultivars in mixtures was frequency-dependent if the competitive abilities of the companion cultivars were similar. If, however, competitive interactions were asymmetrical, frequency-dependence was not evident (Finckh and Mundt, 1992b). There may also be 'feed-back' effects from the pathogen that may alter competitive interactions among the hosts. Experiments with near-isogenic lines that differ only for resistance genes may be helpful, for example, in confirming frequency dependent effects.
4. The **diversity of the pathogen population** itself can be important as shown by Dileone and Mundt (1994), who found that disease decreased with increasing pathogen diversity.
5. Where pathogen specialisation for host genotypes does occur, **resistance reactions induced** by avirulent spores may prevent or delay infection by adjacent virulent spores. The effect was shown to be important in cultivar mixtures in the field for powdery mildew of barley (Chin and Wolfe, 1984a) and for yellow rust (caused by *P. striiformis*) of wheat (Lannou *et al.*, 1995; Calonnec *et al.*, 1996) with a similar magnitude, roughly 20% of the total disease restriction obtained in the mixture.
6. **Interactions among pathogen races** (e.g. competition for available host tissue) may reduce disease severity.
7. **Barrier effects are reciprocal**, i.e. plants of one host genotype will act as a barrier for the pathogen specialised to a different genotype and plants of the latter will act as a barrier for the pathogen specialised to the first genotype.

Because of the universality of these mechanisms with respect to air-borne, splash-borne and some soil-borne diseases, mixtures of host genotypes that vary in response to a range of plant diseases will tend to show a simultaneous response to those diseases, and this is correlated with the disease levels of the components that are most resistant to those diseases. In addition, where particular components are affected by disease, there is a tendency for less affected components to compensate for them in terms of yield (Finckh *et al.*, 2000).

Mixtures are most effective in restricting disease early in the epidemic because the pathogen is relatively slow in establishing on previously uninfected plants. Later, when auto-infection occurs on more plants, the rate of infection in the mixture increases. This may occur at a time when the rate of infection is slowing down on the components grown as pure stands as the disease approaches the mean carrying capacity on the components. Depending on the carrying capacity of the mixture and the duration of the epidemic, disease on the mixture may eventually equal the mean of that on the pure stands as Sitch and Whittington (1983) showed for powdery mildew (caused by *Erysiphe polygoni*) of swede (*Brassica napus*). However, the early delay in epidemic development in the swede mixture, at a crucial time for plant development, was sufficient to limit yield loss due to powdery mildew.

All diseases start from initial foci but they are often divided into 'focal' and 'non-focal' depending on the rate of new infections and dispersal relative to symptom development, the number and proximity of the original foci, and the source of the initial inoculum e.g. volunteers, crop residues or an incoming spore cloud.

Using focally-inoculated oat crown rust as an example, Mundt and Browning (1985b) investigated different sizes of 'host genotype units' (GU) to determine whether blocks of pure cultivars would be as effective as an intimate mixture in containing the initial infection and spread of the pathogen. Experiments (Mundt and Browning, 1985b; Mundt and Leonard, 1986) and computer simulations (Mundt *et al.*, 1986; Mundt and Brophy, 1988) indicated that the number and the size of the GUs are important parameters but intimate mixing may not be essential for optimal restriction of disease. Where several different pathogens are expected, the best planting arrangement will depend on the pathogen with the shallowest dispersal gradient.

Species mixtures have the added advantage that the pathogens involved are specialised to one of the components with virtually no possibility of selection of races able to attack more than one. Some species mixtures have been used widely for many years, for example in temperate and tropical, short-term and long-term pastures. Cereal species mixtures for feed production were grown on more than 1.4 Mio ha in Poland and have been shown consistently to restrict diseases (Czembor and Gacek, 1996). More recent experiments in Poland also demonstrate possible positive effects on weeds in cereal and cereal-legume species mixtures (Gacek, personal communication; see also Bulson *et al.*, 1990). In Switzerland, the 'maize-ley' system (i.e. maize planted without tillage into established leys), which is promoted to reduce soil losses and nutrient leaching, has been shown to reduce smut (caused by *Ustilago maydis)* and attacks by European stem borer (*Ostrinia nubilalis)* and aphids (Bigler *et al.*, 1995).

Examples at different scales

Substantial reductions of several air-borne foliar diseases have been reported in experiments with multilines, cultivar and species mixtures of cereals (Wolfe and Barrett, 1980; Chin and Husin, 1982; Alexander *et al.*, 1986; Koizumi and Kato, 1987; Wolfe, 1987; McDonald *et al.*, 1989; Finckh and Mundt, 1992a,b; Czembor and Gacek, 1996, among others, see Finckh *et al.,* 2000 for review). Cultivar mixtures and multilines are being used on a commercial scale in the US, Denmark, Finland, Poland, Switzerland (Table 10.2) and Colombia (coffee), to control,

Table 10.2. Some examples of diversified production in Europe and the US and its effects on targeted and non-targeted diseases. Results from an informal survey conducted by Finckh, Merz and Schaerer (unpublished)

Country	*Crop*	*Area cropped with mixtures*	*Type of mixtures*[h]	*Number of components in mixtures*	*Pathogens targeted and reduced*	*Non-target pathogens also affected*	*Production system*[i]
Switzerland[a]	Barley/wheat, Grass/clover	150,000 ha	V/ S	2-6	*Blumeria graminis, Puccinia recondita*		
Denmark[b]	Barley	64,000 ha	V	3 or 4	*Blumeria graminis*	Undocumented but generally assumed	[i]
Poland[c]	Barley/wheat, Cereal sp/legume	1.5 Mha	V/ S	3	*Blumeria graminis*	*Rhynchosporium, Drechslera, Puccinia spp., Gaeumannomyces graminis, Pyrenophora teres, Pseudocercosporella*	[e]
Finland[d]	Cereal/legume	20,000 ha	V/ S/ I	2	*Botrytis cinerea, Stagonospora nodorum*	Undocumented but assumed	[i]
Pacific Northwest, USA[e]	Wheat	> 100,000 ha	V/ ML	2 or 3	*Puccinia striiformis, Stagonospora nodorum*	Yield loss to *Cephalosporium gramineum* is reduced	[e]
Iowa, USA[f]	Forage cereals	unknown	S	> 4	*Puccinia coronata*		
Southern great plains, USA[g]	Forage cereals	>>10,000 ha	V/ S	2-6			

[a-g] personal communication from [a]Merz, [b]Pedersen, [c]Gacek, [d]Helenius, [e]Mundt, [f]Holland, [g]Marshall
[h] V = variety mixture, ML = multiline, S = species mixture, I = intercropping
[i] e = extensive, i = intensive production

respectively, wheat yellow rust, oat crown rust, (caused by *P. coronata*), barley mildew and coffee rust (caused by *Hemileia vastatrix*) (Wolfe and Finckh, 1997). When used on more than 300,000 ha in the former German Democratic Republic (GDR), powdery mildew of barley was reduced by 80% in barley cultivar mixtures within five years (see 10.6.3; Wolfe, 1992). Up to 92% reduction in rice blast, caused by *Pyricularia grisea,* was observed on susceptible rice cultivars (*Oryza sativa*) when grown as single rows separated by strips of resistant cultivars in large areas in China (Leung *et al.*, 2003; Zhu *et al.*, 2000; Wolfe, 2000). Cultivar mixtures have also been shown to reduce splash-borne foliar pathogens such as *Stagonospora nodorum* (Jeger *et al.*, 1981) and *Septoria tritici* (Cowger and Mundt, 2002) in cereals and *Bremia lactucae* in lettuce (Maisonneuve, *et al.,* 2004; G. Davies, pers. comm.). Mixtures may even reduce the incidence of soil-borne wheat mosaic virus which is transmitted by *Polymyxa graminis* (Hariri *et al.,* 2001).

In contrast, cultivar mixtures often had moderate to insignificant effects in reducing focal and general epidemics of potato late blight (caused by *Phytophthora infestans)* (Garrett and Mundt, 2000; Stolz *et al.*, 2003) although some combinations produced large reductions (Phillips, 2004). Mixture effects were generally greater under moderate natural inoculum pressure than under high natural inoculum pressure (Garrett *et al.*, 2001; Pilet, 2003). Similarly, when planting potatoes in alternating rows, the best results were obtained for the slowest epidemics (Andrivon *et al.*, 2003). Because *P. infestans* is highly variable at least in Europe, most race-specific resistances have been overcome. This is part of the explanation for this limited success of mixtures. However, intercropping potatoes with strips of non-hosts led to significant disease reductions in field experiments over two years (Finckh *et al.,* 2004; Bouws-Beuermann, 2005). The reductions were due to reduced initial inoculum, reduced infection rates, changes in microclimatic conditions and, most importantly, loss of inoculum from strips planted perpendicular to the prevailing wind direction (Bouws-Beuermann, 2005).

For regional deployment, host genotypes with different resistances are planted in different geographic areas. This is most effective against pathogens that are transported from areas where they survive the off-season to areas where they survive only for the growing season. Thus, an epidemic in the latter area is dependent on immigrant inoculum which carries the appropriate virulence factors. If the resistances in the host populations differ between the regions, then a pathogen race dispersed along its usual seasonal pathway will be stopped when it meets hosts that are resistant to it.

Regional gene deployment strategies have been proposed for potato late blight (Van der Plank, 1963), for breaking the 'Puccinia path' of oat (*Avena sativa*) crown rust in the central United States (Browning *et al.*, 1969) and for wheat stem rust in North America (Van der Plank, 1968; Knott, 1972), and to control barley powdery mildew in the UK and Europe (Wolfe and Barrett, 1977; Wolfe, 1992). One incidental strategy was the introduction of stem rust-resistant wheat varieties into the Mediterranean region which has led to the virtual disappearance of the disease from Europe since the 1960s because of the broken migration route.

10.3.2 Diversity and insect pest control

Mechanisms

Several surveys (Vandermeer, 1989; Andow, 1991) indicate that intercropping can be beneficial for insect pest control. The mechanisms affecting insect infestation in diverse systems are more complex than for pathogen-plant interactions because, in addition to the host plant and the insect, predators and parasites of the insects have to be considered (hyperparasites of pathogens occur but they are not considered, generally, to be important). Some hypotheses for the mechanisms of interaction have been proposed (Altieri and Letourneau, 1982; Andow, 1991; Letourneau, 1997):

The **natural enemies hypothesis** states that predators and parasites are more effective in complex environments because diversity of prey, alternative food sources and microhabitats are greater, allowing for the persistence of generalized natural enemies. Specialized natural enemies, moreover, are less likely to fluctuate because their prey can more easily escape extinction and should therefore provide a more continuous food source even if less abundant (Root, 1973; Risch, 1981; Andow, 1991).

The **resource concentration hypothesis** (Root, 1973) was formulated because insects are more likely to find and remain on their hosts in uniform and concentrated patches than in mixed stands, confirmed in experimental studies (Bach, 1980; Risch, 1981; Risch *et al.*, 1983). In general, it appears that lack of resource concentration is more important than are natural enemies in polycultures (Andow, 1991). However, there are cases where the mechanisms work in opposite directions (Andow, 1991).

In addition, a **plant's apparency** is greatly increased in a monoculture and defence mechanisms that might have been adequate in a more diverse setting may be insufficient. (Price *et al.*, 1980; Andow, 1991).

Interactions among the host components of a species mixture may affect, physiologically, the suitability of the host plants as a food source (Vandermeer, 1989).

Associational resistance refers to the reduced herbivore attack that a plant experiences in association with genetically or taxonomically diverse plants and is part of a plant's defensive system (Price *et al.*, 1980; Andow, 1991). Examples are simple barrier effects of resistant plants or cover crops impeding access to susceptible plant parts, olfactory effects, diversion of the pest to other plants and effects of shading of susceptible plants on their attractiveness to the pest (Perrin, 1977; Price *et al.*, 1980; Risch, 1981). Thus, the concept of diversification for the control of insect pests must go beyond the mixing of resistant cultivars or isogenic lines. Deliberately retaining weeds in fields or planting borders to plant species that support natural enemies, as proposed by Altieri and Letourneau (1982) and others, could thus be understood as the deployment of associational resistance. Reduction of disease or insect pests should reduce pesticide inputs and improve conditions for natural enemies thus reducing the need for pesticides even further. The improvement of natural insect control through insecticide reduction is well documented (Heinrichs *et al.*, 1982, 1986; Gallagher *et al.*, 1994; Heinrichs, 1994).

The required reduction in insect populations for effective reduction of insect transmitted diseases may be beyond that which can be achieved by diversity alone (Power, 1988). However, simultaneous reduction of insect vectors and disease inoculum can be effective as has been shown with the beet western yellows disease and its aphid vector *Myzus persicae* (Andow, 1991). Recent results have also shown that there is a potential to reduce insect transmitted viruses such as PVY using straw mulch (Saucke and Döring, 2004). This result is relevant because the mechanisms involved can be made use of in diversified systems. It is commonly thought that reduced visual apparency through the use of coloured or reflective materials is the reason for the reduced aphid and white fly attacks in mulched stands. However, experiments with straw mulch in potatoes have shown that it is likely that the surface structure of the straw induces a long distance take-off flight in aphids rather than the change in visual cues (Döring *et al.,* 2005). Thus, structural cues may also be of importance here and it might be possible to achieve better virus control by selecting plants that induce long-distance flight.

Examples at different scales

Intrafield diversification for the management of insect pests has to a large extent focused on the use of polycultures or intercropping (e.g., Bach, 1980; Risch, 1983; for reviews see Altieri and Letourneau, 1982; Risch *et al.*, 1983; Andow, 1991). One successful example is the restriction of carrot root fly by growing the host crop in a legume background (Finch and Kienegger, 1997). The wholly green crop comprising 60% of non-host confused the pest which was unable to find and/or settle on its normal host.

A more complex, but highly successful example based on semiochemical effects is the 'push-pull' strategy developed in Kenya for control of stem borer in maize (Khan *et al.*, 1997). Two intercrop plants repel the adult stem borers while two edge or trap crop plants attract the stem borers. It also emerged that one of the 'push' plants, the legume silverleaf (*Desmodium uncinatum*), used as an intercrop, also gave excellent control of the parasitic witchweed, *Striga hermonthica* (Pickett, 1999).

Natural enemies increased greatly and soil erosion was reduced by the deliberate planting or maintenance of flowering weeds and grass in established vineyards in Switzerland and southern Germany (Boller *et al.*, 1997). This practice is increasingly popular in Californian wine-growing areas and in apple production (Finckh, personal observation; Crowder, 1996). A low-input system for growing wheat with a permanent understorey of white clover (*Trifolium repens)* can reduce the major pest aphid species and slugs (Jones and Clements, 1993); in addition, reductions in splash-dispersed diseases, such as those caused by *Septoria* spp. were shown by Bannon and Cooke (1998), because of the clover covering the soil surface.

There may be beneficial effects of greater intra-varietal diversity in the oat-frit fly (*Oscinella frit*) system (van Emden, 1966). The flies can attack the host plants only at a particular growth stage and a higher degree of variability within an oat crop could allow for escape from attack and subsequent compensation. Varietal mixtures have been tested in many systems and results from such experiments have been variable with reductions in pest attacks in some cases and increases in others (see Andow, 1991).

At larger scales among insect pests, regional deployment strategies can also be considered. For example, the brown planthopper (*Nilaparvata lugens*) cannot survive the winter in Japan and temperate Korea and it re-immigrates into these areas annually from subtropical and tropical Asia. Despite this migration pattern, regional deployment would be unlikely to be successful, however, because the migrating populations are extremely diverse and able to adapt to the host genetic backgrounds they encounter within one cropping season (Roderick, 1994). This underlines the dependence of regional deployment strategies on relatively low diversity and consequent slow build-up of the migrating pest population in each season.

10.3.3 Diversity and weed control

Species mixtures have been shown often to reduce weeds both when the mixtures are intended to reduce weeds and as a side-effect in intercropping systems (Altieri and Liebman, 1986; Bulson *et al.*, 1990; Liebman, 1995; Liebmann and Gallandt, 1997). Variation in patterns of competition, allelopathic interference, soil disturbance and mechanical damage due to crop rotation make the environment generally less hospitable for weeds (Table 10.3; Liebman and Dyck, 1993). One of the most important factors affecting weed suppression is crop density (Liebman, 1995) which is generally greater in polyculture than in monoculture. In addition, advantage can be taken of different times at which plants may be competitive against weeds. In peas, there has been a trend towards semi-leafless or leafless varieties to improve the microclimate and thus reduce disease and lodging. However, this leaves the ground bare and open to weeds early in the season. In organic farming, where no herbicides are used, a new system is now being tested using *Camelina sativa*, a brassica that develops early as a dense rosette suppressing all other plants around it. Later in spring, however, only thin stalks are sent up so there is no competition for light with the developing peas. The seed of the brassica and peas can be harvested together and easily separated as the peas are much larger. The brassica is an oil seed that can be used for the production of biodiesel, for example. First experiments have shown no yield reductions in peas but almost complete suppression of weeds and an additional yield of oil seeds (Ackermann *et al.,* 2005). An example of pre-emptive weed control is the sequential cropping system of winter peas and maize (Grass, 2003; Grass and Scheffer, 2003). By using leafy winter pea varieties, early in the spring the field is covered with dense growth suppressing all weeds. Just before the maize is sown, the biomass is harvested for silage or biogas, leaving most of the fixed nitrogen in the soil. Maize can be sown without tillage in a clean soil which in addition is protected from erosion through the residual pea roots that cannot regrow.

Table 10.3. Effects of crop rotation and intercropping on weed infestation. The number of studies in which a given effect was observed is listed (data adapted from Liebman and Dyck, 1993)

(a) Effects of crop rotation

With crop rotation	*Weed density*	*Seed density*	*Crop yield*
Lower than without	21	9	0
Equal than without	5	3	3
Higher than without	1	0	11

(b) Effects of intercropping

Weed biomass in intercropping was	
Less than in sole-crop	47
Variable	3
Higher	4

Planting patterns can play an important role particularly if the crop plant has a large seed size. In this sense, Weiner (1990a,b) used the term 'size-asymmetric competition' with respect to planting patterns with wheat. Because wheat grains are considerably larger than many weed seeds, Weiner *et al.* (2001) found in field trials that regular spacing of wheat seed was more effective than conventional row spacing in ensuring strong competition of wheat against weeds.

If pathogens, insects or weeds depend on surviving the off-season in the crop area, a reduction in their severity should reduce the amount of pathogen inoculum, number of insects or seeds surviving until the next season. This will translate into lower disease, insect or weed pressure early in the following season and consequently will delay epidemic development (Kennedy *et al.*, 1987; Wolfe, 1992; Liebman and Dyck, 1993; see 10.5). If initial inoculum or insect populations migrate from other regions, however, these benefits can be reaped only if gene deployment is practised on the appropriate scale (Kennedy *et al.*, 1987; Wolfe, 1992; see 10.4).

10.3.4 Diversity and abiotic stress

There is little information in the literature on the ability of mixtures to reduce abiotic stress, although this may well be common. Mechanisms could be through niche differentiation and protection with compensation. For example, we observed such effects in Poland, where a cold sensitive winter barley variety had markedly improved survival when mixed with a cold tolerant variety than when grown alone (Finckh *et al.*, 2000). In Pakistan, wheat is intercropped with sugar cane to protect the latter from cold injury (Aslam, personal communication) and variety mixtures of wheat are used for the same reason among others by farmers in Oregon, US (Mundt, personal communication).

An interesting example is row-mixtures of rice in China (Zhu *et al.,* 2000). In pure stands neck blast severities of about 35% were reported that were reduced to 2-5% in the mixtures. At the same time, yields in mixtures increased by 100-200%. As 35% severity of neck blast should not result in more than 35% disease loss (Thinlay *et al.,* 2000a) there must be additional benefits in this system. To understand this it is important to know that in Chinese rice production, 200 kg or more of mineral N are added routinely per ha and year. These fertiliser dosages invariably result in lodging of the old traditional varieties which in addition are susceptible to blast (M.R. Finckh, personal observation). Growing these varieties as single rows between the much smaller and stiffer hybrids effectively prevents them from lodging and allows them to make use of the large amounts of fertiliser. In addition, the hybrids probably compete effectively for the fertiliser thus reducing the dosage available to the landraces relative to the pure stands. To fully understand these interactions, plants in pure stands and mixtures need to be analyzed for nutrient content against a control treatment at lower fertiliser levels appropriate for the traditional varieties.

10.3.5 Diversity and yield

The yield of cultivar mixtures in the absence of disease in most cases varies around the mean of the components, with a tendency to be greater than the mean. Such increases (e.g. Allard, 1960; Allard and Adams, 1969; Norrington-Davies and Hutto, 1972; Nitzsche and Hesselbach, 1983; Finckh and Mundt, 1992a,b) may be explained at least partially by niche differentiation among the components. Allelopathy and synergisms of unknown origin might also play a role.

In species mixtures, niche differentiation is an even more prominent feature. Indeed, in almost all cases reported, the overall yield of species in mixtures exceeds the mean of the species grown in monocultures, i.e. the **land use efficiency** is increased (Liebman, 1995; see also Tilman *et al.,* 2001).

Two kinds of competition act in mixtures: intra-genotypic and inter-genotypic (Jolliffe *et al.*, 1984). This differentiation is important: if the average inter-genotypic competition is less than the average intra-genotypic competition, the mixture will yield more than the pure stand average. Because of the different competitive abilities and density response thresholds of different genotypes, competition is usually asymmetrical and difficult to measure (Spitters, 1983; Firbank and Watkinson, 1985). Competitive relationships among plants may also change over time, for example, if intra-genotypic competition becomes greater than inter-genotypic competition beyond a certain plant density, frequency, or developmental stage (Finckh and Mundt, 1996). Frequency-dependent selection has been reported in wheat cultivar mixtures in the presence and absence of disease but may not be common (Finckh and Mundt, 1993). An example of negative yield interactions was reported for cassava (*Manihot esculenta*) cultivar mixtures. There, inter-genotypic competition was greater than intra-genotypic competition resulting in yield depression in complete mixtures (Daellenbach *et al.*, 2005).

It is important to underline that strong competitive ability may not improve a genotype's performance if it affects its harvest index adversely relative to other mixture components (Hamblin, 1975; Hamblin and Rowell, 1975), or if it is more disease susceptible than the others (Suneson, 1949). Highly skewed competitive interactions may also be disadvantageous if seed from the crop is to be kept for further multiplication (Harlan and Martini, 1938; Jennings and de Jesus, 1968; Wolfe, 1991; Finckh and Mundt, 1993).

In the presence of disease, mixtures of cultivars frequently yield more than the mean of the components grown as pure stands (Table 10.4; Baumer, 1983; Wolfe, 1987; Finckh and Mundt, 1992a; Czembor and Gacek, 1996; Gacek *et al.*, 1996a,c; Finckh *et al.*, 2000: see Wolfe and Finckh, 1997 and Finckh *et al.*, 2000 for reviews). Although the correlation between disease severity and yield is often clear in pure stands, it is not always so in mixtures. This is because the correlation between disease severity and yield on individual mixture components is often poor (Brophy and Mundt, 1991; Finckh and Mundt, 1992b; Finckh *et al.*, 2000). Thus, there may be underlying interactions among component cultivars that can be modified depending on the amount of disease on those components.

Table 10.4. Summary of yield (t ha^{-1}) comparisons for a large number of field trials in England over 11 years comparing the yields of different mixtures of three varieties of barley with their components grown as pure stands. The positive effects of the mixtures were considered to be due largely to restriction of powdery mildew because resistance was largely effective and infection was consistently severe in the area (data from Wolfe, 1987)

(a) Mean yields

No. of trial comparisons	*Mean yield of mixtures*	*Mean yield of components*	*Per cent increase in mixtures ($P<0.001$)*
122	5.61	5.20	7.88

(b) Ranking summary

Number of times mixture yielded more than only:			
0 components	*1 component*	*2 components*	*All 3 components*
0	12	50	60

Generally, in contemporary use of cultivar mixtures, the components have not been selected for performance in mixtures. It is unlikely, therefore, that they will perform as well as lines selected for mixture use, although it is worth noting that even in pedigree line breeding programmes, the F1 and F2 generations are usually grown as populations. However, recent trials have shown that tests of two-way mixtures can provide useful indications of specific and general mixing ability for predicting performance of more complex mixtures of wheat (Knott and Mundt, 1990; Lopez and Mundt, 2000; Mille *et al.*, 2005), barley (Gacek *et al.*, 1996a,b) and potato (Phillips, 2004; Phillips and Wolfe, unpublished) in terms of both yield and disease restriction.

A different approach was described by Suneson (1956) as evolutionary plant breeding, focusing on composite cross populations of barley. Research on these populations and others was reviewed by Phillips and Wolfe (2005) as a background to the development of composite cross populations of wheat in the UK. One objective in this programme is simply to develop a genetic resource for breeders. A more speculative objective, however, is to allow population samples to undergo natural selection under different environmental conditions. One question is whether or not this will provide a rapid means of selecting segregating lines that 'nick' well together to improve yield and yield stability under organic conditions.

10.3.6 Diversity and stability

Yield stability is considered to be one of the main advantages of cultivar mixtures (Wolfe and Barrett, 1980). In practice, the definition of **yield stability** is *high* yield over a range of environments (Eberhart and Russell, 1966) and this accounts for the difficulty of finding an appropriate form of statistical analysis for stability (Crossa, 1988; Dubin and Wolfe, 1994). From the six analyses considered by Dubin and Wolfe, the Westcott geometric analysis (Westcott, 1987) in combination with regression analysis (Eberhart and Russell, 1966, modified by Mundt *et al.*, 1995) gave a consistent and clear picture of the comparative performance of mixtures and their components (Finckh *et al.*, 2000).

Since different cultivars and species tolerate different ranges of environmental variation, it is logical to expect that a mixture of cultivars or species will be more stable in yield than any of the components. For example, the overall better and more stable performance of soybean (Schutz and Brim, 1971) and wheat (Finckh and Mundt, 1996) cultivar mixtures when compared with the mean of their pure stands in different locations, was due to variable interactions over environments. Within large data sets, the variance of the mixture is usually less than that of most if not all of the components (e.g. Table 10.5; Allard, 1960; Dubin and Wolfe, 1994; Finckh *et al.*, 2000).

Such gains are not, of course, automatic. The environment plays a crucial role together with the appropriate choice of system and variety. The choice of system depends on being able to minimise competition and maximise facilitation. An example of facilitation is the benefit often found from intercropping grass species with legume species where the latter may increase the nitrogen available to the former.

These considerations relating to yield are relevant also to stability of quality. For example, Baumer (1983) in Bavaria found that the malting quality of barley variety mixtures grown in different environments was more uniform than that of the components grown in the same environments. Baking quality parameters (falling numbers, e.g. Zeleny values) are also influenced in different ways; available reports indicate that mixtures show no negative effects or slight increases in baking volume (Bayerische Landesanstalt für Bodenkultur und Pflanzenbau, 1987; Zoschke, 1987; Jackson and Wenning, 1997).

Table 10.5. Yield, powdery mildew severity and the yield stability of barley cultivar mixtures relative to their pure stands in five experiments (each with 6 to 11 repetitions) between 1987 and 1995 in Poland. Three-way mixtures and pure stands of four to seven cultivars were compared and ranked within each environment. Ranks were regressed on the mean yield per environment. The mean square error (MSE) of the rank regression of mixtures is expressed relative to the MSE of the rank regression of pure stands. An MSE < 1 indicates higher yield stability than the pure stands (data from Finckh et al.*, 1997)*

Cereal	*cv.*	*Year*	*Envts*	*Rel. yield*	*Rel. dis.**	*Rel. MSE*
Spring Barley (f)[a]	7	87-89	7	1.03	0.81	0.56
Spring Barley (m)	6	88-89	6	1.03	0.85	0.48
Spring Barley (f)	5	91-93	11	1.02	0.70	0.79
Spring Barley (m)	4	91-93	11	1.02	0.60	0.67
Winter Barley (f)	5	93-95	10	1.02	-	0.97
Spring Barley (f)	6	94-96	9	1.01	0.77	0.82
Spring Barley (m)	4	94-96	9	0.99	0.74	0.52
Spring Barley (f)	5	94-96	9	1.01	0.68	1.07

* Disease data not from all sites

[a] f = feed barley; m = malting barley

10.4 BENEFITS OF DIVERSIFICATION IN TIME (CROP ROTATION)

Diversification can be implemented in time by sequential or rotational cropping of species and varieties, which can have highly beneficial effects in controlling weeds, diseases and pests. Crop rotations are fundamental in improving crop health in various ways (Finckh, 2003). These can be divided into: time effects reducing pathogen propagules in the soil or on crop residues; indirect effects via soil microbial activity; direct suppressive effects of certain crops on certain pathogens. While the presence of a pathogen is required to cause disease, the absence of a pathogen is not necessarily required for a healthy crop. Rather, the balance between beneficial and detrimental organisms usually determines the outcome. Rotations are the key to reduced pesticide inputs and may enable ecological methods such as no or reduced tillage to reduce erosion but, at the same time, may increase certain disease problems. This can be dealt with by appropriate crop rotations and green manure treatments (e.g. Teich, 1994; Davis *et al.*, 1996). For example, mulches and minimum tillage are known to increase earthworm populations and these have been shown to reduce Rhizoctonia bare patch (caused by *Rhizoctonia cerealis*) and take-all (caused by *Gaeumannomyces graminis*) of wheat, two diseases for which no genetic resistance is available (Stephens *et al.*, 1994 a,b,c). In cereal production in

Germany, Odoerfer *et al.* (1994) showed that the use of pesticides alone was not sufficient to achieve maximum yield; this was possible only when rotations were used, indicating that factors in addition to recognised diseases play a crucial role in determining yield.

Due to the prohibition of methyl bromide, the search for alternatives such as rotations and suppressive effects of certain plant species and varieties has recently received much attention (Martin, 2003). By depriving pathogens of their hosts for one to several seasons, inoculum can be reduced substantially. In addition, green manure and rotational crops may have direct negative effects on pathogens. For example, the beneficial effects of growing certain *Brassica* species between cereal crops are due to volatile compounds released by the brassica plants that have direct inhibitory effects on the growth of pathogens such as *Gaeumannomyces graminis* and *Rhizoctonia spp.* (Kirkegaard *et al.*, 1998; Kirkegaard and Sarwar, 1999; Mayton *et al.*, 1996) or *Verticillium spp.* (Xiao *et al.*, 1998; Subbarao *et al.*, 1999; Shetty *et al.*, 2000) and *Sclerotinia* (Hao *et al.*, 2003); they may also inhibit some weeds. Amending soil with hairy vetch (*Vicia villosa*) residues can suppress Fusarium wilt in watermelon (Zhou and Everts, 2004) while different oat cover crops in combination with ammonium sulphate as a fertiliser greatly reduced nematode (*Pratylenchus penetrans)* and black rot (caused by *Rhizoctonia fragariae*) infestation of strawberries (Elmer and LaMondia, 1999). Similar considerations apply to a number of weed species; rotations can help in reducing the soil seed bank.

Deployment in time or rotation of resistance to the green leafhopper (*Nephotettix virescens*), the vector of Tungro disease of rice (*Oryza sativa*), contributes to the control of the disease in South Sulawesi in Indonesia. Rice varieties with different green leafhopper resistance genes are rotated between seasons to prevent the build-up of hopper populations highly virulent on any one variety (Manwan and Sama, 1985; Sama *et al.*, 1991).

The sequential release of single gene resistance (i.e. once a gene becomes ineffective it is replaced with another gene) has been practised successfully for the control of Hessian fly (*Mayeticola destructor)* (Gallun, 1977; Hare, 1994). The strategy has also been used for brown planthopper (*Nilaparvata lugens*) control but the success was more short-lived (Heinrichs, 1994). In Korea, gene rotations have been designed and implemented for the control of rice blast, caused by *Pyricularia grisea* (Crill *et al.*, 1981).

A proposal for diversification in time is strongly held by pathologists in Europe for the use of the *mlo*-resistance in barley against powdery mildew. Since its introduction in 1979 and, despite wide-scale use on more than 40% of the area grown to spring barley, this resistance has been durable. So far, it has been used only in spring varieties and it is absent from the somewhat larger winter barley crop. Atzema *et al.* (1996) have shown that pathogen virulence against *mlo* is due to a combination of several genes. It can be argued therefore, that successful recombination and selection of virulent genotypes will be delayed if selection for virulence is relaxed through the absence of *mlo* in the winter barley crop. The potential importance of this proposal is emphasised by the observations of Caffier *et al.* (1996), who showed that populations of the mildew pathogen on winter and spring barley crops can be distinguished easily when there are no common selective factors between the crops. Combining the winter and spring pathogen pools by common selection for *mlo*-virulence would probably have a disastrous effect in increasing selection pressure and the overall frequency of *mlo*-virulent inoculum. Our recommendation for Europe is therefore not to use *mlo* in winter barley or to transfer the gene into other cereals (see 10.7).

This recommendation is difficult to implement because breeders may want to use spring *mlo*-varieties in their winter barley programmes for reasons other than mildew resistance. The decision to exclude such varieties from cultivation then has to rest with the authorities concerned with varietal recognition and regulation. In principle, this is the fate of all regional diversification strategies (see 10.7). Even if they can be shown to be potentially useful from a pathological point of view, there may be too many other factors affecting variety distribution on a regional scale. The exceptions are, of course, the known examples where disease is the overriding consideration from a quarantine standpoint.

There is a potential possibility for 'recycling' defeated resistance genes after a certain time has elapsed and virulent pathogen or insect strains have decreased in frequency (Wolfe, 1992). However, this will work only if, in the absence of host plant resistance, there is selection against the corresponding virulence. This is the case for the Hessian fly (Hare, 1994); however, such selection does not exist in the brown planthopper (Roderick, 1994).

10.5 DIVERSITY AND INTERACTIONS

The tendency in the literature is to concentrate on the effects of diversity on diseases or pests or weeds, or on particular species within these three main groups. Few studies have tried to disentangle the many interactions that probably occur in all of the examples analysed. For example, in our own trials, we have noted a tendency to reduced weed infestation in variety mixtures being analysed for disease restriction. Various mechanisms may be involved (see 10.3.3), but one factor may be that disease restriction at a critical growth period improves the vigour of the crop plants relative to their vigour in pure stands: this could provide a critical advantage in terms of competition with weeds. The converse may also be true; a diversification strategy that has a main effect in reducing weed competition may improve the disease (and pest) resistance of the crop.

A further consideration is that plants may be weeds only during certain phases of crop development. At other stages, the presence of the same 'weeds' may be beneficial because they provide food and habitat for beneficial insects and also erosion control. The weeds may also be infected with their 'own' pathogens which may have some advantage where spores are produced that induce resistance in neighbouring crop plants. This feature may be significant under organic conditions where weeds are more numerous and are free from fungicide treatment.

Even a single crop cultivar grown in monoculture may be subject to a wide range of interactions with its biotic and abiotic environment, which are little understood. As the level of crop diversification increases there is likely to be a multiplicative increase in the range of interactions, from which it may be difficult to determine the major factors with any confidence. This is partly because of the numbers of interactions involved and partly because of their dynamic nature. For these reasons, it may be more useful in practice to analyse the gross effects of different levels of diversification in relation to a range of major biotic and abiotic stresses, rather than to concentrate on a particular form of diversification and a single effect.

10.6 RESPONSES OF PEST AND PATHOGEN POPULATIONS TO DIVERSIFICATION STRATEGIES

For crop diseases, host selection is the most important of the evolutionary forces affecting the structure of pathogen populations so that mixtures may be expected to have a strong within-crop influence on the evolution of pathogen populations relative to pure stands.

A pure cultivar selects strongly for a pathogen race able to overcome it (and thus for pathogen uniformity within the crop). However, there is relatively weak selection across different monocultures for races able to overcome more than one resistance (Hovmøller *et al.,* 1993). Thus, in a crop monoculture system, simple races of the pathogen, able to attack cultivars carrying only one resistance (or none at all), are able to survive in large numbers (Hovmøller *et al.,* 1993; Wolfe and McDermott, 1994; Caffier *et al.*, 1996). Under these conditions, gametic disequilibrium (i.e. non-random association of gene combinations) may be common in the pathogen population even if sexual recombination occurs (Wolfe and Knott, 1982; Wolfe and McDermott, 1994). Within a cultivar mixture, diversity may also decline but more slowly because of selection for complex (able to attack more than one component) or super-races (able to attack all components), caused by intimacy of the host components. If the mixture is common, then gametic disequilibrium generally may decline. It is clearly important to determine whether this form of selection is important and, if so, how it might be avoided or reduced (see section 10.7).

10.6.1 Theoretical analyses and models

The development of models dealing with the effects of selection by host populations on pathogen populations has helped to reveal some of the important factors. Simple models have been based on many assumptions among which is a fitness cost for the pathogen if it possesses unnecessary virulence. The reason for a strong adherence to this assumption (and a corresponding cost of resistance) was mainly that such a cost was needed to prevent selection of complex or super-races in mixture models (e.g. Leonard, 1977; Leonard and Czochor, 1980; Lannou and Mundt, 1995). However, such fitness differences are not observed in practice and, in asexual or partly asexual organisms, they would be small relative to selection for associated characters in common genotypes (Brown, 1995).

Nevertheless, even in the absence of a cost of virulence, a number of factors tend to favour polymorphism and to delay or prevent unidirectional selection for complex races. For example, simple gene-for-gene relationships as assumed in most models are complicated in reality by epistasis, pleiotropy, polygenic traits, linkage and other factors (May and Anderson, 1983). Furthermore, Wolfe *et al.* (1983) suggested that particular resistance genes may influence selection of non-corresponding virulence genes (hitchhiking); this was confirmed by Hovmøller *et al.,* (1993) and Huang *et al.* (1995). In addition, infection by more than one pathogen genotype on the same host may lead to changes in pathogenic fitness (Dileone and Mundt, 1994). Other factors that could result in equilibria are density and frequency-dependent selection and selection in heterogeneous environments (Mundt, 1994); migration, founder and drift effects could also be of major importance (Brändle, 1994).

Theoretical population genetics models have included many parameters, such as effects of multiple resistances (resistance gene pyramids) and specific and non-specific resistance (see Marshall, 1989 for review). Important epidemiological considerations were included by Barrett (1978, 1980, 1988) who showed the importance of auto- and allo-infection as determinants for the amount of disease in a mixture and for selection of complex genotypes. For example, early in the epidemic, maximal restriction of disease implies maximum selection for pathogen phenotypes able to grow on more than one component. However, as the epidemic progresses and particularly if available leaf space becomes limiting, then simple races sporulating close to available uninfected spaces have improved competitive ability. This may have been why, after an initial increase in the mixture, complex races were sometimes seen to decrease in frequency (Barrett and Wolfe, 1980). Such a process will be dependent on the initial success of the complex races and the number of pathogen generations, which can vary with the season (Schaerer and Wolfe, 1996).

A related consideration is that spore immigration into the mixture is probably continuous. The impact of immigrant spores is likely to be more important on the mixture relative to that on the mean of the pure stands, because the pathogen population on the mixture may be smaller, particularly in the early stages of the epidemic.

More recent modelling studies have included spatio-temporal and competitive interactions between pathogen races and induced resistance allowing for stability in the absence of selection against unnecessary virulence (see Finckh *et al.,* 2000 and Mundt, 2002 for reviews)

Gould (1986a,b) was the first to develop a detailed mathematical model of insect-host interactions, using the Hessian fly as a model system. The model (Gould, 1986a) explored the effects of sequential release of hosts possessing single resistance genes, two resistance genes (gene pyramids) and mixtures with or without susceptible hosts. The basic assumptions were first, that there was either no difference in preference for the insect between resistant and susceptible hosts or that the susceptible hosts were preferred; second, the insects were assumed to mate randomly.

If no susceptible hosts were used in the system (high selection pressure), mixtures and sequential release were predicted to last four times as long as pyramids if selection pressure was high. If selection pressure was low, the predicted difference was only two-fold. A number of possible genetic factors such as dominance, epistasis, or linkage of virulence were shown to have a major impact on the relative effectiveness of the three different strategies. Gene pyramids could be as long-lasting as mixtures and both could be much more durable if susceptible hosts were mixed into the population and insects that had been selected on resistant and susceptible hosts mated randomly (Gould, 1986b). Thus, fine-scale diversity slowed down adaptation.

When simulating the planting of resistant and susceptible hosts in adjacent fields rather than in mixtures, Gould (1986a,b) found that the critical factors influencing the effectiveness of these strategies were the insects' migration rates and field size. The rotation of resistant and susceptible hosts over time was predicted to speed up insect adaptation because it allowed for the build-up of half-adapted insects every other year.

Natural enemies potentially also influence pest adaptation to resistance factors in host plants. Depending on the kind of host resistance and the timing of attack by the natural enemies, these can either increase or decrease the rate of pest adaptation. Adaptation to a resistant plant type should be slower if the pest suppression is due to a combination of natural enemies and resistance rather than to resistance alone (Gould *et al.,* 1991).

It is at the interface of analytical population genetics and epidemiological simulation models where studies have provided insights into population dynamic processes in pathogen populations. For example, Luo and Zadoks (1992) found that maximum stabilising effects and maximum disease reduction could not be achieved simultaneously. Lannou and Mundt (1995) explored different parameters and showed that maximising control of disease means maximising control of simple races which leads to maximisation of selection for the complex race. These results indicate that, in a systems approach, one cannot and must not aim simply to minimise disease but rather to consider short-term disease control in the context of long-term effects on the system as a whole.

10.6.2 Results from small scale field experiments

In the field, attempts to follow changes in pathogen population structure on mixed hosts relative to those on their pure components have produced inconclusive results. For example, Chin and Wolfe (1984b) found that barley mixtures selected for more complex powdery mildew races than did the pure stands, but, because of epidemic delay in the mixtures, complex genotypes did not necessarily increase in absolute number relative to the numbers on pure stands. Moreover, they found that disruptive selection by different barley cultivars containing the same race-specific resistance, split races of the pathogen into sub-races that were differentially adapted to the genetic background of the cultivars. Villareal and Lannou (2000) also found that the rate of progress towards complexity in a pathogen population being selected on a cultivar mixture was significantly slower than that predicted from simple models. Chuke and Bonman (1988) found little evidence for an increase in race complexity in rice blast on rice among isolates from mixtures, relative to those from pure stands. In other trials, complex pathogen genotypes increased in frequency on host mixtures relative to pure stands (Wolfe, 1984; Wolfe *et al.*, 1984; Dileone and Mundt, 1994; Huang *et al.*, 1994).

Unfortunately, small-scale, short-term field plot experiments may be inadequate for answering questions about field selection because of different factors including plot interference and a limited range of genotypes in the starting inoculum. For example, Dileone and Mundt (1994) found that the increase of complex pathotypes in mixtures was related inversely to the number of other pathotypes occurring in the mixture.

Rates of change in population structure varied in different experiments and are obviously dependent on many factors. One of the most important is the genetic composition of the pathogen population surviving between seasons (Schaffner *et al.*, 1992). This cannot be tested in small trials because the populations produced have little influence on pathogen survival between crops, except, perhaps, on an extremely small area. Consequently, a sequence of trials, even if they are carried out on the same plots, cannot provide an estimate of selection over years on a specific mixture used in agriculture.

10.6.3 Results from large-scale mixture production

Experimental evidence that the useful lifetime of resistance genes can be extended through diversification, is practically impossible to gather. However, Mundt (1994) observed that a wheat cultivar released as a component of a multiline maintained its resistance to yellow rust for several years. When released as a pure line cultivar, the resistance was overcome within two or three seasons.

A more detailed analysis became possible during the 1980s in the former GDR (Schaffner *et al.*, 1992; Wolfe and McDermott, 1994). Mixtures of malting quality barley cultivars were introduced in 1984 and, by 1988/9, the whole of the spring barley crop (ca 300,000 ha) was grown in this way. The strategy was effective, with mildew held consistently at a low level and fungicide use considerably reduced. There was also a restriction of leaf rust, even though this was not a target disease in the choice of mixture components. Crop yields remained high and the grain was used for high quality malt and beer production (Wolfe *et al.*, 1991; Wolfe, 1992).

These results were obtained despite little effective diversity: only three resistance genes, *mlo, Mla12* and *Mla13* were distributed among most of the cultivars used and field size was large, often 50 to 100 hectares. Thus, the pathogen needed to recombine only three virulences to produce a super race. The virulence against the *mlo*-gene has not yet occurred in Europe (see 10.5) but an increase in the recombinant genotype with *Va12* and *Va13,* occurred eventually in the GDR in 1990. However, common pathogen isolates with this gene combination were not generated locally but had immigrated from Poland and the former Czechoslovakia (Wolfe *et al.*, 1992) and complex races did not dominate the pathogen population every year (Schaffner, 1993). The important inference from these observations is that, despite the limited mixture diversity in this large-scale application, progress in the pathogen population towards dominance of complex or super races was remarkably slow (see 10.6.1 and 10.7).

A second large-scale investigation followed the pathogen response to the use of cultivar mixtures of spring barley in Poland, which increased to a significant scale by 1996 (approx. 80,000 ha, Gacek *et al.*, 1996d). Here, complex races were already common in the pathogen population due to the widespread use of the corresponding resistances at that time. Probably for this reason, it was difficult to detect large or consistent differences between pathogen populations from mixtures or pure stands of the components (Schaerer and Wolfe, 1996). In other words, there was little effective diversity of resistance in the mixtures which probably explains why the yields of the mixtures were not much higher than the means of the components, although they did reduce disease and exhibited valuable stability (Gacek *et al.*, 1996a,c; Table 10.5).

In China, rice cultivar mixtures are now grown on more than 1 mio ha (Y. Zhu, personal communication) principally to protect susceptible land races from rice blast. The success of the strategy is due to a combination of factors. First, the farmers no longer grow the susceptible landraces in pure stands because the yields in the mixtures are much increased (Zhu, *et al.,* 2000; Leung *et al.*, 2003). This therefore reduces the overall inoculum pressure. In addition, different blast populations are associated with the landraces and the hybrids and the taller land races are exposed to a dry microclimate in the mixtures which is not conducive to infection.

The indications from the small-scale field experiments, models and other data, are that complex races can be selected more or less quickly leading to a reduction in effectiveness of mixtures, though this is likely to be slower than the dramatic 'breakdowns' that are common in monoculture. Mundt (1994) concluded that despite the theoretical considerations, complex races do not tend to dominate the pathogen population in diverse systems. The rates of change that do occur suggest that normal shifts in the range of new varieties coming into the market, or, better still, planned changes in the composition of host mixtures (see 10.7), should slow down the rate of selection and increase of undesirable complex races; in other words, durability can be managed.

10.6.4 Interactive effects of host and pathogen populations

Diseases affect plant-plant interactions by altering competitive interactions among plant genotypes within a season (Burdon *et al.*, 1984; Alexander *et al.*, 1986; Paul and Ayres, 1986a, 1987b; Paul, 1989; Finckh and Mundt, 1992a,b; Boudreau and Mundt, 1997) and by affecting the survival and fitness of hosts differentially (Alexander, 1984; Alexander and Burdon, 1984; Paul and Ayres, 1986b,c, 1987a; Jarosz *et al.*, 1989; Finckh and Mundt, 1993). In the shorter term, the frequency of resistant host plants can be increased significantly by disease pressure (Wahl, 1970; Burdon *et al.*, 1981; Webster *et al.*, 1986; Kilen and Keeling, 1990; Finckh and Mundt, 1993). If resistance is linked to unfavourable traits, however, it may be selected against even if disease pressure is strong (Parker, 1991). If one component does become frequent because of its resistance, a virulent pathogen may later spread easily on that component, thus reducing its competitive ability and hence its frequency (Chilvers and Brittain, 1972; Wills, 1996). However, the geographic or time-scales usually studied may be insufficient to allow clear observation of such interactive effects of host and pathogen populations (see Burdon and Jarosz, 1992).

A major problem in analysing plant-plant interactions in mixtures in the presence of disease is that comparisons often have to be made between treatments in which the same genotype suffers different levels of disease. Among the consequences of this are: i) fitnesses and competitive abilities of differentially susceptible genotypes may change in different ways in the presence of disease; ii) depending on the frequency of a genotype in the mixture, the disease severity on it, and thus its fitness, will vary in a complex way. For example, when variable mixture compositions are considered, the frequency and density of the genotypes change simultaneously, affecting disease and competitive relations. Changing frequency and density may have opposite effects on disease severity due to changes in host nutritional status and a lack of barriers in reduced density stands, (e.g. Finckh *et al.,* 1999); iii) effects of disease on host fitness may be influenced by the companion genotypes (Brophy and Mundt, 1991); conversely, disease severity on a genotype may be affected by plant-plant interactions (Finckh and Mundt, 1992b) and by induced resistance (Calonnec *et al.,* 1996); iv) comparisons between diseased and non-diseased treatments are often questionable because of possible direct and perhaps genotype-specific effects of the fungicide used (Paul *et al.*, 1989).

10.7 DIVERSIFICATION STRATEGIES IN PRACTICE

Basic biological considerations indicate that different forms of crop diversification can deliver many benefits relative to monoculture. There are also increasing numbers of large-scale demonstrations in practice that confirm this view.

10.7.1 Recorded examples

Intra-specific diversity: In Europe and the US, barley and wheat variety mixtures are being used on a large scale (Table 10.2) and barley mixtures have been successful in the former GDR (see 10.6.3) with considerable benefits to the farmer and the environment. Mixtures of winter wheat and barley became popular rapidly in Switzerland through a system of financial encouragement to reduce pesticide inputs (Merz and Wolfe, 1996). Intra-specific diversity is the rule rather than the exception in landraces of many species and much of this diversity is related to disease and pest control (Jarvis *et al.*, 2005). A notable example of functional diversity against rice blast is the high diversity within land races of rice in Bhutan. High pathogen pressure is correlated with high diversity for virulence and this, in turn, is positively correlated with diversity for resistance within and among landraces. Consequently, smaller losses were observed in areas of continuous high pathogen pressure (measured by the use of trap crops) than in areas where disease outbreaks are rare (Thinlay *et al.*, 2000a,b; Finckh, 2003).

Outside the cereal crop, one successful application of mixtures was in Colombia from 1982 on, where coffee mixtures were grown on some 400,000 ha to produce beans of high quality (the components are selected for uniformity in this character) from components that had defined variation for rust resistance (Moreno-Ruiz and Castillo-Zapata, 1990; Browning, 1997). Mixtures of different genotypes are common for *Phaseolus* beans in many developing countries and efforts have been undertaken to improve such diverse varieties through partial replacement rather than by changing them into monocultures (Trutmann and Pyndji, 1994).

There are now developments in progress to try to improve, with the help of local farmers, the effectiveness of mixtures through the use of diversity functional against rice blast in blast-prone areas in Vietnam (Mundt and Nelson, personal communication). Efforts are also currently under way to reintroduce variability into crops such as faba beans (Ghaouti *et al.,* 2005) or wheat (see www.efrc.com, and www.cost860.dk) in an attempt to increase yield stability and adaptability of the crop.

Inter-specific diversity. Again, in the developing world, intercropping is common and is being investigated and encouraged on a scale larger than ever. In Europe, there is increasing recognition of the value of species mixtures involving legumes, in addition to the well-established and widely-used forage grass-legume mixtures. For example, in Switzerland, in areas of adequate rainfall, maize is planted into a grass-legume ley (see 10.3.2; Bigler *et al.*, 1995) and in the UK, winter wheat has been successfully planted into white clover (Jones and Clements, 1993). There are also promising results with vegetable-clover combinations (Baumann *et al.*, 2001).

In vegetable production, diversified production (e.g. strip planting) is commonly practised. Reasons for this include the need to supply markets continuously with fresh produce and to reduce risks from crop losses. Interestingly, in recent years, the extension of diversity among salad leaf crops, encouraged by the consumer, has enabled the grower to spread his/her risk among a wider range of species than previously, thus reducing dependence on lettuce (*Lactuca sativa*) with its associated problem of lettuce downy mildew (*Bremia lactucae*), which is expensive and difficult to deal with because of the genetic variation in the pathogen which allows it to overcome resistant varieties and fungicides in intensive production (Michelmore *et al.,* 1984). Another example is the strip intercropping of potatoes with grass/clover described earlier (see 10.3.1).

Despite these examples and the positive research results, application of diversification strategies is slow to develop in practice. This is often because of the large-scale inertia and conservatism of mainstream production. There are also criticisms about the difficulties of planting, managing and harvesting crop mixtures. However, many of these difficulties can be dealt with by simple technical measures such as appropriate modification of planting and harvesting equipment. For example, mechanical separation of wheat and beans is practised on organic farms in Switzerland using screens (Finckh, personal observation).

There is also controversy surrounding selection of complex and super-races (see 10.6). However, the evolution of super-races can also be slowed down or avoided by developing a planned approach to diversification strategies and the composition of mixtures for commercial use. With species mixtures, of course, there is no problem of pathogen adaptation to more than one component.

Theoretically, transgenic resistance could be used to increase functional diversity in crops but it is likely that only few genotypes will be supplied with transgenes and that they would be marketed to occupy even larger areas per variety than before. The process would be driven by the current system of patenting and plant varietal protection which is inappropriate to the development and exploitation of biodiversity (Busch, 1995). As an example, we may consider the *mlo*-resistance gene of barley to powdery mildew which has never been overcome by a virulent pathotype in the field and which has been cloned (Büschges *et al.*, 1997). This gene might be regarded as a prime candidate for genetic engineering to introduce it into other cereals such as oats and wheat. We have already pointed out the danger that *mlo*-resistant winter barley would greatly increase selection on the pathogen, increasing the frequency of the virulence genes and thus the chances for clones with the appropriate virulences to emerge (see 10.5). The additional use of the *mlo*-resistance in other cereals would dramatically increase the genetic uniformity for mildew resistance and might greatly increase the danger of gene-exchange between the *formae speciales* of powdery mildew since host species specificity in mildew may be based on only one or a few genes (Tosa, 1994).

10.7.2 System diversity

Until recently, numerical success in the development of the human population has been dependent on various systems of ecological agriculture and agroforestry. Major historical departures from this theme towards monoculture resulted in the loss of many civilisations (Ponting, 1991). Our recent ability to develop vast monocultures has depended on major developments in synthetic chemistry and engineering technology. It is now clear that these modern applications have many negative impacts; attempts to overcome these by further technological developments lead often to more and novel problems. In our view, there is an overwhelming argument to bring the power of modern biological sciences to bear on the understanding and development of the wide range of ecological processes and interactions that allowed the human population to increase without major impacts on the environment as a whole. This means developing and understanding agricultural systems that are highly diversified at many different levels. As pointed out earlier, however, the diversification needs to be functional in relation to the required outcomes from the systems. As also pointed out earlier, the use of different forms of functional diversity simultaneously can lead to a multiplicative increase in interactions among the many components.

One example of complex systems, although there are still many deficiencies, is the development of modern organic agriculture. Organic systems throughout the world are based on temporal diversity (more or less long rotations) which necessarily involves a useful degree of crop diversity. The use of composts and manures represents another form of diversity within these systems by using a range of crops to generate microbial diversity; many of the microbes involved have positive effects on diseases (Hoitink and Fahy, 1986) and nutrient use efficiency (Maeder *et al.*, 2002). There is also growing interest and application of intercropping systems, particularly in organic agriculture. Agroforestry systems, integrating tree and crop management, represents the highest level of system complexity; these can incorporate a wide range of intra- and inter-specific mixtures and intercropping together with temporal diversity.

However, in addition to the biological inputs necessary for the success of highly diversified systems, there is a need to consider the market aspect. Developments towards highly diversified systems are best served by appropriate local marketing strategies. We also believe that development of systems in these directions will probably be the best answer to both reducing the pressure towards global climate change and to surviving the effects of such changes as they become more obvious.

10.8 CONCLUSIONS

Crop diversification can influence the agricultural ecosystem in many ways, from abiotic effects on microclimate, erosion and soil structure, to biotic effects on pathogens, insects, weeds and the crops themselves. We need to seek the positive effects that emerge from the use of what we have termed functional diversity.

There has been a tendency to concentrate on specific aspects of the use of diversification, for example, in the large body of work concerned with cultivar mixtures and disease restriction. One advantage of this focused approach is that it has facilitated the development of large-scale examples of simple changes (wheat mixtures in the USA, barley mixtures in the GDR, rice mixtures in China) that can be introduced easily with the existing technology (the Chinese harvest by hand...) leading to reduced costs, increased production and less stress on the environment. Progress towards increased exploitation of functional diversity in developing countries is accelerating rapidly and on a large scale. For example, there is considerable interest in rice-pasture rotation systems for the development of large areas of South American savannahs. At the same time, care will be needed, of course to ensure that the natural biodiversity of those areas is not lost.

Although it seems clear that more complex levels of diversification will lead to increases in interactions among the components involved, relatively little work has been attempted in these directions, despite their likely importance. Indeed, because of the difficulties of analysis at high levels of complexity, it may be possible only to follow and analyse the major end effects.

A major deficiency in the development of diversification strategies is that there are often few cultivars of species available that have been actively selected for good performance in diversification strategies and systems. This is highlighted by examples of diversified systems that do not work (e.g. Daellenbach *et al.*, 2005).

It also raises a more general question for mixtures generally of whether the literature gives an adequate picture of reality: is there a tendency for positive results to be published more than negative? We are also aware from our own experience that negative results are sometimes obtained from inadequate trial designs. It has to be accepted also that the positive returns from diversification strategies may sometimes be inadequate. For example, rotations that are adequate for controlling persistent weeds or pathogens may be too long in relation to other needs in a farm plan.

The most important development in our view is the need to consider diversification strategies not in terms of control of an individual pest or pathogen but rather to consider diversification in terms of the whole range of abiotic and biotic challenges facing crop production in a particular locality (see Weiner, 2003). Where the most important strategy becomes avoidance or prevention of problems through system design, the nature of pest, disease and weed problems may then be no longer well-defined, because the system is appropriately buffered and sustainable (see also Bird *et al.*, 1990).

REFERENCES

Ackermann, K. and Saucke, H. (2005) Einfluss des Gemengepartners Leindotter (*Camelina sativa* L.) auf Beikrautbesatz, Schädlingsbefall und Ertrag in Körnererbsen [Effect of the companion crop linseed dodder (*Camelina sativa* L.) on weed development, pest incidence and yield in grain peas], in *Ende der Nische, Beiträge zur 8. Wissenschaftstagung Ökologischer Landbau. 1.3-4.3.2005*, http://orgprints.org/3788/

Alexander, H.M. (1984) Spatial patterns of disease induced by *Fusarium moniliforme* var. *subglutinans* in a population of *Plantago lanceolata*. *Oecologia,* **62**, 141-143.

Alexander, H.M. and Burdon, J.J. (1984) The effect of disease induced by *Albugo candida* (white rust) and *Peronospora parasitica* on the survival and reproduction of *Capsella bursa-pastoris*. *Oecologia,* **64**, 314-318.

Alexander, H.M., Roelfs, A.P. and Cobbs, G. (1986) Effects of disease and plant competition in monocultures and mixtures of two wheat cultivars. *Plant Pathology,* **35**, 457-465.

Allard, R.W. (1960) Relationship between genetic diversity and consistency of performance in different environments. *Crop Science,* **1**, 127-133.

Allard, R.W. and Adams, J. (1969) Population studies in predominantly self-pollinating species XIII. Intergenotypic competition and population structure in barley and wheat. *American Naturalist,* **103**, 620-645.

Altieri, M.A. and Letourneau, D.K. (1982) Vegetation management and biological control in agroecosystems. *Crop Protection,* **1**, 405-430.

Altieri, M.A. and Liebman, M. (1986) Insect, weed and plant disease management in multiple cropping systems, in *Multiple Cropping Systems* (Ed. C.A.Francis). Macmillan: New York, pp. 183-218.

Altieri, M.A. (1995) *Agroecology. The Science of Sustainable Agriculture.*, 2nd Edition. Westview Press, IT Publications: Boulder, Colorado, London.

Altieri, M.A., Nicholls, C.I. and Wolfe, M.S. (1996) Biodiversity – a central concept in organic agriculture: Restraining pests and diseases, in *Fundamentals of Organic Agriculture. Vol. 1.* (Ed. T.V.Ostergaard). IFOAM: Ökozentrum Imsbach, D-66636 Tholey-Theley, pp. 91-112.

Andow, D.A. (1991) Vegetational diversity and arthropod population responses. *Annual Review of Entomology,* **36**, 561-586.

Andrivon, D., Lucas, J.M.A. and Ellisseche, D. (2003) Development of natural late blight epidemics in pure and mixed plots of potato cultivars with different levels of partial resistance. *Plant Pathology,* **52**, 586-594.

Atzema, J.L., Finckh, M.R. and Wolfe, M.S. (1996) Genetics of the response of *Erysiphe graminis* f.sp. *hordei* to *mlo*-resistance in barley, in *Proceedings of the 9th European and Mediterranean Cereal Rusts & Powdery Mildew Conference, 2-6 Sept. 1996, Lunteren, The Netherlands* (Eds. G.H.J.Kema, R.E.Niks and R.A.Daamen). Research Institute for Plant Protection: Wageningen, NL, pp. 55-57.

Bach, C.E. (1980) Effects of plant density and diversity on a specialist herbivore, the striped cucumber beetle, *Acalymna vittata* (Fab.). *Ecology,* **61**, 1515-1530.

Bannon, F.J. and Cooke, B.M. (1998) Studies on dispersal of *Septoria tritici* pycnidiospores in wheat-clover intercrops. *Plant Pathology,* **47**, 49-56.

Barrett, J.A. (1978) A model of epidemic development in variety mixtures, in *Plant Disease Epidemiology* (eds P.R.Scott and A.Bainbridge). Blackwell: Oxford, pp. 129-137.

Barrett, J.A. (1980) Pathogen evolution in multilines and variety mixtures. *Zeitschrift fur Pflanzenkrankheiten und Pflanzenschutz,* **87**, 383-396.

Barrett, J.A. and Wolfe, M.S. (1980) Pathogen response to host resistance and its implication in breeding programmes. *EPPO Bulletin,* **10**, 341-347.

Barrett, J.A. (1988) Frequency-dependent selection in plant-fungal interactions. *Philosophical Transactions of the Royal Society London, Series B,* **319**, 473-483.

Baumann, D.T., Bastiaans, L. and Kropff, M.J. (2001) Competition and Crop Performance in a Leek-Celery Intercropping System. *Crop Science,* **47**, 764-774.

Baumer, M. (1983) Neue Ergebnisse mit Sortenmischungen bei Sommergerste. *Top Agrar,* **2**, 82-86.

Bayerische Landesanstalt für Bodenkultur und Pflanzenbau, (1987) *Versuchsergebnisse Winterweizen, Ertragsstrukturdaten.* Bayerische Landesanstalt für Bodenkultur und Pflanzenbau: Freising, Germany.

Biffen, R.H. (1905) Mendel's law of inheritance and wheat breeding. *Journal of Agricultural Sciences,* **1**, 4-48.

Bigler, F., Waldburger, M. and Frei, G. (1995) Vier Maisanbauverfahren 1990 bis 1993. Krankheiten und Schaedlinge. *Agrarforschung,* **2**, 380-382.

Bird, G.W., Edens, T., Drummond, F. and Groden, E. (1990) Design of pest management systems for sustainable agriculture, in *Sustainable Agriculture in Temperate Zones* (eds C.A.Francis, C.B.Flora and L.D.King). Wiley & Sons, Inc.: New York, pp. 55-110.

Boller, E., Gut, D. and Remund, V. (1997) Biodiversity in three trophic levels: The ecosystem vineyard, in *Vertical food web interactions: evolutionary patterns and driving forces* (eds K.Dettner, G.Bauer and W.Völkl). Springer.

Boudreau, M.A. and Mundt, C.C. (1992) Mechanisms of alteration in bean rust epidemiology due to intercropping with maize. *Phytopathology,* **82**, 1051-1060.

Boudreau, M.A. and Mundt, C.C. (1997) Ecological approaches to disease control, in *Environmentally Safe Approaches to Disease Control* (eds J.Rechcigl and N.Rechcigl). CRC Press: Boca Raton, LA, pp. 33-62.

Bouws-Beuermann, H. (2005) Effects of strip intercropping on late blight severity, yields of potatoes (*Solanum tuberosum* Lindl.) and on population structure of *Phytophthora infestans.* Dissertation, University of Kassel, Faculty of Organic Agricultural Sciences.

Braendle, U.E. (1994) Studies on the genetic structure of local populations of *Erysiphe graminis* f.sp. *hordei* Marchal. Swiss Federal Institute of Technology (Diss. ETH Nr. 10859): Ph.D. Thesis.

Brophy, L.S. and Mundt, C.C. (1991) Influence of plant spatial patterns on disease dynamics, plant competition and grain yield in genetically diverse wheat populations. *Agriculture, Ecosystems and Environment,* **35**, 1-12.

Brown, J.K.M. (1995) Pathogen's responses to the management of disease resistance genes. *Advances in Plant Pathology,* **11**, 75-102.

Browning, J.A., Simons, M.D., Frey, K.J. and Murphy, H.C. (1969) Regional deployment for conservation of oat crown rust resistance genes. *Special Report Iowa Agriculture and Home Economics Experiment Station,* **64**, 49-56.

Browning, J.A. and Frey, K.J. (1969) Multiline cultivars as a means of disease control. *Annual Review of Phytopathology,* **7**, 355-382.

Browning, J.A. (1997) A unifying theory of the genetic protection of crop plant populations from diseases, in *Disease Resistance from Crop Progenitors and Other Wild Relatives* (eds I.Wahl, G.Fischbeck and J.A.Browning). Springer Verlag: Berlin, Heidelberg, New York, Tokyo.

Bulson, H.A.J., Snaydon, R.W. and Stopes, C.E. (1990) Intercropping autumn-sown field beans and wheat: effects on weeds under organic farming conditions, in *Crop Protection in Organic and Low Input Agriculture* (ed. R.J. Unwin). The British Crop Protection Council: Farnham, Surrey GU9 7PH, pp. 55-62.

Burdon, J.J. and Chilvers, G.A. (1976) Controlled environment experiments on epidemics of barley mildew in different density host stands. *Oecologia,* **26**, 61-72.

Burdon, J.J., Groves, R.H. and Cullen, J.M. (1981) The impact of biological control on the distribution and abundance of *Chondrilla juncea* in south-eastern Australia. *Journal of Applied Ecology,* **18**, 957-966.

Burdon, J.J. and Chilvers, G.A. (1982) Host density as a factor in disease ecology. *Annual Review of Phytopathology,* **20**, 143-166.

Burdon, J.J., Groves, R.H., Kaye, P.E. and Speer, S.S. (1984) Competition in mixtures of susceptible and resistant genotypes of *Chondrilla juncaea* differentially infected with rust. *Oecologia,* **64**, 199-203.

Burdon, J.J. and Jarosz, A.M. (1992) Temporal variation in the racial structure of flax rust (*Melampsora lini*) populations growing on natural stands of wild flax (*Linum marginale*): local versus metapopulation dynamics. *Plant Pathology,* **41**, 165-179.

Busch, L., Gunter, V., Mentele, T. *et al.* (1994) Socializing nature: Technoscience and the transformation of rapeseed into canola. *Crop Science,* **34**, 607-614.

Busch, L. (1995) Eight reasons why patents should not be extended to plants and animals. *Biotechnology and Development Monitor,* **No 24**, 24.

Büschges, R., Hollricher, K., Panstruga, R. *et al.* (1997) The barley *mlo*-gene: A novel control element of plant pathogen resistance. *Cell,* **88**, 695-705.

Caffier, V., Hoffstadt, T., Leconte, M. and de Vallavieille-Pope, C. (1996) Seasonal changes in French populations of barley powdery mildew. *Plant Pathology,* **45**, 454-468.

Calonnec, A., Goyeau, H. and Devallavieillepope, C. (1996) Effects of induced resistance on infection efficiency and sporulation of *Puccinia striiformis* on seedlings in varietal mixtures and on field epidemics in pure stands. *European Journal of Plant Pathology,* **102**, 733-741.

Chilvers, G.A. and Brittain, E.G. (1972) Plant competition mediated by host-specific parasites. A simple model. *Australian Journal of Biological Sciences,* **25**, 749-756.

Chin, K.M. and Husin, A.N. (1982) Rice variety mixtures in disease control, in *Proceedings of the International Conference on Disease Control in the Tropics*, pp. 241-246.

Chin, K.M. and Wolfe, M.S. (1984a) The spread of *Erysiphe graminis* f. sp. *hordei* in mixtures of barley varieties. *Plant Pathology,* **33**, 89-100.

Chin, K.M. and Wolfe, M.S. (1984b) Selection on *Erysiphe graminis* in pure and mixed stands of barley. *Plant Pathology,* **33**, 535-546.

Chuke, K.C. and Bonman, J.M. (1988) Changes in virulence frequencies of *Pyricularia oryzae* in pure and mixed stands of rice. *Journal of Plant Protection in the Tropics,* **5**, 23-29.

Clunies-Ross, T. and Hildyard, N. (1992) *The Politics of Industrial Agriculture*. Earthscan Publications Ltd.: London.

Colbourn, T., Dumanoski, D. and Myers, J.P. (1996) *Our Stolen Future: Are we threatening our fertility, intelligence and survival?- A scientific detective story*. Penguin: New York, London, Victoria, Ontario, Auckland.

Cowger, C. and Mundt, C. (2002) Effects of wheat cultivar mixtures on epidemic progression of *Septoria tritici* blotch and pathogenicity of *Mycosphaerella graminicola. Phytopathology,* **92**, 617-623.

Crill, P., Ham, Y.S. and Beachell, H.M. (1981) The rice blast disease in Korea and its control with race prediction and gene rotation. *Korean Journal of Plant Breeding,* **13**, 106-114.

Crossa, J. (1988) A comparison of results obtained with two methods for assessing yield stability. *Theoretical and Applied Genetics,* **75**, 460-467.

Crowder, R. (1996) Education in organics: Practising that which is preached, in *Fundamentals of Organic Agriculture. Vol. 1* (ed. T.V.Ostergaard). IFOAM: Ökozentrum ImsbachD-66636 Tholey-Theley, pp. 240-252.

Czembor, H.J. and Gacek, E.S. (1996) The use of cultivar and species mixtures to control diseases and for yield improvement in cereals in Poland, in *Proceedings of the 3rd Workshop on Integrated Control of Cereal Mildews across Europe, Nov. 5-10 1994, Kappel a. Albis, Switzerland* (eds E. Limpert, M.R. Finckh and M.S. Wolfe). Office for Official Publications of the EC: Brussels, Belgium, pp. 177-184.

Daellenbach, G.C., Kerridge, P.C., Wolfe, M.S. *et al.* (2005) Plant productivity in cassava-based mixed cropping systems in Colombian hillside farms. *Agriculture, Ecosystems and Environment,* **105**, 595-614.

Davis, J.R., Huisman, O.C., Westermann, D.T. *et al.* (1996) Effects of green manures on Verticillium wilt of potato. *Phytopathology,* **86**, 444-453.

Dileone, J.A. and Mundt, C.C. (1994) Effect of wheat cultivar mixtures on populations of *Puccinia striiformis* races. *Plant Pathology,* **43**, 917-930.

Döring, T.F., Brandt, M., Heß, J. *et al.* (2005) *Field Crop Research*, in press.

Dubin, H.J. and Wolfe, M.S. (1994) Comparative behaviour of three wheat cultivars and their mixture in India, Nepal and Pakistan. *Field Crop Research,* **39**, 71-83.

Eberhart, S.A. and Russell, W.A. (1966) Stability parameters for comparing yield. *Crop Science,* **6**, 36-40.

Elmer, W.H. and LaMondia, J.L. (1999) Influence of ammonium sulfate and rotation crops on strawberry black root rot. *Plant Disease,* **83**, 119-123.

Finch, S. and Kienegger, M. (1997) A behavioural study to help clarify how undersowing with clover affects host plant selection by pest insects of brassica crops. *Experimental and Applied Entomology*, **84**, 165-172.

Finckh, M.R. and Mundt, C.C. (1992a) Stripe rust, yield, and plant competition in wheat cultivar mixtures. *Phytopathology*, **82**, 905-913.

Finckh, M.R. and Mundt, C.C. (1992b) Plant competition and disease in genetically diverse wheat populations. *Oecologia*, **91**, 82-92.

Finckh, M.R. and Mundt, C.C. (1993) Effects of stripe rust on the evolution of genetically diverse wheat populations. *Theoretical and Applied Genetics*, **85**, 809-821.

Finckh, M.R. and Mundt, C.C. (1996) Temporal dynamics of plant competition in genetically diverse wheat populations in the presence and absence of stripe rust. *Journal of Applied Ecology*, **33**, 1041-1052.

Finckh, M.R. and Wolfe, M.S. (1997) The use of biodiversity to restrict plant diseases and some consequences for farmers and society, in *Ecology in Agriculture* (ed. L.E.Jackson). Academic Press: San Diego, pp. 199-233.

Finckh, M.R., Gacek, E.S., Czembor, H.J. and Wolfe, M.S. (1999) Host frequency and density effects on disease and yield in mixtures of barley. *Plant Pathology*, **48**, 807-816.

Finckh, M.R., Gacek, E.S., Goyeau, H. *et al.* (2000) Cereal variety and species mixtures in practice, with emphasis on disease resistance. *Agronomie*, **20**, 813-837.

Finckh, M.R. (2003) Ecological benefits of diversification, in *Rice Science: innovations and impact for livelihood* (eds T.W. Mew, D.S. Brar, S. Peng, D. Dawe and B. Hardy). Internatona Rice Research Institute: Los Banos, Philippines, pp. 549-564.

Finckh, M.R., Bouws-Beuermann, H., Piepho, H.P. and Büchse, A. (2004) Auswirkungen von Streifenanbau und Ausrichtung zum Wind auf die räumliche Verteilung und epidemiologische Parameter der Kraut-und Knollenfäule. *Mitt.Bio.Bundesanst.Land-Forstwirtsch*, **396**, 515-516.

Firbank, L.G. and Watkinson, A.R. (1985) On the analysis of competition within two-species mixtures of plants. *Journal of Applied Ecology*, **22**, 503-517.

Gacek, E.S., Czembor, H.J. and Nadziak, J. (1996a) Disease restriction, grain yield and its stability in winter barley cultivar mixtures, in *Proceedings of the Third Workshop on Integrated Control of Cereal Mildews Across Europe. Kappel a. Albis, Switzerland, 5-9 Nov. 1994* (eds E. Limpert, M.R. Finckh and M.S. Wolfe). Office for Official Publications of the EC: Brussels, Belgium, pp. 185-190.

Gacek, E.S., Finckh, M.R. and Wolfe, M.S. (1996b) Disease control and yield effects in spring feed and malting barley mixtures in Poland, in *Proceedings of the Third Workshop on Integrated Control of Cereal Mildews Across Europe. Kappel a. Albis, Switzerland, 5-9 Nov. 1994* (eds E. Limpert, M.R. Finckh and M.S. Wolfe). Office for Official Publications of the EC: Brussels, Belgium, pp. 203-207.

Gacek, E.S., Strzembicka, H. and Wegrzyn, S. (1996c) Mixtures of spring wheat: their influence on powdery mildew and grain yield, in *Proceedings of the Third Workshop on Integrated Control of Cereal Mildews Across Europe. Kappel a. Albis, Switzerland, 5-9 Nov. 1994* (eds E. Limpert, M.R. Finckh and M.S. Wolfe). Office for Official Publications of the EC: Brussels, Belgium, pp. 193-196.

Gacek, E.S., Finckh, M.R., Hurej, M. *et al.* (1996d) The use of cultivar and species mixtures for restriction of diseases, weed infestation and other pests, in *Proceedings of the 9th European and Mediterranean Cereal Rusts and Powdery Mildews Conference, 2-6 September 1996, Lunteren, The Netherlands* (eds G.H.J. Kema, R.E. Niks and R.A. Daamen). Research Institute for Plant Protection: Wageningen, NL, p.302.

Gallagher, K.D., Kenmore, P.E. and Sogawa, K. (1994) Judicial use of insecticides deter planthopper outbreaks and extend the life of resistant varieties in southeast Asian rice, in *Planthoppers: Their Ecology and Management* (eds R.F. Denno and T.J. Perfect). New York, London: Chapman & Hall, pp. 599-614.

Gallun, R.L. (1977) Genetic basis of Hessian fly epidemics. *Annals of the New York Academy of Sciences*, **287**, 223-229.

Garrett, K.A. and Mundt, C.C. (2000) Host diversity can reduce potato late blight severity for focal and general patterns of primary inoculum. *Phytopathology*, **90**, 1307-1312.

Garrett, K.A., Nelson, R.J., Mundt, C.C. *et al.* (2001) The effects of host diversity and other management components on epidemics of potato late blight in the humid highland tropics. *Phytopathology*, **91**, 993-1000.

Ghaouti, L., Vogt-Kaute, W. and Link, W. (2005) Entwicklung ökologischer Regionalsorten bei Ackerbohnen [Development of region-specific organic cultivars in faba bean], in *8. Wissenschaftstagung Ökologischer Landbau - Ende der Nische, Kassel 01.0.-04.3.2005*. pp. 61-62.
Gould, F. (1986a) Simulation models for predicting durability of insect resistant germ plasm: A deterministic diploid, two locus model. *Environmental Entomology,* **15**, 1-10.
Gould, F. (1986b) Simulation models for predicting durability of insect resistant germ plasm: Hessian Fly (Diptera:Cecidomyiidae) resistant winter wheat. *Environmental Entomology,* **15**, 11-23.
Gould, F., Kennedy, G.G. and Johnson, M.T. (1991) Effects of natural enemies on the rate of Herbivore adaptation to resistant host plants. *Experimental and Applied Entomology,* **58**, 1-14.
Graß, R. and Scheffer, K. (2003) Kombinierter Anbau von Energie- und Futterpflanzen im Rahmen eines Fruchtfolgegliedes – Beispiel Direckt- und Spätsaat von Silomais nach Wintererbsenvorfurcht. *Mitteilungen der Gesellschaft für Pflanzenbauwissenschaften,* **15**, 106-109.
Graß, R. (2003) *Direkt- und Spätsaat von Silomais- Ein neues Anbausystem zur Reduzierung von Umweltgefährdungen und Anbauproblemen bei Opimierung der Erträge. Dissertation Universität Kassel.* Cuvillier-Verlag: Göttingen.
Hamblin, J. (1975) Effect of environment, seed size, and competitive ability on yield and survival of *Phaseolus vulgaris* (L.) genotypes in mixtures. *Euphytica,* **24**, 435-445.
Hamblin, J. and Rowell, J.G. (1975) Breeding implications of the relationship between competitive ability and pure culture yield in self-pollinated grain crops. *Euphytica,* **24**, 221-228.
Hao, J., Subbarao, K.V. and Koike, S.T. (2003) Effects of broccoli rotation on lettuce drop caused by *Sclerotinia* minor and on the population density of sclerotia in soil. *Plant Disease,* **87**, 159-166.
Hare, J.D. (1994) Status and prospects for an integrated approach to the control of rice planthoppers, in *Planthoppers: Their Ecology and Management* (eds R.F. Denno and T.J. Perfect). Chapman & Hall: New York, London, pp. 614-632.
Hariri, D., Fouchard, M. and Prud'homme, H. (2001) Incidence of soil-borne wheat mosaic virus in mixtures of susceptible and resistant wheat cultivars. *European Journal of Plant Pathology,* **107**, 625-631.
Harlan, H.V. and Martini, M.L. (1938) The effect of natural selection in a mixture of barley varieties. *Journal of Agricultural Research,* **57**, 189-199.
Heinrichs, E.A., Aquino, G.B., Chelliah, S. *et al.* (1982) Resurgence of *Nilaparvata lugens* (Stal) populations as influenced by method and timing of insecticide applications in lowland rice. *Environmental Entomology,* **11**, 78-84.
Heinrichs, E.A., Aquino, G.B., Valencia, S.L. *et al.* (1986) Management of the brown plant hopper, *Nilaparvata lugens* (Homoptera: Delphacidae) with early maturing rice cultivars. *Environmental Entomology,* **15**, 93-95.
Heinrichs, E.A. (1994) Impact of insecticide resistance and resurgence of rice planthoppers, in *Planthoppers: Their Ecology and Management* (eds R.F. Denno and T.J. Perfect). Chapman & Hall: New York, London, pp. 571-598.
Hoitink, H.A. and Fahy, P.C. (1986) Basis for the control of soilborne plant pathogens with composts. *Annual Review of Phytopathology,* **24**, 93-114.
Hovmoller, M.S., Munk, L. and Ostergaard, H. (1993) Observed and predicted changes in virulence gene frequencies at 11 loci in a local barley powdery mildew population. *Phytopathology,* **83**, 253-260.
Huang, R., Kranz, J. and Welz, H.G. (1994) Selection of pathotypes of *Erysiphe graminis* f. sp. *hordei* in pure and mixed stands of spring barley. *Plant Pathology,* **43**, 458-470.
Huang, R., Kranz, J. and Welz, H.G. (1995) Virulence of *Erysiphe graminis* f.sp. *hordei* isolates collected from barley genotypes with different resistance genes. *Journal of Phytopathology,* **143**, 287-294.
Jackson, L.F. and Wenning, R.W. (1997) Use of wheat cultivar blends to improve grain yield and quality and reduce disease and lodging. *Field Crop Research,* **52**, 261-269.
Jackson, W. (1980) *New roots for agriculture*. Friends of the Earth: San Francisco.
Jarosz, A.M., Burdon, J.J. and Mueller, W.J. (1989) Longterm effects of disease epidemics. *Journal of Applied Ecology,* **26**, 725-733.
Jarvis, D.I., Brown, A.H.D., Imbruce, V. *et al.* (2005) Managing Crop Disease in Traditional Agroecosystems:the Benefits and Hazards of Genetic Diversity, in *Managing Biodiversity in Agricultural Ecosystems* (eds D.I. Jarvis, C. Padoch and D. Cooper). Columbia University Press: New York.

Jeger, M.J., Jones, D.G. and Griffiths, E. (1981) Disease progress of non-specialized fungal pathogens in intraspecific mixed stands of cereal cultivars. II. Field experiments. *Annals of Applied Biology*, **98**, 199-210.

Jennings, P.R. and de Jesus, J.J. (1968) Studies on competition in rice. I. Competition in mixtures of varieties. *Evolution*, **22**, 119-124.

Johnson, R. (1984) A critical analysis of durable resistance. *Annual Review of Phytopathology*, **22**, 309-330.

Johnson, R. (1993) Durability of disease resistance in crops: some closing remarks about the topic and the symposium, in *Durability of Disease Resistance* (eds T. Jacobs and J.E. Parlevliet). Kluwer Academic Publishers: Dordrecht, pp. 283-300.

Jolliffe, P.A., Minjas, A.N. and Runeckles, V.C. (1984) A reinterpretation of yield relationships in replacement series experiments. *Journal of Applied Ecology*, **21**, 227-243.

Jones, L. and Clements, R.O. (1993) Development of a low-input system for growing wheat (*Triticum vulgare)* in a permanent understorey of white clover (*Trifolium repens*). *Annals of Applied Biology*, **123**, 109-119.

Juska, A., Busch, L. and Tanaka, K. (1997) The blackleg epidemic in Canadian rapeseed as a 'normal agricultural accident'. *Ecological Applications*, **7**, 1350-1356.

Kennedy, G.G., Gould, F., dePonti, O.M.B. and Stinner, R.E. (1987) Ecological, agricultural, genetic, and commercial considerations in the deployment of insect-resistant germplasm. *Environmental Entomology*, **16**, 327-338.

Khan, Z.R., Ampong-Nyarko, K., Chiliswa, P. *et al.* (1997) Intercropping increases parasitism of pests. *Nature*, **388**, 631-632.

Kilen, T.C. and Keeling, B.L. (1990) Gene frequency changes in soybean bulk populations exposed to Phytophthora rot. *Crop Science*, **30**, 575-578.

Kirkegaard, J.A., Wong, P.T.W. and Desmarchelier, J.M. (1996) In *vitro* suppression of fungal root pathogens of cereals by *Brassica* tissues. *Plant Pathology*, **45**, 593-603.

Kirkegaard, J.A., Sarwar, M., Wong, P.T.W. and Mead, A. (1998) Biofumigation by brassicas reduces take-all infection, in *Proceedings of the 9th Australian Agronomy Conference, Wagga Wagga, 1998.* pp. 465-468.

Kirkegaard, J.A. and Sarwar, M. (1999) Glucosinolate profiles of Australian canola (*Brassica napus annua* L.) and Indian mustard *(Brassica juncea* L.) cultivars: implications for biofumigation. *Australian Journal of Agricultural Research*, **50**, 315-324.

Knott, D.R. (1972) Using race-specific resistance to manage the evolution of plant pathogens. *Journal of Environmental Quality*, **1**, 227-231.

Knott, E.A. and Mundt, C.C. (1990) Mixing ability analysis of wheat cultivar mixtures under diseased and non-diseased conditions. *Theoretical and Applied Genetics*, **80**, 313-320.

Koizumi, S. and Kato, H. (1987) Effect of mixed plantings of susceptible and resistant rice cultivars on leaf blast development. *Annals of the Phytopathological Society of Japan*, **53**, 28-38.

Lannou, C. and Mundt, C.C. (1995) Evolution of a pathogen population in host mixtures: I: Study of the simple race-complex race competition. *Plant Pathology*, **45**, 440-453.

Lannou, C., de Vallavieille-Pope, C. and Goyeau, H. (1995) Induced resistance in host mixtures and its effect on disease control in computer-simulated epidemics. *Plant Pathology*, **44**, 478-489.

Leonard, K.J. (1977) Selection pressures and plant pathogens. *Annals of the New York Academy of Sciences*, **287**, 207-222.

Leonard, K.J. and Czochor, R.J. (1980) Theory of genetic interactions among populations of plants and their pathogens. *Annual Review of Phytopathology*, **18**, 337-358.

Letourneau, D.K. (1997) Plant-arthropod interactions in agroecosystems, in *Ecology in Agriculture* (Ed. L.E. Jackson). Academic Press: London, New York, pp. 239-290.

Leung, H., Zhu, Y., Revilla-Molina, I. *et al.* (2003) Using genetic diversity to achieve sustainable rice disease management. *Plant Disease*, **87**, 1156-1169.

Liebman, M.L. and Dyck, E. (1993) Crop rotation and intercropping strategies for weed management. *Ecological Applications*, **3**, 92-122.

Liebman, M.L. (1995) Polyculture cropping systems, in *Acroecology: The Science of Sustainable Agriculture. Second Edition* (ed. M.A. Altieri). Westview Press: Boulder, Colorado, pp. 205-218.

Liebman, M.L. and Gallandt, E.R. (1997) Many little hammers: Ecological management of crop-weed interactions, in *Ecology in Agriculture* (ed. L.E. Jackson). Academic Press: San Diego, New York, London, pp. 291-343.

Lopez, C.G. and Mundt, C.C. (2000) Using mixing ability analysis from two-way cultivar mixtures to predict the performance of cultivars in complex mixtures. *Field Crops Research,* **68**, 121-132.

Luo, Y. and Zadoks, J.C. (1992) A decision model for variety mixtures to control yellow rust on winter wheat. *Agricultural Systems,* **38**, 17-33.

Maeder, P., Fliessbach, A., Dubois, D. *et al.* (2002) Soil Fertility and Biodiversity in Organic Farming. *Science,* **296**, 1694-1697.

Maisonneuve, B., Martin, E., de Vallavieille-Pope, C. and Pitrat, M. (2004) Développement de *Bremia lactucae* sur une variéte de laitue cultivée en association avec une résistance introgressée de *Lactuca virosa. Rencontres de Mycologie - Phytopathologie, Aussois*, 13-17 Jan. 2004.

Manwan, I. and Sama, S. (1985) Use of varietal rotation in the management of tungro disease in Indonesia. *Indonesian Agriculture, Research and Development Journal,* **7**, 43-48.

Marshall, D.R. (1989) Modeling the effects of multiline varieties on the population genetics of plant pathogens, in *Plant Disease Epidemiology*, Vol II (eds K.J. Leonard and W.E. Fry). McGraw Hill, pp. 284-317.

Martin, F.N. (2003) Development of alternative strategies for management of soilborne pathogens currently controlled with methyl bromide. *Annual Review of Phytopathology,* **41**, 325-350.

May, R.M. and Anderson, R.M. (1983) Epidemiology and genetics in the coevolution of parasites and hosts. *Philosophical Transactions of the Royal Society London, Series B,* **219**, 281-313.

Mayton, H.S., Oliviar, C., Vaughn, S.F. and Loria, R. (1996) Correlation of fungicidal activity of *Brassica* species with allyl isothiocyanate production in macerated leaf tissue. *Phytopathology,* **86**, 267-271.

McCann, K.S. (2000) The diversity-stability debate. *Science,* **405**, 228-233.

McDonald, B.A., McDermott, J.M., Goodwin, S.B. and Allard, R.W. (1989) The population biology of host-parasite interactions. *Annual Review of Phytopathology,* **27**, 77-94.

Merz, U. and Wolfe, M.S. (1996) Barley and Wheat Mixtures in Switzerland: Resume and Outlook, in *Proceedings of the 3rd Workshop on Integrated Control of Cereal Mildews across Europe, Nov. 5-10, 1994, Kappel a. Albis, Switzerland* (eds E. Limpert, M.R. Finckh and M.S. Wolfe). Office for Official Publications of the EC: Brussels, Belgium, pp. 191-192.

Michelmore, R.W., Norwood, J.M., Ingram, D.S. *et al.* (1984) The inheritance of virulence in *Bremia lactucae* to match resistance factors 3, 4, 5, 6, 8, 9, 10, and 11 in lettuce (*Lactuca sativa*). *Plant Pathology,* **33**, 301-315.

Mille, B., Belhaj Fraj, M., Monod, H. and de Vallavieille-Pope, C. (2005) Assessing four-way mixtures of winter bread wheat for disease resistance, yield, and grain quality using the performance of their two-way and individual cultivar components. *European Journal of Plant Pathology,* (in press).

Moreno-Ruiz, G. and Castillo-Zapata, J. (1990) The variety Colombia: A variety of coffee with resistance to rust (*Hemileia vastatrix* Berk. & Br.). *Cenicafe Chinchiná-Caldas-Colombia Technical Bulletin,* **9**, 1-27.

Mundt, C.C. and Browning, J.A. (1985a) Genetic diversity and cereal rust management, in *The Cereal Rusts, Vol. II* (eds A.P. Roelfs and W.R. Bushnell). Academic Press: Orlando, pp. 527-559.

Mundt, C.C. and Browning, J.A. (1985b) Development of crown rust epidemics in genetically diverse oat populations: effect of genotype unit area. *Phytopathology,* **75**, 607-610.

Mundt, C.C. and Leonard, K.J. (1986) Effect of host genotype unit area on development of focal epidemics of bean rust and common maize rust in mixtures of resistant and susceptible plants. *Phytopathology,* **76**, 895-900.

Mundt, C.C., Leonard, K.J., Thal, W.M. and Fulton, J.H. (1986) Computerized simulation of crown rust epidemics in mixtures of immune and susceptible oat plants with different genotype unit areas and spatial distribution of initial disease. *Phytopathology,* **76**, 590-598.

Mundt, C.C. and Brophy, L.S. (1988) Influence of host genotype units on the effectiveness of host mixtures for disease control: A modeling approach. *Phytopathology,* **78**, 1087-1094.

Mundt, C.C. (1994) Techniques for managing pathogen co-evolution with host plants to prolong resistance, in *Rice Pest Science and Management* (eds P.S. Teng, K.L. Heong and K. Moody). International Rice Research Institute: P.O. Box 933, 1099 Manila, Philippines, pp. 193-205.

Mundt, C.C., Brophy, L.S. and Schmitt, M.S. (1995) Disease severity and yield of pure-line wheat cultivars and mixtures in the presence of eyespot, yellow rust, and their combination. *Plant Pathology,* **44**, 173-182.

Mundt, C.C. (2002) Use of multiline cultivars and cultivar mixtures for disease management. *Annual Review of Phytopathology,* **40**, 381-410.

Nair, P.K.R. (1993) *An Introduction to Agroforestry*. Kluwer Academic Publishers: Dordrecht.
Nitzsche, W. and Hesselbach, J. (1983) Sortenmischungen statt Vielliniensorten. *Zeitschrift fur Pflanzenzuchtung,* **90**, 68-74.
Norrington-Davies, J. and Hutto, J.M. (1972) Diallel analysis of competition between diploid and tetraploid genotypes of *Secale cereale* grown at two densities. *Journal of Agricultural Sciences Cambridge,* **78**, 251-256.
Odoerfer, A., Obst, A. and Pommer, G. (1994) The Effects of Different Leaf Crops in a Long Lasting Monoculture with Winter Wheat. 2. Disease Development and Effects of Phytosanitary Measures. *Agribiology Research,* **47**, 56-66.
Parker, M.E. (1991) Nonadaptive evolution of disease resistance in an annual legume. *Evolution,* **45**, 1209-1217.
Paul, N.D. and Ayres, P.G. (1986a) Interference between healthy and rusted groundsel within mixed populations of different densities and proportions. *New Phytologist,* **104**, 257-269.
Paul, N.D. and Ayres, P.G. (1986b) The impact of a pathogen (*Puccinia lagenophorae*) on populations of groundsel (*Senecio vulgaris*) overwintering in the field. I. Mortality, vegetative growth and the development of size hierarchies. *Journal of Ecology,* **74**, 1069-1084.
Paul, N.D. and Ayres, P.G. (1986c) The impact of a pathogen (*Puccinia lagenophorae*) on populations of groundsel (*Senecio vulgaris*) overwintering in the field. II. Reproduction. *Journal of Ecology,* **74**, 1085-1094.
Paul, N.D. and Ayres, P.G. (1987a) Effects of rust infection of *Senecio vulgaris* on competition with lettuce. *Weed Research,* **27**, 431-441.
Paul, N.D. and Ayres, P.G. (1987b) Survival growth and reproduction of groundsel (*Senecio vulgaris*) infected by rust (*Puccinia lagenophorae*) in the field during summer. *Journal of Ecology,* **75**, 61-71.
Paul, N.D. (1989) The effects of *Puccinia lagenophorae* on *Senecio vulgaris* in competition with *Euphorbia peplus*. *Journal of Ecology,* **77**, 552-564.
Paul, N.D., Ayres, P.G. and Wyness, L.E. (1989) On the use of fungicides for experimentation in natural vegetation. *Functional Ecology,* **3**, 759-769.
Perrin, R.M. (1977) Pest management in multiple agroecosystems. *Agro-ecosystems,* **3**, 93-118.
Pickett, J.A. (1999) Pest control that helps control weeds at the same time. *BBSRC Business Report, April 1999.*
Phillips, S. (2004) The ecology and epidemiology of potato variety mixtures in organic production. Dissertation, University of Reading, UK.
Phillips, S.L. and Wolfe, M.S. (2005) Centenary review: evolutionary plant breeding for low input systems. *Journal of Agricultural Sciences Cambridge,* **140**, 1-10.
Pilet, F. Epidémiologie et biologie adaptative des populations de *Phytophthora infestans* dans des cultures pures et hétérogènes de variétés de pomme de terre. 2003. Unité de Mixte Recherche ENSA-INRA, Rennes - Le Rheu, France. 11-7-2003.
Ponting, C. (1991) *A Green History of the World.* Sinclair Stevenson Ltd.: London.
Power, A.G. (1988) Leaf hopper response to genetically diverse maize stands. *Experimental and Applied Entomology,* **49**, 213-219.
Price, P.W., Bouton, C.E., Gross, P. *et al.* (1980) Interactions among three trophic levels: influence of plants on interactions between herbivores and natural enemies. *Annual Review of Ecology and Systematics,* **11**, 41-65.
Risch, S.J. (1981) Insect herbivore abundance in tropical monocultures and polycultures: an experimental test of two hypotheses. *Ecology,* **62**, 1325-1340.
Risch, S.J., Andow, D. and Altieri, M.A. (1983) Agroecosystem diversity and pest control: Data, tentative conclusions and new research directions. *Environmental Entomology,* **12**, 625-629.
Roderick, G.K. (1994) Genetics of host plant adaptation in delphacid planthoppers, in *Planthoppers: Their Ecology and Management* (eds R.F. Denno and T.J. Perfect). Chapman & Hall: New York, London, pp. 551-570.
Root, R.B. (1973) Organization of a plant-arthropod association in simple and diverse habitats: the fauna of collards (*Brassica oleraceae*). *Ecological Monographs,* **43**, 95-124.
Sama, S., Hasanuddin, A., Manwan, I. *et al.* (1991) Integrated management of rice tungro disease in South Sulawesi, Indonesia. *Crop Protection,* **10**, 34-40.
Saucke, H. and Döring, T.F. (2004) Potato virus Y reduction by straw mulch in organic potatoes. *Annals of Applied Biology,* **144**, 347-355.

Schaerer, H.J. and Wolfe, M.S. (1996) Virulence complexity of field populations of *Erysiphe graminis* f. sp. *hordei* in variety mixtures and pure stands of barley in Poland, in *Proceedings of the Third Workshop on Integrated Control of Cereal Mildews Across Europe. Kappel a. Albis, Switzerland, 5-9 Nov. 1994* (eds E. Limpert, M.R. Finckh and M.S. Wolfe). Office for Official Publications of the EC: Brussels, Belgium, pp. 173-175.

Schaffner, D., Koller, B., Mueller, K. and Wolfe, M.S. (1992) Response of populations of *Erysiphe graminis* f.sp. *hordei* to large-scale use of variety mixtures. *Vortraege fuer Pflanzenzuechtung,* **24**, 317-319.

Schaffner, D. (1993) Reaktion von Populationen de Gerstenmehltaus, *Erysiphe graminis* DC f.sp. *hordei* Marchal, auf Grossraeumigen Einsatz von Sortenmischungen. Dissertation, Swiss Federal Institute of Technology No 10376: Zuerich, Switzerland.

Schmidt, R.A. (1978) Diseases in forest ecosystems: The importance of functional diversity, in *Plant Disease: An Advanced Treatise, Vol 2* (eds J.G. Horsfall and E.B. Cowling). Academic Press: New York, pp. 287-315.

Schutz, W.M. and Brim, C.A. (1971) Inter-genotypic competition in Soybeans. III. An evaluation of stability in multiline mixtures. *Crop Science,* **11**, 684-689.

Scott, P.R., Benedikz, P.W. and Cox, C.J. (1982) A genetic study of the relationship between height, time of ear emergence and resistance to *Septoria nodorum* in wheat. *Plant Pathology,* **31**, 45-60.

Shetty, K.G., Subbarao, K.V., Huisman, O.C. and Hubbard, J.C. (2000) Mechanism of broccoli-mediated Verticillium wilt reduction in cauliflower. *Phytopathology,* **90**, 305-310.

Sitch, L. and Whittington, W.J. (1983) The effect of variety mixtures on the development of swede powdery mildew. *Plant Pathology,* **32**, 41-46.

Soule, J.D. and Piper, J.K. (1992) *Farming in Nature's Image: an Ecological Approach to Agriculture.* Island Press: Washington D.C.

Spitters, C.J.T. (1983) An alternative approach to the analysis of mixed cropping experiments. 1. Estimation of competition effects. *Netherlands Journal of Agricultural Sciences,* **31**, 1-11.

Stephens, P.M., Davoren, C.W., Doube, B.M. and Ryder, M.H. (1994a) Ability of the lumbricid earthworms *Aporrectodea rosea* and *Aporrectodea trapezoides* to reduce the severity of take-all under greenhouse and field conditions. *Soil Biology and Biochemistry,* **26**, 1291-1297.

Stephens, P.M., Davoren, C.W., Ryder, M.H. and Doube, B.M. (1994b) Influence of the Earthworms *Aporrectodea rosea* and *Aporrectodea trapezoides* on *Rhizoctonia solani* disease of wheat seedlings and the interaction with a surface mulch of cereal-pea straw. *Soil Biology and Biochemistry,* **26**, 1285-1287.

Stephens, P.M., Davoren, C.W., Ryder, M.H. *et al.* (1994c) Field evidence for reduced severity of Rhizoctonia bare-patch disease of wheat, due to the presence of the earthworms *Aporrectodea rosea* and *Aporrectodea trapezoides. Soil Biology and Biochemistry,* **26**, 1495-1500.

Stolz, H., Bruns, C. and Finckh, M.R. (2003) Einfluß genetischer Vielfalt auf den Befall mit *Phytophthora infestans* und auf die Ertragsbildung in Kartoffelbeständen, in *Beiträge zur 7. wissenschaftstagung zum Ökologischen Landbau. Ökologischer Landbau der Zukunft, 24.26.2.2003, Wien* (ed. B. Freyer). Universität für Bodenkultur: Vienna, pp. 569-570.

Subbarao, K.V., Hubbard, J.C. and Koike, S.T. (1999) Evaluation of broccoli residue incorporation into field soil for verticillium wilt control in cauliflower. *Plant Disease,* **83**, 124-129.

Suneson, C.A. (1949) Survival of four barley varieties in mixture. *Agronomy Journal,* **41**, 459-461.

Suneson, C.A. (1956) An evolutionary plant breeding method. *Agronomy Journal,* **48**, 188-191.

Teich, A.H. (1994) Disease control in wheat using ecological principles. *Genetika Polski,* **35B**, 127-135.

Thinlay, Finckh, M.R., Bordeos, A.C. and Zeigler, R.S. (2000a) Effects and possible causes of an unprecedented rice blast epidemic on the traditional farming system of Bhutan. *Agriculture, Ecosystems and Environment,* **78**, 237-248.

Thinlay, Zeigler, R.S. and Finckh, M.R. (2000b) Pathogenic variability of *Pyricularia grisea* from the high- and mid-elevation zones of Bhutan. *Phytopathology,* **90**, 621-628.

Tilman, D., Reich, P.B., Knops, J. *et al.* (2001) Diversity and productivity in a long-term grassland experiment. *Science,* **294**, 843-845.

Tosa, Y. (1994) Gene-for-gene interactions between the rye mildew fungus and wheat cultivars. *Genome,* **37**, 758-762.

Tozzetti, T.G. (1767) *True Nature and Causes and Sad Effects of the Rust, The Bunt, the Smut, and other Maladies of Wheat and of Oats in the Field* (translated from Italian by R. A. Tehon, Phytopathology Classics No. 9, 1952). American Phytopathology Society, St. Paul, Minnesota.

Trenbath, B.R. (1977) Interactions among diverse hosts and diverse parasites. *Annals of the New York Academy of Sciences,* **287**, 124-150.

Trutmann, P. and Pyndji, M.M. (1994) Partial replacement of local common bean mixtures by high yielding angular leaf spot resistant varieties to conserve local genetic diversity while increasing yield. *Annals of Applied Biology,* **125**, 45-52.

Van der Plank J.E. (1963) *Plant Diseases: Epidemics and Control.* Academic Press, New York, 349pp.

Van der Plank, J.E. (1968) *Disease Resistance in Plants.* Academic Press: New York.

van Emden, H.F. (1966) Plant insect relationships and pest control. *World Review of Pest Control,* **5**, 115-123.

Vandermeer, J.H. (1989) *The ecology of intercropping.* Cambridge University Press: Cambridge, New York, Melbourne.

Villareal, L.M.M.A. and Lannou, C. (2000) Selection for increased spore efficacy by host genetic background in a wheat powdery mildew population. *Phytopathology,* **90**, 1300-1306.

Wahl, I. (1970) Prevalence and geographic distribution of resistance to crown rust in *Avena sterilis. Phytopathology,* **60**, 746-749.

Webster, R.K., Saghai-Maroof, M.A. and Allard, R.W. (1986) Evolutionary response of barley composite cross II to *Rhynchosporium secalis* analyzed by pathogenic complexity and by gene-by-race relationships. *Phytopathology,* **76**, 661-668.

Weiner, J. (1990a) Plant population ecology in agriculture, in *Agroecology* (eds J.H. Carroll, J.H. Vandermeer and P. Rosset). McGraw-Hill: New York, pp. 235-262.

Weiner, J. (1990b) Asymmetric competition in plant populations. *Trends in Ecology and Evolution,* **5**, 360-364.

Weiner, J. (2003) Ecology – the science of agriculture in the 21st century. *Journal of Agricultural Science,* **141**, 371-377.

Weiner, J., Griepentrog, H.W. and Kristensen, L. (2001) Suppression of weeds by spring wheat (*Triticum aestivum*) increases with crop density and spatial uniformity. *Journal of Applied Ecology,* **38**, 784-790.

Westcott, B. (1987) A method of assessing the yield stability of crop genotypes. *Journal of Agricultural Sciences Cambridge,* **108**, 267-274.

Wills, C. (1996) Safety in diversity. *New Scientist,* 2022.

Wolfe, M.S. and Barrett, J.A. (1977) Population genetics of powdery mildew epidemics. *Annals of the New York Academy of Sciences,* **287**, 151-163.

Wolfe, M.S. and Barrett, J.A. (1980) Can we lead the pathogen astray? *Plant Disease,* **64**, 148-155.

Wolfe, M.S. and Knott, D.R. (1982) Populations of plant pathogens: Some constraints on analysis of variation in pathogenicity. *Plant Pathology,* **31**, 79-90.

Wolfe, M.S., Barrett, J.A. and Slater, S.E. (1983) Pathogen fitness in cereal mildews, in *Durable resistance in crops* (eds F. Lamberti, J.M. Waller and N.A. Van den Graaff). Plenum Press: New York, pp. 81-100.

Wolfe, M.S., Minchin, P.N. and Barrett, J.A. (1984) Some aspects of the development of heterogeneous cropping, in *Cereal Production, Proceedings of the 2nd International Summer School, Royal Dublin Society* (ed. E.J. Gallagher). Butterworths: London, pp. 95-104.

Wolfe, M.S. (1984) Trying to understand and control powdery mildew. *Plant Pathology,* **33**, 451-466.

Wolfe, M.S. (1985) The current status and prospects of multiline cultivars and variety mixtures for disease resistance. *Annual Review of Phytopathology,* **23**, 251-273.

Wolfe, M.S. (1987) The use of variety mixtures to control diseases and stabilize yield, in *Breeding Strategies for Resistance to the Rusts of Wheat.* CIMMYT: Mexico D.F., pp. 91-99.

Wolfe, M.S., Hartleb, H., Sachs, E. and Zimmermann, H. (1991) Sortenmischungen von Braugerste sind gesuender. *Pflanzenschutz-Praxis,* **2**, 33-35.

Wolfe, M.S. (1991) Recent developments in using variety mixtures to control powdery mildew of barley, in *Integrated control of cereal mildews: virulence patterns and their change. Proceedings of Second European Workshop on Integrated Control of Cereal Mildews, Roskilde, Denmark, 23-25 January 1990* (ed. J.H. Jorgensen). Riso National Laboratory: Roskilde, Denmark, pp. 235-243.

Wolfe, M.S., Braendle, U.E., Koller, B. *et al.* (1992) Barley mildew in Europe: population biology and host resistance. *Euphytica,* **63**, 125-139.

Wolfe, M.S. (1992) Barley diseases: Maintaining the value of our varieties, in *Barley Genetics VI* (ed. L. Munk). Munksgaard International Publishers, Ltd.: Copenhagen, pp. 1055-1067.

Wolfe, M.S. and McDermott, J.M. (1994) Population genetics of plant pathogen interactions: The example of the *Erysiphe graminis-Hordeum vulgare* pathosystem. *Annual Review of Phytopathology,* **32**, 89-113.

Wolfe, M.S. and Finckh, M.R. (1997) Diversity of host resistance within the crop: effects on host, pathogen and disease, in *Plant resistance to fungal diseases* (eds H. Hartleb, R. Heitefuss and H.H. Hoppe). G. Fischer Verlag: Jena, pp. 378-400.

Wolfe, M.S. (2000) Crop strength through diversity. *Nature,* **406**, 681-682.

Xiao, C.L., Subbarao, K.V., Schulbach, K.F. and Koike, S.T. (1998) Effects of crop rotation and irrigation on *Verticillium dahliae* microsclerotia in soil and wilt in cauliflower. *Phytopathology,* **88**, 1046-1055.

Zhou, X.G. and Everts, K.L. (2004) Suppression of Fusarium wilt of watermelon by soil amendment with hairy vetch. *Plant Disease,* **88**, 1357-1365.

Zhu, Y., Chen, H., Fan, J. *et al.* (2000) Genetic Diversity and Disease Control in Rice. *Nature,* **406**, 718-722.

Zoschke, M. (1987) Die Mischkultur als Anbaumethode im Blick auf Resistenz, Ertrag und Qualität. *Ergebnisse landwirtschaftlicher Forschung,* **18**, 57-66.

CHAPTER 11

EPIDEMIOLOGY IN SUSTAINABLE SYSTEMS

R.J. COOK AND D.J. YARHAM

11.1 INTRODUCTION

Knowledge of plant disease epidemiology has had increasing impact in the production-based industry of both the developed and developing world. In the last 50 years European agriculture has been associated with a move towards the simplification of systems, as farms have tended to specialize in arable or livestock production, largely determined by their soil or climatic conditions. Although cereal monoculture is no longer such a common practice as during the 1970s and 80s, the production of autumn-sown combinable crops still dominates large areas. The adverse effect of this on the biodiversity of the countryside has been accentuated by advances in the control of weeds and pests – with consequent direct and indirect effects on non-target species and thus on the food chains dependent on them. Fungicides have had far less deleterious environmental impact, though the vermicidal effects of benomyl and the insecticidal properties of pyrazophos gave cause for concern when these were widely used.

In recent years there has also been an increasing awareness that modern arable agriculture is dependent on non-renewable resources. While the same criticism could be levelled at most other industries, agriculture is one of the few which, since it is by its very nature producing renewable resources, could use this ability to reduce its reliance on external inputs.

Sustainability means different things to different people. In this chapter, we have adopted the definition used by the UK Government (Anon., 1994), the agricultural elements of which are:

- To provide an adequate supply of good quality food and other products in an efficient manner.
- To minimize consumption of non-renewable and other resources including by recycling.
- To safeguard the quality of soil, water and air.
- To preserve and where feasible enhance biodiversity and the appearance of the landscape including the UK's archaeological heritage.
- To encourage environmentally sensitive agriculture.

The aim must be to maintain the economic viability of individual farms while avoiding the unnecessary use of resources and deleterious effects on the agro-ecosystem. Sustainability does not necessarily imply low inputs; it does depend on their optimal use. Disease control, for example, will rely heavily on husbandry practices that reduce

B. M. Cooke, D. Gareth Jones and B. Kaye (eds.), The Epidemiology of Plant Diseases, 2nd edition, 309–334.

disease. The definition of sustainability used here would embrace the integrated crop management (ICM) systems, which have become the key elements of recent attempts in western Europe to produce a more environmentally benign agriculture, and would also include 'organic' systems of production. In this chapter we use the term 'organic' to describe crops grown "without artificial chemicals or genetic modification" (HRH Prince of Wales, Soil Association literature).

Disease control in both ICM and organic systems involves the sound application of epidemiological principles and may be seen as the practical application of the concepts expressed by Van der Plank (1960) as:

$$\Delta t = (230/r)\log(I_0/I'_0) \qquad (11.1)$$

where the expected delay (Δt) in the development of an epidemic is dependent upon the reduction of the initial inoculum from I_0 to I'_0 and the rate of disease increase (r). This chapter considers how such epidemiological information can be used in conjunction with cultural and crop management to reduce the initial inoculum (I'_0) and the rate of disease development (r).

11.2 INOCULUM

The first phase of a disease epidemic depends on the level of inoculum present when the crop is first exposed to infection and on the ability of the pathogen to take advantage of this initial vulnerability of the host to become established in the crop. This ability is usually dependent upon climatic (or micro-climatic) conditions at the time. The importance of these factors relative to those influencing the later stages of epidemic development varies widely from disease to disease, as does the feasibility of reducing inoculum potential as a method of disease control in a sustainable arable system. This may be illustrated by reference to some of the more common diseases of arable crops.

11.2.1 Seedborne inoculum

There are many examples of seedborne inoculum initiating epidemics and this is the primary infection source for a number of major crop diseases (see Chapter 13). Bunt of wheat (caused by *Tilletia tritici*), may be considered as an example. The control of bunt has long depended primarily on the control of inoculum. The disease is usually introduced into a crop on the seed sown, although infection from soilborne inoculum may occur. A low level of infection in a crop may go unnoticed but, if seed is saved from such a crop, a much higher level of disease is likely to occur in the following year. The millions of spores released from bunted ears during threshing become dusted onto the surrounding healthy seeds and thus act as a potent source of inoculum for the following season's epidemic. Dillon Weston and Engledow (1933) calculated the potential rate of increase in infection levels that could occur in a stock of wheat from which seed was saved, but not treated, over a three-year period. If one ear in 8500 is affected by bunt in year 1, one in 450 could be infected in year 2 and one in four in year 3. The proportion of ears developing the disease will be primarily determined by the level of inoculum in the seed, though conditions at the time of crop emergence will have some influence on the number of seedlings becoming infected.

Seedborne inoculum of *T. tritici* is effectively controlled by fungicidal seed treatments. Reliance on routine application of such treatments can be questioned in sustainable systems and is not allowed in organic production. If, however, seed crops are grown from seeds treated with a material which also affords protection against soilborne inoculum it should be unnecessary to treat seed of the subsequent commercial crop, unless it is to be sown in a field suspected to have been contaminated by spores from a recently harvested, infected crop. This approach, although acceptable in ICM systems, is no longer available to producers of organic wheat as current regulations preclude the use of seed from crops that have themselves been treated with a synthetic fungicide (unless suitable seed is unavailable).

Recent research has explored the use of more 'organically acceptable' seed treatments such as acetic acid (Borgen and Nielsen, 2001) and mustard flour (Borgen and Kristensen, 2001). At the time of writing, however, control of bunt in organic wheat is dependent on the reduction in inoculum in the early years of production of a seed stock and on the assumption that the inoculum has not subsequently built up to potentially damaging levels. Oxley and Cockerell (1996) found bunt in fields when the level of contamination in the seed sown exceeded 100 spores/seed. Testing of seed to be used in organic production is strongly recommended to ensure that contaminated stocks are not sown. Because of the difficulty of sourcing organically produced seed, organic farmers are sometimes forced to use seed from crops which themselves were grown from conventionally produced seed. The long-term availability of organically produced seed is a challenge for the organic cereal industry.

The approaches outlined for the control of bunt are also applicable to the control of diseases such as the cereal loose smuts (caused by *Ustilago* spp.) and leaf stripe of barley (caused by *Pyrenophora graminea*). Treatment of seed crops does not, however, preclude the contamination of the seed by pathogens arising from other sources of inoculum. The ubiquitous *Microdochium nivale* (which can cause serious seedling blight) can infect cereal seeds during grain filling, whether or not seed treatment was used on the parent crop. After a wet summer, it is advisable to have wheat seed tested for the presence of this pathogen before deciding (even in an ICM system) whether or not to use untreated seed.

The interest in reducing chemical inputs in arable agriculture has led to increasing interest in the possible use of biological control agents (BCAs) to suppress seed-borne inoculum of plant pathogens (see section 11.4.2 below).

11.2.2 Soil as a source of inoculum

Minimization of soil cultivations should be a primary objective of all sustainable systems to conserve energy, soil and water. Minimal cultivation systems can, however, be counter-productive in the context of disease control as they may leave inoculum at or near the surface of the soil into which the crop is to be sown. Control of the soil-borne phase of ergot (caused by *Claviceps purpurea*), for example, relies on ploughing to bury the sclerotia to a depth of at least 50 mm to prevent the clavae from reaching the surface when germination occurs. However, cultivations have varying effects on crop pathogens and there are often complex interactions.

Delaying crop establishment will allow more time for the inoculum of the pathogen to be destroyed by the activity of other soil microorganisms; but the most effective method for controlling most soilborne diseases is sound rotational practice. A few examples are given here to illustrate the often complex effects of rotation and other husbandry practices on soil-borne pathogens.

(a) Bunt

Soilborne inoculum of *Tilletia tritici* is readily controlled by rotation because the host range of the pathogen is limited to a single species (wheat) and it has limited saprophytic ability. The soilborne inoculum does not produce airborne spores, nor does the fungus produce long-lived resting bodies. However, recent work suggests that its spores can remain viable for much longer than was previously thought possible (Johnsson, 1990) and may be more important as an inoculum source than was previously believed to be the case. Delayed sowing allows inoculum levels in the soil to decline before the seeds germinate, but the benefits of this are partially offset by the fact that in the colder conditions to which late sown winter wheat plants are subjected, it takes the seedlings longer to grow through the coleoptile stage during which they are most susceptible to infection. The initial level of inoculum is by far the most important factor in the development of a bunt epidemic, but even with this disease other factors must be taken into consideration when devising a control strategy.

(b) Take-all

Gaeumannomyces graminis, the cause of take-all of cereals (see also Chapter 14), has a wider host range and a greater ability to survive in the absence of a host than has *T. tritici.* It is, however, restricted to the *Gramineae* and its saprophytic ability is so limited that it may usually be controlled by a one-year break from cereals. The saprophytic survival of *G. graminis* is enhanced by the ready availability of nitrogen and any treatment that reduces nitrogen levels during the break year may be expected to enhance its effectiveness in reducing inoculum levels. In a trial at ADAS Drayton, UK, for example, a sequence of wheat crops was interrupted by a set-aside break. In the following wheat crop take-all levels were lower where the set-aside land had been summer cropped with *Phacelia* sp. or more particularly, with white mustard than where it had been summer fallowed (ADAS unpublished data). Uptake of nitrogen by the mustard and *Phacelia* may have helped to 'starve out' the pathogen – an effect of 'catch crops' long ago noted by Garrett and Buddin (1947). In the case of the mustard the effect is likely to have been enhanced by the fungitoxic isothiocyanates released by the roots of the mustard plants and acting as a natural biofumigant (see 11.2.3. below). When diseases are controlled by rotational practice, any treatment which can accelerate the decline in inoculum levels during 'break' years is likely to reduce disease levels when a susceptible crop is again grown in the field.

Although a single year's break will not completely eliminate inoculum of *G. graminis*, it will normally reduce it to acceptable levels, unless perennial rhizomatous species such as couch *(Elymus repens)* or Yorkshire fog *(Holcus lanatus)* are present (Nilsson and Drew Smith, 1981). Infected rhizomes of such grasses can provide a substantial reservoir of inoculum of the fungus, and even in an ICM system, the use of herbicides for their control is often likely to be required (Yarham, 1981). The herbicide should be applied before the break crop to allow time for the rhizomes that carry the inoculum to rot away. In 'organic' production systems (where the use of any herbicide is obviously precluded) every effort must be expended to prevent the build-up of infestations of such weed grasses.

The benefits of a break crop on the incidence of take-all seldom persist beyond the first one or two subsequent wheat crops. Often after only a single crop, and certainly after two, enough inoculum will again have built up in the field to initiate a severe epidemic should conditions favour the development of the disease.

In the UK, it has often been noted that take-all in second successive wheat crops is often more severe after rape than after other break crops – a phenomenon which well illustrates the complex interaction of factors which influence the epidemiology of plant diseases. Where the break crop is oilseed rape, inoculum can survive on the debris of self-sown cereals killed by herbicide during the winter, its survival being facilitated by the high levels of nitrogen normally applied to the rape crop. Added to this, early harvesting of the rape allows the early sowing of a following wheat crop and this facilitates the build-up of inoculum in the first year of a wheat sequence (see below). It has also been suggested (in Hornby, 1998) that the naturally-occurring biofumigation associated with brassica crops may result in a decrease in certain components of the soil microflora which are antagonistic to *G. graminis*. If these were more active against the fungus in its active pathogenic, rather than in its saprophytic phase their absence might have little effect on its survival during the break year but make it easier for it to build up in the following wheat. Although rotation is of prime importance as a disease control practice in sustainable systems, its effect on the epidemiology of plant pathogens is seldom as simple as it at first appears.

As with bunt, conditions in the early stages of the epidemic can significantly affect its severity. Take-all tends to be worse in early-sown winter crops, as the fungus can become well established on the roots in the autumn before falling temperatures prevent further infection (Clare *et al.*, 1986; Prew and Beane, 1986). Some delay in sowing date can therefore be beneficial as can any treatment (such as early cultivation) that accelerates the breakdown of debris carrying inoculum of the pathogen between the harvesting of one crop and the sowing of the next.

(c) Facultative pathogens

Unlike *T. tritici and G. graminis,* some pathogens (e.g. some *Fusarium* species) can survive very effectively as saprophytes in the soil. In such instances rotational practices offer less effective methods of disease control. However, to prevent a dangerous increase in inoculum levels, intensive cropping with a susceptible species should be avoided.

(d) Sclerotinia

Sclerotinia sclerotiorum differs from *G. graminis* in its ability to produce very long-lived resting bodies, in its production of copious airborne spores and in its wide host range amongst dicotyledonous crops. Once again, rotational practices are unlikely to provide effective control of this fungus and other methods must be sought to reduce inoculum levels. It has been suggested that, after a severe attack, a cereal crop should be established by using minimal cultivation (graminaceous crops are immune to infection). Sclerotia thus left on or near the soil surface will produce apothecia, most of the spores from which will be deposited harmlessly on the cereal leaves or on the soil, though some will be caught up by air currents to act as a source of infection for other susceptible crops in the area. Eradication of inoculum from an individual field will not protect a crop grown in that field from infection by such airborne spores, but it will reduce the risk of a severe attack which is most likely to occur close to the source where inoculum pressure is highest.

Adjacent susceptible crops and weeds can also increase the risk of infection. Jerusalem artichoke *(Helianthus tuberosus),* for example, is very susceptible to sclerotinia and can provide an inoculum source for other subsequent susceptible crops grown in the same field or in neighbouring fields. In southern Lincolnshire, UK, in 1991 (a year particularly conducive to the disease), severe infection occurred in two rape crops adjacent to a field where a strip of Jerusalem artichokes had been ploughed in five years previously, although other local crops were not affected (J.M.Ll. Davies, ADAS Terrington, UK, personal communication). An example of the importance of weeds was provided by Hims (1979). Severe sclerotinia in a crop of oilseed rape was confined to a corner of a field adjacent to a wood where hogweed (*Heracleum sphondylium*) and cow parsley (*Anthriscus sylvestris*) were dying prematurely as a result of infection by the pathogen.

(e) Use of trap-crops

For some diseases, the effect of rotation can be enhanced by stimulation of resting spore germination by growing a ‘trap crop’ host before the commercial crop. This approach was used by White (1954) to reduce powdery scab (caused by *Spongospora subterranea*) in potatoes by planting *Datura stramonium* to stimulate germination of resting spores. However, such trap crops are unlikely to be used in practice unless they provide a profitable crop in their own right.

(f) Soil sterilisation

While the eradication of soilborne inoculum of plant pathogens is very difficult in agricultural systems, it may more effectively be achieved in commercial horticulture by partial soil sterilization. However, the use of chemical sterilants will be unacceptable in a sustainable system, and steam sterilization (once widely practised in the glasshouse industry) is today too costly for most situations. At least in warmer climes a possible alternative is offered by soil solarization – the use of the sun's heat to achieve partial sterilization of soil covered by a polyethylene sheet. The process, which has been successfully used in parts of southern Europe since the late 1970s (Tjamos and Faridis, 1980), has been reviewed by De Vray (1995) and Katan (1994) and has been the subject of international conferences, for example The Second International Conference on Soil Solarization and Integrated Management of Soilborne Pests, ICARDA, Aleppo, Syria, 1997. The technique not only reduces pathogen populations by the direct effect of heat but also stimulates biological processes which contribute to their control.

11.2.3 Crop residues as a source of inoculum

Infected stem and leaf debris act as a potent source of inoculum for many diseases. Ploughing can reduce inoculum by burying debris but can bring its own problems. Advice to leave rape stubble unploughed for sclerotinia control contradicts the once standard recommendation that such stubble should be ploughed to reduce the inoculum of *Leptosphaeria maculans* (teleomorph of *Phoma lingam).* Ascospores of *L. maculans,* produced in the autumn from cankered stems left on the soil surface are more likely to introduce disease into nearby crops than are those of *S. sclerotiorum* which are released in the spring when a covering cereal crop is likely to be present to restrict their movements in air currents. Moreover, ploughing will reduce inoculum of *L. maculans* more effectively than that of *S. sclerotiorum* as the mycelium of the former fungus in the buried stem bases will not survive as long as the buried sclerotia of the latter. A case could thus be made for ploughing rape stubble where canker has been a problem but sclerotinia is absent, and for leaving it on the surface after a severe attack of sclerotinia (in the former case the aim is to reduce the amount of inoculum produced, in the latter it is to restrict the dissemination of the inoculum.) Recently, however, there has been a move towards leaving rape stubble unploughed until seeds shed during harvest have germinated; this prevents ungerminated seeds being buried and providing a weed problem in future years. By the time the volunteer plants have emerged and the field has been ploughed, ascospores of *L. maculans* will already have been produced on the stubble and could have infected young plants of the next season's crop in fields nearby. This well illustrates the way in which an ICM practice employed to solve one problem (weed control) can exacerbate another (disease).

The stem base pathogens, *Oculimacula yallundae* and *O. acuformis* (anamorphs *Pseudocercosporella herpotrichoides* f. spp.), the causes of eyespot of cereals, provide an interesting example of the way in which the effects of inoculum level can be over-ridden by other factors in the early development of an epidemic. Where cereal follows cereal, splash-borne conidia, produced in abundance on stem bases left after harvest, are likely to be of primary importance in infecting the subsequent year's crop. It might be expected that the disease would be less severe after ploughing than after direct drilling, but numerous studies have shown that the opposite is more usually the case (e.g. Yarham and Norton, 1981). Various factors are likely to be involved in this apparent anomaly: a thick layer of straw on the soil surface can impede the splash dispersal of the conidia; the more prostrate habit of the usually more shallowly sown direct-drilled plants makes it less likely that splashed spores will be funnelled down to the stem base; and the greater availability of nitrogen (a result of higher rates of soil nitrification) in ploughed soils in the autumn increases the susceptibility of the young plants to infection (another example of excess nitrogen increasing disease risk; see below).

Observations in the UK have indicated that yield responses to fungicides applied for eyespot control are lower in wheat crops following oilseed rape than where the previous crop has been either a cereal or a non-cereal crop other than rape (Cook and Thomas, 1990). Infected debris from the wheat crop prior to a break crop is normally buried by ploughing and brought to the surface again immediately prior to the following wheat crop. Rape, however, is frequently sown into unploughed cereal stubble, so infected straws from the previous crop will be left on the soil surface during the break from cereals and the fungus will have exhausted its capacity to produce more spores (and, therefore, its inoculum potential) before the next cereal crop is taken, thus reducing disease incidence in that crop.

Moreover, it has been suggested that the rape crop may itself contribute to the partial sterilization of the soil in which it grows. Increases in yields of wheat following rape, far greater than those observed after crops such as linseed or oats, have been observed in Australia, the United States and Chile (e.g. Kirkegaard *et al.*, 1994; Kirkegaard, 1996). The evidence suggests that some form of biofumigation may be responsible; phenyl ethyl glucosinolate released from rape roots is converted by microbial enzyme activity to isothiocyanates which are known to be toxic to many pest and pathogen species.

Leaf debris from previous crops is more ephemeral than stem base debris but can provide a direct inoculum source for many diseases, as for example when fallen leaves affected by apple scab (caused by *Venturia inaequalis*) release ascospores into the emerging leaf canopy during the spring (see also Chapter 18). Wind-blown spores may cause infection when they are discharged from debris remote from the crop, as in the case of septoria tritici blotch of wheat (caused by *Mycosphaerella graminicola*, teleomorph of *Septoria tritici*). This disease is notable for the uniform infection that occurs in wheat crops wherever they are grown. Ascospore infection is the most likely source of disease, straw debris providing a source of long-distance ascospores (Scott *et al.*, 1988; Shaw and Royle, 1989). Dissemination of *M. graminicola* inoculum as ascospores is widespread through the countryside so that every susceptible crop might be expected to become infected sooner or later. Even so, the presence of inoculum either within the field or in close proximity to it can affect disease development. Studies by ADAS and Rothamsted Research, UK have demonstrated that ploughing or straw burning can reduce the severity of infection, at least in the early stages of an epidemic, and some work shows early gradients in the incidence of *M. graminicola* in crops adjacent to unploughed set-aside fields (Yarham and Gladders, 1994). Later, the effects of proximity to within-field or nearby inoculum will often be lost, as in the logarithmic stage of epidemic development, the severity of the disease becomes more determined by environmental conditions than by the initial level of inoculum.

Accumulation of debris from successive crops, owing to reduced cultivations, is probably responsible for increases in tan spot (caused by *Pyrenophora tritici-repentis*) in the United States and Australia (Adee and Pfender, 1989; Summerell and Burgess, 1989). The disease has also become more common and severe in Europe in recent years, probably as a result of the increasing popularity of minimum cultivation techniques (Bill Clark, ADAS Boxworth, UK, personal communication). Similarly, the related net blotch disease of barley (caused by *Pyrenophora teres*) became a serious disease in Britain in the late 1970s and early 1980s when such reduced cultivation techniques coincided with widespread popularity of a very susceptible cultivar.

Infected cereal debris can be effectively reduced by stubble burning, which in many cases reduces rather than eradicates inoculum. The technique is effective against rice stem rot (caused by *Sclerotium oryzae*), for example (Webster *et al.*, 1981), but this practice is illegal in many western countries and may not, therefore, be used to reduce inoculum of diseases of field crops. In marked contrast to these examples, Doupnik and Boosalis (1980) showed that planting sorghum directly into wheat stubble decreased the incidence of stalk rot (caused by *Fusarium moniliforme*) compared with normal cultivations. Moisture stress is an important component in the epidemiology of fusarium diseases in cereals and it is likely that increased water conservation owing to the mulch of stubble decreases subsequent susceptibility of sorghum to fusarium stalk rot.

11.2.4 Living plants as inoculum sources

Obligate pathogens may produce resting structures or survive between crops on living plants (either growing plants or storage organs) that often provide abundant inoculum (survival on seeds has already been considered in section 11.2.1). In cereal production, for example, late developing tillers and volunteer plants provide a 'green bridge' which enables such pathogens to survive the gap between the harvesting of one crop and the emergence of the next. Infected weeds can also act as strong sources of inoculum for nearby crop plants that share their susceptibility to particular diseases.

Blumeria graminis (cause of powdery mildew of cereals) provides a good example of such a pathogen. Cleistothecia can survive on dead leaf debris but, for emerging autumn-sown crops, these are likely to be less important as a source of inoculum than are the windborne conidia produced in such generous abundance on self-sown plants. Disease gradients can be found in crops adjacent to such sources of infection. Similarly, early season mildew gradients can be found in spring barley crops adjacent to an overwintering source of inoculum in winter barley (Yarham and Pye, 1969).

The short generation time and prolific sporulation of the fungus ensures such rapid build-up of mildew that severity across the field soon comes to reflect environmental factors such as topography (and hence shelter/exposure) and soil nutrient status, that influence the susceptibility of the plants (Yarham and Gladders, 1994).

Inoculum sources should not, however, be ignored in an integrated control strategy. When spring barley is sown close to a winter-sown crop, care should be taken to ensure that winter and spring varieties have different genetic sources of host partial resistance to mildew so that they are not susceptible to the same strains of the pathogen (see also Chapter 10). A similar tactic should be adopted if it is necessary to sow winter wheat adjacent to a set-aside field carrying self-sown plants of a wheat cultivar susceptible to yellow rust (*Puccinia striiformis*).

The practical significance of the *r* and *Io* values in Van der Plank's equation is well illustrated by reference to potato late blight (caused by *Phytophthora infestans*), and discussed further by Mizubuti and Fry in Chapter 17. Careless husbandry, which allows dense growth of infected haulm to develop on dumps of discarded tubers, will provide so strong a source of initial inoculum (high *Io*) that the pathogen will be able to take full advantage of the first period of favourable (warm and humid) weather and an early attack of blight may be expected.

Complete elimination of inoculum will be impossible in any intensive potato growing area and even isolated crops are likely to be infected by windblown spores sooner or later. Careful hygiene on the part of the grower (and immediate neighbours) may nevertheless so reduce the initial level of inoculum that the lag phase of epidemic development is lengthened and two or three infection periods may be necessary before the disease builds up to damaging levels. However, if there is a long period of weather favourable to the development of the disease, the value of *r* in Van der Plank's equation will be so increased that a damaging epidemic will soon develop, even if initial inoculum levels were low.

Because of the strain specificity of most obligate pathogens, volunteer plants of the same species (and particularly those of the same cultivar) as the crop are most likely to act as potential sources of inoculum to initiate an epidemic within that crop. However, weeds can also play an important role as inoculum sources and effective weed control has a part to play in the control of diseases in sustainable systems The closer the botanical relationship between weed and crop, the greater the risk that disease will spread from the one to the other. The spread of ergot from black-grass (*Alopecurus myosuroides*) to wheat provides a good example of an obligate pathogen spreading between species within the same family and much wider taxonomic gaps can sometimes be crossed, especially by virus pathogens.

In a case encountered by one of the authors (DY) a very severe attack of cucumber mosaic virus occurred in a newly built glasshouse on land that had never before carried a cucumber crop. The grower, assuming that the 'virgin' land would be disease-free, had omitted the usual soil-sterilisation practices. Unfortunately, the soil happened to contain a large population of chickweed seeds, some of which carried the virus. When the weed seeds germinated aphids very soon transmitted the virus to the crop plants and a serious epidemic ensued. Given that steam sterilization is now prohibitively expensive, any attempt to control such an epidemic in ICM or organic production would encounter obvious difficulties. The use of chemical soil sterilants would be frowned on in ICM, and prohibited in organic systems as the latter would preclude the use of herbicides and insecticides. Mechanical control of the weeds at a very early stage of growth would be the only feasible alternative.

Other examples of obligate parasites crossing taxonomic boundaries are provided by the heteroecious rusts which often require alternate hosts in completely different taxa to complete their life cycles. Successful campaigns to eradicate barberry bushes in hedgerows greatly reduced the incidence of black stem rust (caused by *Puccinia graminis)* in wheat crops in the 18th and 19th centuries and provide an excellent example of the eradication of inoculum by non-chemical means (all be it by a method which reduced the bio-diversity of the countryside!).

11.3 DISEASE DEVELOPMENT

While reduction in inoculum can play an important part in the control of some diseases, for the majority of crop diseases it is likely to be effective as a disease control measure only if steps are taken to slow down the rate of increase of the pathogen during the log phase of the epidemic development, i.e. to reduce the value of r in Van der Plank's equation. This can be achieved by:

- increasing the resistance of the host to infection by the pathogen;
- manipulation of the within-crop environment in such a way as to slow the development of the epidemic;
- ensuring that the crop is not exposed at its most vulnerable stages to weather favourable to the epidemic development of the pathogen;
- direct intervention in the development of the epidemic by the judicious use of fungicide (this can be an effective part of the disease management strategy in ICM systems, but is unacceptable in organic systems);
- manipulation of the balance between the pathogen and its microbial antagonists and competitors to encourage processes of biological control.

11.3.1 Genetic resistance and crop sensitivity

Host resistance is a critical factor determining epidemic development and it influences the need to intervene with fungicides. Brown (Chapter 3) considers surveys of pathogen virulences in more detail and the application of such data, and Deadman (Chapter 5) considers the epidemiological implications of different resistance strategies.

In the UK, the relationship between the frequency of virulence genes in cereal pathogen populations has been monitored since the 1960s by the UK Cereal Pathogen Virulence Survey (UKCPVS) to provide early warnings of shifts in pathotype virulences. Throughout this period, there has been a steady and relentless increase in the complexity of pathotypes of yellow rust on wheat and powdery mildew on both wheat and barley. This has mirrored attempts by plant breeders to develop more sophisticated combinations of host resistance.

Within a wheat-growing area, the epidemic development of (for example) a new yellow rust pathotype will be determined by the popularity of cultivars employing those resistance genes for which the new pathotype has developed virulence. Cultivar selection is based primarily upon agronomic characteristics, rather than disease susceptibility. Since inception of the survey there has been a sequence of often devastating epidemics of this disease as a succession of new pathotypes developed to overcome a previously resistant combination of genes. Bayles *et al.* (1997) discuss these interactions, using data from the UKCPVS for both yellow rust on wheat and powdery mildew on barley. In both cases, the complexity of pathogen races reduces the flexibility of conventional breeding techniques to provide resistance. The changes illustrate the ability of pathogens to overcome the best endeavours of plant breeders who have exploited most of the readily available resistance for these diseases.

One of the UKCPVS's most important products has been diversification schemes (Priestley, 1981) developed to help farmers to identify the risk of disease. Cultivars are assigned to individual diversification groups (e.g., Anon., 1997a) selected on the basis of genetic resistance to pathogens and the frequency of matching virulences in the pathogen population. Cultivar selection, with the aim of reducing disease pressure, is an illustration of the difficulty of disease control in sustainable systems.

The cultivation of suitable mixtures of cultivars, chosen from different diversification groups, can reduce disease. However, in most areas of western Europe, pathogen races are now so complex that it is difficult to combine appropriate resistance gene combinations with suitable agronomic or market quality requirements. The additional complication that it is often easier to market pure stands of single cultivars also reduces the uptake of this approach; it is far easier to meet the grain specifications of different customers by mixing discrete lots of known cultivars as necessary. Wolfe *et al.* (1981) discussed the benefits of cultivar mixtures which is also considered in more detail by Finckh and Wolfe in Chapter 10 of this book. The chosen mixture should comprise of cultivars with different sources of resistance, though this does not necessarily guarantee greater yield stability than that provided by pure stands (Akanda and Mundt, 1997).

An interesting manifestation of crop sensitivity to disease occurred in maize during the 1970s, when male sterile lines were being used for seed production. One strain of the common disease southern corn leaf blight (caused by *Helminthosporium maydis*) became especially virulent on hybrids bred using the Texas form of cytoplasmic male sterility. The disease was effectively controlled by reverting to mechanical detasseling of the male parents.

Despite such difficulties as those outlined above, effective use of genetic resistance to specific diseases or tolerance of disease remain essential elements of sustainable agriculture A range of plant breeding approaches is now being explored throughout the world to improve the robustness of crop plant resistance. These involve the use of molecular markers to identify and more precisely to characterise resistance genes, and the technique commonly known as genetic modification; this offers the prospect of durable disease resistance with associated opportunities to reduce significantly or even obviate the need for fungicide treatment. This would be of substantial benefit to sustainable and organic systems. However, genetic modification involving 'genetic engineering' techniques to transfer genes between unrelated species would, at present, be unacceptable to the organic movement.

Knowledge of the background resistance of potential cultivars provides the basic understanding necessary for the design of crop rotations and fungicide programmes that minimise inputs. This knowledge is put to use by specialist crop advisers to help them to select the least-cost options that maximise disease control and yield in the field.

11.3.2 Modification of the crop environment

The rate of development of a disease epidemic is highly sensitive to environmental conditions as they affect spore production and dispersal, the infection process and the length of the latent period. However, pathogens have evolved to take advantage of any favourable interlude in the microclimate. In some instances (e.g. *M. graminicola*) the result has been the evolution of a well-adapted pathogen capable of causing severe yield losses in conditions as diverse as the US mid-west, the Mediterranean basin, Australasia and north-west Europe. Such adaptation renders it almost impossible to identify simple sets of infection conditions which might be exploited either in the field or in prediction systems (Shaw, 1991).

In protected horticulture, manipulation of the environment is more readily achieved than under field conditions and the control of ventilation and temperature has long been used as a means of delaying the build-up of diseases such as those caused by the grey mould pathogen (*Botrytis cinerea*) and the downy mildews (Peronosporales).

Environmental control also plays an important part in controlling the diseases of stored products. The management of potato stores to prevent the development of tuber rots provides an example – and also illustrates the need for a good understanding of the epidemiological requirements of the pathogens concerned. Cooling to retard the development of bacterial soft rots, for example, can exacerbate losses caused by gangrene (causal agent *Phoma exigua* f. *foveata*) if it is not preceded by a high temperature 'curing period'. Such a warm period facilitates the healing of wounds through which, even at low temperatures, the *Phoma* pathogen can advance into the flesh of the tubers.

Although actual modification of the above-ground environment is difficult under field conditions it is not completely impossible. The epidemic development of many pathogens (e.g. *O. yallundae)* is favoured by the high humidities maintained within very thick crops, and reduction of seed rate and or increasing row width has long been advocated as a means of delaying their development. Orientation of rows of phaseolus beans to the direction of the prevailing wind has been advocated to maximise airflow through the crops and the availability of sunlight to the plants to reduce the incidence of sclerotinia (Haas and Bolwyn, 1972). Similarly, Steadman (1983) cites the increased susceptibility of closed canopy cultivars as opposed to open canopy types, which are more resistant.

Many splash-dispersed diseases rely on the ability of rain droplets to move freely between individual plants and leaves within the canopy. In these situations, crop density can have a significant influence on the epidemic development. One of the best examples is septoria tritici blotch, for which disease the interaction with canopy structure has been explored (Lovell *et al.*, 1997). The combination of proximity of inoculum to susceptible leaves and the coincidence of rainfall able to disperse spores onto the upper leaves is a requirement for severe disease. Short and thin crops are therefore often more severely affected than taller crops with dense canopies. Disease spread can also occur by dew or by wind movement of leaves (Lovell and Parker, 1997). These mechanisms allow disease to spread from one leaf layer to another when layers overlap in short-internode cultivars.

Manipulation of the soil environment can also affect disease development. Liming and irrigation, for example, greatly affect the physico-chemical and microbiological environment of the soil and thus influence the risk of disease from soilborne pathogens. Common scab (caused by *Streptomyces scabies*) of potatoes, for example, is favoured by a high pH, and club root (caused by *Plasmodiophora brassicae*) of brassicas by low pH. In a rotation containing both potatoes and brassicas, it would be advisable to lime before the latter crop and not before the former.

Fungi that produce motile zoospores are likely to be favoured by irrigation. This leads to a problem for potato growers wishing to control tuber scab. Common scab prefers dry soil conditions and can be controlled by irrigation during the period four to six weeks from the time of tuber initiation. This, however, favours the development of powdery scab (causal agent *Spongospora subterranea*). Farmers trying to control the former by irrigation can often reduce the benefits of their strategy by aggravating the latter.

Sclerotium rolfsii, the cause of stem rot of peanuts, is favoured less by continuous wetness than by the alternation of wet and dry conditions that irrigation provides. In such a situation, irrigation can greatly impair the disease control achieved by fungicides (Davis, *et al.,* 1996). However, not to irrigate would reduce yields and so jeopardize the profitability of the crop. The grower seeking to operate an ICM system thus faces the dilemma of balancing the irrigation needs of the crop with the need to reduce disease by limiting soil moisture.

Where it is not possible to modify the environment, reducing crop exposure to risk is a valid alternative. Delay of sowing date can reduce the risk of over-wintering yellow rust in winter wheat – temperatures usually being too low for infection to occur by the time the crop emerges. With mildew on spring barley, on the other hand, delayed sowing increases the risks of a damaging attack on the young plants which emerge when temperatures are rising in late spring.

Where a mixture of pathogens is present, attempts to adjust the timing of husbandry operations to reduce disease incidence may be made more difficult by the different epidemiological requirements of the pathogens concerned. In potatoes, for example, watery wound rot (caused by *Pythium ultimum*) is favoured by high temperatures, and gangrene (caused by *Phoma exigua* f. *foveata*) by low ones. Lifting of the crop too early (when the weather is too warm) can increase risks of losses due to the former disease; lifting too late (when it is too cold) can aggravate the latter.

Husbandry practices employed for purposes other than disease control can sometimes increase the risks of disease and sometimes reduce it. The use of growth regulators on cereal crops, for example, can increase disease incidence making it easier for spores of *Septoria* spp. in wheat or *Rhynchosporium secalis* (the cause of leaf blotch) in barley to be splashed on to the upper leaves (by shortening the crop) and producing denser leaf canopies (hence, higher within-crop humidities), which favour the development of diseases. At the same time, a growth regulator may, by shortening and strengthening the straw, reduce the risks of lodging caused by eyespot.

11.3.3 Nutrient status

Just as the rate of epidemic development of a disease may be determined by the inherent susceptibility of the plant to infection, so it can be influenced by variation in susceptibility consequent upon the nutrient status of the host. The rate of build-up of barley powdery mildew, for example, is likely to be more rapid in a crop suffering from manganese deficiency, and net blotch is more likely to be severe on plants deficient in potash. Nitrogen deficiency, on the other hand, tends to retard the epidemic development of most foliar diseases. Avoidance of synthetic sources of nitrogen and the subsequently lower levels of N available to organic crops is likely to be one of the reasons why such crops show lower levels of foliar diseases, by comparison with conventionally-grown crops where nitrogen is rarely limiting.

Increases in fungal diseases in relation to increases in nitrogen are well documented. In cereals, for example, mildew is known to be associated with high levels of nitrogen. Similarly, brown and yellow rust (causal agents *Puccinia recondita* and *P. striiformis* respectively) are both exacerbated by increased nitrogen (Clare *et al.*, 1990; Nuttall and Stevens, 1991). Daamen *et al.* (1989) found that while yellow rust and mildew were increased on wheat in Holland, eyespot and fusarium foot rot were unaffected. The development of an epidemic may also be influenced by the growth stage at which nutrients are applied. For example, under high disease pressure as is often experienced in south-west Britain, applications of nitrogen earlier than are normally advised increase the risk of severe disease on winter barley (Jordan *et al.*, 1989).

The increase in use of nitrogen during the 1970s increased both crop output and the impact of diseases in UK wheat crops. This relationship was initially suggested as a cause of the epidemics of septoria leaf blotch on winter wheat, although early studies and observations on nitrogen response experiments proved to be inconclusive. Leitch and Jenkins (1995) have indicated that the expansion of green leaf area associated with increased nitrogen allows greater opportunity for disease development. The precise mechanisms for this relationship are unclear but it is likely that increased leaf nitrogen levels and increased canopy owing to the higher nitrogen will both increase disease. Both effects allow more potential for greater disease-induced yield losses associated with decreases in green leaf area caused by damage from pathogens (Bryson *et al.*, 1995). These effects on temperate cereal crops are also reflected in the effect of nitrogen in tropical crops. Cu *et al.* (1996) showed that although high rates of applied nitrogen fertilizer are needed to maintain yields of rice, they increase sheath blight (caused by *Rhizoctonia solani*). Yield reductions of between 20 and 42% were recorded at the highest rates of nitrogen (up to 200 kg ha^{-1}).

A further example of the effect of nitrogen on disease severity may be seen in fruit: Utkhede and Smith (1995) showed that nitrogen as ammonium nitrate (240 g per tree per year), applied as a single or split dose, increased the incidence of crown and root rot caused by *Phytophthora cactorum*. By contrast, foliar sprays of urea or ground applications of nitrogen as monoammonium phosphate had no effect on disease incidence.

Source of nitrogen has also been shown to influence the severity of take-all in cereals. The application of ammonium nitrogen reduces infection by the pathogen by way of its effect on other soil micro-organisms. Nitrate nitrogen, on the other hand, increases the susceptibility of roots to infection. However, nitrate N also increases the number of roots produced by the plant and this generally more than offsets its negative effects. This provides a good example of a treatment which favours the epidemic development of a pathogen but, by also increasing the resilience of the plant, can actually reduce the damage caused by the disease.

11.4 CONTROL STRATEGIES

11.4.1 General considerations

Good farm hygiene (e.g. destruction of haulm on dumps of discarded potatoes), the use of tested, disease-free seeds (or other organs of propagation) and sound rotational practice that should be an integral part of any sustainable system will all help to keep inoculum levels to a minimum and thus reduce disease risk. Complete eradication of inoculum is possible only for a very few diseases, so action is also necessary to retard the development of epidemics (the '*r*' of Van der Plank's equation) within growing crops. The most effective way of achieving this is by the use of genetically resistant cultivars – but these cannot be wholly relied on as resistance is seldom complete and, when it is, selection pressure on the fungal population usually results in the eventual emergence of resistance-breaking strains of the pathogen. Exploitation of genetic resistance is thus a key element in any integrated disease control strategy, but until genetic manipulation produces more durable resistance it must be enhanced by other measures to reduce the crop's susceptibility to infection. Any inherent resistance shown by a crop variety can be augmented by ensuring balanced nutrition, avoiding the deficiencies and excesses that render plants susceptible to infection.

An exciting development for the future is the advent of chemicals which, though themselves are not biocidal, have the ability to enhance the plant's own resistance to pathogen attack. This advance could revolutionize our attitude to the control of fungal diseases. It will, however, necessitate treatment before disease is present and may thus lead to a reappraisal of risk assessment criteria for disease control. The latter problem would be avoided by the use of non-chemical methods of stimulating the production of phytoalexins. Percival, Karim and Dixon (1998), for example, have shown if potato tubers are exposed to light before infection an increase in their glycoalkaloid content increases their resistance to infection by *Fusarium sulphureum* and *F. solani* var. *coeruleum*.

In addition to enhancing the disease resistance of the plants themselves, steps can be taken to modify the environment so that conditions within the crop are less conducive to infection. Lower seed rates, increased row widths, optimisation of nutrients and even careful row orientation (to maximize airflow and the drying effects of sunshine) have all been used to create intracrop microclimates less conducive to the development of foliar disease. Similarly, careful soil management (avoidance of compaction, water-logging and excessively loose seed beds) and careful management of irrigation and soil pH will help to create conditions less favourable to infection by soilborne pathogens. Some aspects of these interactions have been reviewed by Conway (1996). There is an obvious need to understand the conditions favouring the epidemic development of pathogens and to be aware that conditions which reduce one disease may favour another.

Epidemic development may thus be slowed by enhancing host resistance or by making the environment less conducive to the pathogen (or more conducive to its microbial antagonists; see also 11.4.2 below). A further tactic that may be employed is to adjust the cropping calendar to avoid crops being at a vulnerable stage when conditions are most favourable for infection. The early development of many diseases on young plants can, for example, be reduced if the sowing of winter crops is delayed. They will then be less likely to be affected by autumn epidemics of foliar pathogens. It should be noted, however, that the robust plants produced in the autumn as a result of early sowing will often tolerate quite high levels of disease without suffering serious harm and that occasionally (as in the case of mildew on winter barley) plants which have suffered a heavy attack in the autumn are less susceptible to potentially more damaging spring epidemics of the same pathogen. Moreover, delay in sowing can itself cause reductions in yield and can increase the risks of pest (e.g. wheat bulb fly) attack. A careful balance must therefore be struck between that which favours the plant, that which favours its pathogens and that which favours its pests – a further example of the problems inherent in the development of integrated crop protection strategies.

11.4.2 Biological control

The introduction into crop production systems of microorganisms known to be competitive with, or antagonistic to, pathogens of the crop being grown is an attractive concept and is already being used to check the epidemic spread of a few pathogens. Successful examples of the use of such biological control agents (BCAs) are the inoculation of the stumps of felled trees with *Peniophora gigantea* (Greig, 1984) to prevent the build-up on them of *Helicobasidium annosus* (the cause of butt rot of pines), and the dipping of Australian peach cuttings in a cell suspension of a non-pathogenic strain of *Agrobacterium radiobacter* to protect them against crown gall (caused by *Agrobacterium tumefasciens*) (Htay and Kerr, 1974). Inoculation of fruit trees with *Trichoderma viride* (Corke, 1974) to control silver leaf (caused by *Chondrosporium purpureum*) has given less satisfactory results but commercial preparations of the BCA have been marketed for this purpose and the same fungus has also been shown to be effective against a number of other pathogens (D'Allbra and Mutto, 1986).

The bacterium *Pseudomonas chlororaphis* has recently been approved as a seed treatment in Denmark for control of bunt on wheat. Isolates of the same bacterium have been tested (along with strains of *P. fluorescens* and *Pantoea* sp.) for the control of *Fusarium* and *Microdochium* seedling blights in Sweden (Johansson *et al.*, 2003). There has also been interest shown in the possible use of fungal BCAs to replace chemical seed treatments. Knudsen *et al.* (1995), for example, found that strains of *Gliocladium roseum* were antagonistic to *Fusarium culmorum* on wheat seeds and significantly reduced seedling blight in the field. Using an isolate of the same fungal antagonist to treat the seeds, Burgess and Keane (1997) were able to reduce the soft rot of chickpea seedlings caused by seed-borne *B. cinerea.* Commenting on their findings, the latter authors suggested that genetic manipulation might enhance the efficiency of *G. roseum* as a biocontrol agent. If genetic engineering does provide an effective way of improving the efficacy of BCAs it could greatly assist the move away from chemical control agents in sustainable production systems. It is doubtful, however, if such genetically modified control agents would find acceptance in organic agriculture. Amongst many other BCAs not yet exploited commercially, *Coniothyrium minitans* can reduce *Sclerotinia* spp. on a number of crops, including sunflower (Huang, 1980).

Further investigation of the interactions between the epidemiology of the pathogen and the antagonist offers a fruitful field for further study. Attempts to control the diseases of arable crops using this approach have generally given disappointing results. Considerable research effort has, for example, been directed at the development of a BCA for the control of take-all (Hornby, 1998), fluorescent pseudomonads having been the most favoured organisms. By and large, however, these attempts have met with very limited success. Part of the problem is that, of necessity, the BCA has been introduced into an extremely complex ecosystem and has had difficulty in maintaining itself in an environment that includes many of its own competitors and antagonists.

In protected crops, biological control is made easier by the increased control which the grower can have over the growing medium. The use of composted bark, for example, has given excellent control of carnation wilt and certain other diseases – the bark itself containing a complex of microorganisms antagonistic to the pathogen. In this case, success does not depend on the introduction of a single BCA but on the provision of an environment in which naturally occurring BCAs can flourish. Such an approach is obviously more difficult in field crops, but an example is provided by the old advice that grass clippings should be dug into the soil to control common scab of potatoes (Actinomycetes developing on the clippings being antagonistic to *S. scabies).*

Cropping sequences can sometimes be modified so as to favour the development of naturally-occurring microbial antagonists. The use of grass leys to build up the root-colonizing saprophyte *Phialophora radicicola* (which successfully competes for infection sites with *G. graminis)* has been advocated for the control of take-all (Deacon, 1973), and the development under conditions of continuous cereal cropping of a microflora antagonistic to *G. graminis* has long been recognized, and exploited, as the cause of take-all decline (Chapter 14).

There are other methods of biological control that do not involve direct use of different biological agents. In horticultural crops the use of grafting techniques to control crop vigour, by using, for example, a dwarfing rootstock, with a scion of a cultivar with the desired characteristics is well established. The approach has been used for disease control in high value crops, e.g. for control of the corky root rot complex of protected tomatoes during the 1970s. In this case plants of the required, but disease susceptible, cultivar were grafted onto disease-resistant rootstocks. The technique was used in western Europe for a short while, but it is very labour-intensive, requires propagation of double the number of plants needed and could not be justified where financial margins are tight.

The principles of immunisation have also been used in some high value crops. For example, in glasshouse tomatoes control of the highly contagious tomato mosaic virus was achieved with varying degrees of success, by inoculating young plants with an attenuated strain of the pathogen, which protected them against attack by more aggressive strains in later life. In practice, both these last two techniques were only used until effective resistant cultivars became available.

Organic farmers would claim that their method of growing crops, more particularly their attention to conserving the organic content of the soil, is itself an effective biocontrol technique as it enables them to maintain the natural microbial balance which will itself suppress the build up of pathogens. This approach is not new. Sandford (1926) hypothesised that:

- saprophytic micro-organisms can control the activities of plant pathogens;
- the microbiological balance of the soil can be changed by altering soil conditions; in particular the addition of fresh organic material will promote the activity and multiplication of saprophytes, which by their competition for nutrients and oxygen, and by their excretions, will depress the activity and multiplication of the pathogens.

More recently, integrated farming systems relying on minimum cultivations (instead of ploughing) and incorporation of crop debris have developed in response to economic pressures and have taken advantage of these principles. A healthy soil with diverse biomass is not only a tenet of organic farming; it is sound practice in any farming system, and a consequence of non-inversion tillage (Fox, 2000). The finding by Rodgers and Shaw (2000) that straw incorporation reduces some cereal diseases affirms the validity of such claims and highlights the need for further research.

11.4.3 Fungicides for epidemic management

In sustainable arable systems an understanding of epidemiology shows the need to delay infection to reduce the rate of disease development. Where this cannot be achieved by manipulating husbandry, fungicides may be needed. Epidemiological principles then become important in helping to time fungicide applications with greater precision in order to optimize dose, minimise applications and maximise effectiveness. Several of the chapters in this volume explore opportunities for improving the precision of fungicide application by forecasting when disease epidemics are likely to occur. For example, Hardwick (Chapter 9) considers some commercial approaches in Europe, including potato blight, which are explored in more detail by Mitzubuti and Fry in Chapter 17. Some of the approaches, in cereals for example, rely on empirical rules which are based on practical observations. Others depend on more fundamental understanding of the causal relationships between the pathogen and the crop environment. Other chapters with a bearing on the potential for forecasting to improve the precision of fungicide application include those by MacCartney, Fitt and West (Chapter 6), Shaw (Chapter 7) and Maude (Chapter 19). Simple grower-focused guides (Harris and Scott, 1989) to crop protection would be of benefit for all sustainable cropping systems and many have already been produced.

As an example of the often complex interactions which need to be considered, disease management in winter wheat provides a good example of the approaches necessary to achieve sustainable disease control. When cereal fungicides were introduced in western Europe in the 1970s, first principles dictated that they should be applied before the log phase of disease development. As a result, the concept of thresholds became established. Work on powdery mildew in spring barley showed that optimum control was obtained when 3-5% of the area of the lowest leaves was affected at the start of stem extension (Jenkins and Storey, 1975). Similarly, for eyespot in winter wheat, an empirically derived threshold of 20% stems affected during the spring became the norm, Scott and Hollins (1974) having shown that only when more than this proportion of stems were affected at GS 75 (Tottman, 1987) did the disease cause yield losses in the field. Eyespot control is about optimal when fungicides are applied between GS 31 and 32. Most farmers are risk averse and, as subsequent disease development is almost impossible to predict, they are under pressure to apply fungicides during this period, mixing the product with other crop treatments applied at this time (e.g. growth regulators or herbicides) to reduce production costs. This is not necessarily sustainable disease management. Some of the difficulties of making precise predictions as to how disease will develop, and the challenges of forecasting, are discussed in Chapter 9.

Severe infection, likely to cause serious yield loss, is relatively infrequent in practice, except in continuous cropping, very early-sown situations or when susceptible cultivars are widely grown. Winter wheat is the most important arable crop over large areas of eastern England and has been the subject of extensive work on dry matter accumulation and disease control. The priority is to control foliar disease on the upper two and, less essentially, third leaves. The lower leaves contribute relatively little to final dry matter but can act as potent sources of inoculum for the more important upper foliage. Guides for farmers (Anon., 2000; Clark and Paveley, 2004) illustrate when it is necessary to control disease and how fungicides should be used, and provide guidelines for reducing the risks of fungicide resistance.

It is essential that individual assessments are made for all disease risks. Although, for example, about 20% of winter wheat crops in England would benefit from treatment against eyespot (Bill Clark, ADAS, UK, personal communication) there are seasonal and rotational variations. In some areas routine treatments for eyespot control were economic only in three of the 11 years between 1985 and 1995 (M. Nuttall, Morley Research Centre, UK, unpublished). This run of data was collected from experiments within commercial second wheat crops sown during late September or early October (i.e. those supposedly at high risk of eyespot infection), in three or four-year rotations including combinable crops and sugar beet. In these situations, sowing date of the first wheat crop is often delayed by late harvest of the root crop. This reduces the risk of severe eyespot developing in the first crop and, thus, restricts the amount of inoculum available to infect the second crop.

Application of epidemiological principles also improves assessment of risk at the time when a fungicide spray decision has to be taken. Wet conditions over the previous six weeks will increase the likelihood that a high level of latent eyespot infection is present in the crop and will thus make it more likely that a spray will be required. In practice, information of this type has been used in western Europe to modify simple threshold-based decisions and determine spray need and timing for eyespot control.

For powdery mildew and yellow rust, there are compelling reasons why fungicides should be applied to winter cereals to reduce inoculum at the start of stem extension during the spring. When conditions are favourable, these diseases often have a latent period of less than 10 days. This is substantially less than the phyllochron (time between emergence of successive leaves) so that impetus of the epidemic builds up on each successive leaf layer. *Septoria tritici* on the other hand, not only has a long incubation period (typically at least three weeks – well over the phyllochron) but also relies on rain or wet conditions before spore dissemination can occur. The length of the pathochron (Beresford and Royle, 1988) reduces the potential for the disease to build up on upper leaves until stem extension is complete and final leaves are available for infection.

This knowledge of disease epidemiology indicates that there is little point in applying fungicides for septoria control during early stem extension unless a period of wet weather occurs and suggests that disease control on the lowest (3rd and 4th) leaves may only have benefit in open-canopied crops (Cook, 1997). Leaf 3 does not usually emerge to become infected until late April at the earliest in the UK and it is thus the development of disease from this time onwards which is of most importance. Data from disease control experiments and annual surveys of wheat crops in the UK indicate that sprays applied in May to protect upper leaves from late stem extension onwards, rather than earlier, provide the best control of septoria tritici blotch (e.g. Thomas *et al.,* 1989; Cook *et al.,* 1991). Under European conditions, there is little benefit to be obtained by reducing inoculum of the pathogen during the winter as it can never be completely eliminated and, if conditions are favourable, even a low level of inoculum can result in a very high rate of disease development. Farmers often attempt to reduce the inoculum of *S. tritici* by applying fungicides earlier than necessary. Epidemiological knowledge confirms that weather conditions during and immediately after stem extension are critical for the epidemic. Additionally, ascospore infection can occur over a long period of time into the spring. Emphasis must be placed, therefore, on slowing epidemic development on the upper leaves which contribute to the filling of the grain, a strategy which is in contrast to that adopted to control mildew, a short generation pathogen.

The uncertainty of how diseases will develop poses some fundamental questions for disease management in any sustainable system. It is unlikely that we will be able to predict weather for more than six days in advance with any certainty. Attempts to minimise fungicide use by reducing the number or dose of fungicide sprays thus depend on judgements of disease risk during critical periods in the development of both the crop and the disease. These approaches may sometimes be as misguided in terms of developing a sustainable disease control strategy as the routine calendar-based treatments adopted in some intensive systems, simply because the fungicide dose may be inadequate for the level of control required.

Fungicide treatment in organic production is limited to the use of a few naturally occurring substances such as elemental sulphur and certain plant products (e.g. Stopes *et al.,* 2000). These materials are less fungitoxic than synthetic fungicides and are considerably more effective as protectants than eradicants. However, good organic practice discourages the prophylactic use even of the permitted materials so a disease should be present, or a definite risk identified before they are applied. A good knowledge of disease epidemiology, and very careful monitoring of crops to pick up the first signs of disease, are thus required if the use of appropriate 'organic' fungicides is to be successful.

11.5 CONCLUSIONS

This chapter has reviewed some of the cases where disease may be reduced by an understanding of crop disease epidemiology and given some examples of how the identification and exploitation of the weak links in pathogen life cycles may be used to improve control strategies. Complete control is, however, also governed by climatic and microclimatic conditions beyond the control of the practitioner. Understanding these interactions can minimize pesticide use and lead to a more environmentally benign agriculture, thus improving the potential of sustainability to maintain food supplies. Although some chemical disease control is likely to be necessary for the foreseeable future, understanding the epidemiological principles for each disease in each crop can reduce fungicides to a level acceptable to sustainable agriculture. Where yield or quality-threatening infection occurs we anticipate that for the foreseeable future application of fungicides will be necessary. Our conclusions may be summarised as follows:

- Seed-borne fungi are the easiest of pathogens to control by the reduction of inoculum. We can ensure the health of seed stocks and we can also test for the presence of disease in the seed we are to use, rejecting it if it is found to be contaminated or, if necessary, making judicious use of a fungicidal seed treatment. In this way chemical use may be, if not eliminated, then at least minimized.

- Sound rotational practice can have an important part to play in the control of soilborne diseases and the high priority given to it in ICM systems is certainly justified. It does not, of course, provide a complete answer to such diseases – especially where the pathogen can be introduced from surrounding areas by means of airborne spores. It must, moreover, be linked with effective weed control. In an ICM system, mechanical methods may form the basis of the weed control programme but the development of certain infestations (e.g. of couch grass) may necessitate the use of a herbicide to reduce build up of take-all in wheat. In the warmer regions of the world, solarization can provide a very effective and environmentally acceptable method of reducing the inoculum of soilborne pathogens of horticultural crops.
- Phoma canker and eyespot illustrate the need for a sound understanding of the biology of crop protection problems in ICM. A tactic designed to reduce inoculum levels (such as ploughing) can be counter-productive if it increases susceptibility to the disease (as in eyespot) or exacerbates another crop protection problem (sclerotinia stem rot). Disease control strategies which integrate rotation and cultivations can make a substantial contribution to the success of disease control.
- Sound hygiene practices to reduce debris-borne inoculum in individual fields can be of value in ICM but are likely to be less effective if inoculum is also present in the surrounding area.
- Effective hygiene will reduce inoculum and may delay the onset of an epidemic. However, abundant production of airborne spores provides the potential for small amounts of inoculum to cause significant crop damage if weather conditions are favourable to individual diseases at key stages during the growing season.
- Avoiding excess nitrogen application is likely to restrict disease development and reduce waste.
- Biological control offers considerable potential in sustainable systems. The introduction of specific biological control agents is likely to be of limited application. Manipulation of the substrate to favour naturally occurring pathogen antagonists offers a better prospect. At present, attempts to control diseases by introducing specific BCAs have met with little success in agriculture but successes in both horticulture and forestry suggest that further research in this area is fully justified. The potential will only be achieved by understanding the ecological interactions between pathogens and antagonists.
- Genetic resistance offers the best prospect for non-chemical disease control. It is essential that we maintain the broadest possible base of genetic resistance in crops. Advances in gene technology provide the prospect of durable crop resistance to disease by importing external resistance sources.

ACKNOWLEDGEMENT

We are indebted to Bill Clark, ADAS Boxworth for helpful comment.

REFERENCES

Adee, E.A. and Pfender, W.F. (1989) The effect of primary inoculum level of *Pyrenophora tritici-repentis* on tan spot epidemic development in wheat. *Phytopathology,* **79**, 873-877.

Akanda, S.I. and Mundt, C.C. (1997) Effect of two-component cultivar mixtures and yellow rust on yield and yield components of wheat. *Plant Pathology,* **46**, 566-580.

Anon. (1994) *Sustainable Development: the UK strategy,* Cmnd 2426, HMSO London.

Anon. (1997a) *Cereal Variety Handbook,* NIAB, Cambridge.

Anon. (2000) *The wheat disease management guide* HGCA Research and Development, Caledonia House, Pentonville Road, London.

Bayles, R.A., Clarkson, J.D.S. and Slater, S. (1997) The UK Cereal Pathogen Virulence Survey, in *The Gene-for-gene Relationship,* CABI, Wallingford, Oxon, pp. 103-118.

Beresford, R.M. and Royle, D.J. (1988) Relationships between leaf emergence and latent period for leaf rust *(Puccinia hordei)* on spring barley and their significance for disease monitoring. *Zeitschrift fur Planzenkrankeiten und Pflanzenschutz,* **95**, 361-371.

Borgen, A.and Kristensen, L (2001) Use of mustard flour and milk powder to control common bunt *(Tilletia tritici)* in wheat and stem smut *(Urocystis occilta)* in rye in organic agrriculture. *Seed Treatment: Challenges and Opportunities. BCPC Symposium Proceedings No. 76*, 141-150.

Borgen, A. and Nielsen, B.J. (2001) Effect of seed treatment with acetic acid for the control of seed borne diseases. *Seed Treatment: Challenges and Opportunities. BCPC Symposium Proceedings No.* **76**, 135-140.

Bryson, R.J., Sylvester-Bradley, R., Scott, R.K. and Paveley, W. (1995) Reconciling the effects of yellow rust on yield of winter wheat through measurements of green leaf area and radiation interceptions. *Aspects of Applied Biology,* **42**, 9-18.

Burgess, D.R. and Keane, P.J. (1997) Biological control of *Botrytis cinerea* on chickpea seed with *Trichoderma* spp. and *Gliocladium roseum*: indigenous versus non-indigenous isolates. *Plant Pathology,* **46**, 910-918.

Clare, R.W., ap Dewi, I. and Madge, W.E.R. (1986) The effect of autumn nitrogen, insecticide and fungicide on winter wheat sown at two dates and with three levels of spring nitrogen. *Proceedings of the 1986 British Crop Protection Conference – Pests and Diseases,* 165-172.

Clare, R.W., Jordan, V.W.L., Smith, S.P. *et al.* (1990) The effect of nitrogen and fungicide treatment on the yield and grain quality of Avalon winter wheat. *Aspects of Applied Biology,* **25**, 375-386.

Clark, W.S. and Paveley, N.K. (2004) *Wheat disease management – 2004 update.* HGCA Research and Development, Caledonia House, Pentonville Road, London.

Conway, K.E. (1996) An overview of the influence of sustainable agricultural systems on plant diseases. *Crop Protection,* **15**, 223-228.

Cook, R.J. (1999) Management by chemicals, in Septoria *on Cereals: a study of Pathosystems.* Edited by J.A. Lucas, P. Bowyer and H.M. Anderson, CABI Publishing, UK.

Cook, R.J. and Thomas, M.R. (1990) Influence of site factors on yield response of winter wheat to fungicide programmes in England and Wales, 1979-1987. *Plant Pathology,* **39**, 540-557.

Cook, R.J. Polley, R.W. and Thomas, M.R. (1991) Disease-induced losses in winter wheat in England and Wales 1985-1989. *Crop Protection,* **10**, 504-508.

Corke, A.T.M. (1974) The prospect for biotherapy in trees infected by silver leaf. *Journal of Horticultural Science,* **49**, (4) 391-394.

Cu, R.M., Mew, TW., Cassman, K.G. and Teng, P.S. (1996) Effect of sheath blight on yield in tropical, intensive rice production systems. *Plant Disease,* **80**, 1103-1108.

D' Allbra, V. and Mutto, S. (1986) Parasitism of *Trichoderma herzianum* on *Polymyxa betae Journal of Phytopathology,* **115**, 61-72.

Daamen, R.A., Wijnands, F.G. and Vliet, G., van der (1989) Epidemics and pests of winter wheat at different levels of agrochemical input; a study on the possibilities for designing an integrated cropping system. *Journal of Phytopathology,* **125**, 3105-3109.

Davis, R.F., Smith, F.D., Brenneman, T.B. and McLean, H. (1996) Effect of irrigation on expression of stem rot of peanut and comparison of above ground and below ground disease ratings. *Plant Disease,* **80**, 1155-1159.

Deacon, J.W. (1973) Control of the take-all fungus by grass leys in intensive cereal cropping. *Plant Pathology,* **22**, 149-155.

De Vray, J.E. (1995) Solarization; an environment-friendly technology for pest management, in *New Advances in Integrated Pest Management of Cultivated Crops and Forests, Fes, Morocco, 27 Nov.-2 Dec.* 1994, *Arab Journal of Plant Protection,* **13**, 61-56.

Dillon Weston, W.A.R. and Engledow, F.L. (1933) The 'money' side of bunted wheat. *Essex County Farmers Union Yearbook,* 1930, 1-6.

Doupnik, B. and Boosalis, M.G. (1980) Eco fallow – a reduced tillage system and plant diseases. *Plant Disease,* **64**, 31-35.

Fox, M.A. (1990) Changes in basal respiration and soil microbial biomass under LIFE management. *Aspects of Applied Biology,* **62**, 194-204.

Garrett, S.D. and Buddin,W. (1947) Control of take-all under the Chamberlain System of intensive barley growing. *Agriculture, London,* **54**, 425-426.

Greig, B.J.W. (1984) Management of East England pine plantations affected by *Helicobasidion annosum. European Journal of Forest Research,* **14** (7), 392-397.

Haas, J.H and Bolwyn, B. (1972) Ecology and epidemiology of sclerotinia wilt of white beans in Ontario. *Canadian Journal of Plant Science,* **52**, 525-533.

Harris, K.M. and Scott, P.R. (1989) *Crop Protection Information – an international perspective.* CAB International, Wallingford, UK.

Hims, M.J. (1979) Wild plants as a source of *Sclerotinia sclerotiorum* infecting oilseed rape. *Plant Pathology,* **28**, 197-198.

Hornby, D. (1998) *Take-all Disease of Cereals – a Regional Perspective.* CAB International, Wallingford, UK.

Htay, K. and Kerr, A. (1974) Biological control of Crown Gall: seed and root inoculation. *Journal of Applied Bacteriology,* **37**, 525-530.

Huang, H.C. (1980) Control of *Sclerotinia* wilt of sunflower by hyperparasites. *Canadian Journal of Plant Pathology,* **2,** 26-32.

Jenkins, J.E.E. and Storey, I. (1975) Influence of spray timing for the control of powdery mildew on the yield of spring barley. *Plant Pathology,* **24**, 125-134.

Johansson, P.M., Johnsson, L. and Gerhardson, B. (2003) Suppression of wheat seedling diseases caused by *Fusarium culmorum* and *Microdochium nivale* using bacterial seed treatment. *Plant Pathology,* **52**, 219-227.

Johnsson, L. (1990) Survival of common bunt (*Tilletia tritici* (DC) Tul.) in soil and manure. *Zeitschrift fur Pflanzenkrankheiten und Pflanzenschultz,* **97**, 502-507.

Jordan, V.W.L., Stinchcombe, G.R. and Hutcheon, J.A. (1989) Fungicide and nitrogen applications in relation to the improvement of disease control and yield of winter barley *Plant Pathology,* **38**, 26-34.

Katan, J. (1994) Soil solarisation, in *13th International Congress on Plastics in Agriculture, Proceedings of a conference held in Verona, Italy 8-11 March 1994, Volume* 1, Ente Antonomo Fiere di Verona, Verona, 8 pp.

Kirkegaard, J. (1996) *In vitro* suppression of fungal root pathogens of cereals by brassica tissue. *Plant Pathology,* **45**, 593-603.

Kirkegaard, J., Gardner, P.A., Angus, J.F. and Koetz, E. (1994) Effect of brassica crops on the growth and yield of wheat. *Australian Journal of Agricultural Research,* **45**, 529-545.

Knudsen, I.M.B., Hockenhull, D.F. and Jensen, D.F. (1995) Biocontrol of seedling diseases of barley and wheat caused by *Fusarium culmorum* and *Bipolaris sorokiniana*: effects of selected fungal antogonists on growth and yield components. *Plant Pathology,* **44**, 467-477.

Leitch, M.H. and Jenkins, P.D. (1995) Influence of nitrogen on the development of *Septoria tritici* epidemics in winter wheat. *Journal of Agricultural Science,* **124**, 361-368.

Lovell, D.J. and Parker, S.R. (1997) Influence of crop growth and structure on the risk of epidemics of *Mycophaerella graminicola (Septoria tritici)* in winter wheat. *Plant Pathology,* **46**, 126-138.

Lovell, DJ., Parker, SR., Hunter, T. *et al.* (1997) Influence of crop growth and canopy structure on the risk of epidemics by *Mycosphaerella graminicola* (*Septoria tritici*) in winter wheat. *Plant Pathology,* **46**, 126-138.

Nilsson, H.E. and Drew Smith, J. (1981) Take-all of grasses, in *Biology and Control of Take-all,* (eds M.J.C. Asher and P. Shipton), Academic Press, London, pp. 433-448.

Nuttall, M. and Stevens, D.B. (1991) Winter wheat nitrogen and fungicide interaction in breadmaking wheat. Experiment A - 1988-90, in *83rd Annual Report,* Morley Research Centre, pp. 81-95.

Oxley S.J.P and Cockerell, V. (1996) Transmission of bunt (*Tilletia caries*) in winter wheat with reference to seed-borne inoculum, site and climate. *Proceedings of Crop Protection in Northern Britain,* 1996, pp. 97-102.
Percival, G.C., Karim, M.S, and Dixon, G.R. (1998) Influence of light-enhanced glycoalkaloids on resistance of potato tubers to *Fusarium sulphureum* and *Fusarium solani* var. *coeruleum. Plant Pathology,* **47**, 665-670.
Prew, R.D. and Beane, J. (1986) Some factors affecting the growth and yield of winter wheat grown as a third cereal with much or negligible take-all. *Journal of Agriculture Science, Cambridge,* **107**, 639-671.
Priestley, R.H. (1981) Choice and deployment of resistant cultivars for cereal disease control, in *Strategies for the Control of Cereal Disease,* (eds J.F. Jenkyn and R.T. Plumb), Blackwell Scientific Publications, Oxford, 65-72.
Rodgers-Gray, B.S. and Shaw, M. (2000) Substantial reductions in winter wheat disease caused by addition of straw but not manure to soil. *Plant Pathology,* **49**, 590-599.
Scott, P.R. and Hollins, T.W. (1974) Effects of eyespot on the yield of winter wheat. *Annals of Applied Biology* **78**, 269-279.
Sandford, G.B. (1926) Factors affecting the pathogenicity of *A. scabies*. *Phytopathology,* **6**, 125-147.
Scott, P.R., Sanderson, P.R. and Benedikz, P.W. (1988) Occurrence of *Mycosphaerella graminicola,* teleomorph of *Septoria tritici,* on wheat debris in the UK. *Plant Pathology,* **3**, 285-290.
Shaw, M.W. and Royle, D.J. (1989) Airborne inoculum as a major source of *Septoria tritici (Mycosphaerella graminicola)* infections in winter wheat crops in the UK. *Plant Pathology,* **38**, 35-48.
Shaw, M.W. (1991) Interacting effects of interrupted humid periods and light on infection of wheat leaves by *Mycosphaerella graminicola (Septoria tritici)*. *Plant Pathology,* **40**, 595-607.
Steadman, J.R. (1983) White mould – a serious yield limiting disease of bean. *Plant Disease,* **67**, 346-350.
Stopes, C., Pearce, B., Clarkson, J. and Kennedy, R. (2000) What is the scope for using organic acceptable biocides in organic plant production. *BCPC Conference Pests & Diseases* **3**, 177-182.
Summerell, B.A. and Burgess, L.W. (1989) Factors influencing survival of *Pyrenophora tritici-repentis* stubble management. *Mycological Research,* **93**, 38-40.
Tjamos, E.C. and Faradis, B. (1980) Control of soilborne pathogens by solar heating in plastic houses. In: *Proceedings of the Fifth Congress of the Mediterranean Phytopathological Union,* Patras, Greece, 82-84.
Thomas, M.R., Cook, R.J. and King, J.E. (1988) Factors affecting development of *Septoria tritici* in winter wheat and its effect on yield. *Plant Pathology,* **38**, 246-257.
Tottman, D.R. (1987) The decimal code for the growth stages of cereals, with illustrations. (Drawings by Hilary Broad). *Annals of Applied Biology,* **110**, 441-454.
Utkhede, R.S. and Smith, E.M. (1995) Effect of nitrogen form and application method on incidence and severity of phytophthora crown rot of apple trees. *European Journal of Plant Pathology,* **101**, 283-289.
Van der Plank, JE. (1960). Analysis of epidemics, in *Plant Pathology,* Vol. III, (eds J.G. Horsfall and A.E. Dimmond), Academic Press, London, New York, pp. 230-287.
Webster, R.K., Wick, C.M. and Brandon, D.M. (1981) Epidemiology of stem rot of rice: effects of burning vs soil incorporation of rice residue. *Hilgardia,* **49**, 1-12.
White, N.A. (1954) The use of decoy crops in the eradication of certain soil-borne plant diseases *Australian Journal of Science,* **17**, 18-19.
Wolfe, M.S., Barrett, J.A. and Jenkins, J.E.E. (1981) The use of cultivar mixtures for disease control, in *Strategies for the Control of Cereal Disease,* (eds J.F. Jenkyn and R.T. Plumb), Blackwell Scientific Publications, Oxford, 73-80.
Yarham, D.J. (1981) Practical aspects of epidemics and controls, in *Biology and Control of Take-all,* (eds M.J.C. Asher and P. Shipton), Academic Press, London, pp. 417-432.
Yarham, D.J. and Gladders, P. (1994) The influence of set-aside on crop diseases. *Aspects of Applied Biology,* **40**, 277-284.
Yarham, D.J. and Norton, J. (1981) Effects of cultivation methods on disease, in *Strategies for the Control of Cereal Disease,* (eds J.F. Jenkyn and R.T. Plumb), Blackwell Scientific Publications, Oxford, pp. 157-188.
Yarham, D.J. and Pye, D. (1969) Winter barley as a source of mildew for spring barley crops. *Proceedings of the British Insecticide and Fungicide Conference,* 24-33.

CHAPTER 12

INFORMATION TECHNOLOGY IN PLANT DISEASE EPIDEMIOLOGY

ADRIAN NEWTON, NEIL McROBERTS AND GARETH HUGHES

12.1 INTRODUCTION

> "The fundamental problem of communication is that of reproducing at one point either exactly or approximately a message selected at another point." (Shannon, 1948).

Plant disease epidemiology is a discipline that is very often concerned with communicating a message: When is a crop at risk from disease? What are the spatio-temporal dynamics of a new disease? What is the optimum time to take action to limit epidemic development? What are the weather conditions that indicate a high risk of disease development? The answers to these sorts of questions (and many others) contain the messages that epidemiologists routinely attempt to communicate. In order to send meaningful messages epidemiologists are required frequently to deal with large volumes of data generated by, and describing, complex biophysical processes. Given this demand for communication of messages derived from complex processes it is not surprising that there is a long-standing interest in the use of Information Technology (IT) among epidemiologists; the first computer-based simulators of plant disease, for example, were produced in the late 1960s (Waggoner, 1968; Waggoner and Horsfall, 1969). As in many other disciplines, the complexity of use of IT in plant disease epidemiology has grown in step with the opportunity to add complexity offered by advances in personal computer hardware and software over the last 40 years. Whether the increase in available complexity has led to a general increase in the quality of information provided to farmers and growers remains a moot point (Clark, 2000; Henley, 2000). The question here is whether the increasingly sophisticated IT infrastructure available to plant disease epidemiologists has actually been focused on addressing Shannon's (1948) *fundamental problem*.

In this chapter we will discuss two areas of relevance in the application of IT to plant disease epidemiology: (1) data acquisition, organisation and retrieval; (2) the application of knowledge (derived from data) in constructing decision tools in epidemiology. Our discussion will concern the role that IT has played in overcoming, and sometimes contributing to, the fundamental problem as it occurs in plant disease epidemiology. We also look forward, in the light of lessons learned to date, and highlight some areas that require attention in on-going efforts to get the important messages of plant disease epidemiology from their points of origin to points of delivery. At a strategic level we examine some issues concerning the

B. M. Cooke, D. Gareth Jones and B. Kaye (eds.), The Epidemiology of Plant Diseases, 2nd edition, 335–356.

availability and accessibility of information relevant to epidemiologists in the light of an exponentially-increasing volume and diversity of data and data sources. In this discussion we consider both empirical data collected from experiments, surveys etc., and information in publications and databases which has already been processed to some extent. Much of what we say applies equally to both. At the more practical end of epidemiology, in forecasting disease (see also Chapter 9), we highlight recent uses of Bayesian methods to evaluate the performance of decision tools (modern examples of which often have a heavy reliance on IT infrastructure). One of the main advantages of these methods, as we will show, is that they allow epidemiologists to identify those diseases for which the potential exists to change grower behaviour by providing disease or disease risk forecasts. The unifying theme across these areas of discussion is the use and flow of information in plant disease epidemiology under uncertainty.

12.2 DEFINITION OF INFORMATION TECHNOLOGY IN PLANT DISEASE EPIDEMIOLOGY

The definition we have adopted is 'the use of computer-based technology to collate, process and disseminate information and knowledge for application to the study of plant disease epidemiology'. This includes existing technologies such as computer databases, statistical and field trial design packages, (e.g. Genstat, SAS, S-Plus, Agrobase), automated quantitative diagnostic systems, Laboratory Information Management Systems (LIMS) which may utilise bar-coding tools, and the newer ones referred to below, but the definition is broad enough to allow the inclusion of new developments not yet envisaged. Fig. 12.1 summarises the processes involved in the use of IT to develop epidemiological knowledge from data. Collation, processing and dissemination of information may operate separately or in an integrated manner depending upon the purpose. It is important to note that empirical data will consist of a mixture of facts (i.e. things that are true) and errors (i.e. things that are not true). The extent to which the information comprises only facts will depend on the quality of the mechanism by which the data are collected and the quality of the processing (if any) that is applied to the data to filter out errors. Some databases may simply collate information such as disease incidence or pathotype distribution. Both knowledge arising from the analysis of information (for example by statistical methods) and the information itself might be made available for practical use through some IT medium. In either case validation by successful use will ultimately separate those IT-based systems which are useful from those which are not. In theory, the process of validation-by-use should provide a means for further filtering of errors from facts and an improvement (if that is possible in the particular context) in the quality of available information and knowledge. Clearly, the content of databases and the way in which they are accessed have important consequences for the quality of information and knowledge that results from their use.

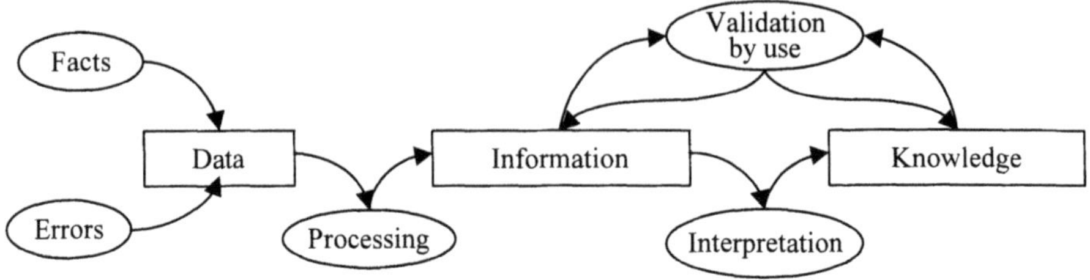

Figure 12.1. The process of knowledge development from data in plant disease epidemiology.

12.3 THE WORLD ACCORDING TO 'GOOGLE'

> "Knowledge is of two kinds. We know a subject ourselves, or we know where we can find information upon it. When we enquire into any subject, the first thing we have to do is to know what books have treated of it. This leads us to look at catalogues and the backs of books in libraries" (Boswell, 1960, p 558).

Increasingly, people approach acquiring information and knowledge by searching the internet. Search engines are powerful and valuable tools. However, like most powerful tools, they can be dangerous in inexperienced hands for the following reasons: (1) searching databases is a skill in which few people have training or experience, and requires an understanding of the database content and search software to be used efficiently (for reasons noted above); (2) the information in the search engine database is likely to be a biased sample of the information available, if for no other reason than because; (3) it is limited to that available on public access servers on the world-wide-web, and (4) there will be a variable level of quality control over on the information, ranging from none to excellent, but probably with little or no indication of what level of credence the user should give to the source. There are many more reasons why information obtained from web searches should be treated cautiously, but we hope that it is apparent that search engines, by themselves, are not a way to obtain an objective assessment of any subject. Perhaps most importantly, whether information is available from direct web searches will depend on a complex set of interacting factors which balance the institutional and personal objectives of the responsible scientist(s) for knowledge transfer with the precedence of peer-reviewed publication and the need for the costs of publication to be recovered.

Much epidemiologically valuable data is made available through the institutions of scientific publishing. Whilst the knowledge published in many epidemiology papers will have been paid for by public funding, and a scientist's usual aim is to disseminate the information free to as wide a constituency as possible, the publishing process has costs which must be met. Furthermore, the requirement to protect originality and precedence in the primary scientific literature leads to time-lags between data being collected and being made widely available. Of course, some of these restrictions are overcome by the use of subscription services for journal access and/or database access. However, even in cases where such access is available, users should bear in mind that reservations, (1) and (2), about public access databases still apply, and further that: (3) the information will still be limited but this time to specific sources, and (4) data quality is defined by the data source.

To epidemiologists, available on-line databases of real-time and historical epidemiological information such as disease incidence, severity, control treatment, host genotype, crop history and other such details are of great value. Similarly, crop advisory information, linked efficiently with its data source and resulting recommendations allows real-world pictures of disease management to be re-constructed, linking theoretical epidemiology with practice. As with research data, crop survey and management databases can be either subscription-access resources or open access. The cautionary comments above notwithstanding, web searches will generate many useful starting points for epidemiologists looking for sources of data. For example, entering the terms 'crop, scouting, disease, weekly, data, status' into Google in early 2005 resulted in approximately 39,500 hits. Some selective searching through the first 100 links resulted in the following data sources that give an impression of the sorts of information that can be obtained. Readers are reminded that the sample is selective and intended for illustrative purposes only: www.nysipm.cornell.edu/lfc/index.html – Cornell University's IPM programme giving access to current and historical information on crop and disease status in the New York State area; www.aragriculture.org/news/cropreport/ – regular reports on crop and disease status published by the Arkansas Cooperative Extension Service; www.oardc.ohio-state.edu/wheat2004/ – web pages reporting results, including large tables of mean results for agronomic and disease resistance scores, of the wheat variety screening trials carried out by Ohio State University; www.sbrusa.net/ – the USDA web site for monitoring soybean rust risk giving up-to-date results from scouting bouts aimed at detecting this new (for the USA) disease. The bias towards USA sites, reflected in the small sample given here should be noted. We suggest two main reasons for this. First, the size and technological sophistication of the USA agricultural science infrastructure is likely to lead to a high representation of USA sites in the searchable population of sites. Second, in the particular context of publicly accessible information on plant disease, the Land Grant mission of relevant USA universities means that there is not the singular emphasis on peer-reviewed publication and impact factors that exists elsewhere, and a corresponding effort on other forms of knowledge transfer (such as web-sites) is made.

Crop Monitor – www.cropmonitor.co.uk/decisiontools/decision.cfm – is a regularly-updated crop disease reporting service for England and Wales which also links to online disease decision tools and risk forecasting systems. In the UK a levy raised on cereal crops is used to support research and knowledge transfer for the benefit of the industry. The levy body (Home Grown Cereals Authority, HGCA) web site has several useful open-access resources including a searchable database of variety performance data – www.hgca.com/varieties/rl-plus/index.html – and historical results from a UK-wide network of trials recording aspects of barley crop development www.hgca.com/BGS/. In Scotland, SAC provides access to these and other similar resources via its web site – www.sac.ac.uk/consultancy/cropclinic/clinic/Adoptacrop. These types of resources are typical of those available in many countries. Most web sites dedicated to plant pathology are run by national plant pathology societies listed by the International Society for Plant Pathology (ISPP) – www.isppweb.org – seeking to serve the needs of their members. These sites could, via the ISPP, serve as a focus for the integration of information technology resources in plant pathology. There are already efforts to provide integrated access to multiple data resources. One example is Plant Management Network International (PMNI) – www.plantmanagementnetwork.org – which is a co-operative body, involving industry, universities and professional bodies such as APS and the Canadian Phytopathological Society. The PMNI web site offers, among other things, the capacity to search a database, dedicated to crop science, constructed from the pooled resources of the partner organisations. Simple searches are free to all users; more detailed searching and full access to the site requires a subscription.

While using on-line search engines does not allow direct access to the primary epidemiology literature, it will, of course, provide the IT-aware epidemiologist with the web sites for journals where the literature can be found. In addition to the major international plant pathology research journals such as *Australasian Plant Pathology*, *European Journal of Plant Pathology*, *Journal of Phytopathology*, *Phytopathology*, *Plant Disease* and *Plant Pathology*, epidemiologists publish their work in a wide range of other journals too numerous to list here. On-line resources such as the ISI Web of Knowledge provide searchable databases of the primary literature allowing information to be retrieved by using keywords independently of the source of the publication. Having promoted the use of on-line databases, we raise the point (often made and often ignored) that they do not generally include information on literature that is more than, typically, 30 years old. Information dating from more than 30 years ago must be obtained by more manual searches, and in this context the final sentence of Samuel Johnson's words remains as true today as when Boswell noted them down in 1775.

For those who may wonder exactly what such a manual search involves, and to highlight the dispersed nature of information relevant to a particular topic, we offer the following quotation, taken from a review of the genetics of horizontal resistance carried out by the late Norman Simmonds (Simmonds, 1991):

> "...it will be appreciated that the literature search needed for the present review presented formidable problems. Some titles were available from reviews but my own reading showed that some valuable papers went uncited. Trials with abstracting journals were not encouraging because of the idiosyncratic and confused terminology adopted and, sometimes, even the apparent failure of authors to understand just what they had done. In the outcome I adopted what might be called a 'brute force' approach. I scanned runs of journals back to about 1965 (about 300 volumes in all) and trusted to secondary sources to identify papers I had failed to find. I might, with advantage, have scanned runs back to the 1930s, but with a declining probability of identifying useful works."

12.4 REAL WORLD DATA CAPTURE

The two crucial factors which initiated the explosion in use of IT are the wide availability of powerful personal computers at very low cost and the growth of the 'internet'. All the IT developments above require standard computers, normally connected to the internet for full operation. This is getting easier with small notebook/laptop computers, wireless connection to the internet, linkage through mobile phones, Personal Digital Assistants (PDAs) etc. Linkage with Global Positioning System (GPS) technology and Geographic Information Systems (GIS) which can assemble multi-layered facts, and associated software such as 'FarmWorks', provide easier access to on-line data in the field allowing synthesis of historical data and current data in real-time rather than relying on going back to an office. Linked to these developments there are now many electronic aids for recording field data automatically with greater speed, volume or accuracy. Examples of instruments include:

- Multi spectral scans (e.g. GreenSeeker and Cropscan) and continuous spectrum scans.
- Remote sensing, e.g. rapid estimation of crop areas, rapid disease level assessment over large areas by satellite spectral/x-ray/infrared etc. analysis.
- Weather stations (for forecasting).
- Photosynthetic measurement systems.
- Digital plant canopy imager and other portable area meters.
- Discriminatory image analysis, e.g. lesion/colony area, space fill.
- Weighing and moisture measurements on harvesters.

In general, digital data capture and storage is now routine. Furthermore, software standardisation on Microsoft® compatible or Open Source formats (and improved transparent translation between these formats) has also increased the speed of integration. The use of remote sensing in phytopathometry is discussed in Chapter 2.

12.5 INFORMATION ACCUMULATION OR DISSEMINATION?

The unplanned use of IT presents the same hazards as the misuse of other technologies. Powerful IT has the potential to allow users to do more than was previously possible. However, this is not a good thing if we do more of the wrong sorts of things. In particular, the ease with which information gets stored rather than being read, analysed, interpreted and then put to further use is a frequent problem associated with easy access to advanced IT facilities, as is unnecessary use of graphics 'because they are there'. Good information systems should always allow easy initial sorting of data into 'discard', 'store for later appraisal' or 'analyse or make full appraisal' categories. Information in the latter category requires more to be revealed for further assessment. However, information classified as 'store for later appraisal' needs to be very carefully labelled, sorted and stored. One point of view is that developments in IT have made it too easy to collect and store data which are then put to no use. Our contention is that this state of affairs is not an argument for not collecting the data, but rather an argument for improving the quality and use of meta-data (i.e. data about data – keywords are a widely used example of meta-data) so that stored data can be easily retrieved and used if the need arises in the future. Particularly in the case of field data, the world does not allow us the chance to make exact repetitions of conditions so the data storage capacity of modern IT should be exploited to allow us to 'go back in time' whenever the need arises. However, we stress that the value in adopting such an approach depends on the quality of the meta-data which describe the contents of the archives of primary data, in much the same way that the quality of databases of publications depends on the quality of the indexing and searching tools which are employed in their construction and interrogation.

Earlier comments notwithstanding, the power of IT for data storage and recovery in the context of publications can lead to the impression that academic publication libraries where hard-copy journals are classified and filed on shelves are obsolete. Hard copy journals are available on-line but retain their page format using 'Portable Document Format' (.pdf) files and must be accessed by viewers such as 'Acrobat'. Ideally, the advent of on-line publishing should have enhanced the value of publications considerably, by allowing much greater access to: libraries of colour graphics; data generated on demand; interactive software tools; active forward and backward links to other information and databases; video images or even sound. However, the vested interests and established system of the big publishing houses simply means more colour images, easier on-line access, and occasionally additional information on the publisher's web site. Indeed on-line publishing rarely even results in faster publication times, which could be from 5 to 9 months less than for 'conventional' journals since individual papers could be available almost immediately they are accepted and formatted. The British Society for Plant Pathology (BSPP) does publish 'New and Unusual Records' of plant diseases on-line as soon as they are accepted, followed by annual hard-copy publication in Plant Pathology. A few other journals are doing likewise. However, the best 'deals' often involve site licences which have to be organised more formally, such as for access to the major bibliographic databases.

Forward citations require the journal to regularly update from citation indices that refer to the article, thereby positioning the publication within a network of information allowing highly flexible citation. Furthermore, software tools can be applied to data published in electronic format which enable each reader to read the paper effectively in a different way by applying further data analyses, informed by more recent data than the author had access to at the time of writing. The use of such tools might lead to cultural changes in publishing allowing the concept of original publication to be preserved, but enabling rapid re-interpretation in the light of more recent data. Raw, or partly processed, data are rarely attached to published articles except theses, but the new technologies open up opportunities to do so in appropriate cases. This would allow further use of the data to the benefit of the original and other authors.

Will textbooks (like this one) be obsolete soon? Should there be a new type of 'electronic text book'? (by 'text book' we mean an assimilating tool giving access to a subject from its fundamental principles; dict: 'manual of instruction, standard book in a branch of study'). Take, for example, someone requiring information on one of the best known plant pathogens, *Phytophthora infestans*, the cause of epidemics of potato late blight. Potential sources of information and resources of interest may be: books (like this), scientific papers, bulletins, crop intelligence reports, conferences, published statistics and surveys, agrochemical control measures available, practical and 'anecdotal' information from agricultural advisors, agricultural policy, germplasm collections of potato, its wild relatives and *P. infestans* and its relatives etc., genetic maps/databases of host and pathogen such as the EUCABLIGHT consortium database of characterised isolates across Europe (www.eucablight.org) scientists working in the field or current work being carried out on research grants. Clearly, to make optimal use of this diversity of information there is the need for a structured information system whereby the various sources can be linked together in a logical and helpful format. However, this demands 'good old-fashioned' experts to invest time and effort to create such resources and this means that the economic case for such resource investments has to be made.

12.6 BRINGING TOGETHER DISCIPLINES

Although the examples above largely bring together knowledge within a discipline, the construction of such web sites involves many challenges. More challenging still, and another dimension where the attributes of IT can be effectively exploited in epidemiology, is in bringing together related, disparate, but relevant, disciplines and making their resources accessible to new users. For example, an important aspect of understanding the epidemiology of plant-microbe interactions is to understand how abiotic stress caused by climate, e.g. drought and heat stresses, can affect the susceptibility of a plant to infection. A web resource called DRASTIC (Database Resource for the Analysis of Signal Transduction in Cells; www.drastic.org.uk) may not at first seem like an epidemiologist's favourite site. Likewise, one of its main resources, a database of plant gene expression data which provides valuable information on the potential interaction between biotic and abiotic stresses, may be difficult at first to relate to epidemiology. However, DRASTIC does not simply provide data from microarrays. The database contains information from a wide range of published papers on whether plant genes are up- or down-regulated in response to various biotic and abiotic stresses. Much of the information is based on experiments with the model plant *Arabidopsis thaliana* and because the database uses Arabidopsis Genome Initiative (AGI) numbers it is possible to be confident about which gene within a family of genes is actually being regulated.

On the DRASTIC web site is a good example of how new knowledge can be obtained by assembling data from disparate sources visually, namely the 'Metabolic pathways of diseased potato' wallchart. Although not an epidemiological example, this was assembled to assist in understanding the complex biochemical changes that can take place in diseased potatoes, but as a wallchart it is fixed both spatially and temporally. Whilst the chart itself is useful it still only touches the surface of making the information within it readily accessible and more importantly, more easily understood by non-specialists. To exploit the power of IT, ideally the information it contains should be stored in a database and drawn on demand from a set of queries. The graphics should not only demonstrate in an accessible and attractive style associations not hitherto observed, but also allow further interaction through an intuitive interface. This is still technologically challenging but becoming practical.

The DRASTIC database contains many examples of *Arabidopsis* genes that are regulated by both abiotic stresses (drought, cold etc.) as well as by infection with a range of fungal, bacterial and virus pathogens. It thereby provides valuable clues as to the molecular basis of how such interactions might occur. The site also brings together information on resistance-inducing compounds that are, or could be, exploited as crop protectants. Current developments centre around text mining as a very powerful tool capable of finding and relating documents on specific subjects to bring together ideas and information from diverse sources in answer to a specific query. We predict that the use of such tools will increase over the next decade as the drive to synthesise diverse sources of information increases. Of course, powerful as such software is, it can only work on data that have been placed on a computer somewhere. Not all of the issues raised by Simmonds (1991) will disappear with new software tools for data mining and management. However, the challenges in the future will be to understand how to construct useful queries and how complex information can be presented in an easily understood format. Indeed, the task of simplifying the presentation of complex information is at the heart of many problems related to biology in general, and epidemiology in particular and is one of the drivers of the development of simple forecasting tools described below, and of the use of methods for the evaluation of those tools.

Making knowledge useful requires synthesis and presentation. For example, the wheat breeding community have done this for marker-assisted breeding tools (maswheat.ucdavis.edu) and a similar site is planned for barley. Another example of more epidemiological relevance is the Cereal Pathogen Resistance Allele Database (CPRAD – www.scri.ac.uk/cprad). This catalogues information on the pathogen resistance alleles reported to be found in genotypes of barley and, in future, wheat and oats. It is fully searchable on a range of criteria and initial query results are displayed in a table. From this table there are links to the source reference or to a pedigree chart with all that is known about the inheritance of the selected allele. It can also show all that is known about all pathogen resistance alleles in any of the genotypes in the pedigree chart. The selected genotype is hypertext linked to display the probabilities that these and other alleles present in the pedigrees are expressed in the genotype. In this way, probabilities that these genes are present in the selected variety but are masked by other genes, for example the *mlo* allele in barley, are displayed. When this information is associated with virulence frequency and virulence combination (race) data, predictions about the likely durability of a variety can be made.

12.7 MODELS, EXPERT SYSTEMS AND DECISION SUPPORT SYSTEMS

There are many Decision Support Systems (DSSs) or software products that integrate diverse information into a decision-making process. The Management Advisory Package for Potato (MAPP: www.potatomapp.co.uk), for example, is a software package that advises growers on seed rates, irrigation strategies and harvest times in order to optimise financial returns based on graded yields. Other examples are ‘PC-Plant Protection’ in Denmark, ‘Plant Diagnosis’, ‘MoreCrop’, ‘Orchard

2000' and 'California Pestcast'. There are also many plant and disease models for canopy and root architecture, virtual plants, and virtual crops. Probabilistic systems can mix elements from different systems, and there are many models built to aid understanding of epidemics. Older established examples include Epimul, Epipre (temporal data) and resistan (fungicide resistance dynamics). There are also many disease/pest risk assessment programmes and libraries of electronic images to aid identification (CABI, 'CyberPest' etc.). There are computer-based systems for identification of organisms that use multi-access keys, for example the Biolog bacterial identification system. Many statistical analyses require IT visualisation tools before an understanding of the processes can be gained. For example, techniques such as wavelet transform, degrade or enhance images to define significant features. There are many cladistics/phylogenetic packages such as Fillip, Genstat routines and Bionumerics. Similarly, there are many mapping packages which allow the use of geostatistical methods for interpolation of data from sparse samples and compositor indexing of GIS data by, for example, soil moisture, elevation, phosphorous or nitrogen.

Some resources are freely available as teaching aids. For example 'LateBlight' or 'Blitecast' which enable the effects of the major epidemiological parameters to be simply simulated and represented graphically, demonstrating the principles of epidemiology, interactively and effectively (Fry, 1990). 'DISTRAIN' is a programme for teaching people to assess disease accurately by asking them to estimate the percentage cover then giving feedback (Tomerlin and Howell, 1988; see also Chapter 2).

Decision support systems highlight a problem with many IT tools. As a knowledge delivery mechanism they can be: inflexible, prone to go rapidly out-of-date, expensive, and often insufficiently user-friendly, and therefore frequently fail to deliver effective solutions or live up to users' expectations (see below). Essentially, developers of IT tools face a choice between one of two general aims in developing a new tool: (1) they can opt for multi-objective tools which will integrate large amounts of information, relevant to one or more important management decisions, and act as comprehensive sources of information (traditional DSSs would fall into this category); while such tools provide the user with potentially a single integrated resource for helping with decision-making, they generally have high development and maintenance requirements, for example, to keep lists of approved pesticide products, application rates and costs up to date, (2) alternatively, they can opt for simple tools which focus on individual decisions and utilise only (or mostly) simple generic information; examples of this approach are disease risk predictors based on statistical decision rules (see below, also Yuen and Hughes, 2002). The pros and cons of this option are, roughly speaking, the inverse of those for option (1). Decision tools of type (2) tend to provide less in the way of background or supporting information, but have lower development and maintenance requirements and ought to be simpler to use.

One aspect of integrating IT into the decision-making process that is sometimes overlooked is that decisions are often made collectively by more than one person, or after the main decision-maker has consulted several other people. These people might collectively be termed the decision-maker's Decision Support Network (DSN). As an illustration of the diversity of the constituents of a DSN, McRoberts *et al.* (2000) found that up to eight different types of person (spouse, business partner, children, accountant, advisor, commercial representative, lawyer, employee) were consulted *occasionally* by a group of Scottish arable farmers when making crop protection decisions. No farmer who provided information for the study *never* consulted anyone when making a crop protection decision. The three most frequently reported types of person who were *always* consulted were: advisors (36% of respondents), commercial representatives (24%) and business partners (22%). Whether a new IT-based tool is of type (1) or type (2) it is likely to have to compliment, rather than replace, the activities of the DSN. Within the context of the DSN, DSSs might be better thought of as Discussion Support Systems. On this basis future developments might most effectively be targeted on type (2) tools which are intuitive to use and which focus on the key decision to be made, leaving the DSN to supply the additional, background information often contained in a DSS. These comments notwithstanding, there are, of course, successful examples of type (1) tools in use. We briefly describe some of these in the next section, before dealing with some generic issues of the efficacy of tools, using examples of type (2) tools in the final section.

12.8 SOME EXAMPLES OF DSS

So far in this chapter we have dealt in broad terms with sources and presentation of information relevant to plant disease epidemiology. In attempting to raise some issues concerning the availability and quality of information that can be obtained directly on the desktop of the modern epidemiologist we have largely concentrated on sources of information of use to epidemiologists and on the use of IT simply in making information available. In this section we report some proven or historical examples of IT and DSSs. In the following section we consider recent developments in epidemiology for situations where IT is used in attempts to put epidemiology into practice for the purpose of forecasting disease and making crop management decisions. Disease forecasting is also dealt with in Chapter 9.

In Denmark a DSS for crop protection called 'PC-Plant Protection' was devised motivated by a decision made in 1986 to reduce pesticide use in Denmark by 50%. The Danish Institute of Plant and Soil Science and the Danish Agricultural Advisory Centre chose to implement their research findings with a detailed use of threshold values to support decisions on treatment need, choice of pesticides and the appropriate dosage for the actual problem using a PC programme. This is not available on the internet as it is licensed to users. Even as far back as 1996, 2800 licences were in use at agricultural schools and with advisers and farmers. It was judged to be user-friendly and the model was considered to be reliable and to meet the requirements of advisers. In 1996, a farm survey of 488 farmers who had used the system in 1995 showed that the system has been well accepted by farmers, not only because of its reliable recommendations but also because of the good profitability that resulted from its use. It also enabled Denmark to make substantial progress in pesticide input reductions. Development continued with a weather module, taking into account detailed weather information and an environmental module, giving farmers the opportunity to choose pesticides according to their environmental risk, integration of the DSS with site-specific (precision) farming and multimedia presentation of biological and pesticide information.

The same developers, have produced an internet-based platform for the dissemination of up-to-date crop protection information (www.planteinfo.dk). This aims to disseminate new and reliable information on the development of pests and diseases in crops during the growing season and is of great value for farmers and advisory services. Up-to-date calculated risks are based on weather data for *Septoria* species on winter wheat, *Drechslera teres* and *Rhynchosporium secalis* on barley, *Oscinella frit*, *Dasineura brassica*, *Phytophthora infestans* and records available from national pest and disease surveys of all major pests and diseases on winter wheat, winter barley and spring barley. Information can be generated automatically and illustrative maps can be presented such as weather-based risk calculations on maps of Denmark divided into 40 x 40 km grids. The grids are coloured according to the calculated risk: green if no risk, yellow if close to risk, and red if at risk. In each grid a comparison with the developments in previous years can be retrieved in a sub-layer. Linked to the maps is further information on the pathogen or pest and comments from crop protection specialists.

Other examples include apple scab forecasting, a component of 'ADEM', a user-friendly programme for growers and advisors developed by HRI at East Malling, Kent, UK. It also models mildew, canker and fruit rot and fireblight, integrating weather data into simultaneous simulations. Cultivar-specific effects are included and it has been used to achieve mildew and scab control levels similar to those with conventional fungicide programmes but with less fungicide application.

Two products primarily aimed at researchers are the Descriptions of Plant Viruses (www.rothamsted.ac.uk/dpv), formerly 'Plant Virus Notebook', which displays the classification of plant virus genera and families as well as genome structure and organization. The features of virus sequences can be displayed in different ways, graphically or textually, so that the positions and relationships of the features can be examined more easily. A complete plant virus classification is included, together with sample genome maps of the type members of each genus (where available). A related plant virus IT resource is the VIDE (Virus Identification Data Exchange – image.fs.uidaho.edu/vide/refs.htm – database).

12.9 DISEASE FORECASTING AND DECISION MAKING IN AN INFORMATION THEORY FRAMEWORK

Figure 1 in Shannon (1948) presents the fundamental problem of communication in diagrammatic form. Shannon's (1948) figure is reproduced, with adaptations, as Fig. 12.2 here. Shannon (1948) was primarily concerned with the mechanical aspects of communication (although he did explicitly state that "*The* destination *is the person (or thing) for which the message is intended*"). We have increased the emphasis on the human-mechanical interaction implied in his Figure 1 here to highlight the human elements in communication. Fig. 12.2 shows half of what is now generally called the Knowledge Transfer (KT) cycle. The missing half would mirror the first by illustrating the flow of information back from the receiver to the information source. The part of the KT cycle which we have not shown is often concerned with defining a problem which is to be solved and thus defining what the message is to be in the part which we have shown. For example, the problem may be a practical difficulty in knowing when to take action to control disease and the desired message is information/advice that resolves the problem. Clearly, it is possible for uncertainty (or noise) to enter the cycle of problem definition, message formulation and delivery at several points and from several sources. We have indicated some of these in Fig. 12.2, particularly the role that those acting as the Information Source and Destination play. As will become clear in the following discussion, the human element and uncertainty are important factors in the success of any IT-based forecasting system.

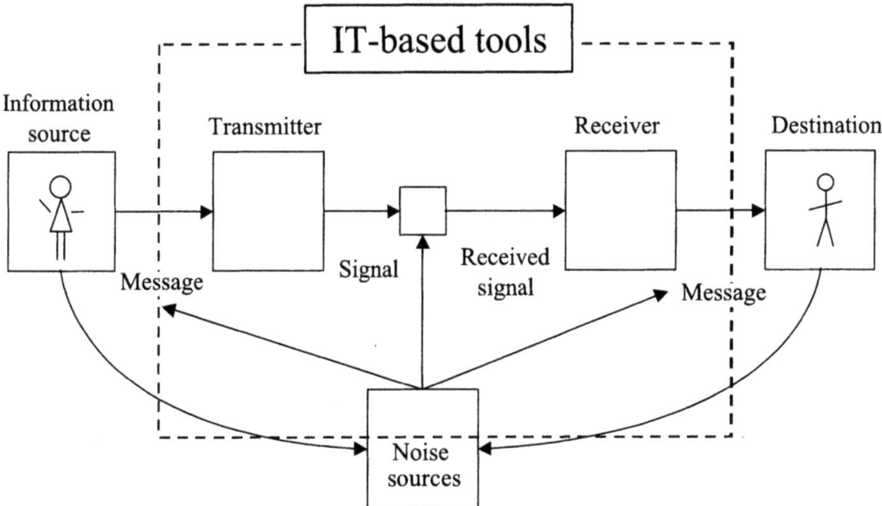

Figure 12.2. An adaptation of Shannon's (1948) schematic diagram of the process of communication illustrating the steps during which uncertainty might arise.

For practical purposes, a forecasting system must do two jobs. First, it must make a useful forecast of disease outbreaks. Secondly, it must present the results of the forecast in a useful format. Early in the development of IT-based forecasters the primary objective was most often to develop tools that would give growers warnings of disease outbreaks so that preventative action could be taken. More recently, concern over the damaging effects of overuse of pesticides on the environment, and a steady reduction in the profitability of crop production, have seen the avoidance of unnecessary pesticide applications become an important objective for disease forecasters. Irrespective of whether the aim of a forecasting system is to provide warnings of disease outbreaks (what might be called positive predictions) or assurances that outbreaks are unlikely (what might, correspondingly, be called negative predictions) the basic performance criterion that any system should meet was spelled out by Campbell and Madden (1990):

> "For a forecasting system to be successful, it must be adopted and implemented by growers. There must be the **perception** that the grower can realize specific, tangible benefits from using the forecasting system **that could not be realized in its absence**." (Campbell and Madden, 1990, p 424; our emphasis).

The quotation from Campbell and Madden (1990) highlights the important relationship that any forecasting system (IT-based or not) must have with the existing knowledge and decision making system of its intended users. The basic point underlying Campbell and Madden's (1990) perceptive statement is that in order to be useful a forecasting system must either: (1) provide new information to the user, or (2) extract previously unrecognized information from already available knowledge and data. Tools developed with these aims in mind might be better described as judgement-assistance or risk assessment tools than DSSs. The objectives for such tools would prioritise user attitudes and preferences, and therefore increase likelihood of not only use, but also feedback, leading to further improvement through active participation of the users in development. Such tools guide growers to a solution, which takes account of the conclusions from the research but which also makes use of growers' experience and their perceptions of risk. It is worth noting the obvious (but sometimes overlooked) point that the combination of objectives (1) and (2) above, constitutes an exhaustive summary of what it is possible for any forecasting system, constructed from empirical data and mathematical, statistical and logical relationships, to do; no such system can contain (or deliver) more information than the content of its components (Doucet and Sloep, 1992). Medawar (1972) made the same point more generally: *"No process of reasoning whatsoever can, **with logical certainty**, enlarge the empirical content of the statements out of which it issues*" (our emphasis).

This epistemic limitation of forecasters, tools, or DSSs holds, no matter how sophisticated the IT infrastructure is that is used in their development or deployment. Medawar's comment is particularly worth bearing in mind when the role of the IT-based forecaster or DSS in disease management is considered in more detail. In addition to the obvious first point (concerning the empirical limitation to the knowledge content of forecasters) it also makes a more subtle point about forecasting. Forecasting entails the loss of logical certainty because forecasts always seek to make statements beyond the empirical content of the information on which they are based. It is thus logically unavoidable that predictions are uncertain, even if we construct deterministic predictors of disease outbreaks that ignore that uncertainty. In the terminology of logical analysis we can say that forecasters are tools for *inductive* reasoning.

However, acceptance of the inductive nature of forecasts and forecasters offers the possibility to evaluate them within a coherent analytical framework using Bayesian analysis within an information theoretical framework. This approach leads to a simple and quantitative assessment of the conditions under which any given forecaster or DSS is likely to meet Campbell and Madden's (1990) '*success*' criterion. (i.e., the extent to which a grower is likely to perceive it as offering something tangibly informative). Evaluation of forecasters in this context is likely to result in more efficient use of IT. It could prevent pointless spending on the development of IT tools which will be useful only to a small proportion of the intended user-base or on tools for problems which cannot usefully be addressed through provision of forecasting systems or DSSs. In this way it might allow plant disease epidemiologists to get closer to solving Shannon's (1948) fundamental problem. The approach also explicitly includes users' perceptions within its analytical scope and thus offers some potential for uniting the technical and human elements of message delivery into one package.

The basic component of the analytical approach – the application of Bayes's theorem to calculate conditional probabilities of events (Howson and Urbach, 1989) – is widely used in medical diagnosis and decision-making. It has been discussed in a plant disease epidemiology context by Yuen and Hughes (2002), McRoberts *et al.* (2003), Yuen and Mila (2003) and Madden (2006). Bayes's theorem provides the means to calculate how the perceived probability of an event (say, the occurrence of a disease outbreak) will change in the light of new information (coming from a forecaster, for example). One version of Bayes's theorem useful in the evaluation of decision tools for control of plant disease, when what we have is an initial assessment of the need for action, is shown in equation 12.1:

$$\textit{Posterior odds}(D+ \textit{(given prediction)}) = LR_x \cdot \textit{Prior odds}(D+) \qquad (12.1)$$

Before describing the terms in equation 12.1, readers are reminded that the odds of an event, E, [odds(E)], is calculated as [P(E)/(1-P(E))]; i.e. the probability that the event occurs divided by the probability that it does not. From this it follows that the probability of an event can be found from the odds: P(E) = [odds(E)/(1+odds(E))]. Equation 12.1, then, states that the odds of need for action, D+ given the prediction, obtained by use of a forecaster with a known LR_x (where x is either + or -), is the product of LR_x and the odds(D+) before the forecast was made (based on whatever information was already available to the decision maker).

The ability of any forecasting system to discriminate between situations where action is needed, or not, can be captured in the likelihood ratio (LR) for the event in question. The LR can be calculated from data describing the frequency with which the forecaster makes correct and incorrect predictions (either positive or negative). This is most easily shown for the case where there is a simple binary choice (action needed or not, for example) and an equivalent binary outcome from the forecaster (prediction of need for action or no need for action). In this case a 2 x 2 decision table gives an exhaustive summary of the results from a set of situations in which forecasts are required and obtained. The frequencies of the four possible outcomes are shown in Table 12.1.

Table 12.1. Possible outcomes of applying a forecaster in a set of situations in which action to control disease is either needed or not needed

	Action actually needed (cases)	*Action actually not needed (controls)*
Forecaster predicts need (+)	a	b
Forecaster predicts no need (-)	c	d

The **sensitivity** of the forecaster is defined as a/(a+c); that is, the frequency of true positive predictions divided by the total number of situations where action was actually needed (so called 'cases'). The **specificity** of the forecaster is defined as d/(b+d); that is, the frequency of true negative predictions divided by the total number of situations in which action was not needed (so called 'controls'). The likelihood ratio for a positive prediction (LR_+) is defined as the sensitivity/[1–specificity] (*i.e.* the true positive proportion divided by the false positive proportion). The likelihood ratio for a negative decision (LR_-) is defined as [1–sensitivity]/specificity (*i.e.* the false negative proportion divided by the true negative proportion). In order to construct Table 12.1 it is necessary to have an independent means of distinguishing cases from controls; i.e., the forecaster must be assessed against an independent 'gold standard'. Murtaugh (1996) provides a useful discussion of this issue in the wider context of ecological indicators. The version of Bayes's theorem given in equation 12.1 allows the LR_x for a forecaster to be combined with an initial assessment of the need for action (the 'prior probability') to produce an updated assessment (the 'posterior probability').

In practical applications of Bayes's theorem in the evaluation of plant disease forecasters, the prior odds have often been based on the long-term prevalence of known cases, and decision makers' judgements of risk are accommodated by specifying different criteria for allocating individual tests to the categories of positive or negative predictions (Hughes *et al.*, 1999; Yuen and Hughes, 2002). However, we note that the prior and posterior odds to which equation 12.1 refers may be based on a variety of formal and informal pieces of information and will, in almost every practical situation involving a real decision maker, contain some element of subjective judgement. Indeed, Howson and Urbach (1989; p39) argue that all probabilities (and hence odds) "*should be understood as subjective assessments of credibility, regulated by the requirement that they be overall consistent*". Some scientists may feel uncomfortable with the idea of allowing a definition of probability which uses the word 'subjective' but it should not be a difficult definition to accept for situations in which people are asked to make estimates of probabilities (or odds) without recourse to formal calculation. It should not, therefore, be too difficult to accept as the relevant description of probability for the context of the current discussion, where decision makers are using and assessing IT-based disease forecasters. We note that this view of probabilities as subjective judgements is in keeping with Campbell and Madden's (1990) comment on the importance of the *perceptions* of growers about the usefulness of forecasters in their adoption.

Consider the potential applications of a forecaster for a grower who has already obtained an initial assessment of the need for action. A positive prediction ought to increase the chance that action will be needed. In this case, LR_+ will be the relevant indicator of performance. A negative prediction ought to decrease the chance that action will be needed, in which case LR_- is of interest. It is not a reasonable expectation that any forecaster will give perfect performance. Equation 12.1 allows the limits for what should be expected from forecasters (given an initial assessment of the need for action) to be calculated for either positive or negative predictions (Yuen and Hughes, 2002; Yuen and Mila, 2003).

The following illustrative discussion combines related examples presented in Yuen and Hughes (2002), McRoberts *et al.* (2003) and Yuen and Mila (2003). Imagine a forecaster that has both specificity and sensitivity equal to 0.9. That is, it correctly distinguishes true positives from false negatives 90% of the time (or, in other words, only one in ten 'cases' would (wrongly) not receive treatment in the long run) and true negatives from false positives 90% of the time (only one in ten 'controls' would (wrongly) receive treatment in the long run). In this case, we have $LR_+ = 0.9/(1\text{-}0.9) = 9.0$ and $LR_- = (1\text{-}0.9)/0.9 = 0.11$. Referring back to Fig. 12.2, everything we have said so far about likelihood ratios would fall inside the central box. That is, the message about requirement for action to control disease is captured in the likelihood ratios, which summarise 'the empirical content' of the data, and it is this message that would be contained within the IT-based tool. We can also see that the development and evaluation of tools in this way involves all of the steps in Fig. 12.1 in an explicit manner that makes the development process transparent.

Consider a grower, A, who, based on previous experience, including personal subjective factors and the known long-term prevalence of disease, has a prior probability of 20% for the need for treatment, written as $P_{prior\_A}(D+) = 0.2$ (that is, $odds_{prior\_A}(D+) = 0.25$). Assume that in using the forecaster, A obtains a positive prediction of need for treatment. According to equation 12.1, A's posterior odds, $odds_{post\_A}(D+)$, are $LR_+ \cdot odds_{prior\_A}(D+) = 9 \times 0.25 = 2.25$, corresponding to a $P_{post\_A}(D+) = 0.692$. The conclusion from this is that A will have moved from a position of 'thinking' there was a 20% chance of need for treatment, to thinking that the chance is just under 70%. A second grower, B, has a prior probability $P_{prior\_B}(D+) = 0.8$ (corresponding to $odds_{prior\_B}(D+) = 4$). Assume that in using the forecaster, B obtains a negative prediction of need for treatment. According to equation 12.1, $odds_{post\_B} = LR_- \cdot odds_{prior\_B}(D+) = 0.11 \times 4 = 0.44$, corresponding to $P_{post\_B}(D+) \cong 0.31$. Summarising, we can say that B's position has changed from one in which the perceived chance of need for treatment was 80% to one in which it was just over 30%.

Returning to Shannon's (1948) fundamental problem, the messages that the forecaster can deliver are summarised in equation 12.1. We can paraphrase them in natural language as follows: (i) given a prediction of need for treatment, multiply your personal odds of need for treatment by LR_+ to find out what the odds of need for treatment are now, in the light of the message contained in the (positive) prediction; (ii) given a prediction of no need for treatment, multiply your personal odds of need for treatment by LR_- to find out what the odds of need for treatment are now, in the light of this (negative) prediction. Messages (i) and (ii) are quite simple and although this, in itself, is no guarantee of them being reproduced '*exactly or approximately*' at the point of delivery, it will not decrease the chances of this happening. Shannon (1948) pointed out that delivering a message with high fidelity says nothing about what that message means. A little thought quickly leads to the conclusion that the meaning of messages (i) and (ii) will depend on the user, for at least two important reasons. First, different users might start with different prior probabilities of need for treatment and will therefore have different posterior probabilities of need for treatment even when they obtain the same forecast (receive the same message). Secondly, even if we remove the variation in prior probabilities among users by using the known long-term prevalence of disease as a universal estimate of the prior probability (see Yuen and Hughes, 2002), the message will still mean different things to different people because different people interpret probabilities (and odds) in different ways, depending on their attitudes to risk.

Looking at the examples used above in relation to Campbell and Madden's (1990) success criterion, there is an intuitive sense in which both A and B have been given new information. This results from the fact that the forecaster in both cases suggested a state of affairs which was contrary to their expectations (based on their prior probabilities of need for treatment). In the case of A $P_{prior\_A}(D+)<0.5<P_{post\_A}(D+)$ and for B, $P_{piror\_B}(D+)>0.5>P_{post\_B}(D+)$. A probability of 0.5 is an intuitive point of interest in decision making because it represents the point at which, for an event with two possible outcomes, either outcome is as likely as the other. Thus, if a forecaster is able to move the user's perceived probability of need for treatment from one side of 0.5 to the other it is likely to offer the user something truly informative. This idea also provides us with the means to specify the required performance of our IT tools ahead of constructing them, thus potentially saving time, effort and money. We illustrate the approach by using the case where a negative prediction (i.e., one of no need for action) is of interest.

By fixing the value of the required posterior probability not to take action and re-arranging equation 12.1 (after substituting the relationship between odds and probability) we obtain the required values for LR_- (written LR_{-req}) which will convert prior probabilities to the required posterior probability, as shown in equation 12.2.

$$LR_{-req} = \left(\frac{P_{post\_req}(D+)}{1-P_{post\_req}(D+)}\right) \Big/ \left(\frac{P_{prior}(D+)}{1-P_{prior}(D+)}\right) \tag{12.2}$$

Two examples are shown in Fig. 12.3, where the required posterior probability (P_{post_req}) of need for action is either ≤ 0.2 or ≤ 0.1. Some general points about Fig. 12.3 are worth highlighting. First, note that the vertical axis in Fig. 12.3 shows the values as $\ln(LR_{-req})$. The curves in Fig. 12.3 are essentially a pair of calibration curves that translate a prior probability into a required LR given either of the two required (target) posterior probabilities. Note that at the P_{post_req} values (0.1 and 0.2 respectively) the $\ln(LR_{-req})$ value is 0, corresponding to a LR_- of 1 ($e^0 = 1$). Referring back to equation 12.1 we can see that this makes sense. If the prior probability already equals the target posterior probability (and therefore the prior and posterior odds are equal) then the value of LR_- in equation 12.1 must be 1. The behaviour of the $\ln(LR_{-req})$ curves at high P_{prior} illustrates the point, made on several occasions (Yuen and Hughes, 2002; Yuen and Mila, 2003; Madden, 2006), that forecasters must have very good performance in order to bring about large changes in perceptions (i.e. differences between posterior and prior probabilities), particularly when the prior is especially small or especially large. Note that the line corresponding to required posterior probability of 0.1 is lower than that for 0.2 across the whole range of prior probabilities, indicating that, for any P_{prior}, LR_{-req} must be smaller to achieve a P_{post_req} of 0.1 than 0.2. An example, based on an assumed P_{prior} of 0.9 is illustrated in Fig. 12.3. To obtain the value of LR_{-req} we begin by projecting vertically upwards from horizontal axis at 0.9, until the projected line intersects with the LR_{-req} curves. Reading across from the point of intersection to the vertical axis provides the desired value. In the example it can be seen that the $\ln(LR_{-req})$ values for P_{post_req} of 0.1 and 0.2 are, respectively, ≅ -4.5 and ≅ -3.5. The actual values LR_{-req} obtained from the mathematical relationship are, respectively, 0.012 and 0.028 to 3 decimal places ($0.012 = e^{-4.39}$ and $0.028 = e^{-3.58}$).

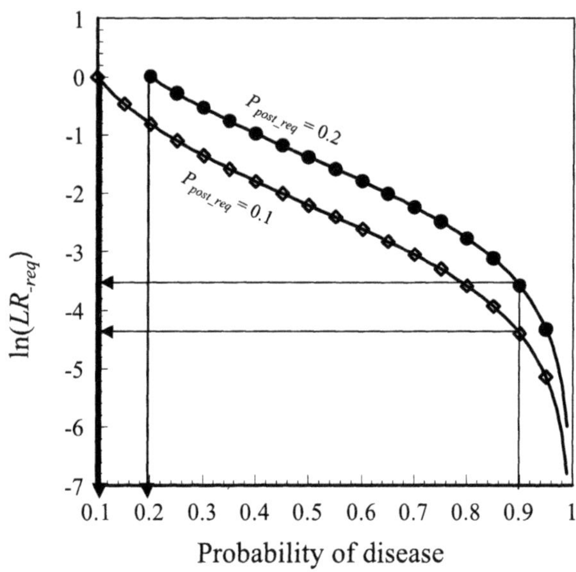

Figure 12.3. Likelihood ratios for the prediction of no disease (LR-) required to achieve fixed posterior probabilities of disease occurrence as a function of the prior probability of disease. The curves for posterior probabilities of 0.2 and 0.1 are shown.

Recalling the definition of LR_- given above, (i.e. the false negative proportion divided by the true negative proportion) we can translate these values of LR_{-req} into an indication of what we are expecting from our IT-based forecaster in terms of correct and incorrect predictions. For $P_{post\_req} = 0.2$ ($LR_- = 0.028$), for example, the frequency of false negative decisions is (approximately) 3% of the frequency of true negative decisions; i.e., for every 100 correct predictions not to take action, only three incorrect decisions not to take action can be tolerated. The more stringent criterion of having the $P_{post\_req} = 0.1$ ($LR_- = 0.012$) requires the forecaster to make only one false negative prediction for every 100 true negative predictions. How do these standards compare with forecasters that are actually in use? Yuen and Hughes (2002) reported three alternative LR_- values for a *Sclerotinia* stem rot forecaster for use in oilseed rape in Sweden: 0.130, 0.274, 0.684, depending on the choice of risk point threshold for action. De Wolf *et al.* (2003) (see also Madden, 2006) estimated a LR_- of 0.200 for their risk prediction system for *Fusarium* head blight of wheat in the north-east USA. A web-based prediction system (Burnett and Hughes, 2004) for the need for stem-extension treatment for eyespot (caused by *Oculimacula* spp.) in winter wheat developed by SAC and the University of Edinburgh, UK has LR_{-} values of approximately 0.3 and 0.45 offering growers two levels of risk acceptance. Comparing the values of LR_{-} for actual forecasters with the value for our hypothetical system, we can see that none of the real systems mentioned has as good performance as our hypothetical system, and, consequently, none of them is capable of generating such large differences between P_{prior} and P_{post}.

12.10 WHERE NEXT?

Bottlenecks in the development of operating systems, computer processing speeds and memory, data storage, and most importantly, communications speed represent potential limitations for applications of IT in plant disease epidemiology. The technology and infrastructure has improved vastly in the last few years, demanding considerable investment by IT users to keep up with developments. The dilemma is now that some control of the quality and appropriateness of material on the internet seems desirable, for the sake of search efficiency. However, control is contrary to the conceptual assumptions of making information available. There is a strong drive towards accreditation of methods and facilities and thereby the data produced, i.e. a move towards Quality Assurance (QA) of data by standardised calibration of instrumentation according to international standards. Security of data and data accreditation and electronic signatures are major issues particularly for protecting intellectual property that can include raw data. As data become more readily available internationally, the same dissemination network should facilitate information validation. The danger lies in the most powerful providers dictating the standards.

The vast improvement in the availability, accessibility and quality of graphics on electronic information systems is one of their great attributes. Although demand for large files increases institutional costs for storage, transmission and processing, efficiencies are also gained by replacing one type of information with others that are more appropriate. For example, images can allow multiple levels of technical description to be bypassed, enabling access by the non-expert into highly specialist fields. This is especially valuable for pest and pathogen identification. Linked with verification data such as geographic location, host etc., many of the pitfalls of such 'picture book' pathology can be avoided. This represents a new or enhanced use of IT, where the combination of information results in a better or a new product, not simply the same ones accessed in a different manner. It is such enhanced usages that represent the greatest potential of IT.

A conceptual change is required in order to best utilise IT. The volume of quality information can and should be maximised as the constraints of storage and access are removed. The major proviso is that all information should be appropriately tagged and it must include verification/authentication details to establish its status. Using text mining software can compensate for lack of appropriate meta-tagging. However, many find, for example, editing and proof reading on-screen far more difficult and error prone than on hard copy. Software to aid such processes should be used with discretion as it is seldom 'smart' enough. Similarly, authoring software is rarely as user-friendly as paper spread out on a desk even with today's large, high-resolution flat screen technology. Perhaps the next major visualisation developments that replace screens with true 'electronic books', i.e. paper-thin flexible media, will herald the next major advance in IT application. Our attitude to IT should be to adapt our methodology and working practices to use it most effectively and to use established low-technology methods if and when they do the job better or complement the IT.

12.11 CONCLUSIONS

One of the biggest problems in developing information networks will be compatibility and accessibility of information resources. Web interfaces already allow hitherto incompatible data storage formats to be made accessible without having to access the software directly. The user is shielded from the complexities of variations in operating system, application software and hardware platform by the use of a common user interface. Options, preferences and permissions can therefore be tailored to users' requirements, allowing different levels of access for different needs. There are several examples of information systems that go some way towards the objectives of a common epidemiology information system and point in the direction that such a development might go: the CABI Crop Protection Compendium; veterinary systems such as 'EqWise' and human medicine where UMLS (NLM) and Galan (EU) have considered conceptual integration of diverse information types.

However, broad information systems relevant to specific subject areas, such as plant disease epidemiology, are difficult to implement because of lack of funding for ends that are difficult to quantify. Furthermore, sustained development is essential since users' needs and available resources will change over time. The key to success for such a project is likely to be the active involvement of users in the development of such systems, perhaps borrowing the ethos of the global user community effort in the development of the Linux operating system and associated Open Source software.

If IT is used in a 'people-centred' manner, it will add value to our work and increase our understanding of epidemiology. Use of evaluation methods for IT-based forecasting and decision systems should lead to a reduction in reliance of validation by end-use, and may lead in the long-term to a higher adoption rate for such tools among growers and extension workers.

ACKNOWLEDGEMENTS

We thank the Scottish Executive Environment and Rural Affairs Department for financial support and the British Society for Plant Pathology for when one of us (ACN) was actively involved in its IT activities. We thank Simon Oxley (SAC, Edinburgh) and Gary Lyon (SCRI) for valuable discussion and ideas.

REFERENCES

Boswell, J. (1960) *Life of Johnson.* J M Dent & Sons, New York (Originally published in 1793 in Vol II of Boswell's *Life of Dr Johnson*).

Burnett, F.J. and Hughes, G. (2004) *The development of a risk assessment method to identify wheat crops at risk from eyespot.* HGCA Project 2382 Final Report, HGCA, London, UK. www.sac.ac.uk/consultancy/cropclinic/cropadvice/hgcaresearch/eyespotmodel.

Campbell, C.L. and Madden L.V. (1990) *Introduction to Plant Disease Epidemiology.* John Wiley & Sons, New York.

Clarke, W.S. (2000) *'Problems of communication and technology transfer in crop protection': A practitioner's perspective.* Proceedings, The BCPC Conference – Pests & Diseases 2000, pp. 1185-1192, BCPC Publications, Farnham, UK.

De Wolf, E.D., Madden, L.V. and Lipps, P.E. (2002). Risk assessment models for wheat Fusarium head blight epidemics based on within-season weather data. *Phytopathology*, **93**, 428-435. – www. wheatscab.psu.edu/riskTool.html.

Doucet, P. and Sloep, P. (1992) *Mathematical Modelling in the Life Sciences.* Ellis Horwood, New York, NY.

Fry, W.E., Milgroom, M.G., Doster, M.A. *et al.* (1990) 'LateBlight': A disease management game. Cornell University, Ithica, New York, USA. www.apsnet.org/online/feature/lateblit/software.htm.

Henley, P.M. (2000) *Issues in the design and delivery of commercial software applications for crop protection.* Proceedings, The BCPC Conference – Pests & Diseases 2000, pp. 1199-1202, BCPC Publications, Farnham, UK.

Howson, C. and Urbach, P. (1989) *Understanding Scientific Reasoning: the Bayesian Approach.* Open Court, La Salle, IL.

Hughes, G., McRoberts, N. and Burnett, F.J. (1999) Decision-making and diagnosis in disease management. *Plant Pathology,* **48**, 147-151.

Madden, L.V. (2006) Botanical epidemiology: some key advances and its continuing role in disease management. *European Journal of Plant Pathology* (in press).

McRoberts, N., Foster, G.N., Sutherland, A. *et al.* (2000) *Do hunter-gatherers make good farmers?* Proceedings, The BCPC Conference – Pests & Diseases 2000, pp. 11935-1198, BCPC Publications, Farnham, UK.

McRoberts, N., Hughes, G. and Savary, S. (2003) Integrated approaches to understanding and control of diseases and pests in field crops. *Australasian Plant Pathology*, **23**, 167-180.

Medawar, P.B. (1972) *Induction and Intuition in Scientific Thought.* (3rd Edition). Methuen & Co. London, UK.

Murtaugh, P.A. (1996) The statistical evaluation of ecological indicators. *Ecological Applications,* **6**, 132-139.

Shannon, C.E. (1948) A mathematical theory of communication. *Bell Systems Technical Journal*, **27**, 37-423. – www.math.psu.edu/gunesch/Entropy/Infcode.html.

Simmonds, N.W. (1991) Genetics of horizontal resistance to diseases of crops. *Biological Reviews*, **66**, 189-241.

Tomerlin, J.R. and Howell, T.A. (1988) DISTRAIN: A computer program for training people to estimate disease severity on cereal leaves. *Plant Disease,* **72**, 455-459.

Waggoner, P.E. (1968) Weather and the rise and fall of fungi, in *Biometeorology* (ed. W P Lowry) Oregon State University Press, Corvallis, pp. 45-60.

Waggoner, P.E. and Horsfall, J.G. (1969) EPIDEM, a simulator of plant disease written for computer. *Connecticut Agricultural Experimental Station Bulletin*, 698, 80 pp.

Yuen, J.E. and Hughes, G. (2002) Bayesian analysis of plant disease prediction. *Plant Pathology*, **51**, 407-412. – krakatau.evp.slu.se/~evat/cgi-bin/sclerot.pl?first=1.

Yuen, J.E. and Mila, A. (2003) *Are Bayesian approaches useful in plant pathology?* Proceedings, Bayesian Statistics and Quality Modelling in the Agro-Food Production Chain, 95-103, Wageningen, The Netherlands – library.wur.nl/frontis/bayes/index.html.

PART TWO

Case Examples

CHAPTER 13

SEEDBORNE DISEASES

W.J. RENNIE AND VALERIE COCKERELL

13.1 INTRODUCTION

Almost 90% of all the world's food crops are grown from seeds (Schwinn, 1994) and seeds are widely distributed in national and international trade. Germplasm is also distributed and exchanged in the form of seeds in breeding programmes. However, many plant pathogens can be seed transmitted and seed distribution is a very efficient means of introducing plant pathogens into new areas as well as a means of survival of the pathogen between growing seasons.

Disease-causing organisms (in plants: usually fungi, bacteria, viruses and nematodes) may be carried with, on or in seeds and, in suitable environmental conditions, may be transmitted to cause disease in developing seedlings or plants. With some diseases, the pathogen attacks the germinating seedling and affects seedling establishment and hence plant populations; with others, disease symptoms are not seen until a later stage of growth.

Studies on the epidemiology of seedborne pathogens are important to understand how the diseases they cause can be controlled most effectively. With seedborne infection, optimum control is often achieved most cost-effectively, and usually with least disruption to trade, by the application of seed treatment chemicals, or through avoiding infection by producing seed in areas or environments that are unfavourable for disease development. Maude (1996) gives a comprehensive review of how seedborne pathogens may be controlled through exclusion and elimination. In some special circumstances, it may be appropriate to test seed before marketing or sowing and infected seed may be rejected or treated with an appropriate chemical. Effective control through seed testing requires not only sensitive and accurate test methods but also appropriate and adequate sampling of the seed lots to be examined (Cockerell *et al.*, 2003; Kruse, 2004).

A point sometimes not appreciated is that seed which is traded nationally or internationally is not necessarily disease-free. Many importing countries have legislation that aims to prevent or limit the introduction of non-indigenous seedborne pathogens (Ebbels, 2003) and national seed certification schemes may have standards for the most damaging seedborne diseases, especially those that cannot easily be controlled by seed treatment chemicals (Rennie, 1993). Frequently a distinction has to be made between diseases that are already widely distributed but which can have an effect on crop performance and those that are of limited distribution but have the potential to cause serious crop losses if introduced into new areas. The former are usually controlled through tolerances or standards in domestic

B. M. Cooke, D. Gareth Jones and B. Kaye (eds.), The Epidemiology of Plant Diseases, 2nd edition, 357–372.

seeds regulations while the latter are often the subject of strict import or quarantine regulations.

Before considering control strategies for specific seedborne diseases, the following epidemiological information must be available.

13.2 EPIDEMIOLOGY

(a) What quantity of inoculum must be present on seeds, and in what form, to transmit the pathogen to developing seedlings and cause disease?

The large majority of seedborne pathogens are carried passively on seed surfaces or are established as hyphae or spores in or under the seed coat. Successful transmission of infection to developing seedlings often depends on the location of infection on the seed and on the amount of inoculum present. Theoretically, one spore of *Tilletia tritici* on a single wheat seed can result in a bunted plant but several workers have shown that, in practice, bunt does not occur in the growing crop until the level of contamination is much higher (40-100 spores per seed); seed contamination needs to be even higher before infection occurs in some less susceptible cultivars (Heald, 1921; Oxley and Cockerell, 1996). The relationship between the amount of inoculum on the seed and disease transmission is usually strongly influenced by environmental conditions and cultivar susceptibility.

(b) How is seed-to-seedling transmission influenced by environmental conditions during germination and subsequent plant development?

Where inoculum is superficial on the seed surface, or is established only in the tissues of the seed coat, successful transmission of the pathogen to the germinating seedling depends not only on the amount and location of inoculum on the seed but also on environmental conditions during germination and seedling establishment. *Microdochium nivale*, which can cause seedling blight in cereals, is often present on seed in the form of mycelium in the pericarp. If germination is rapid, in conditions where moisture is adequate and the temperature is above 10°C, there may be relatively little transfer of infection to developing seedlings and the pathogen has little or no effect on germination or seedling establishment (Hewett, 1983). However, where the soil temperature is between 5 and 10°C and germination proceeds slowly, the fungus readily establishes within the tissues of young seedlings and significant seedling losses can occur (Hewett, 1983; Richards, 1990). Disease transmission is frequently reduced at low soil water contents. Infection of young pea seedlings by *Pseudomonas syringae* pv. *pisi* is greatly reduced in dry seed beds (Holloway *et al.*, 1996).

On the other hand, where inoculum is deep-seated and the embryonic axis of the seed is infected, environmental factors may have relatively little effect on pathogen transmission. Barley seed infected with *Ustilago nuda*, where inoculum becomes established in the scutellum tissues of the seed embryo, is likely to give rise to plants showing symptoms of loose smut over a very wide range of environmental conditions.

(c) Is seedborne inoculum the principal source of infection or are other inoculum sources more important?

For some plant diseases, seedborne inoculum is the only, or principal, source of infection: if healthy seed is sown, disease does not develop in the growing crop. For example, barley leaf stripe occurs only when seeds infected with the fungus *Pyrenophora graminea* are sown. There is no further spread of infection between plants during the growing season and the disease can, therefore, be controlled very effectively by sowing healthy seed, or by applying an effective seed treatment fungicide to infected seed.

In contrast, a closely related fungus, *Pyrenophora teres*, can also be seedborne and infected barley seeds can give rise to seedlings that show symptoms of net blotch, but sowing disease-free seeds will not necessarily result in a crop free from net blotch, since inoculum from other sources can establish the disease at any stage of growth. In practice, *P. teres* present on old stubble, on volunteer barley plants and in neighbouring crops is a much more important source of net blotch infection than seedborne inoculum. Unlike leaf stripe, net blotch can spread between plants during crop growth and, beyond the early seedling stage, the amount of net blotch in a barley crop is rarely related to the initial level of inoculum in the seed sown. Controlling seed infection, therefore, has little or no effect on net blotch at later stages of crop growth. Control of seedborne inoculum is likely to be effective only in situations where inoculum from other sources is relatively insignificant.

(d) Is the disease monocyclic or polycyclic?

Some seedborne pathogens, for example, *Ustilago nuda*, which causes loose smut disease of barley, have relatively uncomplicated life cycles and there is no spread between plants during crop growth. The relationship between seed infection and plant infection is relatively constant and it is only at flowering, when seed reinfection takes place, that environmental conditions influence disease development. With loose smut, cool moist conditions at flowering can lead to higher rates of reinfection because the flowers remain open for a longer period (Hewett, 1978). With monocyclic diseases, reducing or eliminating seedborne inoculum (for example, through appropriate seed treatment) usually gives very effective control of the disease.

Polycyclic diseases, on the other hand, are those that, once established within a crop, will multiply and spread during suitable conditions. They frequently cause local lesions on infected plants and, as the disease develops, these lesions increase in size and produce increasing quantities of inoculum, which is spread to other parts of the infected plant, to neighbouring plants and sometimes to plants in neighbouring crops. The relationship between seedborne inoculum and plant infection is complex. Environmental conditions play an important role and inoculum from other sources, including volunteer plants and old crop debris, complicate the seed-to-plant disease relationship. Many economically important seedborne pathogens have a polycyclic disease pattern. Limiting the effect of polycyclic diseases on crop performance by controlling seedborne infection will only be effective if seedborne disease levels are kept very low and inoculum is not introduced from other sources.

(e) Will reducing seedborne inoculum (for example by selecting relatively healthy seed, or through seed treatment) lead to reduced levels of disease in the growing crop?

Reducing seedborne inoculum, through chemical seed treatment or by selecting healthy seed for sowing, will be an effective disease control measure in situations where inoculum from other sources is not significant. For example, control of monocyclic diseases such as loose smut of barley or wheat by seed treatment is usually very effective. On the other hand, barley net blotch (*P. teres*) or septoria glume blotch on wheat (caused by *Stagonospora nodorum*) cannot usually be controlled by seed treatment or by sowing healthy seed, since susceptible crops very quickly become infected from inoculum originating from old stubble and diseased volunteer plants.

Some polycyclic diseases may be kept in check by reducing the amount of seedborne inoculum. This usually requires any residual seed infection to be very low and is probably best achieved by producing seed in areas where the pathogen does not occur or where environmental conditions are unfavourable for disease development, so limiting or avoiding seed infection. Hewett (1973) showed that if seedborne inoculum of *Didymella fabae* was kept below one infected seed in 200, ascochyta leaf blight did not develop in plots of field beans in England. Grogan (1980) established that seedborne inoculum of the aphid-transmitted lettuce mosaic virus had to be below 0.022% to ensure that disease epidemics did not develop in commercial fields of lettuce in California.

(f) What conditions favour the establishment of infection in developing seeds?

In seed multiplication systems, developing seeds usually become infected from inoculum present within the growing crop. For some diseases, infection may be systemic and seeds developing on diseased plants may become infected as they are formed. In these cases, environmental conditions will have little or no influence on the rate of seed infection.

For many diseases, the pathogen is often carried from infected plants to developing seeds in wind eddies or by rain splash. Fungal conidia and bacteria often develop very readily on diseased plants in humid or wet conditions and seed infection usually depends on the availability of adequate moisture. For many diseases, therefore, seed infection levels are highest when high humidity or rainfall occurs during seed development. Hewett (1965) has shown how *Microdochium nivale* levels on wheat seed tend to be highest in the wetter western and northern parts of the United Kingdom. Commercial cabbage seed production in the United States is concentrated in western Washington state because the dry conditions there result in a low incidence of seedborne inoculum of *Leptosphaeria maculans* (cause of blackleg) and *Xanthononas campestris* pv. *campestris* (cause of black rot) (Gabrielson, 1983). Seed production in areas of low disease pressure, or in environmental conditions unfavourable for seed infection, is a very effective means of producing healthy seeds and is likely to be most cost-effective when seeds are small and transport costs low, or when the seed is of sufficiently high value to justify high transport costs. With low value, high volume cereal seed, isolating seed production may not be practical or economic and healthy crops may become infected during seed production if high levels of inoculum are present in neighbouring crops. Rennie and Cockerell (in Yarham and Jones, 1992) have shown how barley seed may become infected with *P. graminea* 200 m from an inoculum source.

13.3 CASE STUDIES

13.3.1 Bunt of wheat (causal agent Tilletia tritici*)*

Teliospores of the pathogen are present on the surface of contaminated wheat grains. Many lodge in the dorsal crease. When the seed is sown, the teliospores germinate to produce a promycelium. Infection hyphae then penetrate between and through the cells of the coleoptile in the germinating seedling, eventually reaching and invading the young growing point. Mycelium of the fungus is then carried upwards in the inflorescence as the wheat plant grows and, at flowering, the fungus multiplies rapidly to produce a mass of teliospores within the pericarp. The resulting teliospores are protected by the pericarp which is usually broken when the crop is harvested, releasing spores over other healthy seeds to complete the cycle of infection.

Establishment of bunt disease within wheat fields from contaminated seeds depends on the number of spores per seed, the susceptibility of the cultivar to coleoptile infection and environmental conditions during seed germination. Heald (1921) found that the cultivar Marquis needed 542-5000 spores per grain to transmit the disease, compared with only around 100 spores per grain in the susceptible cultivar Jenkins Club. Oxley and Cockerell (1996) also recorded bunt symptoms in the field when the level of contamination exceeded 100 spores per seed.

Seed contamination was generally considered to be the only way in which the pathogen survived between cereal crops and, since seed infection was in the form of seed surface contamination, control of the disease was relatively easily achieved by the application of appropriate seed treatment chemicals. From the 1950s to the 1980s, when almost all UK wheat seed was routinely treated with organomercury seed treatment fungicides, bunt was very rare. There were occasional reports of the disease in wheat crops in England in the 1980s and these were usually associated with the sowing of untreated farm-saved seed. However, Yarham (1993) reported bunt in crops where soilborne inoculum seemed to be the source of infection and this was associated with dry soil conditions occurring in the short period between the harvesting of a heavily diseased crop and sowing a following winter wheat crop. The teliospores appeared to survive between harvest and sowing in the dry soil conditions. Similar reports of occasional soilborne infection of wheat by *T. tritici* have been made in Denmark (Nielsen and Jorgensen, 1994). In conditions favourable for spore germination, teliospores of *T. tritici* are thought to be relatively short-lived (Weltzein, 1957). In these conditions, seedborne inoculum will be of greatest epidemiological significance but spore balls shed by heavily diseased crops can probably protect spores at least until a following wheat crop is sown (Paveley *et al.*, 1996). Recent work suggests that in the UK soil contamination by bunt spores may be more important than was previously thought and the use of seed treatment chemicals where soilborne infection occurs is advocated (Cockerell *et al.*, 2004).

Cockerell and Rennie (1996), in a survey of the health of UK cereal seed in the three years 1992 to 1994, found that *T. tritici* contamination of winter wheat seed was not uncommon. Up to 60% of seed samples were contaminated but contamination levels were generally low, with only 2-8% of the samples examined each year having more than one spore per seed. This widespread, low-level contamination probably reflects the ease with which spores of *T. tritici* can be spread between seed stocks in harvesting, cleaning and storage equipment. *Tilletia tritici* is known to be able to increase rapidly during successive generations in seed production (Dillon-Weston and Engledow, 1930; Oxley and Cockerell, 1996) and for this reason, and because contamination is currently so widespread in UK seed, effective fungicide seed treatments are usually applied to all seed stocks intended for further multiplication.

13.3.2 Loose smut of barley (causal agent Ustilago nuda)

Loose smut of barley is similar to bunt of wheat in that infection becomes established in the young inflorescence at an early stage of seedling growth and the flowering parts are replaced at ear emergence by a mass of fungal spores. Loose smut, like bunt, is a monocyclic disease – the relationship between seed infection and plant infection is constant and there is no increase in plant infection during crop growth. However, there are important epidemiological differences between the two diseases. The mycelium of *U. nuda* is located in the scutellum tissue of infected seeds (Popp, 1951). When an infected seed germinates, the mycelium is also stimulated to grow and it quickly moves into the crown node to become established in the growing point of the young tillers. The future inflorescence is well established at this stage and is already infected by the fungus. As the barley plant grows and eventually 'shoots' to produce its ears, the fungus is carried upwards in the inflorescence.

Because the loose smut fungus is internally borne in the embryo, there is a high probability that each infected seed will produce a diseased plant. The relationship between seed infection and plant infection is good and the environmental influence on disease transmission is slight (Rennie and Seaton, 1975). However, infected barley plants may produce some tillers that avoid infection. This is probably related to the amount and location of mycelium in infected embryos. In some barley varieties that exhibit physiological resistance, the fungus may become established in seed embryos but diseased plants do not develop (Hewett, 1979).

In diseased barley plants, the emerging ears are covered in a mass of black spores which, in contrast to bunt in wheat, are not contained within the pericarp but are easily dislodged at maturity by air currents which carry them to the open inflorescences of neighbouring healthy plants. These spores germinate and the resulting hyphae penetrate the developing seed to complete the life cycle. Infected seeds cannot be distinguished from healthy seeds by visual inspection. The rate of re-infection of barley seeds is determined by the quantity of inoculum available in the crop during flowering and the extent to which the flowers remain open during pollination. High levels of seed infection are associated with varieties that display an open flowering habit. The flowering period may be extended in cool, moist conditions, leading to a potential increase in seed infection. Some cultivars of barley display a 'closed' flowering habit and so avoid infection. Loose smut was rarely recorded in the closed flowering cultivar Golden Promise, even during years of relatively high loose smut infection (Rennie and Seaton, 1975).

Most inoculum for loose smut reinfection probably originates within diseased crops and, given suitable conditions at flowering, the disease tends to multiply over successive generations. However, infection can also spread between neighbouring crops and Rennie (unpublished observations) found that seeds developing in healthy barley crops could become infected by spores released in diseased crops 200 m upwind. Infected volunteer barley plants within seed crops can also act as an important inoculum source for developing seeds.

Because seed infection is deep seated, loose smut was not controlled by chemical seed treatments until the introduction of the systemic fungicide carboxin. Prior to its introduction, elite seed stocks were multiplied in isolation or, occasionally, high value infected stocks intended for further multiplication were subjected to a hot water treatment at a temperature that killed the fungus but not the seed. Because loose smut was not controlled by conventional seed treatments, including those based on organomercury, the European Union required action to be taken against the disease during seed multiplication and standards for loose smut were included in Cereal Seeds Regulations in the UK (Anon., 1985).

In practice, the disease is controlled by treating seed of susceptible cultivars intended for multiplication with an effective systemic seed treatment fungicide (Reeves *et al.,* 1994; Soper, 1995). In the early 1980s, the fungus developed resistance to carboxin and loose smut levels increased in stocks of susceptible cultivars until effective alternative seed treatment fungicides were used on infected stocks (Wray and Pickett, 1985). Seed treatments that are effective in controlling loose smut tend to be more expensive than those that offer more limited control of seedborne diseases and, for this reason, they are more likely to be used on high value seed intended for further multiplication. Loose smut is usually effectively controlled in certified cereal seed but infection can build up to damaging levels in stocks of farm-saved seed of susceptible barley cultivars, especially where the same seed stock is multiplied over several years, in the absence of an effective seed treatment (Rennie, 1987; Cockerell and Rennie, 1996). Surveys of certified and farm-saved seed have shown that the incidence of loose smut varies between seasons and cultivars. High seed infection levels tend to be associated with increasing popularity of susceptible cultivars (Cockerell *et al.,* 2005).

13.3.3 Lettuce mosaic virus

Lettuce mosaic virus (LMV) was first reported by Jagger in Florida (Jagger, 1921) and subsequently shown to be seed transmitted by Newhall (1923). Significant crop losses have been reported in the United States (Grogan *et al.*, 1952; Raid *et al.*, 1996) and in Europe (Broadbent, 1951).

Within growing crops, the virus is spread by aphids in a non-persistent manner, with *Myzus persicae* and *Macrosiphum euphorbiae* among the most common vectors. A small but significant proportion of infected plants (usually up to 10%) produce infected seed and this results in plants that act as a primary source of inoculum in the following season. Diseased plants that develop from infected seeds are often dwarfed and may fail to form marketable heads.

The virus may survive in lettuce-growing areas within susceptible weeds, and Phatak (1974) has reported seed transmission in *Senecio vulgaris*. Effective weed control, particularly in areas where virus-transmitting aphids are active, is important in disease control. Insecticide control of aphids, especially using quick-acting 'knock down' insecticides may limit virus spread. Some cultivars show tolerance to the virus; symptom development in these cases may be limited but diseased plants remain an important source of inoculum for crops of susceptible cultivars. However, seed is the primary source of infection in commercial lettuce crops and the principal means of spread of the virus (Grogan, 1980; Raid *et al.*, 1996).

Seed production schemes usually aim to produce lettuce seed that is either free from LMV or that has a low level of infection, and certified seed will usually meet standards for maximum permitted infection. Grogan *et al.* (1952) reported observations on LMV in California which suggested that typical seed infection of 1-3% would result in around one infected plant in each 1 m row before thinning. Where aphid activity was high and between-plant transmission occurred, up to 30% of the plants could become infected before thinning took place. The rate of seed transmission in lettuce varies from 0.1% to 37% (Shukla *et al.*, 1994) and depends on the time of infection of the mother plant (Couch, 1955), the virus strain (Dinant and Lot, 1992), the lettuce cultivar (Falk and Guzman, 1984) and the temperature at which the infected plants are grown.

Grogan was able to show that virus-free seedlings grown in isolation from other commercial lettuce crops could produce virus-free seed. In further studies, he demonstrated the benefits of using virus-free seed to produce commercial lettuce crops (Grogan, 1980). Plants grown from virus-free seed in a 55 m^2 plot initially showed no symptoms of LMV, whereas 2% of plants grown from the untested seed showed mosaic symptoms. Immediately prior to harvest, approximately 2% of plants within the experimental block showed symptoms whereas infection in the commercial crop had increased to 12%.

These studies also showed that aphid movement was from plant to plant within rows rather than across furrows, encouraging the view that virus transmission between neighbouring crops grown from different seed sources would be lower than spread within crops. In California, seed infection above 0.1% usually gave inadequate disease control in commercial crops: where aphid numbers and activity were high even this low level of seed infection resulted in 23% LMV infected plants in one trial. It became apparent that to ensure effective, consistent control of LMV, seed had to have less than 0.022% seedborne infection (Grogan, 1983). This was achieved by testing 30,000 seeds and accepting only those seed lots that showed no infection. Additional control measures in support of this programme included eliminating potential weed hosts in and near lettuce field borders, avoiding the planting of new lettuce fields adjacent to old fields and ploughing in lettuce fields immediately after harvest. The seed certification programme in Australia includes growing seed crops of lettuce in geographic isolation from commercial crops, in areas of low aphid activity, and the careful inspection of plants to remove any with disease symptoms. Outbreaks of disease in commercial lettuce crops in North America have been associated with breaches in seed certification rules (Raid *et al.*, 1996).

In northern Europe, where aphid pressure may be lower than in California and where commercial lettuce production is probably less intensive, a standard of no LMV in 2200 seeds has been shown to give effective practical control (Van Vuurde and Maat, 1981; Dinant and Lot, 1992). These high standards for seed health have required the introduction of seed testing procedures, usually based on ELISA techniques, that can accommodate a high throughput of samples of 2000-30000 seeds (Zerbini *et al.,* 1995).

13.3.4 Ascochyta leaf and pod spot of field beans (causal agent Didymella fabae)

Field bean (*Vicia faba*) seed may become infected by *Didymella fabae* from diseased pods in infected crops. Seed infection is in the form of mycelium, largely in the seed testa rather than the cotyledons. Heavily infected seeds of susceptible cultivars may show purple-brown discolouration and, on imbibition, dark pycnidia of the pathogen may be seen on infected seeds. Lesions may develop on the primary foliage leaves of seedlings grown from infected seeds. Seed infection appears not to affect germination or seedling establishment (Hewett, 1973) and the proportion of infected seeds that produce seedlings with primary lesions varies from nil to 50% depending on growing conditions (Hewett, 1973; Wallen and Galway, 1977; Gaunt and Liew, 1981). Transmission is often higher in winter-sown crops.

Short-range disease spread occurs by rain splash transmitting the spores, which ooze from pycnidia in leaf lesions in wet weather. Disease development depends very much on environmental conditions and high seed infection is associated with wet seasons. During dry periods, disease development and spread is halted (Hewett, 1973; Biddle, 1994).

In conditions suitable for disease development, Hewett (1973) found that lesions occurred at a height of 30 cm towards the end of May in south-east England and by the end of June leaves 60 cm above ground bore lesions. His observations suggested a rate of spread through the crop of approximately 1 m per month. For successful seed re-infection, there must be rainfall early in the development of a crop so that the pathogen is carried sufficiently high in the leaf canopy to re-infect the pods and seeds as they develop. Hewett (1973) considered that above-average rainfall in May and June was required in south-east England for successful re-infection of developing seeds. In his trials, the maximum increase in seed infection from that on the seed sown to that on the harvested seed was eight times.

Ascochyta leaf and pod spot, like many polycyclic seedborne diseases, develops relatively slowly (Hewett, 1978). Disease increase with time is logarithmic but the maximum infection rate is usually low (Hewett, 1978). The infection rate (*r*), the rate at which the population of the pathogen increases (Van der Plank, 1963), is expressed as the unit increase of infected plants per day and is influenced by climatic factors and the susceptibility of the host plant. *Didymella fabae* was estimated to spread at a rate of 0.07 units per day; over the course of a season, the disease could spread up to 5-10 m from a single focus of infection (Hewett, 1973).

The relatively low levels of seed infection in many seed-producing areas (Hewett, 1973; Wallen and Galway, 1977) and the modest rates of disease development in arable areas suggested that disease control could be relatively easily achieved by limiting seed production to areas of low rainfall and setting a standard of around 1% for commercial seed (Hewett, 1973). For further seed production, lower initial seed infection was necessary, especially in early generations and in the absence of effective seed treatments. Control of disease through standards in seed certification schemes is likely to be most effective where field beans are not widely grown and where there is a rotational interval between crops.

Levels of seed infection vary between years and between varieties grown in the same year and are strongly influenced by weather conditions during seed production (Hewett, 1973; Biddle, 1994). Hewett (1966) found that of 180 seed samples from commercial crops grown in England in 1964–65 only a small proportion were infected whereas the following year over a third of samples tested were infected at levels of 3 to 10%. In 1992 a survey of 451 seed lots of winter beans produced in the UK showed that 31% contained levels of seedborne infection which exceeded 1% and 3% had more than 10% infection. Seed treatment of infected seed stocks often gives only partial control and, in practice, a combination of low seed infection and appropriate chemical treatment is most effective (Knott *et al.,* 1994).

Bond and Pope (1980) found that infected volunteer plants from old field bean crops could act as an important source of inoculum for crops in neighbouring fields. They reported spread up to 120 m into neighbouring fields. They also highlighted the risks from minimum cultivation techniques and drew attention to the risk of spread between crops where spring and winter cultivars were grown in the same area. These additional influences on disease development resulted in a recommendation that seed crops should be isolated by at least 50 m, not only from other bean crops but also from fields in which a bean crop had been grown during the previous year. These actions were in addition to a requirement that certified seed of the first generation should have no infected seeds in a sample of 600. The discovery of the teliomorph *D. fabae,* produced over winter on infected bean straw in stubble fields, suggests that the pathogen can also be spread through the dispersal of airborne sexual spores (Jellis and Punithalingam, 1991).

13.3.5 Pea bacterial blight (causal agent Pseudomonas syringae *pv.* pisi)

Bacterial blight is a widely distributed and potentially damaging disease of peas. The disease is seedborne and primary disease foci develop principally on plants grown from infected, or infested, seed. Water-soaked lesions develop on stems, leaflets and stipules of infected plants. Infection often runs along the veins, causing characteristic fan-shaped lesions. Stem lesions may develop to the extent that they cause shoot death and this can delay and reduce seed production, so causing yield loss.

The disease progresses during wet weather and bacteria oozing from disease lesions are readily spread to healthy tissues or to neighbouring plants in wind-driven rain. Bacteria enter tissues through stomata and damaged cells, so infection is readily spread by machinery moving within and between crops and by animals (including birds) feeding in diseased crops.

Seeds usually become infected directly from pod lesions during wet weather but they may also become contaminated by bacteria-laden dust during harvesting, or from cleaning and grading equipment. Dry conditions and high temperatures during crop growth lead to a reduction in disease development and seed infection. Seedborne bacteria may remain viable for several seasons but populations decline during storage. Seed infection may occur in the apparent absence of field symptoms. The pathogen has a well defined race structure and races can be determined by stem inoculation of the pathogen into a defined set of differential pea varieties (Taylor *et al.,* 1989) or using DNA sequence analysis (Cournoyer *et al.,* 1996).

Pea blight was not known in the UK until 1985, when it was found in a seed crop of Belinda combining peas during a routine inspection (Stead and Pemberton, 1987). Prior to 1985 the disease had been subject to import (quarantine) regulations and had been detected only occasionally in imported seed lots of combining and vining peas.

Following the 1985 outbreak, an action plan was drawn up to obtain general control of the disease and, specifically, to exclude it from certified seed (Stead and Pemberton, 1987). Samples of all breeders' seed, pre-basic and basic seed lots (i.e. seed intended for further multiplication) had to be laboratory tested for *P. syringae* pv. *pisi*. In addition, all seed crops were inspected for symptoms of pea blight during growing crop inspections. Where there was evidence that a particular seed line was infected, steps were taken to eliminate that line from further seed multiplication.

In the period 1986-1990, over 5500 seed samples were tested in four UK seed testing stations (Roberts *et al.,* 1991). The results showed that *P. syringae* pv. *pisi* was relatively widely distributed in UK pea seed, with around 10% or more of samples giving positive results in each of the five years. Infection fluctuated from year to year and seed that was not subject to statutory testing (the final year of certified seed and all farm-saved seed) showed a general increase in infection to reach relatively high levels (more than 40% of samples infected) by 1988.

There was a sharp increase in the frequency of infected stocks between 1987 and 1989 and this suggested that some seed stocks were contaminated from sources other than the parent seed (for example, by harvesting, cleaning or grading equipment) or that the testing programme failed to detect some infected stocks. The testing programme may have failed to identify infection in early generation seed for two reasons. First, only a single sample was required from each crop and this may not have been sufficiently representative when some crops consisted of several hundred tonnes of seed. Second, tests on the required sample size of 1 kg (approximately 3000 seeds) would not have detected infection in some stocks where fewer than 0.1% of seeds were infected.

The epidemiological significance of very low levels of seedborne infection is not well known. Roberts (1992) found that soil water content had a significant effect on transmission of *P. syringae* pv. *pisi* from seed to seedling and seed transmission increased with increasing soil moisture. Hollaway *et al.* (1996) estimated that approximately 10% of infected seeds would give rise to infected seedlings, even in relatively dry seed beds, and that a seed lot with 0.1% infection would, therefore, result in a primary infection of 0.01%, which would be sufficient to cause a field epidemic (Taylor and Dye, 1976).

In a series of glasshouse and field trials, Roberts (1993) and Roberts *et al.* (1995) showed that pea blight could result in significant yield loss. Yield reductions of up to 71% were recorded in glasshouse trials, following inoculation with the pathogen, and a model constructed from field trials data suggested that a yield loss of 0.98 tonnes ha^{-1} would be expected where 5% of the leaf area was diseased at growth stage 208. A feature of the field trials was the considerable spread of infection into plots sown with healthy seed and the interaction between disease and bird damage. In retrospect, the seed sampling and testing procedures introduced in 1986 did not significantly reduce the proportion of seed samples infected with pea blight, though they probably limited the incidence of infection in certified seed. However, despite initial concerns, pea blight did not appear to develop to epidemic levels in spring-sown crops of combining peas and there were few field reports of serious yield loss.

The statutory controls on certified seed have now been removed and, while a voluntary Code of Practice aimed at minimizing the occurrence of blight in pea crops was introduced in 1994 (Anon., 1994), there is currently relatively little testing of spring-sown seed. In contrast, *P. syringae* pv. *pisi* has been shown to have a damaging effect on autumn-sown combining peas and a combination of frost damage in the spring and pea blight can lead to significant plant loss. For this reason, growers of autumn-sown combining peas are advised to have seed tested for pea blight and to avoid sowing seed in which infection is found (A.J. Biddle, personal communication).

The pathogen can survive on or within seeds between seasons and seed infection remains the most important source of disease outbreaks. However, diseased pea debris and infected volunteer plants can also play a role in overwintering and crop rotations are usually considered as part of a strategy in controlling pea blight (Hollaway and Bretag, 1997).

13.4 FUTURE DEVELOPMENTS

The seed industry is becoming increasingly international and there are pressures to relax, or remove, trade barriers between countries. One of the challenges for the future, therefore, is to encourage and facilitate trade in seeds while, at the same time, ensuring acceptable control of damaging seedborne pathogens. Increasingly, import restrictions will focus on those pathogens that, following scientifically sound risk assessments, are shown to present the greatest threat (Ebbels, 2003). Seed production, especially of high value seed, such as hybrid and genetically modified varieties, will be concentrated in countries, or regions, in which specific seedborne pathogens are absent, or which, because of climatic conditions, are conducive to healthy seed multiplication.

For high value seed, cost effective laboratory tests will be required that are sufficiently sensitive to detect very low levels of damaging pathogens. The well established incubation tests used in domestic testing programmes are increasingly being replaced by immunoassays using standard, internationally accepted antibodies or by tests based on nucleic acid technology which offer rapid identification of seed lots with low levels of infection. Tests based on PCR technology are now available for the routine detection of a number of seedborne pathogens (Cockerell *et al.,* 2001; Kenyon *et al.,* 2001; Cockerell *et al.,* 2004).

Tests for seedborne pathogens will increasingly be made 'in-house' by the producer, using methods validated and approved by the International Seed Testing Association (Meijerink, 1997). As seed testing technology develops it may be possible for seed merchants, or even growers who aim to save their own seed, to use diagnostic tests in kit form, perhaps on seed taken directly from the combine, to determine whether it meets acceptable health standards and as a guide to which seed treatment to use, if any (Paveley *et al.,* 1996). A recent development in the UK has been the move away from routine chemical treatment of all cereal seed lots towards a system of treating infected seed lots according to need and directing treatment at specific disease problems (Cockerell *et al.,* 2004). This change has largely been driven by the need to reduce inputs to maintain profitability, and an accompanying increase in the use of farm-saved seed, but it has been facilitated by a better understanding of seedborne disease epidemiology and the increased availability of rapid, reliable test methods. This has shown that a significant proportion of UK cereal seed could safely be sown without treatment, resulting in potential annual savings to growers of up to £8 million. However, in order to avoid disease build-up during seed multiplication, seed to be sown for further seed production should be treated routinely.

REFERENCES

Anon. (1985) *The Cereal Seeds Regulations 1985*, Statutory Instruments 1985 No. 976, Her Majesty's Stationery Office, London.

Anon. (1994) *Mind Your Peas*, Ministry of Agriculture, Fisheries and Food, Cambridge. Ball, S. and Reeves, J.C. (1992) Application of rapid techniques to seed health testing – prospects and potential, in Techniques for the Rapid Detection of Plant Pathogens, (eds J.M. Duncan and L. Torrance), Blackwell Scientific Publications, Oxford, pp. 193-207.

Biddle, A.J. (1994) *Seed treatment usage on peas and beans in the UK*. BCPC Monograph No. 57. Seed Treatment: Progress and Prospects. British Crop Protection Council, Farnham, UK, 143-149.

Bond, D.A. and Pope, M. (1980) *Ascochyta fabae* on winter beans *Vicia faba*: pathogen spread and variation in host import resistance. *Plant Pathology*, **29**, 59-65.

Broadbent, L. (1951) Lettuce mosaic in the field. *Agriculture*, **57**, 578-582.

Cockerell, V. and Rennie, W.J. (1996) Survey of seed-borne pathogens in certified and farm-saved seed in Britain between 1992 and 1994. *Home-Grown Cereals Authority Project Report No. 124*, Home-Grown Cereals Authority, London.

Cockerell, V., Mulholland, V., McEwan, M.M. *et al.* (2001) Seed treatment according to need in winter wheat. BCPC Symposium Proceedings No. 76. *Seed Treatment:Challenges and Opportunities*. British Crop Protection Council, Farnham, UK, 111-116.

Cockerell, V., Thomas, J.T. and Clark, W.S. (2003) H-GCA Topic Sheet No.72. Sampling Wheat Seed for Seed-Borne Diseases. Home-Grown Cereals Authority, London.

Cockerell, V., Paveley, N.D., Clark, W.S. *et al.* (2004) Cereal seed health and seed treatment strategies: exploiting new seed testing technology to optimise seed health decisions for wheat. Home-Grown Cereals Authority Project Report No. 340, Home-Grown Cereals Authority, London.

Cockerell, V., Anderson, M. and Jacks, M. (2005) The incidence of loose smut (*Ustilago nuda*) in Scottish barley seed lots. Scientific Review 2000-2003. Scottish Agricultural Science Agency. Edinburgh, UK.

Couch, H.B.A. (1955) Studies on seed transmission of lettuce mosaic virus. *Phytopathology*, **45**, 63-70.

Cournoyer, B., Arnold, D., Jackson, R. and Vivian, A. (1996) Phylogenetic evidence for diversification of *Pseudomonas syringae* pv. *pisi* race 4 strains into two distinct lineages. *Phytopathology*, **86**(10), 1051-1056.

Dillon-Weston, W.A.R. and Engledow, F.C. (1930) The 'money' side of bunted wheat. Essex County Farmers Union Yearbook 1930, 1-6.

Dinant, S. and Lot, H. (1992) Lettuce mosaic virus. *Plant Pathology*, **41**(5), 528-542.

Ebbels, D.L. (2003) Principles of Plant Health and Quarantine, CABI Publishing, Wallingford, UK.

Falk, B.W. and Guzman, V.L. (1984) Differential detection of seedborne lettuce mosaic virus (LMV) in LMV-susceptible and resistant lettuce breeding lines. *Proceedings of the Florida State Horticulture Society*, **97**, 179-181.

Gabrielson, R.L. (1983) Black leg disease of crucifers caused by *Lepstosphaeria maculans* (*Phoma lingam*) and its control. *Seed Science and Technology*, **11**, 749-780.

Gaunt, R.E. and Liew, R.S.S. (1981) Control strategies for *Ascochyta fabae* in New Zealand field and broad bean crops. *Seed Science and Technology*, **9**(3), 707-715.

Grogan, R.G. (1980) Control of lettuce mosaic with virus-free seed. *Plant Disease*, **64**, 446-449.

Grogan, R.G. (1983) Lettuce mosaic virus control by use of virus-indexed seed. *Seed Science and Technology*, **11**, 1043-1049.

Grogan, R.G., Welch, J.E. and Bardin, R. (1952) Common lettuce mosaic and its control by the use of mosaic-free seed. *Phytopathology*, **42**, 573-578.

Heald, F.D. (1921) The relation of spore load to the per cent of stinking smut appearing in the crop. *Phytopathology*, **11**, 269-278.

Hewett, P.D. (1965) A survey of seed-borne fungi of wheat. The incidence of *Leptosphaeria nodorum* and *Griphosphaeria nivalis*. *Transactions of the British Mycological Society*, **48**, 59-72.

Hewett, P.D. (1966) *Ascochyta fabae* Speg. on tick bean seed. *Plant Pathology*, **15**, 161-163.

Hewett, P.D. (1973) The field behaviour of seed-borne *Ascochyta fabae* and disease control in field beans. *Annals of Applied Biology*, **74**, 287-295.

Hewett, P.D. (1978) Epidemiology of seed-borne disease, in *Plant Disease Epidemiology*, (eds P.R. Scott and A. Bainbridge), Blackwell Scientific Publications, Oxon, UK, pp. 167-176.

Hewett, P.D. (1979) Reaction of selected spring barley cultivars to inoculation with loose smut. *Plant Pathology*, **28**, 77-80.

Hewett, P.D. (1983) Seed-borne *Gerlachia nivalis* (*Fusarium nivale*) and reduced establishment of winter wheat. *Transactions of the British Mycological Society*, **80**, 185-186.

Hollaway, G.J. and Bretag, T.W. (1997) Survival of *Pseudomonas syringae* pv. *pisi* in soil and in pea trash and their importance as a source of inoculum for a following field crop. *Australian Journal of Experimental Agriculture*, **37**(3), 369-375.

Hollaway, G.B., Bretag, T.W., Gooden, J.M. and Hannah, M.C. (1996) Effect of soil water content and temperature on the transmission of *Pseudomonas syringae* pv. *pisi* from pea seed (*Pisum sativum*) to seedling. *Australasian Plant Pathology*, **25**, 26-30.

Jagger, I.C. (1921) A transmissable mosaic disease of lettuce. *Journal of Agricultural Research*, **20**, 737-739.

Jellis, G.J. and Punithalingam, E. (1991) Discovery of *Didymella fabae* sp.nov., the teliomorph of *Ascochyta fabae*, on faba bean straw. *Plant Pathology*, **40**(1), 150-157.

Kenyon, D.M., Thomas, J.E., Bates, J.A. and Taylor, E.J.A. (2001) Quantitative and qualitative detection of *Pyrenophora* species on barley seed using PCR in advisory seed health testing. BCPC Symposium Proceedings No. 76. *Seed Treatment: Challenges and Opportunities*. British Crop Protection Council, Farnham, UK, 63-68.

Knott, C.M., Biddle, A.J. and McKeown, B.M. (1994) PGRO Field Bean Handbook. Processors and Growers Research Organisation, Peterborough, UK.

Kruse, M. (2004) ISTA Handbook on Seed Sampling (second edition). International Seed Testing Association, Basserdorf, Switzerland.

Maude, R.B. (1996) Seed-borne Diseases and their Control. Principles and Practice, CAB International, Wallingford, Oxon, UK.

Meijerink, G. (1997) The International Seed Health Initiative, in *Seed Health Testing: Progress Towards the 21st Century*, (eds J.D. Hutchins and J.C. Reeves), CAB International, Wallingford, Oxon, UK.

Newhall, A.G. (1923) Seed transmission of lettuce mosaic. *Phytopathology*, **13**, 104-106.

Nielsen, B.J. and Jorgensen, C.N. (1994) Control of common bunt (*Tilletia caries* (DC) TULL) in Denmark, in *Seed Treatments: Progress and Prospects*, (ed. T.J. Martin), BCPC Monograph 57, British Crop Protection Council, Farnham, Surrey, UK, pp. 47-52.

Oxley, S.J.P. and Cockerell, V. (1996) Transmission of bunt (*Tilletia caries*) in wheat with reference to seed-borne inoculum, site and climate. *Proceedings Crop Protection Northern Britain 1996*, pp. 97-102.

Paveley, N.D., Rennie, W.J., Reeves, J.C. *et al.* (1996) Cereal Seed Health and Seed Treatment Strategies, Home-Grown Cereals Authority Research Review No. 34, Home-Grown Cereals Authority, London.

Phatak, H.C. (1974) Seed-borne plant viruses – Identification and diagnoses in seed testing. *Seed Science and Technology*, **2**, 3-155.

Popp, W. (1951) Infection in seeds and seedlings of wheat and barley in relation to development of loose smut. *Phytopathology*, **41**, 261-275.

Raid, R.N., Nagata, R.T. and Brown, L.G. (1996) A recent outbreak of lettuce mosaic potyvirus in commercial lettuce production in Florida. *Plant Disease*, **80**(3), 343.

Reeves, J.C., Wray, M.W. and Martin, T. (1994) *Seed Treatment: Progress and Prospects*. Proceedings of a Symposium Held at the University of Kent, Canterbury, 5-7 January, 1994. BCPC Monograph No. 57. British Crop Protection Council, Farnham, Surrey.

Rennie, W.J. (1987) Incidence and control of loose smut in Scottish barley 1980-1985. *Proceedings Crop Protection in Northern Britain 1987*, pp. 85-90.

Rennie, W.J. (1993) Control of seed-borne pathogens in certification schemes, in *Plant Health and the European Single Market*, (ed. D. Ebbels), BCPC Monograph No. 54, British Crop Protection Council, Farnham, Surrey, UK, pp. 61-68.

Rennie, W.J. and Seaton, R.D. (1975) Loose smut of barley. The embryo test as a means of assessing loose smut infection in seed stocks. *Seed Science and Technology*, **3**, 697-709.

Richards, M.C. (1990) Crop protection constraints and choice of variety for organic cereals. *Proceedings Crop Protection in Northern Britain 1990*, pp. 159-164.

Roberts, S.J. (1992) Effect of soil moisture on the transmission of pea bacterial blight (*Pseudomonas syringae* pv. *pisi*) from seed to seedling. *Plant Pathology*, **41**, 136-140.

Roberts, S.J. (1993) Effect of bacterial blight (*Pseudomonas syringae* pv. *pisi*) on the growth and yield of single pea (*Pisum sativum*) plants under glasshouse conditions. *Plant Pathology*, **142**, 568-576.

Roberts, S.J., Reeves, J.C., Biddle, A.J. *et al.* (1991) Prevalence of pea bacterial blight in UK seed stocks, 1986-1990, in *Aspects of Applied Biology*, **27**, Protection of Legumes, (ed. R.J. Froud-Williams, P Gladders, M.C. Heath *et al.*), Association of Applied Biologists, Wellesbourne, pp. 327-332.

Roberts, S.J., Phelps, K., McKeown, B.M. *et al.* (1995) Effect of pea bacterial blight (*Pseudomonas syringae* pv. *pisi*) on the yield of spring sown combining peas (*Pisum sativum*). *Annals of Applied Biology*, **126**, 61-73.

Shukla, D.D., Ward, C.W. and Brunt, A.A. (1994) The Potyviridae. CAB International, Wallingford UK.

Schwinn, F. (1994) Seed treatment – a panacea for plant protection? in *Seed Treatment: Progress and Prospects*, (ed. T.J. Martin), BCPC Monograph No. 57, British Crop Protection Council, Farnham, Surrey, UK, pp. 3-14.

Soper, D. (1995) *A Guide to Seed Treatments in the UK*, British Crop Protection Council, Farnham, Surrey, UK.

Stead, D.E. and Pemberton, A.W. (1987) Recent problems with *Pseudomonas syringae* pv. *pisi* in UK. *EPPO Bulletin*, **17**, 291-294.

Taylor, J.D. and Dye, D.W. (1976) Evaluation of streptomycin seed treatments for the control of bacterial blight of peas (*Pseudomonas pisi* Sackett 1916). *New Zealand Journal of Agricultural Research*, **91**, 91-95.

Taylor, J.D., Bevan, J.R., Crute, I.R. and Reader, S.L. (1989) Genetic relationships between races of *Pseudomonas syringae* pv. *pisi* and cultivars of *Pisum sativum*. *Plant Pathology*, **38**(3), 364-375.

Van der Plank, J.E. (1963) *Plant Diseases: Epidemics and Control*, Academic Press, New York.

Van Vuurde, J.W.L. and Maat, D.Z. (1981) Application of ELISA for the routine detection of lettuce mosaic virus, in *Report on the 17th International Workshop on Seed Pathology*, International Seed Testing Association, Zurich, pp. 26-27.

Wallen, V.R. and Galway, D.A. (1977) Studies on the biology and control of *Ascochyta fabae* on faba bean. *Canadian Plant Disease Survey*, **57**, 31-34.

Weltzein, H.C. (1957) Untersuchungen uber den Befall von Winterwizen durch Tilletia tritici (Bjerk). Winter unter be donderer Berucksichtingung der Frage der Beizmittelresistenz. *Phytopathologische Zeitschrift*, **29**, 121-150.

Wray, M.N. and Pickett, A.A. (1985) Trends in loose smut (*Ustilago nuda*) infections in certified seed of barley in England and Wales 1976-1983. *Journal of the National Institute of Agricultural Botany*, **XVII**, 31-40.

Yarham, D.J. (1993) Soil-borne spores as a source of inoculum for wheat bunt (*Tilletia caries*). *Plant Pathology*, **42**, 654-656.

Yarham, D.J. and Jones, D.R. (1992) The forgotten diseases: why we should remember them. *Proceedings of the Brighton Crop Protection Conference, Pests and Diseases, 1992*, British Crop Protection Council, Farnham, Surrey, UK, pp. 1117-1126.

Zerbini, F.M., Koike, S.T. and Gilbertson, R.L. (1995) Biological and molecular characterisation of lettuce mosaic potyvirus isolates from the Salinas Valley of California. *Phytopathology*, **85**(7), 746-752.

CHAPTER 14

DISEASES CAUSED BY SOIL-BORNE PATHOGENS

P. LUCAS

14.1 INTRODUCTION

The soil is a favourable habitat for microorganisms and is inhabited by a wide range of bacteria, fungi, algae, viruses and protozoa. Soils contain large numbers of microorganisms – usually between one and ten million per gram of soil – with bacteria and fungi the most prevalent. Some microorganisms present in soil are also able to infect plants. These so-called soil-borne plant pathogens may complete their life cycle in the soil, or may spend part of it on the aerial parts of the plant (Bruehl, 1987).

Plant roots take up the mineral nutrients and water essential for plant growth, but they also release a wide range of organic compounds into the surrounding soil. Thus, the area of soil in contact with the plant root, the rhizosphere, is a site of intense microbial activity. Not surprisingly, many microorganisms are more frequent on the surface of plant roots and in the rhizosphere than in the bulk soil not influenced by the presence of roots. The rhizosphere is therefore a key soil habitat, in which numerous interactions occur between plant roots and soil microorganisms. These interactions determine growth conditions for both the plant and the microorganisms in the rhizosphere.

Soil-borne pathogens require a susceptible plant for the development of their parasitic phase, but they may persist in the soil as saprophytes on residues, or as resistant, dormant forms, from several weeks to several years, depending on their biology. Both parasitic and saprophytic phases may be affected by the physico-chemical and biological characteristics of the soil. Soil-borne pathogens generally affect the root system of plants or the base of the stem (foot), in some cases developing on upper parts of the plant through aerial dispersal from soil inoculum or via transport and/or growth in the vessels, leading to vascular diseases.

Such pathogens may cause extensive damage to crops by limiting water and nutrient uptake (root necrosis) and/or transfer towards the upper parts of the plant (vascular disease), or by reducing the quality of crop products developing underground (root or tuber rot, gall, proliferation, etc.). This damage has led to the focusing of considerable effort on improving our understanding of the biology and ecology of these diseases, with the aim of developing control methods.

This chapter, after having identified specific characteristics of soil-borne pathogens, will provide an overview of the research on these pathogens carried out to date. It will then deal with recent advances in epidemiology, bearing in mind that although soil microbiology has been an area of intense research, epidemiological studies have been developed to a lesser extent for soil-borne than for foliar diseases (McDonald, 1994).

B. M. Cooke, D. Gareth Jones and B. Kaye (eds.), The Epidemiology of Plant Diseases, 2nd edition, 373–386.

14.2 THE SOIL-BORNE DISEASE EPIDEMIC

Soil-borne plant pathogens affect crops throughout the world and have been extensively studied. Research efforts have been justified by the economic impact of these diseases on crop production and by specific difficulties associated with studying and controlling soil-borne diseases, mostly due to the complex environment in which these diseases occur (Lucas and Sarniguet, 1998).

14.2.1 A closed environment limiting dispersal

Soil is a closed environment in which propagules capable of initiating epidemics (e.g. spores, sclerotia, mycelia and hyphae) cannot disperse over long distances, with the exception of certain spores or bacteria transported in run-off water or in soil flowing within the soil matrix. It is rare for horizontal dispersal to extend beyond the field margins.

For some soil-borne pathogens, infection is also transmitted by the growth of the pathogen on or through the soil, from a source of inoculum to a susceptible host. This situation mostly applies to fungi, which form mycelia capable of growing through a heterogeneous medium of pores, cracks and aggregates, although this growth is affected by many other physical, biological and chemical factors (Otten and Gilligan, 1998).

Thus, during the crop cycle, soil-borne pathogen propagules are naturally dispersed over short distances (from a few centimetres to a few decimetres). For this reason, diseased plants show up as patches within a field at the start of epidemics (e.g. take-all of winter wheat caused by *Gaeumannomyces graminis* var. *tritici*) (Hornby *et al.*, 1998; Cook, 2003). For soil-borne pathogens with an aerial phase, disease may extend to the whole field during the cropping season if climatic conditions are favourable for the disease (e.g. spore production and dispersal) as is the case for eyespot on wheat, caused by *Tapesia yallundae*.

Most soil-borne pathogens require oxygen. They are therefore mostly located towards the top of the soil profile and their vertical dispersal depends largely on water infiltration pathways and root progression.

14.2.2 A complex, opaque environment with intense biotic and abiotic interactions

Soil is a complex substance with solid, liquid and gaseous components. The organisation and interconnection of these components depend on soil texture, soil structure and external factors, such as climate. Soil structure, and its effects on the relationship between the liquid and gaseous phases, is a major feature determining microbial survival and development in soil. High soil moisture content generally favours microbial activity, but too much water may result in a high prevalence of water-filled pores, resulting in changes in the concentration of O_2, CO_2 or other gases, with consequences for the aerobic or anaerobic microbial communities of the soil microflora (McDonald, 1994).

The effects of the physicochemical characteristics of the soil on the behaviour of microorganisms have been investigated in detail but are still only imperfectly understood. One problem with most of these studies is that the effects of single factors (pH, nutrients, oligoelements, etc.) are often assessed *in vitro* on culture media, making it difficult to account for interactions between these factors. Furthermore, measurements of the physico-chemical characteristics of the soil may lead to the calculation of mean values for a soil sample that mask the great variability between the niches in which specific microbial communities live. It is difficult to observe the microbe in its natural habitat whilst evaluating the environment, without causing a disturbance. New methods for the micro-scale measurement of soil characteristics, based on the use of microsensors, are becoming available (Meyer *et al.*, 2002), as are new molecular techniques for identifying microorganisms and functions (Anderson and Cairney, 2004), and combinations of the two (Lüdemann *et al.*, 2000). These advances should provide us with answers to some of our questions, improving our understanding of what occurs in the niches in which soil-borne pathogens survive before they reach the root cells; this work should also benefit from concepts used in ecology (Griffin, 1985; Reynolds *et al.*, 2003).

The biological characteristics of the soil in relation to plant diseases have been extensively investigated over the last 40 years. One of the first major works published was a book on the ecology of soil-borne pathogens by Baker and Snyder (1965). It was later followed by a book on the biology and control of soil-borne plant pathogens edited by Bruehl (1975). More recently, Hornby (1990) published some of the contributions to the Soil-borne Plant Pathogens Section of the 5th International Congress on Plant Pathology. Making use of the diverse, high level of microbial activity in soil has been seen as a potential means of promoting the biological control of plant diseases. Many studies have concentrated on the identification, selection and application of biocontrol agents and few methods are currently available. Nevertheless, efforts are being made to increase the efficacy of candidate disease antagonists. These include genetic engineering to improve antibiotic production and exploring mechanisms that are important for their establishment in the courts or potential courts of infection by pathogens (Cook, 1993). Another, less well-studied approach is to manipulate soil management techniques such that naturally-occurring biological controls are conserved and can be exploited. Lucas and Sarniguet (1998) discuss these two approaches and suggest that managing the environment by stimulating naturally-occurring microorganisms and then enhancing their efficacy (if necessary and economically acceptable) by introducing specific biocontrol agents (into a more receptive environment) would be an effective complementary strategy (see also chapter 11).

14.2.3 An environment under human influence

Crop production is affected by a number of primary factors. Climate (e.g. sunshine and rainfall, conditioning light interception for photosynthesis and water uptake by the plant, and temperature, which drives crop growth and development) is difficult to modify and farmers simply have to deal with it. In contrast, soil (as a nutrient reservoir and matrix providing the root system with an ideal matrix for its development) is subject to a number of different management practices, from tillage and fertiliser application to sowing or planting. These practices, which are designed to optimise plant growth, also have an impact on soil microbial activity. They therefore have a direct or indirect effect on soil-borne pathogens and may be considered to be a means of managing plant health, in addition to plant growth.

The first objective of soil fertilisation is to satisfy crop nutrition demands. Deficiencies in major and minor nutrients may affect plant physiology, increasing infection levels and exacerbating yield losses caused by the disease. It may also have direct effects on soil-borne pathogens and on the biological and physico-chemical characteristics of the soil. The impact of nitrogen fertilisation on take-all of wheat provides a good illustration of these complex interactions: the application of a source of ammonium reduces take-all in most situations whereas nitrate applications do not have the same effect (Huber *et al.*, 1968). The uptake of NH_4^+ by roots decreases the pH of the rhizosphere. Smiley and Cook (1973) suggested that decreasing pH indirectly inhibits take-all by modifying the rhizosphere microflora at pH values between 5 and 7, and directly below pH 5. Smiley (1978a,b) subsequently reported that the application of a source of NH_4^+ increases the proportion of rhizosphere pseudomonads antagonistic to *G. graminis* var. *tritici in vitro* to a greater extent than the application of NO_3^-. Sarniguet *et al.* (1992a) showed, in pot bioassays and studies of fields cropped with take-all-infected winter wheat, that applications of ammonium-based fertiliser made the soil less receptive to take-all than applications of a nitrate or mixed (NH_4NO_3) fertiliser. The mixed fertiliser had an intermediate effect. Sarniguet *et al.* (1992b) demonstrated that the frequency of *in vivo* antagonistic fluorescent pseudomonads was higher in the NH_4^+-treated soil than in the NO_3^--treated soil. This work also demonstrated that the presence of rhizosphere pseudomonads can increase disease severity. These deleterious bacteria were more frequent in the nitrate-treated than in the ammonium-treated soil. Thus, antagonism observed *in situ* results from the overall effect of antagonistic and deleterious microorganisms, and nitrogen fertilisation (the form of nitrogen applied) affect these two biological components of soil receptivity to take-all.

Soil tillage affects soil structure, thereby affecting the behaviour of microorganisms. It also affects the distribution of crop residues in the soil profile. These residues remain in the top layers in no-tillage systems, but are buried by ploughing. This factor is important if successful infection requires the presence of infectious crop residues close to the soil surface, as is the case for eyespot on winter wheat, disseminated by spores carried over short distances by wind and rain drops (Colbach and Meynard, 1995). Soil tillage and other cultivation practices, including sowing, may also disperse inoculum within the field and even between fields. According to Truscott and Gilligan (2001), the observation that transmission distances within existing patches are frequently smaller than the expansion of patches between seasons suggests that there is a high level of mechanical inoculum dispersal during harvest and cultivation.

Disease can only occur if susceptible crop plants are grown. Successful infection from soil inoculum is more likely to occur with high inoculum and plant densities. Disease propagation is then favoured by short distances between plants. However, in the case of strictly soil-borne diseases, dispersal within a crop has been shown to be very limited, and build-up of the disease to epidemic levels requires several successive susceptible crops, and therefore several years in many cases. Such epidemics are described as polyetic (Zadoks, 1999). Crop rotation is recognised as the best way to keep levels of soil-borne diseases low, although it is not always acceptable to farmers for economic reasons. The mode of action of crop rotation was long thought to involve only the breakdown of pathogen inoculum build-up associated with decay of the inoculum during the cropping of non-host plants. However, different break crops have been shown to have different impacts on disease levels in the following host crop. In a study on the influence of crop rotation on foot and root diseases (take-all, sharp eyespot and eyespot) of wheat, Colbach *et al.* (1994) showed that host crops (wheat and barley) tended to increase the risk of the diseases whereas some non-host crops (alfalfa, peas, sunflower) decreased disease risk and others (maize, and sorghum for sharp eyespot) had an intermediate effect. These findings are consistent with those of Lucas *et al.* (1989), who showed that a soil cropped continuously with maize or cultivated under a wheat-maize rotation was far more conducive to take-all than the same soil cultivated under a wheat-beet rotation. The soil inhibiting disease development most strongly was wheat monoculture, providing evidence for take-all decline, which is known to be due to changes in soil microbial populations. The plant species grown is therefore a significant factor determining the composition of microbial soil communities living in soils and the rhizosphere. This applies not only to pathogens, but also to antagonistic and deleterious microbes. Lemanceau *et al.* (1996) demonstrated that two plant species, flax and tomato, modified in different ways the genetic and phenotypic diversity of the fluorescent pseudomonad community resident in the soil.

The cropping of resistant plants is limited by the fact that, curiously, cultivar selection has produced abundant examples of useful genetic resistance to above-ground but not to below-ground pathogens. Cook *et al.* (1995) suggested that the selection imposed by soil-borne pathogens may favour a different defence strategy which is for the plants to support and respond to populations of rhizosphere micro-organisms antagonistic to their pathogens. Attempts have been made to breed wheat cultivars able to react to hypovirulent strains of *G. graminis* var. *tritici* used as a biocontrol agent against take-all (Lemaire *et al.*, 1982). However, more attention has been paid to selecting bacteria displaying a combination of efficient root colonisation and beneficial effects on the activity of a given plant (Kuiper *et al.*, 2001) than to breeding plants able to exert beneficial selection pressure on microbial communities, although there are some reports of genotype-specific induction of soil microbial communities inhibiting soil-borne diseases such as rhizoctonia root rot in winter wheat cultivars (Mazzola and Gu, 2002; Mazzola, 2004). This approach is supported by studies in other areas demonstrating that the sensitivity of wheat (Rengel, 1997) and oat (Timonin, 1965) genotypes to manganese deficiency, for example, depends on the number of Mn-reducing microorganisms in the rhizosphere.

14.2.4 Consequences for the epidemiology of soil-borne diseases

The restricted dispersal of inoculum is a characteristic of soil-borne pathogens, accounting for the patchy distribution of diseased plants at the start of epidemics. Depending on whether these soil-borne pathogens have an aerial phase of development in their cycle, this patchiness may be observed only in part of the annual cycle or over several years of cultivation, with patches becoming larger before merging (Fig. 14.1). Truscott and Gilligan (2001) described this dynamic as a two-step process: (i) local amplification due to parasitic activity on the plants initially infected and transmission of the disease to neighbouring plants, (ii) dispersal of inoculum by water, wind or humans.

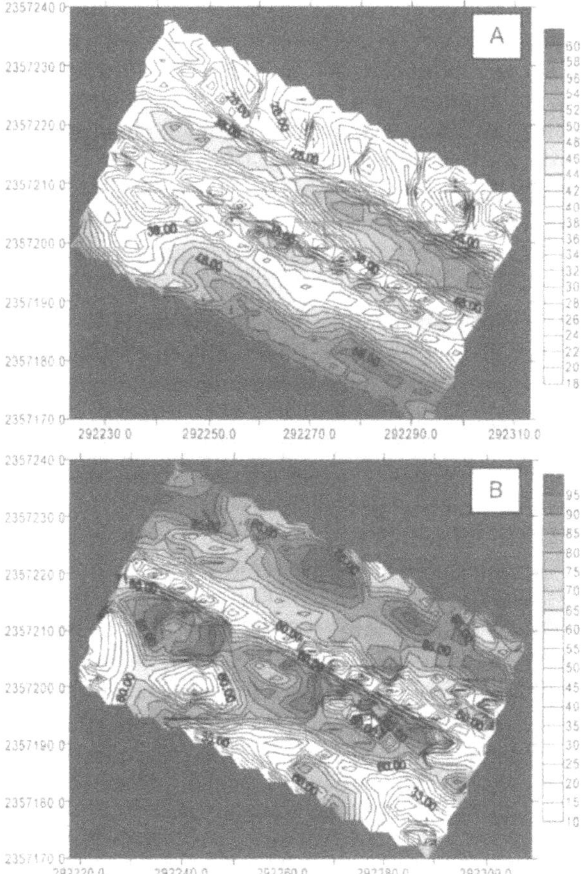

Figure 14.1. Yield (A, $10^{-1}t\ ha^{-1}$) and take-all disease incidence (B, per 100 diseased plants) maps of a 3rd winter wheat field, showing patchiness of the disease (Le Rheu, France, 1999; from Lamkadmi et al.*, 2000).*

Local amplification depends on complex interactions between plants and soil-borne pathogens, regulated by the physicochemical and microbial characteristics of the soil. It is difficult to measure these interactions and characteristics due to soil 'opacity'. One way of relating the epidemiology of a soil-borne disease (take-all of winter wheat) to cultural practices known or thought to have an impact on these interactions and characteristics is described below.

The aim of this work was to propose cropping strategies limiting disease risk. In order to evaluate these strategies based on their ability to decrease epidemics and to increase yield, a dynamic approach to the relationship between the kinetics of the disease and crop growth and development is also presented.

Inoculum dispersal has received less attention, especially that occurring between the harvesting of one crop and the sowing of the subsequent crop. Cultivation leads to the redistribution of inoculum. This may increase the likelihood of invasion, but it may also dilute the inoculum to levels below the threshold required for symptom development in the next crop (Truscott and Gilligan, 2001). The third part of the next section, will illustrate how current or possible management of this intercropping period can affect disease incidence in the next crop.

14.3 MODELLING SOIL-BORNE DISEASE EPIDEMIOLOGY

14.3.1 Relationship between cropping practices and disease dynamics

Much attention has been given to the use of non-linear models to describe the temporal progress of disease (Madden, 1980; Gilligan, 1985). Brasset and Gilligan (1989) compared the use of several non-linear models to describe the increase in the absolute number of diseased roots infected with *G. graminis* var. *tritici* in first and second wheat crops. They concluded that a model incorporating components of primary and secondary infection, together with inoculum decay, described the data in a manner consistent with biological constraints.

Colbach *et al.* (1997a) simplified one of these models and used it to assess the impact of crop management on the primary and secondary infection cycles of take-all epidemics. Origin of inoculum and infection rates are the central elements of this model. Inocula may be found in soils, on plant debris or on the roots of the living plant. Each inoculum is associated with an infection rate. Rate (c_1) corresponds to the capacity of the soil reservoir inoculum to cause infection and disease. The rate of secondary infection (c_2) is a measure of the capacity of infected roots to spread disease to other roots (or from a diseased plant to other plants). The percentage of diseased plants is given by the following equation, where time t is expressed as cumulative degree days (basis 0°C) since sowing:

$$y = \frac{1 - e^{-(c_1+c_2)t}}{1 + \frac{c_2}{c_1} e^{-(c_1+c_2)t}} \tag{14.1}$$

This equation was first successfully tested ($r^2 = 0.99$) on a plot assessed every two weeks after growth stage 30 (Zadoks *et al.*, 1974). It was then fitted to data concerning take-all build-up for each experimental treatment at three sites (three regions of France), where different cultural practices (sowing date, sowing density, total nitrogen dose, nitrogen fertiliser form, burial or removal of preceding crop residue) were tested. The parameters c_1 and c_2 were estimated for each experimental treatment at each site.

A linear model was tested to interpret c_1 and c_2 for each set of estimates at each site as a function of the factors analysed and co-variables measured. Sowing date always affected c_1 (e.g. primary infections) whereas c_2 (e.g. secondary infections) was influenced by sowing date only at the most favourable sites for disease (e.g. those with the highest infection rates, due to favourable climatic conditions). Early sowing systematically increased c_1. This is consistent with previous results (Hornby *et al.*, 1990) and the fact that early sowing provides a longer period for infection before winter. The effect of early sowing on c_2 was variable, positive for one experimental site, negative for another.

A positive correlation was found between plants m^{-2} and parameter c_1, but only at the most favourable sites. Plant density, like sowing date, had a variable effect on c_2. A high plant density at early stages, when the roots are still few in number and short, probably increases the chance of contact between the soil inoculum and living roots, whereas it has a less predictable effect when the root system is well developed.

High levels of nitrogen application increased c_1 and decreased c_2 but both these parameters were decreased by applications of nitrogen in the form of ammonium. As reported by Sarniguet *et al.* (1992a,b), nitrogen can stimulate both the pathogenic and the antagonistic microflora. Increases in the antagonistic microflora early in the infection of seminal roots facilitate the development of fluorescent pseudomonads on necrotic tissue. These pseudomonads then interfere with pathogen expansion, particularly if nitrogen fertiliser is applied in the form of ammonium.

The hierarchy of and interactions between various factors were shown to be important. Factors other than sowing date were generally significant only if sowing date was also significant. Sowing date may therefore be considered the dominant factor, and its interactions with other factors as the most important. This type of interaction is very similar to that observed for site: several factors had a stronger influence or were only significant if the site was favourable for disease development. Each factor seemed to amplify the risk due to the other effects and factors with a weak effect influenced disease only if factors with a strong effect were also present.

This model was used to assess the efficacy of new methods of control, such as the use of fungicidal seed treatments (Fig. 14.2). It was found that, in an early epidemic, the fungicide significantly reduced take-all incidence during all or most of the cropping season whereas, in late epidemics, it decreased incidence only moderately. Seed treatment was shown to reduce incidence by delaying primary infection (Schoeny and Lucas, 1999).

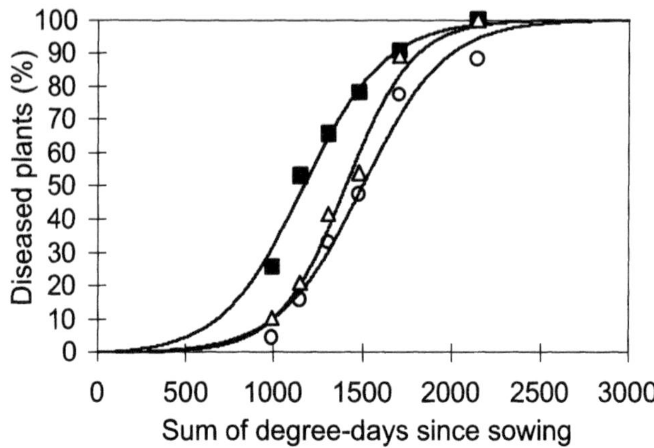

Figure 14.2. Illustration of take-all progress curves for various levels of an experimental seed treatment in a field experiment conducted in 1995, at Le Rheu, France. Symbols represent observed means for each experimental treatment (control ■ and two rates of fungicide Δ, O). (From Schoeny and Lucas, 1999).

Similar approaches, using a similar type of model, have been applied to other soil-borne diseases, including some with an aerial phase, such as eyespot (caused by *T. yallundae*) and sharp eyespot (caused by *Rhizoctonia cerealis*) on winter wheat (Colbach and Saur, 1998; Colbach *et al.,* 1997b). For both diseases, the impact of the major components of cropping systems (crop rotation, soil tillage, wheat management) was assessed by means of field trials, with several disease assessments during wheat growth. Early sowing consistently increased disease incidence through the primary infection cycle, whatever the disease. The frequencies of eyespot and sharp eyespot were increased by high plant density and/or small numbers of shoots per plant during the primary infection cycle. In contrast, in the secondary infection cycle, the frequency of these diseases was decreased by small shoot number per plant, that reduced late disease development at high density. For both diseases, high doses of nitrogen increased disease levels through both infection cycles. However, nitrogen fertiliser in the form of ammonium (*vs.* ammonium nitrate, i.e. 'mixed' fertiliser) decreased eyespot levels as it does for take-all, but had the opposite effect on sharp eyespot.

For eyespot, the model was also used to analyse the influence of crop residue distribution on disease development and infection cycles (Colbach and Meynard, 1995). Differences in the amount and placement of crop residues were achieved by varying crop rotation and soil tillage before the assessed winter wheat crop. When the previous crop was a host crop preceded by a non-host crop, soil inversion resulted in the burial of host residues, thereby decreasing primary infection risk. However, if the previous crop was a non-host crop preceded by a host crop, soil inversion carried the host residues back to soil surface, thereby increasing primary infection risk. Secondary infection was not correlated with crop succession or soil tillage.

14.3.2 Relationship between disease dynamics and yield components

Disease-yield loss relationships (see also Chapter 2) must be determined to assess the agronomic efficacy and economic benefits of control methods. Few studies have focused on the relationships between soil-borne disease progress curves and crop response, for three main reasons. First, studies on the development of soil-borne disease epidemics must include a large number of representative samplings, because of the patchy distribution of most of these diseases. Second, disease assessment for soil-borne pathogens involves destructive sampling, which leads to discontinuities in the dynamic representations of both disease and crop growth. Third, it is not always possible to compare, in equivalent conditions, a healthy situation with various levels of disease in the crop.

The question of yield losses due to take-all has been addressed by comparing situations involving different crop rotations (Slope and Etheridge, 1971), sowing dates (Bateman *et al.*, 1990), or artificial inoculations with different amounts of fungal inoculum (Rothrock, 1988) or with the same amount of inoculum incorporated at different depths (Hornby and Bateman, 1990), in order to generate differences in epidemic patterns. These approaches have generally focused on total yield at harvest, but rarely, the various yield components formed successively during the wheat cropping season (Meynard and Sebillotte, 1994) have been investigated: ear number per square meter (sowing to mid-stem elongation), grain number per ear (floral initiation to flowering), grain number per square meter (sowing to flowering), and 1,000 grain weight (flowering to maturity). The impact of the disease is likely to depend on when disease occurs and, consequently, on the nature of the yield components affected. However, most studies have simply established correlations between damage (i.e. yield reduction) and disease level at flowering (Bateman *et al.*, 1990), grain filling (Slope and Etheridge, 1971; Hornby and Bateman, 1990), or harvest (McNish and Dodman, 1973), and have taken no account of the link between disease dynamics and crop growth dynamics.

Schoeny and Lucas (1999) carried out a series of experiments in which a fungicidal seed treatment was used to generate different disease incidence and severity progress curves at a single location, with identical cultural practices and climatic conditions. Schoeny *et al.* (2001) then investigated the effects of various take-all epidemics on yield formation as a function of disease progression. Simple linear regression models involving various disease variables were compared and their ability to account for and predict the losses of yield components was assessed. Yield losses at harvest were strongly linked to the area under the disease progress curve (AUDPC) for disease incidence calculated between sowing and flowering (Fig. 14.3). The observed losses were larger for plots to which low rates of fertiliser were applied than for plots to which high rates of fertiliser were applied. Losses in terms of ear number per square meter, grain number per ear, and grain number per square meter were mainly related to cumulative disease incidence, calculated as AUDPC, during periods corresponding to yield component formation (sowing to mid-stem elongation, floral initiation to flowering, and sowing to flowering, respectively). In contrast, 1,000 grain weight losses were linked to disease incidence at mid-stem elongation (i.e. at a growth stage before the formation of this yield component, grain filling). This relationship is particularly interesting because of its predictive nature. It can be interpreted as an early effect of take-all on nitrogen and carbon assimilates, limiting re-mobilisation from stems and leaves during grain filling.

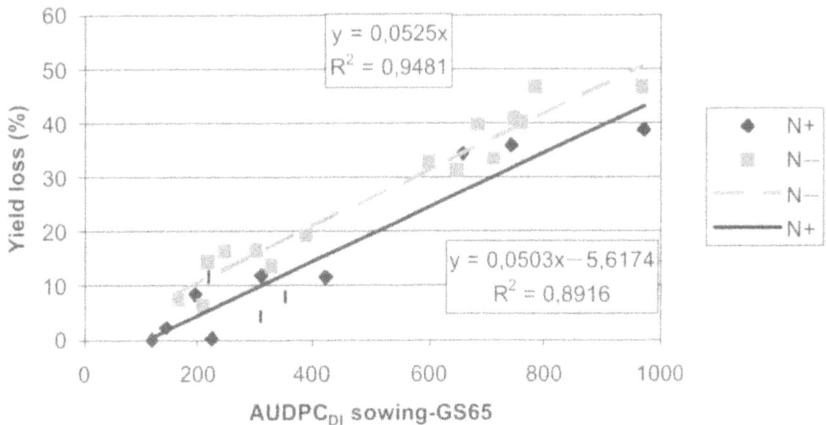

Figure 14.3. Disease-yield loss relationships established from field experiments on winter wheat crops infected with take-all. Yield loss function of cumulative disease incidence between sowing and flowering is established for low (N–) and high (N+) levels of nitrogen fertilisation. (From Schoeny et al., *2001).*

This was confirmed in a subsequent study showing that although wheat plants with severe take-all infection took up more nitrogen per unit of efficient root than uninfected plants, this compensatory response was insufficient to give nitrogen accumulation levels equivalent to those in healthy plants (Schoeny *et al.,* 2003). Thus, split applications of nitrogen with the amount of nitrogen adapted to the lower capacity of infected root systems to absorb nitrogen, as proposed by Lucas *et al.* (1997), might be of value.

14.3.3 Importance of cultivation management between crops

The role played by the intercrop period has not received adequate attention from plant pathologists in the analysis of soil-borne pathogen epidemics. The severity of rhizoctonia root rot on wheat has been linked to the presence of volunteers and weeds growing in the field between harvest and planting of the subsequent crop. These plants act as a 'green bridge', maintaining or increasing the potential inoculum of many plant pathogens, such as *R. solani* AG-8 in particular (Smiley *et al.,* 1992). Dulout *et al.* (1997) compared the effects of wheat volunteers, blackgrass (*Alopecurus myosuroides*) and bare soil on soil infectivity and soil conduciveness to take-all. They showed that both wheat volunteers and blackgrass maintained soil infectivity at a higher level than bare soil. Soil conduciveness was reduced by wheat volunteers whereas bare soil and blackgrass were highly conducive to the disease. Intermediate cropping (e.g. growing of a crop between harvest of the previous cash crop and sowing of the subsequent crop), already recommended for limiting soil erosion and nitrate leaching, should therefore also be considered as a way of improving soil health (Ennaïfar *et al.,* 2005).

14.4 CONCLUSION

Soil-borne pathogens are difficult to control with pesticides because it is difficult to target the niches in which they are found without treating the whole soil profile. Soil disinfection is no longer acceptable, for environmental reasons, and never was acceptable for some crops for economic reasons. Most alternative methods are only partly effective at controlling these diseases, but may act on different phases of the disease development cycle. As soil-borne diseases often develop more slowly than air-borne diseases, this provides an opportunity to time actions throughout the crop cycle, and even between crops, which may be useful given the polyetic characteristic of these epidemics.

This approach requires accurate description of the processes underlying disease epidemics, damage and resulting yield losses, comprehensive organisation of these processes in time and space, identification of the determinants affecting these processes and the identification of possible ways to control epidemics and minimise yield losses. Epidemiology and modelling are central to this kind of approach, (i) describing the behaviour, dynamics and damage of soil-borne pathogens at a range of ecological scales and (ii) making this information available to farmers and advisors to enable them to implement integrated crop protection strategies.

REFERENCES

Anderson, I.C. and Cairney, J.W.G. (2004) Diversity and ecology of soil fungal communities: increased understanding through the application of molecular techniques. *Environmental Microbiology,* **6**(8), 769-779.

Baker, R.R. and Snyder, W.C. (1965) *Ecology of soil-borne plant pathogens – Prelude to biological control.* University of California Press, Berkeley, Los Angeles. 571 pp.

Bateman, G.L., Hornby, D. and Gutteridge, R.J. (1990) Effects of take-all on some aspects of grain quality of winter wheat. *Aspects of Applied Biology,* **25**, 339-348.

Brasset, P.R. and Gilligan, C.A. (1989) Fitting of single models for field disease progress data for the take-all fungus. *Plant Pathology,* **38**, 397-407.

Bruehl, G.W. (1975) *Biology and control of soil-borne plant pathogens.* The American Phytopathological Society, St Paul, Minnesota. 216 pp.

Bruehl, G.W. (1987) *Soilborne plant pathogens,* Macmillan Publishing Company, New York.

Colbach, N., Lucas, P. and Cavelier, N. (1994) Influence des successions culturales sur les maladies du pied et des racines du blé d'hiver. *Agronomie,* **14**, 525-540.

Colbach, N. and Meynard, J.M. (1995) Soil tillage and eyespot: influence of crop residue distribution on disease development and infection cycles. *European Journal of Plant Pathology,* **101**, 601-611.

Colbach, N., Lucas P. and Meynard J.M. (1997a) Influence of crop management on take-all development and disease cycles on winter wheat. *Phytopathology,* **87**, 26-32.

Colbach, N., Lucas, P., Cavelier, N. *et al.* (1997b) Influence of cropping system on sharp eyespot in winter wheat. *Crop Protection,* **16**, 415-422.

Colbach, N. and Saur, L. (1998) Influence of crop management on eyespot development and infection cycles of winter wheat. *European Journal of Plant Pathology,* **104**, 37-48.

Cook, R.J. (1993) Making greater use of introduced microorganisms for biological control of plant pathogens. *Annual Review of Phytopathology,* **31**, 53-80.

Cook, R.J., Tomashow, L.S., Weller, D.W. *et al.* (1995) Molecular mechanisms of defense by rhizobacteria against root disease. *Proceedings of the National Academy of Sciences*, **92**, 4197-4201.

Cook, R.J. (2003) Take-all of wheat. *Physiological and Molecular Plant Pathology,* **62**, 73-86.

Dulout, A., Lucas, P., Sarniguet, A. *et al.* (1997) Effects of wheat volunteers and blackgrass in set-aside following a winter wheat crop on soil infectivity and soil conduciveness to take-all. *Plant and Soil,* **197**, 149-155.

Ennaïfar, S., Lucas, P., Meynard, J-M. *et al.* (2005) Effects of summer fallow management on take-all of winter wheat caused by *Gaeumannomyces graminis* var. *tritici. European Journal of Plant Pathology,* **112**, 167-181.

Gilligan, C.A. (1985) Construction of temporal models: III. Disease progress of soil-borne pathogens, in *Advances in Plant Pathology, Volume 3, Mathematical Modelling of Crop Disease,* (ed. C.A. Gilligan), Academic Press, London, pp. 67-105.

Griffin, D.M. (1985) Soil as an environment for the growth of root pathogens, in *Ecology and Management of soilborne plant pathogens* (eds C.A. Parker, A.D. Rovira, K.J. Moore, P.T.W. Wong and J.F. Kollmorgen), American Phytopathological Society, St Paul, Minnesota, pp. 187-190.

Hornby, D. (1990) *Biological Control of Soil-borne Plant Pathogens.* C.A.B. International, Wallingford, 479 pp.

Hornby, D. and Bateman, G.L. (1990) Artificial infestation of soil with *Gaeumannomyces graminis* var. *tritici* to study the relationship between take-all and wheat yields in field experiments. *Soil Use Management,* **6**, 209-217.

Hornby, D., Bateman, G.L., Gutteridge, R.J. *et al.* (1990) Experiments in England and France on fertilisers, fungicides and agronomic practices to decrease take-all. *Brighton Crop Protection Conference – Pests and Diseases* 1990, **2**, 771-776.

Hornby, D., Bateman, G.L., Gutteridge, R.J., *et al.* (1998) *Take-all Disease of Cereals: A Regional Perspective.* CAB International, London, 400 pp.

Huber, D.M., Painter, C.C., McKay, H.C., *et al.* (1968) Effect of nitrogen fertilization on take-all of winter wheat. *Phytopathology,* **58**, 1470-1472.

Kuiper, I., Bloemberg, G.V. and Lugtenberg, B.J. (2001) Selection of a plant-bacterium pair as a novel tool for rhizostimulation of polycyclic aromatic hydrocarbon-degrading bacteria. *Molecular Plant Microbe Interactions*, **14**, 1197-1205.

Lamkadmi, Z., Fouad, Y., Nicolas, H. *et al.* (2000) Potentialités de la télédétection rapprochée, dans les domaines optique et thermique pour la cartographie d'attaques parasitaires sur culture, in *Agriculture de précision: avancées de la recherche technologique et industrielle. Actes du colloque CEMAGREF – ENESAD, Dijon, 29 et 30 mai 2000,* Educagri Editions, pp. 223-234.

Lemaire J.M., Doussinault G., Lucas P., *et al.* (1982) Possibilités de sélection pour l'aptitude à la prémunition dans le cas du piétin échaudage des céréales, *Gaeumannomyces graminis. Cryptogamie Mycologie,* **3**, 347-359.

Lemanceau, P., Corberand, T., Laguerre, G. *et al.* (1996) The composition of fluorescent pseudomonad populations associated with roots is influenced by plant and soil type. *Applied and Environmental Microbiology,* **62**, 2449-2456.

Lucas, P., Sarniguet, A., Collet, A. *et al.* (1989) Réceptivité du sol au piétin-échaudage (*Gaeumannomyces graminis* var. *tritici*): influence de certaines techniques culturales. *Soil Biology and Biochemistry,* **21**, 1073-1078.

Lucas, P., Jeuffroy, M.-H., Schoeny, A. *et al.* (1997) Basis for nitrogen fertilisation management of winter wheat crops infected with take-all. *Aspects of Applied Biology,* **50**, 255-262.

Lucas, P. and Sarniguet, A., (1998) Chapter 9: Biological control of soil-borne pathogens with resident versus introduced antagonists: Should diverging approaches become strategic convergence? in *Conservation Biological Control,* (ed. P. Barbosa), Academic Press New-York, pp. 351-370.

Lüdemann, H., Arth, I. and Liesack, W. (2000) Spatial changes in the bacterial community structure along a vertical oxygen gradient in flooded paddy soil cores. *Applied and Environmental Microbiology,* **66**, 754-762.

MacDonald J.D. (1994) The Soil Environment, in *Epidemiology and Management of Root Diseases,* (eds C. Lee Campbell and D. Michael Benson), Springer-Verlag Berlin Heidelberg pp. 82-116.

MacNish, G.C., and Dodman, R.L. (1973) Relation between incidence of *Gaeumannomyces graminis* var. *tritici* and grain yield. *Australian Journal of Biological Sciences,* **26**, 1289-1299.

Madden, L.V. (1980) Quantification of disease progression. *Protection Ecology,* **2**, 159-176.
Mazzola, M. and Gu, Y.-H. (2002) Wheat genotype-specific induction of soil microbial communities suppressive to to *Rhizoctonia solani* AG-5 and AG-8. *Phytopathology,* **92**, 1300-1307.
Mazzola, M. (2004) Assessment and management of soil microbial community structure for disease suppression. *Annual Review of Phytopathology,* **42**, 35-59.
Meyer, R.L., Kjaer, T. and Revsbech, N.P. (2002) Nitrification and denitrification near a soil-manure interface studied with a nitrate-nitrite biosensor. *Soil Science Society of America Journal,* **66**, 498-506.
Meynard, J.-M. and Sebillotte, M. (1994) L'élaboration du rendement du blé, base pour l'étude des autres céréales à talles, in *Un point sur... Elaboration du rendement des principales cultures annuelles*, (eds L. Combe and D. Picard). INRA Editions, Paris, pp. 31-51.
Otten, W. and Gilligan, C.A. (1998) Effect of physical conditions on the spatial and temporal dynamics of the soil-borne fungal pathogen *Rhizoctonia solani. New Phytologist,* **138**, 629-637.
Rengel, Z. (1997) Root exudation and microflora populations in the rhizosphaere of crop genotypes differing in tolerance to micronutrient deficiency. *Plant and Soil,* **196**, 255-260.
Reynolds, H.L., Packer, A., Bever, J.D., *et al.* (2003) Grassroots ecology: Plant-microbe-soil interactions as drivers of plant community structure and dynamics. *Ecology,* **84**, 2281-2291.
Rothrock, C.S. (1988). Relative susceptibility of small grains to take-all. *Plant Disease,* **72**, 883-886.
Sarniguet, A., Lucas, P. and Lucas, M. (1992a) Relationships between take-all, soil conduciveness to the disease, populations of fluorescent pseudomonads and nitrogen fertilizers. *Plant and Soil,* **145**, 17-27.
Sarniguet, A., Lucas, P., Lucas, M. *et al.* (1992b) Soil conduciveness to take-all of wheat: Influence of the nitrogen fertilizers on the structure of populations of fluorescent pseudomonads. *Plant and Soil,* **145**, 29-36.
Schoeny, A. and Lucas, P. (1999) Modeling of take-all epidemics to evaluate the efficacy of a new seed-treatment fungicide on wheat. *Phytopathology,* **89**, 954-961.
Schoeny, A, Jeuffroy, M.H. and Lucas, P. (2001) Influence of take-all epidemics on winter wheat yield formation and yield loss. *Phytopathology,* **91**, 694-701.
Schoeny, A., Devienne-Barret, F., Jeuffroy, M.H. *et al.* (2003) Effect of take-all root infections on nitrate uptake in winter wheat. *Plant Pathology,* **52**, 52-59.
Slope, D.B. and Etheridge, J. (1971) Grain yield and incidence of take-all (*Ophiobolus graminis* Sacc.) in wheat grown in different crop sequences. *Annals of Applied Biology,* **67**, 13-22.
Smiley, R.W. and Cook, R.J. (1973) Relationships between take-all of wheat and rhizosphere pH in soils fertilized with ammonium vs nitrate nitrogen. *Phytopathology,* **63**, 882-890.
Smiley, R.W. (1978a) Antagonists of *Gaeumannomyces graminis* from the rhizoplane of wheat in soils fertilized with ammonium or nitrate nitrogen. *Soil Biology and Biochemistry,* **10**, 169-174.
Smiley, R.W. (1978b) Colonization of wheat roots by *Gaeumannomyces graminis* inhibited by specific soils, microorganisms and ammonium-nitrogen. *Soil Biology and Biochemistry,* **10**, 175-179.
Smiley, R.W., Ogg, A.G. and Cook, R.J. (1992) Influence of glyphosate on Rhizoctonia root rot, growth, and yield of barley. *Plant Disease*, **76**, 937-942.
Timonin, M.I. (1965) Interaction of higher plants and soil microorganisms, in *Microbiology and Soil Fertility*, (eds C.M. Gilmore and O.N. Allen). Oregon State University Press, Corvallis, pp. 135-138.
Truscott, J.E. and Gilligan, C.A. (2001) The effect of cultivation on the size, shape, and persistence of disease patches in fields. *Proceedings of the National Academy of Sciences,* **98**, 7128-7133.
Zadoks, J.C., Chang, T.T. and Konzak, C.F. (1974) A decimal code for the growth stages of cereals. *Weed Research*, **14**, 415-421.
Zadoks, J.C. (1999) Reflections on space, time, and diversity. *Annual Review of Phytopathology,* **37**, 1-17.

CHAPTER 15

WIND-DISPERSED DISEASES

B. HAU AND C. DE VALLAVIEILLE-POPE

15.1 INTRODUCTION

Dispersal by wind is important in many plant pathosystems but this chapter is restricted to two groups of important wind-dispersed diseases: the powdery mildews and the rusts. Both groups are of special interest in epidemiology because they are highly sensitive to the environment and their presence can vary from season to season. The fungal pathogens causing these diseases are biotrophic obligate parasites creating well known diseases on many crop plants. Here, the discussion concentrates on cereals, which can be seriously affected by powdery mildew and rust fungi, and particularly on the powdery mildews of wheat and barley and the rusts of wheat.

Powdery mildew of cereals is caused by *Blumeria graminis* (syn. *Erysiphe graminis*) which is specialized into *formae speciales* on cereals in temperate climates, i.e. on wheat, barley, oats, rye and their hybrids. According to Oerke *et al.* (1994), mildew on wheat is an important disease world-wide but more serious in Europe. The biology and epidemiology of cereal powdery mildews have been discussed by Jenkyn and Bainbridge (1978), Aust and von Hoyningen-Huene (1986), Jørgensen (1988) and Wolfe and McDermott (1994), the latter two with emphasis on genetics.

Cereal rusts are divided into five main species, which are also specialized into *formae speciales* (i.e. on wheat, barley, rye, oats), but the separations between the *formae speciales* are not always strict. The three wheat rust species are adapted to particular environmental niches, that correspond to their specific requirements during the infection stages. Stem (black) rust, caused by *Puccinia graminis* f.sp. *tritici* is adapted to the warmest conditions; stripe (yellow) rust caused by *Puccinia striiformis* is adapted to the coolest conditions, while leaf (brown) rust caused by *Puccinia triticina* = *P. recondita* f.sp. *tritici* (Anikster *et al.,* 1997) is intermediate. Stripe rust is well adapted to cool maritime or high altitude climates or where wheat is grown in winter seasons in countries that have warm summers, such as north India, Australia and the US Pacific northwest (Johnson, 1992). Severe stripe rust epidemics are sporadic although they have recently occurred in Australia in 2003, in the United States in 2000 (Chen *et al.,* 2002), and in China in 2002 (Wan *et al.*, 2004). Leaf rust occurs almost annually in most areas and is considered to be the most widely distributed of the three rusts. Cereal rust losses in Europe are primarily associated with stripe and leaf rusts. The importance of stem rust declined due to

B. M. Cooke, D. Gareth Jones and B. Kaye (eds.), The Epidemiology of Plant Diseases, 2nd edition, 387–416.

barberry eradication in the early 20th century (which enabled the survival of *P. graminis* f.sp. *tritici* in the aecidial stage) and from the infrequent occurrence of favourable temperatures. Stem rust occurs in North America, Australia and Argentina, but because of the use of resistant cultivars, its incidence is usually very low nowadays. Selection for earliness in European cultivars contributed to their escape from stem rust. The epidemiology of stripe rust has been reviewed by Rapilly (1979), Stubbs (1985), and Line (2002), that of leaf rust by Samborski (1985) and Eversmeyer and Kramer (2000), and stem rust by Roelfs (1985a,b) and Eversmeyer and Kramer (2000), and the genes for resistance to rust fungi in wheat by McIntosh *et al.* (1995b).

The aim of this section is to compare cereal powdery mildews and rusts which have features in common, such as wide distribution, rapid development within or on host tissue, massive production of spores, the ability to remain viable after long-distance dispersal and a high capacity to become virulent on previously resistant cultivars. The first part compares monocyclic components of the fungal infection cycle, including spore dispersal, influenced by meteorological and biotic factors (e.g. de Vallavieille-Pope *et al.*, 2000). This knowledge forms the basis for the development of models describing epidemics of mildews (e.g. Hau, 1985) and of rusts (e.g. Shrum, 1975). The chapter will then continue with survival strategies in connection with the sexual state and, finally, with the dynamics of sub-populations defined by virulences or fungicide resistance.

15.2 METEOROLOGICAL AND BIOTIC EFFECTS ON THE PHASES OF THE ASEXUAL LIFE CYCLE

Mildews and rusts are polycyclic diseases, which can complete the asexual cycle several times within a season. For mildews, the asexual cycle is the production of haploid conidia while the ascospores, which are the result of the sexual cycle, can occasionally initiate epidemics. For rust fungi, the life cycles are more complicated because macrocyclic heteroecious rusts can develop five types of spores: urediniospores (binucleate), teliospores (in which meiosis occurs), basidiospores (uninucleate and haploid) on the cereal host, and pycniospores (uninucleate, with plasmogamy between different mating types) and aeciospores (binucleate) on the alternate host. The exception is *P. striiformis* f.sp. *tritici*, for which the alternate host is unknown. Rust epidemics on cereals are mainly caused by the dikaryotic urediniospores.

For cereal mildews, it is generally assumed that the different *formae speciales* react identically to environmental conditions so that powdery mildews on wheat and on barley are driven by the same factors. The different rust species on wheat, however, have different environmental requirements.

15.2.1 Infection

The infection process of powdery mildew and rust fungi can be subdivided into germination, elongation of the germ tube, appressorium formation, penetration and haustorium formation. Mildew conidia penetrate the host plant with an infection peg through the epidermis. Rust urediniospores need to be hydrated to germinate and the germ tube tends to grow at right angles to the long axis of the leaf and then to reach closest stomata through which penetration takes place. The germ tube terminates in a swelling to form an appressorium in the cases of *B. graminis, P. triticina* and *P. graminis* f.sp. *tritici; P. striiformis* f.sp. *tritici* does not differentiate appressoria. The establishment of a haustorium in the plant cell, which serves to take up nutrients, terminates the metabolically independent phase of spore infection. *Blumeria graminis* forms haustoria in the epidermal cell and *Puccinia* spp. in adjacent mesophyll cells. The mycelium is exophytic, i.e. on the leaf surface, for powdery mildews and endophytic (in the intercellular spaces within the leaf) for cereal rusts. Average infection efficiency under optimal conditions is similar (20-45%) for these pathogens (Table 15.1).

The germination and penetration of *B. graminis* and *Puccinia* spp. spores are mainly influenced by temperature, relative humidity and wetness conditions. The cardinal temperatures for infection (Table 15.1) reflect the relatively wide adaptation of *B. graminis* and *P. triticina* compared with that of *P. striiformis* f.sp. *tritici* (which is restricted to cooler temperatures) and *P. graminis* f.sp. *tritici* (warmer temperatures). Light is not required for substomatal vesicle formation for *B. graminis*, *P. triticina* and *P. striiformis* f.sp. *tritici* but is reported to be necessary for *P. graminis* f.sp. *tritici* (Burrage, 1970).

Blumeria graminis germination can take place over a wide range of relative humidities, even near 0%. Germination rate increases with relative humidity, reaching the highest levels at 97-100% (Friedrich and Boyle, 1993). Very high vapour pressure deficits of the air, which also reflect dry conditions, have negative effects on mildew infection (Friedrich, 1995c). The high water content of the conidia (63-70%) is considered as a reason for the primary germ tube forming and attaching itself onto the leaf in the absence of external water (Friedrich and Boyle, 1993). For the growth of the appressorial germ tube in dry conditions, the water has to be taken up from the host by the primary germ tube (Carver and Bushnell, 1983). The role of adhesion to the cereal surface in the infection process has been described by Nicholson (1996).

Many contradictory statements exist concerning the effect of liquid water on the mildew germination process. While several workers have reported a deleterious effect of water on conidia (e.g. Pauvert and de la Tullaye, 1977), Merchán and Kranz (1986a) stated that leaf wetness up to 72 h did not affect germination and appressorial formation. The negative effect of rain on infection was mainly due to the washing-off of conidia and not so much due to the inhibition of germination (Merchán and Kranz, 1986b). Conidia of *B. graminis* f.sp. *hordei* can retain their ability to grow normally on leaves after a period on or in water (Sivapalan, 1993). This may be why inoculation of barley plants by spraying conidia suspended in water is possible (Lumbroso *et al.,* 1982). A distinct inhibition of infection was, however, observed in the presence of guttation droplets and liquid excretions of leaf pieces (Merchán and Kranz, 1986a).

Free water on the leaf surface in the form of dew droplets (usually formed during the night) is essential for the germination of rust urediniospores. The minimum

Table 15.1 Assessement of monocyclic parameters for Blumeria graminis, Puccinia striiformis *f. sp.* tritici, Puccinia triticina *and* Puccinia graminis *f. sp.* tritici *under near-optimal conditions*

Species	*Latent period (days)*	*Infectious period (days)*	*Daily sporulation (spores/lesion)*	*Infection efficiency (lesions/spore) (%)*	*Cardinal temperatures for infection (°C)*		
					minimal	*optimal*	*maximal*
B. graminis	5[a]	13-23[d]	$15 \cdot 10^{3}$[d]	33[a]	0-2	12-24[a]	<30[d]
P. striiformis	10-14[b]	30[b]	$15 \cdot 10^{3}$[b]	>40[e,f]	0[f]	5-14[e,f]	20[e,f]
P. triticina	7-8[b]	20[b]	$2 \cdot 10^{3}$[b]	40[f]	2[f]	15[f]	<30[f]
P. graminis tritici	7-9[c]	26[c]	$20 \cdot 10^{3}$[c]	20[g]-45[c]	4[h]	23.5[h]	29[h]

[a]Eckhardt *et al.* (1984b), [b]Sache and de Vallavieille-Pope (1993), [c]Pande *et al.* (1978), [d]Aust (1981), [e]de Vallavieille-Pope and Leconte (unpublished data under conditions of high light intensity received by the plants before inoculation), [f]de Vallavieille-Pope *et al.* (1995, 2002), [g]Wiese and Ravenscroft (1979), [h]Burrage (1970).

continuous dew period necessary for penetration, 2-4 h for *P. graminis* f.sp. *tritici* (Burrage, 1970) and 4-6 h for *P. striiformis* f.sp. *tritici* and *P. triticina* (de Vallavieille-Pope *et al.*, 1995), increases to at least 16 h at sub-optimal temperatures.

An interruption of the wet period by a dry period does not affect ungerminated spores, which are able to infect during a subsequent period (de Vallavieille-Pope *et al.*, 1995). *Puccinia triticina* and *P. striiformis* f.sp. *tritici* are both unable to survive if a dry period occurs between urediniospore germination and penetration. When appressoria are already formed but not substomatal vesicles, *P. triticina* is sensitive to wetness interruption. Resistance to dryness is no greater for *P. triticina* than for *P. striiformis* f.sp. *tritici,* which does not differentiate appressoria. After the formation of substomatal vesicles, the pathogens are not affected by wetness conditions on the leaf surface.

The light quantity received by the plants before inoculation is a major factor in modulating the infection efficiency of *P. striiformis* f.sp. *tritici* but has no effect on that of *P. triticina*. Under both controlled and natural conditions, values of *P. striiformis* f.sp. *tritici* infection efficiency were in the range of 0.4 to 40%, and increased as a function of light quantity received by the plants on the day before inoculation (de Vallavieille-Pope *et al.*, 2002). For stripe rust, three environmental variables – pre-inoculation light quantity received by the plants, post-inoculation temperature, and post-inoculation dew period – were needed to predict the infection efficiency in the field.

Apart from the conditions prevailing during the infection process, *B. graminis* infection efficiency depends also on the conditions at conidiogenesis; it decreases, for example, with higher temperature (Aust, 1981). Moreover, infection efficiency of spores produced during the infectious period changes for both mildews and rusts, even when the temperature remains constant. For instance, the infectivity of mildew conidia at 14°C is highest 5 days after the start of the sporulation (60%) and decreases to 20% after 20 days (Aust, 1981). Infection efficiency of the *P. triticina* urediniospores fluctuates in the range of 6-40% during the infectious period (Sache and de Vallavieille-Pope, 1993). Variation in germination rate of *P. graminis* f.sp. *tritici* urediniospores during the infectious period has been reported (Burrage, 1970) but germination rate is not an estimate of infection efficiency.

The differentiation of infection structures can be induced *in vitro* when appropriate chemical or physical signals are provided. A volatile leaf alcohol acts synergistically with topographical signals mimicking gramineaceous stomata for inducing *P. graminis* f.sp. *tritici* appressoria (Collins *et al.*, 2001). Under a humid atmosphere, a physical signal (mild heat shock) combined with the volatile inductor leads to the differentiation of haustorial mother cells in *P. graminis* f.sp. *tritici* (Wiethölter *et al.*, 2003). The morphogenetic programme of sequential infection structure differentiation is triggered by a number of host-derived signals.

15.2.2 Incubation and latent periods

The speed of a polycylic epidemic is largely influenced by incubation and latent periods, which are the times between host infection and appearance of the first symptoms and between host infection and the start of sporulation, respectively. Even under constant conditions, the delay between appearance of the first and last sporulating lesions can be 10 days for the rusts and 20 days for powdery mildew (Rapilly, 1991). However, this variation is only important for the start of an epidemic. The mean duration of latent period permits classification of *P. striiformis* f.sp. *tritici* as a pathogen having a slow infection cycle, *B. graminis* as having a fast infection cycle, and *P. triticina* and *P. graminis* f.sp. *tritici* with intermediate infection speeds (Table 15.1).

Under field conditions, the *B. graminis* incubation period on one cultivar can vary over a wide range, for instance, on spring barley between 4 and 12 days (Aust *et al.*, 1978) and on winter wheat between 6 and 14 days (Friedrich, 1995b). This is mainly influenced by temperature, which accounted for 74% of the variability in the experiments of Aust *et al.* (1978). The effect of temperature on incubation and latent periods follows an optimum curve, which can be described by non-symmetrical functions (Hau *et al.*, 1985). The shortest latent period occurred at different temperatures for the fungi: 20°C for *B. graminis* and *P. striiformis* f.sp. *tritici*, but about 26°C for *P. triticina* and 26-29°C for *P. graminis* f.sp. *tritici* (Fig. 15.1). Roelfs (1985a) mentioned that the shortest latent period for stem rust occurred at 30°C. For *P. striiformis* f.sp. *tritici*, the fraction of the latent period accomplished per day is a linear relation with the mean daily temperature for 4-20°C (Zadoks, 1961). For 0-4°C, the fraction of the latent period is zero, and for −10 to 0°C, fractions of the latent period are negative: they are substracted from the previously accomplished fractions of latent period and slow down the epidemic (Rapilly, 1991). Stripe rust latent period is not influenced by relative humidity; on the other hand, strong light may affect latency by suppressing sporulation due to a modification of the host reaction type.

Variation in latent periods has been found among isolates. In the US states east of the Rocky Mountains, new *P. striiformis* f.sp. *tritici* races completely replaced the old races that were found before 2000. All new isolates differed from the old isolates by showing shorter latent periods at 18°C than at 12°C (Milus and Seyran, 2004).

15.2.3 Lesion growth

After the latent period, superficial mildew colonies will produce conidia. The size of a lesion increases with time up to 25 mm^2. The lesion expansion rate varies from 0.07 to 2.8 $mm^2 day^{-1}$ (Berger *et al.*, 1995). The curve of lesion growth is linear over time (e.g. Stephan, 1980) at the beginning, but S-shaped when longer periods are taken into consideration (Aust, 1981).

As in the other phases of the life cycle of powdery mildew, temperature is the key factor influencing lesion growth (e.g. Pauvert, 1976). The effect of temperature again follows an optimum curve, with a maximum lesion size at 18-21°C (Stephan, 1980) or 22°C (Aust, 1981). The smaller growth rates at low temperature are partly compensated by a longer duration of colony growth (Eckhardt *et al.*, 1984a).

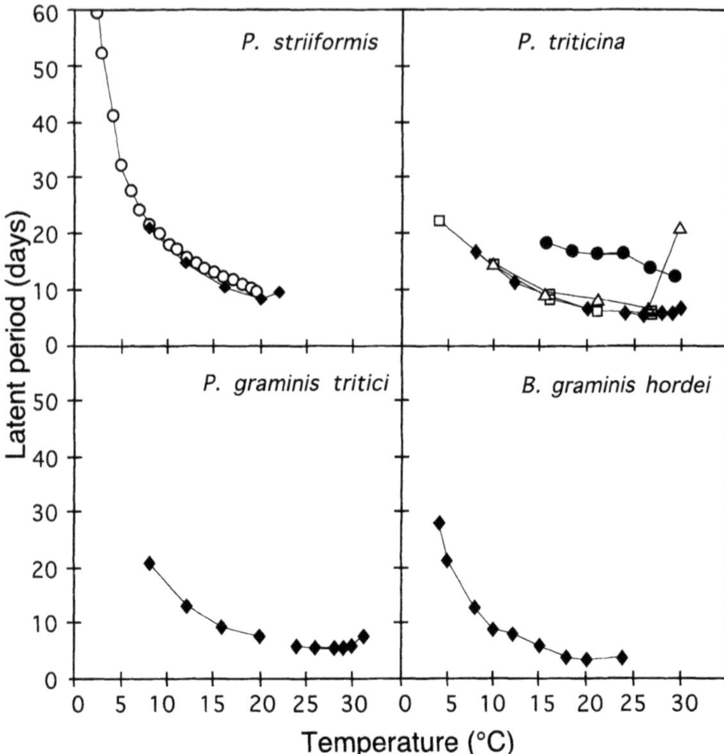

Figure 15.1. Relationships between latent period (in days) and temperature (°C) for Puccinia striiformis *f.sp.* tritici *on wheat (–♦– Tollenaar, 1985; –○– Zadoks, 1961),* Puccinia triticina *on wheat (–♦– Tollenaar, 1985; –●– Tomerlin* et al.*, 1983; –■– Johnson, 1980; –Δ– Eversmeyer* et al.*, 1980),* Puccinia graminis *f.sp.* tritici *on wheat (–♦– Tollenaar, 1985) and* Blumeria graminis *f.sp.* hordei *on barley (–♦– Pauvert and de la Tullaye, 1977).*

A second factor affecting colony growth is precipitation. While the development of colonies is not markedly affected by small droplets, colony growth is impaired partially or totally by heavy precipitation due to the high terminal velocity of large droplets causing destruction of conidiophores (Merchán and Kranz, 1986c).

The three rusts differ by their lesion size, stem rust being intermediate between stripe rust and leaf rust. Leaf rust induces small restricted pustules with an expansion rate of 0.007-0.04 mm^2 day^{-1} (Berger *et al.*, 1995); stem rust differentiates larger ones, but generally colony growth can be neglected for these two rusts. When lesion density is low, *P. triticina* produces concentric rings of minute secondary lesions around the sporulating primary lesions. *Puccinia striiformis* is unique among the biotrophic pathogens because of its semi-systemic growth, i.e. the ability of a lesion to grow within a whole leaf. The sporulation zone increases continually over time. The growth of the lesion was found to be linear (8.8-18.8 $mm^2 day^{-1}$), but the rate of growth of the lesion on certain cultivars appears to decrease with maturation (Emge *et al.*, 1975). After the tillering stage of the host, the growth of the lesion is restricted by leaf veins, causing the characteristic stripe lesions.

15.2.4 Spore production

When the latent period is complete, spore production starts and continues during the infectious period. The spore production of mildews and rusts increases rapidly to a maximum and then decreases gradually to zero, with the typical asymmetrical bell-shaped sporulation curves (e.g. Aust, 1981). To withdraw nutrients from the host over an extended time requires maintenance of integrity of the infected host cells and tissues, which is reflected in the green islands commonly observed around lesions.

Temperature influences the duration of the infectious period. For *B. graminis*, for instance, the infectious period decreases from 23 days at 10°C to 13 days at 26°C (Aust, 1981). On average, the infectious period tends to be longer for *P. striiformis* f.sp. *tritici* than for the other three fungi (Table 15.1). An exception is *P. triticina*, in which sporulation continued for up to two months under optimum conditions and low lesion density (Mehta and Zadoks, 1970). This long sporulation period is interpreted as a survival mechanism.

The sporulation intensity of the pathogens depends also on temperature. For example, barley mildew sporulation intensity increases with temperature up to the optimum at about 18-22°C (e.g. Pauvert, 1976), but decreases for higher temperatures, so that at 27°C only 40% of the maximum value can be reached (Stephan, 1980). On average, the sporulation intensity per lesion of *P. triticina* is one order of magnitude less than that of the three other fungi (Table 15.1).

The number of spores produced is reduced by rain droplets. Heavy precipitation with more than 1 mm of rain caused *B. graminis* sporulation to decline by more than 70%. Friedrich (1995a) showed that sporulation also decreases with high vapour pressure deficit that characterizes dry conditions.

For the rusts, sporulation is a function of relative humidity. For *P. striiformis* f.sp. *tritici*, for instance, the relative humidity must exceed 50% to initiate sporulation. The number of spores formed per unit of time is exponentially related to the increase in relative humidity. However, liquid water stops sporulation (Rapilly, 1979). The rate of lesion development and spore production is curtailed by low light intensity for *P. triticina* (Clifford and Harris, 1981). This is probably an indirect effect of light on host nutritional status, which governs the growth of the biotrophic rust fungi.

Host nitrogen content affects spore production. In leaves with a low nitrogen content, spore production per lesion was reduced compared to medium and high nitrogen content, but lesion size of wheat leaf rust was unaffected (Robert *et al.*, 2004). Nitrogen regime also affects yellow rust through changes of nitrogeneous substrates in wheat leaves, resulting in significant effects on the upper asymptote of *P. striiformis* f.sp. *tritici* epidemics in winter wheat (Neumann *et al.*, 2004).

Infection efficiency and sporulation rate can also be modulated by induced resistance, as shown by Calonnec *et al.* (1996) for *P. striiformis* f.sp. *tritici*.

15.2.5 Biotic factors affecting the monocycle

(a) Pathogen density

Besides the meteorological conditions, the amount of inoculum or the density of lesions can influence the phases of the life cycle. The infection efficiency of *B. graminis* decreases drastically with higher inoculum density (Damgaard and Østergård, 1996). Similarly, the germination rate of *P. striiformis* is reduced with increasing urediniospore concentration, due to self-inhibiting substances produced by germinating urediniospores (Macko *et al.*, 1977).

For *B. graminis*, incubation period (Aust *et al.*, 1978) and latent period (Damgaard and Østergård, 1996) are negatively correlated with inoculum density. Increasing infection density also shortens the latent period of *P. triticina* (Baart *et al.*, 1991).

High lesion density was found to be detrimental to lesion size for *P. graminis* f.sp. *tritici* (Leonard, 1969) and *P. triticina* (Baart *et al.*, 1991).

Spore production per mildew colony decreases exponentially as the number of colonies per leaf increases (Rouse *et al.*, 1984). For *B. graminis* f.sp. *tritici*, the dependence of sporulation on lesion density was explained by the competition for nutrients between lesions (Leonard, 1969). However, total production of *P. triticina* urediniospores per leaf seems to be relatively independent of lesion density, the main limiting factor being the photosynthetic apparatus (Mehta and Zadoks, 1970).

The density of leaf rust lesions on wheat seedlings affects spore production mostly by influencing lesion size. A high density of lesions results in smaller lesions but the number of spores per unit of sporulating area remains approximately constant at a given temperature on wheat seedlings (Robert *et al.,* 2002) and adult plants (Robert *et al.,* 2004).

(b) Host resistance

Characteristics of the host plants can also change the monocyclic parameters due to partial resistance, expressed after the seedling stage or at the adult plant stage.

For *B. graminis*, the monocyclic parameters such as infection efficiency (Royer *et al.*, 1984), latent period (e.g. Asher and Thomas, 1984), colony number cm^{-2} leaf area (e.g. Heun, 1986), lesion size (e.g. Forche, 1981), sporulation intensity (e.g. Dutzmann, 1985) and sporulation capacity (e.g. Royer *et al.*, 1984) can differ between cultivars at the same growth stage.

Similar effects have been detected for the rusts. For instance, the latent periods observed for different wheat genotype and stripe rust race associations varied between 11.0 and 14.8 days (Ghannadha *et al.*, 1995).

Cultivars expressing partial resistance that results in a decrease of the rate of the epidemic are called slow-mildewing or slow-rusting cultivars.

The susceptibility of cereal plants to powdery mildew or rusts can decrease during plant development. The adult plant resistance is especially expressed on leaves formed later in the season, i.e. on the flag leaf and on the flag-1 leaf. Adult plant resistance can reduce the development of mildew and rusts in all phases of the asexual life cycle. For mildew, for instance, the infection efficiency is drastically reduced (e.g. Wright and Heale, 1984), the incubation (e.g. Hau, 1985) and latent periods (Jones, 1978) are prolonged, the lesion size is smaller (e.g. Eckhardt *et al.*, 1984a) and spore production lower (e.g. Russell *et al.*, 1976) on the upper leaves. These components result in reduced disease severities on the upper leaves (Shaner, 1973; Russell *et al.*, 1976).

The expression of adult plant resistance can be influenced by the environment. A barley plant well supplied with water will produce upper leaves with higher susceptibility (Aust and von Hoyningen-Huene, 1986). Adult plant resistance to stripe rust in several winter wheats increases at high post-inoculation temperatures and decreases when subsequently transferred to a lower temperature (Qayoum and Line, 1985; Schultz and Line, 1992). However, high-temperature adult plant resistance expressed under US Pacific northwest conditions may not be adequately expressed under cool temperatures, as under UK conditions (Johnson, 1992). Minor resistance genes to wheat stripe rust are a function of the thermal conditions affecting the plant during the post-inoculation or pre-inoculation phase, which can modify the infection type (Brown and Sharp, 1969).

15.2.6 Spore dispersal

The aerial dispersal of plant pathogens by wind is mainly a passive movement involving three events: liberation, transport and deposition.

(a) Spore liberation

The conidia of *B. graminis* are attached to each other in a chain, whereas the rust urediniospores, produced singly on pedicels, are detached and accumulate in sori erupting through the plant cuticle before dispersal.

To liberate spores, the wind speed must be higher than a certain threshold, which decreases with the duration of calm conditions. For *B. graminis*, the wind speed must exceed 1 m/s after 2 h, and 0.2 m/s after 12 h in calm air. For *P. striiformis* f.sp. *tritici*, 0.6 m/s is necessary after a quiet air period (Rapilly, 1991). The cumulative number of liberated spores is related to the wind speed by an S-shaped curve for both pathogens. However, rapid increase of the wind speed is more important than continuous high wind velocities (e.g. Aylor, 1990).

The maximum spore removal was observed at 2.3 and 2.8 m/s for *P. triticina* and *P. striiformis* f.sp. *tritici*, respectively, in experimental conditions (Geagea *et al.*, 1997). Forces necessary to remove the rust spores are two to three orders of magnitude larger than those required to remove *B. graminis* conidia (Bainbridge and Legg, 1976; Geagea *et al.*, 1997). The difference may be related to the more external position of the conidia of *B. graminis* and the size of the spores.

High relative humidity can cause free water on the leaf surface, which prevents spore liberation (Pauvert, 1984). In the experiments of Adams *et al.* (1986), conidia of *B. graminis* were liberated in darkness when an abrupt change, especially decrease, in relative humidity occurred; this may be interpreted as an active discharge.

The spore surface of *P. striiformis* f.sp. *tritici* is covered with a mucilaginous layer, which becomes thicker when relative humidity increases (Rapilly, 1991). Therefore, under high relative humidity, spores stick together and are dispersed mostly as clusters of 2-10 spores. The size of the dispersal unit is thus larger in *P. striiformis* f.sp. *tritici* than in *P. triticina* and *P. graminis* f.sp. *tritici*, which are mainly dispersed as single spores. Spore removal of *P. striiformis* f.sp. *tritici* requires stronger forces than that of *P. triticina* (Table 15.2).

Table 15.2. Assessement of spore removal parameters in Puccinia striiformis *f.sp.* tritici *and* Puccinia triticina *under controlled conditions*[a]

	P. triticina	*P. striiformis*
Wind threshold (m/s)	1	1.5
Drag force (N)	2.7×10^{-8}	4×10^{-8}
Centrifugal force (N)	0.72×10^{-8}	3.4×10^{-8}
Drop kinetic energy	0.69	0.40

[a]Geagea *et al.* (1997; 1999).

(b) Short-distance transport and gradients

Wheat rusts have long been thought to be released only by wind but rain can play a significant role in the dispersal of urediniospores, either by direct impact or by splashing (Geagea *et al.*, 1999; Sache, 2000). Dispersal by rain, though limited in distance, can be very efficient because the spores have a very high germination potential under wet conditions (Rapilly, 1979). The number of spores removed by single drops and the disease severity on trap plants after rain simulation are proportional to the total kinetic energy of incident rain (Geagea *et al.*, 1999, 2000). Thunder storms remove spores in a very short time period, but also exhaust sporulating lesions and wash off deposited spores. Light rains are less efficient for spore removal but more conducive for the spread of disease. Intermittent rain events of light intensity are the most efficient, especially if they are associated with high wind speeds.

When the spores are deposited from a spore cloud after transport by wind, dispersal gradients can be observed. The gradients of airborne spores are less steep than those of splash-dispersed spores (McCartney and Fitt, 1987). Nevertheless, the number of spores caught with trap plants decreases quickly with distance from the source. For example, compared to the amount of *B. graminis* f.sp. *hordei* spores that landed at a distance of 0.5 m, Hovmøller (1996) trapped only 16% at a distance of 50 m and 0.6% at 200 m. This led to the conclusion that, in samples of airborne spores, distant sources contribute only a few spores in comparison to nearby sources.

For wheat powdery mildew, Fried *et al.* (1979) found that the deposition gradients from a point source are better described with the power law model than with the exponential model. The slope of the gradients is correlated with the wind speed (Fried *et al.,* 1979). In addition, the growth habit of the variety influences the trap efficiency of plants (Russell, 1975).

When the spores land on compatible host plants, disease gradients can be observed as a consequence of deposition gradients, as can occur in spring barley due to neighbouring winter barley (Koch, 1980). Deposition and disease gradients can have the same slope (Fried *et al.,* 1979). Observed disease gradients may not be the result of dispersal gradients but can also be caused by environmental gradients reflecting changes in the microclimate (Koch, 1980).

The immigration of spores into a field is important for the start of an epidemic. Once the epidemic is established, almost all new infections are caused by spores produced within the field, as shown for *B. graminis* f.sp. *hordei* by O'Hara and Brown (1996) and Bousset *et al.* (2001).

Foci are common for stripe rust but rare for leaf and stem rusts as well as for powdery mildew, the epidemics tending to be generalized after a short period of time (about three weeks) after the start of primary foci. The dispersal of *P. triticina, P. graminis* f.sp. *tritici* and *B. graminis* over longer distances than that of *P. striiformis* f.sp. *tritici* prevents the detection of primary foci. The larger dispersal units of *P. striiformis* f.sp. *tritici* require stronger forces for removal and can travel only shorter distances. Therefore, dispersal gradients are usually steeper in stripe rust than in leaf and stem rusts. The dispersal gradients of the three rust fungi can be described equally well with the power law model and with the exponential model (Rapilly, 1991). For the exponential model, the half-distance of the dispersal gradient for *P. striiformis* f.sp. *tritici* is between 2.4 and 2.7 m, which is only one-tenth of the half-distance of *P. graminis* f.sp. *tritici* (25 m) or *P. triticina* (28 m), while the half-distance of *B. graminis* is intermediate (5.8 m).

The primary focus from the initial source of inoculum expands when conditions are favourable, and secondary foci initiated from the primary focus appear in the same field. Foci, almost circular at the beginning, usually become V-shaped because of the effect of the predominant wind direction.

Two theories are proposed to describe the focal expansion: one showing a constant rate of focal expansion (e.g. van den Bosch *et al.,* 1988a) and the other an increasing rate of focal expansion over time (Ferrandino, 1993). The radial speed was assessed as about 10 cm day^{-1} for stripe rust (van den Bosch *et al.,* 1988b), 9.6-61.5 cm day^{-1} for leaf rust (Subba Rao *et al.,* 1990) and about 20 cm day^{-1} for stem rust (Schmitt *et al.,* 1959). The focus expansion rate is correlated with the daily multiplication factor (i.e. the product of spore production and infection efficiency, equivalent to the number of daughter lesions produced per mother lesion) and inversely related to the latent period.

(c) Long-distance transport

Transport by wind can enable pathogens to bridge much longer distances than could be achieved by splash dispersal. The evidence of aerial dispersal was based on the sampling of spores and trajectory analysis of disease progression (Roelfs, 1985a,b;

Eversmeyer and Kramer, 2000). More recently, other characteristics like unnecessary virulence genes (Limpert, 1987; Limpert *et al.,* 1999), isozyme patterns (e.g. Watson and de Sousa, 1982) and molecular markers (Wolfe *et al.*, 1992; Wolfe and McDermott, 1994; Brown and Hovmøller, 2002) have been used to assess genotype flow.

For the transport, a cloud of propagules is lifted up and travels at high altitudes (up to 3,000 m, as shown for leaf and stem rust urediniospores by slide exposures on aercraft; Roelfs, 1985a), following the trajectory of the air mass. Most spores are caught in dry weather during the day but deposition may also be by rain. The probability of viable spores being capable of causing infection after long-distance dispersal is very low but not zero.

The dispersal of barley powdery mildew is enhanced in the main wind direction, i.e. in Europe from west to east, and reduced against it. An example of the transport from west to east is the crossing of the North Sea from Britain to Denmark (Hermansen *et al.*, 1978). Based on high frequencies of the unnecessary virulence gene Va6 in France, Denmark and Austria, Limpert (1987) calculated that *B. graminis* f.sp. *hordei* populations can travel 110 km year^{-1} in the main wind direction. Dispersal from east to west is also possible, as shown by the migration of the most common haplotypes virulent on *Mla13*, characterized for virulence and RAPD (randomly amplified polymorphic DNA) marker spectra, that were first detected in the former Czechoslovakia and later in Austria, the Netherlands, Belgium, Germany and Switzerland (Wolfe *et al.,* 1992). Centrally located regions in Europe receive immigrant spores from numerous neighbouring regions, while the peripheral areas, like Scotland and Spain, have only a low import of spores from distant populations (Müller *et al.*, 1996). Mountains can be a barrier to spore transport. The Alps, for example, reduce the dissemination of powdery mildew spores into Italy and in the opposite direction (Limpert *et al.,* 1990).

Several paths of the long-distance dispersal of cereal rust pathogens are known (Nagarajan and Singh, 1990). The north American Puccinia path is the route taken in the back-and-forth movement of *P. graminis* f.sp. *tritici* urediniospores between northern Mexico/Texas and the US/Canadian border, where stem rust does not overwinter in the uredinial stage. The seasonal movement in sweeps and jumps is northward in the spring and southward in the autumn (Roelfs and Long, 1987). *Puccinia triticina* and *P. striiformis* f.sp. *tritici* have also been implicated in the northward dispersal but leaf rust overwintering in the Great Plains is of major importance. Wheat is grown nearly continuously in the Great Plains from Texas to Minnesota and rust can advance in a series of relatively short, successive jumps. In 1983, stem rust advanced northward at the rate of about 54 km day^{-1} and the longest well-documented jump of *P. graminis* f.sp. *tritici* urediniospores in North America was about 680 km between two wheat-growing regions in Canada separated by forests and lakes (Roelfs, 1985a,b). Even if the inoculum source is a small area, the epidemic in the target zone can be severe, as shown for the stem rust epidemic in the Southern Great Plains in 1986, that resulted from inoculum generated in an area along the Texas Gulf Coast (Roelfs and Long, 1987).

In Europe, the two main routes for stem rust are the east European path from Turkey/Romania to Scandinavia and the west European path from Morocco/Iberia to

Scandinavia (Zadoks, 1967). Further evidence of long-distance dispersal of rust is given by the population structure of *P. triticina* in Western Europe (Park *et al.*, 1996, 2001). The isolates of the four most common pathotypes generally have the same RAPD phenotypes, supporting the hypothesis of a clonal migration over considerable distances. The wind dispersal of urediniospores of *P. striiformis* f.sp. *tritici* from Great Britain to Denmark, and later to Germany and France allowed an explanation of the pathotype distribution in western Europe after the outbreak of the *Yr17* virulent pathotype (Brown and Hovmøller, 2002; Hovmøller *et al.*, 2002).

Another path for *P. graminis* f.sp. *tritici* and *P. triticina* is from the South Indian Hills to the Central Indian Plains (Nagarajan and Singh, 1990). Urediniospores survive throughout the year in south India (in the Nilgiri Hills) and are dispersed to central and northern India by autumn tropical cyclones (following a recurrent pattern of synoptic situations). Unlike in North America, there is no feedback of inoculum to the south.

The possible South African origin of several strains of *P. graminis* f.sp. *tritici* found in Australia in 1968 was speculated because of the similarity in both pathogenicity and isozyme patterns between isolates found in the two continents (Watson and de Sousa, 1982). Stripe rust development in Australia since 1979 illustrates rapidity and long-distance dispersal of the pathogen.

A continental *Puccinia* pathway has also been observed from western China to the main wheat belt, southern Gansu and northern Sichuan being a source for the dispersal of diverse *P. striiformis* f.sp. *tritici* pathotypes (Shan *et al.*, 1998; Brown and Hovmøller, 2002).

This long-distance dispersal has prompted the suggestion to organize the introduction of resistance genes in crops against the main wind direction, for example, Limpert (1987) for barley powdery mildew in Europe, and Bahadur *et al.* (1994) for stem and leaf rusts in India. However, long-distance transport may include diverse populations. Using virulence polymorphism and RAPD markers, Kolmer *et al.* (1995) hypothesized that the two major groups of *P. triticina* isolates currently found in eastern and western Canada originated from different introductions. No relationship between pathogenic and molecular variation was found for *P. striiformis* f.sp. *tritici* in the US Pacific northwest by Chen *et al.* (1993).

15.2.7 Consequences of climatic factors on polycyclic epidemics

Environmental factors can affect different phases of the life cycle differently so that it is not clear what conditions will lead to severe epidemics. Moreover, the progress of an epidemic is heavily influenced by the susceptibility of the crop plant, which is also affected by environmental conditions, for instance, the expression of adult plant resistance against powdery mildew in barley plants is more pronounced in dry and hot weather (Aust and von Hoyningen-Huene, 1986).

For epidemics of mildew, development is optimal between 15 and 22°C and it is retarded at temperatures above 25°C (Mathre, 1982). Epidemics can reach high disease levels if the initial disease level is high and the temperature is relatively warm, for instance, more than 10°C from the beginning of the three-leaf stage until the formation of the flag-2 leaf in spring barley (Stephan, 1984). It is generally assumed (Aust and von Hoyningen-Huene, 1986) that powdery mildews are favoured by relatively dry atmospheric conditions and moderate temperatures if susceptible host tissue is available. A combination of wet and dry conditions, such as rain interrupted by sunny intervals, may favour powdery mildew epidemics (Aust and von Hoyningen-Huene, 1986). No general conclusions can be drawn concerning the environmental factors promoting powdery mildew epidemics, either directly or through the host plant, but it is clear that the apparent infection rates of the disease progress curves are strongly correlated with the initial disease levels on wheat (Rouse *et al.*, 1981) and barley (Scholze, 1985).

Although temperature affects every stage of the disease cycle, viable inoculum, susceptible hosts and duration of free moisture are all factors influencing rust development. Stem rust develops at warmer temperatures (30°C optimum) than do the other wheat rust diseases; thus it is most frequently a disease caused by the reproductive portion of the life cycle (Roelfs, 1985a). Stripe rust is favoured by cool temperatures and, unlike the other cereal rusts, can even function at relatively low temperatures, in particular under the snow. Higher winter temperatures and lower spring temperatures contribute to epidemics. Development of stripe rust epidemics after booting is positively correlated with the mean temperature for the range of 12.9-16.2°C. From 16.2-20.3°C the epidemic rate is negatively correlated with temperature (Ellison and Murray, 1992). Constant or mean temperatures above 22 or 25°C inhibit and may even eliminate the stripe rust fungus. Years with severe epidemics had warmer than normal temperatures in January allowing the survival of fall-infected wheat foliage (Coakley *et al.*, 1988). The precipitation frequency for the month of June was significantly correlated with stripe rust development.

In general, severe stripe rust epidemics are most likely when late summer and fall infections occur, when abundant mycelium survives the winter and when cool nights, warm days and heavy dews occur during spring and summer growing seasons (Coakley *et al.*, 1988). *Puccinia striiformis* f.sp. *tritici* urediniospores produced between 5 and 10°C germinated best, whereas spores produced at 30°C and above were unable to germinate (Rapilly, 1979). This observation may help to explain the correlations observed between high temperatures and the decrease in epidemics. The viability of urediniospores of the stripe rust fungus decreases more rapidly in sunlight than the viability of urediniospores of other cereal rust fungi (Maddison and Manners, 1972).

Stripe rust foci can appear during autumn or winter, when the relative humidity is often very high, which allows only a short distance spread of urediniospores. A generalized epidemic, which usually occurs only in spring, can take place if numerous foci that are difficult to distinguish are present. It seems that endemic multiple foci corresponding to a few infected plants, or even a few sori, initiate epidemics rather than long-distance transport.

For leaf rust, a temperature variable (either the minimum temperature or the hourly temperature equivalence function) was the most important factor explaining variation in overwintering (Eversmeyer and Kramer, 1996). The inclusion of a precipitation variable for summer and autumn indicates the importance of moisture in establishment on volunteer wheat. Re-infection periods during winter and early spring are extremely important in the survival of *P. triticina* when snow cover is not present. Positive average deviations of the fungal temperature equivalent function and the cumulative rainfall indicate periods during which re-infections may occur in the overwintering phase of the epidemic. The study of winter climatic conditions permits evaluation of the possibilities of pathogen multiplication and partially enables forecasting of epidemic changes in spring. Afterwards, leaf rust severity is highly correlated with cumulative degree-days >20°C following inoculation (Subba Rao *et al.,* 1990).

15.3 SURVIVAL AND SEXUAL STATE

15.3.1 Oversummering

Both powdery mildew and cereal rusts oversummer on volunteer crops in the asexual stage, infect the autumn-sown crop and, eventually, overwinter on the volunteers to infect the crops in spring (Zadoks, 1961). Furthermore, stripe rust can oversummer in high altitudes, as in the Alps, where harvest is late and the wind may carry the inoculum to volunteer crops (Zadoks, 1961). *Puccinia striiformis* can survive the non-crop season in mild climates on volunteer cereal plants, or on other graminaceous hosts, although at a very low rate (Sharp and Hehn, 1963; Shaner and Powelson, 1973; Stubbs 1985; Dennis and Brown, 1986; Line, 2002). In these mild areas, volunteer wheat plants are abundant because of the relatively frequent rainfalls in summer that promote the built-up of rust inoculum and are responsible for the recurrence of epidemics (Park, 1990). In dry regions (Iran, South Africa), oversummering can depend on grass species such as *Bromus* spp. *(catharticus, arenarius, oxydon)*, *Hordeum murinum* and *Poa trivialis* (Nazari *et al.*, 1996; Boshoff *et al.,* 2002). Survival of *P. graminis* f.sp. *tritici* is generally difficult during the non-cereal growing season (Roelfs, 1985a). However, a few surviving local uredinia can produce local inoculum.

Powdery mildew differentiates a sexual stage, which contributes to oversummering. With the beginning of senescence of the lower leaves of cereal plants in early summer, the formation of generative mycelium and cleistothecia of *B. graminis* is initiated starting on the lower leaves. This process can take place as early as mid-May in wheat, especially in years with hot dry springs (Götz *et al.*, 1996). In the cleistothecia, 15 to 20 asci develop, each containing eight haploid ascospores which are dispersed by wind, even under high humidity after rain, i.e. in conditions which would be unfavourable for conidial dispersal (Götz *et al.*, 1996). On wheat, ascospores can develop at any time during the second half of the year; as they are released at night and in the early morning hours with favourable infection conditions, they can also contribute to the epidemic during the season (Felsenstein, 1996; Götz *et al.*, 1996). Therefore, the sexual reproduction is more important for powdery mildew on wheat than on barley.

Apart from ascospores, conidia from the summer crop can also infect volunteer plants so that a mixture of ascospores and conidia forms the inoculum for the winter crop. The spore population initiating the autumn epidemic originating from ascospores was estimated to be 25% and 1.5-15% by Brown and Wolfe (1990) and

Brändle *et al.* (1997), respectively. Considering the significant contribution of the ascospores to the infection of winter wheat, Felsenstein (1996) concluded that the distribution of the pathogen by wind is not so important in this period and the effects on neighbouring populations are restricted.

On artificially inoculated barley field plots, the overall ranking of *B. graminis* f.sp. *hordei* pathotype frequencies was the same before and after summer (Bousset and de Vallavieille-Pope, 2003b). However, the variability in pathotype frequencies on volunteers was neither correlated with the frequencies in the populations of conidia on the crop before summer nor with the expected frequencies in the ascospore population. Chance events such as bottlenecks might have a large influence on the pathotype frequencies during summer survival. A pathotype dominant at the end of the summer might possibly have oversummered through asexual reproduction, or alternatively it may have originated from sexual reproduction when the frequency of this pathotype was high and both mating types were represented in the population of conidia before summer (Bousset and de Vallavieille-Pope, 2003a).

15.3.2 Overwintering

The mildew population grows during autumn on the winter crop until the lower cardinal temperature is reached, restricting further development. Mildew can survive the cold period in winter in the vegetative stage on overwintering green plants. The size of the population can be drastically reduced in strong winters, causing severe wilting of leaves.

Puccinia triticina can survive as sporulating or dormant mycelium and/or as viable urediniospores in uredinia on dead leaves (Eversmeyer *et al.*, 1988) under quite severe winter conditions, the limiting factor being the survival of host tissue for re-infection that can occur at relatively low temperatures. Studies in controlled environment chambers indicated that a few hours of freezing are extremely detrimental to the survival of single, unprotected urediniospores of *P. triticina* and *P. graminis* f.sp. *tritici* (Eversmeyer and Kramer, 1995). However, exposure to a wide range of temperatures above 0°C does not significantly reduce viability.

Puccinia triticina can survive during the sexual stage on the alternate hosts *Thalictrum* spp. (the most common); *Isopyrum* spp. in Siberia, *Anchusa* spp. in Portugal and *Clematis* spp. in Italy. In most parts of the world, however, *P. triticina* rarely infects the alternate host and is highly successful in its survival by the uredinial stage alone.

Puccinia graminis f.sp. *tritici* can also survive in the sexual stage on the alternate hosts, for instance, the barberry bush (*Berberis vulgaris*) (the most common) and *Mahonia* spp. After the sexual stage, the spread of stem rust from barberry bushes is local, though there is no reason to suspect that aeciospores are not transported as far as urediniospores (Roelfs, 1985a). Since the widespread eradication of *Berberis*, which, however, can never be complete (Peterson *et al.*, 2005), the occurrence of the aecidial stage is rare and the pathogen survives without the sexual stage in most areas of the world.

Puccinia striiformis, which is endemic in northern Europe, the US Pacific northwest, west Asia, China, South Africa and South America, has a microcyclic life cycle with no known alternate host.

In the north-central and north-eastern United States and adjacent Canada, winter survival of the rusts is rare; the primary inoculum usually originates from wheat in the southern United States and arrives in the region in late spring, when temperatures are too high for stripe rust epidemics (Line, 1995).

15.4 POPULATION DYNAMICS

15.4.1 Dynamics of virulences

Biotrophic pathogens are under selection pressure at all times to infect more hosts and are notorious for their ability to adapt to newly-introduced resistances. Genetic studies have demonstrated that race-specific resistance fits the gene-for-gene theory for *B. graminis*, *P. graminis* f.sp. *tritici* and *P. triticina.* The patterns for resistance genes and races of *P. striiformis* f.sp. *tritici* are also consistent with a gene-for-gene system. Eighty-five different resistance/avirulence gene pairs have been defined for the interaction between barley and *B. graminis* f.sp. *hordei* (Jørgensen, 1994; Caffier *et al.,* 1996a), more than 60 pairs between wheat and *P. graminis* f.sp. *tritici,* over 50 pairs for *P. triticina* and over 40 for *P. striiformis* (Johnson, 1992; McIntosh *et al.*, 1995a; Line and Chen, 1996; McIntosh *et al.*, 2003).

On a regional basis, the composition of the fungal populations is largely determined by the cultivars grown and their respective acreage. When new host resistance genes are introduced, the population of *B. graminis* and *Puccinia* spp. can adapt rapidly by selecting pathotypes with matching virulence gene combinations (e.g. Andrivon and de Vallavieille-Pope, 1993; Bayles *et al.,* 2000; Wan *et al.,* 2004). The changes in virulence frequencies can be caused by strong direct selection as well as by indirect (hitch-hiking) selection (Hovmøller *et al.*, 1993; Huang *et al.*, 1995). The mean number of virulences per isolate can be rather stable (Andrivon and de Vallavieille-Pope, 1995), but in Europe the tendency was towards an increase in pathotype complexity over the years (Wolfe and McDermott, 1994). The tendency for increase in race complexity in rust pathogens like *P. triticina* and *P. striiformis* f.sp. *tritici* (Stubbs, 1985; de Vallavieille-Pope *et al.*, 1990; Line and Qayoum, 1991; Bayles *et al.* 2000; Line, 2002), and for stability of complexity at an intermediate level in powdery mildew fungi when several specific resistance genes are introduced in the host population, is possibly related to the biology of the pathogens, in particular, the presence of a sexual phase in the powdery mildew fungi (Andrivon and de Vallavieille-Pope, 1995). Diversity is greater in sexually reproducing populations of *P. graminis* f.sp. *tritici* (Groth and Roelfs, 1982) than in asexual populations where a single pathotype often predominates.

Races with the highest complexity are found in clonally reproducing pathogens long faced with race-specific host resistance genes, such as *P. graminis* f.sp. *tritici* in the United States or *P. striiformis* f.sp. *tritici* in Europe. Conversely, fungi that may undergo sexual reproduction at frequent intervals, such as *B. graminis* and *Puccinia coronata* f.sp. *avenae*, or that have been only recently combated by means of resistant cultivars, such as *P. coronata* f.sp. *avenae* in Canada (Chong, 1985, 1986), have less complex races and often a greater diversity of virulence phenotypes. However, there is a large local effect of selection by resistance genes on race diversity and complexity (Andrivon and de Vallavieille-Pope, 1995). In the US south-central area, a new race has predominated since 2000, distinguishable by molecular markers and by a narrow virulence spectrum from previous races in the United States (Chen *et al.*, 2002; Markell *et al.*, 2004).

Diversity is higher in populations of cereal mildews than in populations of cereal rusts. *Puccinia striiformis* f.sp. *tritici* is strictly clonal, evolving mainly through the mutation and selection of the fittest clones (Steele *et al.*, 2001; Hovmøller *et al.*, 2002; Roose-Amsaleg *et al.*, 2002; Villaréal *et al.*, 2002; Keipfer *et al.*, 2003), except in western Chinese populations, which are more diverse (Shan *et al.*, 1998; Enjalbert *et al.*, 2002, 2004). There are also cases of clonal populations of *B. graminis* f.sp. *hordei*, as shown during the first years of cultivation of barley cultivars carrying a new resistance gene, for example, *Mla7* and *Ml(La)* (Brown *et al.*, 1993). Furthermore, a *B. graminis* f.sp. *hordei* clone of a race with a single virulence dominated in epidemics on French winter barley cultivars carrying few specific resistance genes to powdery mildew (Caffier *et al.*, 1996b, 1999). Selection solely due to host resistance genes, without assuming any cost of virulence, might lead to such results (Bousset *et al.*, 2002).

A second factor that can change the composition of the fungal populations is the migration of pathogens by wind. Limpert *et al.* (1999), for instance, assumed that the observed increase in race complexity of *B. graminis* from western to eastern Europe was due to spore transport in the main wind direction. Similarly, new *P. striiformis* races occurred in Denmark due to spore migration from England (Hovmøller *et al.*, 2002).

Within a field population, the proportion of races changes with time, depending on the fitness of the pathotypes. Over a long period, the dynamics of *B. graminis* races changed cyclically over seasons with a high diversity of races after the oversummering period of the pathogen due to recombination (Welz and Kranz, 1987) and migration and a decrease until the following summer due to selection (Welz, 1988). Temperature seems to be an important factor that influences the composition of the pathotypes (Eckhardt, 1987) by changing the fitness of races. Furthermore, different races are selected in different zones of the world; for instance, five different groups of *P. striiformis* races were found in Europe (Stubbs, 1988), a main division being between north-western Europe and southern and eastern Europe, due to resistance gene deployment and possibly to climatic conditions.

15.4.2 Dynamics of fungicide resistant subpopulations

Besides the pathotypes which are characterized by their virulence genes, subpopulations within the mildew population can be defined with respect to fungicide resistance. Powdery mildew isolates may differ in a wide range of sensitivity to fungicides that can be characterized by the respective ED_{50} value. The application of fungicides changes the composition of the fungal populations. Under selection pressure, the competitive ability of resistant isolates is higher than that of the more sensitive isolates, which increases the proportion of the resistant sub-population and reduces the control of the disease by the fungicide. For powdery mildew, the population dynamics of fungicide resistance was investigated mainly for ethirimol, triadimenol, fenpropimorph and strobilurin (e.g. Godet and Limpert, 1998). A chemical inducer of host resistance (BTH) induced a further selection pressure in a barley powdery mildew laboratory population in addition to that exerted by ethirimol when both were applied together (Bousset and Pons-Kühnemann, 2003). It has been assumed that the fungicide responses were under polygenic control, but Brown *et al.* (1992) showed that the responses to ethirimol and triadimenol were largely controlled by major genes.

On a regional basis, the level and frequency of resistance against a certain fungicide is generally correlated with the intensity of fungicide use. For example, resistance against triadimenol was high in countries in which this fungicide was used for a long time, as in northern Europe and the UK, while sensitive populations were found in north-eastern Spain and northern Italy with comparatively little fungicide use (Koller *et al.*, 1992). Besides the direct selection of fungicide-resistant sub-populations, indirect selection through hitch-hiking has been assumed in cases in which associations between virulence genes and fungicide resistance exist, for instance, the association of *Va6* with high levels of resistance against triadimenol (Wolfe, 1985). As *B. graminis* is wind-dispersed, a regional population is also affected by the influx of spores from distant areas that may be under strong selection by a certain fungicide. Therefore, Limpert (1987) suggested that fungicides with a specific action should not be introduced in the main wind direction.

The population dynamics within fields depends on the frequency of fungicide applications and the strategy used (Hau and Pons, 1996), such as fungicide mixtures or alternate applications.

For rusts, studies on decreases in efficiency of fungicides are rather scarce. Bayles *et al.* (2001) documented a shift in sensitivity of wheat yellow rust to DMI fungicides in the UK, but there was no evidence of strobilurin resistance.

Fungicide resistance surveys are dealt with in Chapter 3.

15.5 CONCLUDING REMARKS

Cereal powdery mildews and rusts are polycyclic diseases that are important in all cereal-growing areas of the world because of their ability to travel long distances easily by wind dispersal, and only a low level of inoculum density is required for initiating epidemics (e.g., one infection per hectare can cause a stripe rust epidemic in the following year; Line, 2002). However, they differ in their dispersal abilities. A quick development of a stripe rust epidemic requires a random distribution of inoculum foci. This characteristic has never been observed with stem and leaf rusts and powdery mildews, for which dispersal gradients are flat. Although the four pathogens are similar in driving epidemics by repeated asexual infection cycles, their survival strategies differ in the use of a sexual stage: annually for *B. graminis*, rarely for *P. graminis* f.sp. *tritici* and *P. triticina*, or not at all for *P. striiformis* f.sp. *tritici*. The asexual reproduction (by urediniospores or conidia) decreases and stops as sexual reproduction begins (with teliospores and cleistothecia), which is triggered by rising temperature or by host senescence.

The epidemic strategy of *B. graminis*, *P. graminis* f.sp. *tritici* and *P. triticina* is mainly based on a fast infection cycle, a large sporulation capacity and a high infection efficiency. *Puccinia striiformis* f.sp. *tritici* differs by a slower infection cycle compensated for by a systemic growth of the lesion within the leaf. In general, the increase in the disease severity for mildew and stripe rust is due not only to new infections but, more importantly, to lesion growth. For leaf rust and stem rust, colony growth can be neglected.

Blumeria graminis differs from the rust pathogens by its ability to withstand dry conditions, while high humidity and even wetness are necessary for *Puccinia* spp. infection. The strategy of the rusts seems to be adapted to prevailing climates. The cereal rusts adapted to an unstable temperate environment exploit the available resources during short periods; other tropical rusts, adapted to wet and warm conditions in a relatively stable environment with a nearly permanent availability of susceptible host tissue, have a longer infectious period, a later sporulation peak and a lower infection efficiency than some temperate rusts (Sache and de Vallavieille-Pope, 1995). The consequences for the strategy of breeding for resistant cultivars against these temperate pathogens are to recommend selection for genotypes with increased latent periods so delaying epidemics. Partial resistance associated with the frequent failure of haustorium formation is also a source of potentially durable resistance (Niks and Rubiales, 2002). Necrotrophs such as *Septoria* can be controlled mainly by reducing lesion growth and consequently sporulation capacity.

The early stages of infection by these obligate biotrophs are governed by a strict morphogenetic programme and climatic constraints. Their life cycles are synchronized with those of their hosts for both food supply and reproduction. For example, for the asexual stage of *Puccinia* spp., penetration coincides with night conditions, and colonisation of the host tissue coincides with higher temperatures occurring during daytime. These fungi appear to be less flexible than most necrotrophs. The tight morphogenetic control of infection structures (appressoria, haustoria) may also determine the narrow range of hosts that these fungi are able to parasitize and can be seen as a means of preserving the exclusiveness of habitat.

The speed of epidemics of polycyclic diseases depends on the actual values of monocyclic parameters, influenced by meteorological conditions and host susceptibility. A greater understanding of airborne diseases has been achieved by the quantification of these monocyclic parameters. Because of the great number and complexity of the influencing variables, no single technique can assess these epidemiological parameters (de Vallavieille-Pope *et al.*, 2000). For example, the development of standardised and simplified methods enables the quantification of latent periods. Wall sterol content can provide an objective tool to assess pathogen biomass. The quantification of effects of influencing variables forms the basis for the prediction of epidemic progress. Such studies will have more and more useful applications because of the increasing number of weather stations measuring temperature and wetness on short time-scales (one hour or less) and the development of forecasting models.

The monocyclic phases of all the pathogens considered have been modelled, for example, the effect of temperature and wetness period on rust infection (de Vallavieille-Pope *et al.*, 1995, 2002), or the influence of temperature on latent period (Zadoks, 1961). Also, the effect of inoculum or lesion density on monocyclic parameters has been investigated (e.g. Damgaard and Østergård, 1996; Robert *et al.*, 2004), as well as global epidemics as functions of time under the influence of climatic factors (e.g. Coakley *et al.*, 1988). It seems that it would now be useful to combine qualitative and quantitative parameters and to analyse the effects of climatic factors on the whole epidemic using analytical models (Hau, 1990) or simulation models, combined with a growth model of the host. A first step in combining qualitative and quantitative parameters has been achieved by taking into account aggressiveness parameters as well as virulence frequencies (Villaréal and Lannou, 2000; Lannou, 2001). The next step would be to include the effects of climatic factors on each phase of the infection cycle and to integrate the spatial spread of epidemics.

The quantification of disease parameters has made it possible to perform prospective studies to predict disease risks. Forecasting is now possible to optimize fungicide treatments, but another purpose is to suggest methods to select for disease resistance and to manage resistance genes. For example, such advances permitted simulation of slow rusting or slow mildewing epidemics and proposals for exploiting resistance gene diversity in geographic areas and variety mixtures.

REFERENCES

Adams, G.C. Jr., Gottwald, T.R. and Leach, C.M. (1986) Environmental factors initiating liberation of conidia of powdery mildew. *Phytopathology*, **76**, 1239-1245.

Andrivon, D. and de Vallavieille-Pope, C. (1993) Racial diversity and complexity in *Erysiphe graminis* f.sp. *hordei* in France over a 5-year period. *Plant Pathology*, **42**, 443-464.

Andrivon, D. and de Vallavieille-Pope, C. (1995) Race diversity and complexity in selected populations of fungal biotrophic pathogens of cereals. *Phytopathology*, **85**, 897-905.

Anikster, Y., Bushnell, W.R., Eilam, T. *et al.* (1997) *Puccinia recondita* causing leaf rust on cultivated wheats, wild wheats, and rye. *Canadian Journal of Botany*, **75**, 2082-2096.

Asher, M.J.C. and Thomas, C.E. (1984) Components of partial resistance to *Erysiphe graminis* in spring barley. *Plant Pathology*, **33**, 123-130.

Aust, H.-J. (1981) Über den Verlauf von Mehltauepidemien innerhalb des Agro-Ökosystems Gerstenfeld. *Acta Phytomedica*, **7**, 1-76.

Aust, H.-J. and von Hoyningen-Huene, J. (1986) Microclimate in relation to epidemics of powdery mildew. *Annual Review of Phytopathology*, **24**, 491-510.

Aust., H.-J., Hau, B. and Mogk, M. (1978) Wirkung von Temperatur und Konidiendichte auf die Variabilität der Inkubationszeit des Gerstenmehltaues. *Zeitschrift für Pflanzenkrankheiten und Pflanzenschutz*, **85**, 581-585.

Aylor, D.E. (1990) The role of intermittent wind in the dispersal of fungal pathogens. *Annual Review of Phytopathology*, **28**, 73-92.

Baart, P.G., Parlevliet, J.E. and Limburg, H. (1991) Effects of infection density on the size of barley and wheat leaf rust colonies before and on the size of uredia after the start of sporulation. *Journal of Phytopathology*, **131**, 59-64.

Bahadur, P., Singh, D.V. and Srivastava, K.D. (1994) Management of wheat rusts – a revised strategy of gene deployment. *Indian Phytopathology*, **47**, 41-47.

Bainbridge, A. and Legg, B.J. (1976) Release of barley-mildew conidia from shaken leaves. *Transactions of the British Mycological Society*, **66**, 495-498.

Bayles, R.A., Flath, K., Hovmøller, M.S. and de Vallavieille-Pope, C. (2000) Breakdown of the *Yr 17* resistance to yellow rust of wheat in northern Europe – a case study by the yellow rust sub-group of COST 817. *Agronomie*, **20**, 805-811.

Bayles, R., Burnett, F. and Napier, B. (2001) Designing fungicide programmes to minimise the risk of resistance in wheat and barley mildew and wheat yellow rust. *HGCA-Project-Report*, **243**, 1-60.

Berger, R.D., Bergamin Filho, A. and Amorim, L. (1995) Lesion expansion as an epidemic component. *Phytopathology*, **87**, 1005-1013.

Bousset, L., Schaeffer, B. and de Vallavieille-Pope, C. (2001) Effect of early infection on pathotype frequencies in barley powdery mildew *(Blumeria graminis* f.sp. *hordei)* populations in field plots. *Plant Pathology*, **50**, 1-9.

Bousset, L., Hovmøller, M.S., Caffier, V. *et al.* (2002) Changes in frequencies of genotypes avirulent to spring-sown barley in aerial powdery mildew populations from winter barley in France and Denmark. *Plant Pathology*, **51**, 33-44.

Bousset, L. and de Vallavieille-Pope, C. (2003a) Effect of sexual recombination on pathotype frequencies in barley powdery mildew populations of artificially inoculated field plots. *European Journal of Plant Pathology*, **109**, 13-24.

Bousset, L. and de Vallavieille-Pope, C. (2003b) Barley powdery mildew populations on volunteers and changes in pathotype frequencies during summer on artificially inoculated field plots. *European Journal of Plant Pathology*, **109**, 25-33.

Bousset, L. and Pons-Kühnemann, J. (2003) Effects of acibenzolar-S-methyl and ethirimol on the composition of a laboratory population of barley powdery mildew. *Phytopathology*, **93**, 305-315.

Boshoff, W.H.P., Pretorius, Z.A. and Niekerk, B.D.V. (2002) Establishment, distribution, and pathogenicity of *Puccinia striiformis* f.sp. *tritici* in South Africa. *Plant Disease*, **86**, 485-492.

Brändle, U.E., Haemmerlin, A.A., McDermott, J.M.,Wolfe, and M.S. (1997) Interpreting population genetic data with the help of genetic linkage maps, in *The gene-for-gene relationship in plant-parasite interactions* (eds I.R.Crute, E.B.HOLUB, and J.J.Burdon), CAB International, Wallingford, pp. 157-171.

Brown, J.F. and Sharp, E.L. (1969) Interaction of minor host genes for resistance to *P. striiformis* with changing temperature regimes. *Phytopathology*, **59**, 999-1001.

Brown, J.K.M. and Wolfe, M.S. (1990) Structure and evolution of a population of *Erysiphe graminis* f.sp. *hordei*. *Plant Pathology*, **39**, 376-390.

Brown, J.K.M., Jessop, A.C., Thomas, S. and Rezanoor, H.N. (1992) Genetic control of the response of *Erysiphe graminis* f.sp. *hordei* to ethirimiol and triadimenol. *Plant Pathology*, **41**, 126-135.

Brown, J.K.M. and Hovmøller, M.S. (2002) Aerial dispersal of pathogens on the global and continental scales and its impact on plant disease. *Science*, **297**, 537-541.

Brown, J.K.M., Simpson, C.G. and Wolfe, M.S. (1993) Adaptation of barley powdery mildew populations in England to varieties with two resistance genes. *Plant Pathology*, **42**, 108-115.

Burrage, S.W. (1970) Environmental factors influencing the infection of wheat by *Puccinia graminis*. *Annals of Applied Biology*, **66**, 429-440.

Caffier, V., de Vallavieille-Pope, C. and Brown, J.K.M. (1996a) Segregation of avirulences and genetic basis of infection types in *Erysiphe graminis* f.sp. *hordei*. *Phytopathology*, **86**, 1112-1121.

Caffier, V., Hoffstadt, T., Leconte, M. and de Vallavieille-Pope, C. (1996b) Seasonal changes in pathotype complexity in French populations of barley powdery mildew. *Plant Pathology*, **45**, 454-468.

Caffier, V., Brändle, U. and Wolfe, M.S. (1999) Genotypic diversity in barley powdery mildew populations in northern France. *Plant Pathology*, **48**, 582-587.

Calonnec, A., Goyeau, H. and de Vallavieille-Pope, C. (1996) Effects of induced resistance on infection efficiency and sporulation of *Puccinia striiformis* on seedlings in varietal mixtures and on field epidemics in pure stands. *European Journal of Plant Pathology*, **102**, 733-741.

Carver, T.L.W. and Bushnell, W.R. (1983) The probable role of primary germ tubes in water uptake before infection by *Erysiphe graminis*. *Physiological Plant Pathology*, **23**, 229-240.

Chen, X., Line, R.F. and Leung, H. (1993) Relationship between virulence variation and DNA polymorphism in *Puccinia striiformis*. *Phytopathology*, **83**, 1489-1497.

Chen, X.M., Moore, M., Milus E.A. *et al.* (2002) Wheat stripe rust epidemics and races of *Puccinia striiformis* f.sp. *tritici* in the United States. *Plant Disease*, **86**, 39-46.

Chong, J. (1985) Virulence and distribution of *Puccinia coronata* in Canada in 1984. *Canadian Journal of Plant Pathology*, 7, 424-427.
Chong, J. (1986) Virulence and distribution of *Puccinia coronata* in Canada in 1985. *Canadian Journal of Plant Pathology*, **8**, 443-446.
Clifford, B.C. and Harris, R.G. (1981) Controlled environment studies of the epidemic potential of *Puccinia recondita* f.sp. *tritici* on wheat in Britain. *Transactions of the British Mycological Society*, **77**, 351-358.
Coakley, S.M., Line, R.F. and McDaniel, L.R. (1988) Predicting stripe rust severity on winter wheat using an improved method for analyzing meteorological and rust data. *Phytopathology*, **78**, 543-550.
Collins, T.J., Moerschbacher, B.M. and Read, N.D. (2001) Synergistic induction of wheat stem rust appressoria by chemical and topographical signals. *Physiological and Molecular Plant Pathology*, **58**, 259-266.
Damgaard, C. and Østergård, H. (1996) Density dependent growth of barley powdery mildew on a partial resistant barley variety: infection efficiency and spore production, in *COST 817 – Population Studies of Airborne Pathogens on Cereals as a Means of Improving Strategies for Disease Control – Integrated Control of Cereal Mildews and Rusts: Towards Coordination of Research Across Europe* (eds E.Limpert, M.R.Finckh and M.S.Wolfe), Office for Official Publications of the European Communities, Luxembourg, pp. 241-245.
Dennis, J.I. and Brown, J.S. (1986) Summer survival of *Puccinia striiformis* f.sp. *tritici* in Victoria, Australia. *Australian Journal of Plant Pathology*, **15**, 57-60.
Dutzmann, S. (1985) Ein Vergleich verschiedener Methoden zum Messen der Quantität und Qualität der Sporulation von *Erysiphe graminis* f.sp. *hordei*. *Zeitschrift für Pflanzenkrankheiten und Pflanzenschutz*, **92**, 618-628.
Eckhardt, H. (1987) Pathotypendynamik von *Erysiphe graminis* DC. f.sp. *hordei* Marchal in Abhängigkeit von biotischen und abiotischen Faktoren. *Zeitschrift für Pflanzenkrankheiten und Pflanzenschutz*, **94**, 169-177.
Eckhardt, H., Steubing, L. and Kranz, J. (1984a) Das Koloniewachstum von *Erysiphe graminis* DC. f.sp. *hordei* Marchal in Abhängigkeit von Temperatur und Insertionshöhe der Blätter. *Angewandte Botanik*, **58**, 433-443.
Eckhardt, H., Steubing, L. and Kranz, J. (1984b) Untersuchungen zur Infektionseffizienz, Inkubationszeit und Latenzzeit beim Gerstenmehltau *Erysiphe graminis* DC. f.sp. *hordei*. *Zeitschrift für Pflanzenkrankheiten und Pflanzenschutz*, **91**, 590-600.
Enjalbert, J., Duan, X., Giraud, T. *et al.* (2002) Isolation of twelve microsatellite loci, using an enrichment protocol, in the phytopathogenic fungus *Puccinia striiformis* f.sp. *tritici*. *Molecular Ecology Notes*, **2**, 563-565.
Enjalbert, J., Duan, X., Wan, A. *et al.* (2004) Clonality of wheat yellow rust populations in France and high diversity in China. *Proceedings of the 11th Cereal Rusts and Powdery Mildews Conference*, Norwich, England, 22-27 August 2004.
Ellison, P.J. and Murray, G.M. (1992) Epidemiology of *Puccinia striiformis* f.sp. *tritici* on wheat in southern New South Wales. *Australian Journal of Agricultural Research*, **43**, 29-41.
Emge, R.G., Kingsolver, C.H. and Johnson, D.R. (1975) Growth of the sporulating zone of *Puccinia striiformis* and its relationship to stripe rust epiphytology. *Phytopathology*, **65**, 679-681.
Eversmeyer, M.G. and Kramer, C.L. (1995) Survival of *Puccinia recondita* and *P. graminis* urediniospores exposed to temperatures from subfreezing to 35 C. *Phytopathology*, **85**, 161-164.
Eversmeyer, M.G. and Kramer, C.L. (1996) Modeling leaf rust epidemics in the Central Great Plains of the USA, in *Proceedings Ninth European and Mediterranean Cereal Rusts & Powdery Mildews Conference*, Lunteren, The Netherlands, pp. 294-296.
Eversmeyer, M.G. and Kramer, C.L. (2000) Epidemiology of wheat leaf and stem rust in the central Great Plains of the USA. *Annual Review of Phytopathology*, **38**, 491-513.
Eversmeyer, M.G., Kramer, C.L. and Browder, I.E. (1980) Effect of temperature and host:parasite combination on the latent period of *Puccinia recondita* in seedling wheat plants. *Phytopathology*, **70**, 938-941.
Eversmeyer, M.G., Kramer, C.L. and Browder, I.E. (1988) Winter and early spring survival of *Puccinia recondita* on Kansas wheat during 1980-1986. *Plant Disease*, **72**, 1074-1076.
Felsenstein, F.G. (1996) Effects of sexual reproduction on the population dynamics of wheat and barley powdery mildew in Europe, in *COST 817 – Population Studies of Airborne Pathogens on Cereals as a Means of Improving Strategies for Disease Control – Integrated Control of Cereal Mildews and*

Rusts: Towards Coordination of Research Across Europe (eds E.Limpert, M.R.Finckh and M.S.Wolfe), Office for Official Publications of the European Communities, Luxembourg, pp. 91-96.

Ferrandino, F.J. (1993) Dispersive epidemic waves: I. Focus expansion within a linear planting. *Phytopathology*, **83**, 795-802.

Forche, S. (1981) Anzahl und Größe der Kolonien von *Erysiphe graminis* f.sp. *hordei* als Kriterien für die Resistenz von Sommergerstensorten. *Zeitschrift für Pflanzenkrankheiten und Pflanzenschutz*, **88**, 435-441.

Fried, P.M., MacKenzie, D.R. and Nelson, R.R. (1979) Dispersal gradients from a point source of *Erysiphe graminis* f.sp. *tritici* on Chancellor winter wheat and four multilines. *Phytopathologische Zeitschrift*, **95**, 140-150.

Friedrich, S. (1995a) Zur Berechnung der Konidienverbreitung von *Erysiphe graminis* in natürlich befallenen Beständen anhand stündlicher meteorologischer Parameter. *Zeitschrift für Pflanzenkrankheiten und Pflanzenschutz*, **102**, 337-347.

Friedrich, S. (1995b) Berechnung der Inkubationszeit des Echten Mehltaus unter Freilandbedingumgen. *Zeitschrift für Pflanzenkrankheiten und Pflanzenschutz*, **102**, 348-353.

Friedrich, S. (1995c) Modell zur Berechnung der Infektionswahrscheinlichkeit durch Echten Mehltau an Winterweizen anhand meteorologischer Eingangsparameter. *Zeitschrift für Pflanzenkrankheiten und Pflanzenschutz*, **102**, 354-365.

Friedrich, S. and Boyle, S. (1993) Wirkung unterschiedlicher Luftfeuchte auf die Produktion und Keimung der Konidien von *Erysiphe graminis* f.sp. *tritici* in vitro. *Zeitschrift für Pflanzenkrankheiten und Pflanzenschutz*, **100**, 180-188.

Geagea, L., Huber, L. and Sache, I. (1997) Removal of urediniospores of brown (*Puccinia recondita* f.sp. *tritici*) and yellow (*P. striiformis*) rusts of wheat from infected leaves submitted to a mechanical stress. *European Journal of Plant Pathology*, **103**, 785-793.

Geagea, L., Huber, L. and Sache, I. (1999) Dry-dispersal and rain-splash of brown (*Puccinia recondita* f.sp. *tritici*) and yellow (*P. striiformis*) rust spores from infected wheat leaves exposed to simulated raindrops. *Plant Pathology*, **48**, 472-482.

Geagea, L., Huber, L., Sache, I. *et al.* (2000) Influence of simulated rain on dispersal of rust spores from infected wheat seedlings. *Agricultural and Forest Meteorology*, **101**, 53-66.

Ghannadha, M.R., Gordon, I.L., Cromey, M.G. and McEwan, J.M. (1995) Diallel analysis of the latent period of stripe rust in wheat. *Theoretical and Applied Genetics*, **90**, 471-476.

Godet, F. and Limpert, E. (1998) Recent evolution of multiple resistance of *Blumeria (Erysiphe) graminis* f.sp. *tritici* to selected DMI and Morpholine fungicides in France. *Pesticide Science*, **54**, 244-252.

Götz, M., Friedrich, S. and Boyle, S. (1996) Development of cleistothecia and early ascospore release of *Erysiphe graminis* DC. f.sp. *tritici* in winter wheat in relation to host age and climatic conditions. *Zeitschrift für Pflanzenkrankheiten und Pflanzenschutz*, **103**, 134-141.

Groth, J.V. and Roelfs, A.P. (1982) The effect of sexual and asexual reproduction on race abundance in cereal rust fungus populations. *Phytopathology*, **72**, 1503-1507.

Hau, B. (1985) Epidemiologische Simulatoren als Instrumente der Systemanalyse mit besonderer Berücksichtigung eines Modells des Gerstenmehltaus. *Acta Phytomedica*, **9**, 1-101.

Hau, B. (1990) Analytic models of plant disease in a changing environment. *Annual Review of Phytopathology*, **28**, 221-245.

Hau, B. and Pons, J. (1996) Selection of populations of barley powdery mildew influenced by fungicide strategies, in *Modern Fungicides and Antifungal Compounds* (eds H. Lyr, P.E. Russell and H.D. Sisler), Intercept, Andover, pp. 357-364.

Hau, B., Eisensmith, S.P. and Kranz, J. (1985) Construction of temporal models: II Simulation of aerial epidemics, in *Mathematical Modelling of Crop Disease (Advances in Plant Pathology Vol. 3)* (ed C.A. Gilligan), Academic Press, London, pp. 31-65.

Hermansen, J.E., Torp U. and Prahm L.P. (1978) Studies of transport of live spores of cereal mildew and rust fungi across the North Sea. *Grana*, **17**, 41-46.

Heun, M. (1986) Quantitative differences in powdery mildew resistance among spring barley cultivars. *Journal of Phytopathology*, **115**, 222-228.

Hovmøller, M.S. (1996) Powdery mildew spore dispersal and its implications for spore sampling techniques in virulence surveys, in *COST 817—Population Studies of Airborne Pathogens on Cereals as a Means of Improving Strategies for Disease Control—Integrated Control of Cereal Mildews and Rusts: Towards Coordination of Research Across Europe* (eds E.Limpert, M.R.Finckh and M.S.Wolfe), Office for Official Publications of the European Communities, Luxembourg, pp. 81-83.

Hovmøller, M.S., Justesen, A.F. and Brown, J.K.M. (2002) Clonality and long-distance migration of *Puccinia striiformis* f.sp. *tritici* in north-west Europe. *Plant Pathology*, **51**, 24-32.
Hovmøller, M.S., Munk, L. and Østergård, H. (1993) Observed and predicted changes in virulence gene frequencies at 11 loci in a local barley powdery mildew population. *Phytopathology*, **83**, 253-260.
Huang, R., Kranz, J. and Welz, H.G. (1995) Virulence gene frequency change in *Erysiphe graminis* f.sp. *hordei* due to selection by non-corresponding barley mildew resistance genes and hitchhiking. *Journal of Phytopathology*, **143**, 287-294.
Jenkyn, J.F. and Bainbridge, A. (1978) Biology and pathology of cereal powdery mildews, in *The Powdery Mildews* (ed D.M. Spencer), Academic Press, London, pp. 284-321.
Johnson, D.A. (1980) Effect of low temperature on the latent period of slow and fast rusting winter wheat genotypes. *Plant Disease*, **64**, 1006-1008.
Johnson, R. (1992) Past, present and future opportunities in breeding for disease resistance, with examples from wheat. *Euphytica*, **63**, 3-22.
Jones, I.T. (1978) Components of adult plant resistance to powdery mildew (*Erysiphe graminis* f.sp. *avenae*) in oats. *Annals of Applied Biology*, **90**, 233-239.
Jørgensen, J.H. (1988) *Erysiphe graminis*, powdery mildew of cereals and grasses. *Advances in Plant Pathology*, **6**, 137-157.
Jørgensen, J.H. (1994) Genetics of powdery mildew resistance in barley. *Critical Review in Plant Sciences*, **13**, 97-119.
Keiper, F.J., Hayden, M.J., Park, R.F. and Wellings, C.R. (2003) Molecular genetic variability of Australian isolates of five cereal rust pathogens. *Mycological Research*, **107**, 545-556.
Koch, H. (1980) Räumliche Befallsmuster beim echten Mehltau der Gerste. *Zeitschrift für Pflanzenkrankheiten und Pflanzenschutz*, **87**, 731-737.
Koller, B., Müller, K., Limpert, E. and Wolfe, M.S. (1992) Response of populations of *Erysiphe graminis* f.sp. *hordei* to large-scale use of fungicides. *Vorträge Pflanzenzüchtung*, **24**, 335-337.
Kolmer, J.A., Liu, J.Q. and Sies, M. (1995) Virulence and molecular polymorphism in *Puccinia recondita* f.sp. *tritici* in Canada. *Phytopathology*, **85**, 276-285.
Lannou, C. (2001) Intrapathotype diversity for aggressiveness and pathogen evolution in cultivar mixtures. *Phytopathology*, **91**, 500-510.
Leonard, K.J (1969) Factors affecting rates of stem rust increase in mixed plantings of susceptible and resistant oat varieties. *Phytopathology*, **59**, 1845-1850.
Limpert, E. (1987) Barley mildew in Europe: Evidence of wind-dispersal of the pathogen and its implications for improved use of host resistance and of fungicides for mildew control, in *Integrated Control of Cereal Mildews: Monitoring the Pathogen* (eds M.S. Wolfe and E. Limpert), Martinus Nijhof Publishers, Dortrecht, pp. 31-33.
Limpert, E., Andrivon, D. and Fischbeck, G. (1990) Virulence patterns in populations of *Erysiphe graminis* f.sp. *hordei* in Europe in 1986. *Plant Pathology*, **39**, 402-415.
Limpert, E., Godet, F. and Müller, K. (1999) Dispersal of cereal mildews across Europe. *Agricultural and Forest Meteorology*, **97**, 293-308.
Line, R.F. (1995) Successes in breeding for and managing durable resistance to wheat rusts. *Plant Disease*, **79**, 1254-1255.
Line, R.F. and Chen, X. (1996) Wheat and barley stripe rust in North America, in *Proceedings Ninth European and Mediterranean Cereal Rusts & Powdery Mildews Conference,* Lunteren, The Netherlands, pp. 101-104.
Line, R.F. and Qayoum, A. (1991) Virulence, aggressiveness, evolution, and distribution of races of *Puccinia striiformis* (the cause of stripe rust of wheat) in North America, 1968-87. *U.S. Department of Agriculture Technical Bulletin*, **1788**, 44pp.
Line, R.F. (2002) Stripe rust of wheat and barley in North America: a retrospective historical review. *Annual Review of Phytopathology*, **40**, 75-118.
Lumbroso, E., Fischbeck, G. and Wahl, I. (1982) Infection of barley with conidia suspensions of *Erysiphe graminis* f.sp. *hordei*. *Phytopathologische Zeitschrift*, **104**, 222-233.
Macko, V., Trione, E.J. and Young, S.A. (1977) Identification of the germination self-inhibitor from uredospores of *Puccinia striiformis*. *Phytopathology*, **67**, 1473-1474.
Maddison, A.C. and Manners, J.G. (1972) Sunlight and viability of cereal rust uredospores. *Transactions of the British Mycological Society*, **59**, 429-443.

Markell, S.G., Milus, E.A. and Chen, X. (2004) Genetic diversity of *Puccinia striiformis* f.sp. *tritici* in the United States. *Proceedings of* the *11th Cereal Rusts and Powdery Mildews Conference*, Norwich, UK, 22-27 August 2004.

Mathre, D.E. (1982) *Compendium of Barley Diseases*, APS, St. Paul.

McCartney, H.A. and Fitt, B.D.L. (1987) Spore dispersal gradients and disease development, in *Populations of Plant Pathogens: Their Dynamics and Genetics* (eds. M.S.Wolfe and C.E.Caten), Blackwell Scientific Publications, Oxford, pp. 109-118.

McIntosh, R.A., Hart, G.E. and Gale, M.D. (1995a) Catalogue of wheat symbols for wheat, in *Proceedings of the Eighth International Wheat Genetics Symposium,* 1993, Beijing, China, pp. 1333-1451.

McIntosh, R.A., Wellings, C.R. and Park, R.F. (1995b) *An atlas of resistance genes*. CSIRO Australia, Kluwer Academic Publishers.

McIntosh, R.A., Yamasaki, Y., Devos, K.M. *et al.* (2003) Catalogue of wheat symbols for wheat, in *Proceedings of the tenth International Wheat Genetics Symposium,* ed. N.E. Pogna, M. Romano, E.A. Pogna, G. Galterio, 2003, Paestum, Italy, pp. 58-77.

Mehta, Y.R. and Zadoks, J.C. (1970) Uredospore production and sporulation period of *Puccinia recondita* f.sp. *triticina* on primary leaves of wheat. *Netherlands Journal of Plant Pathology*, **76**, 267-276.

Merchán, V.M. and Kranz, J. (1986a) Wirkung der Blattnässe auf den asexuellen Zyklus des Weizenmehltaus *Erysiphe graminis* DC. f.sp. *tritici* Marchal. *Zeitschrift für Pflanzenkrankheiten und Pflanzenschutz*, **93**, 246-254.

Merchán, V.M. and Kranz, J. (1986b) Untersuchungen über den Einfluß des Regens auf die Infektion des Weizens durch *Erysiphe graminis* DC. f.sp. *tritici* Marchal. *Zeitschrift für Pflanzenkrankheiten und Pflanzenschutz*, **93**, 255-261.

Merchán, V.M. and Kranz, J. (1986c) Die Wirkung des Regens auf die Entwicklung des Weizenmehltaus (*Erysiphe graminis* DC. f.sp. *tritici* Marchal). *Zeitschrift für Pflanzenkrankheiten und Pflanzenschutz*, **93**, 262-270.

Milus E.A. and Seyran E., 2004. New races of *Puccinia striiformis* f.sp. *tritici* more aggressive than older races at 18°C. *Proceedings of the 11th Cereal Rusts and Powdery Mildews Conference*, Norwich, UK, 22-27 August 2004.

Müller, K., McDermott, J.M., Wolfe, M.S. and Limpert, E. (1996) Analysis of diversity in populations of plant pathogens: the barley powdery mildew pathogen across Europe. *European Journal of Plant Pathology*, **102**, 385-395.

Nagarajan, S. and Singh, D.V. (1990) Long-distance dispersion of rust pathogens. *Annual Review of Phytopathology*, **28**, 139-153.

Nazari, K., Torabi, M. and Mardoukhi, V. (1996) Wild grass species as oversummering hosts of *Puccinia striiformis* f.sp. *tritici* in Iran. *Cereal Rusts and Powdery Mildews Bulletin*, **24**, 105-107.

Neumann, S., Paveley, N.D., Beed, F.D. and Sylvester-Bradley, R. (2004) Nitrogen per unit leaf area affects the upper asymptote of *Puccinia striiformis* f.sp. *tritici* epidemics in winter wheat. *Plant Pathology*, **53**, 725-732.

Nicholson, R.L. (1996) Adhesion of fungal propagules, in *Histology, Ultrastructure and Molecular Cytology of Plant-Microorganism Interactions* (eds. M. Nicole and V. Gianinazzi-Pearson), Kluwer, Dordrecht, pp. 117-134.

Niks, R.E. and Rubiales, D. (2002) Potentially durable resistance mechanisms in plants to specialized fungal pathogens. *Euphytica*, **124**, 201-216.

Oerke, E.-C., Dehne, H.-W., Schönbeck, F. and Weber, A. (1994) *Crop Production and Crop Protection. Estimated Losses in Major Food and Cash Crops*. Elsevier, Amsterdam.

O'Hara, R.B. and Brown, J.K.M. (1996) Immigration of the barley mildew pathogen into field plots of barley. *Plant Pathology*, **45**, 1071-1076.

Pande, S., Joshi, L.M. and Nagarajan, S. (1978) Quantifying infection and sporulation as possible parameters for measuring host resistance, in *Proceedings of the 5th International Wheat Genetics Symposium,* (ed. S. Rajaram), Indian Society of Genetics and Plant Breeding, New Dehli, pp. 1087-1097.

Park, R.F. (1990) The role of temperature and rainfall in the epidemiology of *Puccinia-striiformis* f.sp. *tritici* in the summer rainfall area of Eastern Australia. *Plant Pathology*, **39**, 416-423.

Park, R.F., Jahoor, A. and Felseinstein, F.G. (1996) Genetic variation in European populations of *Puccinia recondita* f.sp. *tritici* using pathogenicity and molecular markers, in *Proceedings Ninth*

European and Mediterranean Cereal Rusts & Powdery Mildews Conference, Lunteren, The Netherlands, pp. 92-94.

Park, R.F., Goyeau, H., Felsenstein, F.G. *et al.* (2001) Regional phenotypic diversity of *Puccinia triticina* and wheat host resistance in western Europe, 1995. *Euphytica*, **122**, 113-127.

Pauvert, P. (1976) Variations quantitatives de la sporulation d'*Erysiphe graminis* f.sp. *hordei*. *Annales de Phytopathologie*, **8**, 131-140.

Pauvert, P. (1984) Etude expérimentale de la libération des conidies d'*Erysiphe graminis* DC. f.sp. *hordei* sous l'effet du vent. *Agronomie*, **4**, 195-198.

Pauvert, P. and de la Tullaye, B. (1977) Etude des conditions de contamination de l'orge par l'oïdium et de la période de latence chez *Erysiphe graminis* f.sp. *hordei*. *Annales de Phytopathologie*, **9**, 495-501.

Peterson, P.D., Leonard, K.J., Miller, J.D. *et al.* (2005) Prevalence and distribution of common barberry, the alternate host of *Puccinia graminis*, in Minnesota. *Plant Disease*, **89**, 159-163.

Qayoum, A. and Line, R.F. (1985) High-temperature, adult-plant resistance to stripe rust of wheat. *Phytopathology*, **75**, 1121-1125.

Rapilly, F. (1979) Yellow rust epidemiology. *Annual Review of Phytopathology*, **17**, 59-73.

Rapilly, F. (1991) *L'épidémiologie en pathologie végétale : mycoses aériennes*, INRA Paris.

Robert, C., Bancal, M.O. and Lannou, C. (2002) Wheat leaf rust uredospore production and carbon and nitrogen export in relation to lesion size and density. *Phytopathology*, **92**, 762-768.

Robert, C., Bancal, M.O. and Lannou, C. (2004) Wheat leaf rust uredospore production on adult plants: influence of leaf nitrogen content and *Septoria tritici* blotch. *Phytopathology*, **94**, 712-721.

Roelfs, A.P. (1985a) Wheat and rye stem rust, in *The Cereal Rusts. Vol 2, Diseases, Distribution, Epidemiology, and Control* (eds A.P. Roelfs and W.R. Bushnell), Academic Press, Orlando, pp. 1-37.

Roelfs, A.P. (1985b) Epidemiology in North America, in *The Cereal Rusts. Vol 2, Diseases, Distribution, Epidemiology, and Control* (eds A.P. Roelfs and W.R. Bushnell), Academic Press, Orlando, pp. 403-434.

Roelfs, A.P. and Long, D.L. (1987) *Puccinia graminis* development in North America during 1986. *Plant Disease*, **71**, 1089-1093.

Roose-Amsaleg, C., de Vallavieille-Pope, C., Brygoo, Y. and Levis, C. (2002) Characterisation of a length polymorphism in the two intergenic spacers of ribosomal RNA in *Puccinia striiformis* f.sp. *tritici*, the causal agent of wheat yellow rust. *Mycological Research*, **106**, 918-924.

Rouse, D.I., MacKenzie, D.R. and Nelson, R.R. (1981) A relationship between initial inoculum and apparent infection rate in a set of disease progress data for powdery mildew on wheat. *Phytopathologische Zeitschrift*, **100**, 143-149.

Rouse, D.I., MacKenzie, D.R. and Nelson, R.R. (1984) Density dependent sporulation of *Erysiphe graminis* f.sp. *tritici*. *Phytopathology*, **74**, 1176-1180.

Royer, M.H., Nelson, R.R., MacKenzie, D.R. and Diehle, D.A. (1984) Partial resistance of near-isogenic wheat lines compatible with *Erysiphe graminis* f.sp. *tritici*. *Phytopathology*, **74**, 1001-1006.

Russell, G.E. (1975) Deposition of *Erysiphe graminis* f.sp. *hordei* conidia on barley varieties of differing growth habit. *Phytopathologische Zeitschrift*, **84**, 316-321.

Russell, G.E., Andrews, C.R. and Bishop, C.D. (1976) Development of powdery mildew on leaves of several barley varieties at different growth stages. *Annals of Applied Biology*, **82**, 467-476.

Sache, I. (2000) Short-distance dispersal of wheat rust spores by rain and wind. *Agronomie* **20**, 757-767.

Sache, I. and de Vallavieille-Pope, C. (1993) Comparison of the wheat brown and yellow rusts for monocyclic sporulation and infection processes, and their polycyclic consequences. *Journal of Phytopathology*, **138**, 55-65.

Sache, I. and de Vallavieille-Pope, C. (1995) Classification of airborne plant pathogens based on sporulation and infection characteristics. *Canadian Journal of Botany*, **73**, 1186-1195.

Samborski, D.J. (1985) Wheat leaf rust, in: *The Cereal Rusts. Vol 2, Diseases, Distribution, Epidemiology, and Control*. (eds A.P. Roelfs and W.R. Bushnell), Academic Press, Orlando, pp. 39-59.

Schmitt, C.G., Kingsolver, C.H. and Underwood, J.F. (1959) Epidemiology of stem rust of wheat: I. Wheat stem rust development from inoculation foci of different concentration and spatial arrangement. *Plant Disease Reporter*, **43**, 601-606.

Scholze, P. (1985) Zur epidemiologischen Bedeutung der Abgängigkeitsbeziehung zwischen Anfangsbefallsgrad und Befallsprogression beim Gerstenmehltau (*Erysiphe graminis* DC. f.sp. *hordei* March.). *Archiv für Phytopathologie und Pflanzenschutz*, **21**, 131-142.

Schultz, T.R. and Line, R.F. (1992) High-temperature, adult-plant resistance to wheat stripe rust and effects on yield components. *Agronomy Journal*, **84**, 170-175.

Shan, W.X., Chen, S.Y., Kang, L.R. *et al.* (1998) Genetic diversity in *Puccinia striiformis* Westend. f.sp. *tritici* revealed by pathogen genome-specific repetitive sequence. *Canadian Journal of Botany*, **76**, 587-595.

Shaner, G. (1973) Evaluation of slow-mildewing resistance of Know wheat in the field. *Phytopathology*, **63**, 867-872.

Shaner G. and Powelson, R.L. (1973) The oversummering and dispersal of inoculum of *Puccinia striiformis* in Oregon. *Phytopathology*, **63**, 13-17.

Sharp, E.L. and Hehn, E.R. (1963) Overwintering of stripe rust in winter wheat in Montana. *Phytopathology*, **53**, 1239-1240.

Shrum, R.D. (1975) Simulation of wheat stripe rust (*Puccinia striiformis* West) using EPIDEMIC, a flexible plant disease simulator. *Pennsylvania State University Agricultural Experiment Station Progress Report 347.*

Sivapalan, A. (1993) Effects of water on germination of powdery mildew conidia. *Mycological Research*, **97**, 71-76.

Steele, K.A., Humphreys, E., Wellings, C.R. and Dickinson, M.J. (2001) Support for a stepwise mutation model for pathogen evolution in Australasian *Puccinia striiformis* f.sp. *tritici* by use of molecular markers. *Plant Pathology*, **50**, 174-180.

Stephan, S. (1980) Inkubationszeit und Sporulation des Gerstenmehltaus (*Erysiphe graminis* DC.) in Abhängigkeit von meteorologischen Faktoren. *Archiv für Phytopathologie und Pflanzenschutz*, **16**, 173-181.

Stephan, S. (1984) Untersuchungen zum Epidemieverlauf des Gerstenmehltaus. *Archiv für Phytopathologie und Pflanzenschutz*, **20**, 39-52.

Stubbs, R.W. (1985) Stripe rust, in *The Cereal Rusts. Vol 2, Diseases, Distribution, Epidemiology, and Control.* (eds A.P. Roelfs and W.R. Bushnell), Academic Press, Orlando, pp. 61-101.

Stubbs, R.W. (1988) Pathogenicity analysis of yellow (stripe) rust of wheat and its significance in a global context, in *Breeding Strategies for Resistance to the Rusts of Wheat* (eds. N.W. Simmonds and S. Rajaram), CIMMYT, México, D.F. CIMMYT, pp. 23-38.

Subba Rao, K.V., Berggren, G.T., and Snow, J.P. (1990) Characterization of wheat leaf rust epidemics in Lousiana. *Phytopathology*, **80**, 402-410.

Tollenaar, H. (1985) Uredospore germination and development of some cereal rusts from south-central Chile at constant temperatures. *Phytopathologische Zeitschrift*, **114**, 118-125.

Tomerlin, J.R., Eversmeyer, M.G., Kramer, C.L. and Browder, L.E. (1983) Temperature and host effects on latent and infectious periods and on urediniospore production of *Puccinia recondita* f.sp. *tritici. Phytopathology*, **73**, 414-419.

Vallavieille-Pope (de), C., Picard-Formery, H., Radulovic, S. and Johnson, R. (1990) Specific resistance factors to yellow rust in seedlings of some French wheat varieties and races of *Puccinia striiformis* Westend in France. *Agronomie,* **2**, 103-113.

Vallavieille-Pope (de), C., Huber, L., Leconte, M. and Goyeau, H. (1995) Comparative effects of temperature and interrupted wet periods on germination, penetration, and infection of *Puccinia recondita* f.sp. *tritici* and *P. striiformis* urediniospores on wheat seedlings. *Phytopathology*, **85**, 409-415.

Vallavieille-Pope, (de) C., Giosue, S., Munk, L. *et al.* (2000) Assessment of epidemiological parameters and their use in epidemiological and forecasting models of cereal airborne diseases. *Agronomie*, **20**, 715-727.

Vallavieille-Pope (de), C., Huber, L., Leconte, M. and Bethenod, O. (2002) Preinoculation effects of light quantity on wheat seedlings on infection efficiency of *Puccinia striiformis* and *P. triticina. Phytopathology*, **92**,1308-1314.

van den Bosch, F., Zadoks, J.C. and Metz, J.A.J. (1988a) Focus expansion in plant disease. I. The constant rate of focus expansion. *Phytopathology*, **78**, 54-58.

van den Bosch, F., Frinking, H.D., Metz, J.A.J. and Zadoks, J.C. (1988b) Focus expansion in plant disease. III. Two experimental examples. *Phytopathology*, **78**, 919-925.

Villaréal, L.M.M.A. and Lannou, C. (2000) Selection for increased spore efficacy by host genetic background in a wheat powdery mildew population. *Phytopathology*, **90**,1300-1306.

Villaréal, L., Lannou C., de Vallavieille-Pope, C. and Neema, C. (2002) Genetic variability in *Puccinia striiformis* f.sp. *tritici* populations sampled on a local scale during natural epidemics. *Applied and Environmental Microbiology*, **68**, 6138-6145.

Wan, A., Zhao, Z., Chen, X. *et al.* (2004) Wheat stripe rust epidemic and virulence of *Puccinia striiformis* f.sp. *tritici* in China in 2002, *Plant Disease*, **88**, 896-904.

Watson, I.A. and de Sousa, C.N.A. (1982) Long distance transfer of spores of *Puccinia graminis tritici* in the southern hemisphere. *Proceedings of the Linnean Society of New South Wales*, **106**, 311-321.

Welz, G. (1988) Virulenzdynamik einer Population von *Erysiphe graminis* f.sp. *hordei* auf anfälligen Gerstensorten. *Zeitschrift für Pflanzenkrankheiten und Pflanzenschutz*, **95**, 124-137.

Welz, G. and Kranz, J. (1987) Effects of recombination on races of a barley powdery mildew population. *Plant Pathology*, **36**, 107-113.

Wiethölter, N., Horn, S., Reisige, K. *et al.* (2003) *In vitro* differentiation of haustorial mother cells of the wheat stem rust fungus, *Puccinia graminis* f.sp. *tritici*, triggered by the synergistic action of chemical and physical signals. *Fungal Genetics and Biology*, **38**, 320-326.

Wiese, M.V. and Ravenscroft, A.V. (1979) Environmental effects on inoculum quality of dormant rust uredospores. *Phytopathology*, **69**, 1106-1108.

Wolfe, M.S. (1985) Dynamics of the response of barley mildew to the use of sterol synthesis inhibitors. *EPPO Bulletin*, **15**, 451-457.

Wolfe, M.S. and McDermott, J.M. (1994) Population genetics of plant pathogen interactions: the example of *Erysiphe graminis – Hordeum vulgare* pathosystem. *Annual Review of Phytopathology*, **32**, 89-113.

Wolfe, M.S., Brändle, U., Koller, B. *et al.* (1992) Barley mildew in Europe: population biology and host resistance. *Euphytica*, **63**, 125-139.

Wright, A.J. and Heale, J.B. (1984) Adult plant resistance to powdery mildew (*Erysiphe graminis*) in three barley cultivars. *Plant Pathology*, **33**, 493-502.

Zadoks, J.C. (1961) Yellow rust on wheat, studies in epidemiology and physiologic specialization. *Tijdschrift over Plantenziekten*, **67**, 69-256.

Zadoks, J.C. (1967) International dispersal of fungi. *Netherlands Journal of Plant Pathology*, **73**, suppl. 1, 61-80.

CHAPTER 16

ENVIRONMENTAL BIOPHYSICS APPLIED TO THE DISPERSAL OF FUNGAL SPORES BY RAIN-SPLASH

L. HUBER, L. MADDEN AND B.D.L. FITT

16.1 INTRODUCTION

In the past two decades, there has been considerable work on spore dispersal by rain-splash and review articles have been published (Fitt and McCartney, 1986; Fitt *et al.*, 1989; Madden, 1992) which discuss spore dispersal from a biological point of view. Topics covered included the characteristics of splash-dispersed fungi, mechanisms of splash dispersal and methods for studying them, the dispersal of specific pathogen spores by different types of rain under controlled conditions or in field situations and recent spatial distribution and modelling studies.

By contrast, this chapter aims to provide a physical/meteorological perspective on inoculum dispersal by rain-splash and to summarise articles published since the review of Madden (1992). It considers information from plant pathology, epidemiology, agricultural meteorology and soil science that is relevant to understanding splash dispersal and suggests further investigation and application of splash dispersal from the physical/meteorological point of view. It reviews 1. the splash process as a biophysical process; 2. the dispersal of splash droplets or inoculum in relation to physical characteristics of incident drops; 3. the influence of target characteristics on splash efficiency and dispersal; 4. the relevant properties of rainfall above, inside and below the crop canopy in relation to splash dispersal; 5. meteorological instrumentation in relation to rainfall and methodology to quantify splash potential of rain. Dispersal by rain is also discussed in Chapter 6.

16.2 REMOVAL OF SPORES BY SPLASH OF SINGLE INCIDENT DROPS

Beside studies describing spatial patterns of splash-dispersed fungal spores (Fitt *et al.*, 1989; Jenkinson and Parry, 1994; Hörberg, 2002), attempts have been made to quantify the number of spores (or rarely bacteria, Butterworth and McCartney, 1991) removed in splash droplets, by establishing relationships between incident drop properties and subsequent numbers of spores removed for given target characteristics (e.g. leaf structure and lesion pattern). Although seed dispersal by rain-splash has been known for more than 100 years, it has received little attention (Nakanishi, 2002). Occasionally, efforts based on classical or newly-developed techniques (Saint-Jean *et al.*, 2005) have been made to understand the mechanisms of incorporation of spores into the population of splash droplets in relation to their

B. M. Cooke, D. Gareth Jones and B. Kaye (eds.), The Epidemiology of Plant Diseases, 2nd edition, 417–444.

dispersal. Droplet size distributions suggested (Fig. 16.1) were log-normal for droplets obtained by splashing of drops falling onto solid surfaces or leaves (Levin and Hobbs, 1971; Macdonald and McCartney, 1988) or Weibull for droplets produced by drops impacting on a strawberry fruit surface (Yang *et al.*, 1991b).

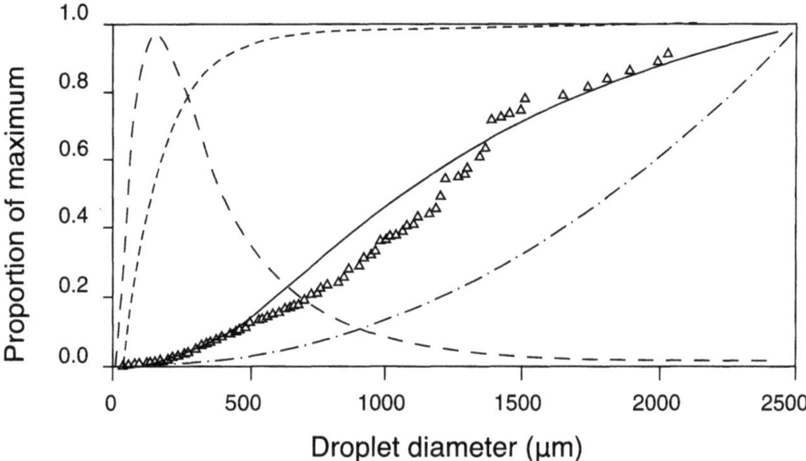

Figure 16.1. Comparison between measured cumulative numbers (Δ) of Pyrenopeziza brassicae *spores dispersed from spore suspensions by falling drops with increasing splash droplet diameter and numbers predicted (—) by a spore incorporation model using the droplet diameter frequency distribution (— —) expressed as a percentage of the mode, the proportion of spore-carrying droplets (---) and the mean spore concentration (—·—) in different droplet diameter categories (modified from Huber* et al., *1996).*

For water drops falling onto thin water films, the log-normal distribution was slightly better than the Weibull distribution; it also has properties that can be used to calculate statistical moments of variables related to droplet diameter based on a power law (Huber *et al.*, 1996). With both models, the droplet size distribution is positively skewed, with more small droplets than large droplets. This description is important because droplets of different sizes travel different distances. Measurements and modelling showed that 100 μm diameter droplets travel 10 cm, whereas 600 μm diameter droplets can travel as far as 1 m (although mean distances are less) (Macdonald and McCartney, 1987). Since this difference has important consequences for the dispersal of spores and subsequent epidemic development, more work is needed on the incorporation of spores into splash droplets using a joint distribution describing both the droplet diameter and the number of spores per droplet. This can be done by a modelling approach (Huber *et al.*, 1996) based on three functions of droplet diameter: diameter distribution of droplets (log-normal), proportion of droplets carrying spores (exponential function) and number of spores per spore-carrying droplet (power law). Fig. 16.1 illustrates the shape of these three functions for an experiment on dispersal of *Pyrenopeziza brassicae* conidia by splash of simulated water drops impacting onto a 0.5 mm deep spore suspension (Fatemi and Fitt, 1983). The parameters for each function depend on characteristics such as spore type, target type and orientation and incident drop diameter. The application of results obtained in spore incorporation experiments in controlled conditions to natural situations is difficult because many different targets are encountered. Even in controlled conditions, lesion morphology and structure influence the source depletion rate; the time until half-depletion was 2-12 times shorter for spores of *Monilia fructigena*, *Gloeosporium fructigenum*, *Fusarium roseum* or *Helminthosporium gramineum* than for spores of *Puccinia striiformis* (Pauvert *et al.*, 1970). Since a source of spores does not remain constant over time, theoretical spore incorporation models require further work to include the biological variation in spore production by lesions.

The interaction between rainfall and crop targets is central in water and energy transfer during a splash event and thus in the removal and subsequent dispersal of fungal spores. Since rainfall is independent of the target (canopy, plant or leaf), experiments using simulated raindrops produced by a drop/rain generator in laboratory conditions can be done with constant targets to derive empirical relationships relating spore removal to raindrop characteristics. The configuration of the target which depends on factors such as surface type, geometry and roughness, also greatly influences the dispersal process.

16.3 FROM A SINGLE IMPACTING RAINDROP TO SPLASH DROPLETS

16.3.1 Drop characteristics

Impacting drops are characterised by their shape and drag coefficient. Possible differences in shape originate from the mechanical equilibrium of forces acting on the drop: the spherical pressure term, hydraulic and dynamic pressure terms, drag force, and weight (force of gravity). The relationship between drag force (F_D) and drag coefficient (C_D) is expressed as:

$$F_D = {}^1/_2\, C_D\, \rho_a\, \pi D^2/4\, V^2 \tag{16.1}$$

where ρ_a is the air density (kg m^{-3}), D the equivalent drop diameter (m) and V the drop fall velocity (m s^{-1}).

Drag coefficients depend on both drop diameter D and Reynolds number ($Re = VD\, \nu^{-1}$ where ν is the kinematic viscosity in m^{2}s^{-1}). The motion equation of a drop falling vertically can be solved numerically to estimate velocities of drops falling from different heights (Fig. 16.2a). Terminal velocities are generally calculated using experimental relationships (Fig. 16.2b).

In the diameter range 0 – 3 mm, the curves of Madden (1992) and Ulbrich (1983) seem similar and provide an adequate basis for calculating velocity-dependent parameters of drops falling at terminal velocity. The fall velocity of drip drops, released at heights ≤ 0.5 – 1 m in most agricultural crops, must be calculated separately for each diameter x release height combination.

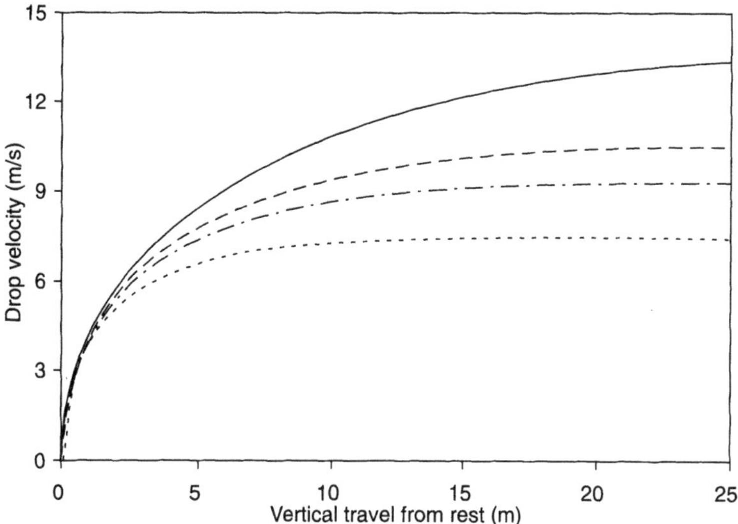

Figure 16.2a. Velocities of drops falling from different heights from rest to terminal velocity. Incident drop diameters 2.2 mm (-----), 2.9 mm (—·—·—), 3.4 mm (— — —), 4.9 mm (—). The velocity values were obained by numerically solving the motion equation resulting from the balance of forces affecting a spherical drop falling in still air.

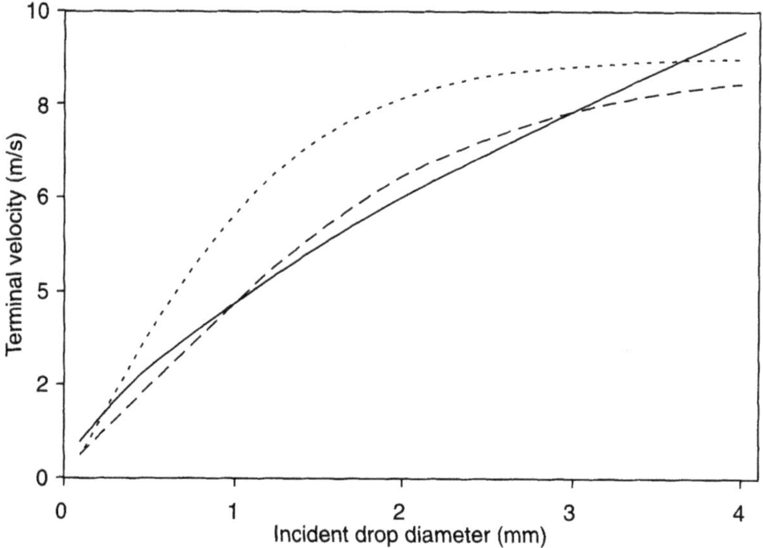

Figure 16.2b. Relationship between incident drop diameter and terminal velocity as described by three formulae (Englemann, 1968: -----; Ulbrich, 1983: —; Madden, 1992: — — —).

Once both diameter D and velocity V of impacting drops are known, any physical quantity X, such as the momentum, impact force or kinetic energy of these drops can be calculated with the expression:

$$X = \alpha D^{\beta} V^{\gamma} \tag{16.2}$$

For drop momentum, $\alpha = \rho_w \pi/6$, $\beta = 2$, $\gamma = 1$ with ρ_w water density in kg m^{-3}; for the impact force, α of the order of $\rho_w \pi/12$, $\beta = 2$, $\gamma = 2$; for kinetic energy, $\alpha = \rho_w \pi/12$, $\beta = 3$, $\gamma = 2$. At terminal velocity, using Ulbrich's function for $V(D)$ (Fig. 16.2b), this expression becomes $X = \alpha' D^{\beta'}$ with $\alpha' = 0.81^{\gamma}\ \alpha$ and $\beta' = \beta + 0.67\gamma$. These parameters can be used for scaling variables characterising splash droplet production (volume, horizontal or vertical distance, kinetic energy of droplets, etc).

16.3.2 Mechanics of splashing and trajectories of splash droplets

By solving the equation of motion (Macdonald and McCartney, 1987; Allen, 1988), the trajectories of splash droplets can be modelled, provided that the relationship $C_D(Re)$ is known and the initial conditions of the trajectory (ejection angle and velocity) are given. Classical Newtonian dynamics can provide an adequate conceptual framework to simulate distances travelled by splash droplets, with most of the discrepancies arising from the variability of initial conditions. For example, in strictly defined conditions with drops falling vertically onto a 1 mm-deep film of water, a narrow range of droplet ejection angles between 45 and 75° was observed, but there was a wide range of ejection speeds between 1 and 4 m s^{-1} (Allen, 1988). Excellent experimental work done by Macdonald and McCartney (1988) illustrates the relationship existing between initial ejection speed and angle of ejection for splash droplets on bean leaves. Since the pioneering work of Gregory *et al.* (1959), most work on the horizontal distance travelled by splash droplets has been done under windless rain conditions. Field experiment results on splash dispersal of conidia of *Mycocentrospora acerina* on caraway show that turbulence contributes to the transportation of splash-dispersed spores at least up to about 10 m (Evenhuis *et al.*,1997). In the field with a prevailing wind direction, distances travelled by splash-dispersed spores can be greater by an order of magnitude than in controlled conditions. In the spread of the anthracnose disease on the tropical pasture legume *Stylosanthes scabra*,, dispersal of *Colletotrichum gloeosporioides* conidia by rain-splash including both primary splashes and re-splash is characterized by half-distances of about 1-10 m in the field in comparison to 5-15 cm in still air (Pangga *et al.*, 2004). Efforts to study wind-driven rain were made by soil physicists (Erpul *et al.*, 2004) investigating the splash saltation process in which particles are lifted by impacting raindrops and subsequently transported by wind shear. Knowing that the average velocity of the splash saltation of soil particles ranges between 3 and 7 ms^{-1} depending on wind velocity, such physical processes are expected to be of crucial importance for spore dispersal by rain-splash.

Although the physics determining trajectories of splash droplets in the air is understood, little is known about the processes controlling variability in the initial conditions of droplet trajectories. Thus, it is very difficult to proceed further in studying the mechanics of splash on a small scale, because modelling the splash process is physically complex and computer-intensive and the upscaling of physical mechanisms from single splash events to canopy level is complicated. Therefore, most efforts to predict the characteristics of a series of splash events for use in plant disease epidemiology and ecology, for given rain/target conditions, are made without analyses of each individual event. The splash process is considered as a black box, ignoring the impact stage, and the outcome is described macroscopically using mean or cumulative variables. For example, water splash by single incident drops can be characterised by considering the overall efficiency of the splash in transferring mass or energy from the incident drops to splash droplets.

Alongside these macroscale studies a new physical approach was developed by Saint-Jean *et al.* (2004) to model water transfer by rain-splash in a 3-D canopy using Monte-Carlo integration. In the framework of this mechanistic and probabilistic modelling work, each individual splash event at the plant-atmosphere interface and the interception of splash droplets by plant organs is simulated. The first results show the potential of such a model as a research tool for studying effects of obstruction patterns on splash droplet trajectories. By assuming that in-flight evaporation of splash droplets can be ignored (which makes sense under high relative humidity of the air during rainfall) and drag force can be ignored, a more operational 2-D parabolic trajectory model was developed to predict rain splash height (Pietravalle *et al.*, 2001).

16.3.3 Water and energy transfer during splash events

The splash mechanism can be studied experimentally and numerically by investigating the physical properties (surface tension, viscosity) of the water drops and the target. In epidemiology, splash events can be studied as exchange systems, using knowledge of empirical relationships between impacting drop characteristics (e.g. size and velocity) and splash parameters (e.g. number and mass of droplets, flight distance, kinetic energy). The concept of distance travelled by a splash droplet during a single splash event is different from that of distance travelled by a spore. A spore may move by a series of successive 'jumps' in several splashes but droplet movement and consequently the distance travelled depend only on the initial conditions of droplet ejection, atmospheric turbulence characteristics and obstruction patterns.

In still air with no obstructions, deposition gradients from a point source have been studied extensively (Fitt *et al.*, 1989). Most average or median distances travelled by splash droplets from point sources are $\leq$15 cm and such droplets rarely travel beyond 1 m (Fitt and McCartney, 1986). Because the spore-carrying droplets of greatest importance are the large ones travelling close to the source (Fig. 16.1), splash dispersal is a short-range phenomenon.

The limited distances travelled by splash droplets and steep spore droplet deposition gradients have been confirmed in different experiments (Fitt and McCartney, 1986), with targets ranging from liquid spore suspensions (Fernandez-Garcia and Fitt, 1993) to cow pats (Granvold, 1984). Negative exponential models generally fitted these deposition gradients better than inverse power models. Half-distances of pathogen dispersion in splash droplets produced by simulated drops are in the range of 5 to 15 cm (Fitt *et al.*, 1989). A diffusion model closely related to the exponential model and based on physical hypotheses also described deposition gradients of spores well (Yang *et al.*, 1991a; Madden *et al.*, 1996).

In an extensive analysis of the drop impaction and splash droplet movement from a strawberry fruit surface (Yang *et al.*, 1991b), mass and kinetic energy reflection of impacting drops (0.5-4 mm) falling from release heights between 0.25 and 1.5 m were determined for different combinations of drop diameter and release height. Frequency distributions and cumulative distributions of droplet diameter and droplet distance of travel were positively skewed and fitted well by the Weibull distribution. On a log-log scale, total kinetic energy and total mass of splash droplets were linearly related to kinetic energy of drops impacting on the strawberry fruit surface (Fig. 16.3a). On a log-linear scale, mass and kinetic energy reflective factors were expressed as quadratic functions of droplet velocity and asymptotically limited by a constant value (Fig. 16.3b). By comparison, the percentages of drop mass and kinetic energy transferred to droplets increased with increasing drop velocity to reach maxima of *c.* 70% for mass and 10% for kinetic energy. Essentially, most of the mass or kinetic energy of splash droplets is derived from high velocity, high mass, high energy incident drops.

The few experimental data sets available show that external variables, such as drop size and impact velocity, cannot always explain splash variability. At low release heights, total kinetic energy or mass of splash droplets per incident drop can be expressed as a power law of the kinetic energy (proportional to D^3V^2) of the drop falling on a surface (Yang *et al.*, 1991b). For a wide range of release heights (1, 2, 11 m), the maximum splash droplet height was predicted well using $\log(DV)$ (Walklate *et al.*, 1989). Using a weighing technique to determine the mass balance between incident drops (falling from 1.5 m or 11 m) and splash droplets, Huber *et al.* (1997) directly obtained the mass reflective factor (proportions) and total mass during splash events for various target configurations. For given impacting surface and target characteristics, Pietravalle *et al.* (2001) confirmed the potential use of the kinetic energy of impacting raindrops for estimating maximum splash height and total number of splashes on a vertical cylindrical splash meter (Shaw, 1987, 1991). The impacting kinetic energy of individual drops or simulated rainfall was a good indicator of rust spore dispersal from wheat leaves or seedlings, respectively (Geagea *et al.*, 1999, 2000).

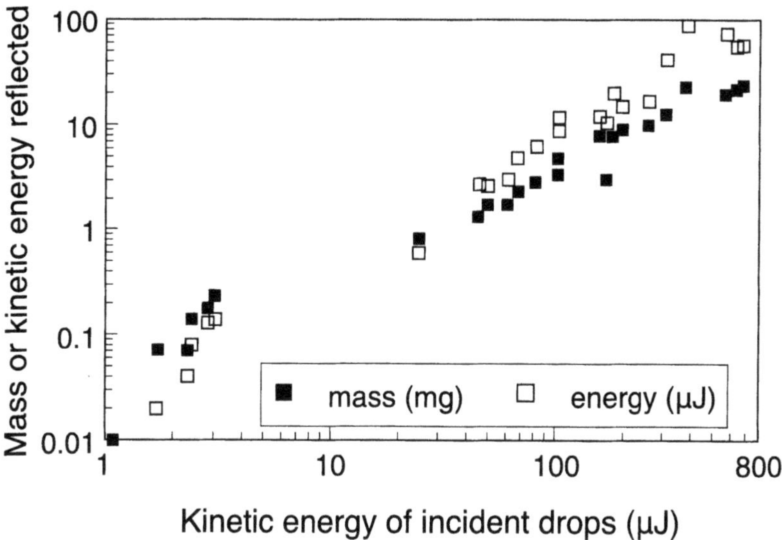

Figure 16.3a. Relationship between the kinetic energy of simulated drops impacting on a strawberry fruit surface and the mass and kinetic energy of water droplets produced during the impact (redrawn from Yang et al., *1991b).*

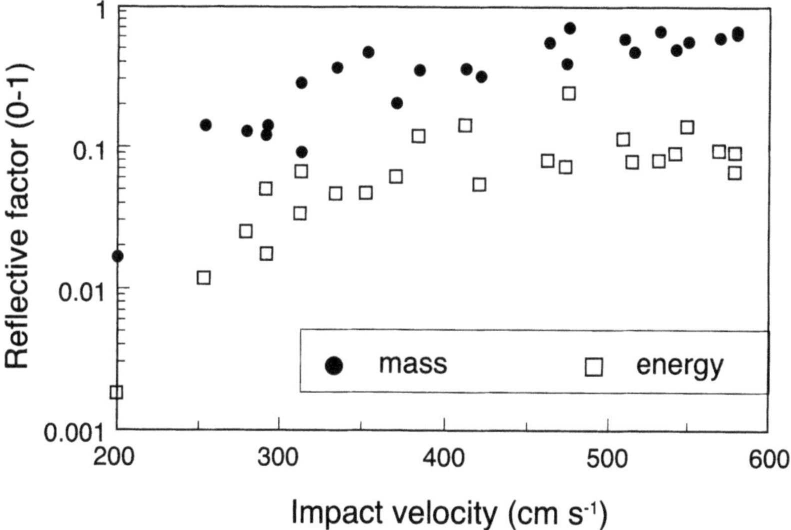

Figure 16.3b. Relationship between the fall speed of simulated drops impacting on a strawberry fruit surface and the mass and kinetic energy reflective factors of water droplets produced during the impact (redrawn from Yang et al., *1991b).*

16.4 INFLUENCE OF TARGET CHARACTERISTICS ON SPLASH PARAMETERS

It is difficult to quantify the influence of target characteristics, such as target type (artificial, leaf or fruit surface), superficial properties (wetness, rugosity, presence of fungal lesions), internal and mechanical properties (turgidity, flexibility), on the splash process. The scale of complexity increases from that of an artificial rigid target to that of a plant surface and ultimately to that of a three-dimensional crop canopy. Of particular interest is the leaf scale because of its importance in the quantitative approach to the study of rain-splash.

16.4.1 Inert rigid target

An artificial rigid surface can be used as a reference target to scale mass and energy transfer during splash under repeatable conditions, so that mass and energy transfer from real leaves, which are neither inert nor rigid, can be studied by comparison with transfer from the artificial leaves. Important characteristics of the surface are the presence or absence of water and the thickness of the water film on a wet target. The variability in the mass of water splashed from dry glass targets is much larger than that from wet targets and decreases as incident drop size increases. The angle of inclination of the target also greatly affects the splash process, especially with large incident drops (>3 mm); the mass of water splashed by large drops (impacting onto a glass target wetted with a water film 0.2 mm deep) decreased by about 50% as the angle from horizontal increased from 0° to 15° (Huber *et al.*, 1997).

Fig. 16.4a illustrates the power law relationship between drop diameter and the mass of water splashed per incident drop for drops falling from a height of 11 m onto artificial rigid glass targets of various types (dry, wet with water films of thickness 0.1 to 0.6 mm). Walklate *et al.* (1989) compared splash from various target types and developed the model:

$$\Delta z = z^* \ln(DV/C^*) \tag{16.3}$$

where Δz is the maximal splash height above the impact level, and z^* and C^* are splash parameters; estimates of z^* were directly affected by target surface properties. For a rigid metallic target, the critical impact velocity above which splashing occurs is a power law of drop diameter D ($1.4 < D < 4.5$ mm) (Stow and Hadfield, 1980), which decreases (in a range less than 3-4 ms^{-1}) with increasing surface roughness.

16.4.2 Plant surfaces

Experiments using leaf or fruit surfaces have indicated that the variability in splash variables is much greater for artificial targets. For example, the pattern of mass transfer of splash droplets was different between healthy or diseased (white leaf spot, caused by *Mycosphaerella capsellae*) sub-horizontal oilseed rape leaves at three stages of maturity and tobacco leaves, showing that leaf surface characteristics

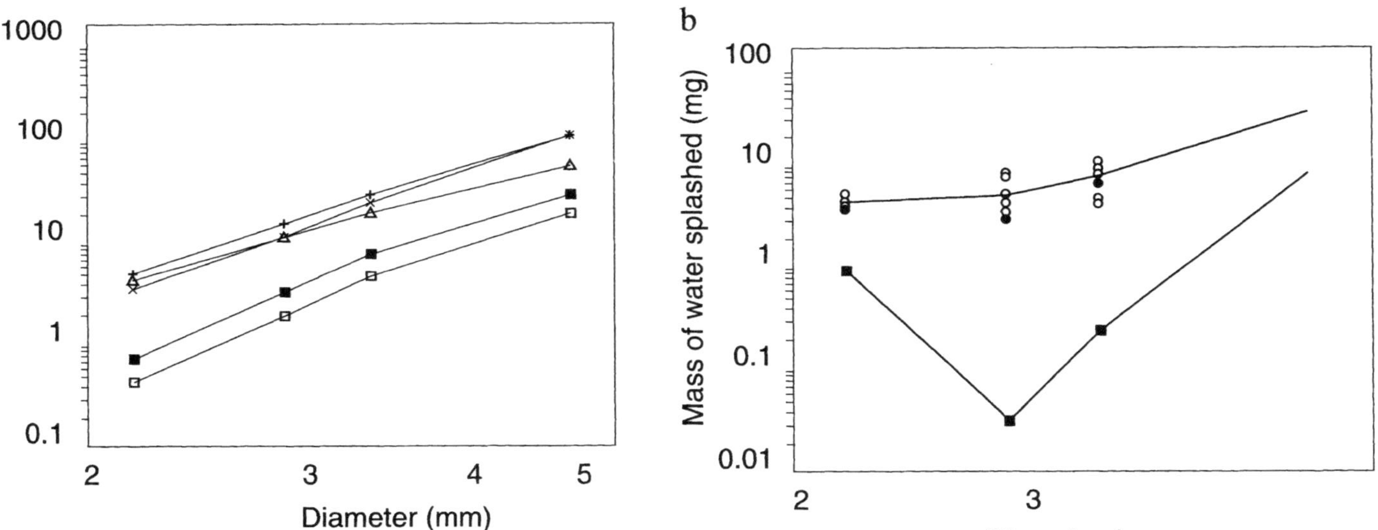

Figure 16.4. Comparison between the amounts of water splashed per drop from horizontal glass targets (a) and from oilseed rape and tobacco leaves (b) plotted against incident drop diameter. Glass targets (a): dry wettable, ■; dry non-wettable, □; 0.1 mm water film, +; 0.4 mm water film, ×; 0.6 mm water film, Δ. Leaf targets (b): healthy oilseed rape leaves of different ages, O; oilseed rape leaves infected with Mycosphaerella capsellae, *●; healthy tobacco leaves, ■ (redrawn from Huber* et al., *1977).*

play an important role in splash formation (Fig. 16.4b). The power law relationship between drop diameter and mass of water splashed established experimentally for horizontal dry glass plates with $2 < D < 5$ mm appears to apply for incident drops impacting on tobacco leaves when $D > 3$ mm (Huber *et al.*, 1997). The quantities of water splashed from tobacco leaves were much smaller than from oilseed rape leaves (by a factor of 6), probably because the hairy surface of the tobacco leaves absorbed a much larger proportion of the drop momentum than the smooth waxy surface of the oilseed rape leaves. Surface wetness must also be taken into account because large differences exist between plant species (Brewer and Smith, 1997).

The physical analysis of drop impaction on a strawberry fruit surface provides another example (Yang *et al.*, 1991b) in which the droplet trajectory vectors were related to fruit surface inclination. The angle of reflection (measured from impact surface normal) was close to the angle of impact (also measured from impact surface normal) and the percentage of droplets moving in the same direction as that of impact surface normal was a linear function of the sine of the impact angle (Fig. 16.5); this relationship was independent of both incident drop diameter and release height (and thus velocity).

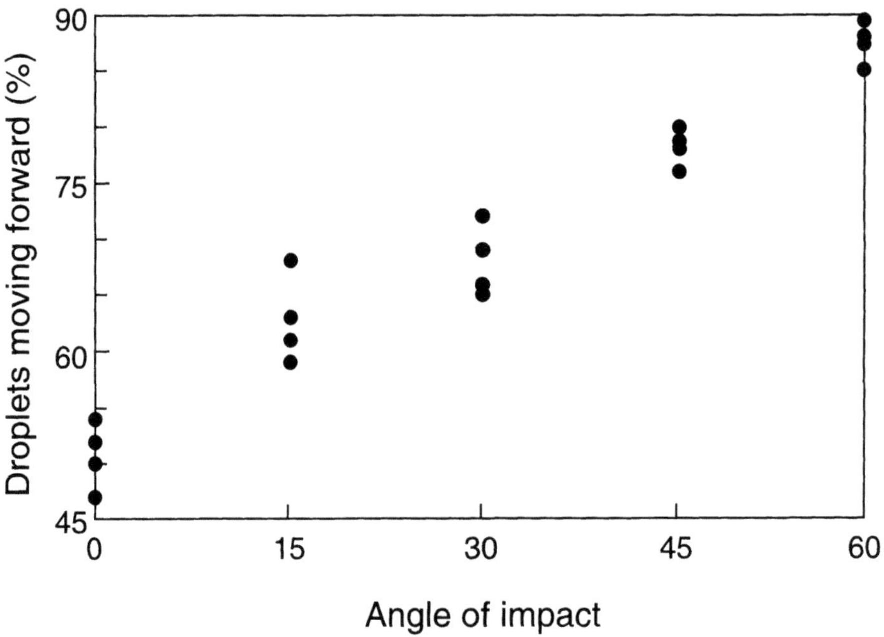

Figure 16.5. Relation between impact angle of a falling drop (defined as the angle between the surface normal and the trajectory direction) and the percentage of droplets moving in the direction of the surface normal. Points are for four drop sizes (1, 2, 3, or 4 mm) and are means of replicates for different heights of drop release (redrawn from Yang et al., *1991b).*

The influence of leaf flexibility on splash was examined by comparing Brussels sprout leaves that were either rigid (taped onto a rigid holder) or normally flexible (Stedman, 1979). There was more splash from rigid leaves because flexible leaves absorb a proportion of the drop momentum so that it is not all available for transfer into splash droplet production, and maximum splash height values for oilseed rape and wheat leaves were much less than those obtained with artificial targets (horizontal surface with thin or accumulated water film) (Walklate *et al.*, 1989). Except in a few cases (Fitt *et al.*, 1992), experimental results on drop splash efficiency have been obtained with plant targets whose flexibility was not taken into account.

16.4.3 Crop targets

Splash dispersal is greatly modified, and generally decreased, in a crop canopy by comparison with splash from an artificial rigid target or an individual plant target. Surface topography (including ground cover) and plant obstruction both have a large influence on splash (Stedman, 1979, 1980; Yang *et al.*, 1991b; Soleimani *et al.*, 1996; Huber *et al.*, 1997).

(a) Ground cover and surface topography

A controlled experiment with uniform-sized drops falling near terminal velocity on barley straw, placed on a horizontal plane surface or on an open nylon grid showed the effects of ground cover on maximum upward splash of water droplets (Walklate *et al.*, 1989). Dispersal is greatly affected by surface roughness, quantified as the standard deviation of elevation; the flux density of conidia (assessed as a number of colonies growing on a selective medium) was greatest with a polyethylene ground cover and smallest with straw ground cover (Yang *et al.*, 1990).

The effects of ground cover/topography on water splash were studied extensively by Yang and Madden (1993) using simulated rainfall impacting on three types of ground cover (soil surface, straw, plastic mulch) or strawberry plants. Flux densities of droplet number and mass over horizontal distance were described well by exponential models, and ground cover affected the steepness of gradients in both numbers and mass of splash droplets; the steepest gradient was observed with the straw ground cover and the shallowest with plastic mulch. The mass reflective factor over a time period of uniform rainfall, defined here as the estimated total mass of splashed water divided by the total mass of impacting raindrops on the intercepting area, was directly influenced by the type of ground cover.

(b) Spatial organisation and temporal changes in plant canopies

The effects of the strawberry plant canopy on splash of water were assessed in the experiments of Yang and Madden (1993). For three values of leaf area index (LAI = 1.4, 1.9, 2.2) and a plant free control (LAI = 0), droplet size distributions were characterised by Weibull functions. The gradients of flux densities of droplet number and mass were significantly shallower with plants than without them. Furthermore, the total reflected mass of water and the mass reflective factor (percentage of incident mass) were considerably decreased by the presence of strawberry plants (Fig.16.6). An inverse relationship between strawberry canopy density and splash dispersal of *Colletotrichum acutatum* spores by simulated rainfall has been observed (Boudreau and Madden, 1995). Three possible mechanisms for decline in spore deposition were suggested: (i) decrease in the numbers of spores removed from lesions, by splash or re-splash, because fewer raindrops reached impaction sites; (ii) decrease in spore dispersal by secondary splash due to local plant obstructions; (iii) direct interception of spore-carrying droplets by plant surfaces.

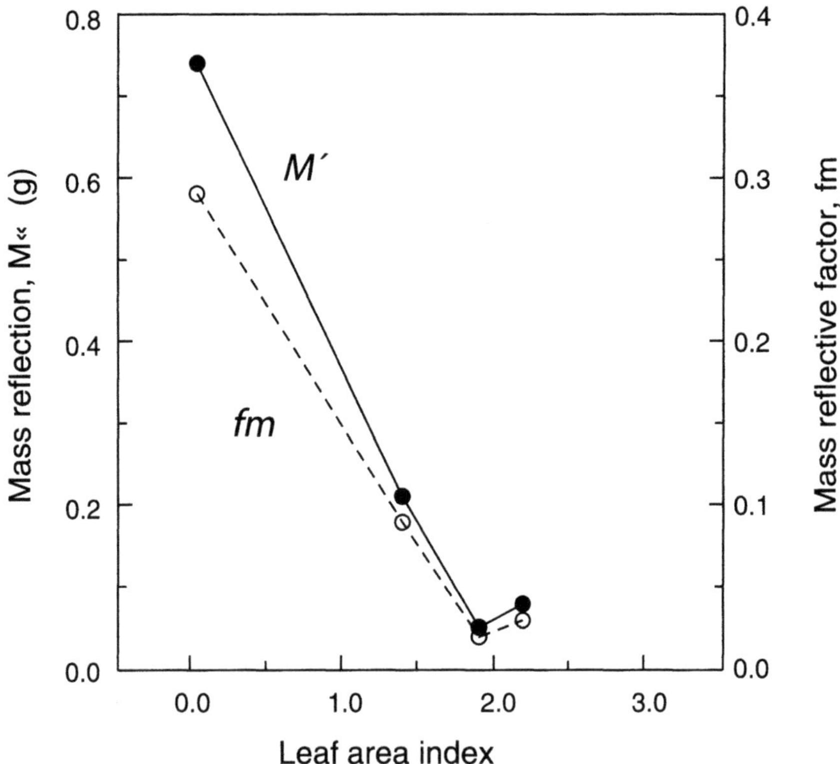

Figure 16.6. Influence of leaf area index on mass reflection and mass reflective factor (percentage of incident mass) of simulated rainfall (intensity = 15 mm h^{-1}) falling across the window of a rain shelter (see Fig. 16.5) on strawberry plants with soil ground cover. Points are means of eight simulated rain events of 2 min each and correspond to three values of LAI (from Yang and Madden, 1993).

16.5 RELEVANT CHARACTERISTICS OF RAINFALL-CANOPY INTERACTIONS

To understand spore dispersal by rain-splash, a more specific description of the properties of rainfall is required. From experimental studies with single incident drops, it is known that drop size is an important factor affecting the splash process on a target (Fitt *et al.*, 1989) because the terminal velocity and hence the kinetic energy of a drop is a power function of diameter (Ulbrich, 1983: $V(D)$ proportional to $D^{0.67}$). Although large drops are a small proportion of the total number of drops in rain, they contribute a large proportion of total rainfall volume and play a decisive role in splash dispersal. Therefore, the size distribution of raindrops is important because a given rainfall intensity and the drop size distribution can change significantly during a rain event (Ulbrich, 1994). The type of rain (e.g. drizzle, shower, widespread and stratiform, orographic, thunderstorm, etc.) is known to affect the frequency distribution of drop sizes (Mason and Andrews, 1960; Ulbrich, 1983). A description of drop size distribution, with the assumption of spherical raindrops at terminal velocity, is needed. Drip drops within a canopy travel at speeds less than the terminal velocity because of the limited fall height and an analysis of rainfall modification by the canopy is a prerequisite to studies on spore dispersal by drip drops (also named secondary drops or gravity drops).

16.5.1 Characteristics of rainfall

Rainfall is composed of discrete drops scattered in a volume of air and several physical and chemical processes contribute to the raindrop size distribution (DSD) in the free atmosphere. Evaporation and drop break-up increase the number of small drops, whereas condensation and coalescence increase the number of large drops. These processes combine to produce a drop size range from 0.2 to 5.5 mm. The temporal variability in DSD during a rain event can also been considered. For instance, Smith and DeVaux (1992) showed that during a one-hour storm, the mean drop diameter ranged from 0.5 to 2 mm and the standard deviation of diameter varied from 1 to 6 times the mean diameter. Moreover, the number of drop impacts on the ground ranged between 50 and 5000 drops $m^{-2} s^{-1}$.

Classical characteristics of rain distribution were discussed by Madden (1992). The size distribution was described initially by Marshall and Palmer (1948), using a negative exponential function of drop diameter: $N(D) = N_0 \exp(-\Lambda D)$, where $N(D)$ is the number of drops in a unit volume (m^3) of air per length unit (cm) of drop diameter D (cm); N_0 and Λ are parameters dependent on the type of drop size distribution and rain intensity, respectively. This function was used in some epidemiological studies (Aylor and Sutton, 1992; Madden *et al.*, 1996). The Marshall-Palmer equation was generalised as a gamma distribution by Ulbrich (1983):

$$N(D) = N_0 D^{\mu} \exp(-\Lambda D) \qquad (16.4)$$

where μ can vary from -2 to +2 (or more rarely +6), depending on the rain type (Table 2 in Ulbrich, 1983). The gamma distribution was used in a simulation study to quantify amounts of rain-splash from leaves (Huber *et al.*, 1997).

The gamma distribution takes account of short time variations better but the Marshall-Palmer function may be satisfactory for long time and space averaging. If the 'instant' spectrum (less than 10 min) is a gamma distribution, the time-averaged spectrum (over a few hours) is a Marshall-Palmer distribution (i.e. negative exponential) (Yangang, 1993). The Marshall-Palmer distribution can generally be used to represent drop size variability if the time scale of the epidemiological processes studied is longer than 20-30 min but the gamma distribution should be used if the time scale is shorter.

From the DSD, integral parameters can be calculated using the generalized expression:

$$P = a_P \int_0^{\infty} D^p N(D)\, dD \tag{16.5}$$

in which a_p and p depend on the integral parameter of interest (Madden *et al.*, 1998; Ulbrich, 1983). With D in cm and $N(D)$ in m^{-3} cm^{-1}, the values of a_p and p can be easily calculated for any integral parameter: for drop number flux density (m^{-2} s^{-1}), $a_p = 17.67$ and $p = 0.67$; for rainfall intensity (mm h^{-1}), $a_p = 33.31$ and $p = 3.67$; for rainfall power (W m^{-2}), $a_p = 1.44$ and $p = 5.01$; and for the back-scattering coefficient (mm^6 m^{-3}), $a_p = 10^6$ and $p = 6$.

Although the DSD can be measured directly with an accuracy in drop diameter measurements of ≤6% (Salles and Poesen, 1999), the distribution can also be estimated using two integral parameters measured independently (Madden *et al.*, 1998; Torres *et al.*, 1994). For instance, measurements at ground level (rain intensity and rain power) can be used to predict the shape parameter of an assumed gamma distribution before calculating the term of direct interest for rain-splash. The atmospheric measurements used by meteorologists to characterise raindrop size distribution, such as liquid water content and the radar back-scattering coefficient, may be applied to splash dispersal at a regional scale since radar is used routinely for assessing the spatial distribution of rainfall. Drops less than 1-2 mm in diameter are of minor importance for splash dispersal because of their small velocity and kinetic energy. Therefore, if a drop diameter threshold is known (or assumed) for a given type of target or canopy, this lower limit can be used in the integration for P to determine an integrative parameter that relates to splash dispersal (Walklate, 1989).

Since the angle between the target normal and the impact directions greatly influences the splash process, it is important to characterise the variations in both angle and direction of rainfall. This can be done with an instrument consisting of a vertical grooved cylinder with the flow-off partitioned into the four cardinal sectors as designed by Crockford *et al.* (1991). Volumes of water collected in each of the four cardinal sectors can be used to derive rough estimates of rain angle and direction at low wind speeds and without significant gusts or a bimodal wind direction spectrum. Caution is required when using this type of instrument because no standard exists.

16.5.2 Rainfall modification by the canopy and drip drop production in crops

The modification of rainfall by the crop canopy is important in splash dispersal (Huber *et al.*, 1995) since drip drops produced by a crop canopy on which rain is falling may have different effects on spore dispersal by splash in comparison to raindrops that impact on spore-bearing lesions directly. Transformation of rainfall is largely influenced by the canopy structure (e.g. LAI vertical partitioning, leaf angle distribution) in relation to rain angle and direction. For uniform canopies, direct penetration of raindrops through the canopy is limited when LAI is more than 2 (Schottman and Walter, 1982). A two-layer stochastic rainfall interception model was developed to take into consideration both primary raindrops (falling at terminal velocity) and secondary drops dripping from leaves (Calder, 1996). This model predicts the mean numbers of impacts and drops retained per leaf. Such a model, with appropriate modification for the splash process, could be used to quantify the amounts of splash in the canopy and the vertical profile of splash volumes. The kinetic energy of drip drops rainfall underneath a canopy can exceed the kinetic energy of the raindrops above the canopy because the canopy has increased the median drop diameter and the proportion of large drops (Armstrong and Mitchell, 1988; Finney, 1984). The DSD of rainfall transformed by a crop canopy is often bimodal, with a first mode corresponding to the DSD of the untransformed rainfall (slightly shifted and skewed towards large drop sizes) and the second mode corresponding to the 4-7 mm range of drip drop diameters (Armstrong and Mitchell, 1987). Range and variability of drip drop sizes seem quite insensitive to the size distribution of primary drops (Armstrong and Mitchell, 1988; Moss and Green, 1987) but sensitive to vegetation type (Hall and Calder, 1993). Madden (1992) provides a prospective synthesis of the effects of canopy on rainfall. Recent work on 3-D modelling of canopy architecture for simulation of rainfall interception parameters has potential to predict downward spore dispersal by wash-off through stem flow or splash of large drip drops on lower leaves (Bussière *et al.*, 2002; Bassette and Bussière, 2005). This may be important in dispersal of *Mycosphaerella fijiensis* and *M. musicola*, cause of sigatoka diseases of banana.

A good example where secondary drip drops reaching the ground dispersed inoculum from soil and litter is in the spread of cocoa black pod disease (Gregory *et al.*, 1984). Ballistic spore-carrying splash droplets carried spores up from the soil to infected pods located up to a height of 70 cm but aerosol droplets carried upwards by vertical air currents created by natural convection were believed to transport spores to the pods above 70 cm which also developed disease symptoms. Fog is another source of drip drops; when aluminium or plastic 'leaves' were placed in a fog wind tunnel for several hours with a fog intensity of 0.3 mm hr^{-1} (Merriam, 1973), fog drip drops formed could be as much as 1 mm in diameter even though there are no primary drops.

16.5.3 Effects of rain and canopy characteristics on spore dispersal by splash

Because the DSD shifts towards larger drop sizes as rain intensity increases (Ulbrich, 1983) and splash droplet numbers and mass increase with increasing size of the impacting drop (Yang *et al.*, 1991b), increasing rain intensity might be expected to increase spore dispersal and disease spread. However, it has been difficult to show a clear relationship between rain intensity and dispersal (Madden *et al.*, 1996; Madden, 1997). Field studies have been hindered by the considerable variability in rain intensity during rain episodes, differences in rain type (e.g. showers), and by the variation in total rainfall volume and duration between different episodes. Until recently, controlled studies did not generate rain with a great enough range of intensities to determine the functional relationship between intensity and dispersal parameters (e.g. mass of water splashed, number of spores transported).

Madden *et al.* (1996) generated rain with a wide range of intensities that matched the theoretical Marshall-Palmer drop size distributions of natural rain to elucidate the effects of rain intensity on splash dispersal of *Colletotrichum acutatum* spores; increasing rainfall intensity increased the mass of water splashed from the surface (with mass reflective factors of *c.* 2-6%), spore removal from lesions and total number of spores dispersed (measured as colony forming units on a selective growth medium) (Fig. 16.7a). However, infection of susceptible strawberry fruit exposed to rain increased with increasing intensity up to 15-30 mm h^{-1} and then decreased with further increases in intensity (Madden *et al.*, 1996; Figure 16.7b) because dispersed spores were then removed from potential infection sites by wash-off or run-off.

Specifically, the rate of wash-off of spores increased with increasing intensity, presumably because the physical processes involved in spore removal from lesions are also involved in removal of spores from infection sites. This disease result is consistent with the observed spore deposition results (Yang *et al.*, 1991a; Madden *et al.*, 1996), which showed an initial increase in spores per unit area with time but then a decline at longer times. Ntahimpera *et al.* (1997) confirmed these results in work with rain of different drop size distributions (that could not be represented by the simple Marshall-Palmer model) and found analogous results. Specifically, increasing the median diameter of drops (on a volume or mass basis) increased spore dispersal, but the incidence of fruit infection was not affected by changing the median diameter (or drop size distributions). Rain splash dispersal of *Giberella zeae* spores within a wheat field can be described by a power law relationship between spore flux density at a given canopy height and rain intensity, with a power independent from factors such as year and location (Paul *et al.*, 2004). Thus, rain intensity has a complex relationship to splash dispersal.

The complexity depends on the role of the physical 3-D structure. For example in the case of strawberry production, Ntahimpera *et al.* (1998) demonstrated the effects of a Suddangrass cover crop on splash dispersal of *Colletotrichum acutatum* conidia. Effects of cover crop architecture can be complex; for example rain splash intensity might be increased and splash droplets obstructed. In experiments on splash dispersal of *Septoria tritici* conidia in a wheat-clover intercrop (Bannon and Cooke, 1998) in controlled conditions using simulated rain, the cover crop decreased spore dispersal in both horizontal and vertical directions. Field work on septoria tritici blotch in wheat showed how the spatial distribution of lesions influences the risk of inoculum dispersal by rain-splash in canopies (Lovell *et al.*, 2003).

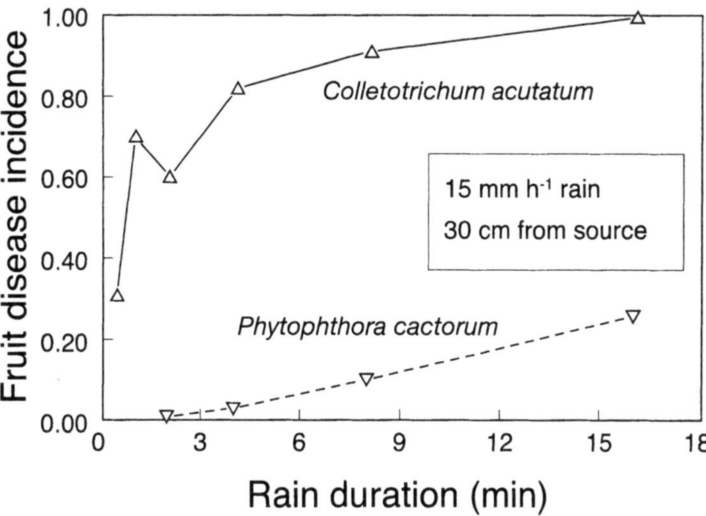

Figure 16.7a. Mean incidence (proportion of fruit affected) of strawberry anthracnose (caused by Colletotrichum acutatum*) and strawberry leather rot (caused by* Phytophthora cactorum*) in relation to duration of simulated rain (redrawn from Madden* et al.*, 1992).*

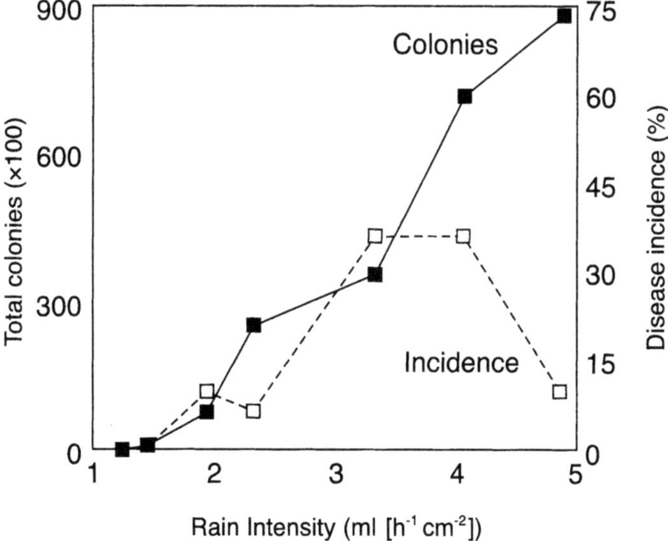

Figure 16.7b. Total number of colonies and disease incidence (% strawberry fruit affected) as a function of rain intensity (redrawn from Madden, 1997).

16.6 CHARACTERIZING RAINFALL IN RELATION TO SPLASH-DISPERSED PATHOGEN DISEASES

Spore dispersal by rainfall is not all by rain-splash; processes such as rain puff, wash-off of spores by stem-flow and dripping water and wash out of air-borne spores by falling raindrops can all disperse spores. These processes may operate in isolation or in combination to increase disease spread if more susceptible hosts are reached by infectious spores or to decrease disease spread if inoculum is lost. One important topic which is rarely investigated is the influence of environmental (particularly physical and chemical) conditions on inoculum during dissemination. However, this influence is probably of much greater importance for air-borne spores dispersed by wind than for splash-dispersed spores. Both spatial and temporal variations in rainfall characteristics influence inoculum movement by rain-splash. The temporal variation in drop size distribution is an external function influencing the physical process of spore dispersal by rain-splash. DSD measurements require expensive and sophisticated instruments (spectropluviometer or disdrometer); however, these instruments are unique in providing the forcing function of rainfall. In most cases, rainfall parameters are measured directly, for example by mechanical integration of drop volume per unit time for different rain intensities. If the DSD is measured, integral parameters of rainfall can be estimated by numerical integration (section 16.5.1).

16.6.1 Rain drop size distribution (DSD) and integral parameters of rainfall

(a) Measurement of DSD

Instruments based on microwave back-scattering by rain have been developed to estimate rainfall intensity over a large scale, nearly instantaneously, for meteorological purposes (Jameson, 1991). To validate this methodology, other instruments have been developed to measure the DSD, which is essential for interpretation of the radar reflection factor. Based on optical and electromechanical principles, these instruments are of two types: spectropluviometers provide the joint statistical distribution of both drop size and drop speed and disdrometers provide only the DSD. One spectropluviometer (Spectropluviomètre Précis Mécanique, Bezons, France) operates by assessing the optical obstruction of a horizontal parallel beam of light by falling raindrops that decrease the light intensity received by a receptor when they pass through the parallelepipedic measurement zone. The electrical signal produced by the photodiode receptor is transformed into positive pulses whose height and duration fluctuations are used to determine both the size and speed of drops. Drop size and velocity histograms can be measured at different sampling intervals (from 15 to 120 s) with 16 drop diameter categories from 0 to 5 mm and 16 drop speed classes between 0 and 10 m/s. Even if this instrument shows limited performance at measuring high rainfall rates (larger than 35 mm h^{-1}), kinetic energy and momentum can be estimated with an error less than 4% (Salles and Poesen, 1999).

The disdrometer described by Hall and Calder (1993) is also based on optical principles. An annular light beam is received by a photodiode detector that generates current pulses whose heights are linearly related to the diameter of drops passing through the sampling volume. There are special algorithms to account for drops that are partially outside the sampling volume or/and partially masking each other. Another disdrometer is based on a transducer exposed to rain (Disdromet, Basel, Switzerland); during rain, the vertical transfer of momentum for each impacting drop produces a pulse whose height is related to the drop diameter by a power law.

(b) Integral parameters and duration characteristics

The kinetic energy (or power) and intensity of rainfall can be calculated using the DSD measurements. The drop velocity is either measured (with a spectropluviometer) or estimated using empirical relationships (Ulbrich, 1983; Madden, 1992). Rain amount and intensity are usually measured directly (Viton, 1990). Based on the two DSD parameters linked to the type of rainfall (convective, stratiform, etc.), a generalized power law relationship between rainfall power (i.e. kinetic energy per unit area and per unit time) and intensity was derived (Salles *et al.*, 2002; see Equation 28 and Table 2). During heavy showers or storms, errors may occur in measuring rainfall intensity that are relevant to rain-splash. Water dumping from one of the two tipping bucket rain-gauge chambers may not be properly recorded since the tipping movement of the vessel is not instantaneous. Tipping-bucket gauges produce electrical pulses which are easily computerized to calculate total amounts of rain, temporal changes in intensity, maximum and mean intensities over different time periods and periods of constant rainfall intensity. When the time interval between rainfall amount measurements is relatively long (rarely <1 hour), the histogram of rainfall intensity can be useful.

A type of rain gauge operates by detecting (rather than by measuring threshold amounts of water) and delivering an analogue voltage output to estimate instantaneous rain intensity. This could be very useful for measuring the duration of periods with rain intensity above given threshold values within a rainfall episode. For studies where limited equipment is available, a direct measurement of rain duration in association with rain amount data could be of use. The rain duration sensors consist of a grid (similar in principle to leaf wetness sensors) composed of two electrodes separated by a space that is bridged by rain water. The electrodes are heated continuously so that fog deposition and dew formation are not recorded (Viton, 1990).

16.6.2 Methodology for measuring potential rain-splash

(a) Measurements using a standard target

A simple instrument for the direct measurement of maximum splash height was developed by Shaw (1987). This 'splash meter' consists of an annular reservoir placed on the ground filled with a UV-fluorescent dye with a vertical cylinder (diameter 5 cm) of chromatography paper in the centre of the annular reservoir. The impact of raindrops on the liquid surface produces splash droplets, some of which are intercepted by the vertical cylinder; the logarithm of the amount of dye coverage (above a threshold height of 5-10 cm) was a linear function of height. Before regression, a correction factor was applied to dye coverage data to account for the decrease in the apparent splash receptor area with distance above the target. The response of this instrument to rainfall is dependent on its design characteristics. The type of target greatly affects the amount of dye splashed; for a liquid target, the depth of the water film affects the amount of water splashed (Huber *et al.*, 1997). Further work is needed to improve target surface and reservoir characteristics to optimize the response of the instrument to rainfall. As suggested by Shaw (1991), a constant level reservoir for the splash meter is useful to avoid the need to refill the reservoir when intense evaporation occurs during days before a rain event (Huber, unpublished data); with a V-shaped aluminium rain-shield at the top of the vertical cylinder, a non-fluorescent dye can be used to permit a visual assessment of the area covered by splash without the need for the sophisticated equipment to quantify coverage by a fluorescent dye. Effects of the angle of rainfall require investigation because the impact force of drops decreases as the angle of rain versus the surface normal increases. In the presence of gusty winds, variability in rain angles produces fluctuations in the quantity and trajectory of splash droplets that may cause significant variation in DSD over a period of a few hours or even of a few minutes. Consequently, both wind and DSD characteristics can limit the accuracy of the splash meter. However, this simple instrument should be used more extensively.

An alternative is the direct measurement of kinetic energy with an electronic sensor. For studying soil erosion, methods based on rainfall intensity/kinetic energy relationships to relate soil transport to rainfall characteristics proved to be inadequate (Kinell, 1981). As suggested by Walklate *et al.* (1989), kinetic energy sensors based on the response of a piezoelectric crystal transducer (producing a single analogue output after signal transformation and integration) appear to be very useful (Madden *et al.*, 1998). Recent work showed the potential use of such rainfall kinetic energy sensors to assess maximal splash height (Lovell *et al.*, 2002), or rainfall power over variable time periods (Madden *et al.*, 1998). A potential problem with sensors based on piezoelectric crystals is the low signal-to-noise ratio for measurements of kinetic energy from impacting small to moderate sized raindrops. The measurement of both rainfall intensity and kinetic energy at the same time is relevant to the characterisation of potential rain-splash, because they vary considerably depending on the time interval of measurements (Smith and deVaux, 1992). Methodologies developed by soil erosion scientists might be of interest for studying spore dispersal by rain-splash or wash-off processes (van Dijk *et al.*, 2003).

(b) Estimates based on DSD measurements or indirect estimates

Two integral parameters (e.g. intensity and kinetic energy at ground level, or back-scattering coefficient and liquid water content) measured at adequate time intervals are sufficient to determine DSD parameters (section 16.5.1). Once the size distribution is known, indices of potential rain-splash can be calculated with or without taking into account some kind of crop target characterization (Fig. 16.8). Based on the existing network of meteorological radars and a specific instrument measuring potential rain-splash at ground level (that remains to be defined), an alternative (Walklate, 1989) is to calibrate relationships between rain-splash indices and back-scattering coefficients in the same way that rainfall intensity is calibrated against radar reflectance (section 16.5.1).

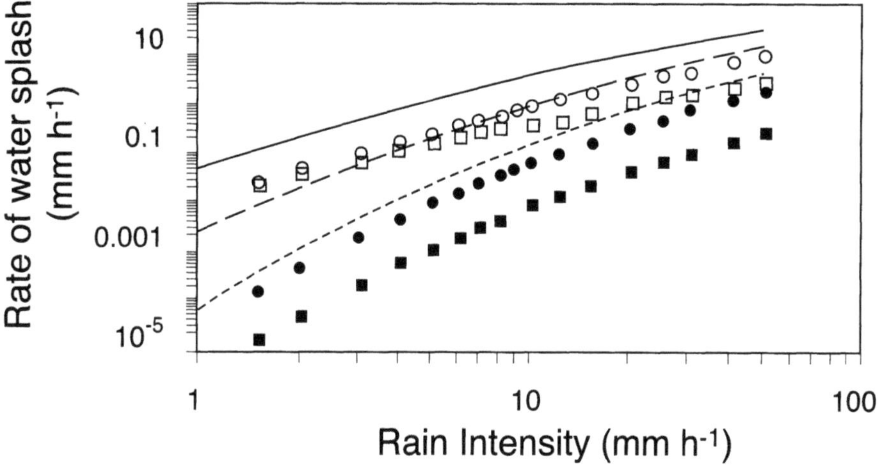

Figure 16.8. Comparison of splash intensity indices based only on rainfall characteristics (lines) (Walklate et al., *1989) or based on both rainfall and target characteristics (symbols) from oilseed rape (circles) and tobacco (squares) leaves and assuming a minimum splash drop size threshold of 2 mm (solid line and empty symbol), 3 mm (dashed line) or 4 mm (dotted line and solid symbol) (redrawn from Huber* et al., *1997).*

16.6.3 Towards monitoring splash-dispersed pathogen diseases in the field

Most experimental studies on splash have been done in controlled conditions, with single drops or simulated rain impacting on individual leaves or a small number of plants. This helps to reduce temporal or spatial variability of input factors (e.g. rain intensity) and avoids the confounding that occurs in nature where many variables are often linked (e.g. rain and leaf wetness). Based on controlled experiments for several air-borne fungal pathogens, it was demonstrated that several concepts (e.g. maximal splash height, mass of water splashed, truncated integration of DSD), and methodologies (e.g. splash meter, measurement of DSD) can be very helpful for assessing spore dispersal by rain-splash (Fitt *et al.*, 1989; Madden, 1992). In the case of *Mycosphaerella graminicola* (*Septoria tritici*) in winter wheat for example, spore dispersal from the bottom leaves by rain-splash is well established, and thresholds for disease control are simply related to daily rainfall amount and plot growth stage (Hansen *et al.,* 1994). However, considerations on penetration of raindrops in the host canopy show that increased levels of nitrogen application might reduce disease development because higher canopy density could lead to the obstruction of spore-carrying droplets (Lovell *et al.,* 1997). Therefore, the rainfall-canopy structure interaction should be considered in detail to improve disease forecasting. Along these lines, the use of overhead irrigation influences both growth of field beans and development of *Ascochyta fabae*; canopy height and LAI on an unprotected irrigated plot were approximately twice the values of those on the reference plot with no irrigation (Huber, 1992). Disease incidence on the irrigated plot was at least twice the value of the non-irrigated plot. In this case, disease development depends on dispersal and infection, the latter being influenced by water persistence. In contrast to *Septoria* spp. in winter wheat, the increase in LAI of the field bean crop favours *Ascochyta* disease development. One can probably assume that dispersal from the bottom leaves decreases as density increases but that frequent overhead irrigation (or rain) leads to substantial and higher infection efficiency, due to prolonged and recurrent wetness. Thus, depending on the disease/environment combination, reduced inoculum movement by rain-splash could limit the epidemic development; but in other situations reduced dispersal is compensated by increased infection or sporulation. Because of the multi-faceted effects of rain (or irrigation) on disease, simple systems such as those developed to forecast or predict septoria tritici blotch on winter wheat would likely be sufficient to assess the risk of disease (Hansen *et al.,* 1994). Otherwise splash dispersal is only one process among other epidemiological processes and hypotheses must be made on how to incorporate it in the full epidemic cycle.

These examples show that splash dispersal events have not been used as field indicators of disease progress and outbreaks to the same extent as infection periods based on wetness and temperature. It is likely that simple forecasting methods applied to pathosystems for splash-dispersed pathogen diseases should be further investigated using rain variables (amount of rainfall, number of days with rainfall greater than a minimal threshold, etc.); field experiments should include some characterisation of rainfall and/or rain-splash. Examples where rain variables have been successfully related to disease development include leather rot and anthracnose of strawberry (Reynolds *et al.*, 1988; Madden *et al.*, 1993). On the other hand, pea anthracnose (caused by *Ascochyta pinodes*) is a disease whose epidemiological components are mostly known (Roger and Tivoli, 1996) to start modelling the host disease pathosystem; however, the dispersal of conidia by rain-splash is still a poorly characterised component whose quantitative influences on inoculum movement and disease progress require further research.

16.7 CONCLUDING REMARKS

The presence, persistence, and movement of liquid surface water greatly affect the epidemiology of plant diseases. Surface wetness is crucial to the infection process of many fungal pathogens and can be modelled using energy balance concepts (Huber and Gillespie, 1992). Rain-splash is second in importance to wind as an agent for inoculum dispersal. However, this essentially small-scale physical process controls spore dispersal of a large number of pathogens (Fitt *et al.*, 1989). Environmental physics and meteorology have introduced several concepts and methods which are very useful for modelling rain-splash and for understanding spore dispersal by splash. Techniques to estimate drop size distributions or integral rainfall parameters over a consistent time scale should be investigated further. Above the crop canopy, a reference splash index could eventually be defined. A simple parameter that changes with rain intensity and rain type, such as the mass median diameter, could be used to predict splash (Ntahimpera *et al.*, 1997). Simple means for adapting this method to different crop targets are needed because mechanical interactions between raindrops and target characteristics are extensive.

Effects of ground cover and plant obstruction on rain-splash have important implications for crop management practices. For example, the advantages of using a thin film of plastic mulch with horticultural crops (e.g. increase in net radiation and soil heat flux and consequently the soil temperature around plant roots) must be balanced against the disadvantage that it favours redispersal of pathogen spores by multiple splashing after they have been removed from infected plant surfaces by wash-off or splash (Madden *et al.*, 1993; Yang *et al.*, 1990). Thus, strategies for disease management require compromise based on the best available information. As the understanding of disease spread by rain-splash is improved through more experimental work with different plant/ground cover combinations and the appropriate modelling, cultural practices can be modified to improve disease management.

Because suitable instruments for measuring rain-splash or characterizing spatial variability of rainfall accumulation have become available, mapping potential rain-splash should make rapid progress (Workneh *et al.*, 2005) and overcome the lack of site-specific weather information. Besides the advantage of measuring rather than predicting the rainfall power, it is still difficult to convince meteorological services to include potential rain-splash in meteorological networks in addition to rain intensity and/or surface wetness duration measurements. Since rain intensity measurements cannot accurately estimate potential rain-splash and the density of the rain gauge network is too low to measure temporal and spatial variability of rainfall, it is likely that microwave imagery to provide maps of rainfall radar measurements will be the technology used to provide spatial information about rain-splash in the long-term.

Improvements in the understanding and prediction of rain-splash will be made by development of conceptual models based on physics and probability theory. There is also a need to improve models of spore dispersal processes from a point source or spatially distributed sources by combining physics and probability theory (Pielaat *et al.*, 1998; Saint-Jean *et al.*, 2004; Yang *et al.*, 1991a). Knowing the crucial role of canopy architecture, one key modelling question concerning disease spread by splash dispersal is the choice between 1-D and 3-D modelling of mass transfer in the canopy. To describe upward spread of light leaf spot on winter oilseed rape, a refined one-dimensional approach taking into account both splash dispersal and stem expansion was developed (Pielaat *et al.*, 2002). Combining water transfer by rain-splash in a 3-D canopy structure (Saint-Jean, 2003) and incorporation of spores in splash droplets may lead to a versatile analysis of canopy architecture effects on the spread of splash-dispersed fungal pathogens. Additionally, since spatial and temporal patterns of rainfall greatly influence spore movement in crop canopies the influence of variability in rainfall characteristics and the effect of the intermittence of rain occurrence on spore dispersal by rain-splash merit further investigation.

REFERENCES

Allen, R.F. (1988) The mechanics of splashing. *Journal of Colloid and Interface Science*, **124**, 309-316.

Armstrong, C.L. and Mitchell, J.K. (1987) Transformations of rainfall by plant canopy. *Transactions of the American Society of Agricultural Engineers*, **30**, 688-696.

Armstrong, C.L. and Mitchell, J.K. (1988) Plant canopy characteristics and processes which affect transformation of rainfall properties. *Transactions of the American Society of Agricultural Engineers*, **31**, 1400-1409.

Aylor, D.E. and Sutton, T.B. (1992) Release of *Venturia inaequalis* ascospores during unsteady rain: relationship to spore transport and deposition. *Phytopathology*, **82**, 532-540.

Bannon F.J. and Cooke, B.M. (1998) Studies on dispersal of *Septoria tritici* pycnidiospores in wheat-clover intercrops. *Plant Pathology*, **47**, 49-56.

Bassette, C. and Bussière, F. (2005) 3-D modelling of the banana architecture for simulation of rainfall interception parameters. *Agricultural and Forest Meteorology*, **129**, 95-100.

Boudreau , M.A. and Madden, L.V. (1995) Effects of strawberry density on dispersal of *Colletotrichum acutatum* by simulated rain. *Phytopathology*, **84**, 934-941.

Brewer, C.A. and Smith, W.K. (1997) Patterns of leaf surface wetness for montane and subalpine plants. *Plant, Cell and Environment*, **20**, 1-11.

Bussière, F., Solmon, F. and Fouéré, A. (2002) Implementation and evaluation of DROP, a model for the simulation of rainfall distribution below plants described in 3D. *Agronomie*, **22**, 93-103.

Butterworth J. and McCartney H.A. (1991) The dispersal of bacteria from leaf surfaces by water splash. *Journal of Applied Bacteriology*, **71**, 484-496.

Calder I.R. (1996) Rainfall interception and drop size-development and calibration of the two-layer stochastic interception model. *Tree Physiology*, **16**, 727-732.

Crockford, R.H., Richardson, D.P., Fleming, P.M. and Kalma, J.D. (1991) A comparison of methods for measuring the angle and direction of rainfall. *Agricultural and Forest Meteorology*, **55**, 213-231.

Englemann, R.J. (1968) The calculation of precipitation scavenging, in *Meteorology and Atomic Energy* (ed. D.H. Slade) U.S. Atomic Energy Commission, Oak Ridge, USA.

Erpul, G., Norton, L.D. and Gabriels, D. (2004) Splash-saltation trajectories of soil particles under wind-driven rain. *Geomorphology*, **59**, 31-42.

Evenhuis, A., Verdam, B. and Zadoks J.C. (1997) Splash dispersal of conidia of *Mycocentrospora acerina* in the field. *Plant Pathology*, **46**, 459-469.

Fatemi, F. and Fitt, B.D.L. (1983) Dispersal of *Pseudocercosporella herpotrichoides* and *Pyrenopeziza brassicae* spores in splash droplets. *Plant Pathology*, **32**, 401-404.

Fernandez-Garcia, E. and Fitt, B.D.L. (1993) Dispersal of the entomopathogen *Hirsutella crypto-sclerotium* by simulated rain. *Journal of Invertebrate Pathology*, **61**, 39-43.

Finney, H.J. (1984) The effects of crop covers on rainfall characteristics and splash detachment. *Journal of Agricultural Engineering Research*, **29**, 337-343.

Fitt, B.D.L. and McCartney, H.A. (1986) Spore dispersal in splash droplets, in *Water, Fungi and Plants* (eds P.G. Ayres, L. Boddy). Cambridge: Cambridge University Press, pp. 87-104.

Fitt, B.D.L., McCartney, H.A. and Walklate, P.J. (1989) The role of rain in dispersal of pathogen inoculum. *Annual Review of Phytopathology*, **27**, 241-270.

Fitt, B.D.L., Inman, A.J., Lacey, M.E. and McCartney, H.A. (1992) Splash dispersal of spores of *Pseudocercosporella capsellae* (white leaf spot) from oilseed rape leaves of different inclination, flexibility and age. *Zeitschrift für Pflanzenkranheiten and Pflanzenschutz*, **99**, 234-244.

Geagea, L., Huber, L. and Sache, I. (1999) Dry-dispersal and rain-splash of brown (*Puccinia recondita*) and yellow (*P. striiformis)* rust spores from infected wheat leaves exposed to simulated raindrops. *Plant Pathology*, **48**, 472-482.

Geagea, L., Huber, L., Sache, I. *et al.* (2000) Influence of simulated rain on dispersal of rust spores from infected wheat seedlings. *Agricultural and Forest Meteorology*, **101**, 53-66.

Granvold, J. (1984) Rain splash dispersal of third-stage larvae of *Cooperia* spp. (Trichostrongylidae). *Journal of Parasitology*, **70**, 924-926.

Gregory, P.H., Guthrie, E.J. and Bunce, M.E. (1959) Experiments on splash dispersal of fungus spores. *Journal of General Microbiology*, **20**, 328-354.

Gregory, P.H., Griffin, M.J., Maddison, A.C. and Ward, M.R. (1984) Cocoa black pod: a reinterpretation. *Cocoa Grower's Bulletin*, **35**, 5-22.

Hall, R.L. and Calder, I.R. (1993) Drop size modification by forest canopies - measurements using a disdrometer. *Journal of Geophysical Research*, **90**, 465-470.

Hansen J.G., Secher B.J.M., Jorgensen L.N. and Welling B. (1994) Thresholds for control of *Septoria* spp. in winter wheat based on precipitation and growth stage. *Plant Pathology*, **43**, 183-189.

Hörberg, M.H. (2002) Patterns of splash dispersed conidia of *Fusarium poae* and *Fusarium culmorum*. *European Journal of Plant Pathology*, **108**, 73-80.

Huber, L. (1992) Déterminisme et mesure de la durée d'humectation en vue de la protection des plantes. *Agronomie*, **12**, 281-295.

Huber, L., Bussière, F., Clastre, P. *et al.* (1995) Interactions entre l'eau liquide et la végétation. Actes de l'Ecole-Chercheurs INRA en Bioclimatologie. Le Croisic 3-7 avril 1995, tome 1: de la plante au couvert végétal (eds P. Cruiziat, J. P. Lagouarde), pp. 533-545.

Huber, L., Fitt, B.D.L. and McCartney, H.A. (1996) The incorporation of pathogen spores into rain-splash droplets: a modelling approach. *Plant Pathology*, **45**, 506-517.

Huber, L. and Itier B. (1990) Leaf wetness duration in a field bean canopy. *Agricultural and Forest Meteorology*, **51**, 281-292.

Huber, L. and Gillespie, T.J. (1992) Modelling leaf wetness in relation to plant disease epidemiology. *Annual Review of Phytopathology*, **30**, 553-577.

Huber, L., McCartney, H.A. and Fitt, B.D.L. (1997) Influence of target characteristics on the amount of water splashed by impacting drops. *Agricultural and Forest Meteorology*, **87**, 201-211.

Jameson, A.R. (1991) A comparison of microwave techniques for measuring rainfall. *Journal of Applied Meteorology*, **30**, 32-54.

Jenkinson, P. and Parry, D.W. (1994) Splash dispersal of conidia of *Fusarium culmorun* and *Fusarium avenaceum*. *Mycological Research*, **98**, 506-510.

Kinnell, P.I.A. (1981) Rainfall intensity-kinetic energy relationships for soil loss prediction. *Soil Science Society American Journal*, **45**, 153-155.

Levin, Z. and Hobbs, P.V. (1971) Splashing of water drops on solid and wetted surfaces : hydrodynamics and charge separation. *Philosophical Transactions of the Royal Society*, **A269**, 555-585.

Lovell, D.J., Parker, S.R., Hunter, T. *et al.* (1997) Influence of crop growth and structure on the risk of epidemics by *Mycosphaerella* (*Septoria tritici*) in winter wheat. *Plant Pathology*, **46**, 126-138.

Lovell, D.J., Parker, S.R., Van Peteghem, P. *et al.* (2002) Quantification of raindrop kinetic energy for improved prediction of splash-dispersed pathogens. *Phytopathology*, **92**, 497-503.

Lovell, D.J., Parker, S.R., Hunter, T. *et al.* (2003) Position of inoculum in the canopy affects the risk of septoria tritici blotch epidemics in winter wheat. *Plant Pathology*, **53**, 11-21.

Macdonald, O.C. and McCartney, H.A. (1987) Calculation of splash droplet trajectories. *Agricultural and Forest Meteorology*, **39**, 95-110.

Macdonald, O.C. and McCartney, H.A. (1988) A photographic technique for investigating the splashing of water drops on leaves. *Annals of Applied Biology*, **113**, 627-638.

Madden L.V. (1992) Rainfall and the dispersal of fungal spores. *Advances in Plant Pathology*, **8**, 39-79.

Madden, L.V., Wilson, L.L., Yang, X. and Ellis, M.A. (1993) Field spread of anthracnose fruit rot of strawberry in relation to ground cover and ambient weather conditions. *Plant Disease*, **77**, 861-866.

Madden, L.V., Wilson, L.L., Yang, X. and Ellis, M.A. (1992) Splash dispersal of *Colletotrichum acutatum* and *Phytophthora cactorum* by short-duration simulated rains. *Plant Pathology*, **41**, 427-436.

Madden, L.V., Yang, X. and Wilson, L.L. (1996) Effects of rain intensity on splash dispersal of *Colletotrichum acutatum*. *Phytopathology*, **86**, 864-874.

Madden, L.V. (1997) Effects of rain on splash dispersal of fungal pathogens. *Canadian Journal of Plant Pathology*, **19**, 225-230.

Madden, L.V., Wilson, L.L. and Ntahimpera, N. (1998) Calibration and evaluation of an electronic sensor for rainfall kinetic energy. *Phytopathology*, **88**, 950-959.

Marshall, J.S. and Palmer, W.M. (1948) The distribution of raindrops with size. *Journal of Meteorology*, **5**, 165-166.

Mason, B.J. and Andrews, J.B. (1960) Drop-size distributions from various types of rain. *Quarterly Journal of the Royal Meteorological Society*, **86**, 346-353.
Merriam, R.A. (1973) Fog drip from artificial leaves in a fog wind tunnel. *Water Resources Research,* **9**, 1591-1598.
Moss, A.J. and Green, T.W. (1987) Erosive effects of the large water drops (gravity drops) that fall from plants. *Australian Journal of Soil Resource*, **25**, 9-20.
Nakanishi, H. (2002) Splash seed dispersal by raindrops. *Ecological Research*, **17**, 663-671.
Ntahimpera, N., Madden, L.V. and Wilson, L.L. (1997) Effect of rain distribution alteration on splash dispersal of *Colletotrichum acutatum*. *Phytopathology*, **87**, 649-655.
Ntahimpera, N., Ellis, M.A., Wilson, L.L. and Madden, L.V. (1998) Effect of a cover crop on splash dispersal of *Colletotrichum acutatum* conidia. *Phytopathology*, **88**, 536-543.
Pangga, L.B., Geagea, L., Huber, L. *et al.* (2004) Dispersal of *Colletotrichum gloeosporioides* conidia, in *High-yielding anthracnose-resistant Stylosanthes for agricultural systems* (ed. S. Chakraborty), Australian Centre for Agricultural Research, pp. 211-222.
Paul, P.A., El-Allaf, S.M., Lipps, P.E. and Madden L.V. (2004) Rain splash dispersal of *Giberella zeae* within wheat canopies in Ohio. *Phytopathology,* **94**, 1342-1349.
Pauvert, P., Fournet , J. and Rapilly, F. (1970). Dissémination d'un inoculum fongique par des gouttes d'eau: influence de la morphologie des fructifications. *Annales de Phytopathologie*, **2**, 43-53.
Pielaat, A., Madden, L.V. and Gort, G. (1998) Spores splashing under different environmental conditions: a modeling approach. *Phytopathology*, **88**, 1131-1140.
Pielaat, A., van den Bosch, F., Fitt, B.D.L. and Jeger M.J. (2002) Simulation of vertical spread of plant diseases in a crop canopy by stem extension and splash dispersal. *Ecological Modelling,* **151**, 195-212.
Pietravalle, S., Van den Bosch, F., Welham, S.J. *et al.* (2001) Modelling of rain splash trajectories and prediction of rain splash height. *Agricultural and Forest Meteorology*, **109**, 171-185.
Reynolds, K.M., Madden, L.V. and Ellis, M.A. (1988) Effect of weather variables on strawberry leather rot epidemics. *Phytopathology*, **78**, 822-827.
Roger, C. and Tivoli, B. (1996) Spatio-temporal development of pycnidia and perithecia and dissemination of spores of *Mycosphaerella pinodes* on pea. *Plant Pathology,* **45**, 518-528.
Saint-Jean, S., Chelle, M. and Huber, L. (2004). Modelling water transfer by rain-splash in a 3D canopy using Monte Carlo integration. *Agricultural and Forest Meteorology*, **121**, 183-196.
Saint-Jean S., Testa A., Kamoun S. and Madden L.V. (2005) Use of a green fluorescent protein marker for studying splash dispersal of sporangia of *Phytophthora infestans*. *European Journal of Plant Pathology*, **112**, 391-394.
Saint-Jean, S. (2003) Étude expérimentale et numérique du mécanisme de transfert d'eau par éclaboussement de gouttes de pluie dans une structure tridimensionnelle. PhD, University of Paris XI, France.
Salles, C. and Poesen J. (1999). Performance of an optical spectro pluviometer in measuring basic rain erosivity characteristics. *Journal of Hydrology,* **218***,* 142-156.
Salles, C., Poesen, J. and Sempere-Torres, D. (2002) Kinetic energy of rain and its functional relationship with intensity. *Journal of Hydrology*, **257**, 256-270.
Schottman, R.W. and Walter, M.F. (1982) Relative erosion potential for coalesing water droplets. *Paper* N° 82-2037. *American Society of Agricultural Engineeres*, St. Joseph, MI 49085.
Shaw, M.W. (1987) Assessment of upward movement of rain splash using a fluorescent tracer method and its application to the epidemiology of cereal pathogens. *Plant Pathology*, **36**, 201-213.
Shaw, M.W. (1991) Variation in the height to which tracer is moved by splash during natural summer rain in the United Kingdom. *Agricultural and Forest Meteorology*, **55**, 1-14.
Smith, J.A. and DeVaux, R.D. (1992) The temporal and spatial variability of rainfall power. *Environmetrics*, **3**, 29-53.
Soleimani, M.J., Deadman, M.L. and McCartney H.A. (1996) Splash dispersal of *Pseudocercosporella herpotrichoides* spores in wheat monocrop and wheat-clover bicrop canopies from simulated rain. *Plant Pathology*, **45**, 1065-1070.
Stedman, O.J. (1979) Patterns of unobstructed splash dispersal. *Annals of Applied Biology*, **91**, 271-285.
Stedman, O.J. (1980) Splash droplet and spore dispersal studies in field beans (*Vicia faba* L). *Agricultural Meteorology*, **21**, 111-127.
Stow, C.D. and Hadfield, M.G. (1980) An investigation of the conditions for splashing of water drops on solid, dry surfaces. *Journal of the Meteorological Society of Japan*, **58**, 59-67.

Torres, D.S., Porra, J.M. and Creutin, J.-D. (1994) A general formulation for raindrop size distribution. *Journal of Applied Meteorology,* **33**, 1494-1502.
Ulbrich, C.W. (1983) Natural variations in the analytical form of the raindrop size distribution. *Journal of Climate and Applied Meteorology*, **22**, 1764-1775.
Ulbrich, C.W. (1994) Corrections to empirical relations derived from rainfall disdrometer data for effects due to drop size distribution truncation. *Atmospheric Research*, **34**, 207-215.
Van Dijk, A.I.J.M., Bruijnzeel, L.A. and Eisma, E.H. (2003) A methodology to study rain splash and wash processes under natural rainfall. *Hydrological Processes*, **17**, 153-167.
Viton, P. (1990) Mesures de la pluie. *La Météorologie*, **33**, 28-33.
Walklate, P.J. (1989) Vertical dispersal of plant pathogens by splashing. Part I: the theoretical relationship between rainfall and upward rain splash. *Plant Pathology*, **38**, 56-63.
Walklate, P.J., McCartney, H.A. and Fitt, B.D.L. (1989) Vertical dispersal of plant pathogens by splashing. Part II: experimental study of the relationship between raindrop size and the maximum splash height. *Plant Pathology*, **38**, 64-70.
Workneh, F., Narasimhan, B., Srinivasan, R., and Rush, C.M. (2005) Potential of radar-estimated rainfall for plant disease risk forecast. *Phytopathology,* **95**, 25-27.
Yang, X., Madden, L.V., Wilson, L.L. and Ellis, M.A. (1990) Effects of surface topography and rain intensity on splash dispersal of *Colletotrichum acutatum. Phytopathology*, **80**, 1115-1120.
Yang, X. and Madden, L.V. (1993) Effect of ground cover, rain intensity and strawberry plants on splash of simulated raindrops. *Agricultural and Forest Meteorology*, **65**, 1-20.
Yang, X., Madden, L.V. and Brazee, R.D. (1991a) Application of the diffusion equation for modelling splash dispersal of point-source pathogens. *The New Phytologist*, **118**, 295-301.
Yang, X., Madden, L.V., Reichard, D.L. *et al.* (1991b) Motion analysis of drop impaction on a strawberry surface. *Agricultural and Forest Meteorology*, **56**, 67-92.
Yangang, L. (1993) Statistical theory of the Marshall-Palmer distribution of raindrops. *Atmospheric Environment*, **274**, 15-19.

CHAPTER 17

POTATO LATE BLIGHT

EDUARDO S.G. MIZUBUTI AND WILLIAM E. FRY

17.1 INTRODUCTION

Potato late blight caused by the oomycete *Phytophthora infestans* (Mont.) de Bary may be the best known, longest studied, and still among the most destructive of all plant diseases. Devastation caused by this plant pathogen in the late 1840s in Europe, led to food shortages throughout Europe, but the effect in Ireland led to the Irish potato famine. The enormity of the famine initiated by this plant disease stimulated much investigation on plant diseases and led to the development of plant pathology as a distinct discipline. However, these studies had much broader impact. Inoculation experiments by de Bary provided evidence for the causal role of this pathogen in the disease and contributed to the later acceptance of Pasteur's more general germ theory of disease.

Late blight impacts humans because the foliar phase limits production of tubers; infections on tubers destroy a potential food source and tremendous effort and resources are employed to prevent both of these phases. The human suffering resulting from the dramatic epidemics in the 1840s was exacerbated by the shortage of potatoes. Early 19th century Ireland had developed a 'potato economy' in which a large, poor population depended almost exclusively on potatoes for their subsistence for most of the year (Bourke, 1993). The high productivity of potatoes permitted a population explosion in Ireland and some have estimated that the Irish population nearly tripled during the 60 years prior to 1840 (Woodham-Smith, 1962). The starvation that resulted from the food shortages caused by the blight epidemics led to the deaths and emigration of an estimated three million people (Large, 1940). A second famine caused by late blight was experienced in Germany during 1916 and 1917. The shortage of copper, the basic constituent of Bordeaux mixture, was caused by the demanding armour industry. Because late blight control was not properly carried out, potato availability for the families of soldiers was short. Many died and this affected the morale of the German troops. Some historians speculate this could have led to the defeat of the Germans in World War I (Horsfall and Cowling, 1978).

Knowledge of the etiology and epidemiology of the disease and discovery and use of fungicides, enabled improved suppression of late blight during the 160 years following the Irish potato famine. However, despite the improved disease suppression, the financial impact remained burdensome. Economists at the International Potato

B. M. Cooke, D. Gareth Jones and B. Kaye (eds.), The Epidemiology of Plant Diseases, 2nd edition, 445–471.

Centre estimated that annual costs of late blight were approximately US $3.2 billion (Raman *et al.*, 2000).

It is logical that the pathogen and the disease have received much attention. There are numerous studies, reviews, book chapters and books devoted to the biology of the pathogen and disease. Some classic studies such as those by Crosier (1934) and Van der Zaag (1956) are still used. Of the many reviews, those by Andrivon (1995, 1996) and Harrison (1992) provide a wealth of references. Some books have been published as proceedings of conferences, but others have been published solely to investigate and illustrate the breadth and depth of knowledge on the disease (Ingram and Williams, 1991; Lucas *et al.*, 1991; Dowley *et al.*, 1995; Erwin and Ribeiro, 1996). With such a rich literature on late blight, this chapter will explore the implications of insights gained from recent studies on the population biology of this very familiar pathogen.

17.2 POPULATION BIOLOGY OF *P. INFESTANS*

Pathogen population biology is a more inclusive term than epidemiology because it encompasses both ecological and genetic aspects of populations (Milgroom and Peever, 2003). In the last 15 years, there has been an explosion of information about the biology of populations of *P. infestans*. Studies were carried out in temperate regions as well as in tropical and subtropical areas of the world. While this knowledge will improve late blight management everywhere, the benefit is expected to be particularly important in the tropics and subtropics.

The life history and evolutionary position of an organism can be very helpful in interpreting population phenomena. Our understanding of the basic biology of *P. infestans* has changed dramatically during the latter part of the 20th century. The organism was first described as *Botrytis infestans* by Jean Montagne in the mid 19th century. Since then our understanding of the phylogenetic relationships of *P. infestans* has been changing. Oomycetes are now generally regarded as existing in a kingdom separate from the true fungi, plants, animals, and prokaryotes. Some suggest that they belong in the kingdom Protoctista but others believe they should be in the kingdom Chromista (discussed in Erwin and Ribeiro, 1996). Regardless of kingdom, evidence is mounting that oomycetes and some algae belong in the same kingdom (Sogin and Silberman, 1998; van West *et al.*, 2003). These organisms share ribosomal ITS similarity, have biflagellate (tinsel and whiplash) zoospores, have similar sexual structures (antheridia and oogonia), are diploid and have cell walls of glucans. *Phytophthora infestans* is heterothallic, with two mating types (A1 or A2), and bisexual. When the organism reproduces sexually, a thick-walled spore – oospore – is formed. However, *P. infestans* can exist quite successfully as an asexual organism; as such it is a near obligate parasite in nature. Because the life history may be essentially asexual in one geographic region, but sexual and asexual in another, the epidemiology of the pathogen can be dramatically different in different regions. For example, some sources of inoculum are important in some regions and completely unimportant in other regions.

17.2.1 Origin and early migrations

Migration has had a dramatically important role in the population biology of *P. infestans*. The first records of the disease in temperate potato production occurred in the mid 19th century. The disease was noticed as a severe problem in northeastern United States in the early 1840s (Stevens, 1933). It was clearly widespread throughout Europe by the end of the growing season in 1845 (Bourke, 1964). Prior to the mid 19th century, late blight had been unknown in North America, Europe, Africa and Asia. After the mid 19th century, the disease became problematic throughout the rest of the world, usually shortly after potato production was established (Cox and Large, 1960).

Evidence has accumulated that most worldwide populations of *P. infestans* had been dominated by single clonal lineage of *P. infestans* prior to the 1980s (Goodwin *et al.*, 1994a). These populations were obviously asexual. A likely explanation for this population genetics structure was that a single clonal lineage had 'colonized' most continents. This could have happened if the pathogen escaped a centre of origin of greater diversity, going through a severe bottleneck such that only one or very few clonal lineages escaped. Through subsequent bottlenecks, a single clonal lineage could have survived and then been distributed throughout much of the world in subsequent migrations. A feasible alternative explanation is difficult to identify.

The current centre of diversity of *P. infestans* and the location from which the primary migrations originated seem to be in the highlands of Central Mexico. It was in this location that isolates with the A2 mating type were first discovered. Other populations throughout the world were asexual and comprised of a single mating type of the organism (Goodwin *et al.*, 1994a). Not only were both mating types found, but they occurred in equal frequencies (Gallegly and Galindo, 1958; Tooley *et al.*, 1985; Niederhauser, 1991). Populations in these highlands are sexual (Tooley *et al.*, 1985; Fernández-Pavía *et al.*, 2004) and remarkably diverse for all neutral markers and for pathotype (Tooley *et al.*, 1986; Rivera-Peña, 1990; Goodwin *et al.*, 1992). No other location has the genetic diversity reported for Central Mexico.

It seems highly likely that migrations led to the occurrences of late blight in the United States and Europe in the mid 19th century. At this point, the mechanism by which *P. infestans* was removed from Mexico and introduced to the United States and Europe is uncertain and the subject of renewed debate (Andrivon, 1996; Abad and Abad, 1997; May and Ristaino, 2004). Three possible theories are proposed: 1. the pathogen migrated from Central Mexico into the USA and then into Europe (Fry *et al.*, 1993); 2. the pathogen was introduced into the USA and Europe from the Andean region (Tooley *et al.*, 1989); and 3. *P. infestans* migrated from Mexico to Peru and then to the USA and Europe (Andrivon, 1996). The chronology of late blight detection suggests that the first migration might have been into the United States/Canada, because the disease was detected first in eastern USA and then in Europe (Fry *et al.,* 2002). Migration from the north eastern United States to Europe could have occurred via transport of infected tubers. Migrations out of Mexico were very limited, because potatoes (*Solanum tuberosum*) were not a common crop in Mexico and certainly were not an export crop until the late 20th century. Movement on infected potato seed tubers is probably the most efficient mechanism for transport of *P. infestans*. Interestingly, analyses of rDNA and mtDNA from herbarium specimens collected during the years of the great famine suggest that late blight epidemics were caused by isolates that had Ia mtDNA type, which is different than the Ib mtDNA of the previously thought widespread lineage US-1 of *P. infestans* (Ristaino *et al.*, 2001; May and Ristaino, 2004). This does not rule out Mexico as the origin of the isolate(s) that migrated to Europe, since the Ia mtDNA type is commonly found in this country (Gavino and Fry, 2002). Further studies based on gene genealogy will certainly contribute to test these theories.

17.2.2 Recent migrations

Evidence for contemporary migrations was first signalled by the discovery of isolates with the A2 mating type in Switzerland in 1981 (Hohl and Iselin, 1984). Prior to the late 20th century, there is no clear evidence of important migrations from Mexico to the rest of the world. The report of isolates of an A2 mating type of *P. infestans* was startling news for persons working with late blight because they had assumed that individuals only of the A1 mating type were present in all worldwide locations outside of Mexico. When other plant pathologists investigated their local populations in response to the report of Hohl and Iselin (Hohl and Iselin, 1984), A2 mating types were discovered throughout Europe (Fry *et al.*, 1993). In northern Europe, immigrant strains have largely displaced the previous indigenous strain (Spielman *et al.*, 1991). The immigrating population contained both A1 and A2 mating types and, as would be predicted, sexual reproduction has now been detected (Drenth *et al.*, 1995; Andersson *et al.*, 1998; Turkensteen *et al.*, 2000).

The pathway by which immigrant strains were introduced recently into Europe is apparently via a large shipment of potatoes from Mexico to Europe in the winter of 1976/77 (Niederhauser, 1991). The importation was necessitated by an underproduction of potatoes in Europe in the summer 1976, due to a drought. Genetic evidence suggests that the ‘new’ population in Europe has again been transported to other continents. There has been significant trade in seed tubers from Europe to Africa and south American countries such as Brazil and Venezuela. The linkage between Africa and South America to Europe seems especially tight, because ‘new’ strains of *P. infestans* with the same *Gpi* alleles as in the European population have been detected in Europe and in Africa and South America (Goodwin *et al.*, 1994a; Forbes *et al.*, 1997; McLeod *et al.*, 2001; Reis *et al.*, 2003).

An apparently separate migration has introduced a new clonal lineage to the Far East (Mosa *et al.*, 1989; Koh *et al.*, 1994). The new clonal lineages in China, Japan, Korea, the Philippines, and Taiwan appear to be displacing the previous clonal lineage that had been distributed worldwide (Koh *et al.*, 1994; Jyan *et al.*, 2004). However, sexual reproduction in Japan or Korea has not yet been detected, perhaps because oospores produced from the A1 and A2 genotypes there fail to germinate (Mosa *et al.*, 1991). Consequently, populations there remain very simple and strictly clonal. In Russia there is also evidence of population displacement, but different from that reported in Japan and Korea; there is also circumstantial evidence of sexual reproduction in Russia (Elansky *et al.*, 2001).

Migrations into the United States and Canada have occurred during the 1980s and the early 1990s. These were first detected in the early 1990s (Deahl *et al.*, 1991; Goodwin *et al.*, 1994b). The first migrations were detected in production regions along the Pacific Coast of Canada and the United States. Subsequently, immigrant strains were detected along the East Coast and then throughout the United States and Canada (Fry and Goodwin, 1997b). The immigrant strains in the United States brought characteristics that were previously not present: A2 as well as A1 mating types; enhanced pathogenicity to either potatoes or tomato; and metalaxyl resistance (Fry and Goodwin, 1997a).

The 1990s migrations into the United States and Canada were detected because of a dramatic increase in the severity of late blight (Johnson *et al.*, 1997; Daayf and Platt, 2000). In locations where late blight became more severe, exotic strains were detected. Often the previous indigenous strain was not detected. In 1994 and 1995, there were locally devastating epidemics that had severe economic repercussions on individual growers throughout the United States (Fry and Goodwin, 1997b; Johnson *et al.*, 1997). Because the migrations were observed as they happened, it was possible to compare immigrant strains directly with the previous indigenous strain. In most of these comparisons, the immigrant strains appeared to have enhanced pathogenicity to tomatoes, to potato foliage and tubers (Legard *et al.*, 1995; Kato *et al.*, 1997; Lambert and Currier, 1997; Miller *et al.*, 1998). Regarding tuber infection, it now appears that there is significant diversity among genotypes of *P. infestans* for pathogenicity on tubers (Lambert and Currier, 1997).

Currently (early 2005) the populations of *P. infestans* worldwide vary from clonal to recombinant (Table 17.1). In countries where it is clonal, there are a single or few lineages which can be more adapted to a particular host species. Host specificity in *P. infestans* is not absolute, i.e. isolates adapted to potato can infect and cause lesions on tomato and the opposite can also occur. There could be differences regarding isolate aggressiveness when inoculated in different hosts (see section 17.3.4). Interestingly, the old lineage US-1 is still present in several places.

Comparison of exotic strains to previous strains in relation to their reactions to abiotic influences suggests that the diverse genotypes may respond somewhat differently to abiotic factors than did previous very simple populations. In an investigation of the germination of sporangia, recent isolates of the US-1 clonal lineage (a representative of the population present in the United States perhaps since the mid 19th century) responded the same as reported by Crosier in the early 20th century (Crosier, 1934). However, germination of sporangia of the US-8 clonal lineage responded significantly differently. Whereas sporangia of US-1 germinated rather well at 20°C sporangia of US-8 did not (Mizubuti and Fry, 1998). Similar responses were obtained for a new lineage (BR-1) associated with potato in Brazil (Maziero, 2001). This observation suggests that plant pathologists throughout the world need to ascertain the responses of new local populations of *P. infestans* to important abiotic factors such as temperature, moisture, and solar radiation.

Table 17.1. Mode of reproduction of Phytophthora infestans *in populations of different countries. Data from Mizubuti and Forbes (2002) and GILB web site (http://gilb.cip.cgiar.org/modules.php?name=News&file=article&sid=415)*

Continent/ Region	*Country*	Mode of reproduction *Clonal (C) or Recombinant (R)*
Africa	Uganda, Ethiopia, South Africa	C
	Morocco	C
Asia	China, Korea, India, Pakistan, Thailand, Nepal, Japan	C
	Bangladesh, Sri Lanka, Philippines, Taiwan, Vietnam, Indonesia	C
	Russia	C (Siberia), R (Moscow)
	Southeast Asia and Israel	C
Europe	Austria, Belgium, Denmark, France, Germany, Ireland, Italy, Spain, Switzerland, United Kingdom - England and Wales	C
	Finland, Hungary, Norway, Poland, Sweden, The Netherlands, UK (Scotland)	R
Latin America	Uruguay, Paraguay, Brazil, Peru, Argentina, Chile, Ecuador, Bolivia, Colombia, Venezuela, Panama, Costa Rica, Honduras, other countries in Central America	C
	Mexico	C and R
North America	Canada	C
	United States	C
Oceania	Australia	C

17.3 PATHOGEN BIOLOGY

The occurrence or not of sexual reproduction has profound implications to population structure and to disease epidemiology. This is reflected in survival and sources of inoculum.

17.3.1 Survival of Phytophthora infestans

The advent of sexual reproduction and production of oospores has stimulated intensive interest in the survival of *P. infestans* in many locations in the world. In locations where sexual reproduction is absent the pathogen is essentially an obligate parasite and cannot survive for long periods in the absence of a susceptible host. However in locations where sexual reproduction is possible, oospores may survive from one season to the next, regardless of host presence.

(a) Survival of asexual structures

Overwintering of free-living sporangia and mycelium is limited. Although sporangia can survive for days or weeks in moist soil (Andrivon, 1995) they do not survive for long periods and especially they do not survive drying conditions (Fernández-Pavía *et al.*, 2004). However, mycelium in infected viable tubers survives well from one season to the next, whether the tubers are in temperature controlled storage, in a pit (clamp), or in soil that does not freeze (Zwankhuizen *et al.*, 1998; Kirk *et al.*, 2001). From November through April, temperatures at the base and at the centre of discarded potato cull piles that varied from 1 to 15 tons rarely dropped below 0°C (Kirk, 2003). Infected tubers that survive from one season to the next can produce plants that become infected (Van Der Zaag, 1956). However, the efficiency of transmission is low – most infected tubers do not initiate epidemics.

Survival of detached *P. infestans* sporangia in the atmosphere is limited (Mizubuti *et al.*, 2000; Sunseri *et al.*, 2002). For airborne sporangia, temperature, relative humidity and solar radiation affect survival. Since the work of Crosier (Crosier, 1934), plant pathologists have questioned the ability of sporangia to survive in a dry atmosphere. While some workers concluded that sporangia could survive for several hours (De Weille, 1963; Rotem and Cohen, 1974) others felt that sporangia could survive only minutes (Crosier, 1934; Glendinning *et al.*, 1963; Warren and Colhoun, 1975). It turned out that a factor leading to different results and different conclusions by different groups of workers was the rehydration rate of dried sporangia. Dried sporangia that were rehydrated rapidly died, whereas dried sporangia that rehydrated slowly were more likely to survive (Minogue and Fry, 1981). Nonetheless, sporangia survived longer at high relative humidities than at low humidities and sporangia survived better at 15 to 20°C than at 30°C. Half-lives of dried sporangia ranged from 2 to 6 hours at temperatures ranging from 15 to 30°C and at relative humidities from 40 to 88% (Minogue and Fry, 1981).

Solar radiation is one of the most important environmental factors that influences sporangia viability. Exposures of field-produced sporangia to direct solar radiation during sunny days caused viability to drop from approximately 70% to 0.3% within one hour. During overcast conditions, however, viability remained practically constant for a period of at least three hours (Mizubuti *et al.*, 2000). Even in surface water normally formed in irrigated fields, solar radiation was detrimental to sporangia (Porter and Johnson, 2004). Survival of detached sporangia under field conditions, varied from 0 to 20 days depending on the amount of solar irradiance (Porter and Johnson, 2004).

Sporangia survive much longer in moist soil than in the atmosphere (Andrivon, 1995). Whereas survival of sporangia in the atmosphere is measured in hours, survival in soil is measured in days and weeks with longer survival in moist than in dry soils. Often, survival of sporangia is determined by 'infectivity' (determining how long a particular soil contains a propagule that can infect a potato tuber or potato leaf). In soil sporangia can survive for approximately 40 days and zoospores for up to 10 days (Zan, 1962). Factors associated with reduced infectivity include low pH and aluminum toxicity as well as moisture (Andrivon, 1995). Additionally, associated soil microbes lessen survival of sporangia in soil.

The presence or absence of sexual reproduction in the pathogen population is especially significant to the epidemiology of tomato late blight. In temperate zones where there is a crop-free period, isolates of *P. infestans* adapted to tomatoes are less certain to survive from one season to the next than are isolates adapted to potatoes. This is because tomato pathogens have no effective means to survive the inter-cropping period. Potato pathogens, on the other hand, survive on tubers (in storage, in fields, or in clamps.) The pathogen appears unable to survive from one season to the next on tomato seeds, and other parts of the tomato plant typically die between seasons. Survival of *P. infestans* in infected tomato tissue (leaflets, stems, and green fruits) either kept on the soil surface or buried 10 cm deep, under field conditions in Brazil, was limited. No viable structures were found after 30 days (Fig. 17.1) (M.A. Lima, L.A. Maffia and E.S.G. Mizubuti, unpublished data).

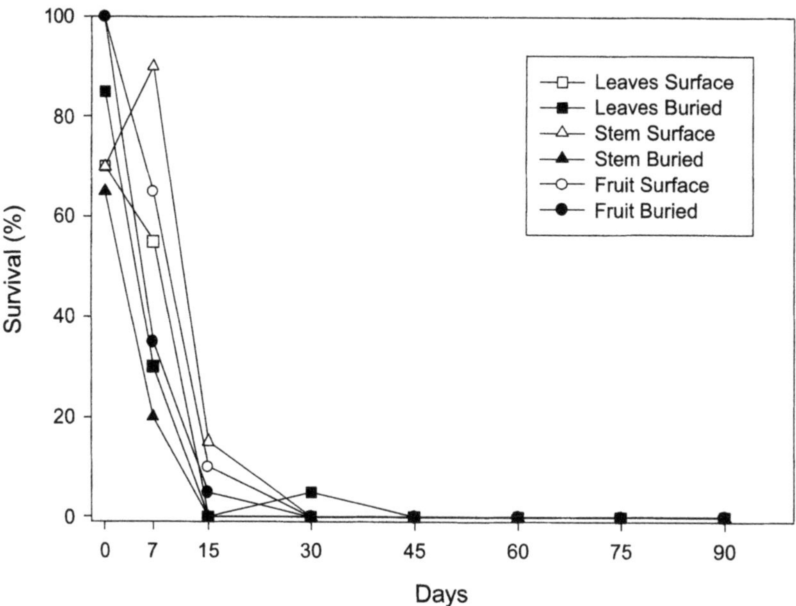

Figure 17.1. Survival of Phytophthora infestans *in infected tomato tissues: leaves, stems or tomato fruit. Samples were kept either buried or on the soil surface, under field conditions (Spring to Summer) in Brazil. Survival was estimated by periodically sampling infected tissue and subjecting samples to a bioassay modified from Drenth* et al. *(1995).*

There are survival mechanisms even if potatoes are not involved. If hosts of *P. infestans* are growing in a region, these can be a source of the pathogen. In temperate regions, some glasshouse or greenhouse production can be a source. Tomatoes are an expected host, but recently it has become clear that petunias and calibrachoas can also be hosts to *P. infestans* (Becktell *et al.*, 2005b). Transplants produced in one region, transported to and subsequently planted in another can certainly be a source of the pathogen. In the highland tropics where potatoes are grown all year round, there can be a constant source of inoculum. Even if potatoes are not grown all year round, there are other solanaceous plants in the Andes that harbour *P. infestans* (Adler *et al.*, 2004) and these can be a source of the pathogen.

(b) Oospore survival

Because sexual reproduction in *P. infestans* is a recent phenomenon in most of the world, there are still many unknowns and much interest in deciphering the details of the role of oospores in the epidemiology of late blight. To date, it is known that oospores certainly survive in soil from one season to the next (Strömberg *et al.*, 1999; Mayton *et al.*, 2000; Fernández-Pavía *et al.*, 2004). Experiments conducted under field conditions in the Toluca Valley, Mexico, suggest that oospores can survive for at least two years in the absence of potato plants (Fernández-Pavía *et al.*, 2004). The direct relationship between late blight intensity in the beginning of the epidemics and the number of oospores per gram of soil together with the presence of stem lesions at the soil interface indicate that oospores are important initial inoculum for late blight epidemics in the field. Similar observations were made by others (Andersson *et al.*, 1998; Medina and Platt, 1999; Turkensteen *et al.*, 2000). Survival of oospores is also reported during the summer when air temperatures reached up to 44°C (Singh *et al.*, 2004). This is of particular interest for potato production areas in tropical and subtropical countries.

Many additional questions remain. We do not yet know whether oospores might give rise to epidemics earlier than infected seed tubers, whether the resulting diverse pathogen population will be more difficult to suppress than clonal populations and whether oospores in soil might give rise to infected tubers regardless of the occurrence of foliar late blight. These questions need to be addressed.

(c) Alternative hosts

The host range of *P. infestans* is somewhat difficult to describe accurately because many studies have identified host responses or lesions in response to inoculations. Erwin and Ribeiro (1996) listed 89 host species but a large minority was included on the basis of artificial inoculation experiments. Throughout most of the world, the two most important crop hosts are potatoes and tomatoes.

While many species have been listed as susceptible to *P. infestans*, the number of weed hosts that might serve as reservoirs of inoculum varies according to regions. In north temperate regions, *Solanum sarachioides*, *S. nigrum*, *S. dulcamara*, and *S. sisymbriifolium* are reported as hosts (Platt, 1999; Cooke *et al.*, 2002; Flier *et al.*, 2003). Plants infected with two strains, one A1 and one A2, may allow oospore formation, and may affect the dynamics of initial inoculum for late blight epidemics. Because *S. sarachioides* (hairy nightshade) is an increasingly important weed in North America, it might serve as either a trap for *P. infestans* or as a source of inoculum. In tropical and subtropical areas, many species of solanaceous as well as of other families are reported to be hosts of *P. infestans*. In South America, a potentially large number of species of *Solanum* are sources of this oomycete pathogen. For example, pear melon

(*Solanum muricatum*) is an agricultural host that can harbour isolates of the A1 and A2 mating types in Ecuador (Adler *et al.*, 2002). The main implication of this finding to the epidemiology of late blight is to emphasize that plants other than potatoes or tomatoes can be a source of the pathogen and can provide opportunity for formation of oospores (Oliva *et al.*, 2002). In Africa, solanaceous and species of Asteraceae were reported as host of *P. infestans* (Fontem *et al.*, 2004).

Despite many reports of potential hosts and reservoir species for *P. infestans*, the epidemiological significance of such plants remains largely unknown. Assessing the contribution of non-economical plants as inoculum sources is not trivial. Population genetics tools such as those reported in a study to detect late blight foci in The Netherlands may be useful to epidemiologists (Zwankhuizen *et al.*, 1998). The ability to trace genotypes will contribute to elucidate the role of weeds, ornamentals, and cultivated plants as inoculum sources.

In north American greenhouse production, late blight is a consideration if petunias and tomatoes are grown in the same house. In general, petunias are not nearly as susceptible as are tomatoes, but they can be infected and they can also be a source of inoculum to initiate an epidemic on tomatoes in the greenhouse (Becktell *et al.*, 2005a). Infections on petunias are not nearly so visible as on tomatoes so infected petunias might not be detected.

17.3.2 Dispersal of P. infestans

Infected tubers can be transported kilometres, hundreds of kilometres, or from one continent to another. Transport of infected tubers is almost certainly the most important mechanism by which global migrations have occurred. Presumably, it is also probable for oospores to be transported long distances. However, the relative importance of oospores in global migration is difficult to ascertain. Oospores may be present in tubers or in soil accompanying tubers. The low germination rates of *P. infestans* oospores (Andrivon, 1995) further challenges our efforts to investigate the role of this spore form in epidemiology.

Although movement of infected seed tubers certainly transports *P. infestans*, the efficiency of subsequent establishment has been reportedly very low (Van Der Zaag, 1956). Typically, many infected tubers die before producing plants and those that produce plants may not be infected. However, a variably low frequency of infected tubers may give rise to infected plants (Inglis *et al.*, 1999). It now appears that specific strains differ from each other in their abilities to infect tubers (Lambert and Currier, 1997) and presumably also in their abilities to initiate new epidemics. A major challenge in late blight epidemiology is to develop quantitative systems to predict the probability and intensity of epidemics starting from infected seed tubers.

Airborne sporangia can also initiate epidemics. However, because free sporangia do not survive long when exposed to sunlight, very long distance transport is rare and highly unlikely. However, shorter range transport can be much more common and very important, especially under cloudy and moist conditions. It is true that while disease gradients indicate that most sporangia are deposited within meters of a source (Paysour and Fry, 1983), some obviously escape the local area and can be dispersed and remain capable of initiating disease after being dispersed for several kilometers. Van der Zaag (1956) concluded that in some cases they are dispersed at least 11 km. Because sporangia are so sensitive to solar radiation, distance aerial transport is very highly unlikely during periods of intense solar radiation. Thus, long distance transport might occur during evening or night or under cloudy conditions.

The basic features of aerial transport have been described (Van Der Zaag, 1956; Hirst and Stedman, 1960). It is generally regarded that sporangia are produced during a wet period of some hours (typically overnight), are dispersed during subsequent drying (in the morning) and then cause infection when the leaves are again wetted (by dew or rain). However, rains during the day such as is typical of some locations in the highland tropics can stimulate production and release of a new crop of sporangia for dispersal later in the day.

Dispersal is comprised of three distinct processes: escape from the canopy, transport through atmosphere, and landing on plant tissues (Aylor, 1986). Escape from the canopy depends on the number of lesions, wind speed, change in relative humidity or solar radiation, and location of lesions in the canopy. Sporangia are typically produced overnight and then are released during change in relative humidity the next morning (Hirst, 1958). The higher in the canopy the lesions are located and the higher the wind speed, the greater the number of sporangia that can escape from the canopy (Aylor *et al.*, 2001). Wind speed of 1 to 2 m s^{-1} can remove considerable numbers of sporangia and also transport those propagules to distances up to 20 km in less than 3 hours (Aylor *et al.*, 2001), a time period during which sporangia can still be viable in cloudy days.

Transport of sporangia has been described using a variety of models. Empirical models such as the power law and the exponential model have been used for very short-range dispersal (Paysour and Fry, 1983). Long-range transport of *P. infestans* is passive and can be modelled either by empirical or by physical models, i.e. *K*-Theory-based models, Gaussian plume (Spijkerboer *et al.*, 2002) or Lagrangian stochastic (Aylor *et al.*, 2001). A more extensive discussion on the advantages and disadvantages of the different modelling approaches is available from several sources (McCartney and Fitt, 1985; Fitt and McCartney, 1986; McCartney and Fitt, 1987). Despite the availability of various models, and the knowledge that sporangia can be transported long distances (Harrison, 1947) application of such models to disease management has not yet occurred. Interestingly, the seemingly similar topic of disease spread over space (Minogue and Fry, 1983a; Ferrandino, 1996a,b) has not yet been linked to models of dispersal over longer ranges.

17.3.3 Inoculation to germination

Sporangia germinate either by releasing zoospores or by producing a germ tube. The mode of germination is heavily influenced by temperature. For most genotypes, temperatures below 15°C cause only the production of zoospores. It appears that when the sporangia are formed they are programmed to produce zoospores because proteins that function in zoospores are present in the mature, uncleaved sporangium (Hardham and Hyde, 1997; Judelson and Blanco, 2005). For some individuals, the proportion of sporangia that produce zoospores declines as temperatures increase above 15°C and some sporangia germinate directly by producing a germ tube. However, the proportion of sporangia that germinate directly rarely rises above ca 20%. The influence of temperature on mode of germination was first described by Crosier (1934). This relationship, described in the first part of the 20th century, appears to be accurate for the US-1 clonal lineage of *P. infestans.* It is presumably this clonal lineage on which the experiments were performed. However, these results do not apply equally to all clonal lineages. For example, the effect of temperature on sporangial germination of US-8 (a clonal lineage introduced from Mexico to the United States) appears to be rather different from the effect on US-1 (Mizubuti and Fry, 1998). Crosier (1934) found that *P. infestans* sporangia (presumably US-1) germinated reasonably well (directly) at 18 to 20°C. In recent experiments, sporangia of US-1 also germinated well at 18 to 20°C, but the situation was different for US-8 (Mizubuti and Fry, 1998). This clonal lineage germinated well at 10 to 15°C, but not at 18 to 20°C. The clonal lineage-dependent response to temperature suggests that predictions concerning the effect of abiotic factors on growth and development of *P. infestans* need to be made with some caution until we learn more about the responses of different genotypes to various abiotic factors.

17.3.4 Interaction with the host: infection, colonization and reproduction

Host plants are of overwhelming importance to the population dynamics of *P. infestans*. Pathogenic specialization within *P. infestans* occurs within a host species as well as among species. Specialization for host genus has recently been documented (Legard *et al.*, 1995; Adler *et al.*, 2004). Some isolates recently introduced into the United States during the 1970s were especially pathogenic to tomatoes, whereas others were especially pathogenic to potatoes. In competitive fitness experiments, tomato-adapted strains rapidly dominated the total population when growing on tomatoes (Legard and Fry, 1996). While many strains can cause lesions on both potatoes and tomatoes, it is primarily the tomato-adapted strains that occur commonly on tomatoes and these typically do not occur commonly on potatoes (Oyarzun *et al.*, 1998; Lebreton *et al.*, 1999; Suassuna *et al.*, 2004).

The intraspecies host-specificity of *P. infestans* results from the compatibility of isolates with resistance genes, R-genes. Some isolates are compatible with cultivars containing diverse R-genes, whereas other isolates may be incompatible with such cultivars. The observed plasticity in populations of *P. infestans* has caused most potato breeders to avoid the use of R-genes (Umaerus *et al.*, 1983; Wastie, 1991). The situation with the cultivar Pentland Dell seems unfortunately common. This cultivar contained genes R1, R2, and R3 and was resistant to late blight when it was first identified. However, during the 'adoption' phase of the cultivar, a compatible pathotype of *P. infestans* appeared (Malcolmson, 1969), and the resistance broke down before widespread adoption. This scenario has occurred many times (Van der Plank, 1968). Goodwin *et al.* (1995a) have suggested that the compatibility loci in *P. infestans* can be highly variable and perhaps are more readily mutable than other loci. Variation in virulence was demonstrated for asexual progenies of *P. infestans* originating from single zoospores cultures (Abu-El Samen *et al.*, 2003). Even when reproduction leads to clonal structure, variation in virulence could occur. However, the recent discovery of an R-gene from *S. bulbocastanum* (Song *et al.*, 2003) may change the dogma concerning variability in populations of *P. infestans.* This R-gene appears, after a few years of testing, to be very durable; no compatible isolates have yet been reported, and this gene has been exposed to highly diverse populations of *P. infestans* in Europe, the USA and in the Toluca Valley in Mexico. This gene is being examined very carefully.

Once inside the plant, colonization of host tissue proceeds rapidly. The latent period can be as short as 3 to 4 days, depending on cultivar resistance and weather conditions (Flier and Turkensteen, 1999; Carlisle *et al.*, 2002; Andrade-Piedra *et al.*, 2005). The formation of sporangiophores in potato tissue usually occurs after tissue necrosis in symptom development; thus, latent period can be completed after the incubation period (time from inoculation to symptom development). Interestingly, the latent period of some isolates infecting tomato is completed before the incubation period, that is sporulation occurs before the tissue is visibly symptomatic (Vega-Sánchez *et al.*, 2000; Smart *et al.*, 2003; Suassuna *et al.*, 2004). These isolates are specialized to tomato and have a prolonged biotrophic colonization phase (Vega-Sánchez *et al.*, 2000; Smart *et al.*, 2003).

Colonization of host tissue is known to be affected by temperature and isolates of distinct lineages can respond differently to temperatures. The optimum temperature for development of isolates of the US-1 lineage, estimated through short latent period and large lesion area and sporulation, was higher than the optimum temperature for isolates of the BR-1, a new lineage associated with potatoes in Brazil. Isolates of BR-1 lineage developed well at low temperatures (Maziero, 2001). At 10°C, the latent period of BR-1 was completed at 135 hours while the latent period of US-1 was not completed. Nevertheless, when leaflets inoculated with US-1 and kept at 10°C were transferred to 22°C, sporangiophores were seen 24 hours after incubation. In a study conducted in the 1980s, probably with an isolate of the US-1 lineage, greater growth of *P. infestans* mycelia was observed at 17°C than at 13°C (Harrison *et al.*, 1990). Unfortunately, there is no detailed study aimed at assessing the effects of temperature on growth of isolates of the new lineages of *P. infestans*.

The effects of humidity on pathogen growth inside plant tissue seem to be less important than the effect of temperature (Harrison *et al.*, 1990), but high humidity is very important for reproduction by sporangial formation. More sporangia are formed in the range of 18 to 22°C in a saturated atmosphere (Harrison, 1992). To date, there is no evidence of a differential response to humidity among different genotypes of *P. infestans*.

17.4 LATE BLIGHT MANAGEMENT

Effective management of late blight requires a comprehensive approach, integrating many strategies and tactics. This is especially important in managing strains that are very aggressive and not especially sensitive to some fungicides. Reliable information to support management strategies comes from experiments done in various environmental conditions and preferably in places where different *P. infestans* genotypes are present. However, exhaustive evaluation of treatments by classical experimental procedures (controlled conditions and field trials) is often of low practical value. Furthermore, in many areas where pathogen populations are not variable, it is not appropriate to introduce new genotypes, especially for field experiments. However, as an inexpensive proxy for exhaustive field experiments, assessment of treatments, including the effects of different genotypes and environmental conditions, can be accomplished using late blight models. Such analyses commonly require only computer time and desktop brainstorming time when analyzing outputs.

17.4.1 Modelling

Potato late blight is a classic polycyclic disease (many generations per season) and has been the subject of analysis via mathematical models. The goals have been to explain phenomena or to investigate outcomes of certain conditions. Nevertheless, the contribution of models extends beyond representation and analysis of epidemics. Models are crucial for a systems analysis approach of evaluating management practices and field experimentation must be conducted to test the predictions of analysis.

Modelling of late blight has utilized both analytical and simulation approaches. Analytical models can provide a general insight, whereas greater specificity is possible from the use of computer simulation models. To date, all models have assumed that the pathogen population is homogeneous and responds uniformly to abiotic and biotic influences.

Analytical models can aid interpretation of temporal and spatial dynamics of late blight epidemics. Progress models such as the exponential, logistic, and Gompertz have been used to describe late blight epidemics and answer epidemiologically relevant questions. Models can be used to compare the effects of diverse control measures: environmental favourability (Gilligan, 1990), efficacy of resistant genotypes (Ojiambo *et al.*, 2000), or resistance deployment (Andrivon *et al.*, 2003) on late blight epidemics. On the spatial side, the spread of potato late blight through 'waves' has been investigated (Minogue and Fry, 1983b; Scherm, 1996). Recently, using geographic information systems (GIS), the temporal and spatial patterns of *P. infestans* genotypes were studied (Jaime-Garcia *et al.*, 2000). It is also possible to assess the likelihood of occurrence of genotypes with important epidemiological features such as fungicide resistance or mating type in an area (Jaime-Garcia *et al.*, 2001). This information can be useful for decision-making regarding, for instance, the choice of proper fungicide to be applied and assessment of high-risk areas where the likelihood of disease development is high and where more intensive fungicide sprays should be considered. From the population genetics point of view, regions containing both A1 and A2 mating types can increase the chances of sexual reproduction; thus more attention should be devoted to late blight control in these areas.

Interpretation of analytical models should be tempered by at least two important considerations. First, all of the assumptions (implicit as well as explicit) used for model construction need to considered when interpreting model results. This is sometimes very difficult because implicit assumptions may be difficult to identify exactly because of their extreme familiarity. For example, use of the logistic model for many purposes assumes a constant environment and constant host susceptibility, assumptions that are probably usually violated when considering real epidemics. The influence of a non-constant environment would have caused incorrect inferences if the logistic model had been used in all cases to compare the effects of fungicide and cultivar resistance (Fry, 1978). Some authors have identified these non-constant influences and have attempted to modify models to reflect changing influences on disease progress (Berger and Jones, 1985; Waggoner, 1986). Second, it is also important to use caution when inferring biological meaning from certain mathematical parameters.

To improve the performance of analytical models, more detailed modelling of the system can be achieved by linked differential equations. This approach can give better representations of some sub-processes of epidemics (Van Oijen, 1992). As an example, a S-I-R (Susceptible-Infectious-Removed) model, with intrinsically linked differential equations, was used to study the importance of different fitness components on breeding programmes. Van Oijen (1989) predicted that infection efficiency and lesion growth rate are important components to look at when breeding for resistance. This hypothesis was tested using field data and indeed lesion growth rate was found to be one of the most important components in breeding for late blight resistance (Colon *et al.*, 1995). These investigations not only illustrate the potential contribution of models from human epidemiology in plant disease but also and more importantly, a clever usage of mathematical models to generate a testable hypothesis.

Complex simulation models allow for more detailed and specific temporal-spatial investigations. Several such models have been constructed (Waggoner and Horsfall, 1969; Bruhn and Fry, 1981; Van Oijen, 1995; Apel *et al.*, 2003). Simulation employs numerical solution to evaluate different treatments and analysis of a considerable range of situations under known conditions. With simulation, a better understanding of sub-processes can be accomplished. Inferences and decision-making based on the analysis of simulation modelling may be more readily acceptable because the relationships between cause and effect can be identified. Such approaches have contributed to devising efficient late blight management strategies: timing initial sprays and subsequent applications; evaluation of the importance of individual sprays in controlling the disease and determining the contribution of host resistance in an integrated disease management programme (Shtienberg *et al.*, 1994; Van Oijen, 1995; Apel *et al.*, 2003). Thus it appears that simulators are useful if properly verified and validated but verification and validation require considerable time and effort. Finally, it must be constantly remembered that simulators are constructed from a given set of conditions and use of the models is only known to be appropriate for that set of conditions. Extrapolation to different conditions should be done with caution.

The late blight model of Bruhn, developed for temperate North America and the US-1 clonal lineage of *P. infestans*, has been modified to reflect: i) the more diverse environmental conditions in the Andes; ii) the disease cycle more accurately; and iii) the characteristics of a strain of *P. infestans* recently introduced to the Andes (Andrade-Piedra *et al.*, 2005a,b). Validation of the model was accomplished with observed data not used in model construction. The observed data came from field trials with three potato cultivars in three diverse locations in Peru for a total of twelve epidemics. Observed and simulated epidemics were compared graphically using disease progress curves and numerically using the area under the disease progress curve in a confidence interval test, an equivalence test, and an envelope of acceptance test. The level of agreement between observed and simulated epidemics was high, and the model was found to be valid according to subjective and objective performance criteria. The model was then evaluated against thirty-two epidemics in diverse locations (Ecuador, Mexico, Israel, USA). If the susceptibility of the cultivar was accurately known, the model provided realistic predictions in these diverse locations.

The rather large effort devoted to the late blight disease has caused this disease to become a model system for modelling activities. The wealth of information on the disease, the significant modelling effort already devoted to it and its continued economic importance, combine to provide continued motivation for improving models of this disease.

17.4.2 Late blight forecasting

Potato late blight has been the subject of numerous forecasting efforts. Most efforts have been directed at improving the efficiency and efficacy of fungicides, but a forecast could be directed at any tactic. To define efficient use of fungicides is to mean that fungicides are not used when they are *not* needed but are used when they *are* needed.

One of the first forecast systems with solid fundamental knowledge was developed in The Netherlands by Van Everdingen (1926) and became known as the 'Dutch Rules'. According to these rules outbreaks of late blight are probable 15 days after the occurrence of two days with two specific characteristics: the first day requires more than 4 hours of dew during night time with a minimum temperature above 10°C and the following day a mean cloudiness above 0.8 with measurable rainfall. Many other late blight forecast systems have been developed over the years (Hyre and Horsfall, 1951; Wallin, 1962; Krause *et al.*, 1975; Johnson *et al.*, 1998; Singh *et al.*, 2000), with most utilizing empirical relations of weather variables with disease development. Usually, the aim of these systems was the timing of the first fungicide application. Despite the empiricism, the forecast systems helped, under some environment conditions, to achieve greater efficiency in controlling late blight compared with weekly fungicide spray applications. It seems that under non-conducive environmental conditions, forecast systems tend to save sprays without significant crop losses compared with weekly fungicide applications (Raposo *et al.*, 1993).

Most forecast systems have been modified after their initial development. Several modifications have been applied to Blitecast, perhaps because computers were just coming into widespread use when Blitecast was first published. Blitecast itself is an integration of two previous forecasts developed by Hyre (1954) and Wallin (1962). The best known modification was to incorporate Blitecast into a comprehensive potato management programme that was available on a diskette (Connell *et al.*, 1991; Stevenson *et al.*, 1994). Attempts to incorporate weather forecasts (Raposo *et al.*, 1993) into practical use of Blitecast have not yet been successful.

For other purposes, such as estimating global severity of late blight based on climate characteristics, analyses using Blitecast integrated with GIS technology was informative (Hijmans *et al.*, 2000). It seems likely that the next improvements in late blight forecasting will be to include host susceptibility, pathogen occurrence and virulence and aggressiveness as variables. To accomplish these tasks, simulation-based forecast models need to become more flexible. Recently, SimCast, a simulation-based forecast system, was modified to incorporate the effects of cultivars with high levels of resistance and evaluated in Central Mexico (Grünwald *et al.*, 2002a). Effective management of late blight was obtained by combining host resistance and fungicide sprays according to SimCast.

Local validation of any forecast system is necessary, because local environmental conditions and the pathogen population may differ from those existing where the forecast system was developed. A forecast may reflect an implicit assumption that is perfectly appropriate for the location in which the forecast was developed, but is not universally applicable. For example the rules of Van Everdingen would not be useful in the Toluca Valley in Mexico, because the night- time temperatures are usually below 10°C.

The assumption (typical of temperate zone potato production) that the pathogen is at a very low population level at the beginning of the season may not be accurate for all agro-ecosystems. In some areas of the tropics inoculum is available all year-round. Using Burkard samplers, the concentration of sporangia in the troposphere (the atmosphere layer that extends from ground level up to approximately 10 km height) was monitored throughout the year (January 2004 to January 2005) in Viçosa, Brasil (20° 45'S, 42° 52'W). Sporangia were collected in 41 of 51 (78%) sampled weeks (Fig. 17.2) (M.A. Lima, L.A. Maffia and E.S.G. Mizubuti, unpublished data).

On the other hand, for countries in temperate climates, there have been periods in which *P. infestans* was essentially absent. For example, some locations in the United States were free of late blight for some years during the 1980s (Goodwin *et al.*, 1995b; Johnson *et al.*, 1997). The incorporation of reliable estimates of incoming inoculum would have contributed to more efficient disease management during those years. In the mid 1990s in the USA, after the introduction of immigrant strains, late blight was widespread and the assumption of inoculum availability was again justified. However, in the future and in some locations we may again return to the situation with rare occurrences of late blight and then information about the probability of inoculum availability will again be useful in disease management. Efforts to forecast the probable occurrence of the pathogen are justified, and when successful, would be much welcomed. Disease forecasting is covered in detail in Chapter 9.

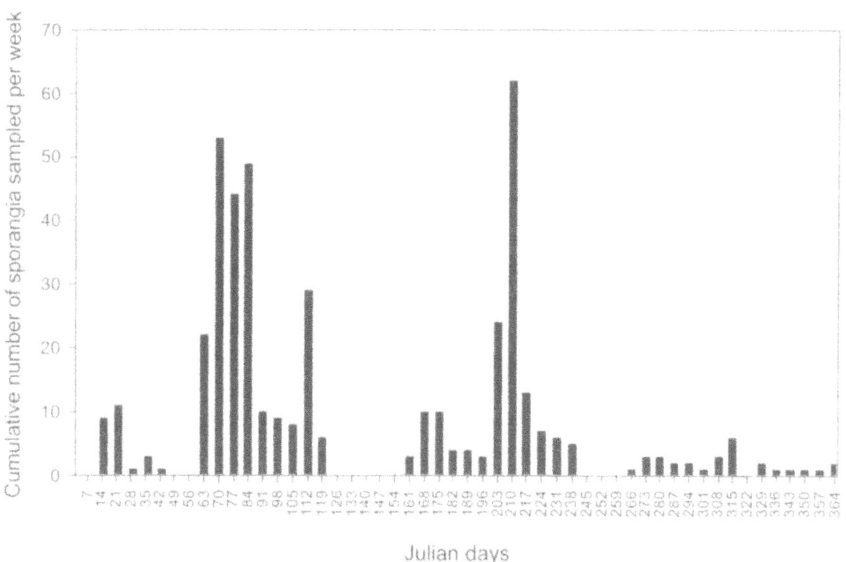

Figure 17.2. Cumulative number of P. infestans *sporangia collected per week throughout 2004 in Viçosa, MG, Brazil, using a Burkard sampler.*

17.4.3 Control measures

Management strategies are conveniently grouped into those that limit or reduce the initial inoculum and those that retard pathogen growth rates.

A variety of techniques can limit or reduce the size of the pathogen population at the beginning of the season. Elimination of infected potatoes can be carried out whether they are in piles as individual tubers left in the field, or used as seed tubers. Thus destruction of cull piles and destruction of volunteers are important components of a comprehensive management system. Planting only healthy seed tubers is also very important. If chemicals that can prevent pathogen growth in seed tubers become available, these will be very beneficial. However, because sporangia of *P. infestans* can be dispersed aerially, late blight is a community phenomenon and, if there are many growers in a region, each of the members of the community needs to take proper precautions to suppress late blight. If some growers do not take these precautions, then their farms become potential sources of inoculum for their neighbours.

Techniques that limit pathogen growth rates include utilization of cultivars with field resistance, and periodic application of protectant or effective systemic fungicides. Against sensitive isolates, the systemic fungicide metalaxyl could halt an established epidemic. However, many populations throughout the world are no longer sensitive to metalaxyl. Currently, many oomycete-specific fungicides such as QoI compounds, dimethomorph, propamocarb, etc. (Köller, 1998) have been commercialized, but currently, we are unaware of any fungicide that could effectively halt epidemics caused by metalaxyl-resistant strains under conditions favourable to *P. infestans* growth and development. Thus it is important to have protectant fungicide on plants prior to establishment of disease. Good coverage is required for adequate suppression, and good coverage is most readily achieved with relatively frequent application, particularly when plants are growing rapidly. It was learned that, on average, not all sprays had an equal contribution to the suppression of foliar disease (Shtienberg, 1989, 1994). Sprays applied early or mid-season had larger effects on foliar disease than sprays applied late season. Management programmes that include host resistance as a variable in addition to weather (Fry, 1977; Fry *et al.*, 1983) can achieve considerable efficiencies in fungicide utilization.

Plants with field resistance slow pathogen growth rates. Unfortunately, most cultivars that have market acceptance are not highly resistant to late blight. Nonetheless there is diversity among commercial potato cultivars in terms of resistance to late blight and these levels can be incorporated into an overall management strategy. Cultivars with higher levels of resistance require less fungicide than cultivars with lower levels of resistance (Fry, 1978).

Durable or polygenic resistance is sometimes interpreted to be synonymous with intermediate resistance levels but cultivars ranging from complete susceptibility to very highly resistant are possible (unpublished results). The problem has been that in most locations these breeding lines have not yet become popular (an exception appears to be Mexico, where more resistant cultivars have some market share) (Grünwald *et al.*, 2002b). Until these more resistant cultivars and breeding lines become accepted, or until breeders create resistant potatoes that have more of the qualities desired by the market place, integrated strategies will need to be used. One possible strategy is to adjust the amount of fungicide to complement the level of resistance in a cultivar (Fry, 1978). Another possible strategy that could be adopted in certain locations is to combine field resistance with other cultivars in cultivar mixtures. Cultivar mixtures have been evaluated for late blight management, but the effects were small and not detectable in all locations (Garrett and Mundt, 2000; Garrett *et al.*, 2001; Andrivon *et al.*, 2003).

The increased importance of late blight during the past few years has stimulated plant molecular biologists to become interested in resistance to late blight. There has been intense effort directed at cloning R-genes and in defining the signalling pathways that lead to resistance. One potentially very important discovery has been an R-gene from *S. bulbocastanum*. This gene has now been cloned, and it appears to be similar in structure to other R-genes. The very significant difference is that this gene seems to provide durable resistance to late blight, even when exposed for seven years to highly variable populations of *P. infestans* (Song *et al.*, 2003). Other efforts have investigated the roles of genes coding for antimicrobial proteins and PR (pathogenesis-related) proteins. When transformed into recipient plants, some of these genes have conferred partial resistance in greenhouse experiments (Liu *et al.*, 1994; Wu *et al.*, 1995).

The role of microbes in the population dynamics and survival of *P. infestans* is largely unknown. Several reports of testing different organisms attempting to biocontrol late blight have been published during the past decade. Biological control is an interesting alternative for the control of late blight, mainly on organic potato crops. Two biocontrol strategies have received most attention: use of phylloplane microorganisms (Ng and Webster, 1997; Garita *et al.*, 1998, 1999; Daayf *et al.*, 2003) and microbial-induced resistance (Yan *et al.*, 2002; Silva *et al.*, 2004). Unfortunately, we are not aware of any study that demonstrated the effectiveness of biocontrol agents against potato late blight under field conditions. There have been limited efforts to thoroughly investigate interactions between *P. infestans* and other microbes on leaf surfaces and efforts are just being initiated to investigate the interactions between oospores of *P. infestans* and the soil microbial community.

17.5 CONCLUDING REMARKS

This chapter has attempted to illustrate how the population genetics of *P. infestans* has influenced the epidemiology of the late blight disease, and therefore why it is important to include population genetics in any consideration of the epidemiology or management of the disease. The transition from an asexual life history to a life history that also includes sexual reproduction is changing how we view both the epidemiology and management. This chapter has also attempted to illustrate that the population dynamics of *P. infestans* in temperate potato production zones can differ significantly from that in some tropical production zones. Ignorance of the pathogen population biology at the local level can lead to failures in disease control, and it is now clear that the pathogen biology can be dramatically different in one location compared to another. Unfortunately, even if we have an excellent understanding of the pathogen biology in a location, the efficacy of available tools to suppress this disease can be inadequate when the environment is particularly favourable to the pathogen. Most notably, cultivars with high levels of durable resistance that are acceptable to the market are needed. Additionally, a curative fungicide that could halt an established epidemic is also required.

Until advances in disease suppression technology are made, epidemiologists need to address questions, the answers to which will lead to improved efficacy of disease management. Some of these are:

- What is the quantitative relationship between oospores in the field and epidemic development?
- What is the quantitative relationship between late blight in the foliage and late blight in the tubers?
- What is the quantitative role in a particular production field of long distance, aerial transport of sporangia?
- What are the most effective integrated management systems for particular production regions?

These questions need to be answered at the local level to be useful, and they need to be answered generically for universal application. Obtaining the answers will be challenging, but important. The past fifteen years has seen an explosion in our understanding of the population dynamics of *P. infestans* and there has also been very successful application of that information to limit the damage caused by this plant pathogen. We expect to see new discoveries during the next fifteen years that provide similar new understanding and improvements in our efforts to manage this very destructive plant disease.

REFERENCES

Abad, Z.G. and Abad, J.A. (1997) Another look at the origin of late blight of potatoes, tomatoes, and pear melon in the Andes of South America. *Plant Disease*, **81**, 682-688.

Abu-El Samen, F.M., Secor, G.A. and Gudmestad, N.C. (2003) Variability in virulence among asexual progenies of *Phytophthora infestans*. *Phytopathology*, **93**, 293-304.

Adler, N.E., Chacon, G., Flier, W.G. *et al.* (2002) The Andean fruit crop, pear melon (*Solanum muricatum*) is a common host for A1 and A2 strains of *Phytophthora infestans* in Ecuador. *Plant Pathology*, **51**, 802-802.

Adler, N.E., Erselius, L.J., Chacon, M.G. *et al.* (2004) Genetic diversity of *Phytophthora infestans sensu lato* in Ecuador provides new insight into the origin of this important plant pathogen. *Phytopathology*, **94**, 154-162.

Andersson, B., Sandström, M. and Strömberg, A. (1998) Indications of soil borne inoculum of *Phytophthora infestans*. *Potato Research*, **41**, 305-310.

Andrade-Piedra, J.L., Hijmans, R.J., Forbes, G.A. *et al.* (2005a) Simulation of potato late blight in the Andes I: Modification and parameterization of the LATEBLIGHT model. *Phytopathology*, **95**, 1191-1199.

Andrade-Piedra, J.L., Hijmans, R.J., Juárez, H.S. *et al.* (2005b) Simulation of potato late blight in the Andes II: Validation of the LATEBLIGHT model. *Phytopathology*, **95**, 1200-1208.

Andrivon, D. (1995) Biology, ecology, and epidemiology of the potato late blight pathogen *Phytophthora infestans* in soil. *Phytopathology*, **85**, 1053-1056.

Andrivon, D. (1996) The origin of *Phytophthora infestans* populations present in Europe in the 1840s: a critical review of historical and scientific evidence. *Plant Pathology*, **45**, 1027-1035.

Andrivon, D., Lucas, J.M. and Ellisseche, D. (2003) Development of natural late blight epidemics in pure and mixed plots of potato cultivars with different levels of partial resistance. *Plant Pathology*, **52**, 586-594.

Apel, H., Paudyal, M.S. and Richter, O. (2003) Evaluation of treatment strategies of the late blight *Phytophthora infestans* in Nepal by population dynamics modelling. *Environmental Modelling & Software*, **18**, 355-364.

Aylor, D.E. (1986) A framework for examining inter-regional aerial transport of fungal spores. *Agricultural and Forest Meteorology*, **38**, 263-288.

Aylor, D.E., Fry, W.E., Mayton, H. *et al.* (2001) Quantifying the rate of release and escape of *Phytophthora infestans* sporangia from a potato canopy. *Phytopathology*, **91**, 1189-1196.

Becktell, M.C., Daughtrey M.L. and Fry W.E. (2005a). Epidemiology and management of petunia and tomato late blight in the greenhouse. *Plant Disease*, **89**, 1000-1008.

Becktell, M.C., Smart, C.D., Haney, C.H., *et al.* (2005b) Host-pathogen interactions between *Phytophthora infestans* and the solanaceous hosts *Calibrachoa* x *hybridus*, *Petunia* x *hybrida* and *Nicotiana benthamiana*. *Plant Disease*, **89**, (in press).

Berger, R.D. and Jones, J.W. (1985) A general model for disease progress with functions for variable latency and lesion expansion on growing host plants. *Phytopathology*, **75**, 792-797.

Bourke, A. (1993) *The visitation of god?*, The Lilliput Press Ltd., Dublin, 230 pp.

Bourke, P.M.A. (1964) Emergence of potato blight, 1843-46. *Nature*, **203**, 805-808.

Bruhn, J.A. and Fry, W.E. (1981) Analysis of potato late blight epidemiology by simulation modeling. *Phytopathology*, **71**, 612-616.

Carlisle, D.J., Cooke, L.R., Watson, S. *et al.* (2002) Foliar aggressiveness of Northern Ireland isolates of *Phytophthora infestans* on detached leaflets of three potato cultivars. *Plant Pathology*, **51**, 424-434.

Colon, L.T., Budding, D.J., Paul Keizer, L.C. *et al.* (1995) Components of resistance to late blight (*Phytophthora infestans*) in eight South American *Solanum* species. *European Journal of Plant Pathology*, **101**, 441-456.

Connell, T.R., Koenig, J.P., Stevenson, W.R. *et al.* (1991) An integrated systems approach to potato crop management. *Journal of Production Agriculture*, **4**, 453-460.

Cooke, L.R., Carlisle, D.J., Wilson, D.G. *et al.* (2002) Natural occurrence of *Phytophthora infestans* on woody nightshade (*Solanum dulcamara*) in Ireland. *Plant Pathology*, **51**, 392-392.
Cox, A.E. and Large, E.C. (1960) *Potato blight epidemics - Throughout the world*, ARS USDA Handbook No. 174, Government Printing Office, Washington, DC., 230 pp.
Crosier, W. (1934) Studies in the biology of *Phytophthora infestans* (Mont.) De Bary. *Cornell University Agricultural Experiment Station*, **Memoir 155**, 40 pp.
Daayf, F. and Platt, H.W. (2000) Changes in metalaxyl resistance among glucose phosphate isomerase genotypes of *Phytophthora infestans* in Canada during 1997-1998. *American Journal of Potato Research*, **77**, 311-318.
Daayf, F., Adam, L. and Fernando, W.G.D. (2003) Comparative screening of bacteria for biological control of potato late blight (strain US-8), using *in vitro*, detached-leaves, and whole-plant testing systems. *Canadian Journal of Plant Pathology*, **25**, 276-284.
De Weille, G.A. (1963) Laboratory results regarding potato blight and their significance in the epidemiology of blight. *European Potato Journal*, **6**, 121-130.
Deahl, K.L., Goth, R.W., Young, R. *et al.* (1991) Occurrence of the A2 mating type of *Phytophthora infestans* in the United States and Canada. *American Potato Journal*, **68**, 717-725.
Dowley, L.J., Bannon, E., Cooke, L.R. *et al.* (1995) *Phytophthora infestans 150*, Boole Press Ltd., Dublin, 382 pp.
Drenth, A., Janssen, E.M. and Govers, F. (1995) Formation and survival of oospores of *Phytophthora infestans* under natural conditions. *Plant Pathology*, **44**, 86-94.
Elansky, S., Smirnov, A., Dyakov, Y. *et al.* (2001) Genotypic analysis of Russian isolates of *Phytophthora infestans* from the Moscow region, Siberia and Far East. *Journal of Phytopathology*, **149**, 605-611.
Erwin, D.C. and Ribeiro, O.K. (1996) *Phytophthora diseases worldwide*, APS Press, St. Paul, 562 pp.
Fernández-Pavía, S.P., Grünwald, N.J., Díaz-Valasis, M. *et al.* (2004) Soilborne oospores of *Phytophthora infestans* in Central Mexico survive winter fallow and infect potato plants in the field. *Plant Disease*, **88**, 29-33.
Ferrandino, F.J. (1996a) Two-dimensional distance class analysis of disease incidence data: Problems and possible solutions. *Phytopathology*, **86**, 685-691.
Ferrandino, F.J. (1996b) Length scale of disease spread: Fact or artifact of experimental geometry. *Phytopathology*, **86**, 806-811.
Fitt, B.D.L. and McCartney, H.A. (1986), Spore dispersal in relation to epidemic models, in *Plant disease epidemiology* (eds K.J. Leonard and W.E. Fry), MacMillan, New York, pp. 311-345.
Flier, W.G. and Turkensteen, L.J. (1999) Foliar aggressiveness of *Phytophthora infestans* in three potato growing regions in the Netherlands. *European Journal of Plant Pathology*, **105**, 381-388.
Flier, W.G., van den Bosch, G.B.M. and Turkensteen, L.J. (2003) Epidemiological importance of *Solanum sisymbriifolium*, *S. nigrum* and *S. dulcamara* as alternative hosts for *Phytophthora infestans*. *Plant Pathology*, **52**, 595-603.
Fontem, D.A., Olanya, O.M. and Njualem, B.F. (2004) Reaction of certain solanaceous and asteraceous plant species to inoculation with *Phytophthora infestans* in Cameroon. *Journal of Phytopathology*, **152**, 331-336.
Forbes, G.A., Escobar, X.C., Ayalla, C.C. *et al.* (1997) Population genetic structure of *Phytophthora infestans* in Ecuador. *Phytopathology*, **87**, 375-380.
Fry, W.E. (1977) Integrated control of potato late blight - Effects of polygenic resistance and techniques of timing fungicide applications. *Phytopathology*, **67**, 415-420.
Fry, W.E. (1978) Quantification of general resistance of potato cultivars and fungicide effects for integrated control of potato late blight. *Phytopathology*, **68**, 1650-1655.
Fry, W.E. and Goodwin, S.B. (1997a) Resurgence of the Irish Potato Famine fungus. *BioScience*, **47**, 363-371.
Fry, W.E. and Goodwin, S.B. (1997b) Re-emergence of potato and tomato late blight in the United States. *Plant Disease*, **81**, 1349-1357.
Fry, W.E., Apple, A.E. and Bruhn, J.A. (1983) Evaluation of potato late blight forecasts modified to incorporate host resistance and fungicide weathering. *Phytopathology*, **73**, 1054-1059.
Fry, W.E., Goodwin, S.B., Dyer, A.T. *et al.* (1993) Historical and recent migrations of *Phytophthora infestans*: Chronology, pathways, and implications. *Plant Disease*, **77**, 653-661.
Gallegly, M.E. and Galindo, J. (1958) Mating types and oospores of *Phytophthora infestans* in nature in Mexico. *Phytopathology*, **48**, 274-277.

Garita, V.S., Bustamante, E. and Shattock, R. (1998) Seleccion de antagonistas para el control biologico de *Phytophthora infestans* en tomate. *Manejo Integrado de Plagas (Costa Rica)*, **no. 48**, 25-34.
Garita, V.S., Bustamante, E. and Shattock, R. (1999) Control microbiologico de *Phytophthora infestans* en tomate. *Manejo Integrado de Plagas (Costa Rica)*, **no. 51**, 47-58.
Garrett, K.A. and Mundt, C.C. (2000) Host diversity can reduce potato late blight severity for focal and general patterns of primary inoculum. *Phytopathology*, **90**, 1307-1312.
Garrett, K.A., Nelson, R.J., Mundt, C.C. *et al.* (2001) The effects of host diversity and other management components on epidemics of potato late blight in the humid highland tropics. *Phytopathology*, **91**, 993-1000.
Gavino, P.D. and Fry, W.E. (2002) Diversity in and evidence for selection on the mitochondrial genome of *Phytophthora infestans*. *Mycologia*, **94**, 781-793.
Gilligan, C.A. (1990) Comparison of disease progress curves. *New Phytologist*, **115**, 223-242.
Glendinning, D., MacDonald, J.A. and Grainger, J. (1963) Factors affecting the germination of sporangia in *Phytophthora infestans*. *Transactions of the British Mycological Society*, **46**, 595-603.
Goodwin, S.B., Cohen, B.A. and Fry, W.E. (1994a) Panglobal distribution of a single clonal lineage of the Irish potato famine fungus. *Proceedings of the National Academy of Sciences of the United States of America*, **91**, 11591-11595.
Goodwin, S.B., Sujkowski, L.S. and Fry, W.E. (1995a) Rapid evolution of pathogenicity within clonal lineages of the potato late blight disease fungus. *Phytopathology*, **85**, 669-676.
Goodwin, S.B., Cohen, B.A., Deahl, K.L. *et al.* (1994b) Migration from Northern Mexico as the probable cause of recent genetic changes in populations of *Phytophthora infestans* in the United States and Canada. *Phytopathology*, **84**, 553-558.
Goodwin, S.B., Spielman, L.J., Matuszak, J.M. *et al.* (1992) Clonal diversity and genetic differentiation of *Phytophthora infestans* populations in Northern and Central Mexico. *Phytopathology*, **82**, 955-961.
Goodwin, S.B., Sujkowski, L.S., Dyer, A.T. *et al.* (1995b) Direct detection of gene flow and probable sexual reproduction of *Phytophthora infestans* in Northern North America. *Phytopathology*, **85**, 473-479.
Grünwald, N.J., Romero Montes, G., Lozoya Saldaña, H. *et al.* (2002a) Potato late blight management in the Toluca Valley: Field validation of SimCast modified for cultivars with high field resistance. *Plant Disease*, **86**, 1163-1168.
Grünwald, N.J., Cadena-Hinojosa, M.A., Rubio-Covarrubias, O. *et al.* (2002b) Potato cultivars from the Mexican national potato program: Sources and durability of resistance against late blight. *Phytopathology*, **92**, 688-693.
Hardham, A.R. and Hyde, G.J. (1997), Asexual sporulation in the Oomycetes, in *Advances in Botanical Research* (eds T.C. Tommerup and J.H. Andrews), Academic Press, pp. 353-398.
Harrison, A.L. (1947) The relation of weathering to epiphytotics of late blight on tomatoes. *Phytopathology*, **37**, 533-538.
Harrison, J.G. (1992) Effects of the aerial environment on late blight of potato foliage - a review. *Plant Pathology*, **41**, 384-416.
Harrison, J.G., Barker, H., Lowe, R. *et al.* (1990) Estimation of amounts of *Phytophthora infestans* mycelium in leaf tissues by enzyme-linked immunosorbent assay. *Plant Pathology*, **39**, 274-277.
Hijmans, R.J., Forbes, G.A. and Walker, T.S. (2000) Estimating the global severity of potato late blight with GIS-linked disease forecast models. *Plant Pathology*, **49**, 697-705.
Hirst, J.M. (1958) New methods for studying plant disease epidemics. *Outlook on Agriculture*, **2**, 16-26.
Hirst, J.M. and Stedman, O.J. (1960) The epidemiology of *Phytophthora infestans* II. The source of inoculum. *Annals of Applied Biology*, **48**, 489-517.
Hohl, H.R. and Iselin, K. (1984) Strains of *Phytophthora infestans* from Switzerland with A2 mating type behaviour. *Transactions of the British Mycological Society*, **83**, 529-531.
Horsfall, J.G. and Cowling, E.B. (1978), Some epidemics man has known, in *Plant disease: An advanced treatise* (eds J.G. Horsfall and E.B. Cowling), Academic Press, New York, pp. 17-32.
Hyre, R. and Horsfall, J.G. (1951) Forecasting potato late blight in Connecticut. *Plant Disease Reporter*, **35**, 423-431.
Hyre, R.A. (1954) Progress in forecasting late blight of potato and tomato. *Plant Disease Reporter*, **38**, 245-253.
Inglis, D.A., Powelson, M.L. and Dorrance, A.E. (1999) Effect of registered potato seed piece fungicides on tuber-borne *Phytophthora infestans*. *Plant Disease*, **83**, 229-234.

Ingram, D.S. and Williams, P.H. (1991) *Phytophthora infestans, the cause of late blight of potato*, Academic Press, London, 273 pp.

Aylor, D.E., Fry, W.E., Mayton, H. and Andrade-Piedra, J.L. (2001) Quantifying the rate of release and escape of *Phytophthora infestans* sporangia from a potato canopy. *Phytopathology*, **91**, 1189-1196.

Jaime-Garcia, R., Trinidad-Correa, R., Felix-Gastelum, R. *et al.* (2000) Temporal and spatial patterns of genetic structure of *Phytophthora infestans* from tomato and potato in the Del Fuerte Valley. *Phytopathology*, **90**, 1188-1195.

Jaime-Garcia, R., Orum, T.V., Felix-Gastelum, R. *et al.* (2001) Spatial analysis of *Phytophthora infestans* genotypes and late blight severity on tomato and potato in the Del Fuerte Valley using geostatistics and geographic information systems. *Phytopathology*, **91**, 1156-1165.

Johnson, D.A., Alldredge, J.R. and Hamm, P.B. (1998) Expansion of potato late blight forecasting models for the Columbia Basin of Washington and Oregon. *Plant Disease*, **82**, 642-645.

Johnson, D.A., Cummings, T.F., Hamm, P.B. *et al.* (1997) Potato late blight in the Columbia basin: An economic analysis of the 1995 epidemic. *Plant Disease*, **81**, 103-106.

Judelson, H.S. and Blanco, F.A. (2005) The spores of *Phytophthora*: Weapons of the plant destroyer. *Nature Reviews - Microbiology*, **3**, 47-58.

Jyan, M.H., Ann, P.J., Tsai, J.N. *et al.* (2004) Recent occurrence of *Phytophthora infestans* US-11 as the cause of severe late blight on potato and tomato in Taiwan. *Canadian Journal of Plant Pathology*, **26**, 188-192.

Kato, M., Mizubuti, E.S., Goodwin, S.B. *et al.* (1997) Sensitivity to protectant fungicides and pathogenic fitness of clonal lineages of *Phytophthora infestans* in the United States. *Phytopathology*, **87**, 973-978.

Kirk, W.W. (2003) Thermal properties of overwintered piles of cull potatoes. *American Journal of Potato Research*, **80**, 145-149.

Kirk, W.W., Niemira, B.A. and Stein, J.M. (2001) Influence of storage temperature on rate of potato tuber tissue infection caused by *Phytophthora infestans* (Mont.) de Bary estimated by digital image analysis. *Potato Research*, **44**, 87-96.

Koh, Y.J., Goodwin, S.B., Dyer, A.T. *et al.* (1994) Migrations and displacements of *Phytophthora infestans* populations in East Asian countries. *Phytopathology*, **84**, 922-927.

Köller, W. (1998), Chemical approaches to managing plant pathogens, in *Handbook of pest management* (ed. J.R. Ruberson), Marcel Dekker, Inc., New York, pp. 337-376.

Krause, R.A., Massie, L.B. and Hyre, R.A. (1975) BLITECAST: A computerized forecast of potato late blight. *Plant Disease Reporter*, **59**, 95-98.

Lambert, D.H. and Currier, A.I. (1997) Differences in tuber rot development for North American clones of *Phytophthora infestans*. *American Potato Journal*, **74**, 39-43.

Large, E.C. (1940) *The advance of the fungi*, Dover Publications Inc., New York, 488 pp.

Lebreton, L., Lucas, J.M. and Andrivon, D. (1999) Aggressiveness and competitive fitness of *Phytophthora infestans* isolates collected from potato and tomato in France. *Phytopathology*, **89**, 679-686.

Legard, D.E. and Fry, W.E. (1996) Evaluation of field experiments by direct allozyme analysis of late blight lesions caused by *Phytophthora infestans*. *Mycologia*, **88**, 608-612.

Legard, D.E., Lee, T.Y. and Fry, W.E. (1995) Pathogenic specialization in *Phytophthora infestans*: Aggressiveness on tomato. *Phytopathology*, **85**, 1356-1361.

Liu, D., Kashchandra, G.R., Hasegawa, P.M. *et al.* (1994) Osmotin overexpression in potato delays development of disease symptoms. *Proceedings of the National Academy of Science USA*, **91**, 1888-1892.

Lucas, J.A., Shattock, R.C., Shaw, D.S. *et al.* (1991) *Phytophthora*, British Mycological Society Cambridge University Press, Cambridge, 447 pp.

Malcolmson, J.F. (1969) Races of *Phytophthora infestans* occurring in Great Britain. *Transactions of the British Mycological Society*, **53**, 417-423.

May, K.J. and Ristaino, J.B. (2004) Identity of the mtDNA haplotype(s) of *Phytophthora infestans* in historical specimens from the Irish Potato Famine. *Mycological Research*, **108**, 471-479.

Mayton, H., Smart, C.D., Moravec, B. *et al.* (2000) Oospore survival and pathogenicity of single oospore recombinant progeny from a cross involving US-17 and US-8 genotypes of *Phytophthora infestans*. *Plant Disease*, **84**, 1190-1196.

Maziero, J.M.N. (2001) *Influência da temperatura e do tempo de molhamento foliar nos componentes epidemiológicos de Phytophthora infestans e validação do simulador Blight no Brasil*. Universidade Federal de Viçosa, Viçosa, 61 pp. (Master's degree thesis).

McCartney, H.A. and Fitt, B.D.L. (1985), Construction of dispersal models, in *Advances in Plant Pathology* (ed. C.A. Gilligan), Academic Press, London, pp. 107-143.

McCartney, H.A. and Fitt, B.D.L. (1987), Spore dispersal gradients and plant disease development, in *Populations of plant pathogens: Their dynamics and genetics* (eds M.S. Wolfe and C.E. Caten), Blackwell Scientific Publications, Oxford, pp. 109-118.

McLeod, A., Denman, S., Sadie, A. *et al.* (2001) Characterization of South African isolates of *Phytophthora infestans*. *Plant Disease*, **85**, 287-291.

Medina, M.V. and Platt, H.W. (1999) Viability of oospores of *Phytophthora infestans* under field conditions in northeastern North America. *Canadian Journal of Plant Pathology*, **21**, 137-143.

Milgroom, M.G. and Peever, T.L. (2003) Population biology of plant pathogens. The synthesis of plant disease epidemiology and population genetics. *Plant Disease*, **87**, 608-617.

Miller, J.S., Johnson, D.A. and Hamm, P.B. (1998) Aggressiveness of isolates of *Phytophthora infestans* from the Columbia Basin of Washington and Oregon. *Phytopathology*, **88**, 190-197.

Minogue, K.P. and Fry, W.E. (1981) Effect of temperature, relative humidity, and rehydration rate on germination of dried sporangia of *Phytophthora infestans*. *Phytopathology*, **71**, 1181-1184.

Minogue, K.P. and Fry, W.E. (1983a) Models for the spread of disease: Model description. *Phytopathology*, **73**, 1168-1173.

Minogue, K.P. and Fry, W.E. (1983b) Models for the spread of plant disease: Some experimental results. *Phytopathology*, **73**, 1173-1176.

Mizubuti, E.S.G. and Fry, W.E. (1998) Temperature effects on developmental stages of isolates from three clonal lineages of *Phytophthora infestans*. *Phytopathology*, **88**, 837-843.

Mizubuti, E.S.G. and Forbes, G.A. (2002). Potato late blight IPM in the developing countries, in *The Global Initiative on Late Blight Conference* (ed. C. Lizárraga), Hamburg: The International Potato Center, pp. 93-97.

Mizubuti, E.S.G., Aylor, D.E. and Fry, W.E. (2000) Survival of *Phytophthora infestans* sporangia exposed to solar radiation. *Phytopathology*, **90**, 78-84.

Mosa, A.A., Kato, M., Sato, N. *et al.* (1989) Occurrence of the A2 mating type of *Phytophthora infestans* on potato in Japan. *Annals of the Phytopathological Society of Japan*, **55**, 615-620.

Mosa, A.A., Kobayashi, K., Ogoshi, A. *et al.* (1991) Formation of oospores by *Phytophthora infestans* in inoculated potato tissues. *Annals of the Phytopathological Society of Japan*, **57**, 334-338.

Ng, K.K. and Webster, J.M. (1997) Antimycotic activity of *Xenorhabdus bovienii* (Enterobacteriaceae) metabolites against *Phytophthora infestans* on potato plants. *Canadian Journal of Plant Pathology*, **19**, 125-132.

Niederhauser, J.S. (1991), *Phytophthora infestans*: The Mexican connection, in *Phytophthora* (eds J.A. Lucas, R.C. Shattock, D.S. Shaw, and L.R. Cooke), Cambridge University Press, Cambridge, pp. 25-45.

Ojiambo, P.S., Nyanapah, J.O., Lung, C. *et al.* (2000) Comparing different epidemiological models in field evaluations of selected genotypes from *Solanum tuberosum* CIP population A for resistance to *Phytophthora infestans* (Mont.) De Bary in Kenya. *Euphytica*, **111**, 211-218.

Oliva, R.F., Erselius, L.J., Adler, N.E. *et al.* (2002) Potential of sexual reproduction among host-adapted populations of *Phytophthora infestans sensu lato* in Ecuador. *Plant Pathology*, **51**, 710-719.

Oyarzun, P.J., Pozo, A., Ordoñez, M.E. *et al.* (1998) Host specificity of *Phytophthora infestans* on tomato and potato in Ecuador. *Phytopathology*, **88**, 265-271.

Paysour, R.E. and Fry, W.E. (1983) Interplot interference: A model for planning field experiments with aerially disseminated pathogens. *Phytopathology*, **73**, 1014-1020.

Platt, H.W. (1999) Response of solanaceous cultivated plants and weed species to inoculation with A1 or A2 mating type strains of *Phytophthora infestans*. *Canadian Journal of Plant Pathology*, **21**, 301-307.

Porter, L.D. and Johnson, D.A. (2004) Survival of *Phytophthora infestans* in surface water. *Phytopathology*, **94**, 380-387.

Raman, K.V., Grünwald, N.J. and Fry, W.E. (2000) Promoting international collaboration for potato late blight disease management. *Pesticide Outlook*, **11**, 181-185.

Raposo, R., Wilks, D.S. and Fry, W.E. (1993) Evaluation of potato late blight forecasts modified to include weather forecasts: A simulation analysis. *Phytopathology*, **83**, 103-108.

Reis, A., Smart, C.D., Fry, W.E. *et al.* (2003) Characterization of *Phytophthora infestans* isolates from Southern and Southeastern Brazil from 1998 to 2000. *Plant Disease*, **87**, 896-900.
Ristaino, J.B., Groves, C.T. and Parra, G.R. (2001) PCR amplification of the Irish potato famine pathogen from historic specimens. *Nature*, **411**, 695-697.
Rivera-Peña, A. (1990) Wild tuber-bearing species of *Solanum* and incidence of *Phytophthora infestans* (Mont.) de Bary on the western slopes of the volcano Nevado de Toluca. 3. Physiological races of *Phytophthora infestans*. *Potato Research*, **33**, 349-355.
Rotem, J. and Cohen, Y. (1974) Epidemiological patterns of *Phytophthora infestans* under semi-arid conditions. *Phytopathology*, **64**, 711-714.
Scherm, H. (1996) On the velocity of epidemic waves in model plant disease epidemics. *Ecological Modelling*, **87**, 217-222.
Shtienberg, D., Doster, M.A., Pelletier, J.R. *et al.* (1989) Use of simulation models to develop a low-risk strategy to suppress early and late blight in potato foliage. *Phytopathology*, **79**, 590-595.
Shtienberg, D., Raposo, R., Bergeron, S.N. *et al.* (1994) Incorporation of cultivar resistance in reduced-sprays strategy to suppress early and late blights on potato. *Plant Disease*, **78**, 23-26.
Silva, H.S.A., Romeiro, R.S., Carrer, R. *et al.* (2004) Induction of systemic resistance by *Bacillus cereus* against tomato foliar diseases under field conditions. *Journal of Phytopathology*, **152**, 371-375.
Singh, B.P., Gupta, J., Roy, S. *et al.* (2004) Production of *Phytophthora infestans* oospores *in planta* and inoculum potential of *in vitro* produced oospores under temperate highlands and sub-tropical plains of India. *Annals of Applied Biology*, **144**, 363-370.
Singh, B.P., Islam, A., Sharma, V.C. *et al.* (2000) JHULSACAST: a computerized forecast of potato late blight in western Uttar Pradesh. *Journal of the Indian Potato Association*, **27**, 25-34.
Smart, C.D., Myers, K.L., Restrepo, S. *et al.* (2003) Partial resistance of tomato to *Phytophthora infestans* is not dependent upon ethylene, jasmonic acid, or salicylic acid signaling pathways. *Molecular Plant-Microbe Interactions*, **16**, 141-148.
Sogin, M.L. and Silberman, J.D. (1998) Evolution of the protists and protistan parasites from the perspective of molecular systematics. *International Journal For Parasitology*, **28**, 11-20.
Song, J., Bradeen, J.M., Naess, S.K. *et al.* (2003) Gene RB cloned from *Solanum bulbocastanum* confers broad spectrum resistance to potato late blight. *Proceedings of the National Academy of Science United States*, **100**, 9128-9133.
Spielman, L.J., Drenth, A., Davidse, L.C. *et al.* (1991) A second world-wide migration and population displacement of *Phytophthora infestans*? *Plant Pathology*, **40**, 422-430.
Spijkerboer, H.P., Beniers, J.E., Jaspers, D. *et al.* (2002) Ability of the Gaussian plume model to predict and describe spore dispersal over a potato crop. *Ecological Modelling*, **155**, 1-18.
Stevens, N.E. (1933) The dark ages in plant pathology in America: 1830-1870. *Journal of the Washington Academy of Sciences*, **23**, 435-446.
Stevenson, W.R., Curwen, D., Kelling, K.A. *et al.* (1994) Wisconsin's IPM program for potato: the developmental process. *HortTech*, **4**, 90-95.
Strömberg, A., Persson, L. and Wikström, M. (1999) Infection of potatoes by oospores of *Phytophthora infestans* in soil. *Plant Disease*, **83**, 876.
Suassuna, N.D., Maffia, L.A. and Mizubuti, E.S.G. (2004) Aggressiveness and host specificity of Brazilian isolates of *Phytophthora infestans*. *Plant Pathology*, **53**, 405-413.
Sunseri, M.A., Johnson, D.A. and Dasgupta, N. (2002) Survival of detached sporangia of *Phytophthora infestans* exposed to ambient, relatively dry atmospheric conditions. *American Journal of Potato Research*, **79**, 443-450.
Tooley, P.W., Fry, W.E. and Villareal-González, M.J. (1985) Isozyme characterization of sexual and asexual *Phytophthora infestans* populations. *Journal of Heredity*, **76**, 431-435.
Tooley, P.W., Sweigard, J.A. and Fry, W.E. (1986) Fitness and virulence of *Phytophthora infestans* isolates from sexual and asexual populations. *Phytopathology*, **76**, 1209-1212.
Tooley, P.W., Therrien, C.D. and Ritch, D.L. (1989) Mating type, race composition, nuclear DNA content, and isozyme analysis of Peruvian isolates of *Phytophthora infestans*. *Phytopathology*, **79**, 478-481.
Turkensteen, L.J., Flier, W.G., Wanningen, R. *et al.* (2000) Production, survival and infectivity of oospores of *Phytophthora infestans*. *Plant Pathology*, **49**, 688-696.
Umaerus, V., Umaerus, M., Erjefalt, L. *et al.* (1983), Control of *Phytophthora* by host resistance: Problems and Progress, in *Phytophthora: Its biology, taxonomy, ecology, and pathology* (eds D.C.

Erwin, S. Bartnicki-Garcia, and P.H. Tsao), American Phytopathological Society, St. Paul, pp. 315-326.

Van der Plank, J.E. (1968) The progress of disease in relation to resistance in the host.

Van Der Zaag, D.E. (1956) Overwintering en epidemiologie van *Phytophthora infestans*, tevens enige nieuwe bestrijdingsmogelijkheden. *Tijdschrift Over Plantenziekten*, **62**, 89-156.

Van Everdingen, E. (1926) Het verband tusschen de weersgesteldheid en de aardappelziekte (*Phytophthora infestans*). *Tijdschrift Over Plantenziekten*, **32**, 129-140.

Van Oijen, M. (1989). On the use of mathematical models from human epidemiology in breeding for resistance to polycyclic fungal leaf diseases of crops, in *Joint Conference of the EAPR Breeding Section and the EUCARPIA Potato Section* (eds K.M. Louwes, H.A.J.M. Toussaint, and L.M.W. Dellaert), Wageningen, The Netherlands: Pudoc Wageningen, pp. 26-37.

Van Oijen, M. (1992) Selection and use of a mathematical model to evaluate components of resistance to *Phytophthora infestans* in potato. *Netherlands Journal of Plant Pathology*, **98**, 192-202.

Van Oijen, M. (1995), Simulation models of potato late blight, in *Potato ecology and modelling of crops under conditions limiting growth* (eds A.J. Haverkort and D.K.L. MacKerron), Kluwer Academic Publishers, Dordrecht, pp. 237-250.

van West, P., Appiah, A.A. and Gow, N.A.R. (2003) Advances in research on oomycete root pathogens. *Physiological and Molecular Plant Pathology*, **62**, 99-113.

Vega-Sánchez, M.E., Erselius, L.J., Rodriguez, A.M. *et al.* (2000) Host adaptation to potato and tomato within the US-1 clonal lineage of *Phytophthora infestans* in Uganda and Kenya. *Plant Pathology*, **49**, 531-539.

Waggoner, P.E. (1986), Progress curves of foliar diseases: their interpretation and use, in *Plant disease epidemiology* (eds K.J. Leonard and W.E. Fry), MacMillan Publishers, New York, pp. 3-37.

Waggoner, P.E. and Horsfall, J.G. (1969) EPIDEM. A simulator of plant disease written for a computer. *Bulletin of the Connecticut Agricultural Experiment Station, New Haven*, **No. 698**.

Wallin, J.R. (1962) Summary of recent progress in predicting late blight epidemics in United States and Canada. *American Potato Journal*, **39**, 306-312.

Warren, R.C. and Colhoun, J. (1975) Viability of sporangia of *Phytophthora infestans* in relation to drying. *Transactions of the British Mycological Society*, **64**, 73-78.

Wastie, R.L. (1991), Breeding for resistance, in *Phytophthora infestans: the cause of late blight of potato. Advances in Plant Pathology* (eds D.S. Ingram and P.H. Williams), Academic Press, San Diego, pp. 193-224.

Woodham-Smith, C. (1962) *The great hunger*, Harper & Row, New York, 510 pp.

Wu, G., Shortt, B.J., Lawrence, E.B. *et al.* (1995) Disease resistance conferred by expression of a gene encoding H_2O_2 - generating glucose oxidase in transgenic potato plants. *The Plant Cell*, **7**, 1357-1368.

Yan, Z., Reddy, M.S., Ryu, C. *et al.* (2002) Induced systemic protection against tomato late blight elicited by plant growth-promoting rhizobacteria. *Phytopathology*, **92**, 1329-1333.

Zan, K. (1962) Activity of *Phytophthora infestans* in soil in relation to tuber infection. *Transactions of the British Mycological Society*, **45**, 205-221.

Zwankhuizen, M.J., Govers, F. and Zadoks, J.C. (1998) Development of potato late blight epidemics: Disease foci, disease gradients, and infection sources. *Phytopathology*, **88**, 754-763.

CHAPTER 18

APPLE SCAB: ROLE OF ENVIRONMENT IN PATHOGEN AND EPIDEMIC DEVELOPMENT

A.L. JONES AND G.W. SUNDIN

18.1 INTRODUCTION

Apple scab (apple black spot) continues to be a serious disease in virtually all apple-producing regions worldwide despite the steady flow of research reports on its biology, epidemiology and management for more than a century; for reviews see MacHardy (1996) and Aylor (1998). Largely due to these efforts, the disease has been satisfactorily managed but lasting solutions have been elusive. Periodic outbreaks of problems with fungicide resistance (Köller, 1994) and the breakdown of resistance in scab-resistant cultivars (Parisi and Lespinasse, 1996) are constant reminders that scientists are only ahead of the pathogen temporarily.

The early research on the epidemiology of apple scab provided a strong foundation for future studies. The use of controlled environmental chambers for defining the environmental factors that favour infection was pioneered by Keitt and Jones (1926). Later, Mills (1944) used data from these studies to develop one of the first practical disease predictive systems. The Mills system established a rational basis for developing management strategies for scab. Much of today's scientific literature on apple scab represents the efforts of many scientists to re-evaluate, embellish and update the research contributions of these early scientists.

The objective of this chapter is to provide a modern assessment of the epidemiology of apple scab and is organized as a running narrative of research on the Mills system, the foundation for many of the epidemiological studies on apple scab in modern times.

18.2 AETIOLOGY OF APPLE SCAB

In most apple-producing regions *Venturia inaequalis* (Cooke) G. Wint., the cause of apple scab, has one sexual infection cycle per year and a series of asexual infection cycles. In these regions *V. inaequalis* may also produce conidial primary inoculum in lesions on wood, shoots, and outer and inner surfaces of bud scales (Becker *et al.*, 1992; Holb *et al.*, 2004; Moosherr and Kennel, 1995; Stensvand *et al.*, 1996). The sexual cycle is highly significant in causing scab epidemics because the number of ascospores potentially available from apple leaf litter is considerably greater than the number of conidia potentially available from diseased wood, shoots and buds, and

B. M. Cooke, D. Gareth Jones and B. Kaye (eds.), The Epidemiology of Plant Diseases, 2nd edition, 473–489.

the conidia often exhibit low viability (Holb *et al.*, 2004). An exception to the sexual infection paradigm occurs in areas of Israel where winter temperatures are unfavourable for sexual reproduction. Here, *V. inaequalis* has asexual infection cycles only; spring epidemics arise from conidia produced in diseased leaves that persist on shoot tips through the mild dormant period (Boehm *et al.*, 2003).

The sexual infection cycle arises from ascospores released from pseudothecia. The distinctive ascospores of *V. inaequalis* are recognized based on their colour, morphology and size. Mature ascospores (11–15 x 5–7 μm) are two-celled, yellowish-green to tan, with smooth walls. The upper cell is shorter and wider than the lower cell. The ease of recognizing these spores has been very useful and important to studies on the epidemiology of apple scab. Mature conidia (12–22 x 6−9 μm) are usually one-celled, yellowish-olive, and ovate to lanceolate.

Venturia inaequalis overwinters in infected fallen leaves on the orchard floor. In late autumn and early spring, pseudothecia are initiated in these dead leaves. Maturation of pseudothecia in late winter and early spring is favoured by alternating periods of wetness and dryness. Normally, some pseudothecia produce mature ascospores, beginning around the green-tip stage of flower bud development (Gilpatrick and Szkolnik, 1978). Maturation and discharge of ascospores usually lasts for 9 to 12 weeks. Peak ascospore maturation and discharge normally occurs around the bloom period (Gilpatrick and Szkolnik, 1978; Villalta *et al.*, 2002). When the leaf litter becomes wet, ascospores are forcibly ejected about 3 mm (range 0.1−8.1 mm) from the leaf surface (Aylor and Anagnostakis, 1991). Aerial concentrations of ascospores are highest close to the ground and decrease rapidly with height above the ground (Aylor, 1995; Aylor and Qiu, 1996).

Germination begins, provided a film of moisture is present, as soon as ascospores and conidia land on the host. About 9–17 days are required from inoculation to the appearance of the olive-green, velvety scab lesions (Mills and LaPlante, 1951; Tomerlin and Jones, 1983). The average temperature after penetration governs the length of the latent period, provided humidity is not limiting (Tomerlin and Jones, 1983).

Conidia within new scab lesions are disseminated by splashing rain and by air. Interestingly, dry removal of conidia from lesions is common, and the frequency of conidia in orchard air on days without rain is often diurnal; more conidia are trapped in a volumetric spore trap in the afternoon and evening than in the morning (Sutton *et al.*, 1976). Conidial germination and infection occurs under similar conditions as those for ascospores (Keitt and Jones, 1926; Moore, 1964; Schwabe, 1980). Secondary infection on fruit can occur in autumn but visible symptoms may not develop until after the fruit has been stored for several months. Due to the loss of ontogenetic resistance, the pathogen can also build up on apple leaves, particularly on the under surface, late in the season (Kollar, 1996; Li and Xu, 2002; Sutton *et al.*, 1976). Infections on old leaves may either develop as small faint lesions with diffuse colonies during the autumn or as symptomless infections consisting of mycelium and conidia when examined microscopically (Li and Xu, 2002). Such lesions may give rise to pseudothecia for the next growing season.

18.3 PREDICTING APPLE SCAB RISK BASED ON THE PHYSICAL ENVIRONMENT

The Mills system for predicting apple scab infection periods is used worldwide in various forms to provide a rational basis for implementing scab control strategies. Developed in New York State, the Mills system was important initially for scheduling applications of sulphur during rain (Mills, 1944; Mills and LaPlante, 1951). Sulphur, the main fungicide available in the 1940s, was effective only when applied immediately ahead of infection events. Unnecessary applications of sulphur dusts and sprays could be avoided by delaying each treatment to within a few hours before infection was anticipated based on Mills. The Mills system was also used to establish the need and timing for emergency post-infection application of lime sulphur. High rates of lime sulphur would control scab when applied within a few hours after a predicted infection event but the treatment was used sparingly to avoid possible reductions in fruiting.

Today, the Mills system is the foundation for most integrated pest management (IPM), integrated fruit production (IFP) and stewardship programmes for apple disease management (Cross, 1994; Jones, 1995; Sutton, 1996). It is used for improving the timing of fungicides with post-infection activity and for establishing when protective spray programmes may need to be backed up with post-infection sprays. Depending on the degree of their kick-back or reach-back activity, post-infection fungicides are applied within 24 to 96 hours after the onset of a predicted infection period. The evolution of this scab prediction system since its introduction in the 1940s provides an excellent case history for students interested in how forecast systems are introduced and then maintained as production systems change. The principles of disease forecasting are discussed in Chapter 9.

Mills's publication of curves (Fig. 18.1) and later a table relating infection by the scab fungus to the number of hours of wetting at various temperatures was a seminal event in apple scab epidemiology. Initial inoculum was assumed to be abundant – a realistic assumption considering that sulphur is not especially efficacious and was the predominant fungicide used at that time. Then, based on correlations with leaf wetness and temperature, predictions were made for slight, moderate, or severe infection based on the duration of continuous wetness at temperatures between 5.5 and 25.5°C. Research in Wisconsin by Keitt and Jones (1926) provided Mills with the baseline for his curves. This baseline was the minimum number of hours of wetting that was necessary for infection of leaves on inoculated apple trees placed in temperature-controlled dew chambers. Because of his uncertainty about the minimum conditions for infection in the field, Mills made the curve for light infection from ascospores much thicker than the curves for moderate and severe infection (Mills, 1944). The ambiguous way in which Mills described light infection has been overlooked in many reproductions of Mills's original curves. The thick line was replaced by the word 'approximate' in the heading for the Mills table. Research aimed at improving the accuracy of this infection curve has been intensive and ongoing in many countries since the 1940s.

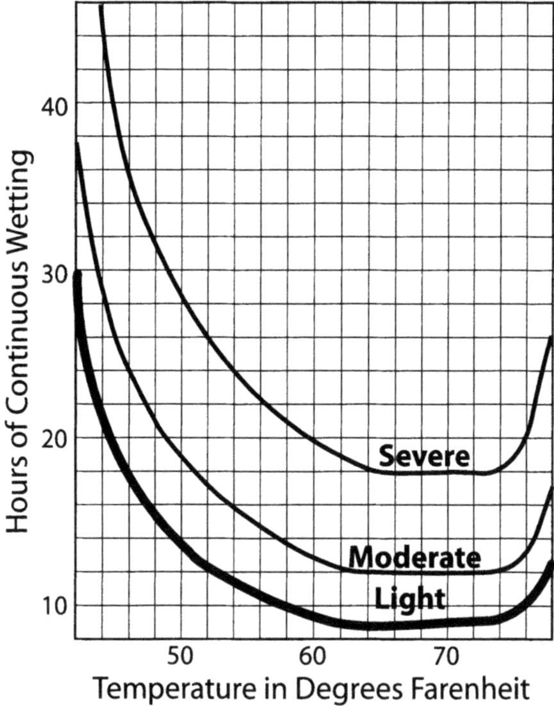

Figure 18.1. The relation of wetness duration, temperature and degree of apple scab infection as originally published by Mills (1944). Scab warning systems based on these criteria have survived several decades.

Mills made two assumptions in drawing up his minimum curves for primary apple scab infection. These assumptions were well known to those who helped him validate his curves in the 1940s but were never presented in refereed journal articles. First, he used the onset of rain as the trigger mechanism for initiating the models because he desired a starting point that could be easily identified (available data at the time indicated that the onset of ascospore discharge, unlike the release of conidia, was often delayed compared to the onset of rain). Second, he assumed it would take at least three hours longer in the field than in growth chambers for infection from ascospores. In the growth chamber studies, ascospores were applied directly to the plants but in the field ascospore release is delayed until the spore discharge mechanisms of the pseudothecia are activated. Mills also reasoned that it would take more time for airborne ascospores than waterborne conidia to reach susceptible tissue in sufficient quantities to lead to a significant level of infection. It was the difference in the time of inoculation, not possible differences in infection rate after the spores reached the host, that concerned Mills and led him to develop two minimum infection curves, one for ascospores and one for conidia.

18.3.1 Using Mills's criteria for scab prediction in practice

The Mills table uses the number of hours of continuous wetness from the beginning of rain and the average temperature during the wet period for predicting scab infection. In practice wet periods are not always continuous resulting in split wetting periods. Several procedures have been suggested for how to deal with dry periods between rains in the Mills system. Mills initially suggested that wetting periods separated by no more than four hours be added together but later he changed it to twelve hours (half a day) or more with sunny weather. Studies in South Africa with ascospores indicated that wetting periods separated by dry intervals of sixteen hours or less should be combined (Schwabe, 1980). Work in The Netherlands indicated that the twelve hour dry interval proposed by Mills should be shortened to eight hours (Rooseje, 1963), an interval used in some warning systems (Jones *et al.*, 1980). With conidial inoculum, several researchers (Schwabe, 1980; Becker and Burr, 1994) have suggested that wetting periods should be added together when dry intervals are shorter than 32 to 48 hours. Becker and Burr (1944) proposed a series of regression equations for determining the proportion of viable conidia remaining at the start of secondary wetting periods. Of these various suggestions, the eight-hour rule has worked reasonably well when predicting infection from ascospores.

Mills's table (Mills and LaPlante, 1951) indicated that predicting infection from conidia would require one-third less wetting time than for ascospores; this has prompted numerous studies on the time required for infection by ascospores and conidia. Results from greenhouse and moist chamber studies conducted in the United States (Keitt and Jones, 1926; Stensvand *et al.*, 1997), The Netherlands (Roosje, 1963), Belgium (Sys and Soenen, 1970), UK (Moore, 1964) and South Africa (Schwabe, 1980) all indicated that infection times for conidia are similar to or slightly longer than those for ascospores when spore suspensions are applied directly onto the host. Although these results suggest that the infection times used for ascospores and conidia in the Mills system should be the same (MacHardy and Gadoury, 1989; Stensvand *et al.*, 1997), none of the studies address the question of whether ascospores require more time than conidia to reach host tissue. Again, it is generally forgotten that Mills had reasoned, based on the spore release data available at that time, that it would take about three hours for a significant number of ascospores to be deposited on susceptible apple tissues while the deposition of conidia would be nearly instantaneous with the onset of rainfall.

The significance of night-time release of ascospores and how to utilize this information in advisory services and predictive models based on Mills has been debated since Brook (1969) first demonstrated that the ascospore release response for *V. inaequalis* was stimulated by far red light and suppressed by darkness. When day-time and night-time rain events were compared, ascospore release was often delayed until sunrise during night-time rain events while release was initiated soon after the onset of day-time rain events. The cumulative number of ascospores trapped at night in orchard air was found to represent a small percentage of the total number trapped each season. Spore trapping studies conducted in several countries indicated that 80–95% of the ascospores were trapped during the daylight hours (Brook, 1966; MacHardy and Gadoury, 1986; Villalta *et al.*, 2002). However, a delay in ascospore release until sunrise does not happen every time. Substantial numbers of ascospores have been trapped on occasion during natural night-time rain events (Aylor and Sutton, 1992; Rossi *et al.*, 2001; Warner and Braun, 1992), and ascospore release during darkness has been confirmed under certain conditions in laboratory tests (Gadoury *et al.*, 1998). Darkness release of ascospores is observed in both the field and laboratory when pseudothecia contain high numbers of mature spores ready for discharge or near the tail end of the release season as pseudothecia become senescent (Gadoury *et al.*, 1998; Rossi *et al.*, 2001).

As night-time ascospore release events cannot always be anticipated; disease outbreaks may occur occasionally when advisory services ignore night-time rain events until sunrise. For example, advisory services in Australia have ignored night-time wetting hours when wetness events are initiated at night for nearly 40 years (Wicks, 1970; Penrose *et al.*, 1985); however, current scab researchers in Australia are suggesting that this practice be modified for orchards where the density of overwintering inoculum is high or unknown (Villalta *et al.*, 2002). The criteria were changed in part because, as in other parts of the world, release of ascospores of *V. inaequalis* was detected during darkness in some Australian orchards (Villalta *et al.*, 2002). Since ascospore densities are rarely known and perhaps conidia are present, it is advisable to calculate infection periods from the beginning of a rain event even when the wetness begins at night.

18.3.2 Modifications and revisions of the Mills criteria

Mills made no further modifications to his model following its introduction in table form, nor was the table modified by other researchers during his lifetime. In the early 1980s, Jones *et al.* (1980) proposed that the number of hours of wetting required for infection at temperatures below 8.3°C be reduced from those used by Mills. The modified Mills table, which was part of an electronic disease predictor (Jones *et al.*, 1980), was validated in multi-year field studies in Michigan, New York and Ohio (Ellis *et al.*, 1984; Jones *et al.*, 1984) and Poland (Nowack and Cimanowski, 1985). However, researchers in Australia concluded that the modified Mills table predicted too many infection periods because no wetting period adjustments were made when rains began at night (Penrose *et al.*, 1985).

MacHardy and Gadoury (1989) recommended a revision of the Mills table after reviewing published reports on its performance in apple regions worldwide. They proposed that infection times given in Mills's curve for infection from ascospores be reduced by three hours at most temperatures (referred to as Mills/a-3 or as revised Mills). Infection periods for ascospores were computed from sunrise for rain periods beginning the previous night and those for conidia from the beginning of all rain periods. Later, infection times at temperatures at or below 8°C were adjusted based on minimum infection times recorded in growth chamber studies at 2° to 8°C (Stensvand *et al.*, 1997). In the finalized revised table the minimum hours of wetness required for infection from ascospores at 1° to 26°C were much shorter than those in the original Mills table while the hours of wetting needed for infection from conidia were longer than in Mills and slightly longer than the minimum for ascospores under greenhouse conditions (Stensvand *et al.*, 1997). It is likely the Mills system has been used for decades without problems, except for temperatures below 8°C, because minimum infection times in the field are longer than minimum infection times in growth chamber experiments. It is probable that when false positives and false negatives are considered, there will be little if any improvement in the identification of scab infection periods by substituting the revised table for the modified Mills table; however, the two systems must be compared rigorously over several growing seasons to establish if revised Mills is a significant improvement.

The revised Mills table proposed by MacHardy and Gadoury (1989) and by Stensvand *et al.* (1997) made no distinction between light, moderate and heavy infection periods. Mills noted that the severity of an infection period would increase as the hours of wetting increased because potentially more inoculum could reach and infect the host. Data from volumetric spore trapping studies in the field indicate that some daytime discharge events are prolonged requiring six to eight hours of rain before 50% of the ascospores are trapped while other discharge events are concentrated requiring just two hours of rain before 50% of the ascospores are trapped (Rossi *et al.*, 2003). As the duration of a wetness period increases beyond minimum infection times, more ascospores, particularly delayed-release ascospores, can infect thereby increasing disease severity. Mills's severity level system is often criticised because a light Mills period may result in more disease than a moderate or severe Mills period. However, inoculum dose often accounts for the difference in disease severity between two apparently similar infection events. It is often forgotten that the severity of a particular infection period is determined not only by favourability of the physical environment as judged by Mills, but also by the potential ascospore dose (PAD; see Gadoury and MacHardy, 1986) which is rarely known. A prediction for light infection means that the minimum conditions for infection have been met and disease will develop provided inoculum is present (inoculum was assumed to be present by Mills) and no control is attempted. Severity predictions are important because, based on circumstances beyond Mills, a grower may respond to a prediction for light infection differently than to a prediction for moderate or heavy infection.

18.3.3 Other scab predictive systems based on physical environment

Trapman (1994) and Xu *et al.* (1995) have developed simulation models for detecting scab infection periods and have incorporated them into personal computer-based warning systems called RIMpro and Ventem™, respectively. Research on these systems was stimulated by work on microprocessor-based instruments for disease prediction in the early 1980s (Jones *et al.*, 1980). Unlike scab warning systems based on the Mills criteria, simulation models can deal with the dynamic nature of weather and give growers real-time predictions of disease. Rainfall, surface wetness, temperature and relative humidity data are used to determine, for each wetting period, the proportion of spores that successfully infect the leaves, referred to as relative infection measure (RIM) by Trapman (1994) and as infection efficiency (IE) by Xu *et al.* (1995). Scab control obtained in 1992 with three curative sprays following periods with high RIM values was similar to control obtained with nine curative sprays following Mills infection periods (Trapman, 1994), while Ventem™ identified infection periods more accurately than the Mills system in each of three years (Xu *et al.*, 1995).

In the same way that many scientists have questioned the accuracy of the Mills criteria for predicting scab infection periods (MacHardy and Gadoury, 1989; Trapman, 1994; Xu *et al.*, 1995), scientists are likely to question the accuracy of these new simulation models. The problems of measuring leaf wetness are well known (Huber and Gillespie, 1992), yet scientists rarely questioned the accuracy of wetness duration measurements. Even study to study variation in the placement of wetness sensors, from within trees to between tree rows to outside the orchard, can make a big difference in the duration of wetness readings. In most of the cases reviewed by MacHardy and Gadoury (1989), it was not possible to determine whether the greater sensitivity of the warning system relative to Mills's table is due to increased sensitivity of weather recording instruments or to a modification of the Mills criteria. Consistently underestimating wetting periods would result in the conclusion that infection actually occurs in fewer hours than Mills's table indicates. Jones *et al.* (1980) and Xu *et al.* (1995) both incorporated provisions in their models to compensate for possible errors in determining the length of the wetting period. Until wetting periods can be evaluated consistently from region to region, models developed in one region may or may not work in another, depending on the amount of error associated with wetting period measurements.

18.3.4 Modifying Mills's criteria for changing fruit susceptibility

In the Mills system, the relationship between leaf wetness, temperature and infection of fruit and of young apple leaves is considered to be identical. However, as apple fruit mature, infection periods that result in infection on young leaves often fail to result in infection on fruit (Tomerlin and Jones, 1983). The hours of continuous wetting needed for fruit infection increases exponentially in the 1–23 week period following full bloom (Schwabe *et al.*, 1984). A model was developed for estimating the percentage of fruit infection based on temperature, wetness duration and fruit age. The percentage of fruit infection (P) can be estimated from the equation:

$$P = (-6.7240 + 0.3407W + 0.7265T - 0.0043W^2 - 0.0228T^2 + 0.0070WT)^2 \quad (18.1)$$

where T is temperature (°C) and W is an estimate of the hours of wetting needed for infection at that stage of fruit development. Estimates of W are from the following equation:

$$W = W/\{(-6.482 + 119.4X^{0.5})/120\} \quad (18.2)$$

where X is the number of weeks past full bloom, *120* is a scalar that adjusts the duration of wetness according to the age of the fruit, and W is the hours of wetting observed during the wet period. Since highly susceptible leaves are present on shoots until leaf emergence has finished, adjustments for changing fruit susceptibility with age are usually not made until shoot growth stops. The model is particularly valuable for evaluating the importance of long continuous wet periods late in the season.

18.3.5 Latent period predictions

The Mills table can also be used to predict the approximate time in days for development of scab lesions on young leaves. This information is a valuable guide to advisers and growers on when to inspect orchards for lesions following predicted infection periods. Lesions with conidia may be visible 9 to 17 days after the onset of a wet period suitable for infection (Mills and LaPlante, 1951). The time required for visible lesions is a function of temperature and relative humidity. The mean temperature for the first five days of the infection period is used to predict days from infection to visible lesions. High relative humidities favour sporulation, low humidities delay lesion development and can extend the latent period for several days (Tomerlin and Jones, 1983).

18.4 PREDICTING APPLE SCAB RISK BASED ON PRIMARY INOCULUM LEVELS

After the Mills system, perhaps the greatest advancement in timing of fungicide-based strategies for controlling apple scab has been the development of both indirect and direct methods for estimating the potential release of ascospores during periods of environmental risk. Fungicide applications for scab control are initiated at bud break (green tip) in the spring to coincide with the beginning of the ascospore release season. Both application frequency and the quantity of fungicide per application are often relaxed at the end of the ascospore release season.

Mills's assumption that primary inoculum is abundant during all environmental risk periods is often incorrect today because of the high level of scab control achieved with modern fungicides. Primary inoculum may not be important at times due to year-to-year variation in when the release season begins or ends, suppression of ascospore maturity, often drought related, during the release season, and variation among orchards in potential ascospore dose. Knowledge of inoculum risk is as important as knowledge of environmental risk in designing scab control programmes aimed at reducing fungicide use on apples.

18.4.1 Estimating ascospore maturity and discharge

Visual assessments of pseudothecia are useful for estimating the beginning and end of the ascospore release season and have been used for nearly sixty years in the United States to time the initiation and the end of control programmes for primary apple scab (Gilpatrick and Szkolnik, 1978; Gadoury *et al.*, 1992). During the ascospore release season, visual assessments are useful for estimating the potential for ascospore release in the event of rain. Although the green tip stage of tree development is often used as a biofix for the beginning of the ascospore release season, maturity assessments can be used to establish in a particular year whether ascospores are maturing earlier or later than green tip.

Microscopic examination of the contents of pseudothecia crushed on glass slides is a widely used method for assessing ascospore maturity (Szkolnik, 1969; Gadoury and MacHardy, 1982a) but such examinations require considerable skill in interpretation (Gadoury *et al.*, 1992). The maturity status of each pseudothecium is judged, based on classification of ascospore maturity in each ascus into five categories. Data from 20 to 25 pseudothecia taken from several freshly collected leaves in each orchard are summarized and provide an estimate of the percentage of mature ascospores. Pseudothecial maturity can also be determined by sectioning fixed specimens (James and Sutton, 1982a) but this method is very time consuming.

Laboratory methods for estimating ascospore maturity are often supplemented with ascospore discharge tests. Several methods have been developed for assessing the potential release of ascospores of *V. inaequalis* from infected leaves or leaf discs freshly collected from the field and include spore release towers and tunnels (Hirst and Stedman, 1962; Brook, 1969; Gilpatrick *et al.*, 1972), aerated or agitated water in bottles or flasks (Hutton and Burchill, 1965; Kollar, 1998), or simply mounting moist leaf tissue on the top inside surface of Petri dishes (Szkolnik, 1969). Ascospore discharge tests are easy to conduct and they provide a more accurate estimate of release potential than pseudothecial assessments. Approximately 10 to 15% of the asci need to be rated as mature before an ascospore release season is initiated (Gadoury *et al.*, 1992) and ascospore discharge tests with spore release towers may overestimate the number of spores likely to be airborne in an orchard late in the season (Aylor, 1996). Also, cumulative ascospore release when measured by microscopic examination of crushed pseudothecia and discharge tests lags behind cumulative estimates made with volumetric spore traps in the field (Gadoury *et al.*, 2004).

The results from ascospore discharge tests indicate that ascospores can be discharged from leaves freshly collected over a 10 to 12-week period but the exact duration of the discharge period is governed by temperature (Gadoury and MacHardy, 1982b; James and Sutton, 1982b) and can be limited by lack of moisture (James and Sutton, 1982a). Peak discharge occurs between the pink and bloom stages of apple flower bud development, with the discharge pattern approximately normally distributed between green tip and four to six weeks after petal fall (Gilpatrick and Szkolnik, 1978; Villalta *et al.*, 2002). The accumulated number of ascospores discharged from leaves approximates a sigmoid curve. Ascospore discharge data, when combined with meteorological data, can provide usable estimates of aerial spore concentrations (Aylor and Qiu, 1996). By knowing the concentration of ascospores in air above an orchard and the number of lesions resulting from deposition of ascospores on susceptible tissue (Aylor and Kiyomoto, 1993), it will become possible to evaluate the need for fungicide applications based on inoculum and environmental risks combined.

18.4.2 Monitoring ascospore release in orchards

Once infection periods are predicted based on the physical environment, the number of ascospores released into the orchard air (dose) largely determines the number of spores subsequently deposited on leaves and fruit. Because of the nature of a particular wetting period or because of changes in ascospore maturation after sampling, actual release of ascospores in the orchard may not always reflect laboratory estimates. To obtain real-time estimates of total spore dose during each wetting period, some advisory programmes use spore samplers to measure the aerial concentration of ascospores in the orchard (Sutton and Jones, 1976).

Burkard volumentric spore traps and Rotorod samplers are instruments used to monitor and estimate airborne ascospores in orchard air during natural rain events (Aylor, 1993; Sutton and Jones, 1976). The Burkard trap, because of its high cost and complex design, is used primarily as a research tool. Rotorod samplers, because of their lower cost, efficiency and ease of operation, can be used to determine total ascospore dose during each wetting period. Due to the highly aggregated pattern of ascospore distribution across an orchard (Charest *et al.*, 2002), spore traps and samplers are normally placed in locations likely to yield high numbers of ascospores. Sometimes overwintering infected leaves are placed at the trapping site to increase the potential for airborne ascospores. Used in this way, spore trapping devices are useful for detecting discharge events to time fungicide applications. In Michigan, field advisers monitor airborne ascospores in orchards with Rotorod samplers and make information on ascospore release available to growers via self-answering telephones, fax networks, electronic mail and radio (Jones, 1995). The main advantage of using spore samplers is that the variable effects of weather and environment on release are reflected in the monitoring data.

18.4.3 Models of pseudothecial development and ascospore maturity

Initial efforts to predict ascospore maturity involved the use of degree-day accumulations starting from the date of 50% leaf fall for a New York model (Massie and Szkolnik, 1974) or from the time when asci began to develop for a North Carolina model (James and Sutton, 1982b). The New York model predicted mature ascospores much earlier in North Carolina than they were observed in nature (Sutton *et al.*, 1981), while predictions with the North Carolina model were in close agreement with observations in Italy on the maturation of ascospores (Mancini *et al.*, 1984). These studies were the first to show that the rate of ascospore maturation in nature was directly related to temperature.

Degree-day accumulations starting with the first appearance of mature ascospores in nature or with the green-tip stage of bud development are very useful for predicting ascospore maturity using daily temperatures. The maturity of ascospores of *V. inaequalis* can be estimated from the following model: $Y = 2.51 + 0.01\ X$, where Y is the probit of proportion of matured ascospores and X is accumulated degree-days (base 0°C) from the first appearance of mature spores (Gadoury and MacHardy, 1982b). Transformation of estimated probit values to percentage values can be accomplished using a graph (Gadoury and MacHardy, 1982b), a table of standard normal curve areas, or the standard normal cumulative distribution function (such as NORMSDIST(Z) in Microsoft's Excel). When using tables or a distribution function, the estimated probit value is reduced by a scaling factor of five. This model uses an easy-to-identify biofix or starting point and predictions are relatively accurate since degree-days are accumulated for a few weeks, compared with a few months in the New York and North Carolina models. However, moisture can limit ascospore maturation (James and Sutton, 1982a) and prolonged dry periods during the spring may temporarily interrupt maturation, thereby reducing the accuracy of this degree-day model (Rossi *et al.*, 1999; Schwabe *et al.*, 1989).

A method for estimating the risk of airborne ascospores in Italian apple orchards was developed by Giosuè *et al.* (2000). The model was designed to replace or augment visual assessments of ascospore maturity and discharge. Daily changes in pseudothecial development were estimated using a model developed by James and Sutton (1982b) and a modification by Mancini *et al.* (1984). Once pseudothecia were predicted to contain the first mature ascospores (stage st = 9.5) a second model was used to predict the number of ascospores expected to be discharged during each wetting period. The model used a cumulative airborne dose measured with volumetric traps in a previous season to quantify the expected discharge. The driving variables, hourly temperature and leaf wetness, were used to accumulate degrees above 0°C when leaves were wet. A Mills table was used to determine the risk of infection on rainy days with predicted ascospore discharge.

18.4.4 Assessing differences in inoculum density among orchards

Perhaps the research on inoculum density by MacHardy and co-workers has had the greatest effect on scab control strategies of any work on scab in the last 30 years. Prior to this research, scab programmes were initiated based on estimates for first ascospore release using leaves taken from locations with high levels of over-wintering inoculum, which meant that spray programmes were normally initiated no later than green tip. Gadoury and MacHardy (1986) developed an assessment method, based on the incidence and severity of apple scab just before leaf fall and the amount of leaf litter present at bud break, for establishing the potential inoculum density (PAD = potential ascospore density) in orchards. In orchards with a low PAD, it was found that the initiation of scab spray programmes in spring could be delayed for two to three weeks after budbreak compared to orchards with high levels of inoculum (Wilcox *et al.*, 1992; MacHardy *et al.*, 1993). Thus, in low inoculum orchards, the first scab sprays coincided with the tight-cluster to pink stage of flower-bud development rather than green tip. Although a model was developed for estimating potential ascospore dose in individual orchards (Gadoury and MacHardy, 1986), it is somewhat complicated and seldom used in practice. To reduce possible risk from unprotected infection periods before the first sprays were applied, Wilcox *et al.* (1992) proposed that a series of sprays with sterol inhibitor fungicides with post-infection control activity be used in orchards where early season sprays had been omitted. However, the selection of scab strains with reduced sensitivity to the sterol inhibitor groups of post-infection fungicides in several areas of the world will complicate and limit the potential use of this scab control strategy (Köller *et al.*, 1997; Schnabel and Jones, 2001).

Overwintering of scab in lesions on apple twigs should also be considered as part of the inoculum assessment process. The risk of overwintering conidial inoculum is greatest in orchards where scab was a problem the previous autumn (Becker *et al.*, 1992). Conidia of *V. inaequalis* are detected in the spring in overwintering scab lesions on shoots and on the interior and exterior of bud scales in some areas of Europe (McKay, 1938; Cook, 1974; Hills, 1975; Moosherr and Kennel, 1995) and the United States (Becker *et al.*, 1992). The viability of the overwintered conidia on the outer surface of twigs and bud scales is very low and therefore this inoculum source is unlikely to cause scab outbreaks (Holb *et al.*, 2004). However, the viability of conidia found on inner bud tissues is much higher and these conidia are the ones likely to cause early season scab outbreaks (Holb *et al.*, 2004). In orchards with overwintering conidial inoculum, the release of conidia would be likely to occur with the onset of rain. Because of the risk of conidial infection at night, the assumption in revised Mills that the release of ascospores from night-time rains begins at sun-rise, should be ignored.

18.4.5 Integrating inoculum and environmental risks into warnings

The Mills table provided extension agents in New York State with a tool for developing scab warnings based on environmental risks, and inoculum assessments were soon incorporated to reinforce these warnings. Extension agents, who were the integrators, worked closely with local meteorologists so that warnings were based on the best available weather information, and with local radio station and post-office personnel to ensure that the warnings were disseminated quickly. As scab warning systems proved successful in reducing problems with apple scab control, they were quickly adapted for use in other countries. Because of grower acceptance, there is still a demand for these systems, but they are harder to maintain because of declining local resources.

Scab warning systems continue to be improved and should enhance the competitiveness of apple growers in the future. Microprocessor technology and personal computers have stimulated the development of new models and instruments for scab prediction (Jones *et al.*, 1984; Trapman, 1994; Xu *et al.*, 1995). In Europe, as part of local IFP efforts, modern scab warning systems are being developed in several countries (Butt, 1994). There are indications that scab warning systems will continue to advance as new knowledge on apple scab and new technologies improve their reliability, affordability, convenience and relevance to current and future scab control techniques.

18.5 SUMMARY

Research on the epidemiology of apple scab during the last thirty years has been dominated by studies aimed at improving the Mills criteria for predicting apple scab infection. It is a tribute to Mills and to the robustness of his table that, after more than sixty years, his research is still being used in predictive models. Studies on inoculum densities in orchards and how inoculum densities vary with each rain through the primary season have been very helpful in understanding differences in scab severity among orchards.

There is, however, a need for more research in several areas. Probably the greatest need is for a standard reproducible method for measuring wetness periods. Direct or indirect methods are needed for monitoring the wetness duration of leaves and fruit and of overwintering leaves on the ground. Then, wetness-driven predictive models could be tested and used reliably in different apple regions. Such a system would be widely applicable because wetness periods govern the progress of many plant diseases. There is also a lack of understanding of the extent that dry weather or prolonged periods of low humidity in spring interrupt ascospore maturity and discharge. Current predictive models for estimating ascospore maturation are based on temperature, while in nature maturation is often limited by the lack of moisture. Also, the revised Mills criteria and other scab forecasting systems need to be validated in several countries using a standard protocol. Additionally, more information is needed on factors that affect the overwintering of inoculum, the relationship between inoculum density and the potential for infection and the relationship between ascospore discharge, deposition and infection. In order to make more use of information on inoculum density, a simple but suitable procedure is needed for pest managers to monitor inoculum levels.

ACKNOWLEDGEMENT

Appreciation is extended to Dr. Turner B. Sutton (Department of Plant Pathology, North Carolina State University, Raleigh) for review of the first edition of this manuscript.

REFERENCES

Aylor, D.E. (1993) Relative collection efficiency of Rotorod and Burkard spore samplers for airborne *Venturia inaequalis* ascospores. *Phytopathology,* **83**, 1116–1119.

Aylor, D.E. (1995) Vertical variation of aerial concentration of *Venturia inaequalis* ascospores in an apple orchard. *Phytopathology,* **85**, 175–181.

Aylor, D.E. (1996) Comparison of the seasonal pattern of airborne *Venturia inaequalis* ascospores with the release potential of *V. inaequalis* ascospores from a source. *Phytopathology,* **86**, 769–776.

Aylor, D.E. (1998) The aerobiology of apple scab. *Plant Disease,* **82**, 838–849.

Aylor, D.E. and Anagnostakis, S.L. (1991) Active discharge distance of ascospores of *Venturia inaequalis*. *Phytopathology,* **81**, 548–551.

Aylor, D.E. and Kiyomoto, R.K. (1993) Relationship between aerial concentration of *Venturia inaequalis* ascospores and development of apple scab. *Agricultural and Forest Meteorology,* **63**, 133–147.

Aylor, D.E. and Qiu, J. (1996) Micrometeorological determination of release rate of *Venturia inaequalis* ascospores from a ground-level source during rain. *Agricultural and Forest Meteorology,* **81**, 157–178.

Aylor, D.E. and Sutton, T.B. (1992) Release of *Venturia inaequalis* ascospores during unsteady rain: relationship to spore transport and deposition. *Phytopathology,* **82**, 532–540.

Becker, C.M. and Burr, T.J. (1994) Discontinuous wetting and survival of conidia of *Venturia inaequalis* on apple leaves. *Phytopathology,* **84**, 372–378.

Becker, C.M., Burr, T.J. and Smith, C.A. (1992) Overwintering of conidia of *Venturia inaequalis* in apple buds in New York orchards. *Plant Disease,* **76**, 121–126.
Boehm, E.W.A., Freeman, S., Shabi, E, *et al.* (2003) Microsatellite primers indicate the presence of asexual populations of *Venturia inaequalis* in coastal Israeli apple orchards. *Phytoparasitica,* **31**, 236–251.
Brook, P.J. (1966) The ascospore production season of *Venturia inaequalis* (Cke.) Wint., the apple black spot fungus. *New Zealand Journal of Agricultural Research,* **9**, 1064–1069.
Brook, P.J. (1969) Effects of light, temperature, and moisture on release of ascospores of *Venturia inaequalis* (Cke.) Wint. *New Zealand Journal of Agricultural Research,* **12**, 214–227.
Butt, D.J. (ed) (1994) *Integrated Control of Pome Fruit Diseases,* Norwegian Journal of Agricultural Sciences, Supplement No. 17, Agricultural University of Norway, Advisory Service, Ås, Norway.
Charest, J., Dewdney, M., Paulitz, T., *et al.* (2002) Spatial distribution of *Venturia inaequalis* airborne ascospores in orchards. *Phytopathology,* **92**, 769–779.
Cook, R.T.A. (1974) Pustules on wood as sources of inoculum in apple scab and their response to chemical treatment. *Annals of Applied Biology,* **77**, 1–9.
Cross, J. (1994) Integrated fruit production – current status in Europe. *Norwegian Journal of Agricultural Sciences,* Supplement **17**, 13–17.
Ellis, M.A., Madden, L.V. and Wilson, LL. (1984) Evaluation of an electronic apple scab predictor for scheduling fungicides with curative activity. *Plant Disease,* **68**, 1055–1057.
Gadoury, D.M. and MacHardy, W.E. (1982a) Preparation and interpretation of squash mounts of pseudothecia of *Venturia inaequalis*. *Phytopathology,* **72**, 92–95.
Gadoury, D.M. and MacHardy, W.E. (1982b) A model to estimate the maturity of ascospores of *Venturia inaequalis*. *Phytopathology,* **72**, 901–904.
Gadoury, D.M. and MacHardy, W.E. (1986) Forecasting ascospore dose of *Venturia inaequalis* in commercial apple orchards. *Phytopathology,* **76**, 112–118.
Gadoury, D.M., Seem, R.C., MacHardy, W.E., *et al.* (2004) A comparison of methods used to estimate the maturity and release of ascospores of *Venturia inaequalis*. *Plant Disease,* **88**, 869–874.
Gadoury, D.M., Seem, R.C., Rosenberger, D.A., *et al.* (1992) Disparity between morphological maturity of ascospores and physiological maturity of asci in *Venturia inaequalis*. *Plant Disease,* **76**, 277–282.
Gadoury, D.M., Stensvand, A. and Seem, R.C. (1998) Influence of light, relative humidity, and maturity of populations on discharge of ascospores of *Venturia inaequalis*. *Phytopathology,* **88**, 902-909.
Gilpatrick, J.D. and Szkolnik, M. (1978) Maturation and discharge of ascospores of the apple scab fungus, in *Apple and Pear Scab Workshop Proceedings,* (eds A.L. Jones and J.D. Gilpatrick), New York State Agricultural Experiment Station Special Report 28, pp. 1–6.
Gilpatrick, J.D., Smith, C.A. and Blowers, D.R. (1972) A method of collecting ascospores of *Venturia inaequalis* for spore germination studies. *Plant Disease Reporter,* **56**, 39–42.
Giosuè, S., Rossi, V., Ponti, I., *et al.* (2000) Estimating the dynamics of airborne ascospores of *Venturia inaequalis*. *Bulletin OEPP/EPPO Bulletin,* **30**, 137–142.
Hills, S.A. (1975) The importance of wood scab caused by *Venturia inaequalis* (Cke.) Wint. as a source of infection for apple leaves in the spring. *Phytopathologische Zeitschrift,* **82**, 216–223.
Hirst, J.M. and Stedman, O.J. (1962) The epidemiology of apple scab (*Venturia inaequalis* (Cke.) Wint.) II. Observations on the liberation of ascospores. *Annals of Applied Biology,* **50**, 525–550.
Holb, I.J, Heijne, B., and Jeger, M.J. (2004) Overwintering of conidia of *Venturia inaequalis* and the contribution to early epidemics of apple scab. *Plant Disease,* **88**, 751–757.
Huber, L. and Gillespie, T.J. (1992) Modelling leaf wetness in relation to plant disease epidemiology. *Annual Review of Phytopathology,* **30**, 553–557.
Hutton, K.E. and Burchill, R.T. (1965) The effect of some fungicides and herbicides on ascospore production of *Venturia inaequalis* (Cke.)Wint. *Annals of Applied Biology,* **56**, 279–284.
James, J.R. and Sutton, T.B. (1982a) Environmental factors influencing pseudothecial development and ascospore maturation of *Venturia inaequalis*. *Phytopathology,* **72**, 1073–1080.
James, J.R. and Sutton, T.B. (1982b) A model for predicting ascospore maturation of *Venturia inaequalis*. *Phytopathology,* **72**, 1081–1085.
Jones, A.L. (1995) A stewardship program for using fungicides and antibiotics in apple disease management programs. *Plant Disease,* **79**, 427–432.
Jones, A.L., Lillevik, S.L., Fisher, P.D., *et al.* (1980) A microcomputer-based instrument to predict primary apple scab infection periods. *Plant Disease,* **64**, 69–72.

Jones, A.L., Fisher, P.D., Seem, R.C., *et al.* (1984) Development and commercialization of an in-field microcomputer delivery system for weather-driven predictive models. *Plant Disease,* **64**, 458–463.

Keitt, G.W. and Jones, L.K. (1926) Studies of the epidemiology and control of apple scab. *Wisconsin Agricultural Experiment Station Research Bulletin,* **73**, 1–104.

Kollar, A. (1998) A simple method to forecast the ascospore discharge of *Venturia inaequalis. Journal of Plant Diseases and Protection,* **105**, 489–495.

Kollar, A. (1996) Evidence for loss of ontogenetic resistance of apple leaves against *Venturia inaequalis. European Journal of Plant Pathology,* **102**, 773–778.

Köller, W. (1994) Chemical control of apple scab – status quo and future. *Norwegian Journal of Agricultural Sciences*, Supplement **17**, 149–170.

Köller, W., Wilcox, W.F., Barnard, J., *et al.* (1997) Detection and quantification of resistance of *Venturia inaequalis* populations to sterol demethylation inhibitors. *Phytopathology,* **87**, 184–190.

Li, B. and Xu, X. (2002) Infection and development of apple scab (*Venturia inaequalis*) on old leaves. *Journal of Phytopathology,* **150**, 687–691.

MacHardy, W.E. (1996) *Apple scab: Biology, Epidemiology and Management,* APS Press, St. Paul, MN.

MacHardy, W.E. and Gadoury, D.M. (1986) Patterns of ascospore discharge by *Venturia inaequalis. Phytopathology,* **76**, 985–990.

MacHardy, W.E. and Gadoury, D.M. (1989) A revision of Mills's criteria for predicting apple scab infection periods. *Phytopathology,* **79**, 304–310.

MacHardy, W.E. and Jeger, M.J. (1983) Integrating control measures for the management of primary apple scab, *Venturia inaequalis* (Cke.) Wint. *Protection Ecology,* **5**, 103–125.

MacHardy, W.E., Gadoury, D.M. and Rosenberger, D.A. (1993) Delaying the onset of fungicide programs for control of apple scab in orchards with low potential ascospore dose of *Venturia inaequalis. Plant Disease,* **77**, 372–275.

Mancini, G., Cotroneo, A. and Galliano, A. (1984) Evaluation of two models for predicting ascospore maturation of *Venturia inaequalis* in Piedmont (NW Italy). *Rivista di Patologia Vegetale, Series IV,* **20**, 25–37.

Massie, L.B. and Szkolnik, M. (1974) Prediction of ascospore maturation of *Venturia inaequalis* utilizing cumulative degree days. (Abstr.) *Proceedings of the American Phytopathological Society,* **1**, 140.

McKay, R. (1938) Conidia from infected budscales and adjacent wood as a main source of primary infection with the apple scab fungus *Venturia inaequalis* (Cooke) Wint. *Science Proceedings Royal Dublin Society n.s.,* **21**, 623–644.

Mills, W.D. (1944) Efficient use of sulphur dusts and sprays during rain to control apple scab. *Cornell University Extension Bulletin,* **630**, 1–4.

Mills, W.D. and LaPlante, A.A. (1951) Diseases and insects in the orchard. Cornell University Extension Bulletin, **711**, 21–27.

Moore, M.H. (1964) Glasshouse experiments on apple scab I. Foliage infection in relation to wet and dry periods. *Annals of Applied Biology,* **53**, 423–435.

Moosherr, W. and Kennel, W. (1995) Investigations on superficial apple scab on apple shoots. *Journal of Plant Diseases and Protection,* **102**, 171–183.

Nowacka, H. and Cimanowski, J. (1985) Evaluation of the methods determining critical periods in scab infection for its control. *Fruit Science Report,* **12**, 35–39.

Parisi, L. and Lespinasse, Y. (1996) Pathogenicity of *Venturia inaequalis* strains of race 6 on apple clones (*Malus* sp.). *Plant Disease,* **80**, 1179–1183.

Penrose, L.J., Heaton, J.B., Washington, W.S. and Wicks, T. (1985) Australian evaluation of an orchard based electronic device to predict primary apple scab infections. *Journal of Austrian Institute of Agricultural Science,* **51**, 74–78.

Roosje, G.S. (1963) Research on apple and pear scab in the Netherlands from 1938 until 1961. *Netherlands Journal of Plant Pathology,* **69**, 132–137.

Rossi, V., Giosuè, S., and Bugiani, R. (2003) Influence of air temperature on the release of ascospores of *Venturia inaequalis. Journal of Phytopathology,* **151**, 50–58.

Rossi, V., Ponti, I., Marinelli, M., *et al.* (2001) Environmental factors influencing the dispersal of *Venturia inaequalis* ascospores in the orchard air. *Journal of Phytopathology,* **149**, 11–19.

Rossi, V., Ponti, I., Marinelli, M., *et al.* (1999) Field evaluation of some models estimating the seasonal pattern of airborne ascospores of *Venturia inaequalis. Journal of Phytopathology,* **147**, 567–575.

Schnabel, G. and Jones, A.L. (2001) The 14α–demethylase (*CYP51A1*) gene is overexpressed in *Venturia inaequalis* strains resistant to myclobutanil. *Phytopathology,* **91**, 102–110.

Schwabe, W.F.S. (1980) Wetting and temperature requirements for apple leaf infections by *Venturia inaequalis* in South Africa. *Phytophylactica,* **12**, 69–80.

Schwabe, W.F.S., Jones, A.L. and Jonker, J.P. (1984) Changes in the susceptibility of developing apple fruit to *Venturia inaequalis. Phytopathology,* **74**, 118–121.

Schwabe, W.F.S., Jones, A.L. and van Blerk, E. (1989) Relation of degree-day accumulations to maturation of ascospores of *Venturia inaequalis* in South Africa. *Phytophylactica,* **21**, 13–16.

Stensvand, A., Amundsen, T., and Semb, L. (1996) Observations on wood scab caused by *Venturia inaequalis* and *V. pirina* in Norway. *Norwegian Journal of Agricultural Science,* **10**, 533–340.

Stensvand, A., Gadoury, D.M., Amundsen, T., *et al.* (1997) Ascospore release and infection of apple leaves by conidia and ascospores of *Venturia inaequalis* at low temperatures. *Phytopathology,* **87**, 1046–1053.

Sutton, T.B. (1996) Changing options for the control of deciduous fruit tree diseases. *Annual Review of Phytopathology,* **34**, 527–547.

Sutton, T.B. and Jones, A.L. (1976) Evaluation of four spore traps for monitoring discharge of ascospores of *Venturia inaequalis. Phytopathology,* **66**, 453–456.

Sutton, T.B., Jones, A.L. and Nelson, L.A. (1976) Factors affecting dispersal of conidia in the apple scab fungus. *Phytopathology,* **66**, 1313–1317.

Sutton, T.B., James, J.R. and Nardacci, J.F. (1981) Evaluation of a New York ascospore maturity model for *Venturia inaequalis* in North Carolina. *Phytopathology,* **71**, 1030–1032.

Sys, S. and Soenen, A. (1970) Investigations on the infection criteria of scab (*Venturia inaequalis* Cke. Wint.) on apples with respect to the table of Mills and LaPlante. *Agricultura,* **18**, 3–8.

Szkolnik, M. (1969) Maturation and discharge of ascospores of *Venturia inaequalis. Plant Disease Reporter,* **53**, 534–537.

Tomerlin, J.R. and Jones, A.L. (1983) Effect of temperature and relative humidity on the latent period of *Venturia inaequalis* in apple leaves. *Phytopathology,* **73**, 51–54.

Trapman, M. (1994) Development and evaluation of a simulation model for ascospore infections of *Venturia inaequalis. Norwegian Journal of Agricultural Sciences,* Supplement **17**, 55–67.

Villalta, O.N., Washington, W.S., Kita, N., *et al.* (2002) The use of weather and ascospore data for forecasting apple and pear scab in Victoria, Australia. *Australasian Plant Pathology*, **31**, 205–215.

Warner, J. and Braun, P.G. (1992) Discharge of *Venturia inaequalis* ascospores during daytime and night-time wetting periods in Ontario and Nova Scotia. *Canadian Journal of Plant Pathology,* **14**, 315–321.

Wicks, T.J. (1970) Our apple black spot warning service. *South Australia Department of Agriculture, Extension Bulletin,* **7.70**, 1–8.

Wilcox, W.E., Wasson, D.I. and Kovach, J. (1992) Development and evaluation of an integrated, reduced-spray program using sterol demethylation inhibitor fungicides for control of primary apple scab. *Plant Disease,* **76**, 669–677.

Xu, X.-M., Butt, D.J. and van Santen, G. (1995) A dynamic model simulating infection of apple leaves by *Venturia inaequalis. Plant Pathology,* **44**, 865–876.

CHAPTER 19

ONION DISEASES

R.B. MAUDE

19.1 INTRODUCTION: WORLD ONIONS

Onions (*Allium cepa*) are grown worldwide as a vegetable for human consumption. They are harvested as dry bulbs that are eaten raw or cooked. In certain countries e.g. Mexico, they may be thinned when young for use in salads. Also, they are grown at high densities as green (salad) onions and as small bulbs for pickling. Shallots (*A. cepa* var. *ascalonicum*) and multiplier or potato onion (*A. cepa* var. *aggregatum*) are small bulb strains of *Allium cepa*.

This chapter concentrates on dry bulb onions that are grown in temperate, tropical and sub-tropical regions of the world (Currah and Proctor, 1990; Rabinowitch and Brewster, 1990; Rabinowitch and Currah, 2002). Globally, 32 million tonnes were produced in 1994 (Anon., 1995; Jones, 1998). Production increased by 63% to 52 million tonnes by 2002 (Anon., 2003; Table 19.1). More than 65% of these crops were grown in Asia, with China (about 45%) and India (some 20%) the major producers of bulb onions (Anon., 2003; Table 19.1). Indian exports of bulb onions have accounted for 70% of its total foreign exchange earnings in fresh vegetables (Gupta *et al.*, 1991). There the crop has been grown on 0.29 million ha with an annual production of 3.14 million tonnes (Gupta *et al.*, 1994). Currently, world bulb onion yields average 17 tonnes per hectare with the more mechanised nations achieving higher yields (Anon., 2003; Table 19.1).

Table 19.1. Dry bulb onion statistics, from FAO Production Yearbook for 2002 (Anon., 2003)

Location	*Area harvested (1,000 ha)*	*Yield (1,000 tonnes)*	*Tonnes ha^{-1}*
World	2972	51914	17
Asia	2014	33453	17
Europe	420	7409	18
Africa	276	3917	14
South America	159	3281	21
* North America	98	3604	37
Australia	5	250	47

* *includes Central America*

B. M. Cooke, D. Gareth Jones and B. Kaye (eds.), The Epidemiology of Plant Diseases, 2nd edition, 491–520.

Onions are valued for their pungency and give flavour (Randle and Lancaster, 2002) and texture to foods. In addition, both onion and garlic (*Allium sativum*) have medicinal effects which may be important in the prevention and amelioration of coronary problems (Augusti, 1990). In a recent review on the effect of *Alliums* (onion and garlic) on health (Keusgen, 2002) it is suggested that consumption of *Allium* vegetables may lower the risk of gastrointestinal cancers. Also, it is suggested that daily intake of garlic (*A. sativum*) significantly reduces the incidence of diseases that are caused by atherosclerosis (Keusgen, 2002).

Physiologically, the onion is a long day plant and a certain length of day must occur before bulbing is induced. Bulbing in high northern latitudes, such as UK, is initiated when days are 14-16 h long. In low latitudes within the tropics and sub-tropics, daylength varies little and onions have adapted to respond to shorter days; these genotypes may initiate and form bulbs under daylengths of 13 h or less (Currah and Proctor, 1990). Onion cultivars may be classified by daylength sensitivity: very long day (VLD) (northern Europe), long day (LD) (as characterised by US 'long day' onions), intermediate day (ID) and short day (SD) types (Bosch *et al.*, 2002).

The onion bulb consists of a series of concentric fleshy leaf bases that are enclosed in thin dry protective wrapper leaves. Because of this anatomy, onion bulbs can be stored for considerable periods.

Brown-skinned onions of the 'Rijnsburger' type predominate in the UK and the Netherlands. The crop is drilled directly in early spring as seeds or planted as sets (small bulbils) to gain crop growth advantage. Two weeks before harvest, crops are sprayed with a sprout suppressant, the onion foliage is removed mechanically and the bulbs are direct harvested and transferred to bulk stores. The onions are surface dried in forced air at 425 m^3 h^{-1} t^{-1} at 30°C and this is followed by gentler drying with recirculated air until the onion necks (pseudostems) are fully dry (Anon., 1977). The bulk stores are maintained overwinter at ambient temperatures for up to five months; the storage life of the crop may be extended further by cold storage. Gubb and MacTavish (2002) discuss the various factors that affect the storage capability of different onion cultivars in different countries.

There are many different types of sub-tropical and tropical onions. Texas Grano and New Mexico Yellow Grano are typical cultivars grown in the southern states of North America (Corgan and Kedar, 1990). Similar cultivars are grown in Israel and hybrid cultivars are widely used. About 80% of crops are direct seeded but sets and transplants are also used. Generally the bulbs are harvested by hand.

In the tropics red-skinned types are common, for example, Red Creole and Bombay Red; brown- and yellow-skinned onions also are grown. Texas Grano is a cultivar common to many of the countries. Seeds are sown into flat beds and the seedlings are transplanted into the field. Sometimes plants are grown on a ridge system to enable irrigation during the growing period (Uzo and Currah, 1990). Bulbs are harvested by hand and may be sold immediately. Also, bulbs may be field-cured at high temperatures, after which some are stored in well-ventilated thatched houses (Uzo and Currah, 1990).

Bosch *et al.* (2002) list the principal commercial types of onions for Europe, Asia, USA, Canada, South America, Australasia, Africa and Russia. Similarly,

Currah (2002) gives lists of onion cultivars grown in Africa, tropical and sub-tropical Asia, the Americas and the Caribbean.

19.2 ONION DISEASES

Onions are attacked by viruses, phytoplasmas, bacteria and fungi (Schwartz and Mohan, 1995). The case histories given in this chapter relate to fungal pathogens of onions and illustrate epidemiological concepts. Johnston and Booth (1983) defined epidemiology as a 'Study of the factors affecting outbreaks of disease and spread of infection' and this definition is followed here. The epidemiological factors, given below, that influence infection are being used by scientists to produce predictive models for the onset of diseases, particularly those caused by foliar pathogens. Some of these models will be considered in this chapter.

Climatic factors, particularly temperature, rainfall, relative humidity (RH), duration of leaf wetness, crop phenology and nutritional status are selective of the range of fungal pathogens which may occur in temperate (cool dry and cool moist conditions) and tropical (hot dry and hot moist conditions) countries. Light influences the sporulation of fungi and the spore form (i.e. ascospores, pycnidiospores, and conidia) and dictates their method of distribution. Inoculum source (seed, soil, debris, etc.) are foci for disease transmission; disease progress depends on inoculum concentration and on the operation (or not) of some of the factors given above.

19.3 CASE HISTORIES: SEEDBORNE DISEASES

19.3.1 Constraints affecting seeds as infection sources

(a) Temperature implications

Temperatures that are optimal for the growth of pathogens of onion in culture (Table 19.2) may be broadly indicative of the climatic temperatures that will favour their development as disease agents in crops.

The microflora of onion seeds comprises a mixture of organisms - bacteria and fungi in particular; some of these are superficial and others are more deeply situated in the tissues of the seeds. The respective locations of organisms on or within seed tissues is a direct result of the available sources of inoculum and of the environmental conditions that obtained when the seed crop was growing and maturing.

The mycofloral composition of seeds is largely determined by the temperatures and moisture conditions (rainfall, duration of leaf wetness and RH) that occurred during seed crop production. Some of these fungi are pathogens of onions; their ability to transmit and cause disease depends on their interactions with other seedborne and soilborne organisms within the temperature profiles of their respective environments.

For example, *Botrytis aclada* (*Botrytis allii*) grows optimally in culture at 21.5°C and is inhibited by temperatures of 30°C (Fig. 19.1). Onion seeds produced under temperate conditions may be infected with *Botrytis aclada*. Infected seeds transmit the fungus to onions grown in cool moist climates, such as that of the UK, causing neck rot (Maude, 1983). In the UK, a drying temperature of 30°C, at which the fungus ceases to grow in culture (Fig. 19.1), is used to inhibit the growth of the fungus on the necks of freshly-topped harvested onions.

Table 19.2. Effects of temperature on growth of fungal pathogens in vitro

Disease	*Pathogen*	*Temp. range and optima for growth °C*	*Reference*
Cool climates			
Downy mildew	*Peronospora destructor*	1-28 (13)	Yarwood, 1943
Smut	*Urocystis cepulae*	12-28 (20)	Dow and Lacy, 1969
White rot	*Sclerotium cepivorum*	< 5-29 (20-24)	Walker, 1926
Neck rot	*Botrytis aclada*	5-25 (22-23)	Maude, 1990a
Leaf rot	*Botrytis squamosa*	9-31 (24)	Shoemaker and Lorbeer, 1977
Hot climates			
Purple blotch	*Alternaria porri*	10-30 (25)	Fahim, 1966
Pink root	*Pyrenochaeta terrestris*	13-32 (26)	Hansen, 1929
Basal plate rot	*F. oxysporum* f.sp. *cepae*	9-36 (24-27)	Abawi and Lorbeer, 1972
Black mould	*Aspergillus niger*	10-40 (31)	Maude, 1990a
Blue/green mould	*Aspergillus fumigatus*	20-45 (37.5)	Maude, 1990a

All in vitro *measurements were made on agar*

High temperature fungi, such as *Aspergillus niger*, grow optimally in culture at 32.5°C (Fig. 19.1); they rarely, if ever, occur on onion seed samples produced under temperate conditions but occur increasingly on seed samples produced under sub-tropical and tropical conditions (Table 19.3) achieving maximum incidence where seeds are produced under hot, dry tropical conditions with temperatures in seed production fields in excess of 30°C. *Aspergillus niger* is seed-transmitted (Hayden and Maude, 1992) and causes black mould of bulb onions produced and stored in hot climates, e.g., Texas (USA), Egypt, Sudan and India (Maude, 1990a). Thus, black mould is rarely a problem of the temperate onion crop; however, in 1982 there were serious outbreaks of this disease in onions in bulk stores in the UK, where bulb drying temperatures in excess of 30°C and relative humidities in excess of 80% RH for long periods elicited infection by *A. niger* (Maude and Burchill, 1988).

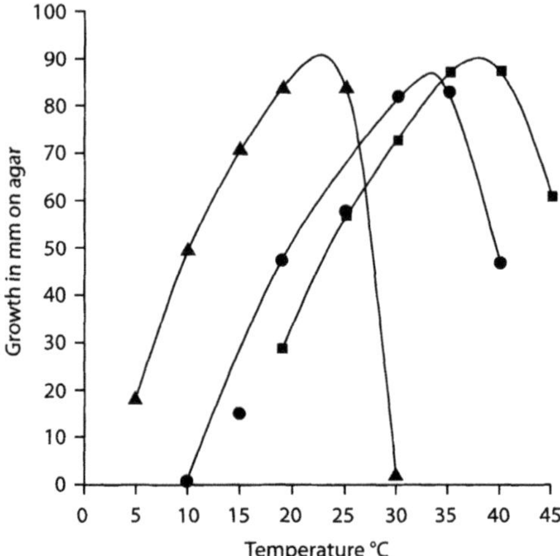

Figure 19.1. Growth response of onion bulb rotting fungi, in vitro. Botrytis allii ▲, Aspergillus niger ●, A. fumigatus ■ *(redrawn from Maude, 1990a).*

Table 19.3. Occurrence of Aspergillus niger *on onion seed collected from different geographical localities (from Hayden and Maude, 1992)*

	Percentage seeds with A. niger		
Climate	*No. of samples*	*Mean*	*Range*
Temperate, humid (rainy with mild winters, warmest month below 22°C)	24	1.9	0-26
Sub-tropical, humid (rainy with dry winters, warmest month above 23°C)	5	16.9	0.5-41
Tropical, humid (mostly rainy, warmest month >18°C, e.g. Savanna)	3	21.0	10.5-40.5
Tropical, dry (hot, semi-arid climate, e.g. Steppes; ambient temperatures generally >20-30°C)	1	23.0	-
Tropical, desert (hot, dry, arid climate; ambient temperatures generally >30°C, 50°C common)	30	88.9	26-100

In addition, the mycofloral balance of *Aspergillus* spp. on a single sample of onion seeds can alter depending on the recovery temperature that is used. Thus, *A. niger* was recovered from naturally infected seeds over an incubation temperature range of 20 to 35°C but at 40 and 45°C, it was replaced by *A. fumigatus* from the same seed sample – presumably because this fungus has a higher growth optimum (see Fig. 19.1) and was more competitive at those temperatures (Hayden and Maude, 1994).

(b) Transmission and transmission rate

Transmission and transmission rate are factors of epidemiological importance in the initiation of spread of disease. Consequential transmission of many seedborne organisms, and of seedborne fungal pathogens of onions in particular, requires that the incidence of infected seeds (i.e. the inoculum threshold) is sufficient to transfer infection to a proportion of the emerging seedlings causing disease outbreaks of economic proportions. Inoculum thresholds have been determined for many seedborne organisms (Maude, 1996). They are of critical importance where infected seeds are the main source of crop disease.

The epidemiological consequences of sowing onions with a threshold level of 1% *B. aclada*-infected seeds are such that, under wet summer conditions, there is considerable spread of disease in the growing crop, resulting in up to 10% of bulbs with neck rot in store (Fig. 19.2). This is the maximum incidence of diseased bulbs that can be handled in commercial grading lines and seed samples with 1% or more infected seeds should be discarded or should be treated before sowing.

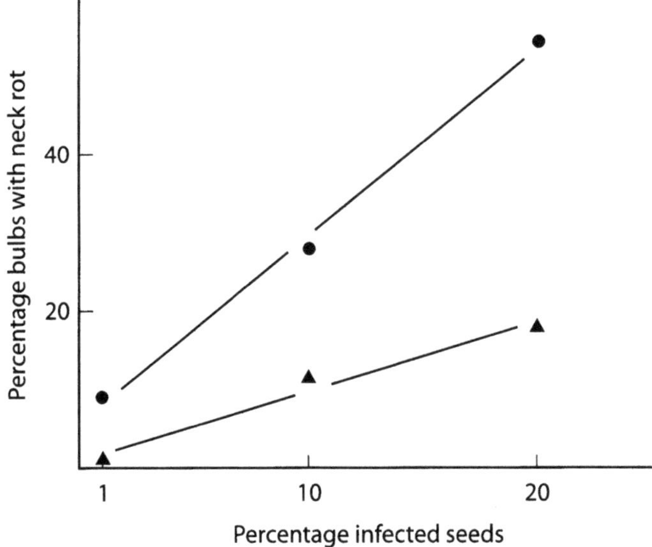

Figure 19.2. Relationship between Botrytis allii-*infected onion seeds and post-harvest neck rot of bulb onions:* ▲, *dry summer (1973);* ●, *wet summer (1974) (redrawn from Maude, 1983).*

Such transmission relationships fit Van der Plank's equation (1963), which states: the amount of initial inoculum *(x_o)*, the average infection rate *(r)* and the time over which infection occurs *(t)* determines the amount of disease that develops *(x)* and therefore $x = x_o e^{rt}$. In the example of *B. aclada*-infected onion seeds, the minimum tolerance level would be based on reducing the initial inoculum (i.e. infected seeds) *(x_o)* so that the disease threshold *(x_d)* is not exceeded, even under conditions favouring the average infection rate *(r)*.

The rate of transmission of seedborne onion pathogens is modified by a number of factors, including fungal type (i.e. biotroph, necrotroph), severity of seed infection, soil conditions (e.g. temperature and moisture) and soil microflora.

For example, the biotroph *Peronospora destructor* (cause of downy mildew of onion) is commonly present in onion seed production fields, where it may invade developing onion seeds. In common with many other downy mildews, its survival on onion seeds is relatively short-lived and so there are opposing views on whether transmission of the disease can occur from a seed source.

Other pathogens (mainly necrotrophic) are probably sufficiently persistent on or in seeds to transmit and cause outbreaks of disease. Of these, *Botrytis aclada* (*B. allii* – cause of neck rot) is the main temperate seedborne pathogen of onions. *Aspergillus niger* (black mould) (Hayden and Maude, 1992) and *Alternaria porri* (cause of purple blotch) (Wu, 1979) are hot climate equivalents (Table 2). The importance of infected seeds in the transmission of either of the latter fungi is difficult to quantify, because field sources of infection exist (Pandotra, 1965; Hayden *et al.*, 1994a,b).

19.4 CASE HISTORIES: FOLIAR DISEASES

Many temperate (*Botrytis aclada, Botrytis squamosa, Peronospora destructor*, etc.) and hot climate (*Alternaria porri, Glomerella cingulata, Aspergillus niger, Cercospora duddiae*, etc.) pathogens, from a variety of sources, infect the foliage of growing onion crops causing major outbreaks of disease. A number of epidemiological factors constrain or facilitate leaf disease development, some of which are reviewed here in relation to specific pathogens.

19.4.1 Botrytis aclada *(cause of onion neck rot)*

(a) Pathogen and disease

Neck rot is the main storage disease of bulb onions in temperate parts of the world (see Maude, 1990a). Munn identified the pathogen as *Botrytis allii* n.sp. causing small sclerotial neck rot disease of onions in Michigan and New York State, USA (Munn, 1917). This identification was confirmed by Walker (1925, 1926). Hennebert (1963, 1973) following examination of herbarium material and the use of adjusted spore measurements concluded that the fungus named *B. aclada* by Fresenius in 1850 (Fresenius, 1850) was the same as *B. allii*.

New research (light microscope - Shirane *et al.* (1989); DNA fingerprinting - Nielsen *et al.*, 2001; Nielsen and Yohalem, 2001; Yohalem *et al.*, 2003) has identified a small-spored 16-chromosome form (type A1) and a larger-spored 32-chromosome form (type A11) of the fungus. These *B. aclada* subgroups (A1 and A11) were genetically different from *B. byssoidea* (the cause of mycelial neck rot) and significantly different from each other (Nielsen *et al.*, 2001). From the use of universally primed (UP) polymerase chain reaction (UP-PCR) DNA fingerprints, it is suggested that subgroup A11 of *B. aclada* is a possible hybrid species between *B. byssoidea* and subgroup A1 of *B. aclada* (Nielsen and Yohalem, 2001). Both the smallest-spored group (*B. aclada*) and the largest-spored group (*B. byssoidea*) each have 16 chromosomes, while the intermediate group (*B. allii*) has 32 (Yohalem *et al.*, 2003). This research shows that the three groups are genetically distinct and that isolates of *B. aclada* and *B. byssoidea* were the possible ancestors of the polyploid *B. allii* (Yohalem *et al.*, 2003). Yohalem *et al.*, (2003) suggest that both *B. aclada* and *B. allii* are valid names and they propose that *B. aclada* is reserved for the small-spored sub-group (A1) and *B. allii* for the larger-spored sub-group (A11) of *B. aclada*.

In this chapter the name *B. aclada* (*sensu* Fresenius) has been used for both subgroups (A1 and A11) replacing *B. allii* (*sensu* Munn). *Botrytis aclada* is the asexual form (anamorph) of an ascomycete fungus. There is no known teliomorph (sexual stage). The fungus is a necrotroph.

The disease only becomes evident when onions have been in store for eight or more weeks. Typically the upper parts of affected bulbs are soft when pressed and removal of the brown wrapper scales (leaves) reveals a black mass of sclerotia enclosing the neck tissues. Individual sclerotia are up to 5 mm in diameter and a grey mould sometimes extends below them on bulbs. The fungus invades downwards in the food scales of bulbs and these have a brown 'cooked' appearance.

The pathogen became important in commercial onion bulb crops in the 1970s in the UK through the use of imported diseased seed (Maude and Presly, 1977 a,b; Maude, 1983). In a recent survey, *B. aclada* has been detected in many seed lots harvested from bulb onion seed crops grown in the semi-arid Columbia basin of Washington, US (du Toit *et al.*, 2004). Preliminary molecular analysis of isolates of *B. aclada* collected during this survey may indicate that both *B. aclada* and *B. allii* (*sensu* Yohalem *et al.*, 2003) were present.

(b) Epidemiology

The epidemiology of most foliar pathogens of onions can be deduced by relating the symptoms produced at various stages of crop development to the environmental conditions that prevailed immediately prior to symptom appearance. In this respect, onion neck rot is different: the disease is virtually symptomless in the growing crop and to record its spread, which occurs by the aerial dissemination of conidia, it is necessary to incubate onion foliage to demonstrate the presence of conidiophores and conidia of the fungus on those tissues. This reveals that there is a direct linear relationship between incidence of seedborne infection, the number of infected plants in onion crops and the number of neck rot infected bulbs in store (Maude and Presly, 1977a,b; Maude, 1983). During dry growing seasons, there is a virtual one-to-one relationship between percentage infected seeds and percentage infected bulbs in store, but in wet humid seasons the storage disease levels are proportionately greater (Fig. 19.2) (Maude, 1983).

Stewart and Franicevic (1994) obtained similar seed/bulb relationships with higher seed infections resulting in greater levels of bulb rot in store in New Zealand. Tylkowska and Dorna (2001) in Poland reported a stronger seed/bulb neck rot relationship from infected versus contaminated seed. There was no apparent relationship between plant infection and *B. aclada* infection of the harvested seed from infected bulb onion seed crops in Washington, USA (du Toit *et al.*, 2004).

The latency of this disease in onion crops increases the difficulty of establishing the critical epidemiological parameters. The most consistent correlation over a period of seven years in the UK was the relationship between the log % neck rot and RH in the month of June; relative humidities over 85% then were correlated (r = 0.97; $P < 0.05$) with high levels of neck rot in stored onions six to seven months later (Maude *et al.*, 1985).

Thus, rainfall as evidenced by high RH appears to be important in determining levels of infection in field crops and consequently of neck rot in stored onion bulbs.

The effects if any, of the crop microflora and mycoflora on *Botrytis aclada* on/in onion leaf tissues may have a further influence on the epidemiology of the disease.

Laboratory experiments on dead onion leaf pieces held under controlled moist/wet conditions (Köhl *et al.*, 1995; Köhl *et al.*, 1997; Nielsen *et al.*, 2000) demonstrated that there were a number of mycofloral antagonists that might adversely affect the sporulation of *B. aclada*. Their modes of action include: antifungal effects causing hyphal plasmolysis, antibiosis, and exclusion by rapid colonisation of necrotic leaf tissues (Köhl *et al.*, 1997). Parasitism was not observed (Köhl *et al.*, 1997). Their role in the onion crop is not known but there may be a potential use for some of them as biocontrol fungi.

19.4.2 Botrytis squamosa *(cause of leaf rot, leaf blight, blast)*

(a) Pathogen and disease

Botrytis squamosa is the asexual state of the ascomycete *Sclerotinia squamosa* (teliomorph) and both asexual spores (conidia) and sexual spores (ascospores) may be important sources of infection for this necrotroph.

This fungus differs from *B. aclada* in that it produces symptoms appearing as greyish white leaf lesions (spots) which are circular to elliptical in shape, 1 to 10 mm long by 1 to 4 mm wide (Hancock and Lorbeer, 1963). Severe infection leads to dieback of onion leaf-tips and sporulation occurs only on this dead tissue (Lorbeer, 1992). The incidence of leaf spots and their density on the leaf surface has been used to monitor the progress of disease outbreaks.

Sources of the pathogen in the United States are represented by conidia from onion debris and cull piles in seed production fields, and ascospores from apothecia produced from sclerotia, or conidia from sclerotia present on the surface of soil which had previously borne onions (Ellerbrock and Lorbeer, 1977; Lorbeer, 1992; Clarkson *et al.*, 2000). Inoculum bridging also occurs in successionally sown onions in a single season or over the span of a year where green, pickling, ware or seed production crops link the seasons into a continuous cycle of onions (Maude, 1990b). The disease is more prevalent on densely produced crops (e.g. green and pickling onions) in the UK but it is very destructive in most onion crops in temperate parts of the world (Lorbeer, 1992).

Primary inoculum from the sources described above and secondary inoculum produced as conidia on the leaves of infected crop plants, spore dispersal and survival, infection and colonisation by and the sporulation capacity of *B. squamosa* are basic requirements for the development of leaf blight in onion crops (Sutton *et al.,* 1986).

(b) Epidemiology

The significance of a primary inoculum source such as sclerotia in initiating disease outbreaks depends on the ability of such a source to produce conidia and this is regulated by temperature and soil water potential (Clarkson *et al.,* 2000). The greatest number of conidia per sclerotium in the field were produced at temperatures of 5 to 10°C and reduction in soil water potential reduced the number of conidia produced. However, conidia were produced at water potentials as low as –2 Mpa, at which sclerotial germination was at least 60% (Clarkson *et al.,* 2000). Water potential was more likely to be the limiting factor for conidiogenic germination between April and September than between October and March, as temperatures were more limiting in these winter months in the UK (Clarkson *et al.,* 2000).

Disease development on onion foliage is dependent on the positive action and interaction of a number of epidemiological factors. For example, *B. squamosa* produces conidia at night on the necrotic tips of onion leaves when mean temperatures during a wet period are between 8 and 22°C. Formation of conidia is greatest when the duration of wetness is more than 12 h and least when there is wetness for 5 h or less. Conidial release is favoured by decreasing or increasing RH and by rain; once spores are detected, release occurs on a daily basis (Shoemaker and Lorbeer, 1977; Sutton *et al.,* 1978). Dispersal follows a diurnal pattern, which peaks between 0900 to 1200h in Canada (Sutton *et al.,* 1978).

When conidia land on onion leaves, temperature and leaf wetness duration are the main variables influencing infection. Although conidia may germinate over a wide temperature range (6 to 33°C), lesions develop optimally at 20°C and are reduced at 15 and 25°C; lesion numbers increase with increasing leaf wetness duration up to 48 h (Alderman and Lacy, 1983; Lorbeer, 1992). Leaf penetration is direct and localised leaf spot lesions develop; these may expand leading to blighting when leaf wetness is prolonged (Alderman and Lacy, 1983). The greatest production of conidia occurs on necrotic parts of leaves and leaf tips and not on the leaf spot lesions that represent a host hypersensitivity reaction to a weak pathogen (Lorbeer, 1992).

Routine applications of fungicides are used globally for the control of onion leaf blight but their number can be reduced by the use of disease forecasting.

The first disease forecasting system known as BOTCAST was developed in Canada (Sutton *et al.,* 1986). Between 1988 and 1992 two forecasting models,

BOTCAST (used to predict the application of the first spray) and SIV (a sporulation model used to time subsequent sprays) were tested in the Netherlands (de Visser, 1996). The application of these models reduced chemical sprays by 54% without an observed yield loss or increase in disease severity. Modifications to BOTCAST are suggested by de Visser (1996) and comparisons have been made with sporulation models such as DINOV (Vincelli and Lorbeer, 1988). BOTCAST is used in practice in the Netherlands to reduce fungicide usage in the control of leaf blight (Meier, 2000).

In the USA integrated pest management programmes have been developed for dry bulb onions, based on careful monitoring of the development of leaf lesions, and these have enabled fungicide spray applications to be delayed until a critical disease level (CDL) is reached (Lorbeer, 1992). In addition, microclimatic measurements have been used to predict daily inoculum incidence and infection density, from which disease severity indices are computed. From this, bioclimatological data thresholds are determined that indicate when fungicide sprays should be applied.

The forecasting system that has been developed is known as BLIGHT-ALERT (Lorbeer *et al.*, 2002) and is available commercially.

Both BOTCAST and BLIGHT-ALERT are weather-based predictive systems. Field monitoring to determine the CDL is used in BLIGHT-ALERT but not in BOTCAST (Lorbeer *et al.*, 2002). Carisse *et al.* (2003) reported that in Canada bulb onion leaf blight epidemics began when airborne conidia of *B. squamosa* reached a concentration of 10 conidia m^{-3} air. Following experimentation a critical threshold of 15 to 20 conidia m^{-3} air was established and in some years, applying this threshold, it was possible to manage the disease and reduce the number of spray applications by 20% to zero (Carisse *et al.*, 2003).

BOTCAST has been tested in conjunction with the application of *Gliocladium roseum* spores for the biological control of leaf blight in bulb onions (James and Sutton, 1996). The fungus was about half as effective as fungicide (chlorothalonil) in reducing the density of *B. squamosa* leaf spots but it was considered that the antagonist had potential for controlling the disease sufficiently to avoid economic yield losses (James and Sutton, 1996).

However, it was noted that germination of *B. squamosa* sclerotia in the soil could be predicted and may be of use in predicting release of primary inoculum (Clarkson *et al.*, 2000). This model might be combined with existing models for leaf blight development such as BOTCAST (Sutton *et al.*, 1986) to provide a more comprehensive system for predicting leaf blight (Clarkson *et al.*, 2000).

19.4.3 Peronospora destructor *(cause of downy mildew)*

(a) Pathogen and disease

Peronospora destructor is a phycomycete fungus. It differs from the *Botrytis* spp. already described in that it is an obligate biotroph. Otherwise, similar epidemiological principles govern the development and spread of downy mildew.

Downy mildew can cause serious losses in bulb and green (salad) onions and in onion seed production crops (Gilles *et al.*, 2004). The authors cited bulb onion yield losses of 60-75%; losses as high as 100% were reported for salad onions, where complete crops may be rejected because of mildew and frequent reductions were caused in seed production onion crops through collapse of flowering stalks and poor germination of harvested seeds (Gilles *et al.,* 2004).

The pathogen causes pale leaf lesions on onions on which sporangiophores bearing infective sporangia are produced which give a greyish-violet colour to the lesions. Decaying onion debris may release oospores into the soil to act as a further source of inoculum (Dixon, 1981; Viranyi, 1988).

The main infection source in seed production fields may be infected mother bulbs.

Factors that are important in the epidemiology of the disease include temperature, light, RH and air-flow.

(b) Epidemiology

Sporangia produced on leaf lesions are mostly released in the morning, as leaves dry and humidity is reduced, and not during the hours of darkness when air is saturated with moisture. Greatest discharge occurs when leaves are vibrated while exposed to brief periods of red-infrared radiation under reducing humidity (RH less than 59%) (Leach *et al.,* 1982). Spores survive under a wide range of weather conditions during the day but are frequently killed at night by alternating wetness and dryness, especially when these cycles are associated with low dew deposition (Hildebrand and Sutton, 1984a).

Production, sporulation and germination of released sporangia and penetration of leaf tissues are favoured by temperatures of 10 to 13°C, high relative humidities (RH) and prolonged periods of leaf wetness (Viranyi, 1988). Sporangia germinate only in free water on dull days.

The infection cycle is characterised by long latent periods (about 9 to 16 days) and short periods (about 1 to 2 days) when the pathogen sporulates (Hildebrand and Sutton, 1982). Factors that contribute to its variability include temperature and inoculum dose (Hildebrand and Sutton, 1984b).

The disease is controlled by the routine applications of fungicides and it was suggested by de Weille (1975) that it might be possible to reduce their number and/or the periodicity by developing a bioclimatological modelling approach to disease forecasting as part of an integrated strategy for disease control.

The first model forecasting downy mildew was given the acronym DOWNCAST and it utilised information for predicting sporulation and infection of onions by *Peronospora destructor* under field conditions in Canada (Jesperson and Sutton, 1987). A sporulation-infection period was predicted when conditions were conducive to sporulation, spore dispersal, spore survival and infection. These criteria, derived from earlier work (Hildebrand and Sutton, 1982; 1984a,b,c) established quantitative and temporal relationships of temperature, rain, high humidity, rate of dew deposition and its duration with the infection cycle of the pathogen (Jesperson and Sutton, 1987). The method correctly predicted sporulation incidence on 111 of 119 nights in two growing seasons (Jesperson and Sutton, 1987). Since then, a number of new models have been developed. These include ONIMIL in Italy (Battilani *et al.*, 1996), a de Visser modified DOWNCAST in Holland (de Visser, 1998), ZWIPERO in Germany (Friedrich *et al.*, 2003) and MILIONCAST (an acronym for 'MILdew on onION foreCAST') in the UK (Gilles *et al.*, 2004).

Australian experience on onion trap plants suggested that DOWNCAST could predict downy mildew infection events (Fitzgerald and O'Brien, 1994) and in New Zealand the use of the model reduced fungicide applications by 40% (Wright *et al.*, 2002).

However, although DOWNCAST predicts sporulation it cannot predict the quantity of sporangia produced (Gilles *et al.*, 2004). The sporulation model within DOWNCAST and DOWNCAST-based models (ONIMIL and de Visser modified DOWNCAST) was found to be inaccurate under Dutch and UK climatic conditions (de Visser, 1998; Gilles *et al.*, 2004). This has led to the development of newer improved models to enable more accurate sporulation and infection forecasts (Friedrich *et al.*, 2003; Gilles *et al.*, 2004).

The most recent of these, MILIONCAST is comprised of three predictive models: (1) predicting sporulation, (2) predicting infection and (3) predicting latent periods (Gilles and Kennedy, 2004). The data are derived from the effects of temperature and relative humidity on the rate of downy mildew sporulation on onion leaves tested in controlled environmental conditions (Gilles *et al.*, 2004). Sporangia were produced most rapidly at 8 to 12°C after 5 h of high humidity during dark periods. The greatest number of sporangia was produced at 100% RH and sporulation decreased to almost nil when relative humidity decreased to 93% RH. The accuracy of this sporulation data was tested by comparative observations made on infected plants in pots outdoors. Latent periods were 9 to 11 days between 15 to 23°C and increased with temperature decreasing below 15°C.

Accuracy of prediction with MILIONCAST was greater than that of the existing models based on DOWNCAST (Gilles *et al.*, 2004). When the complete model was applied to downy mildew disease outbreak data collected over six previous seasons (1996 to 2001) MILIONCAST correctly predicted over 60% of the observed peaks of increase in disease severity (Gilles and Kennedy, 2004). Field trials are needed to compare the accuracy of MILIONCAST with that of the other models (Gilles and Kennedy, 2004).

The individual merits of each model have been reviewed (Gilles and Kennedy, 2004; Gilles *et al.*, 2004).

19.4.4 Alternaria porri *(cause of purple blotch)*

(a) Pathogen and disease

Alternaria porri is the asexual conidial-producing state of an ascomycete fungus. There is no known sexual stage. The fungus is a necrotroph and causes purple blotch, a disease that is important on alliums worldwide (Aveling, 1998) but is most severe on onions where they are grown in areas with hot humid climates (Miller and Lacy, 1995; Lakra, 1999).

The first symptoms on leaves are small water-soaked lesions that develop white centres and enlarge to become zonate and brown to purple in colour. Older leaves are more susceptible than younger ones to infection (Miller, 1983; Lakra, 1999). Severely affected foliage may die back and seed stalks may fail to produce seed heads causing serious losses in onion seed crops (Lakra, 1999).

(b) Epidemiology

The epidemiological factors that control the development of temperate foliar pathogens such as *Botrytis squamosa* and *Peronospora destructor* also apply to *Alternaria porri*. A major difference is that *A. porri* is active over a higher temperature range with an optimum of 25°C (Fahim, 1966; Miller and Lacy, 1995). Otherwise, conidia are formed at night on leaf lesions at relative humidities of 90% or more and are fully mature after 15 h of dew. Sporulation occurs at night at high relative humidities and conidia are released between 0800 and 1400 h as humidity decreases on calm days (Fig. 19.3) (Meredith, 1966). The concentration of conidia in the air increases on windy days and also after rainfall, irrigation or spraying (Meredith, 1966).

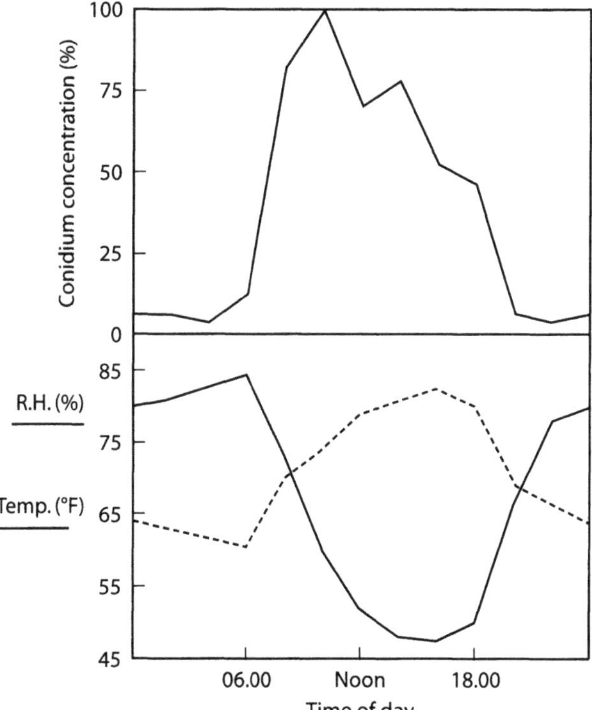

Figure 19.3. Diurnal periodicity of Alternaria porri *conidia in calm dry weather (redrawn from Meredith, 1966).*

With this disease, the duration of conditions favourable for infection and subsequent RH are important in determining the type of lesion that is produced. Thus, conidia formed under dew durations of 12 h or less and low RH cause sterile leaf flecks, while those formed under dew conditions lasting for 16 h or more with high RH cause typical lesions (Fig. 19.4) (Bock, 1964; Everts and Lacy, 1990a). It was suggested that flecks may be caused by immature conidia, which are not sufficiently aggressive to infect (Everts and Lacy, 1990a). However, in onion leaf inoculation studies, germinated conidia were found near the centre of flecks but these flecks did not develop into lesions (Everts and Lacy, 1996).

Aveling *et al.* (1994) found that 96% of normal conidia germinated at 25°C within 24 h of inoculation and that penetration of leaf tissue is direct or through stomata.

Miller *et al.* (1978) reported that the fungal necrotroph *Stemphylium vesicarium* alone or in complex with *A. porri* caused a severe leaf blight of onions in 1976 in Texas, USA. *Stemphylium vesicarium* produced white and purple lesions on commercial garlic crops in Spain (Basallote-Ureba *et al.,* 1999). The symptoms were reproduced when the fungus was inoculated into garlic and onions.

In recent Australian research (Suheri and Price, 2001) either or both pathogens, *Alternaria porri* and *Stemphylium vesicarium* were isolated from purple leaf blotch lesions on field leeks grown in the state of Victoria. In this field situation *S. vesicarium* was the dominant pathogen. However, the diurnal pattern of *A. porri* and *S. vesicarium* conidia in the leek crops was similar to the periodicity of *A. porri* in onion crops (Meredith, 1966; Everts and Lacy, 1990b), but the purple blotch lesions developed under cooler (10 to 13°C) conditions (Suheri and Price, 2001).

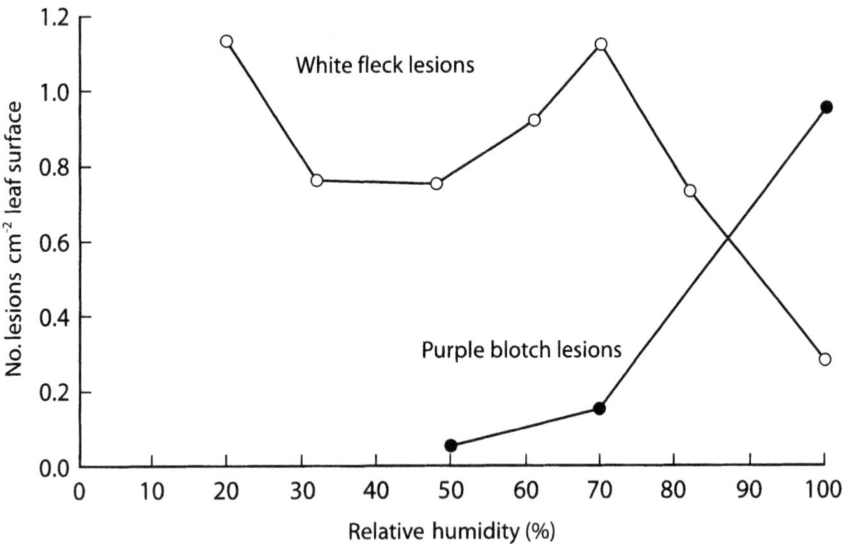

Figure 19.4. Effect of duration of dew and post-infection relative humidity on symptom expression by Alternaria porri *(redrawn from Bock, 1964).*

The pathogen usually associated with purple blotch symptoms is *Alternaria porri*. However, these symptoms are generally indistinguishable from those of *Stemphylium* leaf blight caused by *Stemphylium vesicarium* and purple leaf blotch disease is now considered by Suheri and Price (2000a,b) to be a complex caused by both pathogens.

19.5 CASE HISTORIES: SOILBORNE DISEASES

Soilborne pathogens are located and localised in an environment where their epidemiology is determined by edaphic effects of the soil and its microflora. In the soil, host roots grow to contact pathogens in their immediate locality; above ground and by comparison, leaf pathogens are disseminated in the air to infect new crops over a wide area.

The main temperate soilborne pathogens described here are *Sclerotium cepivorum* (cause of onion white rot), a specialised necrotroph, and *Urocystis cepulae* (cause of onion smut) a biotroph (Table 19.2).

The warm-climate pathogens described are *Pyrenochaeta terrestris* (cause of pink root) and *Fusarium oxysporum* f.sp. *cepae* (cause of basal rot); both function as necrotrophs (Table 19.2).

Soilborne pathogens of onions affect the rooting systems of the plants commonly causing wilting, loss of turgor, stunting and death.

Most root diseases are disseminated by the movement of infected vegetative material, or by any action that moves soil containing the pathogen – for example, in soil clinging to farm machinery, implements and workers' boots (Entwistle, 1990).

19.5.1 Sclerotium cepivorum *(cause of onion white rot)*

(a) Pathogen and disease

Sclerotium cepivorum is an ascomycete fungus, which produces black, nearly spherical sclerotia (about 0.2 to 0.5 mm in diameter); there is no sexual, or conidial state.

The pathogen is specific to *Allium* spp. and in that respect is a specialised necrotroph. White rot occurs in most parts of the world where onions are grown under temperate conditions. It varies in severity from place to place. New outbreaks commonly occur where the disease is prevalent locally but, occasionally, the disease can remain completely localised for considerable periods of time i.e. over 20 years (Entwistle, 1990).

There are no commercial *Allium* varieties with sufficient resistance to *S. cepivorum* (Utkhede *et al.*, 1982; Brix and Zinkernagel, 1992).

Onions may be attacked at all stages of crop growth. Early attacks cause poor crop establishment; later infections produce yellowing and wilting, which in some cases results in the complete collapse of the plant. Affected plants have a grey-white fluffy mycelium on their stem bases, which gives the disease its name: white rot. The disease is caused by the sclerotia of *Sclerotium cepivorum,* which perennate in the soil. Sclerotia germinate producing hyphae that infect the roots and stem plates of onions. New sclerotia are formed in the stem base tissues of onions and these are released into the soil when plants die or are harvested (Entwistle and Munasinghe, 1978).

There is a considerable literature on the epidemiology of the disease (Coley-Smith, 1988; Entwistle, 1990; Metcalfe and Wilson, 1999; Clarkson *et al.,* 2004).

(b) Epidemiology

Host and soil effects are important in the epidemiology of white rot. Firstly, for plant infection to occur, the sclerotia held in check by soil fungistasis have to receive a stimulus from specific factors released as exudates by the roots of onions (Esler and Coley-Smith, 1983). When this occurs sclerotia germinate producing hyphae that may grow 1 to 2 cm through the soil to infect the roots of onions. They penetrate the root epidermis producing appressoria and infection cushions and the hyphae enter between cell wall junctions (Metcalfe and Wilson, 1999). Then they invade the hypodermis and grow into the cortex of the roots. During the early stages of infection, cell death is limited to the cells penetrated by the hyphae of *S. cepivorum*; however, onion cell walls in the path of the leading infection hyphae often dissolve before the hyphal tips reach them. This was shown to be indicative of enzymic action by certain pectinases (Metcalfe and Wilson, 1999). Plant-to-plant spread of infection may occur in densely-sown or closely-spaced crops.

Secondly, soil temperature is a critical factor affecting germination of sclerotia, mycelial growth and root infection of onions. Sclerotial germination is favoured by temperatures between 10 and 20°C (Fig. 19.5); outside this range, germination is slow but returns to normal when temperature is restored to 15°C. Thus, in the UK, sclerotial germination is poor during the winter but improves in spring and early summer (Fig. 19.5); later in the summer, germination may be inhibited by high soil temperatures – sclerotia cannot survive temperatures of 35°C and above (Entwistle and Munasinghe, 1978). As a result, disease losses are high in overwintered crops in the autumn and spring, when temperatures are suitable for infection, (the optimum is between 15 to 18°C – Crowe and Hall, 1980), but white rot may be of minor importance in hot summers (Entwistle and Munasinghe, 1978). In countries with hot climates, white rot may be confined to the cooler summer months (Coley-Smith, 1988).

Sclerotial germination is little affected by pH values varying from 4.8 to 8.5 (Entwistle and Granger, 1977); there was a decrease in germination of sclerotia as soils became drier or wetter than field capacity (-300 millibars) (Crowe and Hall, 1980). Recent studies (Clarkson *et al.,* 2004) showed that over 90% of sclerotia were degraded at high water potentials where the soil was nearly saturated. This confirmed the findings of previous workers (Crowe and Hall, 1980) and those of Leggett and Rahe (1985) and Crowe and Carlson, (1994) who used flooding to degrade sclerotia as a means of controlling the disease.

Sclerotia may persist in the soil from at least four to 20 years (Entwistle and Munasinghe, 1978; Entwistle, 1990). Attempts to forecast white rot based on assessments of the numbers of sclerotia per given weight of soil have met with variable results within and between countries (Entwistle, 1990). Newer models based on inoculum dynamics (Backhouse, 2003) have so far failed because of insufficient data to develop the relationship for the production of new sclerotia based on the size of infected bulbs. The maximum increase in numbers of sclerotia in the soil occurs at the end of the onion growing season and this will affect the following crop - but only if it is an onion crop.

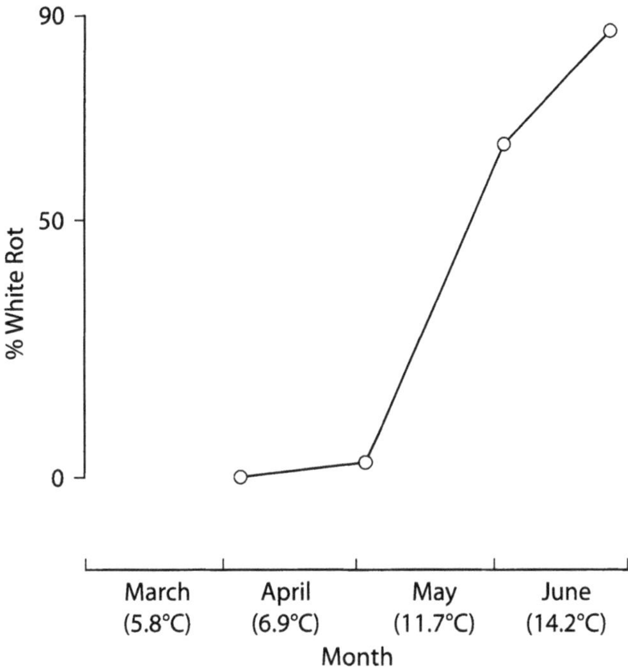

Figure 19.5. Incidence of white rot in overwintered salad onions, Southfleet, UK (1976/1977) (redrawn from unpublished data of A. R. Entwistle). Sampling dates were 5 April, 3 May, 2 June and 27 June, 1977. (°C) = Mean monthly soil temperatures at 100 mm depth.

In this respect the epidemiology of the white rot fungus is unlike that of an airborne pathogen where inoculum increase and disease spread, within and between like crops, can occur throughout the growing season.

Investigations of the epidemiology of this soilborne disease raise the question of how the soil environment might be altered to the disadvantage of the pathogen, for example, its sclerotia. Physical, chemical and biological approaches have been made (Clarkson *et al.,* 2004). The use of fungal antagonists or biocontrol agents (Adams and Ayres, 1981; Ahmed and Tribe, 1977; Coley-Smith, 1987; Clarkson *et al.,* 2002; Kay and Stewart, 1994) has met with varying degrees of success dependent on the limitations of the experimental systems used. The interaction between biocontrol agent and pathogen is complex (Metcalfe and Wilson, 2001) and to model it, strict control of the environment and of the inoculum (pathogen and biocontrol agent) is necessary.

To an extent this has been achieved by Clarkson *et al.*, (2004) who showed in laboratory assays that isolates of *Trichoderma viride* effectively degraded sclerotia of *S. cepivorum* (four isolates) in four onion soil types (silty clay, sand, silt and peat) under a range of temperature and soil moisture conditions.

19.5.2 Urocystis cepulae *(cause of onion smut)*

(a) Pathogen and disease

Urocystis cepulae is a basidiomycete fungus and obligate biotroph that produces infective thick-walled compound teliospores which remain viable in the soil for as long as 25 years (Thaxter, 1890). The pathogen is specific to *Allium* spp., principally affecting onion, leek and shallots (Horst, 2001).

Smut is a most destructive disease of onions (Horst, 2001) and is of considerable economic importance in temperate onion-growing areas of the world (Utkhede and Rahe, 1980; Crevier and Stewart, 1988). Monnet and Thibault (2000) described the symptoms of *Allium* diseases, including smut, on onions and shallots in France. In the UK, Brooks (1953) described disease outbreaks of smut as serious but local in green onions grown from seeds.

Dark lead-coloured lesions appear on the cotyledons of infected onions soon after emergence. Young lesioned leaves become thickened and twist or curl backwards. Lesions rupture, releasing the black powdery teliospores of the fungus back into the soil. Also, the spores may be disseminated by running water, tools, the roots of transplanted vegetables, and by the feet of people and animals (Horst, 2001).

Teliospores comprise a single thick-walled central cell enclosed by a series of smaller, hyaline cells that serve as nurse cells. Teliospores give rise to uninucleate thalli of + or – hyphal mating types which, separately, are not pathogenic. Fusion of hyphae of opposite mating types results in a binucleate mycelium which is pathogenic (Tachibana and Duran, 1961) and which infects the emerging cotyledon leaf below ground level.

Both white rot and onion smut are initiated by fungal propagules, which remain dormant yet infectious in the soil for considerable periods of time. Unlike the white rot fungus, which can attack onions at all stages of crop growth, *Urocystis cepulae* can only infect seedlings and very young plants. These are killed within 3 to 4 weeks after emergence; those that survive are stunted, with brittle, distorted, lesioned leaves.

(b) Epidemiology

Host stimulation of sclerotial germination is a feature of the epidemiology of onion white rot. There appears to be no similar comparison for *U. cepulae* teliospores. Soil temperature is a major factor in the epidemiology of the disease. There is abundant infection of onions at soil temperatures of 10 to 12°C and up to 25°C; there is no infection at 29°C and there is a rapid decline in disease (Walker and Wellman, 1926).

Optimum germination and growth of teliospores and hyphal fragments occurs in temperate soils between 13 and 22°C; above 25°C teliospore germination and above

28°C hyphal growth, respectively, are severely restricted. At these higher temperatures, seedlings grow more quickly through the susceptible state.

Seedlings are susceptible only from the second day after seed germination for a period of 10 to 15 days (Horst, 2001). During this time, infection hyphae may penetrate the cuticle of the cotyledon; if they invade the meristematic tissues, then successive leaves (which are produced basipetally in onion) become infected during the short period of susceptibility.

Thus the ecological limits of smut are determined by soil temperature.

19.5.3 Fusarium oxysporum *f.sp.* cepae (cause of basal plate rot)

(a) Pathogen and disease

Fusarium oxysporum is a common soilborne plant pathogen with a worldwide distribution. As a species, it probably causes more damage to agricultural crops than any other plant pathogen (Correll, 1991). Within the species there is a high level of specificity with over 120 described *formae speciales* and races capable of causing vascular wilt diseases of many agricultural crops (Armstrong and Armstrong, 1981).

Fusarium oxysporum f.sp. *cepae* is one of these and it is the imperfect state of an ascomycete fungus. In this form, the pathogen is a soilborne necrotroph that causes disease outbreaks in onions at high soil temperatures (25 to 35°C). Basal plate rot is prevalent in areas of the world where onions are grown under high temperature conditions (Sherf and Macnab, 1986). Thus, over the years, the disease has been responsible for significant losses in field and stored crops of dry bulb onions in many parts of the world, for example, in Colorado, USA (Swift, *et al.,* 2002), in Turkey (Özer *et al.,* 2003) and in Italy (Fantino and Schiavi, 1987). Also, the fungus attacks shallot and garlic (Horst, 2001).

Infection may occur at any time during the growth of the crop; below ground the roots, and via these the stem base plate (modified stem) from which the leaves are produced, are attacked, resulting in a progressive yellowing (*Fusarium* yellows) or necrosis of the leaves from the tips downward culminating in wilting (*Fusarium* wilt) in some cases (Havey, 1995). The roots ultimately rot; infected stem plates may show a brown discolouration and affected bulbs rot progressively in store (Havey, 1995).

Wilt-resistant, intermediate- and long-day onion cultivars have been developed but resistant short-day cultivars are lacking (Cramer, 2000).

Disease severity during onion tissue colonisation varies depending on variation in production of pectolytic isoenzymes produced by the fungus (Özer *et al.,* 2003).

(b) Epidemiology

The fungus commonly occurs in the soil where long-term survival is by means of thick-walled resting spores (chlamydospores); macroconidia and less often microconidia are also produced. Soil temperatures of 25 to 28°C are optimal for disease development; at 15°C the disease is rarely seen (Havey, 1995). The fungus may be active at any soil moisture that permits plant growth.

In the soil, the propagules (chlamydospores) of the fungus close to or on onion roots germinate to produce limited or more extensive mycelial growth on the root surface (Abawi and Lorbeer, 1971b). From this, hyphal penetration of the root may be direct through uninjured root tissue or via wounds; appressoria may or may not be formed (Abawi and Lorbeer, 1971b). Although the pathogen may remain latent in onion plants and bulbs, crop yield may be reduced and basal rot may develop in the stored bulbs (Stadnik and Dhringa, 1997). Bulb rot begins as soon as the pathogen spreads from the stem base plate to the leaf sheath tissues (Abawi and Lorbeer, 1971b; Holz and Knox-Davies, 1985).

Soil temperature is the main limiting factor in the epidemiology of the disease but inoculum concentration and certain soil factors are also important.

Inoculum was found to be unevenly distributed in organic soils (Abawi and Lorbeer, 1971a), ranging from 300 to 6,500 propagules of *F. oxysporum* f.sp. *cepae* per gram of oven dried soil in the top 15 cm of soil.

Abawi and Lorbeer (1972) demonstrated that a population of 50,000 propagules g^{-1} of oven dried soil was needed before significant disease development could be detected in field soil.

Farms with a history of the disease usually had the highest populations of the fungus, but in some cases there was evidence of biotic and or abiotic effects where fields with high populations of *F. oxysporum* f.sp. *cepae* did not have a high incidence of basal rot (Abawi and Lorbeer, 1971a).

There are suppressive and conducive soils that affect the incidence of diseases caused by *Fusarium* wilt fungi (Mace *et al.,* 1981). General suppression is conferred by overall activity of the microbial community resident in the soil, but specific soil suppressiveness is active against a specific pathogen (Mazzola, 2004). Thus, although microbial interactions in the rhizosphere are extremely complex, specificity in soil-plant-microbe interactions for disease suppression exists (Whipps, 2001).

Wilt-suppressive soils are specific to *Fusarium* wilts and are not effective against diseases caused by non-vascular *Fusarium* species (Weller *et al.,* 2002). Certain soilborne bacterial and fungal genera contribute to suppression. Suppression of pathogenic *F. oxysporum* has been attributed to the activity of non-pathogenic *F. oxysporum* and fluorescent bacteria (*Pseudomonas* spp.) (Weller *et al.,* 2002).

19.5.4 Pyrenochaeta terrestris *(cause of pink root)*

(a) Pathogen and disease

Pyrenochaeta terrestris is the pycnidial asexual state of an ascomycete fungus; there is no teliomorph. It is a soil inhabitant and a weak necrotroph.

Pink root is a devastating disease of onions, especially of those grown in the tropics and sub-tropics (Entwistle, 1990; Sumner, 1995; Pagès and Notteghem, 1996). Also, leeks may be affected (Biesiada *et al.,* 2004).

Unlike the diseases already described, isolates of the fungus are weakly associated with the roots of non-allium crops - for example, pepper, eggplant, soyabean, tomato, peas, barley, wheat, oats, cucumber, cauliflower and many other plant species (Sumner, 1995).

Hyphae of the fungus penetrate the root tips, which become light pink; as infection progresses, roots become deeper pink, turning to deep purple. Affected roots become water-soaked, dry out and disintegrate. Diseased plants exhibit drought symptoms; stunting results and bulb size is reduced.

Unlike *F. oxysporum* f.sp. *cepae* the fungus does not invade the stem base plate or the bulbs of onions. Like *F. oxysporum* f.sp. *cepae* this fungus is most active in the upper 15 cm of soil.

Resistance to *P. terrestris* has been tested for in a yellow short-day onion in New Mexico, USA (Corgan and Holland, 1993) and in Australia, cultivars with prolific root systems were more tolerant of the pathogen than those with weak root systems (Rogers and Henderson, 1989). Yellow Bermuda is the most resistant of commercial onion cultivars (Horst, 2001).

(b) Epidemiology

The pathogen probably survives in soil as chlamydospores, pycnidia or pycnidiospores or in colonised roots or plant debris (Sumner, 1995). Populations of fungus in the soil increase rapidly in the presence of onions (Entwistle, 1990).

Temperature is the main epidemiological constraint that applies to *P. terrestris* as it does to *F. oxysporum* f.sp. *cepae*. The optimum temperatures for the growth of the pathogen and for disease development range from 24 to 28°C; infection is reduced at 20°C and there is little disease at less than 16°C (Sumner, 1995). The fungus is little affected by soil moisture. Thus, the damaging effects of pink root are determined by the seasonal patterns of soil temperature and the growth period of the host (Entwistle, 1990).

Both the above fungi fit the same environmental niche and often have been present together in a disease complex in onion roots; this makes it difficult to decide which of the two is the main cause of damage (Lacy and Roberts, 1982). It is now accepted that the pathogens can act independently to cause disease: *P. terrestris* reduces numbers of plants but not weights of bulbs in organic soils, whereas *F. oxysporum* f.sp. *cepae* reduces plant numbers and weights of bulbs (Lacy and Roberts, 1982).

19.6 CONCLUDING REMARKS

Completely different fungal species occupy similar disease niches in temperate compared with hot climates. All of them, regardless of function (biotrophic or necrotrophic), means of survival (chlamydospores, teliospores, sclerotia, etc.) infection (conidia, sporangia, etc.) and host parts invaded (seed, root or leaf) are subject to epidemiological constraints. Provided inoculum (seedborne, soilborne or airborne) is present at a threshold concentration capable of causing disease, then temperature and moisture (as rainfall, dew, soil moisture and/or relative humidity) are the main epidemiological factors that limit or facilitate the outbreak and the further development of that disease. For example, in moist, warm soils it is mainly temperature that determines infection of roots by and the ecological limits of *Pyrenochaeta terrestris* and *Fusarium oxysporum* f.sp. *cepae*. In moist, cool soils, moderate temperatures are critical for root infection by *Sclerotium cepivorum* and *Urocystis cepulae*. Possibly the interaction of a greater number of epidemiological factors is required for leaf infection of onions (ware and seed production plants) by foliar pathogens. The temperature ranges for conidiation are clearly different for pathogens of temperate and those of hot climates but in either situation the interaction of temperature with long periods of leaf wetness and high relative humidities are similar requirements and are overriding factors controlling spore production and leaf infection, with decreasing relative humidities activating spore release of these pathogens. These factors also apply in the infection of seed production onion plants and some may apply to the transmission and spread of the diseases initiated by the sowing of infected onion seeds.

The critical evaluation of epidemiological factors, and particularly of those involved in foliar disease outbreaks, has resulted in the production of accurate disease forecasting models, some of which have been used successfully in practice to reduce fungicide applications in commercial crops of onions. However, many such models have failed to consider availability of and/or concentration of inoculum in assessment of disease risk (Phillon, 2003). Such information if applied (Phillon, 2003; Bugiani *et al.,* 1995; Carisse, *et al.,* 2003; Gilles *et al.,* 2004) would likely improve disease management (Carisse, *et al.,* 2003), and if linked to newer, rapid methods of inoculum capture and pathogen identification (Kennedy *et al.,* 2000; Wakeham *et al.,* 2004) could produce faster, more accurate assessment of disease risk and thereby may further reduce pesticide applications in field crops.

New molecular-based research has established genetic methods for fungal identification (Nielsen *et al.,* 2001; Nielsen and Yohalem, 2001; Yohalem *et al.,* 2003). Practical research is now required to investigate the respective pathogenicities of *B. aclada* (*sensu* Yohalem, *et al.,* 2003) and of *B. allii* (*sensu* Yohalem, *et al.,* 2003) and to establish their roles in the epidemiology of onion neck rot disease.

ACKNOWLEDGEMENTS

The author wishes to record his thanks to Miss Lindsey Peach, who redrew the graphs from published information, and to Miss Jane Fellows who provided some of the meteorological data. I am grateful to my wife for proof reading the manuscript.

The author is most grateful to Warwick HRI for access to library facilities at Wellesbourne.

REFERENCES

Abawi, G.S. and Lorbeer, J.W. (1971a). Populations of *Fusarium oxysporum* f.sp. *cepae* in organic soils in New York. *Phytopathology,* **61**, 1042-1048.

Abawi, G.S. and Lorbeer, J.W. (1971b). Pathological histology of four onion cultivars infected by *Fusarium oxysporum* f.sp. *cepae*. *Phytopathology,* **61**, 1164-1169.

Abawi, G.S. and Lorbeer, J.W. (1972). Several aspects of the ecology and pathology of *Fusarium oxysporum* f.sp. *cepae*. *Phytopathology,* **62**, 870-876.

Adams, P.B. and Ayres W.A. (1981). *Sporodesmium sclerotivorum*: distribution and function in natural biological control of sclerotial fungi. *Phytopathology*, **71**, 90-93.

Ahmed, A.H.M. and Tribe, H.T. (1977). Biological control of white rot of onion (*Sclerotium cepivorum*) by *Coniothyrium minitans*. *Plant Pathology*, **26**, 75-78.

Alderman, S.C. and Lacy, M.L. (1983). Influence of dew period and temperature on infection of onion leaves by dry conidia of *Botrytis squamosa*. *Phytopathology*, **73**, 1020-1023.

Anon. (1977). Dry bulb onions. Effects of various handling and drying practices on storage v7/40. *Report of Kirton Experimental Horticulture Station*, **14**, 53.

Anon. (1995). *FAO Production Yearbook 1994*, FAO Statistics Series No. 125, Food and Agriculture Organisation of the United Nations, Rome, 243pp.

Anon. (2003). FAO *Production Yearbook 2002*, FAO Statistics Series No. 176, Food and Agriculture Organisation of the United Nations, Rome, 261pp.

Armstrong, G.M. and Armstrong, J.K. (1981). *Formae speciales* and races of *Fusarium oxysporum* causing wilt diseases, in *Fusarium: Diseases, Biology, and Taxonomy*, (eds P. E. Nelson, T. A. Toussoun, and R. J. Cook) Pennsylvania State University Press, University Park, pp. 391-399.

Augusti, K.T. (1990). Therapeutic and medicinal values of onions and garlic, in *Onions and Allied Crops: Volume 111 - Biochemistry, Food Science and Minor Crops*, (eds H. D. Rabinowitch and J. L. Brewster) CRC Press, Inc., Boca Raton, Florida, pp. 93-108.

Aveling, T.A.S. (1998). Purple blotch (*Alternaria porri*) of onion. *Recent Research Developments in Plant Pathology*, **2**, 63-76.

Aveling, T.A.S., Snyman, H.G. and Rijkenberg, F.H J. (1994). Morphology of infection of onion leaves by *Alternaria porri*. *Canadian Journal of Botany*, **72**, 1164-1170.

Backhouse, D. (2003). Modelling inoculum dynamics and epidemiology of onion white rot, in *Eighth International Congress of Plant Pathology*, (eds D. Swain and S. Zydenbos) Christchurch, New Zealand, February 2nd-7th p. 102.

Basallotte-Ureba, M.J., Prados-Ligero, A.M. and Melero-Vara, J.M. (1999). Aetiology of leaf spot of garlic and onion caused by *Stemphylium vesicarium* in Spain. *Plant Pathology*, **48**, 139-145.

Battilani, P., Rossi, V. and Giosue, S. (1996). ONIMIL, a forecaster for primary infection of downy mildew of onion. *EPPO Bulletin*, **26**, 567-576.

Biesiada, A., Kolota, E., Pietr, S. *et al.* (2004). Evaluation of some biological methods of pink root rot control on leek. *Acta Horticulturae*, **635**, 187-193.

Bock, K.R. (1964). Purple blotch (*Alternaria porri*) of onion in Kenya. *Annals of Applied Biology*, **54**, 303-311.

Bosch Serra, A-D. and Currah, L. (2002). Agronomy of onions, in *Allium Crop Science: Recent Advances*, (eds H. D. Rabinowitch and L. Currah) CAB International, Wallingford, Oxon. UK, pp. 187-232.

Brix, H.D. and Zinkernagel, V. (1992). Screening for resistance of *Allium* species to *Sclerotium cepivorum* with special reference to nonstimulatory resistance. *Plant Pathology*, **41**, 13-19.

Brooks, F.T. (1953). *Plant Diseases*. Oxford University Press, 457 pp.

Bugiani, R., Govoni, P., Bottazzi, R. *et al.* (1995). Monitoring airborne concentrations of sporangia of *Phythophthora infestans* in relation to tomato late blight in Emilia Romagna, Italy. *Aerobiologia*, **11**, 41-46.

Carisse, O., Rolland, D., Lefebvre, A. and Talbot, B. (2003). Use of aerobiology data to manage onion blight caused by *Botrytis squamosa*, in *Eighth International Congress of Plant Pathology*, (eds D. Swain and S. Zydenbos) Christchurch, New Zealand, February 2nd-7th p. 5.

Clarkson, J.P., Kennedy, R. and Phelps, K. (2000). The effect of temperature and water potential on the production of conidia by sclerotia of *Botrytis squamosa*. *Plant Pathology*, **49**, 119-128.

Clarkson, J.P., Payne, T., Mead, A. and Whipps, J.M. (2002). Selection of fungal biocontrol agents of *Sclerotium cepivorum* for control of white rot by sclerotial degradation in a UK soil. *Plant Pathology*, **51**, 735-745.

Clarkson, J.P., Mead, A., Payne, T. and Whipps, J.M. (2004). Effect of environmental factors and *Sclerotium cepivorum* isolate on sclerotial degradation and biological control of white rot by *Trichoderma*. *Plant Pathology*, **53**, 353-362.

Coley-Smith, J.R. (1987). Alternative methods of controlling white rot disease of *Allium*, in *Innovative Approaches to Plant Disease Control*, (ed. I. Chet) New York, USA: John Wiley, 161-167.

Coley-Smith, J.R. (1988). *Sclerotium cepivorum* Berk., in *European Handbook of Plant Diseases,* (eds I. M. Smith, J. Dunez, D. H. Phillips, R. A. Lelliot and S. A. Archer) Blackwell Scientific Publications, Oxford, pp. 446-447.

Corgan, J.N. and Kedar, N. (1990). Onion cultivation in subtropical climates, in *Onions and Allied Crops: Volume 11 - Agronomy, Biotic Actions, Pathology, and Crop Protection,* (eds H.D. Rabinowitch and J.L. Brewster) CRC Press, Inc., Boca Raton, Florida, pp. 32-47.

Corgan, J.N. and Holland, M. (1993). 'NuMex Starlite' onion. *HortScience*, **28**, 66-67.

Correll, J.C. (1991). The relationship between *formae speciales*, races, and vegetative compatibility groups in *Fusarium oxysporum. Phytopathology*, **81**, 1061-1064.

Cramer, C.S. (2000). Breeding and genetics of *Fusarium* basal rot resistance in onion. *Euphytica*, **115**, 159-166.

Crevier, H. and Stewart, K.A. (1988). Effect of a systemic fungicide gel mixture on onion growth and smut. *Canadian Journal of Plant Science,* **68**, 557-559.

Crowe, F.J. and Hall, D.H. (1980). Soil temperature and moisture effects on sclerotium germination and infection of onion seedlings by *Sclerotium cepivorum. Phytopathology,* **70**, 74-78.

Crowe, F.J. and Carlson, H. (1994). Continued investigations of flooding as a means of *Allium* white rot control, in *Proceedings of the Fifth International Workshop on Allium White Rot, 1994,* (eds A.R. Entwistle and J.M. Melero-Vara) Cordoba, Spain and Warwick, UK: Instituto de Agricultura Sostenible and Horticulture Research International, pp. 169-183.

Currah, L. (2002). Onions in the Tropics: Cultivars and Country Reports, in *Allium Crop Science: Recent Advances* (eds H. D. Rabinowitch and L. Currah), CAB International, Wallingford, Oxon. UK, pp. 379-407.

Currah, L. and Proctor, F.J. (1990). Onions in tropical regions. *Bulletin No. 35*, Natural Resources Institute, UK. X111 232 pp + X111.

Dixon, G.R. (1981). *Vegetable Crop Diseases.* Macmillan Publishers Ltd, London. 404 pp.

Dow, R.L. and Lacy, M.L. (1969). Factors affecting the growth of *Urocystis colchici* in culture. *Phytopathology,* **59**, 1219-1222.

Ellerbrock, L.A. and Lorbeer, J.W. (1977). Sources of primary inoculum of *Botrytis squamosa. Phytopathology,* **67**, 363-372.

Entwistle, A.R. (1990). Root diseases, in *Onions and Allied Crops: Volume 11 - Agronomy, Biotic Actions, Pathology, and Crop Protection,* (eds H. D. Rabinowitch and J. L. Brewster) CRC Press, Inc., Boca Raton, Florida, pp.103-154.

Entwistle, A.R. and Granger, J. (1977). Sclerotium germination studies. *Report of the National Vegetable Research Station for 1976,* The British Society for the Promotion of Vegetable Research, Wellesbourne, Warwick, CV35 9EF, UK, pp. 97-98.

Entwistle, A.R. and Munasinghe, H.L. (1978). Epidemiology and control of white rot disease of onions, in *Plant Disease Epidemiology,* (eds P.R. Scott and A. Bainbridge) Blackwells Scientific Publications, Oxford, pp. 187-191.

Esler, G. and Coley-Smith, J.R. (1983). Flavour and odour characteristics of species of *Allium* in relation to their capacity to stimulate germination of sclerotia of *Sclerotium cepivorum. Plant Pathology,* **32**, 13-22.

Everts, K.L. and Lacy, M.L. (1990a). The influence of dew duration, relative humidity, and leaf senescence on conidial formation and infection of onion by *Alternaria porri. Phytopathology,* **80**, 1203-1207.

Everts, K.L. and Lacy, M.L. (1990b). Influence of environment on conidial concentration of *Alternaria porri* in air and on purple blotch incidence on onion. *Phytopathology,* **80**, 1387-1391.

Everts, K.L. and Lacy, M.L. (1996). Factors influencing infection of onion leaves by *Alternaria porri* and subsequent lesion expansion. *Plant Disease*, **80**, 276-280.

Fahim, M.M. (1966). The effect of light and other factors on the sporulation of *Alternaria porri. Transactions of the British Mycological Society*, **49**, 73-78.

Fantino, M.G. and Schiavi, M. (1987). Onion breeding for tolerance to *Fusarium oxysporum* f. sp. *cepae,* in Italy. *Phytopathologia Mediterranea*, **26**, 108-112.

Fitzgerald, S.M. and O'Brien, R.G. (1994). Validation of 'Downcast' in the Prediction of sporulation-infection periods of *Peronospora destructor* in the Lockyer Valley. *Australian Journal of Experimental Agriculture,* **34**, 537-539.

Fresenius, G. (1850). *Botrytis aclada* Fresen., *Beitrage zur Mykologie,* 16, Heinrich Ludwig Bronner, Frankfurt, Germany.

Friedrich, S., Leinhos, G.M.E. and Löpmeier, F.-J. (2003). Development of ZWIPERO, a model forecasting sporulation and infection periods of onion downy mildew based on meteorological data. *European Journal of Plant Pathology*, **109**, 35-45.

Gilles, T. and Kennedy, R. (2004). Evaluation of MILIONCAST, a forecaster for onion downy mildew, with historical data, in *Advances in Downy Mildew Research: Volume 2*, (eds P.T.N. Spencer-Phillips and M.J. Jeger) Kluwer Academic Publishers, Dordrecht, Netherlands, pp. 81-90.

Gilles, T., Phelps, K., Clarkson, J.P. and Kennedy, R. (2004). Development of MILIONCAST, an improved model for predicting downy mildew sporulation on onions. *Plant Disease*, **88**, 695-702.

Gubb, I.R. and MacTavish, H.S. (2002). Onion pre- and post-harvest considerations, in *Allium Crop Science: Recent Advances*, (eds H.D. Rabinowitch and L. Currah) CAB International, Wallingford, Oxford, UK, pp. 233-265.

Gupta, R.P., Srivastiva, K.J. and Pandey, U.B. (1991). Management of onion diseases and insect pests in India. *Onion Newsletter for the Tropics*, (eds L. Currah and F.J. Proctor), Natural Resources Institute, Chatham Maritime, Kent, UK, No. 3, 15-17.

Gupta, R.P., Srivastiva, K.J. and Pandey, U.B. (1994). Diseases and insect pests of onions in India. *Acta Horticulturae*, **358**, 265-268.

Hancock, J.G. and Lorbeer, J.W. (1963). Pathogenesis of *Botrytis cinerea*, *B. squamosa* and *B. allii* on onion leaves. *Phytopathology*, **53**, 669-673.

Hansen, H.N. (1929). Etiology of the pink root disease of onions. *Phytopathology*, **19**, 691-704.

Havey, M.J. (1995). *Fusarium* basal plate rot, in *Compendium of Onion and Garlic Disease*, (eds H.F. Schwartz and S.K. Mohan) The American Phytopathological Society, Minnesota, USA, pp. 10-11.

Hayden, N.J. and Maude, R.B. (1992). The role of seedborne *Aspergillus niger* in transmission of black mould of onion. *Plant Pathology*, **41**, 573-581.

Hayden, N.J. and Maude, R.B. (1994). The effect of heat on the growth and recovery of *Aspergillus* spp. from the mycoflora of onion seeds. *Plant Pathology*, **43**, 627 -630.

Hayden, N.J., Maude, R.B. and Proctor, F. J. (1994a). Studies on the biology of black mould (*Aspergillus niger*) on temperate and tropical onions. 1. A comparison of sources of disease in temperate and tropical field crops. *Plant Pathology*, **43**, 562-569.

Hayden, N.J., Maude, R.B., El Hassan, H.S. and Al Magid, A.A. (1994b). Studies on the biology of black mould (*Aspergillus niger*) on temperate and tropical onions. 2. The effect of treatments on the control of seedborne *A. niger*. *Plant Pathology*, **43**, 570-578.

Hennebert, G.L. (1963). *Les Botrytis des Allium, 15th Int. Symp. Phytopharmacie and Phytiatry*, pp. 851-876.

Hennebert, G.L. (1973). *Botrytis* and *Botrytis*-like genera. *Persoonia*, **7**, 183-204.

Hildebrand, P.D. and Sutton, J.C. (1982). Weather variables in relation to an epidemic of downy mildew. *Phytopathology*, **72**, 219-224.

Hildebrand, P.D. and Sutton, J.C. (1984a). Effect of weather variables on spore survival and infection of onion leaves by *Peronospora destructor*. *Canadian Journal of Plant Pathology*, **6**, 119-126.

Hildebrand, P.D. and Sutton, J.C. (1984b). Relationship of temperature, moisture, and inoculum density to the infection cycle of *Peronospora destructor*. *Canadian Journal of Plant Pathology*, **6**, 127-134.

Hildebrand, P.D. and Sutton, J.C. (1984c). Interactive effects of the dark period, humid period, temperature and light on sporulation of *Peronospora destructor*. *Phytopathology*, **74**, 1444-1449.

Holz, G. and Knox-Davies, P.S. (1985). Production of pectic enzymes by *Fusarium oxysporum* f. sp. *cepae* and its involvement in onion bulb rot. *Phytopathologische Zeitschrift*, **112**, 69-80.

Horst, R.K. (ed.) (2001). *Westcott's Plant Disease Handbook (6th edition)*. Van Nostrand Reinhold, New York, 1008 pp.

James, T.D.W. and Sutton, J.C. (1996). Biological control of leaf blight of onion by *Gliocladium roseum* applied as sprays and with fabric applicators. *European Journal of Plant Pathology*, **102**, 265-275.

Jesperson, G.D. and Sutton, J.C. (1987). Evaluation of a forecaster for downy mildew of onion (*Allium cepa* L.). *Crop Protection*, **6**, 95-103.

Johnston, A. and Booth, C. (eds) (1983). *Plant Pathologist's Handbook*, 2nd edn, Commonwealth Agricultural Bureaux, Slough, England, pp. 439.

Jones, D. Gareth (ed) (1998). *The Epidemiology of Plant Diseases*. Kluwer Academic Publishers, London, 460 pp.

Kay, S.J. and Stewart, A. (1994). Evaluation of fungal antagonists for control of onion white rot in soil box trials. *Plant Pathology*, **43**, 371-377.

Kennedy, R., Wakeham, A.J., Byrne, K.G. *et al.* (2000). A new method to monitor airborne inoculum of the fungal plant pathogens *Mycosphaerella brassicicola* and *Botrytis cinerea*. *Applied and Environmental Microbiology*, **66,** 2996-3000.

Keusgen, M. (2002). Health and Alliums, in *Allium Crop Science: Recent Advances,* (eds H.D. Rabinowitch and L. Currah) CAB International, Wallingford, Oxon. UK, pp. 357-378.

Köhl, J., Plas, C.H. van der, Molhoek, W.M.L. and Fokkema, N.J. (1995). Effect of interrupted leaf wetness periods on suppression of sporulation of *Botrytis allii* and *B. cinerea* by antagonists on dead onion leaves. *European Journal of Plant Pathology*, **101**, 627-637.

Köhl, J., Bélanger, R.R. and Fokkema, N.J. (1997). Interaction of four antagonistic fungi with *Botrytis aclada* in dead onion leaves: a comparative microscopic and ultrastructural study. *Phytopathology,* **87**, 634-642.

Lacy, M.L. and Roberts, D.L. (1982). Yields of onion cultivars in Midwestern organic soils infested with *Fusarium oxysporum* f.sp. *cepae* and *Pyrenochaeta terrestris*. *Plant Disease,* **66**, 1003-1006.

Lakra, B.S. (1999). Development of purple blotch incited by *Alternaria porri* and its losses in seed crop of onion (*Allium cepa*). *Indian Journal of Agricultural Sciences*, **69**, 144-146.

Leach, C.M., Hildebrand, P.D. and Sutton, J.C. (1982). Sporangium discharge by *Peronospora destructor*: Influence of humidity, red-infrared radiation and vibration. *Phytopathology,* **72**, 1052-1056.

Leggett, M.E. and Rahe, J.E. (1985). Factors affecting the survival of sclerotia of *Sclerotium cepivorum* in the Fraser Valley of British Columbia. *Annals of Applied Biology,* **106**, 255-263.

Lorbeer, J.W. (1992). *Botrytis* leaf blight of onion, in *Diseases of Vegetables and Oil Seed Rape Crops, Volume 11,* (eds H .S. Chaube, J. Kumar, A. N. Mukhopadhyay and U. S. Singh) Prentice Hall, Englewood Cliffs, New Jersey, pp. 186-211.

Lorbeer, J.W., Kuhar, T.P. and Hoffman, M.P. (2002). Monitoring and forecasting for disease and insect attack in onions and *Allium* crops within IPM strategies, in *Allium Crop Science: Recent Advances,* (eds H. D. Rabinowitch and L. Currah) CAB International, Wallingford, Oxon. UK, pp. 293-309.

Mace, E.M., Bell, A.A. and Beckman, C.H. (eds) (1981). *Fungal Wilt Diseases of Plants,* Academic Press, New York.

Maude, R.B. (1983). The correlation between seed-borne infection by *Botrytis allii* and neck rot development in store. *Seed Science and Technology,* **11**, 829-834.

Maude, R.B. (1990a). Storage diseases of onions, in *Onions and Allied Crops: Volume 11 - Agronomy, Biotic Actions, Pathology, and Crop Protection,* (eds H.D. Rabinowich and J.L. Brewster) CRC Press, Inc., Boca Raton, Florida, pp. 273-296.

Maude, R.B. (1990b). Leaf diseases of onions, in *Onions and Allied Crops: Volume 11 - Agronomy, Biotic Actions, Pathology, and Crop Protection,* (eds H. D. Rabinowitch and J. L. Brewster) CRC Press, Inc., Boca Raton, Florida, pp. 173-189.

Maude, R.B. (1996). *Seedborne Diseases and Their Control - Principles & Practice,* CAB International, Wallingford, Oxon OX10 8DE, UK, 280 pp.

Maude, R.B. and Presly, A.H. (1977a). Neck rot (*Botrytis allii*) of bulb onions. 1. Seed-borne infection and its relationship to the disease in the onion crop. *Annals of Applied Biology,* **86**, 163-180.

Maude, R.B. and Presly, A.H. (1977b). Neck rot (*Botrytis allii*) of bulb onions. 11. Seed-borne infection in relationship to the disease in store and the effect of seed treatment. *Annals of Applied Biology,* **86**, 181-188.

Maude, R.B. and Burchill, R.T. (1988). Black mould (*Aspergillus niger*) storage rot of bulb onions in the UK, in *Abstracts, 5th International Congress of Plant Pathology, Kyoto, Japan,* August 1988, p. 419.

Maude, R.B., Bambridge, J.M., Presly, A.H. and Phelps, K. (1985). Effects of field environment and cultural practices on incidence of *Botrytis allii* (neck rot) in stored bulb onions. *Quaderni della scuola di Specializzazione in Viticoltura ed Enologia 1985,* **9**, 241-243.

Mazzola, M. (2004). Assessment and management of soil microbial community structure for disease suppression. *Annual Review of Phytopathology*, **42**, 35-59.

Meier, R. (2000). Downy mildew and leaf spots in onions. *PAV-Bulletin Akkerbouw,* No. 4, 25-30.

Meredith, D.S. (1966). Spore dispersal in *Alternaria porri* (Ellis) Neerg. on onions in Nebraska. *Annals of Applied Biology,* **57**, 67-73.

Metcalfe, D.A. and Wilson, C.R. (1999). Histology of *Sclerotium cepivorum* infection of onion roots and the spatial relationships of pectinases in the infection process. *Plant Pathology,* **48**, 445-452.

Metcalfe, D.A. and Wilson, C.R. (2001). The process of antagonism of *Sclerotium cepivorum* in white rot infected onion roots by *Trichoderma koningii*. *Plant Pathology,* **50**, 249-257.

Miller, M.E. (1983). Relationships between onion leaf age and susceptibility to *Alternaria porri. Plant Disease*, **67**, 284-286.

Miller, M.E. and Lacy, M.L. (1995). Purple blotch, in *Compendium of Onion and Garlic Diseases,* (eds H. F. Schwartz, and S. K. Mohan) The American Phytopathological Society, Minnesota, USA, pp. 23-24.

Miller, M.E., Taber, R.A. and Amador, J.M. (1978). *Stemphylium* blight of onion in South Texas. *Plant Disease Reporter,* **62**, 851.

Monnet, Y. and Thibault, J. (2000). Maladies de l'oignon et de l'échalote. *PHM Revue Horticole* No. 413, 42-43.

Munn, M.T. (1917). Neck rot of onions. *Bulletin of the New York Agricultural Experimental Station,* No. 437, 363-455.

Nielsen, K. and Yohalem, D.S. (2001). Origin of a polyploid *Botrytis* pathogen through interspecific hybridisation between *Botrytis aclada* and *B. byssoidea. Mycologia*, **93**, 1064-1071.

Nielsen, K., Yohalem, D.S., Green, H. and Jensen, D.F. (2000). Biological control of grey mould in onion, in *18th Danish Plant Protection Conference 111. DJF Rapport, Havebrug,* No. 17, 55-61.

Nielsen, K., Justesen, A.F., Funck Jensen, D. and Yohalem, D.S. (2001). Universally primed polymerase chain reaction alleles and internal transcribed spacer restriction fragment length polymorphisms distinguish two subgroups in *Botrytis aclada* distinct from *B. byssoidea. Phytopathology,* **91**, 527-533.

Özer, N., Köycü, D., Chilosi, G. *et al.* (2003). Pectolytic isoenzymes by *Fusarium oxysporum* f. sp. *cepae* and antifungal compounds in onion cultivars as a response to pathogen infection. *Canadian Journal of Plant Pathology*, **25**, 249-257.

Pagès, J. and Notteghem, J.L. (1996). Effect of soil treatment practices on pink root disease in the Senegalese cultivation system. *International Journal of Pest Management*, **42**, 29-34.

Pandotra, V.R. (1965). Purple blotch disease of onion in Punjab. 11. Studies on the life-history, viability and infectivity of the causal organism *Alternaria porri. Proceedings of the Indian Academy of Science, Section B,* **61**, 326-330.

Phillon, P. (2003). Timing of sprays against potato late blight, based on average spore concentration, in *Eighth International Congress of Plant Pathology,* (eds D. Swain and S. Zydenbos) Christchurch, New Zealand, February 2nd- 7th p. 6.

Rabinowitch, H.D. and Brewster, J.L. (eds) (1990). *Onions and Allied Crops: Volume 11 - Agronomy, Biotic Actions, Pathology, and Crop Protection,* CRC Press, Inc., Boca Raton, Florida, 320 pp.

Rabinowitch, H.D. and Currah, L. (eds) (2002). *Allium Crop Science: Recent Advances,* CAB International, Wallingford, Oxon. UK, 515 pp.

Randle, W.M. and Lancaster, J.E. (2002). Sulphur compounds in Alliums in relation to flavour quality, in *Allium Crop Science: Recent Advances,* (eds H. D. Rabinowitch and L. Currah) CAB International, Wallingford, Oxon. UK, pp. 329-356.

Rogers, I.S. and Henderson, R.D. (1989). Testing suitability of onion cultivars for dehydration. *Acta Horticulturae*, **247**, 157-162.

Schwartz, H.F. and Mohan, S.K. (eds) (1995). *Compendium of Onion and Garlic Diseases,* The American Phytopathological Society, Minnesota, USA, 54 pp.

Sherf, A.F. and Macnab, A.A. (1986). Onions, garlic, leeks and shallots, in *Vegetable Diseases and Their Control,* John Wiley & Sons, NewYork, 433-470.

Shirane, N., Masuko, M. and Hayashi, Y. (1989). Light microscope observation of nuclei and mitotic chromosomes of *Botrytis* species. *Phytopathology*, **79**, 728-730.

Shoemaker, P.B. and Lorbeer, J.C. (1977). The role of dew and temperature in the epidemiology of *Botrytis* leaf blight of onion. *Phytopathology*, **67**, 1267-1272.

Stadnik, M.J. and Dhringa, O.D. (1997). Root infection by *Fusarium oxysporum* f.sp. *cepae* at different growth stages and its relation to the development of onion basal rot. *Phytopathologia Mediterranea,* **36**, 8-11.

Stewart, A. and Franicevic, S.C. (1994). Infected seed as a source of inoculum for *Botrytis* infection of onion bulbs in store. *Australasian Plant Pathology*, **23**, 36-40.

Suheri, H. and Price, T.V. (2000a). Infection by *Alternaria porri* and *Stemphylium vesicarium* on onion leaves and disease development under controlled environments. *Plant Pathology,* **49**, 377-384.

Suheri, H. and Price, T.V. (2000b). *Stemphylium* leaf blight of garlic (*Allium sativum*) in Australia. *Australasian Plant Pathology*, **29**, 192-199.

Suheri, H. and Price, T.V. (2001). The epidemiology of purple leaf blotch on leeks in Victoria, Australia. *European Journal of Plant Pathology,* **107**, 503-510.

Sumner, D.R. (1995). Pink root, in *Compendium of Onion and Garlic Diseases,* (eds H. F. Schwartz and S. K. Mohan), The American Phytopathological Society, Minnesota, USA, pp.12-13.

Sutton, J.C., Swanton, C.J. and Gillespie, T.J. (1978). Relation of weather variables and host factors to the incidence of airborne spores of *Botrytis squamosa. Canadian Journal of Botany,* **56**, 2460-2469.

Sutton, J.C., James, T.D.W. and Rowell, P.M. (1986). BOTCAST: A forecasting system to time the initial fungicide spray for managing *Botrytis* leaf blight of onions. *Agriculture, Ecosystems and Environment,* **18**, 123-143.

Swift, C.E., Wickliffe, E.R. and Schwartz, H.F. (2002). Vegetative compatibility groups of *Fusarium oxysporum* f.sp. *cepae* from onion in Colorado. *Plant Disease*, **86**, 606-610.

Tachibana, H. and Duran, R. (1961). The pathogenicity of monosporic versus polysporic cultures of *Urocystis cepulae* (Schlecht.) Rabenh. *Phytopathology,* **51**, 67.

Thaxter, R. (1890). The smut of onions (*Urocystis cepulae* Frost). *Report of the Connecticut Agricultural Experiment Station for 1889,* 127-154.

Toit, L J. du, Derie, M.L. and Peter, G.Q. (2004). Prevalence of *Botrytis* spp. in onion seed crops in the Columbia Basin of Washington. *Plant Disease*, **88**, 1061-1068.

Tylkowska, K. and Dorna, H. (2001). Onion (*Allium cepa*) seed and plant health with special reference to *Botrytis allii. Phytopatologia Polonica* No. 21, 55-68.

Utkhede, R.S. and Rahe, J.E. (1980). Screening world onion germplasm collection and commercial cultivars for resistance to smut. *Canadian Journal of Plant Science,* **60**, 157-161.

Utkhede, R.S., Rahe J.E., Coley-Smith, J.R. *et al.* (1982). Genotype-environment interactions for resistance to onion white rot. *Canadian Journal of Plant Pathology*, **4**, 269-271.

Uzo, J.O and Currah, L. (1990). Cultural systems and agronomic practices in tropical climates, in *Onions and Allied Crops: Volume 11 - Agronomy, Biotic Actions, Pathology, and Crop Protection,* (eds H. D. Rabinowitch and J. L. Brewster), CRC Press, Inc., Boca Raton, Florida, pp. 49-62.

Van der Plank, J.E. (1963). *Plant Diseases: Epidemics and Control.* Academic Press, New York and London. 349 pp.

Vincelli, P.C. and Lorbeer, J.W. (1988). Forecasting spore episodes of *Botrytis squamosa* in commercial onion fields in New York. *Phytopathology,* **78**, 966-970.

Viranyi, F. (1988). *Peronospora destructor* (Berk.) Caspary, in *European Handbook of Plant Diseases,* (eds I. M. Smith, J. Dunez, D. H. Phillips, R. A. Lelliot and S. A. Archer), Blackwell Scientific Publications, Oxford, pp. 331-333.

Visser, C.L.M de (1996). Field evaluation of a supervised control system for *Botrytis* leaf blight in spring sown onions in the Netherlands. *European Journal of Plant Pathology*, **102**, 795-805.

Visser, C.L.M de (1998). Development of a downy mildew advisory model based on downcast. *European Journal of Plant Pathology,* **104**, 933-943.

Wakeham, A., Kennedy, R. and McCartney, A. (2004). The collection and retention of a range of common airborne spore types trapped directly into microtiter wells for enzyme-linked immunosorbent analysis. *Aerosol Science,* **35**, 835-850.

Walker, J.C. (1925). Two undescribed species of *Botrytis* associated with the neck rot diseases of bulb onions. *Phytopathology,* **15**, 708-713.

Walker, J.C. (1926). *Botrytis* neck rots of onions. *Journal of Agricultural Research,* **33**, 893-928.

Walker, J.C. and Wellman, F.L. (1926). Relation of temperature to spore germination and growth of *Urocystis cepulae. Journal of Agricultural Research,* **32**, 133-146.

Weille, G.A. de (1975). An approach to the possibilities of forecasting downy mildew infection in onion crops. *Mededelingen en Verhandelingen Koninklijk Nederlands Meteorologisch Instituut,* no. 97, 1-83.

Weller, D.M., Raaijmakers, J.M., McSpadden Gardener, B.B. and Thomashow, L.S. (2002) Microbial populations responsible for specific soil suppressiveness to plant pathogens. *Annual Review of Phytopathology*, **40**, 309-348.

Whipps, J.M. (2001). Microbial interactions and biocontrol in the rhizosphere. *Journal of Experimental Botany,* **52**, 487-511.

Wright, P.J., Chynoweth, R.W., Beresford, R.M. and Henshall, W.R. (2002). Comparison of strategies for timing protective and curative fungicides for control of onion downy mildew. *Proceedings of the BCPC Conference 2002 – Pests and Diseases.* The British Crop Protection Council, Surrey, UK. pp. 207-212.

Wu, W.-S. (1979). Survey on seed-borne fungi of vegetables. *Plant Protection Bulletin, Taiwan,* **21**, 206-219.
Yarwood, C.E. (1943). Onion downy mildew. *Hilgardia,* **14**, 595-691.
Yohalem, D.S., Nielsen, K. and Nicolaisen, M. (2003). Taxonomic and nomenclatural clarification of the onion neck rotting *Botrytis* species. *Mycotaxon*, **85**, 175-183.

CHAPTER 20

THE RECENT EPIDEMIC OF CASSAVA MOSAIC VIRUS DISEASE IN UGANDA

G.W. OTIM-NAPE AND J.M. THRESH

20.1 INTRODUCTION

This chapter differs from many of the other contributions to this volume in that it deals with only one specific disease. Cassava mosaic virus disease (CMD) justifies special treatment for several reasons:

- Cassava provides one of the main staple foods in many tropical areas and it is particularly important in several of the poorest countries of sub-Saharan Africa.
- CMD is the most important disease of cassava in Africa and also in Sri Lanka and southern India.
- The disease has a long history in Africa, where research has been in progress for many years. Indeed, CMD is likely to have received more attention than any other vector-borne disease of an African food crop.
- CMD is a prime example of a particularly important group of virus diseases - those that have a dual strategy of dispersal as they are disseminated in vegetative propagules and transmitted by an arthropod vector.
- A very damaging epidemic of CMD in Uganda in the 1990s caused serious food shortages and affected the livelihood and well-being of rural communities in many parts of the country. This necessitated major interventions and numerous cassava improvement projects were mounted by governmental and non-governmental organizations.
- When the first attempts were made to control the epidemic in Uganda, it became apparent that there was insufficient information available to initiate effective management strategies and this led to greatly increased research on CMD, not only in Uganda but also elsewhere in Africa.
- The Ugandan epidemic spread to the adjacent countries of Kenya, Tanzania, Burundi and Rwanda and so attained the status of a pandemic, which currently poses a threat to other important cassava-growing areas in Africa.

B. M. Cooke, D. Gareth Jones and B. Kaye (eds.), The Epidemiology of Plant Diseases, 2nd edition, 521–549.

- The current pandemic provides an example of the way in which the occurrence and control of a disease is closely associated with the vulnerability of the varieties being grown and the availability of resistant varieties.
- Experience with CMD in Uganda demonstrated the advantages of adopting an ecological approach in seeking to explain the main features of the epidemic in the country and of the pandemic in the region and their effects on the amount and type of cassava grown.
- Research on CMD and the viruses responsible has benefited greatly from collaboration between virologists in the tropics and those in developed countries of Europe and North America with access to sophisticated laboratory equipment and techniques.
- The highly successful outcome of the research and extension activities mounted in Uganda in response to the epidemic provides a striking example of the benefits to be gained from such investments in agriculture, as evident from the very favourable benefit/cost ratios reported.

This chapter briefly considers the main features of CMD and its history, distribution and importance in Africa. The recent epidemic in Uganda is then discussed together with the approaches that were adopted to achieve control. Additional information and further details of the control measures introduced and deployed in Uganda are presented elsewhere (Otim-Nape *et al.,* 2000). The emphasis here is on the main features of general epidemiological interest.

20.2 CASSAVA AND CASSAVA MOSAIC DISEASE IN AFRICA

Cassava (*Manihot esculenta:* Euphorbiaceae) is a semi-woody perennial that is not known in the wild. The crop is considered to have evolved from wild *Manihot* species in South/Central America, where it has been cultivated for millennia. Cassava is still widely grown in the neotropics, which currently account for *c.* 18% of total world production - estimated to be *c.* 185 million tonnes per annum (FAO, 2003).

The tropical areas of sub-Saharan Africa now form the main region of cassava production, even though cultivation did not begin there until comparatively recent times following the introduction of the crop to coastal areas of West Africa in the sixteenth century and to coastal East Africa in the eighteenth century (Jones, 1969). Cassava soon spread inland from the coastal areas, but does not seem to have been widely grown until the twentieth century. Production then expanded rapidly in many countries and in very diverse agroecologies. The total area cultivated is currently estimated to be 11.2 million hectares (M. ha) and the leading producers are Nigeria (3.5 M. ha), Democratic Republic of Congo (DRC) (1.8 M. ha), Mozambique (0.9 M. ha) and Ghana (0.8 M. ha). However, the area exceeds 50 000 ha in each of 17 other African countries (FAO, 2003).

In Africa, cassava is grown mainly for the tuberous roots, which are used for human consumption in diverse forms. The leaves are also consumed in DRC and some other countries as an important component of dietary protein and vitamins.

Little cassava is exported, but some roots are used locally to feed livestock or poultry, or to prepare starch or alcoholic products. Production is almost exclusively by small-scale farmers, many of whom are in the poorest sectors of society. The very diverse methods of production and utilization were assessed in ten countries during 1990-94 in the Rockefeller-funded Collaborative Study of Cassava in Africa (Nweke, 1994). This showed big differences between areas in the importance and overall status of cassava. However, in all countries there is great varietal diversity and the crop is usually intercropped with one or more other species.

Cassava is routinely propagated vegetatively using hardwood stem cuttings. Vegetatively propagated crops are particularly prone to virus infection and cassava is no exception. At least 18 different viruses or putative viruses have been described, of which eight are known to occur in Africa (Thottappilly *et al.*, 2003). The most important of these are the whitefly-borne geminiviruses (Family: *Geminiviridae*; Genus: *Begomovirus*), which cause CMD. This was first reported in 1894 in what is now Tanzania. The disease was soon shown to be transmissible by grafts and by whiteflies and, in the absence of a visible pathogen, was attributed to virus infection. However, no virus particles were detected until the 1970s, when sap inoculations to cassava and also to *Nicotiana clevelandii* and several other herbaceous host species were successful. The geminivirus so isolated was later characterized and caused the typical symptoms of CMD when returned to cassava, so fulfilling Koch's postulates (Bock and Woods, 1983). Subsequently, three similar but distinct cassava mosaic geminiviruses (CMGs) were distinguished, of which two occur in Africa and the third in the Indian subcontinent (Swanson and Harrison, 1994). The two African CMGs were at first considered to have distinct and largely non-overlapping distributions, but subsequent studies on a wider range of isolates have shown the situation to be more complex than had been assumed. Moreover, five additional CMGs have been distinguished recently (Fauquet and Stanley, 2003) and a novel recombinant virus or strain has been associated with the current CMD pandemic in East and Central Africa (Deng *et al.,* 1997; Harrison *et al.,* 1997a; Zhou *et al.,* 1997).

There is little information on the early history of CMD in Tanzania or elsewhere in Africa. The disease seems to have become increasingly important in the 1920s and 1930s when cassava production expanded rapidly, and serious problems were reported in many countries and research began in several. During this period, the first attempts were made to select CMD-resistant varieties from those being grown at the time in Africa, or introduced from elsewhere and resistance breeding programmes were initiated (Jennings, 1994). These continue in several countries and, since 1971, the most influential has been at the International Institute of Tropical Agriculture (IITA), Ibadan, Nigeria. The overall research effort on CMD has been considerable, as evident from the sequence of reviews and proceedings (Fauquet and Fargette, 1988, 1990; Jennings, 1994; Thresh and Otim-Nape, 1994; Thresh *et al.,* 1994a; Otim-Nape *et al.,* 1996; Thresh *et al.,* 1998a,b; Legg, 1999; Legg and Thresh, 2000, 2004; Calvert and Thresh, 2002; Thottappilly *et al.,* 2003; Thresh and Cooter, 2005). Indeed, the disease has received more attention than any other virus disease of an African food crop, yet it remains prevalent in many areas and is still regarded as the most important disease of cassava in Africa. Moreover, CMD was rated as the most important vector-borne disease of any African food crop in a detailed economic assessment completed before the occurrence of the current regional pandemic (Geddes, 1990).

Until recently there were few quantitative data to support these assertions, or on which to base definitive estimates of the losses caused by CMD. Nevertheless, various estimates have been made, most recently by Thresh *et al.* (1997), who suggested overall losses in Africa of 15 to 24% of total production. This is equivalent to 13-23 M.t., compared with estimated production at the time of 73 M.t. On these assumptions, annual losses total US$1300 to 2300 million at a conservative value of US$100^{t-1}. In deriving these figures a broad distinction was made between three distinct epidemiological situations, designated 'benign', 'endemic' and 'epidemic'.

The *benign* situation is currently represented by many parts of Tanzania and the mid-altitude areas of Cameroon and Malawi, where the incidence of CMD is generally low and mainly due to the use of partially infected stocks of planting material. In these areas, the symptoms of CMD are not usually very conspicuous, there is little spread by the whitefly vector (*Bemisia tabaci*) and disease control measures are not considered essential. Some loss of crop is incurred, but it is accepted by farmers and researchers as unimportant and there are no compelling reasons to adopt CMD-resistant varieties or phytosanitation. Thus, the situation is stable and there is a dynamic equilibrium between counteracting tendencies. One is for the incidence of infection to increase, albeit slowly, due to a limited amount of spread by vectors. The other is for the incidence to decrease because infected plants tend to be underrepresented on further vegetative propagation. This is because the debilitating effect of CMD on growth decreases the number of stems that are suitable for use as cuttings and farmers tend to select these from the most vigorous unaffected plants available. An additional factor is that infected cuttings grow less vigorously than healthy ones, sustain greater mortality during the early stages of growth and are more likely to be damaged during the first weeding. Moreover, CMD is not fully systemic in infected plants, especially those infected at a late stage of growth, and so they provide at least some uninfected cuttings. This so-called 'reversion' phenomenon has important epidemiological implications in restricting the degeneration that would otherwise occur on successive cycles of vegetative propagation (Fargette *et al.,* 1994). The interactions between the various factors are complex and have been modelled. These studies indicate that dynamic equilibria can develop under different conditions and at incidences of disease determined by the values of the assumptions made on rates of spread, the amount of reversion and the extent to which infected cuttings are under-represented because of vigour/selection effects (Fargette *et al.,* 1994; Fargette and Vié, 1995; Holt *et al.,* 1997).

The *endemic* situation is also stable, but completely different from the benign in that there is such a high incidence of CMD that farmers have only limited access to uninfected cuttings for new plantings. This is the situation in many important cassava-growing areas of West and Central Africa, where farmers and even researchers have become so accustomed to CMD that it is regarded as largely inevitable and of no great significance. Yields are substandard, but they are accepted as normal and specific control measures are seldom recommended or adopted. However, farmers are discriminating in their choice of variety and adopt those that grow and yield satisfactorily, despite the occurrence of CMD (Thresh *et al.,* 1994b).

There is also increasing use of improved varieties, selected at IITA or elsewhere partly for their ability to resist or tolerate infection with CMD. Stocks of these varieties that are known to be free of CMD are seldom available, the removal (roguing) of diseased plants is seldom practised and the benefits to be gained from adopting such phytosanitation measures are unclear.

In ecological terms, the *epidemic* situation is one of acute instability and represents an extreme perturbation of the dynamic equilibrium to be expected between an obligate parasite and its host(s). As in the endemic situation, there is a high incidence of CMD, but symptoms are so severe that plants are stunted and produce little or no yield. Food security and economic livelihoods are threatened and the whole future of the crop in the area is jeopardized. Clearly, this is an unsustainable situation and farmers have little alternative but to decrease the area of cassava grown and switch to other crops, at least until such time as the epidemic abates or effective control is achieved using resistant varieties or by other means.

For these reasons, the epidemic situation is unlikely to be widespread or sustained for long and has seldom been reported. However, this may be because of the limited research on cassava in many parts of Africa. At least some outbreaks may be overlooked or unreported, or their full significance is not appreciated until farmers have worked out their own solutions by changing variety or decreasing the area of cassava grown. There are only brief accounts of the early epidemics associated with the expansion of cassava production in East and West Africa in the 1920s and 1930s and of the more recent epidemics in Cape Verde (Anon., 1992) and Akwa Ibom State, Nigeria (Anon., 1993). The 1940s epidemic in Madagascar is relatively well documented (Cours *et al.,* 1997). The recent epidemic in Uganda has been followed in unprecedented detail and is considered here.

20.3 CASSAVA AND CASSAVA MOSAIC DISEASE IN UGANDA

Cassava was first reported in Uganda during the latter part of the nineteenth century and it was grown on only a limited scale until the 1920s, when production expanded rapidly. As in many other parts of Africa at the time, this expansion was soon followed by reports of CMD, which was first recorded in Uganda in 1928. By 1934, CMD was of considerable importance in many areas, including what are now Soroti and Kumi administrative districts. (Fig. 20.1 locates the different districts of Uganda mentioned in this chapter). There was almost total infection of the main varieties in the worst affected areas. New varieties and breeding lines introduced from Tanzania were evaluated for resistance and concerted breeding efforts led to the selection of CMD-resistant varieties. These were multiplied rapidly in the 1950s and distributed in quantity to farmers (Jameson, 1964). The use of such varieties, and the measures of selection and roguing, were enforced by local government statute and led to the control of the disease for several decades. This satisfactory situation seems to have continued in the 1970s and into the 1980s. However, little detailed information is available on the incidence of CMD during this period and the disease received only limited research attention until new problems were reported in the late 1980s and the latest investigations began.

20.4 THE 1990s EPIDEMIC IN UGANDA

20.4.1 Initial reports from Luwero district: 1988

In 1988 there were reports of crop failure in an important cassava-growing area of northern Luwero district (now Nakasongola following recent boundary changes). An estimated 2000 ha of cassava were severely affected, which was a substantial proportion of the total area being grown in the locality. The damage was at first attributed to the cassava green mite *(Mononychellus tanajoa)*. However, surveys of the affected area and adjacent localities showed that it was due to an unusually

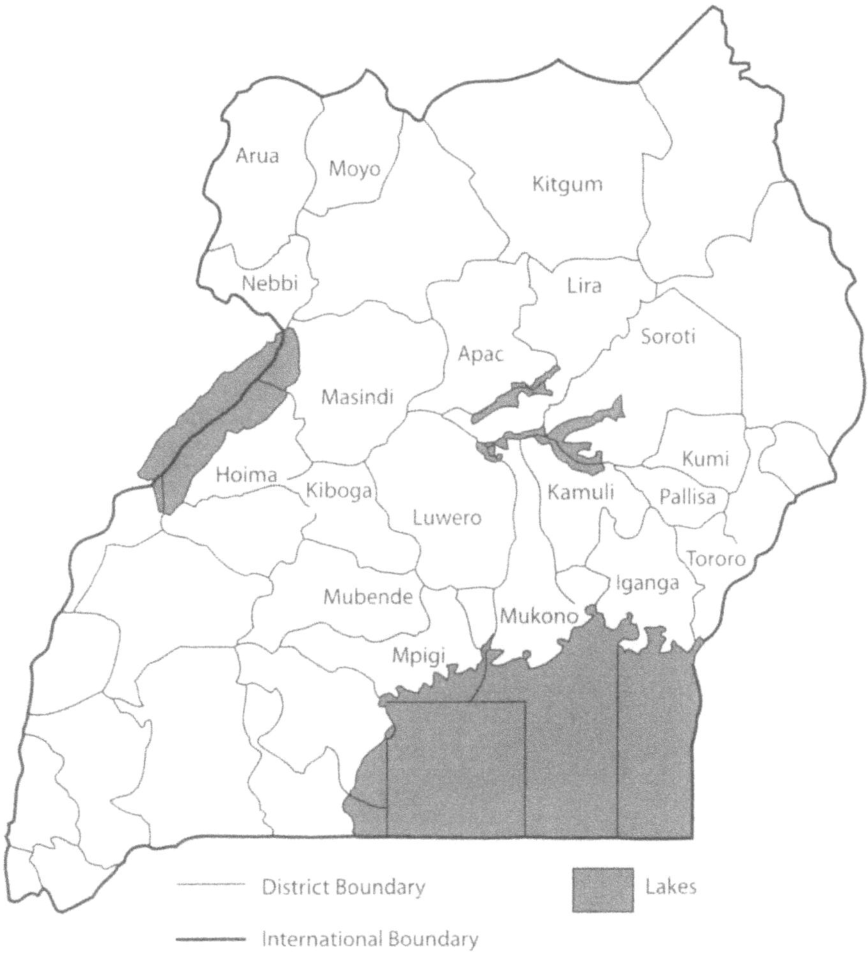

Figure 20.1. Sketch map of Uganda showing the administrative boundaries at the time of the 1990s epidemic and the names of the districts that feature in the text.

severe outbreak of CMD, which hitherto had been relatively unimportant (Otim-Nape *et al.*, 2000). Affected plants developed severe leaf mosaic symptoms and chlorosis. There was much leaf distortion and abscission and little or no yield of tuberous roots, except from the plants recently infected. Plants grown from infected cuttings were the most severely affected and grew so badly that whole plantings were abandoned or removed and replaced by other crops.

Uganda at the time was emerging from a prolonged period of political instability and insecurity. Access to many parts of the country had been impossible or severely curtailed and the work of the National Cassava Programme had been seriously disrupted following the enforced evacuation of all senior staff from the Serere Research Station in Soroti District and their relocation at Namulonge Agricultural Research Institute, near Kampala in 1987. This explains the limited information on the background and emergence of the Luwero epidemic. It soon became apparent from observations in northern Luwero that the affected area was expanding and the progress of the epidemic was monitored by regular assessments of farmers' plantings along the main north-south road through Luwero and on into Mpigi district and Kampala. Each year from 1988, the epidemic moved progressively southwards along a broad front towards Kampala at a rate of *c.* 20-30 km per year (Fig. 20.2). Within the epidemic area, CMD-affected plants developed very severe symptoms (Gibson *et al.*, 1996), and produced such low yields that serious food shortages ensued (Otim-Nape *et al.*, 2000).

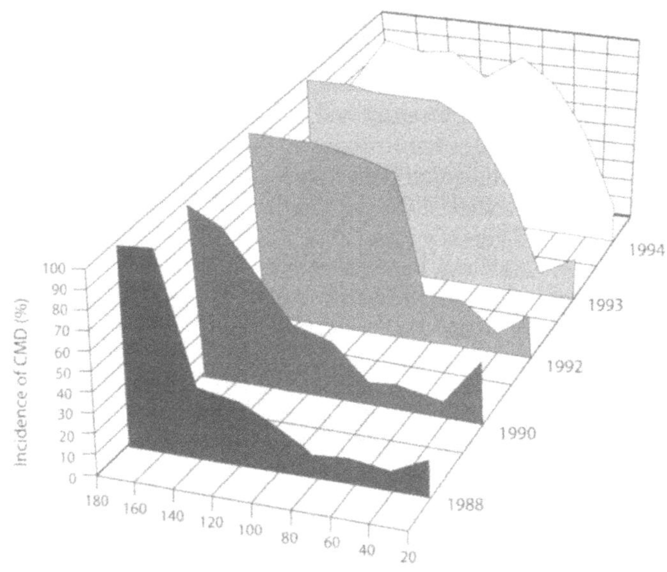

Figure 20.2. The overall incidence of CMD in successive years in representative farmers' plantings along a south-north transect from Kampala into northern Luwero district, illustrating the southward movement of the epidemic front between 1988 and 1994.

Attempts were made to restore production in the areas of Luwero that were first affected by providing farmers with uninfected cuttings of Ugandan varieties obtained from districts to the south, where CMD was not a problem at the time. Almost total infection occurred within a few months of planting, even in locations where farmers had followed the official advice to remove all diseased plants from the neighbourhood before the uninfected cuttings were planted. This emphasizes the high 'infection pressure' prevailing at the time, the mobility of the whitefly vector and the vulnerability of the introduced varieties. There was an obvious need for varieties that were resistant to infection, but these were not available in quantity for distribution to farmers. This led to virus screening and on-farm trials of potential new varieties for release in the area.

While these studies were in progress, farmers in Luwero responded in various ways to the great scarcity of CMD-free cuttings for new plantings. Some farmers sought to maintain production by collecting cuttings from plants in the locality that had escaped infection, or that developed inconspicuous symptoms and continued to produce at least some yield. There was a rapid turnover in varieties as those that were vulnerable soon succumbed and were discarded, whereas relatively tolerant varieties became increasingly prevalent. Nevertheless, many farmers were so badly affected that they ceased to grow cassava and increased the area of other food crops including maize, sorghum, sweet potato and finger millet. There was also continued use of cuttings obtained from relatively unaffected areas to the south and from other districts to the west bordering Lake Albert. This traffic was facilitated by the good communications and the ready availability of transport along main and feeder roads. Cassava production was sustained, although many of the varieties used were susceptible to CMD. They eventually succumbed to the disease, which created the need for further replacement material and CMD-resistant varieties were adopted when they became available.

Initially, the main rehabilitation project in Luwero and also in five other districts of Uganda was funded by the UK Gatsby Charitable Foundation (GCF) (Otim-Nape *et al.,* 1994, 1997). This supported the activities of the National Cassava Programme, extension staff and local government collaborators in multiplying and distributing the improved CMD-resistant varieties that had been shown to be successful in on-farm trials and were sought by farmers. Funds were also provided for on-farm trials and to train farmers and extension staff on CMD and its control. The initial emphasis was on providing cuttings of three improved virus-resistant varieties that originated from IITA, Nigeria, in the 1970s. These and other officially released varieties now account for a substantial proportion of the area grown in Luwero district and predominated in 46% of the fields assessed there in 2003 (Bua *et al.,* 2005). Additional CMD-resistant varieties are being developed to extend the range of genotypes available (Otim-Nape *et al.*, 2000).

20.4.2 Arua, Nebbi and Moyo districts (West Nile region)

Cassava has long been the main staple food in Arua, Nebbi and Moyo districts of the West Nile region of north-west Uganda, near the border with eastern DRC. A severe epidemic of CMD was first noticed in these districts in 1989 and coincided with an unusually severe drought, which lasted five months and affected cassava and other food crops. Food shortages ensued and there were many thefts of tuberous roots from any surviving fields. The epidemic developed rapidly between July and October 1990, and symptoms were very conspicuous. The predominant local varieties were all very susceptible and severely damaged.

The first attempts to restore cassava production in West Nile were in 1991 and they were funded by FAO. Many cuttings were introduced to the area from unaffected districts of southern Uganda. However, the varieties used were not sufficiently resistant to withstand infection and soon succumbed, as had occurred earlier in Luwero. CMD-resistant varieties were then introduced and distributed by several non-governmental organizations (Otim-Nape *et al.,* 2000). The IITA CMD-resistant variety TMS 30572 (released in Uganda as Nase 3 and also known as Migyera or Nigeria) proved to be particularly well adapted to local conditions. It is now widely grown and considerable use has also been made of varieties obtained locally, or from neighbouring areas of DRC, which have at least some degree of resistance to CMD. Production has continued and the effects of the epidemic have been less severe than in many other districts where there is less varietal diversity.

20.4.3 Kumi district

In 1991, attention turned to the Kumi district of eastern Uganda, which had been one of the areas worst affected by the long period of civil unrest and insecurity. Much of the population had been held in detention camps and normal agricultural activity was impossible or badly disrupted. The situation had improved considerably by 1991, when the area again became accessible and rehabilitation and resettlement projects were initiated by relief agencies. The main UK-funded OXFAM project involved the distribution of fishing equipment, farm tools, seed and cuttings of cassava. This had previously been the most important staple food crop and had been grown widely for local consumption and for export to other areas.

The initial rehabilitation effort was greatly impeded by the epidemic of CMD occurring in the area at the time. Initially, it was most damaging in the northern counties, especially in the popular variety Ebwanateraka which predominated throughout the district. Infection was least in counties to the south bordering the relatively fertile high rainfall areas of Mbale district near Mt Elgon on the Kenya border.

A striking feature of the detailed observations made near Kumi township in April 1991 was that the recently planted cuttings had been largely free of CMD, as indicated by the complete absence of symptoms on the first-formed leaves. Nevertheless, the youngest leaves of almost all plants showed conspicuous symptoms of CMD, due to a sudden influx of whitefly-borne inoculum some weeks previously. Whiteflies were exceptionally numerous in the area at the time and up to 60 adults were recorded per shoot tip.

A later survey of Kumi district in April 1992 revealed that the overall disease situation had deteriorated further and the symptoms were very severe. There was a high incidence of infection throughout the whole district, including the southern areas where CMD incidence and severity were relatively low in 1991. Much of the infection in the southern areas had occurred soon after planting and was due to whiteflies, as recorded to the north around Kumi township in 1991. The situation was different there in 1992 as, in the absence of healthy planting material, farmers had collected cuttings from plants that had been infected late the previous year and consequently virtually all plants showed severe symptoms from the outset. Growth was very stunted; many plantings were soon abandoned and became completely overgrown by weeds.

Some use was being made of cuttings obtained from less severely affected districts immediately to the south of Kumi, but movement was restricted and the traffic in planting material was much less than in Luwero district. Consequently, there was a rapid decline in the area of cassava as farmers abandoned the crop after realizing the futility of planting infected cuttings. This had very severe consequences in 1994, when there were widespread food shortages and famine-related deaths were reported to a later Presidential Commission of Inquiry into the causes of the crisis. The areas worst affected were those where the second rains of 1993 had failed and there was little or no cassava available to sustain populations until the 1994 crops of sweet potato, sorghum and other grain crops were harvested.

Various attempts were made to restore cassava production in Kumi using the most appropriate varieties available. In the initial 1991 OXFAM resettlement project, cuttings of the Ugandan varieties Bao and Aladu were obtained from Apac district to the west. However, these varieties were not very resistant to infection and, as in Luwero and West Nile, they were introduced at the height of the epidemic when whiteflies were numerous and CMD was prevalent. Rapid spread occurred to the new plantings and many were soon almost totally infected. This occurred whether or not farmers had heeded the advice to combine the small individual family allocation of cuttings so as to establish larger blocks and to provide these with at least some degree of isolation.

Later rehabilitation projects were much more successful, although some of the varieties used were even more susceptible than those distributed in 1991. This can be explained by the reduction in infection pressure that occurred as a consequence of the big decrease in the area of cassava grown, following earlier losses and the switch to other crops. A particularly successful project was operated by the Ugandan charity Vision Terudo, which introduced Nase 3 (TMS 30572). This and other improved CMD-resistant varieties were also distributed in an Agricultural Development Project in 1995 that had a particularly big impact. A survey in 2003 established that CMD-resistant varieties were being widely grown and they predominated in 88% of the fields assessed. This explains why the overall incidence of infection in Kumi was only 14%, symptoms were generally inconspicuous and there has been a full recovery in cassava production (Bua *et al.*, 2005).

20.4.4 Soroti district

Soroti district was severely disrupted in the 1980s and the security situation did not improve sufficiently to allow ready access to the area until 1991. It was then learnt from discussions with farmers that a severe epidemic of CMD had already occurred and caused a drastic decrease in cassava cultivation. The crop had previously been grown so extensively and profitably for sale locally and for export to northern districts of Uganda and into southern Sudan that it was referred to as 'white gold'. A survey in 1991 revealed that cassava plantings were few, the incidence of CMD in the remaining fields in all counties was almost total and the symptoms were very severe (Otim-Nape *et al.,* 1998).

Ebwanateraka was virtually the only variety being grown in the district in the late 1980s. It was highly susceptible to infection and the failure of this variety due to CMD caused grave food shortages and hardship. A further survey in 1992 confirmed that many farmers had largely abandoned growing cassava and switched to sweet potato and other crops. The cassava area had decreased by 95% as farmers realized the futility of planting infected cuttings. These were the only ones available in quantity due to the insecurity and limited access to material from the relatively unaffected areas south of Lake Kyoga.

Because of the insecurity, no attempts could be made to restore cassava production until 1991, by which time little cassava was being grown and there were few remaining sources of infection, especially in counties where farmers had heeded the official advice to remove any remaining infected stands before planting CMD-free cuttings introduced from elsewhere. This explains why the first attempts to re-stock the area with cuttings of Bao and Aladu obtained from Apac district were more successful than previous ones in Luwero, West Nile and Kumi, which were made earlier at the height of the epidemic. Another important difference is that farmers in Soroti were encouraged to plant the introduced material in large isolated blocks. These were under group management and extension staff provided advice on roguing and other aspects of control (Otim-Nape *et al.,* 1997). Infection was much less than when the same varieties had been introduced several months earlier into Kumi. Other later projects were also successful, especially when Nase 3 and other CMD-resistant varieties became available. By 1996, there had been a considerable recovery in cassava production based on the improved varieties released through the various rehabilitation projects and other varieties introduced by farmers themselves from elsewhere in Uganda. There has been a big increase in varietal diversity and resistant varieties predominated in 69% of the plantings assessed in 2003, when the overall incidence of infection in the district was 27% and even less in the resistant varieties (Bua *et al.,* 2005).

20.4.5 Kitgum, Lira, Apac, Masindi and Kiboga districts

Kitgum, Lira, Apac, Masindi and Kiboga are other districts of Uganda where cassava has long been the traditional staple food crop. From anecdotal and other evidence, it became apparent that many parts of these districts were affected by severe CMD epidemics in the late 1980s and the remaining areas were affected subsequently. CMD incidence and severity were uniformly high in all five districts surveyed between 1990 and 1992 and severe losses were reported (Otim-Nape *et al.*, 1998). However, the decline in production that occurred in Masindi, Kiboga and parts of Apac was not as drastic as in Kumi or Soroti. This was associated with the very wide range of varieties being grown in western districts and, especially in Masindi, there has been a big increase in the area of Tongolo and other locally-selected varieties which have some degree of tolerance. The GCF project has also had an important influence in providing farmers with virus-resistant varieties. Masindi was included in the original 1991 project and operations were extended to Lira and parts of Apac in 1992 (Otim-Nape *et al.*, 1997). Nase 3 and other improved varieties soon accounted for a substantial proportion of the cassava grown in these districts and by 1996 there had been a considerable diffusion of this material into non-project areas (Bua *et al.*, 1997). However, from subsequent surveys it is apparent that many farmers have since reverted to the use of locally selected varieties now that the CMD problem has abated and resistance to CMD is no longer an over-riding requirement. Resistant varieties predominated in only 8% of the plantings assessed in Masindi in 2003 when the overall incidence and severity of CMD were generally intermediate (Bua *et al.*, 2005).

20.4.6 Districts south of Lake Kyoga

The northern areas of Mukono and Kamuli districts to the south of Lake Kyoga received little attention until 1994, when surveys were made of the incidence and severity of CMD in representative areas and assessments were made along main and feeder roads traversing the areas east and west of Kamuli township. It then became possible to relate the epidemiological situation in the area to previous findings elsewhere in Uganda (Otim-Nape *et al.*, 2000). Six distinct zones were distinguished, using an ecological approach, and they were shown to occur in a consistent and predictable sequence from pre-epidemic areas in the south to post-epidemic areas in the north:

- **Zone 1: the pre-epidemic zone,** characterized by few whiteflies on cassava, a low incidence of CMD and little or no evidence of spread by whitefly. Symptoms generally mild and the overall disease situation benign.
- **Zone 2: a transitional zone,** where CMD incidence and spread were variable, but somewhat greater than in Zone 1.
- **Zone 3: the epidemic zone,** characterized by generally high population densities of whiteflies, very rapid spread of CMD and severe symptoms, mainly on the youngest leaves of affected plants.
- **Zone 4: a zone of chronic and severe infection,** where CMD was so prevalent that farmers had little option but to plant infected cuttings; growth and yield were seriously impaired and many plantings were abandoned before harvest.
- **Zone 5: a post-epidemic zone,** in which cassava production had largely been abandoned, except by the few farmers who had introduced uninfected planting material from elsewhere.

- **Zone 6: a zone of recovery,** where production had been restored by farmers introducing planting material from elsewhere or adopting improved virus-resistant varieties.

From the experience in Kamuli and previous observations elsewhere in Uganda, it was possible to generalize and interpret the six zones as forming a *temporal* as well as a *spatial* series. The zones occurred as a temporal series over several years as the epidemic passed through a particular site, or as a spatial series apparent at any one time when travelling south to north from pre-epidemic to post-epidemic areas. An important feature of the zonation in districts where Ebwanateraka or other susceptible varieties of cassava were widely grown was the very obvious and abrupt distinction between the Zone 3/Zone 4 areas and the largely unaffected Zone 1 to the south. This created an obvious disease 'front' that was readily apparent to an experienced observer, even when travelling in a vehicle at speeds of up to 50-60 km h^{-1}.

The approximate position of the front was mapped over substantial distances across southern Uganda in 1994 and assessments continued in subsequent years to record the number and overall health status of the plantings and the varieties grown. Thus, the existence and southward movement of the epidemic was established across Mukono, Kamuli and Iganga districts and into Pallisa and Tororo districts to the east and into Mpigi, Mubende and Kibaale districts to the west (Legg and Ogwal, 1998; Otim-Nape *et al.*, 2000; Colvin *et al.*, 2004).

The situation in the western and south-western districts of Uganda has been difficult to interpret because the topography and patterns of land use are more varied than in the east. Cassava is less widely grown and there are rocky outcrops and considerable areas of coffee, tea, forest and rangeland used for cattle. An important feature of cassava production in the western districts is that it is mainly for local consumption and not for export to other areas. Many different varieties are grown, often within the same plantings, and this has provided a degree of resilience in that the more resistant varieties have been selected preferentially and adopted whenever CMD has become a problem. This was evident in several western districts, including Masindi, Hoima and Mubende, where farmers were seen removing the most severely affected plants from within plantings of mixed varieties and retained plants that were relatively unaffected and still productive. Consequently, the epidemic front was either not apparent or ill-defined and the situation was in marked contrast to that in Iganga, Kamuli, Kumi and Soroti districts to the east. There the vulnerable Ebwanateraka predominated (Otim-Nape *et al.*, 1998, 2001) and there was little scope for the local selection of relatively resistant types.

20.4.7 Subsequent developments and the post-epidemic situation

Views on the nature and magnitude of the CMD problem in Uganda changed considerably during the 1990s as additional areas were affected and further information became available. When the first problems were reported in northern Luwero and in West Nile in 1988-89, they were regarded as localized and unrelated.

The situation changed in the early 1990s when it became apparent that the Kumi epidemic was an extension of the earlier one in Soroti, that much of central Uganda was affected and that the epidemic was continuing to spread southwards. In subsequent years the epidemic and associated whitefly infestations continued to spread southwards into Mpigi, southern Mukono and Iganga and eventually reached Lake Victoria near Jinja township in 1996. Other lake shore areas to the east and west were then affected and the progress of the epidemic was monitored in detail by sequential observations along a north-south transect in southern Mukono district (Colvin *et al.,* 2004). Kalangala Island in Lake Victoria was first affected in 1997 and the epidemic continued to spread into Masaka and Rakai and eventually reached the south-west districts bordering Tanzania and Rwanda. Moreover, the areas that were affected first entered a post-epidemic recovery phase as the problem abated, cassava plantings increased and production was restored.

The extent and rapidity of the changes that have occurred in the incidence and severity of CMD and in the varieties of cassava grown in Uganda since the epidemic was first reported in 1988 are evident from comprehensive surveys of large areas of the country in 1990-92 (Otim-Nape *et al.,* 1998), 1994 (Otim-Nape *et al.,* 2001) and 2003 (Bua *et al.,* 2005). There have also been surveys of a smaller number of representative districts in 1997 and annually between 1998 and 2001. Nevertheless, the information available is incomplete because official access to some northern and western districts has been restricted or impossible because of continuing insecurity. Despite these difficulties, it is apparent that the epidemic spread across all parts of central and southern Uganda during the 1990s and that by the end of the decade virtually all areas were in a post-epidemic phase.

The current situation in post-epidemic areas is variable and related to the overall status of cassava, and the extent to which CMD-resistant varieties have been adopted. The greatest impact of the epidemic has been in Kumi, Soroti and other districts where cassava was particularly important and where there had been an almost total reliance on Ebwanateraka, or one of the other farmer-selected varieties that proved to be extremely vulnerable to CMD. In these districts there was an initial drastic decline in production as the original cassava was largely eliminated. The subsequent recovery was associated with a big increase in varietal diversity as farmers introduced varieties from other parts of Uganda and CMD-resistant varieties were made available through the many different cassava rehabilitation projects. This is apparent from the surveys that also revealed a generally high incidence of infection in the farmer-selected varieties and much use of infected cuttings as planting material. There was a relatively low incidence of infection in the CMD-resistant varieties, especially Nase 3, Nase 4 and the recently released TME 14 (Bua *et al.,* 2005).

Another feature of the most recent survey results is that the symptoms in the local varieties are generally less severe than those encountered earlier at the height of the epidemic. This is apparent even in Ebwanateraka and other vulnerable varieties that have persisted in many areas, or reappeared after almost disappearing for several years. Many of these varieties now grow and yield satisfactorily, although almost totally infected. These developments can be explained by the emergence of somewhat tolerant landraces and by the occurrence of avirulent virus strains or strain combinations that are considerably less damaging than those encountered earlier. This was demonstrated by Pita *et al.* (2001) and also in a recent study of avirulent virus isolates from Ebwanateraka which provided some degree of protection from damaging virulent strains (Owor *et al.*, 2004). Another feature of the post-epidemic phase is that there has been an increase in the proportion of single virus infections and a decrease in the prevalence of the particularly damaging combination of viruses that predominated at the height of the epidemic (Sseruwagi *et al.*, 2004). The emergence of somewhat tolerant landraces and of avirulent virus strains or strain combinations explains why the recovery in cassava production that has occurred in Uganda is not confined to areas where CMD-resistant varieties have been widely adopted.

The 1997 and subsequent surveys show a progressive increase in the adoption of resistant varieties and overall these predominated in about one third of all plantings assessed in 2003 (Bua *et al.*, 2005). However, uptake was variable and the proportion ranged from less than 10% in six of the 21 districts surveyed to more than 50% in six other districts. The highly resistant Nase 3 has been the most successful of the varieties released officially in the 1990s and it predominated in seven of the 21 districts surveyed in 2003. However, in several districts the uptake of resistant varieties was limited by a shortage of planting material, or because the varieties released originally did not meet the requirements of farmers or consumers; especially after the epidemic had abated, the need for resistant varieties had lessened and farmers became more concerned with other attributes. Increasing attention is now being given to the quality and other features of CMD-resistant varieties and TME 14 and some of the other recent selections already feature prominently in some districts (Bua *et al.*, 2005). This suggests that the use of resistant varieties will continue to increase, even though there is likely to be a decrease in the funds available for rehabilitation projects of the type mounted during the 1990s crisis. However, it is unclear whether CMD-resistant varieties will eventually displace the local mainly susceptible ones grown previously, or whether the two types of material will continue to co-exist. Much will depend on the success of the cassava breeding and selection programmes and of any attempts made to improve the current unsatisfactory health status of the many popular local varieties that are still widely grown.

20.4.8 The impact of the epidemic in Uganda

In attempts to assess the impact of the epidemic in Uganda, there is general agreement on:

- the drastic decline in cassava production and the food shortages that occurred in many districts during the 1990s following the onset and spread of the epidemic;
- the big decrease in the area of cassava being grown at the height of the epidemic, especially in the worst-affected areas;
- the high and largely unfulfilled demand for cassava at the height of the epidemic which led to a big increase in theft and in the prices of cassava roots for consumption and of cuttings for new plantings;

- the marked changes that occurred in many areas in the varieties being grown and the big increase in varietal diversity;
- the importance of the CMD-resistant varieties which were first introduced in the early 1990s and predominated in a third of all plantings by 2003;
- the greatly increased traffic in cassava cuttings that occurred within and between districts as a direct consequence of the epidemic;
- the increased plantings of alternative staple food crops that were necessary until cassava production was restored;
- the big allocation of funds for cassava rehabilitation projects made by many governmental and non-governmental organizations.

Only some of these changes and developments have been quantified and there is a dearth of information from which to estimate the overall impact of the epidemic in economic or social terms. Indeed, the official FAO statistics for Uganda give no indication of decreased production as they show progressive increases since 1980 in the area of cassava grown and in the production achieved. This is implausible and indicates basic flaws in the method of data collection.

Indicative estimates of the losses caused by the CMD epidemic were made on the assumption that each year in the mid-1990s an area of Uganda equivalent to four whole districts was virtually out of production (Otim-Nape *et al.,* 2000). This represents an annual total of 60,000 ha, assuming an average of 15,000 ha per district. Assuming productivity at 10 t ha^{-1}, this is equivalent to an annual loss of 600,000 t, worth US$6O million, at a conservative valuation of US$100 t^{-1}. These are huge losses, to which must be added the cost of developing control measures and of mounting the various large-scale cassava rehabilitation projects. It is also necessary to consider the diversion of national, donor and international resources that could otherwise have been used to enhance cassava production, or to mount other agricultural projects in Uganda or elsewhere.

The social and human costs of the epidemic are even more difficult to quantify because some are largely intangible. Nevertheless, it is apparent that the epidemic had a big impact on many communities, especially those relying heavily on cassava as a subsistence crop and source of income to meet family requirements. National and local government officials became gravely concerned at the social and economic implications of the epidemic and were closely involved in the various cassava rehabilitation projects, especially the distribution of CMD-resistant varieties. Inevitably, cassava and CMD attained a high political 'profile' and it was calculated that the disease had decreased overall national GDP.

The increased resources and attention that were given to cassava may eventually be seen as a benefit arising from the epidemic in that it drew attention to the importance of the crop and to the prospects for increased productivity by introducing improved varieties and other innovations. Other very damaging epidemics have ultimately been seen to have had such beneficial effects. For example, the severe epidemic of *Helminthosporium* blight of maize in the USA in the 1970s drew international attention to the dangers of genetic uniformity with several important crops and of an undue dependence on a few genotypes of common origin (Ullstrup, 1972; Day, 1977). Previous tropical examples of such devastating diseases include maize rust (caused by *Puccinia polysora*) and cacao swollen shoot virus disease in West Africa and sugarcane mosaic virus disease, which greatly stimulated research on these important crops and led ultimately to the introduction of improved genotypes and enhanced crop productivity (Thresh, 1990). Such benefits are emerging already with cassava in Uganda and elsewhere and will be even more apparent if the latest CMD-resistant varieties fulfil their early potential.

20.4.9 The situation in adjoining countries

Kenya: As the epidemic spread across Uganda in the 1990s it became inevitable that the neighbouring countries of Kenya, Tanzania and Rwanda would soon be affected. The threat first became a reality in 1995, when the severe effects of the epidemic were evident in Mbale and Tororo districts of eastern Uganda along the border with western Kenya, south of Mt Elgon (R.W. Gibson, unpublished report). Surveys in adjacent parts of Kenya then revealed a high incidence of recent infection associated with unusually large infestations of *B. tabaci*. Western Kenya was at the time the most important cassava-producing area of the country and accounted for *c.* 63% of total national production. CMD had occurred previously in the region, but the incidence and severity of the disease were generally low and there was considerable use of infected cuttings as planting material and little spread by whiteflies.

Subsequent surveys of western Kenya in 1997 and 1998 showed a progressive spread of the epidemic westwards to reach Kakamega and Kisumu on Winam Gulf of Lake Victoria. As in adjacent parts of Uganda, the farmer-selected varieties being grown were extremely vulnerable to infection. Severe damage occurred and there were few suitable cuttings available for new plantings. This led to a drastic decline in production as farmers lacked access to CMD-resistant varieties, except where these were acquired by farmers from adjacent parts of Uganda. Losses in western Kenya in 1998 alone were estimated at more than 140,000 t, worth US$10 million (Legg, 1999).

The eastward spread of the epidemic in western Kenya was restricted by an extensive area of high ground that is mainly forested and unsuitable for cassava cultivation. In contrast, areas south of Kisumu along the eastern side of Lake Victoria were considered to be at serious risk in 1997 to 1998 (Legg *et al.,* 1999). There has since been continued spread southwards towards the Tanzania border. However, spread has been less rapid than expected and seems to have been impeded by the natural barriers of the Winam Gulf and the Nyando Plain that separate north from south Nyanza Province.

Monitoring operations in western Kenya have been funded by grants from the US AID Office for Foreign Disaster Assistance (OFDA), which has also supported similar activities in north-west Tanzania and Burundi. Moreover, OFDA, Gatsby, Rockefeller and other donors have supported attempts to restore production in the areas of western Kenya that were first affected. Initially, Nase 3 and Nase 4 (SS4) were introduced from Uganda and multiplied for release to farmers. Subsequently, other selections were made from CMD-resistant material introduced from the IITA regional cassava breeding programme based at Serere in Soroti district of Uganda.

The latest surveys have shown some recovery in production, but the overall incidence of CMD still exceeds 50%. This is due largely to the continued use of infected cuttings of local varieties and the limited availability of resistant material.

Tanzania: South-west districts of Uganda bordering Tanzania were considered to be in the 'pandemic expansion' zone in 1997 to 1998 (Legg *et al.*, 1999). By 1999 the epidemic was reported in Bukobo district of Kagera region of Tanzania on the western shore of Lake Victoria and the epidemic-associated virus was identified in the area (Legg, 1999). Subsequent spread southwards has been unusually rapid and the epidemic has now reached the important cassava-growing areas in southern parts of Lake Victoria near Mwanze, as established during OFDA-funded monitoring surveys.

Rwanda: Cassava is particularly important in Rwanda which has been at risk since the onset of the epidemic in the late 1990s in adjacent parts of Uganda to the north and Tanzania to the east. A survey in July 2000 provided the first evidence of the epidemic in Rwanda. It was at the time restricted to Umutana province where CMD incidence and severity were greater than elsewhere in the country and the epidemic-associated virus was identified (Legg *et al.*, 2001). There has since been further spread and the epidemic is having particularly serious consequences for a country that has already been severely affected by civil war and a prolonged period of insecurity.

Burundi: Burundi has been at risk from Rwanda to the north and Tanzania to the east. This accounts for the first occurrence of the epidemic in the north-east regions of Kirundu and Muyinga in 2003. The incidence and severity of CMD as recorded in surveys were greater than elsewhere in the country, there was evidence of considerable spread by whitefly and the epidemic-associated virus was identified (Bigirimana *et al.*, 2004).

Sudan: Cassava is an important staple food crop in southern parts of Sudan bordering Kitgum and other districts of northern Uganda that were severely affected by the 1990s epidemic. However, it is unclear whether the epidemic has affected Sudan and if so whether there has been spread from Uganda to Sudan or *vice versa*. Access has been restricted because of the prolonged period of insecurity, although the epidemic-associated Uganda recombinant virus (EACMV-UG) was detected in samples collected from cassava in Sudan and tested in the UK (Harrison *et al.*, 1997b).

Democratic Republic of Congo (DRC): There is similar uncertainty in DRC (formerly Zaire) to the west of Uganda where access has been difficult because of civil war and general insecurity. Cassava crop failure was reported by aid workers in *c.* 1990 in Kivu region near the Ugandan border, but it is not known whether this was due to CMD, cassava bacterial blight, cassava mealybug, cassava green mite or drought. Subsequently, EACMV-UG was detected in samples from around Kisangani *c.* 500 km from the western border with Uganda (S. Winter, unpublished,

1998). The virus has also been detected in samples of severely diseased cassava collected even further to the west near Kinshasa, and also in Congo Republic (Neuenschwander *et al.*, 2002), and most recently in Gabon (Legg *et al.*, 2004). Currently these are the western-most areas where EACMV-UG has been detected, but their significance is uncertain. It is unclear whether the recombinant virus is the one causing the pandemic in eastern Africa and whether it occurs continuously from Kinshasa and the Congo Republic in the west (longitude 15°E) to western Kenya in the east (35°E). It is also not known whether spread has been from east to west or west to east. These uncertainties will be difficult to resolve because of continuing insecurity and poor communications in the region and the dearth of trained crop protectionists and facilities for virus identification. Nevertheless, there is an obvious threat to the important cassava-growing areas of Cameroon and westwards into West Africa. Both parents of the EACMV-UG recombinant are known to occur in these countries and a virulent form could arise at any time, even if not already present. It is also likely that there will be spread westwards in Gabon and from Congo Republic and DRC. This emphasises the importance of regular surveys to monitor the changing situation, as in the OFDA-funded operations in eastern Africa and the Great Lakes region. There are obvious problems of doing surveys in Central Africa because of the difficult terrain, the vast areas at risk and chronic insecurity. It will also be difficult to enforce quarantine controls on the movement of cassava material, to avoid disseminating novel strains of virus or vectors into new areas. Nevertheless, restrictions should be enforced and it is also important to introduce and propagate CMD-resistant varieties so that they are available for distribution to farmers when, or preferably before, the need arises.

20.4.10 Causes of the recent epidemic

The underlying reasons for the recent epidemic in Uganda are undoubtedly complex and explanations have been sought so as to improve the effectiveness of control measures and to facilitate predictions of the course of the subsequent pandemic in the region and the areas most likely to be affected.

Any explanation must account for the main features of the epidemic, and particularly:

- the sudden increase in rates of spread of CMD that have occurred in areas where the disease had previously spread slowly;
- the consistent and predictable progress of the epidemic front southwards across much of Uganda and on into western Kenya, north-west Tanzania, Rwanda and Burundi;
- the marked increase in disease severity and in whitefly infestations that are associated with the onset of the epidemic;
- the ecological characteristics of the epidemic that resemble those of a biological invasion of new territory by a novel virus species or pathogenic variant.

Attention was given from the outset to the possibility that a new and particularly damaging form of cassava mosaic virus was spreading to areas where previously CMD had been relatively benign. However, initial attempts to demonstrate differences between virus isolates from epidemic and pre-epidemic areas were unsuccessful. All isolates behaved similarly and had the properties of *African cassava mosaic virus* (ACMV) in serological tests with monoclonal antibodies (Harrison *et al.,* 1997a). This virus had been detected previously in Uganda and other parts of Africa west of the Great African Rift Valley, whereas *East African cassava mosaic virus* (EACMV) occurred in coastal areas to the east and also in Madagascar (Harrison *et al.,* 1995).

Subsequently, PCR primers were used in Uganda to distinguish between isolates of ACMV that occurred in all districts and those of a distinct virus that occurred in the many districts that were affected by the epidemic, but not in the few remaining unaffected areas. The epidemic-associated virus was referred to initially as the Uganda variant (UgV) (Harrison *et al.,* 1997a) or Ugandan cassava virus (UCV) (Deng *et al.,* 1997). It was shown to have the serological properties of ACMV, but the A component of the DNA genome contains nucleotide sequences of both ACMV and EACMV. Consequently, UgV/UCV is considered to be a recombinant of the two viruses.

Additional evidence on the properties of the UgV/UCV recombinant, including the complete nucleotide sequence of the DNA-B component, was presented by Zhou *et al.* (1997). Moreover, it was shown that plants inoculated with UgV/UCG developed more severe symptoms than ACMV alone and that dual infection with both viruses was even more damaging (Harrison *et al.,* 1997b). Thus the severe effects of the epidemic can be explained by the spread of the virulent recombinant UgV/UCV into areas where previously only ACMV was present and caused relatively mild symptoms.

In further studies using virus isolates collected in different parts of Uganda in 1997, the recombinant was re-designated EACMV-UG (Pita *et al.,* 2001). Additional evidence was obtained on the synergism between EACMV-UG and ACMV and also on the occurrence of mild strains of EACMV-UG (Harrison *et al.,* 1997b). There have also been several other studies on the occurrence and distribution of ACMV and EACMV-UG in different parts of Uganda. These have shown a decrease in the incidence of ACMV and in the prevalence of dual infection, together with the increased occurrence of mild strains (Owor *et al.,* 2004; Sseruwagi *et al.,* 2004). These trends have contributed to the decline in symptom severity that is a feature of the post-epidemic situation in recent years and one that is also associated with the adoption of varieties that are resistant to or tolerant of infection with CMD (Bua *et al.,* 2005).

The occurrence and behaviour of the novel recombinant explain many features of the epidemic in Uganda, but not the big increase in population densities of the whitefly vector. Moreover, this cannot be attributed to the use or misuse of insecticides, as recorded with *B. tabaci* on other crops in many parts of Asia and the Americas (Gerling and Meyer, 1996). This is because insecticides are not widely used on cotton or other field crops in Uganda and they are not applied at all to cassava, except by a few farmers who resorted to chemicals in abortive attempts to control whiteflies at the height of the epidemic. Some other explanation is required for the increased whitefly populations and attention has been given to two possibilities that are not mutually exclusive. One is that the original pre-epidemic population of *B. tabaci* has been supplemented or displaced by a new invader type that is more fecund and may also be more efficient as a virus vector. The other possibility is that there is a complex interaction between virus, host and vector such that whitefly populations on cassava are enhanced as a direct or indirect consequence of virus infection.

Initial studies on adult whiteflies collected from cassava in different parts of Uganda used electrophoresis to detect heterogeneity within, but no consistent differences between populations from epidemic and non-epidemic areas (Legg *et al.*, 1994). Maruthi *et al.* (2001) also collected whiteflies from epidemic and pre-epidemic areas of Uganda and detected no differences in fecundity or development time between the two groups and no evidence of mating incompatibility. Moreover, there were no differences between epidemic and pre-epidemic populations in analyses of DNA extracts using a RAPD-PCR technique. This led to the conclusion that the epidemic and associated increase in whitefly populations were not associated with a novel reproductively isolated biotype.

In parallel studies, Legg *et al.* (2002) used molecular techniques to analyse the DNA sequences of the mitochondrial cytochrome oxidase I gene of whiteflies collected along transects from pre-epidemic to epidemic areas of southern Uganda in 1997 and from two post-epidemic sites in Uganda in 1999. Two distinct haplotypes were distinguished in the 1997 collections. The one designated Ug1 occurred primarily at sites ahead of the epidemic front and was considered to be 'the local vector'. Ug2 was referred to as 'the invader' and occurred at or behind the epidemic front. Only Ug1 was detected in each of the ten samples collected in 1999. Ug2 was regarded as an invader from elsewhere in Africa that is associated with the epidemic in Uganda, even though it was not detected in 1999 and may have integrated with the pre-existing Ug1 population. The significance of these findings is unclear in the absence of any evidence of differences in the fecundity of Ug1 and Ug2 populations, or in their ability to transmit different cassava mosaic viruses. This indicates the need for additional studies in Uganda and also in Kenya, Tanzania and other parts of Africa where the pandemic continues to progress. Meanwhile, there has been an emphasis on the interaction hypothesis in seeking to explain the big increase in whitefly populations that has occurred in epidemic areas (Colvin *et al.,* 1999, 2004; Legg and Thresh, 2000).

Initial field observations provided no evidence to support the interaction hypothesis, as adults and nymphs were fewer on CMD-affected cassava than on equivalent healthy plants (Gibson *et al.,* 1996). It was also observed that nymphs and eggs were far fewer on the chlorotic areas of CMD-affected leaves than on adjacent dark green areas. However, in subsequent studies using growth chambers whitefly fecundity was greater on CMD-infected than on uninfected plants and this was associated with increased concentrations of asparagine and three other amino acids that are known to enhance the fecundity of other sap-sucking insects (Colvin *et al.,* 1999). Moreover, it was noted that the whiteflies on CMD-infected cassava tend to aggregate on the dark green areas of the laminae that leads to crowding and can be expected to promote emigration.

These findings provide an explanation for the increased whitefly activity and virus spread that have been such a marked feature of the epidemic in Uganda and elsewhere. Spread is also likely to be facilitated by the increased concentration of virus that occurs in plants dually infected with ACMV and EACMV-UG (Harrison *et al.*, 1997b). Nevertheless, additional studies are required on the complexities of the interaction between virus, host and vector. It is particularly important to explain how the limited amount of dark green tissue produced by severely diseased plants can generate such large populations of whiteflies. There is also a need to account for the large population of whiteflies and direct feeding damage now being reported on Nase 4 (SS4) and several other recently introduced varieties, even though these are resistant to CMD and seldom become infected and express symptoms.

20.4.11 The rôle of varietal diversity

The recent epidemic in Uganda was undoubtedly exacerbated by the widespread cultivation of Ebwanateraka and several other farmer-selected varieties that proved to be extremely vulnerable to the severe form of CMD. Ebwanateraka is thought to have originated in Soroti district in the 1970s and was first included in official variety trials there in 1983. CMD was not a problem in the area at the time and in these circumstances of low infection pressure, the inherent vulnerability of Ebwanateraka to CMD was not apparent and the variety was soon grown extensively in Soroti and several other districts, where it largely displaced the many other local varieties being grown previously. The history of Bao and other vulnerable varieties is likely to be similar to that of Ebwanateraka, but it is less well documented.

Many other varieties of local origin are grown in Uganda and farmers often grow mixtures of several different varieties, especially in areas to the west and south where cassava is grown mainly for local consumption. There is similar genetic diversity in many other parts of Africa, and a rapid turnover in the varieties grown (Nweke *et al.*, 1994). This occurs as new ones become available from introductions or by selection from self-sown seedlings and as existing varieties fail because of their unacceptable vulnerability to one or more of the prevailing pests or diseases, or for some other defect. Varieties that are severely affected by CMD grow so badly that they are unlikely to survive in areas where a virulent form of the disease is prevalent and, inevitably, there will be a conscious or unconscious trend towards the cultivation of varieties that escape or tolerate infection. Conversely, vulnerable varieties can emerge and may even become dominant in areas of low inoculum pressure where other selection criteria are paramount.

These considerations are consistent with the trends observed in Uganda where the impact of the 1990s epidemic was greatest in Soroti and other areas where cassava was being widely cultivated; here there was limited diversity and the main varieties grown were extremely vulnerable to infection. In contrast, in areas of diversity there were marked shifts in the relative importance of the different varieties being grown. Some vulnerable ones that were widely grown almost completely disappeared, whereas others appeared or increased in importance because they continued to provide acceptable yields, even when infected. These developments have been evident in many districts of Uganda and indicate how farmers can adjust to the occurrence of severe CMD epidemics if there is a sufficient degree of diversity amongst the local varieties being grown, or if they have ready access to such material from elsewhere.

An obvious difficulty with this approach is that the adjustment takes years to achieve, during which time substantial losses occur, farmers have considerable difficulty in maintaining production and food security is jeopardized. Furthermore, the outcome is not entirely satisfactory in that many of the local varieties being grown became totally infected, albeit with a generally mild form of CMD. This emphasises the need to improve the health status of local varieties and also the scope for deploying resistant genotypes of the type available from IITA and elsewhere. These were the main objectives of the various official interventions made in Uganda in response to the epidemic, as discussed in the following section.

20.4.12 Control measures

In 1988, there was only limited research in progress on CMD in Uganda and the development and release of virus-resistant varieties was not a high priority of the National Cassava Programme. Moreover, there was little information from previous research done in Uganda or elsewhere in Africa on which to base effective control measures. This necessitated a greatly increased commitment to research on CMD and its control. Much progress has been made, as is evident from the many publications that have appeared since the first version of this chapter was completed in 1997. Moreover, the experience gained in Uganda has been invaluable in mounting similar control programmes in Kenya, Tanzania and elsewhere.

Because of the massive impact of the CMD epidemic, it was necessary to introduce control measures on a large scale in Uganda before definitive results were obtained and suitably resistant varieties were selected and made available. Accordingly, and since 1989, various organizations funded relief and rehabilitation schemes, but only some were arranged in close collaboration with research staff (Otim-Nape *et al.*, 1994, 2000). This explains why different varieties and approaches were adopted in mounting the various projects and why the results were very variable and not always successful (Otim-Nape *et al.*, 2000). Much of the variation in outcome can be explained by differences in the resistance of the varieties released to farmers and in the circumstances and way in which they were deployed. It soon became evident that plantings made soon after the onset of the epidemic were likely to be unsuccessful unless very resistant varieties were used. In epidemic areas whiteflies were generally numerous, sources of inoculum were abundant, there was seldom much separation between plantings and the inoculum pressure was so high that susceptible varieties soon succumbed. Moreover, because of the mobility of the whitefly vector this occurred even if somewhat isolated sites were selected, roguing was practised and all infected plants were removed from the immediate locality before planting commenced.

In contrast, such an approach achieved at least partial success when adopted during the later post-epidemic stage when there had been a big reduction in the amount of cassava grown, sources of infection were relatively few, planting was done in large blocks and stringent roguing was promoted. Even greater success was achieved where suitably resistant varieties were available, but the problem was to produce sufficient quantities of such material to meet the enormous demand, especially in districts where only vulnerable varieties had been grown previously. This explains the importance of the early Gatsby, Vision Terudo and other NGO-funded projects, in which the emphasis was on CMD-resistant varieties that offered prospects of a lasting solution to the CMD problem.

Cassava production has been restored in Uganda and economic assessments have been made of the great benefits obtained from the substantial investment in agricultural research and development necessitated by the epidemic. However, there is a need to avoid an undue dependence on the few resistant varieties that have been released to date. It is also important to improve the generally unsatisfactory health status of the many local varieties that are still being grown and which are almost totally infected. Only limited attention has been given to this aspect of control and the most appropriate measures to adopt and the benefits to be gained have not been determined. It is likely that much could be achieved by selecting healthy planting material and by adopting other phytosanitation measures. There is also scope for evaluating mild strain protection and appropriate crop disposition and for exploiting biodiversity as a means of control by using intercrops or varietal mixtures of resistant and susceptible varieties (Thresh and Otim-Nape, 1994). Another priority is to obtain the funding required to utilize the experience gained in Uganda and mount effective rehabilitation projects on a suitably large-scale in Kenya, Tanzania and other countries affected more recently. Ideally, stocks of resistant varieties should also be introduced to areas ahead of the epidemic for use when required. This has seldom been possible, because of the limited funds allocated and the understandable emphasis on areas already affected and in greatest need.

20.5 GENERAL EPIDEMIOLOGICAL FEATURES OF CASSAVA MOSAIC DISEASE

CMD is likely to be an unfamiliar disease to many readers of this volume. Nevertheless, it is important in providing an example of general epidemiological features of wide applicability:

- As a 'new encounter' disease *(sensu* Buddenhagen, 1977) arising from the juxtaposition of an exotic crop and an indigenous pathogen. Several other important African virus diseases are of this type, including groundnut rosette, maize streak and cacao swollen shoot (Thresh, 1985).
- As one of the many tropical diseases to have caused serious problems following an intensification of cropping practices (Thresh, 1985; Bos, 1992). Rice tungro in South-East Asia and rice yellow mottle and maize streak in Africa are other well-known virus diseases of this type.

- As a 'vagile' disease that spreads quickly and far and as a consequence is difficult to control by isolation, roguing and other phytosanitary practices. 'Crowd diseases' such as cacao swollen shoot differ in that they do not spread quickly or far in any considerable amount and they are relatively easy to control by such methods (Thresh *et al.,* 1988).
- As one of the many diseases that have become particularly prevalent in times of war or civil strife that facilitate spread and impede the development and adoption of control measures. Cacao swollen shoot disease in Ghana and plum pox disease in Europe during the Second World War (1939-1945) are other examples (Thresh, 1985).
- As a disease that has caused particular problems following the emergence of a novel highly virulent form of the causal pathogen which displaces the form(s) occurring previously.
- Of the way in which disease incidence and severity are closely associated with the vulnerability of the cultivars grown. There are many other examples of disease problems associated with the widespread cultivation of particular genotypes (Thresh, 1985, 1990). Indeed, Simmonds (1962) commented that '*disease patterns are to a great extent a product of our plant breeding and agricultural practices*'.
- Of the way in which a disease can necessitate the introduction of resistant genotypes and provide a powerful incentive for their adoption and use on a large scale and in a shorter period than is usual for agricultural innovations. Sugarbeet curly-top disease in the south-west states of USA and sugarcane mosaic disease in many tropical and sub-tropical countries are other examples of this type (Thresh, 1990).
- Of the way in which farmers exploit host genetic diversity to provide a degree of stability and resilience that is lost when there is an over-reliance on relatively few closely related genotypes (Day, 1977).
- Of the extent and rapidity with which a long-standing, stable equilibrium between a pathogen and its host(s) can be disrupted during a perturbation, as caused by an increase in the abundance of the vector, or in the virulence of the pathogen. Some of the most striking examples of this type have occurred following the first appearance of a new virus strain or vector species or biotype and they have many of the features of a biological invasion. Ecologists have given much attention to such events (Kornberg and Williamson, 1987; Hengeveld, 1989) and an ecological approach has been advantageous in seeking to interpret the Ugandan epidemic.
- Of the focal expansion of disease as a 'travelling wave' progressing uniformly at a rate that approaches a constant value (van den Bosch *et al.,* 1988; Zadoks and van den Bosch, 1994). Spread in this way can be visualized as a disease profile that moves through space at a constant velocity without changing shape; there are other examples in the epidemiological literature (Zadoks and van den Bosch, 1994).
- Of a disease capable of causing a pandemic in the sense of an epidemic that progresses over a period of years to affect very large areas (Gäumann, 1946).

Several of these features have become evident from studies over the last decade in Uganda and elsewhere in eastern and Central Africa, yet many uncertainties remain. This is hardly surprising, because, although considerable, the research effort it has been possible to mount has been inadequate in relation to the magnitude and complexity of the problem. Such a damaging pandemic occurring in Europe or North America would undoubtedly have led to a far greater allocation of funds and resources and to the deployment of teams of researchers supported by extensionists and adequate funding. This is simply not possible in Uganda or elsewhere in sub-Saharan Africa, although considerable assistance has been provided by International Agricultural Research Centres and advanced laboratories in Europe and North America. There has also been substantial financial support from charitable organizations and other donors. Nevertheless, it is quite likely that in some African countries an epidemic of the type that occurred in Uganda would be completely overlooked or considered simply as an unaccountable crop failure. These considerations explain why CMD and many other diseases continue to cause serious losses, food insecurity and hardship in Africa and why information on the epidemiology and control of these diseases lags far behind that available on comparable diseases in developed countries.

REFERENCES

Anon. (1992) Quarantine Implications: Cassava Program: 1987-1991. Working document No. 116, CIAT, Colombia.

Anon. (1993) How Akwa Ibom overcame a crisis in cassava production. *Cassava Newsletter,* **17**, 9-10.

Bigirimana, S., Barumbanze, P., Obonyo, R. and Legg, J.P. (2004) First evidence for the spread of *East African cassava mosaic virus*-Uganda (EACMV-UG) and the pandemic of cassava mosaic disease to Burundi. *Plant Pathology*, **53**, 231.

Bock, K.R. and Woods, R.D. (1983) Etiology of African cassava mosaic disease. *Plant Disease,* **67**, 994-995.

Bos, L. (1992) New plant virus problems in developing countries: a corollary of agricultural modernization. *Advances in Virus Research,* **38**, 349-407.

Bua, A., Otim-Nape, G.W., Acola, G. and Baguma, Y.K. (1997) The adoption, approaches and impact of cassava multiplication in Uganda, in *Progress in Cassava Technology Transfer in Uganda,* (eds G.W. Otim-Nape, A. Bua and J.M. Thresh), Proceedings of the National Workshop on Cassava Multiplication. Masindi 9-12 January 1996, NARO/Gatsbv-NRI Publication.

Bua, A., Sserubombwe, W.S., Alicai, T. *et al.* (2005) The incidence and severity of cassava mosaic virus disease and the varieties of cassava grown in Uganda: 2003. *Roots* (in press).

Buddenhagen, I.W. (1977) Resistance and vulnerability of tropical crops in relation to their evolution and breeding. *Annals New York Academy of Sciences,* **287**, 309-326.

Calvert, L. and Thresh, J.M. (2002) The viruses and virus diseases of cassava, in *Cassava: Biology, Production and Utilization,* (eds R.J. Hillocks, J.M. Thresh and A. Bellotti), CAB International, Wallingford, UK, pp. 237-260.

Colvin, J., Omongo, C.A., Maruthi, M.N. *et al.* (2004) Dual begomovirus infections and high *Bemisia tabaci* populations: two factors driving the spread of a cassava mosaic disease pandemic. *Plant Pathology*, **53**, 577-584.

Colvin, J., Otim-Nape, G.W., Holt, J. *et al.* (1999) Symbiotic interactions drive epidemic of whitefly-borne disease in Uganda, in *Cassava Mosaic Disease Management in Smallholder Cropping Systems*, (eds R.J. Cooter, G.W. Otim-Nape, A. Bua and J.M. Thresh), Natural Resources Institute/NARO, Chatham, UK, pp. 76-86.

Cours, G., Fargette, D., Otim-Nape, G.W. and Thresh, J.M. (1997) The epidemic of cassava mosaic virus disease in Madagascar in the 1930s-1940s: lessons for the current situation in Uganda. *Tropical Science,* **37**, 1-7.

Day, P.R. (ed.) (1977) The Genetic Basis of Epidemics in Agriculture. *Annals of the New York Academy of Sciences,* **287**, 1-386.

Deng, D., Otim-Nape, G.W., Sangare, A. *et al.* (1997) Presence of a new virus closely related to East African cassava mosaic geminivirus associated with cassava mosaic outbreak in Uganda. *African Journal of Root and Tuber Crops,* **2**, 23-28.

Ewald, P.W. (1983) Host-parasite relations, vectors and the evolution of disease severity. *Annual Review of Ecology and Systematics,* **14**, 465-485.

FAO (2003) *The Food and Agricultural Organization of the United Nations. Production Year Book 2002:* FAO Statistics Series No. 176, 56, 102-103. Rome.

Fargette, D. and Vié, K. (1995) Simulation of the effects of host resistance, reversion, and cutting selection on incidence of African cassava mosaic virus and yield losses in cassava. *Phytopathology,* **85**, 370-375.

Fargette, D., Thouvenel, J.-C. and Fauquet, C. (1987) Virus content of leaves of cassava infected by African cassava mosaic virus. *Annals of Applied Biology,* **110**, 65-73.

Fargette, D., Thresh, J.M. and Otim-Nape, G.W. (1994) The epidemiology of African cassava mosaic geminivirus: reversion and the concept of equilibrium. *Tropical Science,* **34**, 123-133.

Fauquet, C. and Fargette, D. (eds) (1988) *Proceedings: The International Seminar on African Cassava Mosaic Disease and its Control,* Yamoussoukro, Côte d'Ivoire, 4-8 May 1987, CTA/FAO /ORSTOM /IITA/IAPC.

Fauquet, C. and Fargette, D. (1990) African cassava mosaic virus: etiology, epidemiology and control. *Plant Disease,* **74**, 404-411.

Fauquet, C.M. and Stanley, J. (2003) Geminivirus classification and nomenclature: progress and problems. *Annals of Applied Biology*, **142**, 165-189.

Gäumann, F. (1946) *Pflanzliche Infektionslehre,* Birkhäuser, Basel, 611 pp.

Geddes, A.M.W. (1990) *The Relative Importance of Crop Pests in Sub-Saharan Africa,* Bulletin No. 36, Natural Resources Institute, Chatham, UK.

Gerling, D. and Meyer, R.I. (eds) (1996) *Bemisia 1995: Taxonomy, Biology, Damage, Control and Management,* Intercept Publishers, Andover, UK, 702 pp.

Gibson, R.W., Legg, J.P. and Otim-Nape, G.W. (1996) Unusually severe symptoms are a characteristic of the current epidemic of mosaic virus disease of cassava in Uganda. *Annals of Applied Biology,* **128**, 479-490.

Harrison, B.D., Swanson, M.M. and Robinson, D.J. (1995) Cassava viruses in the Old World, in *Proceedings: Second International Scientific Meeting of the Cassava Biotechnology Network,* Bogor, Indonesia, 22-26 August 1994, Working Document No. 150, pp. 463-472, CBN/CRJFC/ AARD/CIAT.

Harrison, B.D., Liu, Y.L., Zhou, X. *et al.* (1997a) Properties, differentiation and geographical distribution of geminiviruses that cause cassava mosaic disease. *African Journal of Root and Tuber Crops,* **2**, 19-22.

Harrison, B.D., Zhou, X., Otim-Nape, G.W. *et al.* (1997b) Role of a novel type of double infection in the geminivirus-induced epidemic of severe cassava mosaic in Uganda. *Annals of Applied Biology,* **131**, 437-448.

Hengeveld, R. (1989) *Dynamics of Biological Invasions,* Chapman and Hall, London, 160 pp.

Holt, J., Jeger, M.J., Thresh, J.M. and Otim-Nape, G.W. (1997) An epidemiological model incorporating vector population dynamics applied to African cassava mosaic virus disease. *Journal of Applied Ecology,* **34**, 793-806.

Jameson, J.D. (1964) Cassava mosaic disease in Uganda. *East African Agricultural and Forestry Journal,* **29**, 208-213.

Jennings, D.L. (1994) Breeding for resistance to African cassava mosaic geminivirus in East Africa. *Tropical Science,* **34**, 110-122.

Jones, W.O. (1969) *Manioc in Africa,* Stanford University Press, Stanford, California.

Kornberg, H. and Williamson, M.H. (eds) (1987) *Quantitative Aspects of the Ecology of Biological Invasions,* Transactions of the Royal Society of London, 240 pp.

Legg, J.P. (1999) Emergence, spread and strategies for controlling the pandemic of cassava mosaic virus disease in east and central Africa. *Crop Protection,* **18**, 627-637.

Legg, J.P. and Ogwal, S. (1998) Changes in the incidence of African cassava mosaic virus disease and the abundance of its whitefly vector along south-north transects in Uganda. *Journal of Applied Entomology,* **122**, 169-178.

Legg, J.P. and Thresh, J.M. (2000) Cassava mosaic virus disease in East Africa: a dynamic disease in a changing environment. *Virus Research*, **71**, 135-149.

Legg, J.P. and Thresh J.M. (2004) Cassava virus diseases in Africa, in *Plant virology in sub-Saharan Africa.* (eds Jd'A. Hughes and B.O. Adu), Conference Proceedings, IITA, Ibadan, pp. 517-552.

Legg, J.P., French, R., Rogan, D. *et al.* (2002) A distinct *Bemisia tabaci* (Gennadius) (Hemiptera: Sternorrhyncha: Aleyrodidae) genotype cluster is associated with the epidemic of severe cassava mosaic virus disease in Uganda. *Molecular Ecology*, **11**, 1219-1229.

Legg, J.P., Gibson, R.W. and Otim-Nape, G.W. (1994) Genetic polymorphism amongst Ugandan populations of *Bemisia tabaci* (Gennadius) (Homoptera: Aleyrodidae), vector of African cassava mosaic geminivirus. *Tropical Science*, **34**, 73-81.

Legg, J.P., Ndjelassili, F. and Okao-Okuja, G (2004) First report of cassava mosaic disease and cassava mosaic geminiviruses in Gabon. *Plant Pathology*, **53**, 232.

Legg, J.P., Okao-Okuja, G., Mayala, R. and Muhinyuza, J.-B. (2001) Spread into Rwanda of the severe cassava mosaic virus disease pandemic and the associated Uganda variant of East African cassava mosaic virus (EACMV-Ug). *Plant Pathology*, **50**, 796.

Legg, J.P., Sseruwagi, P., Kamau, J., *et al.* (1999) The pandemic of severe cassava mosaic disease in East Africa: current status and future threats, in *Food Security and Crop Diversification in SADC Countries: The Rôle of Cassava and Sweet Potato*, (eds M.O. Akoroda and J.M. Teri), Proceedings of the Scientific Workshop of the Southern African Root Crops Research Network, SADC/IITA/CIP, pp. 236-251.

Maruthi, M.N., Colvin, J. and Seal, S. (2001) Mating compatibility, life-history traits, and RAPD-PCR variation in *Bemisia tabaci* associated with the cassava mosaic disease pandemic in East Africa. *Entomologia Experimentalis et Applicata*, **99**, 13-23.

Neuenschwander, P., Hughes, J. d'A., Ogbe, F. *et al.* (2002) Occurrence of the Uganda variant of *East African cassava mosaic virus* (EACMV-Ug) in western Democratic Republic of Congo and the Congo Republic defines the westernmost extent of the pandemic in East/Central Africa. *Plant Pathology*, **51**, 385.

Nweke, F.I. (1994) Farm level practices relevant to cassava plant protection. *African Crop Science Journal*, **2**, 563-582.

Nweke, F.I., Dixon, A.G.O., Asiedu, R. and Folayan, S.A. (1994) Cassava varietal needs of farmers and the potential for production growth in Africa. *Collaborative Study of Cassava in Africa: Working Paper No.* 10, IITA, Ibadan, 239 pp.

Otim-Nape, G.W., Alicai, T. and Thresh, J.M. (2001) Changes in the incidence and severity of cassava mosaic virus disease, varietal diversity and cassava production in Uganda. *Annals of Applied Biology*, **138**, 313-327.

Otim-Nape, G.W., Bua, A. and Baguma, Y. (1994) Accelerating the transfer of improved production technologies: controlling African cassava mosaic virus disease epidemics in Uganda. *African Crop Science Journal*, **2**, 479-495.

Otim-Nape, G.W., Bua, A. and Thresh, J.M. (eds) (1997) *Progress in Cassava Technology Transfer in Uganda*, Proceedings of the National Workshop on Cassava Multiplication, Masindi, 9-12 January 1996, NARO/Gatsby/NRI Publication.

Otim-Nape, G.W., Thresh, J.M. and Fargette, D. (1996) *Bemisia tabaci* and cassava mosaic virus disease in Africa, in *Bemisia 1995: Taxonomy, Biology, Damage, Control and Management*, (eds D. Gerling and R.T. Meyer), Intercept Publishers, Andover, UK, pp. 319-350.

Otim-Nape, G.W., Thresh, J.M. and Shaw, M.W. (1998) The incidence and severity of cassava mosaic virus disease in Uganda: 1990-1992. *Tropical Science*, **38**, 25-37.

Otim-Nape, G.W., Bua, A., Thresh, J.M. *et al.* (2000) *The Current Pandemic of Cassava Mosaic Virus Disease in East Africa and its control*, NARO/NRI/DFID publication, Chatham Maritime, UK, 100 pp.

Owor, B. Legg, J.P., Okao-Okuja, G., *et al.* (2004) Field studies of cross protection with cassava mosaic geminiviruses in Uganda. *Journal of Phytopathology*, **152**, 243-249.

Pita, J.S., Fondong, V.N., Sangaré, A., *et al.* (2001) Recombination, pseudo-recombination and synergism of geminiviruses are determinant keys to the epidemic of severe cassava mosaic disease in Uganda. *Journal of General Virology*, **82**, 655-665.

Simmonds, N.W. (1962) Variability in crop plants, its use and conservation. *Biological Reviews*, **37**, 422-465.

Sseruwagi, P., Rey, M.E.C., Brown, J.K. and Legg, J.P. (2004) The cassava mosaic geminiviruses occurring in Uganda following the 1990s epidemic of severe cassava mosaic disease. *Annals of Applied Biology*, **145**, 113-121.

Swanson, M.M. and Harrison, B.D. (1994) Properties, relationships and distribution of cassava mosaic geminiviruses. *Tropical Science*, **34**, 15-25.

Thottappilly, G., Thresh, J.M., Calvert, L.A. and Winter, S. (2003) Cassava, in *Virus and Virus-like Diseases of Major Crops in Developing Countries*, (eds G. Loebenstein and G. Thottappilly), Kluwer, Netherlands, pp. 107-165.

Thresh, J.M. (1985) The origin and epidemiology of some important plant virus diseases. *Applied Biology*, **5**, 1-65.

Thresh, J.M. (1990) Plant virus epidemiology: the battle of the genes, in *Recognition and Response in Plant-Virus Interactions*, (ed. R.S.S. Fraser), NATO AS1 Series H, *Cell Biology*, **41**, 93-121, Springer-Verlag, Berlin/Heidelberg.

Thresh, J.M. and Cooter, R.J. (2005) Strategies for controlling cassava mosaic disease in Africa. *Plant Pathology*, **54**, 587-614.

Thresh, J.M. and Otim-Nape, G.W. (1994) Strategies for controlling African cassava mosaic geminivirus. *Advances in Disease Vector Research*, **10**, 215-236.

Thresh, J.M., Fargette, D. and Otim-Nape, G.W. (1994a) The viruses and virus diseases of cassava in Africa. *African Crop Science Journal*, **2**, 459-478.

Thresh, J.M., Otim-Nape, G.W. and Fargette, D. (1998a) The components and deployment of resistance to cassava mosaic virus disease. *Integrated Pest Management Reviews*, **3**, 209-224.

Thresh, J.M., Otim-Nape, G.W. and Jennings, D.L. (1994b) Exploiting resistance to African cassava mosaic virus. *Aspects of Applied Biology*, **39**, 51-60.

Thresh, J.M., Owusu, G.L.K. and Ollennu, L.A.A. (1988) Cocoa swollen shoot: an archetypal crowd disease. *Journal of Plant Diseases and Protection*, **95**, 428-446.

Thresh, J.M., Otim-Nape, G.W., Legg, J.P. and Fargette, D. (1997) African cassava mosaic virus disease: the magnitude of the problem. *African Journal of Root and Tuber Crops*, **2**, 13-19.

Thresh, J.M., Otim-Nape, G.W., Thankappan, M. and Muniyappa, V. (1998b) The mosaic diseases of cassava in Africa and India caused by whitefly-borne geminiviruses. *Review of Plant Pathology*, **77**, 935-945.

Ullstrup, A.J. (1972) The impacts of the southern corn leaf blight epidemics of 1970-1971. *Annual Review of Phtytopathology*, **10**, 37-50.

van den Bosch, F., Zadoks, J.C. and Metz, J.A.J. (1988) Focus expansion in plant disease. 1. The constant rate of focus expansion. *Phytopathology*, **78**, 54-58.

Zadoks, J.C. and van den Bosch, F. (1994) On the spread of plant disease; a theory on foci. *Annual Review of Phytopathology*, **32**, 503-521.

Zhou, X., Liu, Y., Calvert, L. *et al.* (1997) Evidence that DNA-A of a geminivirus associated with severe cassava mosaic disease in Uganda has arisen by interspecific recombination. *Journal of General Virology*, **78**, 2101-2111.

INDEX

GPSR Compliance
The European Union's (EU) General Product Safety Regulation (GPSR) is a set of rules that requires consumer products to be safe and our obligations to ensure this.

If you have any concerns about our products, you can contact us on

ProductSafety@springernature.com

In case Publisher is established outside the EU, the EU authorized representative is:

Springer Nature Customer Service Center GmbH
Europaplatz 3
69115 Heidelberg, Germany

www.ingramcontent.com/pod-product-compliance
Ingram Content Group UK Ltd.
Pitfield, Milton Keynes, MK11 3LW, UK
UKHW012141240726
13966UKWH00001B/97

9781402045806